Lecture Notes in Networks and Systems

Volume 380

The series "Lecture Notes in Networks and Systems" publishes the latest developments in Networks and Systems—quickly, informally and with high quality. Original research reported in proceedings and post-proceedings represents the core of LNNS.

Volumes published in LNNS embrace all aspects and subfields of, as well as new challenges in, Networks and Systems.

The series contains proceedings and edited volumes in systems and networks, spanning the areas of Cyber-Physical Systems, Autonomous Systems, Sensor Networks, Control Systems, Energy Systems, Automotive Systems, Biological Systems, Vehicular Networking and Connected Vehicles, Aerospace Systems, Automation, Manufacturing, Smart Grids, Nonlinear Systems, Power Systems, Robotics, Social Systems, Economic Systems and other. Of particular value to both the contributors and the readership are the short publication timeframe and the world-wide distribution and exposure which enable both a wide and rapid dissemination of research output.

The series covers the theory, applications, and perspectives on the state of the art and future developments relevant to systems and networks, decision making, control, complex processes and related areas, as embedded in the fields of interdisciplinary and applied sciences, engineering, computer science, physics, economics, social, and life sciences, as well as the paradigms and methodologies behind them.

Indexed by SCOPUS, INSPEC, WTI Frankfurt eG, zbMATH, SCImago.

All books published in the series are submitted for consideration in Web of Science.
For proposals from Asia please contact Aninda Bose (aninda.bose@springer.com).

More information about this series at https://link.springer.com/bookseries/15179

Pavel V. Trifonov · Marina V. Charaeva
Editors

Strategies and Trends in Organizational and Project Management

Springer

Editors
Pavel V. Trifonov
Financial University under the Government of the Russian Federation
Moscow, Russia

Marina V. Charaeva
Faculty of Management
Southern Federal University
Rostov-on-Don, Russia

ISSN 2367-3370 ISSN 2367-3389 (electronic)
Lecture Notes in Networks and Systems
ISBN 978-3-030-94244-1 ISBN 978-3-030-94245-8 (eBook)
https://doi.org/10.1007/978-3-030-94245-8

This Springer imprint is published by the registered company Springer Nature Switzerland AG
The registered company address is: Gewerbestrasse 11, 6330 Cham, Switzerland

Introductory Article

The international scientific conference "Operational and Project Management: Problems, Challenges, Solutions" was held on May 19–21, 2021, organised by Financial University under the Government of the Russian Federation and Southern Federal University, with the participation of professors and scholars from the Russian Federation and eleven neighbouring and other countries—Belarus, Bulgaria, Great Britain, Greece, China, Latvia, Poland, North Macedonia, Serbia, Slovakia and Ukraine.

The purpose of the conference was to discuss the development of operational and project management technologies in Russia and abroad. The scientific agenda of the conference concerned the search for optimal mechanisms, technologies and areas aimed to improve the operational and project activities of modern organisations. It reflected the need to search for new methods in the field of process-based, production and project management, as well as to increase the operational efficiency and digitalisation of management systems.

The conference also featured a diverse analysis of financial economy issues in the context of management, the concepts of prospective economy, regional development and strategising with regard to the socio-economic conditions of different countries, education, digitalisation and globalisation processes.

The actualisation of financial economics as a subject area and its management is not accidental, since special importance in the present-day economic conditions is given to the issues of funding of project activities as the core aspect. In this context, it is important to understand the etymology of financial economics, the management principles in the context of total digitalisation, the transformation of financial methods and levers, as well as the formation of new economic trends. All these issues are important not only from the point of view of fundamental science, but also in practical terms, as well as for identifying priority areas in the educational process.

The conference aroused lively interest in the scientific community and brought together more than 300 foreign and Russian participants at the discussion platforms in the face-to-face and distance–participation format. The conference participants noted that its results are significant and valuable for the development of global and Russian science and economy.

The conference proceedings were split into six sections.

The section **"Risks and opportunities of modern managerial practices"** considered the foreign and domestic experience in implementing lean production management methods ("Lean Production", "Just in Time", "Kaizen Management"); key problems of inculcating project management systems at Russian production enterprises; current trends in the development of business process management methodology in Russia. Much attention was paid to the risks and conditions for the implementation of quality management systems at the enterprise during the crisis period of economic development, the choice of operational strategies for corporate production systems based on loss and cost management analysis in the conditions of restrained growth, as well as the choice of managerial decision-making mechanisms in the field of improvement of production systems management at enterprises during the recessionary period of economic development.

The section **"Imperatives and driving forces of the prospective economy"** was a platform for the discussion of subjects related to the development of social economy in the modern realities, in particular, the transformation of educational space in the system of digitalisation of economy, formation of functional working age model in terms of human capital development. The participants of the discussion came to the conclusion that the development of human capital and improved quality of labour force should be the basis for the modernisation of the national economy, serving as one of the main driving forces of the prospective economy.

The section **"Modern strategies of socio-economic development"** raised the questions connected with prioritisation of the strategic development areas of the global and Russian economy, sustainable socio-economic development, economic growth, income distribution, innovations policy, modelling of strategies for small and medium-size business development, which allowed to form an opinion on the problems in the sphere of digitalisation of economy and development of approaches to assessing the efficiency of national projects.

The sections **"Management under digitalisation: transformation of management paradigm or traditional approaches to management"** and **"Financial economy: priorities, trends, new solutions and areas"** considered exhaustively the issues of management and trends in the financial economy as a separate area. The foreign and Russian scholars and experts–practitioners discussed the results of research in the field of marketing and human resource management obtained with regard to the innovations-related, social and environmental agenda, urging to present comprehensively the core of modern management and digital transformation of the managerial paradigm. The presented studies in the sphere of financial economics aimed to reflect the best practices and the specific tools of financial management, as well as the solutions created on their basis.

The section **"Business education and new economic realities: challenges or opportunities"** paid special attention to the problems of integration of educational levels, relevance and practicality of competences acquired by university graduates upon their entering the international academic environment of the digital economy, when the said factors acquire a new context with the introduction of the new mission of universities as integrators, innovators and commercialisers of know-how into the academic reality, as well as its impact on the educational system.

A number of conclusions were drawn on the outcomes of the conference as follows:

1. A number of directions for development of cross-functional areas of cooperation between project and process specialists were defined, depending on industry affiliation;
2. An assessment of the scope of innovational potential of high-tech industries in Russia was made;
3. A set of key problems aimed to develop professional competencies of project and process managers in the Russian labour market was identified;
4. Some key areas of process management development were formulated for the Russian corporate sector;
5. A number of options for applying "lean manufacturing" technologies in such sectors as education and public administration were considered.

The conference participants prepared scientific publications on the subjects of the presented reports. A participant could deliver a presentation at the conference within the framework of a specified section if his/her scientific report was confirmed by a relative article and was enclosed to the application. Subsequently, the content of the publication was reviewed by representatives of the programme committee according to the "blind principle" (the reviewer could not see the article author's name, while the author did not have any information about the reviewer). After verification of the article for compliance with the requirements for scholarly content, design and conformity with the subject area of the conference, the reviewer gave his/her opinion on the admission of the article for publication in the collection of papers.

The interdisciplinarity and the breadth of coverage of the listed above fundamental and practical problems made this conference attractive for specialists representing different scientific spheres in Russia and foreign countries. The organising committee of the conference hopes that the covered subject area will be further developed by other scientists interested in the problems of operational and project management in the modern environment.

Organisation

Organising Committee of the Conference

Chairman

Mikhail Eskindarov	Financial University, Russian Federation

Members of the Organising Committee

Vladimir Maslennikov	Financial University under the Government of the Russian Federation, Russia
Arkady Trachuk	Financial University under the Government of the Russian Federation, Russia
Elena Arsenova	Financial University under the Government of the Russian Federation, Russia
Yulia Gruzina	Financial University under the Government of the Russian Federation, Russia
Vitaly Zhidikov	Financial University under the Government of the Russian Federation, Russia
Pavel Trifonov	Financial University under the Government of the Russian Federation, Russia
Natalia Linder	Financial University under the Government of the Russian Federation, Russia
Tatiana Popadyuk	Financial University under the Government of the Russian Federation, Russia
Olga Andreeva	Rostov State University of Economics, Russia
Roman Volodin	Southern Federal University, Russia
Natalia Muravyova	Southern Federal University, Russia
Alexandra Novitskaya	Southern Federal University, Russia
Asem Naushabaeva	Bitlis Eren University, Turkey

Allahyarova Tahima	Centre for Social Research, Azerbaijan
Rakhimov Mukhsin	Tajik Technical University named after M.S. Osimi, Tajikistan

Programme Committee of the Conference

Chairperson

Arkady Trachuk	Financial University, Russian Federation

Programme Committee Members

Linder Natalia	Financial University under the Government of the Russian Federation, Russia
Abdikeev Niyaz	Financial University under the Government of the Russian Federation, Russia
Baev Georgy	Production Management Centre, Bauman Moscow State Technical University, Russia
Boshnyak Timur	Coordinator of Lean Manufacturing Programmes, Ruselectronics, JSC Powered by Rostec, Russia
Vasiliev, Vitaly	Director General, Kaizen Institute Russia, Russia
Ibragimov Ruslan	Head of Department of SeverGroup, JSC, Russia
Ildemenov Sergey	Financial University under the Government of the Russian Federation, Russia
Polevoy Sergey	Financial University under the Government of the Russian Federation, Russia
Popadyuk Tatiana	Financial University under the Government of the Russian Federation, Russia
Rzhavin Georgy	Vice-President of ABPMP Russia Association
Trifonov Pavel	Financial University under the Government of the Russian Federation, Russia
Barmuta Karine	Don State Technical University, Russian Federation
Bogdanova Margarita	D.A. Tsenov Academy of Economics, Bulgaria
Natalia Vovchenko	Rostov State University of Economics, Russian Federation
Grishina Olga	Plekhanov Russian University of Economics, Russian Federation
Dimitrovski Robert	Institute of Knowledge Management, Northern Macedonia
Dziwok Ewa	University of Economics in Katowice, Poland
Evstafieva Elena	Rostov State University of Economics, Russian Federation
Zarić Siniša	University of Belgrade, Serbia

Olga Ivanova	Rostov State University of Economics, Russian Federation
Natalia Kazakova	Plekhanov Russian University of Economics, Russian Federation
Klimuk Vladimir	Baranovichi State University, Belarus
Lennerová Ivana	University of Economics in Bratislava, Slovakia
Matusovic Martin	University of Economics in Bratislava, Slovakia
Molchanov Igor	Lomonosov Moscow State University, Financial University under the Government of the Russian Federation, Russian Federation
Rupeika-Apoga Ramona	University of Latvia, Latvia
Stehlíková Beáta	Slovak University of Technology, Slovakia
Sagitova Rosa	University of Portsmouth, UK
Sadovnikova Natalia	Plekhanov Russian University of Economics, Russian Federation
Terziev Venelin	Angel Kanchev University of Ruse, Bulgaria
Trinchuk Victor	National University of Life and Environmental Sciences of Ukraine, Ukraine
Charaeva Marina	Southern Federal University, Russian Federation
Dmitry Shevchenko	Southern Federal University, Russian Federation

Contents

Risks and Opportunities of Modern Managerial Practices

Assessing the Impact of Living Standards on Managing the Development Potential of the Arctic Zones

Tamara P. Danko, Lyubov A. Chaikovskaya, Nikita D. Epstein, Natalia N. Kovalenko, and Kirill E. Chikunov(✉)

Plekhanov Russian University of Economics, Moscow, Russia

Abstract. Environmental safety is becoming increasingly important with the growth of environmental pollution. The Arctic zones (Murmansk Region, Republic of Karelia and Archangelsk Region) potential assessment is of great interest from the point of view of evaluating its impact on innovative development of the Arctic zones in terms of quality of life and environmental safety. It should first be noted in this research that the Arctic is not only a large territory of our country, but also a zone of economic interests of the world at large; today the Arctic presents a challenge and inspires for the invention of innovative ways of managing its expanse in various aspects. In this regard, the assessment of the Russian Arctic potential is necessary and truly important in terms of evaluation of impact of the quality of life on managing the potential of this area. The authors of the research believe that in order to ensure due realisation of the sustainable development concept at the macro level it is necessary to improve the operation of all subsystems of the organisation at the micro level, which requires the development of new technologies, improvement of the quality of end results of relevant activities and introduction of new, more efficient management methods.

Keywords: Arctic zones potential · Innovative methods · Own expanse · Mean per capita income

1 Introduction

Environmental safety is becoming increasingly important in the conditions of the growing environmental pollution; therefore, environmental innovations should replace hazardous ones owing to the development and acquisition of environmentally friendly technologies, development of relevant scientific research as well as waste recycling methods reducing pollution, introduction of necessary innovative machinery, equipment, components, consumables. Such solutions are in demand in industry, power engineering, construction, logistics and water resource management. Innovative activities as an important factor of sustainable development should be taken into account in calculations used for strategic planning.

P. V. Trifonov and M. V. Charaeva (Eds.): DITEM 2021, LNNS 380, pp. 3–11, 2022.
https://doi.org/10.1007/978-3-030-94245-8_1

The authors aimed to assess the impact of the quality of life on the management of innovative development potential of the Arctic zones actors in the conditions of the pandemic and environmental pollution. The number of scientific papers on regional environmental security published in Russia in peer-reviewed journals in the last 2 years amounted to 750 and on those on environmental innovation – about 500, with dozens of PhD theses defended on the above subjects. About 350 articles on the role of innovations in sustainable development of the region were published in the journals indexed in the Web of Science and Scopus databases. The researchers agree that proper environmental regulation promotes technological innovations and is one of the most important means to achieve green transformation [1, 2]. Meanwhile, ecological regulation also differs significantly in mechanisms through which it influences different approaches to technological innovation [3].

The authors of the research believe that in order to ensure due realisation of the sustainable development concept at the macro level, it is necessary to improve the operation of all subsystems of the organisation at the micro level, which requires the development of new technologies, improvement of the quality of end results of relevant activities and introduction of new, more efficient management methods [4]. As to the regional level, it seems necessary to establish the relationship between the activity of enterprises in the sphere of environmental safety and the costs of technological and ecological innovations.

Considering the environmental initiatives, different organisations implement ecological innovations with the purpose to ensure compliance with the modern technical regulations, rules and standards (environmental legislation provisions), as well as the expected tightening of legal norms. In addition, economic innovations by organisations may be driven by the need to meet market (consumers') requirements, the availability of governmental grants, subsidies or other financial incentives for implementing environmental innovations, or may be motivated by voluntary adherence to the general principles of environmental protection [15].

2 Methods

Currently, there exist a number of methodologies used by state authorities and scientists to assess the quality of life, this evaluation being a direct or indirect indicator of per capita income of the population and the living standards in the ecologically safe environment.

For the research, the authors chose per capita income, median income, the first decile, the ninth decile, the decile coefficient of income differentiation, the Gini coefficient. Before assessing the innovative development potential, the information on pandemic trends in the regions under investigation will be presented. The analysis of the situation is shown in Table 1.

Table 1. Characteristics of the pandemic situation in Murmansk Region, Republic of Karelia and Archangelsk Region in May 2021 [5].

	Total cases of contamination, human subjects	Recovered, human subjects	Recovered (%)	Died, human subjects	Died (%)	Infected, human subjects
Murmansk Region	50578	47843	94.6	1176	2.3	1559
Republic of Karelia	46167	44366	96.1	496	1.1	1305
Archangelsk Region	62545	59286	94.8	962	1.5	2297

3 Results and Discussion

The analysis of the presented materials shows the general picture of morbidity in the given areas. The results show that the danger (pandemic) in fact exists. The indicators differ from each other by regions. We can state that the situation is more benign in the Republic of Karelia which has the lowest number of contaminated persons during the pandemic as well as the lowest 2019-nCoV mortality among the compared entities [6]. The highest incidence rate is shown by Archangelsk Region, but the region ranks second in terms of mortality. Murmansk Region holds a middling position in terms of morbidity over the entire period, but the death rate is the highest there. Table 1 shows as well the people registered as ill in the region on the date specified in the table, those who were sick at the time the resource was accessed. After the presentation of these figures, the authors proceed to expound on the content of the given research to assess the impact of the living standards on the management of development potential by the example of the selected territories.

The first thing undertaken within the framework of the research was choosing the methodology through which it was possible to turn to an assessment of the living standards in the conditions of the present ecological and pandemic situation.

After reviewing the literature on the evaluation of living standards, the authors concluded that the Gini coefficient was the most interesting tool at that stage.

The first step was to derive the per capita income using the formula:

$$\overline{X} = \frac{\sum X_i f_i}{\sum f_i}.$$

Obviously, the calculations made the picture more informative. In particular, Murmansk Region has the best result in terms of per capita income in relation to the other areas. However, it is impossible to conclude only on the basis of this calculation; it is necessary to calculate the median income that shows the difference between the low-income and the high-income groups of the population using the formula

$$Me = X_0 + i * \frac{0,5 * \sum f_i - S_{Me-1}}{f_{Me}},$$

where x_0 – the lower limit of the median interval;
i – the value of the median interval;
f_{Me} – the frequency of the median interval;
S_{Me-1} – the cumulative frequency of the intervals preceding the median interval;
$\sum f_i$ – sum of frequencies.

Table 2. Indicators of per capita and median income in Murmansk Region, Republic of Karelia and Archangelsk Region in 2016–2020 [7–13].

	2016	2017	2018	2019	2020
Murmansk Region					
Per capita income – rubles a month	37359	39273	41564	44237	45996
Median income – rubles a month	29895.9	31586.7	33644.7	36481.8	38273.5
Republic of Karelia					
Per capita income – rubles a month	26247.0	27472.7	29149.6	30853.8	32124.1
Median income – rubles a month	21797.7	22771.1	24054.0	25476.1	26536.5
Archangelsk Region					
Per capita income – rubles a month	31394.3	32309.8	33830.5	35693.1	36572.6
Median income – rubles a month	24990.0	25349.1	26447.8	27970.4	29298.1

According to the data in Table 2, one can state that Murmansk Region (per capita income – 45996 rubles a month) has been in the lead for five years compared with the other specified regions; Archangelsk Region comes second and the Republic of Karelia ranks third. It lags behind its nearest partner by 13871.9 rubles. The situation is a little different, as far as the median income is concerned. The first position used to change over the 5 years. Before 2018 it was occupied by Archangelsk Region, after 2018 – by Murmansk Region; the same situation is with the second position: before 2018 Murmansk Region ranked second, after 2018 – Archangelsk Region, the Republic of Karelia was the third. The monthly median income is 26536.5 rubles, which is by 1,737.0 lower.

In addition to the median income, it is very important to estimate the decile group in the analysis, which demonstrates the state of the low-income group in terms of the upper and lower bounds of its earnings. As shown by the calculation materials, the first and the ninth decile characterise different income boundaries. The first decile demonstrates the upper income bound of the low-income group, being calculated by the formula:

$$D_1 = X_0 + i * \frac{0,1 * \sum f_i - S_{D1-1}}{f_{D1}},$$

where x_0 – the lower boundary of the interval containing the first decile;
i – the value of the interval containing the first decile;
f_{D1} – the frequency of the interval containing the first decile;

S_{D1-1} – the cumulative frequency of the intervals preceding the interval containing the first decile;

The ninth decile demonstrates the lower bound of the low-income group, being calculated according to the formula:

$$D_9 = X_0 + i * \frac{0,1 * \sum f_i - S_{D9-1}}{f_{D9}},$$

where x_0 – the lower limit of the interval containing the ninth decile;

i – the value of the interval containing the ninth decile;

f_{D9} – the frequency of the interval containing the ninth decile;

S_{D9-1} – the cumulative frequency of the intervals preceding the interval containing the ninth decile;

(Table with calculated indices). Finally, it was decided to use two more coefficients, the decile coefficient of income differentiation:

$$k_D = \frac{D_9}{D_1},$$

where D_9 – the ninth decile;

D_1 – the first decile;

and the Gini coefficient as such

$$G = 1 - 2 * \sum x_i Cumy_i + \sum x_i y_i;$$

where x_i – the share of population belonging to the i-th social group in the total population;

y_i – the share of income concentrated with the i-th social group;

$Cumy_i$ – cumulative income share.

Table 3. The Gini coefficients over 5 years for 3 constituent entities of the Russian Federation in the years 2016–2020 [14].

	2016	2017	2018	2019	2020
Murmansk Region					
The Gini coefficient	0.363	0.357	0.348	0.342	0.334
Republic of Karelia					
The Gini coefficient	0.336	0.338	0.341	0.341	0.34
Archangelsk Region					
The Gini coefficient	0.369	0.379	0.382	0.38	0.364

The authors had an opportunity to construct the Lorenz curve on the basis of the last calculated Gini coefficient, which is a graphical representation of differentiation of people's incomes in the given regions and which consequently characterises the format of management of the region's marketing potential. The calculations are presented in the chart for the three regions (Figs. 1, 2, 3, 4, 5, and 6).

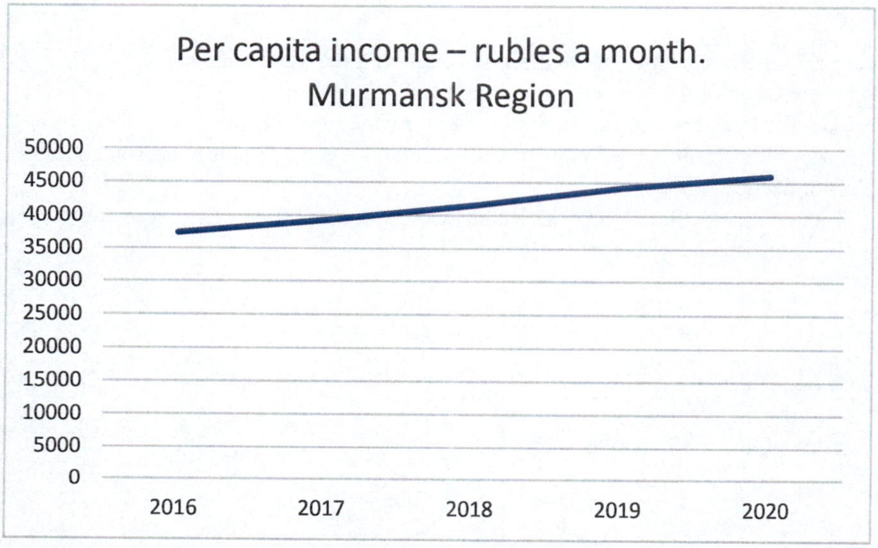

Fig. 1. Per capita income in Murmansk Region over 5 years.

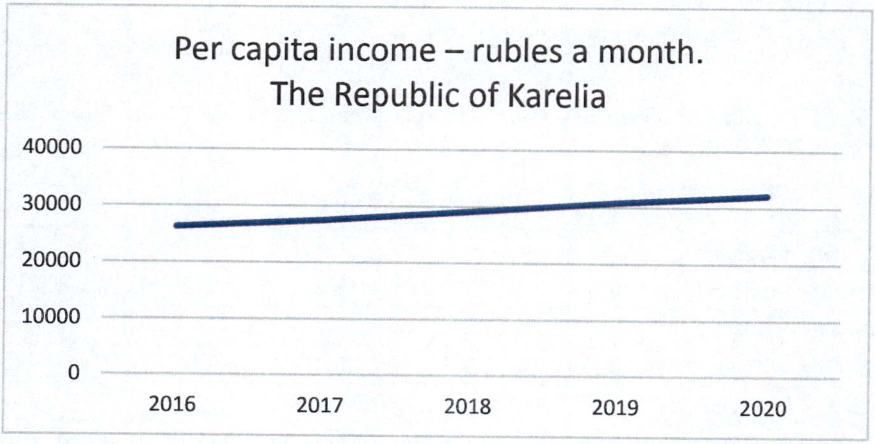

Fig. 2. Per capita income in the Republic of Karelia over 5 years.

The research supposes that the authors had to choose under what conditions the analysis of the quality of life judged by income assumes the analysis of income concentration. The presented calculations and the constructed charts make it possible to assert

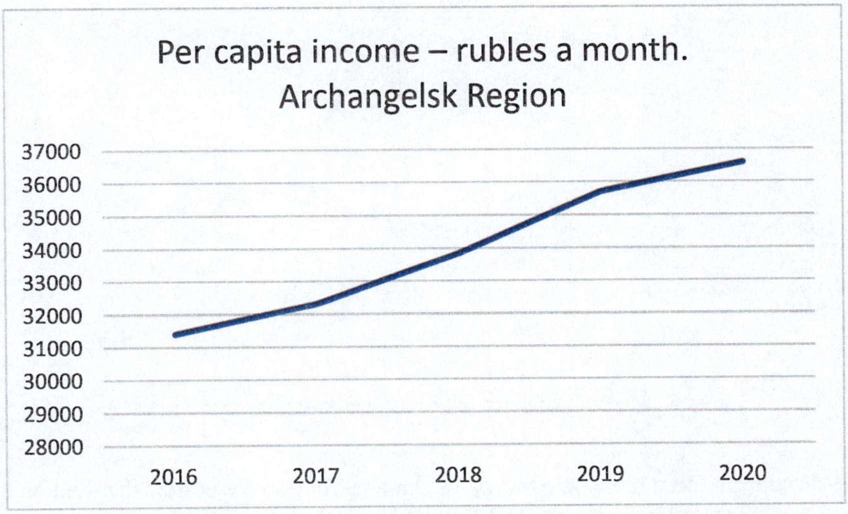

Fig. 3. Per capita income in Archangelsk Region over 5 years.

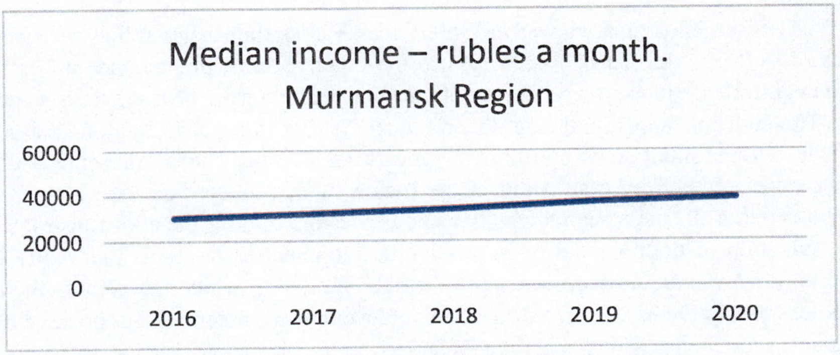

Fig. 4. Median income in Murmansk Region over 5 years.

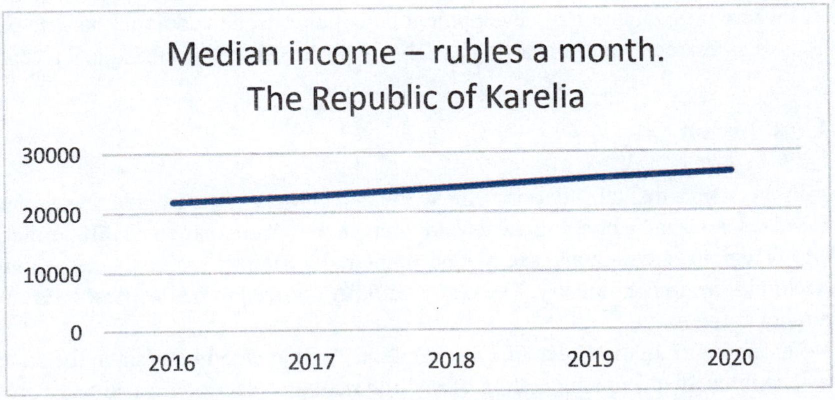

Fig. 5. Median income in the Republic of Karelia over 5 years.

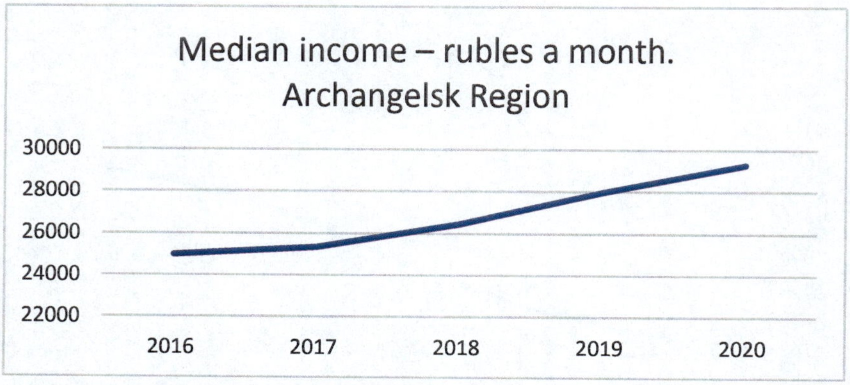

Fig. 6. Median income in the Archangelsk Region over 5 years.

that the concentration of income in a particular area follows the normal distribution law and the Gini coefficient is close to 0. This enables the authors to state that the three areas have different income differentiation patterns and, respectively, different developmental potential.

The Gini coefficient, as viewed in Table 3, makes it possible to assert that the position is close to 0 for Murmansk Region. As to the other regions, one can see a different income differentiation and discrepancy of the obtained data with the normal distribution law. The above data enable the authors to conclude that Murmansk Region shows the most favourable situation with its high income differentiation. The last presented results create an extremely tense situation in the region, which hinders the development of the potential. Thus the research results make it possible to offer the above instruments for evaluation of the potential of innovative development as the tools that really can be used to assess the development pace format. Naturally, we do not exclude the use of other tools in the assessment of innovative development potential, but presently the conducted research might enable the scholars to use the offered values both to assess the development potential and for due development of their prognostication. This does not exclude the possibility of extending the constraints of use of this methodology in other spheres where the evaluation of development potential is used in conjunction with other methods of assessment and forecasting which will be the subject of subsequent research.

4 Conclusion

The current environmental situation in the world and in Russia, in particular, necessitates advanced research and technological development in the sphere of income differentiation in various regions, creation and use of environmentally friendly materials, new production technologies in the industry. The Gini coefficient proved to be the most interesting instrument at this stage.

No groundwork in the given area has been duly undertaken in Russia in the studied areas. The intensified implementation of environmental protection measures by organisations requires the development and practical application of mechanisms intended to draw ecological investments in technologies and infrastructure. The authors addressed the currently realised projects and development strategies in the Russian Federation related to innovations and environmental safety, as well as the activities to stimulate the realisation of environmental protection measures at the federal and regional levels.

References

1. Liu, Y., Zhu, J., Li, E.Y., Meng, Z., Song, Y.: (2020)
2. Ouyang, X., Li, Q., Du, K.: (2020)
3. Liu, J., Xie, J.: (2020)
4. Tsenina, T., Tsenina, E., Tsou, T.: State regulation of logistics processes in accordance with the concept of sustainable development based on digital technologies. In: Proceedings of the International Conference on Digital Technologies in Logistics and Infrastructure (ICDTLI 2019) (2019). https://doi.org/10.2991/icdtli-19.2019.39
5. Russia News. https://covid-stat.com/en/country/rossiya/
6. Chen, N., et al.: Epidemiological and clinical characteristics of 99 cases of 2019 novel coronavirus pneumonia in Wuhan, China: a descriptive study. The Lancet **395**(10223), 507–513 (2020). https://doi.org/10.1016/S0140-6736(20)30211-7
7. Annual data on the regions of Russia. Standard of living of the population. https://regstat.rea.ru/indicator/parent/6754/data
8. The list of the most important indicators of socio-economic development of the Murmansk region. https://gov-murman.ru/region/ser_mo/
9. The average, median and modal level of monetary income of the population as a whole in Russia and in the subjects of the Russian federation for 2016. https://gks.ru/free_doc/new_site/population/bednost/tabl/1-2-6_2016.doc
10. The average, median and modal level of monetary income of the population as a whole in Russia and in the subjects of the Russian federation for 2017. https://gks.ru/free_doc/new_site/population/bednost/tabl/1-2-6_2017.doc
11. The average, median and modal level of monetary income of the population as a whole in Russia and in the subjects of the Russian federation for 2018. https://gks.ru/free_doc/new_site/population/bednost/tabl/1-2-6_2018.doc
12. The average, median and modal level of monetary income of the population as a whole in Russia and in the subjects of the Russian federation for 2019. https://gks.ru/free_doc/new_site/population/bednost/tabl/1-2-6_2019.doc
13. https://rosstat.gov.ru/storage/mediabank/me_1_2013-2020.doc
14. Gini coefficient (Income concentration index). https://www.fedstat.ru/indicator/31165
15. Danko, T.P., et al.: Marketing management of potentials. In: Danko, T.P. (ed.). Regions, Cities, Companies: For University Students of the Academic Field "Economics" and "Management". UNITY-DANA, Moscow (2018)

Digitalization and Industry 4.0-Changes Caused by COVID-19

Mira Sh. Rysalieva[1], Valentina B. Dzobelova[2](✉), Sergey L. Yablochnikov[3], Alisa V. Olisaeva[2], and Alyona E. Salamova[2]

[1] Kyrgyz-Russian Slavic University named after B.N. Yeltsin, Bishkek, Kyrgyzstan
[2] North-Ossetian State University named after K. L. Khetagurov, Vladikavkaz, Russia
[3] Moscow Technical University of Communications and Informatics, Moscow, Russia

Abstract. Today the world community functions under the conditions of the Fourth Industrial Revolution implementation, one of the main factors of which is global digitalization. The rapidly increasing role of digitalization creates new conditions for the functioning and formation of socio-economic all over the world. The main attention in the implementation of the above processes is focused on the fullest satisfaction of the constantly growing needs of citizens in receiving various types of information and information services. This is the goal of business, as well as national and municipal authorities in most countries. However, in the past two years, a number of trends have emerged in the formation of the global information space and the evolution of the information services market. These trends are due to the COVID-2019 pandemic. In the article, the authors investigate such formed trends in Russia and the world, based on the analysis of available statistical data. Conclusions are made regarding the possible scenarios for the further course of events.

Keywords: Digitalization · Industry 4.0 · Covid-19 · Digital technologies · E-Commerce · Global integration

1 Introduction

Digitalization is one of the factors of the Fourth Industrial Revolution realization. Its specific trigger is the cognitive activity of mankind, which created a significant intellectual and technological potential by the beginning of the 21st century. This, in turn, made it possible to make qualitative changes not only in the economy but also in social relations, in mentality, in a person's perception of reality. Once, the Third Industrial Revolution provided people with huge energy resources, expanding the possibilities for the production of a wide range of various goods and services. Current innovations are reorienting production, first of all, to the virtual space, aiming at meeting the constantly growing human need for information and information services, knowledge, improvement of intelligence and cognitive activity. In this regard, the study of various aspects of the realization of the universal digitalization concept and the implementation of the Industry 4.0 is a very urgent task.

P. V. Trifonov and M. V. Charaeva (Eds.): DITEM 2021, LNNS 380, pp. 12–20, 2022.
https://doi.org/10.1007/978-3-030-94245-8_2

2 Materials and Methods

The Fourth Industrial Revolution is focused specifically on man, as a creative person, capable of changing reality purposefully, as well as influencing actively historical processes due to a significant expansion of his opportunities. However, a threat that could create global problems on the way of implementing technological and social reforms, in particular, and the evolution of society as a whole, arose in 2019 for man and humanity as a whole. The aim of the article is the study of trends in the socio-economic sphere caused by the COVID-19 pandemic, based on the analysis of available statistical data. The methodology of this study assumes the determination of possible scenarios for the further course of events based on the above-mentioned statistical analysis of data on the economy and society functioning, as well as the forecasting.

3 Results

The pandemic, Industry 4.0, and global digitalization have influenced the transformation of a set of processes in the socio-economic sphere of all countries of the world, correlating with most spheres of human activity in economically developed and developing countries, undergoing significant changes, adapting to new conditions. Individuals and legal entities are forced to organize remote work of employees using hardware and software of modern information and telecommunications technologies. Companies that invest in the development of digital technologies and have covered the costs incurred for the introduction of innovations have a significant increase in the net profit growth rate by 17%, and companies have more time to develop and apply innovations due to the digitalization of business processes. Such data was confirmed by 84% of companies [1].

Researchers at Capgemini Consulting, conducted an analysis of the value of digital business for enterprises, based on data from more than 400 large companies operating in various fields. According to the results of the conducted research, analysts concluded that the economic indicators of the enterprise are directly dependent on the introduction of the latest technologies and management methods: – companies that have not chosen a new development model, according to the results, have negative economic indicators compared to other market participants-minus 24%; – companies that pay attention to two areas – the introduction of the latest technologies and the application of new management methods, have profit indicators that are 26% higher than those of competing enterprises [2].

In the Russian Federation, the citizens spent a huge sum of 3.22 trillion rubles making online purchases of goods in 2020. This is almost 1.6 times more than in 2019. According to the data of the Association of Internet Trade Companies E-commerce amounted in general to 9.6% of the entire retail market in Russia in 2020 [3].

The ICT adoption index was chosen as a key factor reflecting the level of development of the information and communication technologies sphere [4]. Besides, this gave a significant impulse to the development of the material base for the widespread introduction of Internet technologies, the development of new and modernization of traditional hardware and software. According to Data Insight analysts, due to the COVID-19 pandemic, the growth of online trading in Russia will be on average at least 6% annually until 2024. An additional increase in income due to the coronavirus pandemic over these years will amount to 4.4 trillion rubles, with the total volume of the Russian market of 23.3 trillion rubles (Fig. 1).

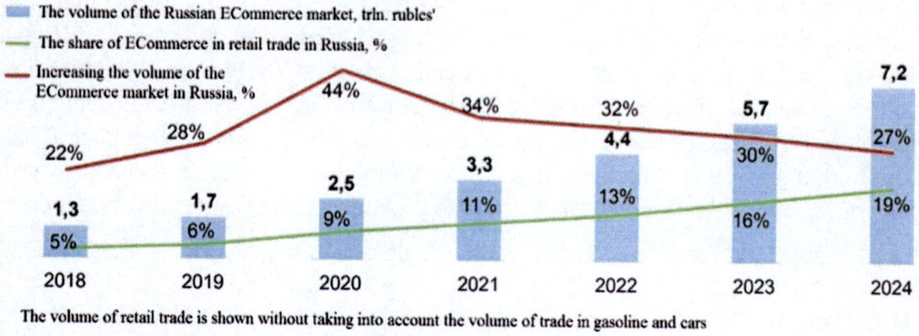

Fig. 1. Real increase of the volumes of the E-commerce market in Russia in 2018–2020 and forecast for 2022–2024.

Small and medium-sized businesses in Russia began to adopt successfully the experience of big market participants, actively moving to the Internet space and cooperating with logistics companies. Enterprises of this type in Europe make up 99% of the total number of all companies, 77% of them have their own web pages, but only 17% are engaged in online sales of goods [5] (Fig. 2).

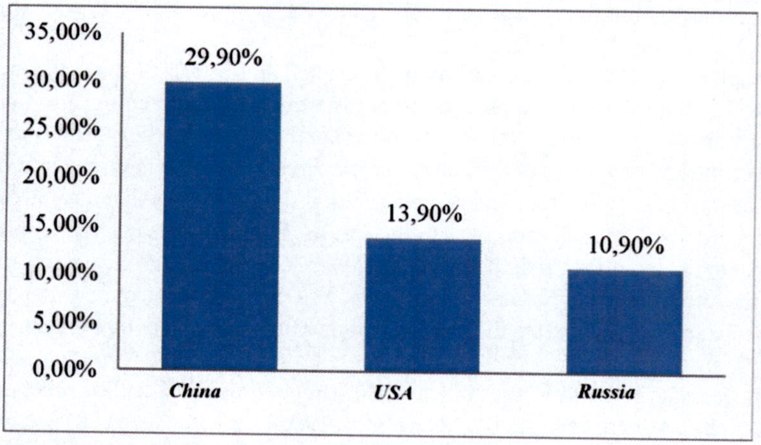

Fig. 2. The share of online sales of the main participants of the E-commerce market (data for June 2020).

The data given above indicate that more than half of the global Internet trading volume (54.7%) refer to China, the United States and Russia. At the same time, China is the leader in this sphere of business (its share of the total volume of commercial transactions carried out in the virtual space is 29.9%). The United States is in the second place (13.9%). In fact, the Russian Federation does not lag behind (10.9%). Herewith, the total volume of online sales amounted to more than US $19 billion [3].

From the beginning of COVID-19 pandemic, not only the digital transformation of society is being implemented, but also the rethinking of the essence of people's production activities. According to the OECD 2019–2020 research on employment prospects, the share of highly qualified workers in the OECD countries increased by about 9% from 1995 to 2018. The share of workers engaged in low-skilled labor also increased by 3% somewhat unexpectedly. At the same time, the share of average skilled labor decreased by 11% [6]. The trigger for Industry 4.0 was the acceleration of social processes due to communication, analytics and automation of operations. During this time, most enterprises have gone through three stages of this kind of transformation, namely: adjustment, restoration, resumption of work under post-crisis conditions. It should be noted that 92% of companies in the McKinsey survey already felt the need to update their business processes in connection with the changes caused by the digital transformation of the economy before the pandemic [7]. The faster the business structure overcomes the first two stages, the more sustainable and adaptive its business policy will be in the future.

The E-commerce market is mainly implemented in the form of two formats, such as online stores and trading platforms.

The dynamics of the global E-commerce market for the period from 2016 to 2020 is presented in Fig. 3.

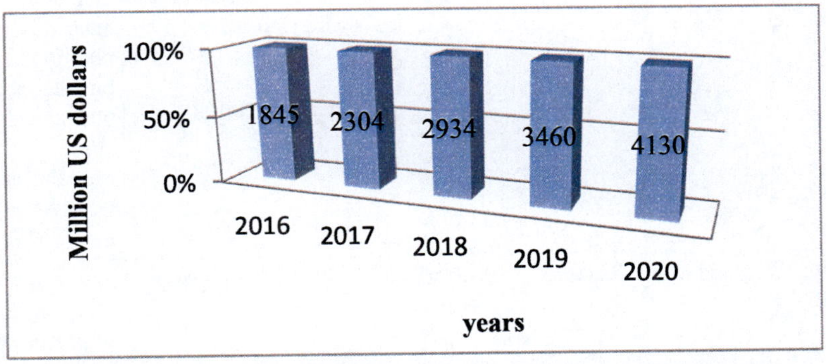

Fig. 3. Dynamics of the global E-commerce market from 2016 to 2020

At the beginning of 2020, demand for such services as WeChat, Zoom, Skype, Tencent, Cisco's Webex, Ding, and Microsoft Teams increased significantly. According to the report of the analytical company "Brand Finance", the leading positions are occupied by global companies that directly or indirectly interact with the digital technology sector. Among the leaders in this rating, companies such as Amazon, Apple, Google, Samsung, Facebook, Microsoft and so on can be distinguished. This fact, undoubtedly, can indicate that the modern world economy has set a course for digital transformation [8]. The use of these online platforms has led to an increasing demand for data storage and analysis. Due to this fact there has been a stable growth in demand for technology products from Microsoft, Amazon Web Services, Alibaba and Tencent. In just three months of 2020, the value of Amazon shares increased by 23.6%, the company set a new record of market capitalization – more than $1.23 trillion. According to Americans for Tax Fairness and the Institute for Policy Studies, representatives of key ICT corporations earned the most from the pandemic: the head of Amazon, J. Bezos – $34.6 billion, Facebook founder M. Zuckerberg – $25.3 billion, Microsoft creator B. Gates – $8 billion and Oracle co-founder L. Ellison – $7 billion [9].

The European Union and the Organization for Economic Cooperation and Development have initiated several digital projects for the further realization of technological perspectives (see Table 1 [10]). There is a stable trend regarding the transformation of modern information and communication technologies into an effective tool for struggling against the pandemic.

Table 1. Digital projects and solutions in the countries of the European Union and the Organization for Economic Cooperation and Development.

Project	Participating countries	Project promoter	Basic framework
Education and Training. Coronavirus: online learning resources	The world	The European Commission	List of online resources for teachers and students
COVID-19 Global Education Coalition	The world	Moodle in association with UNESCO	Online education platform that includes about 50.000 web sites. MoodleCloud has 1.67 million users
Remote-Work-and-Study-Resources	Switzerland	GitHub-HankiDesign	List of articles and plans for employees of companies, as well as for individuals; various tools and resources (in open access) for remote work and study

(*continued*)

Table 1. (*continued*)

Project	Participating countries	Project promoter	Basic framework
GNSS Apps for COVID-19 Response	The world	European Global Navigation Satellite Systems Agency	Applications to support public authorities in understanding the dynamics of the COVID-19 outbreak, as well as the support of citizens in their everyday life
COVIDSafe	Europe	Australian Government Department of Health	Services to prevent the spread of COVID-19. Using Bluetooth technology, the application helps identify people exposed to coronavirus
Open JOGL – Just One Giant Lab	France	Voluntary association	A mobilisation platform operating as a research and innovation laboratory. JOGL helps to solve many problems, as it cooperates with public services, companies and academic labs
AI-Robotics VS COVID-19	Europe	The European Commission	Online platform for collecting and storing information about technological solutions to face the pandemic, as well as ideas about Artificial Intelligence and Robotics

(*continued*)

Table 1. *(continued)*

Project	Participating countries	Project promoter	Basic framework
CoronaHack-Al VS COVID-19	Great Britain	KTN – Knowledge Transfer Network	A virtual platform for biomedical engineering scientists, analysts and entrepreneurs, whose activities are aimed at eliminating the pandemic
Telehealth Clinic during COVID-19	The world	Voluntary association	Online service for receiving medical care or consultation, under the conditions of impossibility of health care providing institution visiting. Video chats with medical staff or volunteers, who can help to resolve all issues

Despite the large number of projects being implemented by different countries and the intensive development of modern technologies, there is a significant dissonance in the field of access to such technologies. First of all, this problem is typical for the least developed countries, where, in particular, problems with the technological provision of information and communication infrastructure exist. This is especially true for education, where real opportunities for switching to a distance format are difficult, since about 65% of the population does not have access to global Internet resources at all. For example, in Indonesia, only 34% of citizens have the opportunity to implement online education, while in such countries as Poland, Lithuania, Norway, Switzerland, the Netherlands and others, this index is 95% [11].

4 Discussion

COVID-19 has significantly influenced the evolution of communications and telecommunications, as the density of information flows has significantly increased. The ICT sector is one of the most dynamically developing areas both in the world and in Russia. According to the analytical company Gartner, in 2019, the volume of the global information technology market was estimated at $3.74 trillion. The volume of the Russian market of information and communication technologies in 2019 reached 47.05 billion US dollars [12]. 5G technology has huge potential to stimulate significant socio-economic development. Real projects of using 5G networks in such industries as engineering, transport, logistics and others already exist. The global Digital IQ study for 2020 also confirms that companies investing in the development of digital technologies and covering the costs incurred for the introduction of innovations have a significant increase in the net profit growth rate by 17%, in addition, companies have more time to develop and apply innovations due to the digitalization of business processes. Such data was confirmed by 84% of companies [1]. The governments of most European countries have created favorable conditions for investment. Besides, a powerful technological and industrial base has been created on the European continent, which contributes to the active use of 5G networks [13].

5 Conclusion

The pandemic has impacted the concept of Industry 4.0 as a whole. The introduction of elements of artificial intelligence in the implementation of industrial and educational activities contributed to the efficient processing of large amounts of information, and the creation of powerful data transmission networks helped to ensure the transformation of many processes into virtual space. Active digital transformation in the context of a pandemic affected, first of all, those industries that ensure the safety and life of society as a whole, namely: online services of authorities, healthcare and education, the economy, finance, processing and storage of personal data, etc. In general, based on digital solutions and technological aspects of their implementation, the authorities of many countries over the past two years have radically revised their telecommunications policy, and this, in turn, has caused a surge of ideas and initiatives from large and small companies, public organizations and ordinary citizens of most countries in Europe and the world. According to the authors, all technological solutions developed and implemented during the COVID-19 pandemic will play a positive role in the future and will be relevant after several decades.

References

1. Department of International and Regional Cooperation of the AC RF. How the coronavirus is changing the work with technologies: the transition to "remote", robots, neural networks and cybersecurity. https://ach.gov.ru/news/kak-koronavirus-menyaet-rabotu-s-tekhnologiyami-perekhod-na-udalenku-roboty-neyroseti-i-kiberbezopas

2. New Research from Capgemini Consulting and MIT Sloan Management Review reveals. https://www.capgemini.com/news/new-research-fromcapgemini-consulting-and-mit-sloan-management-review-reveals-why
3. AECC. E-commerce Market 2020. https://akit.ru/rinok-internet-torgovli-2020/
4. Schwab, K.: World Economic Forum. The Global Competitiveness Report 2019 (2019)
5. IPG.RESEARCH. Global Development of E-Commerce, USA, China, Russia. https://ipg-estate.ru/issledovaniia/ipgresearch-globalnoe-razvitie-e-commerce-ssha-kitai-rossiia
6. OECD Employment Outlook 2020: Worker Security and the COVID-19 Crisis. OECD Publishing, Paris (2020). https://doi.org/10.1787/1686c758-en
7. The Digital-Led Recovery from COVID-19: Five Questions for CEOs. https://www.mckinsey.com/business-functions/mckinsey-digital/our-insights/thedigital-led-recovery-from-covid-19-five-questions-for-ceos
8. Brand Finance Global 500. Annual report on the world's most valuable brands. https://brandfinance.com/wp-content/uploads/1/brand_finance_global_500_2020_preview.pdf
9. Isaev, A.: Razvitie 5G development will seriously slow down due to COVID-19. https://1informer.com/internet/razvitie-5gseryozno-zamedlitsya-iz-za-covid-19-15523
10. HSE: Programme of Fundamental Research of the NRU HSE "Digital Agenda and Initiatives in the Field of Digital Technologies in the Context of COVID-19." NRU HSE, Moscow (2020)
11. OECD. Distance learning during school closure: how well are students and schools prepared (2020). https://www.oecd.org/coronavirus/
12. Gartner. Gartner IT Research. https://www.gartner.com/en/information-technology/research
13. Kuptsov, M.I., et al.: Modeling internet business optimization processes. In: International Conference on Engineering Management of Communication and Technology (EMCTECH), Vienna, 2020, pp. 1–5 (2020). doi:https://doi.org/10.1109/EMCTECH49634.2020.9261507

The Basis of Greening the Tax System of Russia

Nadezhda A. Zatsarnaya(✉)

Plekhanov Russian University of Economics, Moscow, Russia

Abstract. The need for greening the Russian tax system is increasingly determined by the vector of its modernization. However, this popular term has many variations of its interpretation. This article defines "greening" the tax system. In addition, in the course of the study, the author identified the functions and roles performed by environmental taxes for the purpose of greening. It is concluded that the greening of the tax system implies the subordination of its functioning to the goals of ensuring environmental safety. This can be achieved in different ways, including the introduction of the concept of "environmental tax" into the tax code, the transition from indirect to direct taxation, shifting the tax burden on activities that harm the environment, and making the imposed environmental taxes targeted. To write the article, the materials of the scientific works of domestic scientists were used, the analysis of the information obtained was carried out; modeling was carried out using mathematical and econometric methods. In order to rank and prioritize the functions of environmental taxes, a system of equations was drawn up that simulates the situation when taxes perform the function of a price in a market of imperfect competition. However, according to the results of the research carried out in the article, it was concluded that this theory is inconsistent: the transposition of the principles of the functioning of economic entities in the market of imperfect competition cannot be applied to the taxation system.

Keywords: Greening tax system · Functions of taxes · Tax elasticity coefficient

1 Introduction

The need for greening the Russian tax system is increasingly determined by the main vector of its modernization. This is due to the fact that the greening of the economic basis as a set of financial and economic relations is a multidimensional concept that affects the solution of economic, industrial, and social problems. At its core, greening consists of changing the approach to the analysis of the activities of economic entities. It is about focusing not on the characteristics of activity as a process, but on the effects that such activity and its results provoke.

There are several approaches to determining the nature of the interaction of economic and environmental parameters of development, common to which is the establishment of a direct mutual relationship between changes in economic and environmental parameters, based on the synergistic aspect, which is usually represented within the framework of the institutional theory of social development [1–8].

P. V. Trifonov and M. V. Charaeva (Eds.): DITEM 2021, LNNS 380, pp. 21–27, 2022.
https://doi.org/10.1007/978-3-030-94245-8_3

Although, based on the definition of the environmental tax [9], the fiscal function should not be the main one, but can be applied from the position of the tax administration in relation to conditionally non-renewable resources, the use of which cannot be stopped due to their importance for the stable functioning of social-economic systems of the state and society, according to a number of researchers [10–12], modern taxes, conventionally considered environmental in Russia, are only fiscal in nature, without performing compensatory or stimulating functions. In this regard, tax rates are often equated to the price for the right to use natural resources or for the right to exert a negative impact on them, and tax revenues for the state acquire the character of proceeds from the sale of goods and rights, the main goal being their maximization. The article analyzes the possibility of applying to the state tax system the principles of the functioning of monopolists in a market of imperfect competition, which, according to the analysis of information sources, the state appears at the moment in tax relations.

In addition, the objectives of this study are to define the category of "greening" of the tax system and designate the pyritization of the functions of environmental taxes.

2 Materials and Methods

To write the article, materials of scientific works of Russian scientists were used. For the purposes of the study, a number of various sources of information were studied, the analysis and synthesis of the information obtained were carried out, modeling was carried out using mathematical and econometric methods. In particular, the author has built an econometric model consisting of three equations, combined on the basis of applying the laws of the imperfect competition market to tax calculations.

The first equation of the model is a function of maximizing budget revenues of the state as a monopolist in the rights to own and dispose of public goods, which, among other things, include ecology and environmental parameters. The second equation of the model describes the dependence of the volume of environmental impact on changes in the rates of environmental taxes. The third equation describes the relationship for calculating the degree to which the propensity to cause environmental harm responds to a change in the environmental tax as the price of that opportunity. The described relationship is based on the principle that sources of pollution must bear the burden of the cost of pollution in order to reduce its consequences of an amount corresponding to the damage caused to society or exceeding the cost of eliminating pollution. The principle is based on the idea that natural objects are a public good, and everyone who harms them must pay compensation to society. Suppose that the burden of compensating for environmental damage expressed as a tax will fall on the polluter if the demand for the right to pollute is relatively elastic compared to the supply and if the supply in the form of allowing polluting is relatively inelastic compared with demand, the burden of compensation for environmental damage will fall on the state and society.

The elasticity of the tax system shows the relationship between tax rates and tax revenues to the budget. Thus, we obtain by what percentage the volume of demand for pollution, or the propensity to harm environmental safety, will change if the rate of the environmental tax on this impact changes by 1%. Since there is always an inverse relationship between the price (tax) and the volume of demand, the indicator $E_D^P < 0$

must be correct. If an increase in the tax by 1% leads to a greater reduction in pollution ($E_D^P < -1$), then it can be considered elastic, in this case, the negative technogenic impact on environmental safety will decrease, an increase in the tax will lead to a sharp reduction in the tax base, and budget revenues in parts of environmental taxation will decrease, which means that the burden of eliminating environmental problems will fall on the state and society, that is, compensation will be carried out at the expense of other non-environmental taxes. If it turns out that $E_D^P > -1$, then a tax increase will lead to an increase in budget revenues since the increase in the tax will exceed the reduction in the tax base, and the change in the tax will lead to a smaller change in the volume of environmental impact, in which case the principle "polluter pays" can be fully implemented.

Note that in the market of imperfect competition for a monopolist, maximization of its revenue occurs if the elasticity index E = 1. This state corresponds to the Laffer point – the extremum of the parabola (Laffer curve), which occurs at a tax rate at which tax revenues to the budget reach a maximum [13].

3 Results

The greening of the tax system in Russia appears to us as a transformation process within the framework of changing the functions of taxes present in the system, for the implementation of priority goals through the impact both on the entire financial and economic system of the state, and on the activities of its individual subjects, realizing their economic interests within its individual sectors, by encouraging economic entities to level or completely eliminate environmental stress that impedes the smooth functioning and expanded reproduction of financial and economic processes.

This means that environmental taxes should serve as a tool for ensuring sustainable environmental and economic development of the country, which is a long-term priority task, the solution of which can be achieved if environmental taxes present in the tax system meet not only formal criteria, fulfilling the function of filling the budget, but also perform:

- Preventive function – the impact of the power of tax instruments on the behavior of economic agents, aimed at preventing their environmentally unsafe behavior;
- Compensatory function – the restoration of environmental characteristics, including replenishment of the resource base and optimization of natural conditions in which the implementation of economic activities is carried out, by channeling funds accumulated from the proceeds of the environmental tax for these purposes;
- Stimulating function – the impact of the tax burden on economic processes and the incentive to use environmental, resource-saving technologies, the use of which in the process of economic activity costs the taxpayer cheaper than paying taxes;
- Accounting and information function – for collecting and analyzing statistical information in order to determine the indicator of environmental safety for the analyzed region, making a forecast of its development and plans for leveling negative trends;
- Ideological function – the tax is a reflection of the priority goals and directions of development of the state;

- Punitive function – the punishment of persons guilty of causing harm to environmental safety is expressed in imposing a tax burden on them, and the tax, in this case, acts as a price for the right to violate environmental safety.

Based on the foregoing, a system was compiled that simulates the relationship of individual indicators of the greening process, and consists of three equations:

$$\mathrm{TR}' = \mathrm{MR} = \frac{\Delta TR}{\Delta Q} = \frac{TR_1 - TR_0}{Q_1 - Q_0} = \frac{P_1{*}Q_1 - P_0{*}Q_0}{Q_1 - Q_0} = 0 \tag{1}$$

$$Q_1 = \left(\left(\frac{P_1}{P_0} - 1\right) * \mathrm{E} + 1\right) * Q_0 = \left(\frac{P_1{*}E}{P_0} - \mathrm{E} + 1\right) * Q_0 = \frac{P_1{*}E{*}Q_0}{P_0} - \mathrm{E} * Q_0 + Q_0 \tag{2}$$

$$E_D^P = \frac{\frac{\Delta Q}{Q_{cp}^D} \cdot 100\%}{\frac{\Delta P}{P_{cp}} \cdot 100\%} = \frac{Q_1 - Q_0}{\frac{Q_1 + Q_0}{2}} * \frac{\frac{P_1 + P_0}{2}}{P_1 - P_0} = \frac{Q_1 - Q_0}{Q_1 + Q_0} * \frac{P_1 + P_0}{P_1 - P_0} \tag{3}$$

where TR is the total tax revenue of the budget,

MR is the derivative of the TR function, the marginal total budget revenue,

P_0 – existing tax rates (known variable),

P_1 – optimal tax rates at which the negative impact on the environment is reduced (unknown variable),

Q_0 – existing volumes of environmental impact (known variable),

Q_1 – optimally minimized volumes of impact on the environment – volumes of impact at which the environment does not cause irreparable damage (unknown variable).

Note that for each of the taxes the indicators TR and MR should be calculated separately, as a separate item of budget revenues and can be summed up only as the final financial result in monetary form, and maximization of budget revenue is possible if the marginal revenue of the monopolist is zero.

By means of mathematical transformations described in Eqs. (1) and (2), we obtain a new system of Eqs. (4) and (5), respectively:

$$P_1 * Q_1 - P_0 * Q_0 = 0 \tag{4}$$

$$Q_1 = \frac{P_1 * E * Q_0}{P_0} - E * Q_0 + Q_0 \tag{5}$$

Thus, expressing the unknown variable P1 in terms of the unknown variable Q1, we obtain the Eq. (6):

$$P_1 = \frac{P_0 * Q_0}{\frac{P_1{*}E{*}Q_0}{P_0} - E * Q_0 + Q_0} = \frac{P_0}{\frac{P_1{*}E}{P_0} - E + 1} \tag{6}$$

The Eq. (6) we obtained can be transformed into a quadratic Eq. (7), which can be solved as an algebraic equation of the general form $ax^2 + bx + c = 0$ through the discriminant:

$$\frac{E}{P_0} - E * P_1^2 + (1 - E) * P_1 - P_0 = 0 \tag{7}$$

$$D = b^2 - 4 * a * c \quad (8)$$

where D > 0 – 2 solutions: $x_{1,2} = \frac{-b \pm \sqrt{D}}{2a}$

D = 0 – 1 solution: $x = \frac{-b}{2a}$

D < 0 – no solutions.

Thus, the solution of Eq. (7) by mathematical transformations through the discriminant (9) acquires the solutions presented in formulas:

$$D = (1 - E)^2 + 4 * \frac{E}{P_0} * P_0 = 1 - 2E + E^2 + 4E = E^2 + 2E + 1 = (E + 1)^2 \quad (9)$$

$$P_{11,2} = \frac{E - 1 \pm (E + 1)}{2 * \frac{E}{P_0}} \quad (10)$$

$$P_{11} = \frac{(E - 1 + E + 1) * P_0}{2E} = P_0 \quad (11)$$

$$P_{12} = \frac{(E - 1 - E - 1) * P_0}{2E} = \frac{-2P_0}{2E} = \frac{-P_0}{E} \quad (12)$$

This means that the system consisting of Eqs. (1–3) has solutions described in formulas (11–14).

If $P_1 = P_0$, then

$$Q_1 = \frac{P_0 * E * Q_0}{P_0} - E * Q_0 + Q_0 = Q_0 \quad (13)$$

If $P_1 = \frac{-P_0}{E}$, then

$$Q_1 = \frac{\frac{-P_0}{E} * E * Q_0}{P_0} - E * Q_0 + Q_0 = \frac{-P_0 * Q_0}{P_0} - E * Q_0 + Q_0$$
$$= -Q_0 - E * Q_0 + Q_0 = -E * Q_0 \quad (14)$$

In this case, if the modulus of elasticity is equal to one, that is, E = 1 or E = −1, then the considered indicators will be equal either to their current known values or to their opposite values, which is impossible based on the economic meaning of the variables under consideration, since the values tax rates cannot be negative. From which it follows that the applied theory of transposition of the principles of functioning of economic entities in the market of imperfect competition is not applicable to the taxation system.

4 Discussion

Earlier, a number of works were published on the issue considered in the study, including ideas that are similar in their essence but focusing on different aspects of the issue.

In particular, relying on the principles of an environmentally friendly economy, based on the understanding of the impossibility of infinite expansion of the sphere of influence in conditions of limited space, limited resources, the relationship of everything with everything, it is possible to define the greening of the economy as "a set of relationships between the state, society, and enterprises arising from ensuring sustainable harmonization of the interests of the economy and the environment and aimed at reducing the socio-economic costs of economic activity by ensuring the elimination of the negative consequences of the intensification of the use of natural resources" [14].

Another version of the interpretation of this term is the definition, which consists in maximizing the results obtained while maintaining the ecological balance in the environment and preventing its pollution on the basis of the development and implementation of low-waste, environmental, energy, and resource-saving technologies [15].

The concept of greening the tax system in Russian theory is presented by researchers in accordance with the ideas of Benoit Bosquet as the implementation of "environmental" reforms in taxation through taxation of rent received for the right to use natural resources and through taxation of environmental pollution (shifting the tax burden from the tax on personal income and corporate income tax on payments for the emission of pollutants) [16]. In this aspect, the greening of the tax system is often interpreted by modern Russian researchers as:

- Introduction to the tax code of the concept of "environmental tax";
- The transition from indirect to direct taxation;
- Shifting the tax burden on activities that harm the environment;
- Giving targeted environmental taxes to be introduced [17, 18].

Agreeing with the outlined postulates, an author's definition was created that describes the process of greening, taking into account the functions and roles performed by taxes.

5 Conclusion

The study provides an interpretation of the concept of greening the tax system, defines the functions performed by environmental taxes. In order to rank and prioritize the functions of environmental taxes, a system of equations has been drawn up that simulates the situation when taxes perform the function of a price in the market of imperfect competition. It is concluded that the greening of the tax system implies the subordination of its functioning to the goals of ensuring environmental safety. This can be achieved in different ways, including the introduction of the concept of "environmental tax" into the tax code, the transition from indirect to direct taxation, shifting the tax burden on activities that harm the environment, making the imposed environmental taxes targeted. Environmental taxes should meet not only formal criteria, fulfilling a fiscal function. Taxes, conventionally considered environmental in Russia, in current realities are only of fiscal nature, as evidenced by a number of studies. In this regard, tax rates are often equated to the price for the right to use natural resources or for the right to negatively affect them. However, according to the results of the research carried out in the article, it was concluded that the transposition of the principles of the functioning of economic entities in the market of imperfect competition cannot be applied to the taxation system.

References

1. Birmamitova, A.A., Gubachikov, B.A., Dikinov, A.: Bul. Exp. Council **1**(8), 92–99 (2017)
2. Edenhofer, O., Dao, N.T.: J. Macroecon. **55**, 253–273 (2017). https://doi.org/10.1016/j.jmacro.2017.10.007
3. Freire-González, J., Ho, M.S.: Sustainability **10**(2), 501 (2018). https://doi.org/10.3390/su10020501
4. Mantaeva, E.I., Bataeva, B.S., Goldenova, V.S., Avadaeva, I.V.: Bul. AGTU Ser: Econ. **1**, 7–17 (2018). https://doi.org/10.24143/2073-5537-2018-1-7-17
5. Mantaeva, E.I., Goldenova, V.S., Slobodchikova, I.V.: Bul. VolSU. Ser 3: Econ. Ecol. **20**(3), 24–31 (2018). https://doi.org/10.15688/jvolsu3.2018.3.4
6. Plaksunova, T.A., Safronov, O.M., Chernyavskaya, Yu.N.: Bul. VSUIT **4**(74), 385–392 (2017). https://doi.org/10.20914/2310-1202-2017-4-385-392
7. Streimikiene, D., Butkienė, I.S., Zavadskas, E.K., Cavallaro, F.: Energies **11**(5), 1193 (2018). https://doi.org/10.3390/en11051193
8. Valentey, S.D., Dzhanaeva, N.G.: Stat. Econ. **5**, 61–70 (2017). https://doi.org/10.21686/2500-3925-2017-5-61-70
9. Zatsarnaya, N.A.: Fin. Credit **26**(5(797)), 1135–1150 (2020). https://doi.org/10.24891/fc.26.5.1135
10. Bondarenko, T.G., Prodanova, N.A., Zhdanova, O.A., Maksimova, T.P.: Sys. Rev. Pharm. **11**(12), 1495–1506 (2020)
11. Chernyavskaya, N.V., Kleiman, A.V.: Int. Acc. **8**(398), 38–50 (2016)
12. Yashalova, N.N.: Fin Credit **2**(626), 38–46 (2015)
13. Rassadin, B.I.: Service **15**(1), 67–79 (2021). https://doi.org/10.24411/2413-693X-2021-10106
14. Lavrov, V.N., Rychkov, A., Bashorina, O.V.: Bul. Ural. Inst. Econ. Manag. Law **2**(23), 48–54 (2013)
15. Golubev, V.V., Golubev, K.V., Shishkov, Y.I., Ershov, A.K.: Method of ecologization of technologies of agricultural complex. Russian patent. IPC C02F11/04 C12P5/00 C05F3/00 A23K10/0012 (2019)
16. Benoit, B.: Greening the Tax System in Russia. Russian University, Moscow (2001)
17. Altukhova-Nys, Y., Bascourret, J.-M., Ory, J.-F., Petitjean, J.-L.: Mesurer la compétitivité des exploitations agricoles familiales en transition. In: Colloque SFER – Compétitivité, Agriculture et Alimentation. Université de Reims Champagne-Ardenne (2017)
18. Mochalova, L.A., Grinenko, D.A.: Discussion **6**(91), 70–76 (2018). https://doi.org/10.24411/2077-7639-2018-10017

Implementation of Project Management in Russian Agriculture

Dmitry A. Loginov(✉) and Aleksander L. Nosov

Vyatka State Agrotechnological University, Kirov, Russia

Abstract. Agriculture is a complex and conservative branch of the economy. Nevertheless, advanced management solutions are also penetrating it. This also applies to project management. The transition to project management is justified due to the fact that it provides an increase in the efficiency of investment. From this position, the study of the features of project management in agriculture in Russia is an urgent scientific and applied problem. The authors of the study summarized the scientific material on the transition to project management in agriculture in Russia and concluded that the scientific support of this direction has not yet been developed sufficiently. Summarizing the practical experience of project management in the agro-industrial complex of the Kirov region, the authors described the features of the transition to project management in the agro-industrial complex and gave recommendations for improving the management of this process. In particular, it is proposed to strengthen the emphasis on the development of the innovative component in agricultural investment projects.

Keywords: Project management · Agriculture · Project management · Agro-industrial complex · Agricultural policy · Innovation · Investment

1 Introduction

Project management is a widely discussed topic in modern economic and managerial discussions. This is primarily due to the fact that the project approach contributes to a more responsible use and distribution of resources invested in a particular area of economic activity. In addition, the application of the project approach makes it possible to control in more detail the cash, resource and other material flow during the completion of the business plan to the final implementation. In this regard, the interest in project management is quite understandable both in theoretical and practical terms.

Agriculture is one of the most conservative branches of the economy, which is due to a number of reasons, including the peculiarities of the mentality of rural workers, the high level of dependence of the industry on natural, agro-climatic conditions, the duration of the production cycle, the huge level of capital intensity of the industry and many others. The conservative nature of the agricultural sector determines the slow course of changes in it. This also applies to the implementation of project management. Nevertheless, the transition to project management in agriculture is extremely necessary due to the fact that, along with most other branches of the national economy, agribusiness is in a tough

P. V. Trifonov and M. V. Charaeva (Eds.): DITEM 2021, LNNS 380, pp. 28–35, 2022.
https://doi.org/10.1007/978-3-030-94245-8_4

competitive environment, when the survival of economic units can only be ensured by the most efficient use of the entire resource base. Thus, it is necessary to implement project management in agriculture.

Currently, a certain interest in the implementation of project management in agriculture has emerged and is growing in scientific circles. This trend has not spared Russia either.

One of the earliest publications related to the study of the transition to project management in the agro-industrial complex of Russia was made in 2008 [1]. In this work, the expediency of the transition to project management in agriculture is justified, but it is indicated that this process will be quite long and costly, since most agricultural enterprises are not ready for such a transformation technically, technologically and financially.

Some researchers [2] emphasize that the implementation of the project approach in the agro-industrial complex of Russia is extremely difficult due to the insufficient availability of updating fixed assets. In particular, it is emphasized that most of the samples of advanced, progressive machinery and equipment for agricultural production are imported for Russian manufacturers. Accordingly, their acquisition is accompanied by a number of currency restrictions and risks [3, 4]. In such conditions, the acquisition of new fixed assets itself cannot be put on stream and remains, rather, a one-time, separate event, for which it makes no sense to organize a separate project support.

In the current conditions, some authors [5, 6] rely on the development of software for agricultural enterprises in the complex of proposed measures for the development of project management in agriculture. Sharing this idea and supporting the need to focus on the technological side of project management in agribusiness, the authors of this article believe that along with the information technology aspect, the motivational aspect of the transition to project management is also preserved, which is due to the lack of understanding of the feasibility of changes in management on the part of many agricultural entrepreneurs.

As Borisov rightly noted [7], project management is included as an integral part of the project management system and marks the transition from classical management to project management. At the same time, the key benefit from the transition to project management is an increase in the efficiency of the use of investment resources. These processes are important for agriculture. And as a way to achieve such a transition, Borisov suggests a more active use of information and communication technologies in the management of agricultural production at all levels (both microeconomic and macroeconomic). Moreover, the development of project management in the agro-industrial complex can ensure an increase in the efficiency of state support for the industry.

The presented problems and approaches to its solution are widely considered in the works of a number of Russian authors. An in-depth study was conducted on the transition to project management in agriculture based on the use of special information systems [8–10]. These authors emphasize that the transition to project management will not only increase the efficiency of agriculture, but will also help reduce risks in the industry (in particular, investment risks). The authors give recommendations on the use of specific information systems in the transition to project management in agriculture. At the same time, it is stated that the transition from classical management to project management in the agricultural sector will provide a breakthrough in attracting external investment resources to the agro-industrial complex. Indeed, project management, by increasing the transparency of monitoring resource flows, provides a higher level of investment attractiveness of the industry. This advantage means that it is expedient to activate the transition to project management in agriculture with the active support of the state. Thus, state support can be considered not only as a resource for the development of project management in agribusiness, but also as a factor in stimulating the transition to project management in the agricultural sector initiated by the state. In the case of successful implementation of project management in the agro-industrial complex, a higher level of efficiency in management, in making and implementing management decisions will be provided, which is extremely important at the present stage of economic and social development, when most management processes are undergoing significant acceleration. At the same time, it is emphasized that the pace of implementation of project management in agriculture strongly depends on the mentality of entrepreneurs and managers in agriculture.

It is possible to distinguish a separate line in research focused on the fact that project management in agriculture is closely related to the implementation of innovative projects [11, 12]. These authors emphasize a rather significant view on the prospects of the project approach in agriculture, the essence of which is that project management is significant only when (compared to classical management) the project has a significant innovative component. Further, the concept proposed by the authors moves on to the aspects of the implementation of the project approach in agriculture. In this plan, it is emphasized that the implementation of the project approach needs comprehensive state support with an emphasis on stimulating innovative processes. For this purpose, it is proposed to create, with the support of the state, a modern innovative infrastructure of agriculture that contributes to the generation and implementation of innovative ideas in real agricultural production. At the same time, it is proposed to transform the relations between the state and agriculture from the "science – production" scheme to the "science – innovative economy – production" scheme in the macroeconomic plan. Thus, the value of the presented point of view lies in the fact that it is proposed to concretize project management in agriculture methodologically to support innovative projects, which implies a transition to more advanced technologies for the production of agricultural products, and therefore an increase in the efficiency of agricultural production.

There is a number of proofs for this approach. In particular, the results of the analysis of the economic growth of individual foreign countries have been published, confirming that one of the leading factors of economic growth in the 21st century has become high-tech, innovative agriculture, in which innovative projects are implemented overwhelmingly on the basis of initially state-supported scientific and technical development carried out on the basis of one of the national universities [13–15]. Agreeing with the author of this thesis, we consider it necessary to emphasize that without a strong domestic technological basis, without breakthrough ideas and technical solutions created by compatriots, there will really be nothing to implement in Russian agriculture through a project approach, except for standard investment solutions for the purchase and installation of foreign equipment of well-known world brands. But, as practice shows, when implementing such projects, saving resource flows is not in the first place, and therefore project management remains in demand to a small extent.

It is important to emphasize that the project approach is often considered from the point of view of a way to increase the effectiveness of state support measures [16]. In addition, the implementation of the project philosophy in the state regulation of agricultural development should have a positive impact on the support of innovative processes in the agro-industrial complex [17]. Project management becomes a condition for the digital transformation of agricultural management [18] and provides an opportunity to optimize the flow of resources of the agro-industrial complex and agricultural products between regions [19]. It is noted [20] that project management can significantly improve the accuracy, effectiveness and efficiency of planning and forecasting in agriculture.

The purpose of this study is to identify the current state of project management in agriculture on the example of the Kirov region.

Research objectives: to give an overview of scientific publications on the research topic; to analyze the current state of project management in the agro-industrial complex on the example of the Kirov region; to give recommendations for the further development of project management in agriculture in Russia.

2 Materials and Methods

This study is based on a combination of a number of scientific methods. One of them is a monographic method aimed at collecting and summarizing scientific publications on the topic of research. The authors evaluated the collected scientific literature critically, from the standpoint of the research topic and their own professional experience, which made it possible to determine the key methodological directions for the development of project management in the agro-industrial complex of Russia. Another method implemented in research is the method of observation and generalization of one's own professional experience. Working in the field of agricultural science and agricultural education, the authors had the opportunity to observe in a long retrospective the development of agriculture throughout the country and especially in detail – the Kirov region. The participation of the authors in a number of regional meetings and other management events allowed the researchers to summarize extensive material on the features of the implementation of the project method in the management of agricultural production.

3 Results

The introduction of project management in the agro-industrial complex of the Kirov region began around 2002. It was during this period that the interest of private investors who had not previously invested in agriculture increased in placing part or all of their capital in agricultural entrepreneurship. Of interest were both agricultural enterprises and food industry enterprises, as well as economic entities that combine the production of agricultural raw materials and its industrial processing. In a sense, the presence of industrial processing became a kind of insurance for investors due to the fact that it was easier to sell ready-made food products and the risk of the manufacturer's dependence on the manipulation of the purchase price by the wholesale buyer (most often, large food industry enterprises) was excluded.

So the first agricultural holdings began to appear in the Kirov region. It should be noted that the processes of transition to project management were carried out both at the micro-level – in individual enterprises of the agro-industrial complex, and at the meso-level in the regional executive authority responsible for the development of agriculture. In particular, the regional Ministry of Agriculture and Food collected the experience of the best practices in the regions of Russia on the unification of agricultural enterprises and food industry enterprises into effective associations – agricultural firms. However, it should be noted that such an association without investing from the outside did not give any noticeable result in economic terms. Thus, during the implementation of project management, it is important to avoid such a risk as the risk of "changing signage". It is not so important which economic units are united into an agricultural firm, as it is important how the management is organized in the new association and how much investment is invested in the development of the economy.

In 2002–2005, the foundations of project management were laid in the agro-industrial complex of the Kirov region, as a result of which agricultural holdings were formed-associations of agricultural enterprises consisting of the main enterprise, which is the center for making managerial and investment decisions and a kind of management company of the entire holding and affiliated subsidiaries that lost their legal independence under various pretexts and became production units within the holding. In some cases, such farms remained nominally legal entities with the rights of a subsidiary, in other cases they lost the status of a legal entity, becoming a branch or subdivision of an agricultural holding.

The size of such agricultural holdings can be called unlimited within one region. The borders of the leading agricultural holdings quickly expanded beyond the boundaries of a separate municipality, and soon such holdings acquired a territorial intermunicipal character, sometimes covering almost the majority of municipalities in the region.

As criteria of project management in the formation of agricultural holdings in the Kirov region, such as achieving a sufficient degree of diversification of production and achieving the necessary minimum production volume corresponding to the break-even point were used.

Starting from 2005–2006, the activation of project management at the meso-level began, associated with the transition to the implementation of the corresponding sectoral national project. It was the national project for the development of the agro-industrial complex and state support within its framework that extended the project approach to most agricultural enterprises in the region, and not only to agricultural holdings. In 2005–2015, there was a layer of independent agricultural enterprises of the Kirov region that functioned within the historically established scale of activity, but reached a higher level of efficiency due to participation in the national project for the development of the agro-industrial complex.

From 2016 to the present, the newest stage of the development of project management in the agro-industrial complex of the Kirov region is being implemented, connected with the formation of independent regional goals for the development of the agricultural industry. In particular, the leadership of the region has set a goal to take a leading position in milk production in the Volga Federal District. Such a strategic choice made it possible to focus the funds of the state regional support of the agro-industrial complex on the relevant areas related to stimulating milk production, and, indeed, allowed the region to reach new, higher positions in milk production in the Volga Federal District.

In parallel, within this period, project management has spread from large and medium-sized forms of agricultural production to small forms. Farming has been actively supported in the region, including through such rather "generous" forms of state support as grants (gratuitous targeted financing).

Thus, it can be stated that by 2021, project management has spread quite widely in the agro-industrial complex of the Kirov region, covering both all forms of agricultural production according to the scale criterion (large, medium and small production), and by specialization (agricultural production, industrial processing of agricultural raw materials), and by territory (design solutions in the agro-industrial complex are implemented both in the suburbs of the administrative center of the Kirov region, the city of Kirov, and in the peripheral agricultural territories of the region).

4 Discussion

It should be noted that the solutions presented in the scientific literature are generally consistent with the practical experience achieved in the Kirov region on the implementation of project management in agricultural activities.

However, there are distinctive features that need to be noted.

Firstly, the agro-industrial complex of the Kirov region did not use any template solutions for the use of information and communication support for the implementation of investment projects. The issue of information and communication support remained at the micro-level and was solved tactically by each individual farm. This feature did not become an obstacle to the implementation of the project approach in the agro-industrial complex of the Kirov region.

Secondly, the information and consulting service of the region did not play a key role in the implementation of the project approach. The consulting was conducted mainly in the free market space. This also did not prevent the implementation of the entrepreneurial function when investing in agriculture in the region.

Thirdly, there is no close connection between investment projects implemented in the region and advanced innovative solutions. The innovativeness of most of the implemented investment projects in the region is reduced to the fact that the technologies and equipment introduced into production within the framework of the projects are advanced, but already mastered in world practice. This means that the agro-industrial complex of the region adheres mainly to a catch-up strategy in investment development, but does not seek to become an innovative leader.

Fourth, the project approach is focused on those branches of agriculture in the region that are most profitable for the agro-climatic zone of the Kirov region. These are, first of all, traditional branches of animal husbandry.

5 Conclusion

Thus, it can be stated that the transition from traditional to project management in the agro-industrial complex of the Kirov region has taken place. Taking into account the fact that the Kirov Region is a typical agricultural region for its agro-climatic belt, this conclusion (taking into account reasonable reservations) can be extended to agriculture in Russia as a whole.

The state needs to focus now on the transition from simple project management to innovative project management, in every possible way stimulating the processes of activating agricultural science and introducing advanced technological solutions into practice. It is necessary to move away from the strategy of catching up development in agriculture, focused on the mastered foreign technologies, and move to the strategy of industry leadership based on the progress of agar science.

The presented research may be of interest both for the executive authorities responsible for agricultural development at the federal level, and for the relevant bodies at the level of the constituent entities of the Russian Federation.

References

1. Kuprava, T.A.: Bul. Peop. Friendship Univ. Rus. Ser: Agr. Anim. Husb. **3**, 14–19 (2008)
2. Retyakov, I.I., Tkachenko, V.V.: Economic characteristics of the geographical information system for monitoring the machine-tractor fleet in agriculture. In: VI All-Russian Scientific and Practical Conference of young scientists: Scientific support of the agro-industrial complex. Kuban State Agrarian University named after I. T. Trubilin, Krasnodar (2012)
3. Korshikova, M.V., Svistunova, I.G.: Krasnogorya Sci. **8**(3), 45–53 (2019). https://doi.org/10.12731/2070-7568-2019-3-45-53
4. Vecherin, D.S.: Legal regulation of project management as a mechanism for implementing the state program for the development of agriculture and regulating the markets of agricultural products, raw materials and food. In: Collection of Abstracts: Moscow Legal Forum Online 2020, pp. 174–179. Moscow State University of Technologies and Management named after K. G. Razumovsky (The First Cossack University), Moscow (2020)
5. Tkachenko, N.A.: Design and development of a software package for effective management of crop production processes. In: Collection of Articles Based on the Materials of the 71st Scientific and Practical Conference of Teachers on the Results of Research for 2015: Scientific Support of the Agro-industrial Complex. Kuban State Agrarian University, Krasnodar (2016)
6. Okhremenko, I.V., Kozlov, V.I., Prokopov, S.V.: Econ. Entrep. **96**(7), 300–305 (2018)
7. Borisov, N.S.: Fin. Mark. Banks **5**, 80–82 (2020)
8. Vorobyeva, M.A., Tkachenko, V.V., Tkachenko, N.A.: Polyth. Netw. El Sci. J. Kuban State Agr. Univ. **126**, 879–890 (2017). https://doi.org/10.21515/1990-4665-126-062
9. Surovitskaya, G.V., Maslova, I.V.: Innov. Sci. Edu. Bus. **1**, 6–12 (2018)
10. Kholodova, M.A.: Bul. Michurinsky State Agr. Univ. **62**(3), 138–144 (2020)
11. Kibirov, A.Ya., Petrov, S.E.: Agri-Food Pol. Rus. **65**(5), 24–28 (2017)
12. Muratova, E.A.: Int. Legal Courier **3**, 17–22 (2020)
13. Goncharenko, I.Yu.: Formation of an effective model of innovation management in the agricultural sector of the Russian economy. Abstract PhD Thesis in Economic Sciences. State University of Management, Moscow (2016)
14. Tebekin, A.V., Lomakin, O.E., Shafirov, V.G.: Transp. Bus. Rus. **1**, 3–10 (2018)
15. Panteleeva, T.A.: Econ. Entrep. **106**(5), 121–125 (2019)
16. Kholodova, M., Podgorskya, S.: Project management methods in agriculture. In: E3S Web of Conferences: Innovative Technologies in Science and Education. 8th Innovative Technologies in Science and Education, ITSE/2020 Rostov-on-Don, 19–30 August 2020 (2020). https://doi.org/10.1051/e3sconf/202021011007

17. Kyalturina, E.N., Urmuhametov, I.M., Smirnov, A.A., Kreneva, S.G., Vinokurov, I.S.: Methods of applying innovative development processes in agriculture. In: Proceedings of the 32nd International Business Information Management Association Conference, IBIMA 2018. Vision 2020: Sustainable Economic Development and Application of Innovation Management from Regional Expansion to Global Growth 32, Vision 2020: Sustainable Economic Development and Application of Innovation Management from Regional Expansion to Global Growth, Seville, 15–16 November 2018, pp. 7848–7854 (2018)
18. Kotsur, E.V., et al.: J. Phys. Conf. Ser **1399**, 033009 (2019). https://doi.org/10.1088/1742-6596/1399/3/033009
19. Solntseva, O.V., Yashina, M.L.: Solving the placement problem as an element of the information and analytical platform for digital agriculture. In: Bogoviz, A.V. (ed.) Complex Systems: Innovation and Sustainability in the Digital Age. SSDC, vol. 283, pp. 389–398. Springer, Cham (2021). https://doi.org/10.1007/978-3-030-58823-6_43
20. Gorlov, I.F., et al.: IOP Conf. Ser. Earth Environ. Sci. **677**, 032022 (2021). https://doi.org/10.1088/1755-1315/677/3/032022

Usage Feasibility of AI-Based Intellectual Instruments for In-Firm Planning

Fedor A. Zagumennov(✉), Andrey V. Bystrov, and Alexey G. Radaykin

Plekhanov Russian University of Economics, Moscow, Russia

Abstract. Artificial Intelligence features as instruments for in-firm and strategic planning are expanding to markets. This paper provides an overview of modern AI approaches. As AI tools, several instruments with detailed descriptions are listed in this paper. The algorithms allow a more flexible approach for in-firm processes forecasting. Such instruments are several in-firm tools using deep-neural networks as a flexible alternative to regression. They are intended for usage in demand forecasting and production batch size prediction. As a result, a prospective cross-industry SaaS-platform (software as a service) is provided as the most convenient way to implement and use such technologies. This paper provides calculations for such software, gives examples, and what role this software plays in modern business conditions. Integration of such methods can be expensive for customers. Costs and risk estimations of modern ERP and CRM systems are calculated. As a result, it is described why expediency of integration of such systems is substantiated.

Keywords: Justification · In-firm planning · Machine learning · Deep learning · Instruments · AI · Saas

1 Introduction

Modern digital transformation technologies are becoming more and more necessary as a new driver for business growth. There are many solutions, however, each one has its own features and drawbacks. The cost of implementation of such technologies in a particular company remains high. The integration of such systems, however, becomes necessary. The failure of selection of particular software leads to in-firm troubles and budget loss [1].

AI-based approaches in business software, as well as migration of processes to a cloud platform, is the next step of ERP and CRM software. However, it may lead to excessive expenses and incorrect decisions [2]. The risk potential of AI discrimination is still high and may lead to problems for in-firm planning [3, 4].

This paper provides an overview of AI technologies to be implemented and justifies its integration to a particular company in the electronic manufacturing industry. Such software can be combined with an ordinary ERP-based or CRM system. Such combines several approaches:

P. V. Trifonov and M. V. Charaeva (Eds.): DITEM 2021, LNNS 380, pp. 36–42, 2022.
https://doi.org/10.1007/978-3-030-94245-8_5

- Cloud software as a service system
- Ordinary ERP and CRM functions
- Cross-industry peer-to-peer business-to-business (B2B) marketplace functions, which allow companies to exchange illiquid assets, digital products and hardware documentation.
- Proposed artificial intelligence (AI) functions, realized with the help of convolutional and Long-Short term memory neural networks: demand forecasting, income, revenue forecasting and optimization of batch size.

In general, Software as a service differs from conventional applications, because it requires the user to subscribe to the service [5].

The market for these services is constantly growing and has experienced extremely exponential growth over the past many years except the last two or three years, where a trend towards stabilization of growth is visible (Fig. 1).

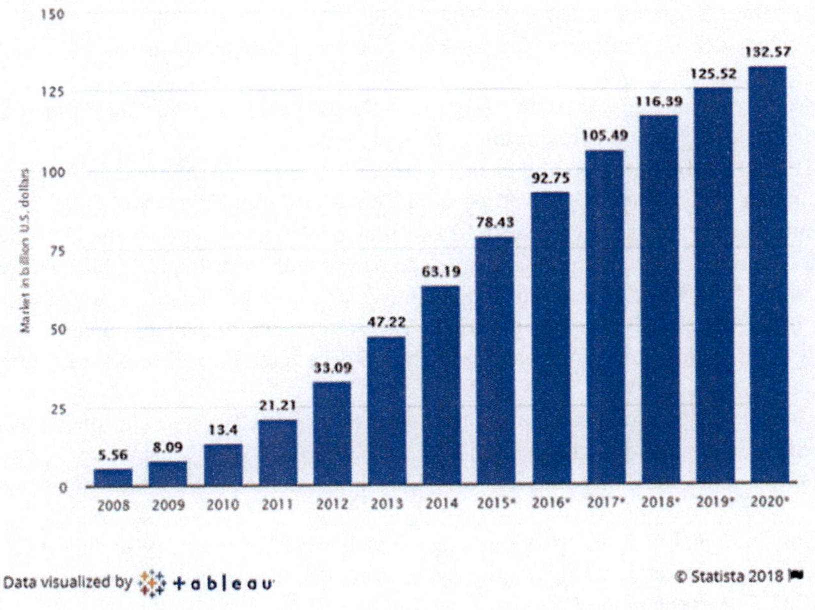

Fig. 1. SaaS applications market growth. *Source*: Statista 2018.

One of the definitions of SaaS is that it is:

- cloud-based;
- has the benefit of faster performance;
- provides access from multiple devices;
- provides subscription model.

SaaS is maintained by the development company and usually does not require updates on the client machine. SaaS applications are both business applications and software for ordinary users.

Once a month, a certain amount is debited from the card or corporate account, and access to this application is prolonged, which can both be a risk factor or a benefit in the case of integration justification.

2 Materials and Methods

The research conducted is devoted to describing the AI and non-AI systems for organizing the in-firm business processes. The goals and tasks of research are:

1. To determine whether it is necessary to implement the proposed instruments for in-firm business systems. Integration of in-firm business features helps to organize business processes. On the other hand, expenses of such integration may be valuable which makes overall integration non-feasible. The proposed research includes calculations of integration expenses.
2. To collect information about modern AI approaches for in-firm planning.
3. To propose new AI tools for in-firm planning.
4. To consider the SaaS approach for ERP systems and to justify its use.

We have gathered our data using publicly available sources and from 4 local companies using digital software for in-firm planning.

The obligation of integration of such software comes when the company grows, and the number of tasks increases disproportionally. An ERP implementation survey showed that mid-scale companies who earn $100 million to $250 million experienced the fastest implementation at just 6.6 months. Large companies with more than $25 billion revenue took more than 12.35 months to implement. Once the implementation was complete, 49% of the organizations stated that ERP improved the entire business process. As the result, 95% of companies found such integration beneficial for in-firm planning after all [6].

Three local enterprises have implemented several ERP systems in their business processes, while the remaining company.

For AI approaches there are 2 technologies implemented and tested:

1. Convolutional neural network connected to Long-Short memory predictor (CNN-LSTM) which is working as a regression component to forecast business processes while in-firm planning [7].
2. A method of batch size calculation, based on a statistical data.

Provided research is dedicated to the electronic manufacturing industry and expenses estimations may be limited for the particular tasks involved in a business process.

The definition and exploratory research design were used [8, 9] as the research methodology.

3 Results

As the result, there is Table 1 provided which presents data collected from the open statistical information about the implementation of intellectual instruments for in-firm business processes [10]. Similar data are reported by G2 Company [11].

Table 1. Justification of intellectual systems integration for in-firm business processes. *Source*: [12].

Benefits:	Value, % of respondents
Is beneficial for in-firm planning	95%
Improved business processes	49%
Speeds up operations	20–26%
Use of cloud ERPs	86%
Manufacturing industry	46%
Drawbacks:	Value, % of respondents
Poorly defined technical requirements	53%
High price of integration	66%
Too large time of implementation	72%

According to the same sources, the cost of integration is 6000–9000 USD, which can be compared to the annual revenue of a small innovative enterprise.

It is possible to calculate the justification coefficient for in-firm planning.

Taking into account that the target market for proposed features is manufacturing, we can construct a benefit coefficient as:

$$k_b = 46\% \cdot 86\% * 95\% = 37.6\%.$$

So, at least 19.4 respondents from the industry are using cloud ERP and are satisfied with the result.

Now we calculate a drawback coefficient:

$$k_d = 53\% * 66\% * 72\% = 25.2\%.$$

According to the provided information, the integration process is successful in most cases, but less than half of the respondents are satisfied with the results.

80% of developers believe that many ERP processes can be completely replaced by machine learning or AI. AI approaches in ERP, MES, CRM and other business software are in an expanding integration state and are intended to automate processes in such in-firm instruments. 44% of AI developers have already implemented AI to their workflows in ERP systems. Intelligent technology was only a part of 4% of ERP solutions. These technologies have already been implemented in several ERP-systems, in particular, in Microsoft ERP, SAP S/4 HANA, as well as, experimentally, in programs based on 1C-Enterprise.

Considering the experience of others, new algorithms for in-firm planning were constructed.

1. The production batch depends on the size of your company and your revenue. Therefore, it means, you cannot choose minimal or maximum possible batch size, and you should make a production profitability curve first. This task can be automated and released as a feature for in-firm planning. Figure 2 illustrates such an algorithm.
2. Component risk management is important. It depends on the industry for your company, but sometimes the risky components level may lower the production costs and increase your income
3. Generally, using linear or even cubic regression models is inefficient for forecasting business processes. Instead, we provide a CNN-LSTM approach [13, 14] which allows predicting several types of business data (income, revenue, product/stock price, demand of products, etc.) The neural network structure as well as experiments with a constructed predictor is presented in Fig. 3.

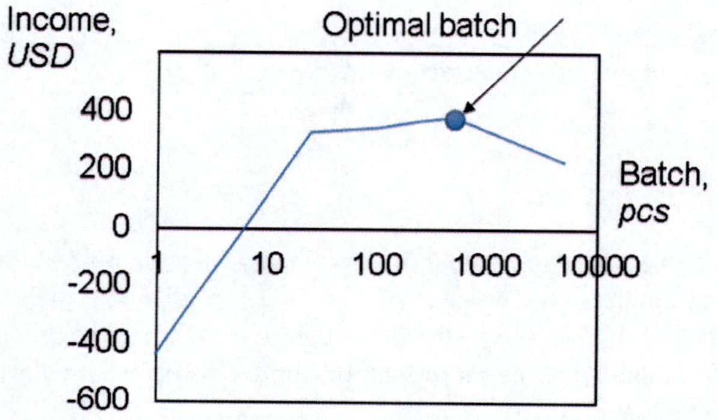

Fig. 2. Calculation of optimal batch size for a particular product.

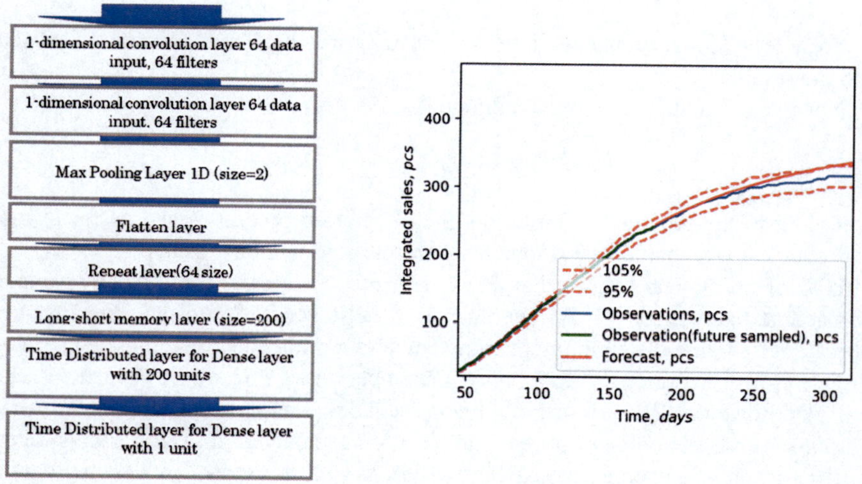

Fig. 3. CNN-LSTM Neural Network (left), Constructed predictor for demand forecasting (with the evaluation of results).

ERP functions united with such new in-firm planning features can be united as a cross-industry cooperation B2B platform, which will help companies to establish their growth.

4 Discussion

The listed results are applicable mainly for companies with a predictive line of growth. For example, if the company has a stable and established line of products, it is easier to produce calculations and forecasting.

It is also applicable to machinery and electronics industries only since the data obtained was only from such enterprises.

The future of enterprises lies in such features as remote job technology advanced and dedicated manufacturing technologies, fabless technologies [15]. That means that a new enterprise concept arises, which is known as a virtual factory [16]. Virtual factories are systems of complex technological solutions that provide, in the shortest possible time, the design and manufacture of globally competitive products. A virtual factory implies the presence of enterprise management information systems that allow developing and using as a single object a virtual model of all organizational, technological, logistics, and other processes at the level of global supply chains and, which is important, by distributed production assets. The usage of the latest technologies in enterprise software, as AI in-firm planning instruments are, is especially suitable for use in virtual factories [17].

5 Conclusion

It should be noted that most respondents report that integration of intellectual systems in their in-firm processes is beneficial (88–95%). High price remains the highest risk factor for most companies. That is why most of the big enterprises have ERP implemented and small companies prefer to have it as an optional system.

This paper provides two innovative approaches for AI and non-AI in-firm planning using digital technologies.

Regarding the feasibility of ERP and other systems integration, it is a good opportunity for businesses to boost their performance, and even for small enterprises, it should be considered for integration.

Acknowledgements. This paper was published with the help of a research program conducted in Plekhanov Russian University of Economics (University Order No. 649 dated 06/07/2021).

References

1. Bystrov, A., Tolstykh, T., Agaeva, A.: Mod. Syst. Netw. Econ. Eng. Nat. Soc. **2022** (2020). https://doi.org/10.21685/2227-8486-2020-2-2
2. Bystrov, A., Tolstykh, T., Radaykin, A.: Econ. Manag. **26**, 564–576 (2020). https://doi.org/10.35854/1998-1627-2020-6-564-576

3. Artificial Intelligence: A Rights-Based Blueprint for Business. https://www.bsr.org/en/our-insights/report-view/artificial-intelligence-a-rights-based-blueprint-for-business
4. Manakhova, I., Bystrov, A., Ignatyeva, G., Alekhina, O.: Personnel security in the conditions of digitalization of the economy. In: Proceedings of the Volgograd State University International Scientific Conference "Competitive, Sustainable and Safe Development of the Regional Economy", pp. 391–395 (2019). https://doi.org/10.2991/cssdre-19.2019.76
5. Radaykin, A.G.: Econ. Entrep. **8–2**(85), 847–850 (2017)
6. Kulyasova, A., Esina, A., Svirchevskiy, V.: Rus. J. Ind. Econ. **12**, 316–322 (2019). https://doi.org/10.17073/2072-1633-2019-3-316-322
7. Brownlee, J.: Multi-Step LSTM Time Series Forecasting Models for Power Usage. https://machinelearningmastery.com/how-to-develop-lstm-models-for-multi-step-time-series-forecasting-of-household-power-consumption/
8. Anastas, J.W.: Research design for social work and the human services. In: Flexible Methods: Descriptive Research, Chap. 5. Columbia University Press, New York (1999). https://doi.org/10.7312/anas11890
9. Cuthill, M.: Sust. Dev. **10**, 79–89 (2002)
10. Nikunj, D.: ERP Facts and Statistics 2020: Market Trends, Data and Analysis Software Suggest (2020). https://www.softwaresuggest.com/blog/erp-facts-and-statistics-market-trends-data-and-analysis/
11. G2 ERP Statistics. https://learn.g2.com/erp-statistics
12. Discover Top Business
13. Software & Service Partners. https://www.softwaresuggest.com/
14. Pedregosa, et al.: JMLR **12**, 2825–2830 (2011)
15. Wadii, B., et al.: A Novel CNN-LSTM-based Approach to Predict Urban Expansion (2021). https://arxiv.org/abs/2103.01695
16. Radaykin, A.G.: Econ. Horiz. **3**(49), 12–18 (2019)
17. Zagumennov, F., Radaykin, A., Dzvinko, R.: IOP Conf. Ser. Earth Environ. Sci. **666**, 062091 (2021). https://doi.org/10.1088/1755-1315/666/6/062091
18. Association "Technet". https://technet-nti.ru/

Development of Tools for Intra-company Planning of Production Systems in the Digital Economy

Ivan I. Tonkih[1(✉)], Marina S. Kuzmina[2], and Kristina A. Logun[3]

[1] Joint-Stock Company "Voronezh Central Design Bureau "Polyus"", Voronezh, Russia
[2] Joint-Stock Company "Concern "Sozvezdie"", Voronezh, Russia
[3] North-Eastern State University, Magadan, Russia

Abstract. Improving the efficiency of production systems is an important part of forming and maintaining a competitive position for an enterprise in any industry. In terms of economic aspects, improving cost-effectiveness can be expressed in cost reduction, increased profitability, a positive image of the product, and increased customer loyalty. Various factors influence the efficiency of production systems, today the digitalization of the economy occupies the unconditional leadership in the external environment factor, among the factors of the internal environment, a special place belongs to intra-company planning, as it forms the standards and norms of the resource usage, tension, and synchronism of the finished products, lay the basis for the profitability of products. The theoretical and practical basis of the study were the works of Russian and foreign scientists related to the problems of enterprise management in the context of digitalization and probe research conducted by the authors. The article considers the stages of digital transformation of intra-company planning, aimed at increasing the speed and flexibility of response of production systems, proposed a methodical approach to studying the level of digitalization of tools for intra-company planning. The results of its testing at the MIC enterprises (military-industrial complex) are presented. The above study is the author's contribution to improving the efficiency of manufacturing enterprises in the digital economy.

Keywords: Digitalization · Flexibility · Responsiveness of production systems · Digitalization level of intra-company planning

1 Introduction

Generalizing the various points of view, we can say that improving tools for intra-company planning and improving the efficiency of production systems is a multidimensional and multifactorial phenomenon [1, 2]. In our opinion, today, one of the main factors in the development of intra-company planning tools is economic digitalization [3, 4]. Examining various statistics, we see significant growth in the digital industry [5]. The most relevant digital technologies include big data, neurotechnologies, and artificial intelligence [6]. Thus, we can say that the digitalization of the economy creates new

P. V. Trifonov and M. V. Charaeva (Eds.): DITEM 2021, LNNS 380, pp. 43–48, 2022.
https://doi.org/10.1007/978-3-030-94245-8_6

opportunities to coordinate production and planning processes [7, 8], which can take it to a new level. We are talking about a new approach to working with information, its collection and transmission efficiency, analysis, and correction [9]. The global nature of digitalization implies a gradual adaptation and raises the question of the content and sequence of work leading to the digitalization of intra-company planning. Despite the increasing urgency, this issue remains poorly developed.

2 Materials and Methods

The purpose of the study is to design and develop methodological provisions to provide a reasonable choice of effective directions for improving the intra-company tools, aimed at increasing the efficiency of production systems, providing increased flexibility and responsiveness in the digitalization of the economy.

The objective is specified in the solution of the following tasks:

- identifying the necessary preliminary steps that lead intra-company planning tools to an effective digital transformation;
- assessment of the current level of digitalization of intra-company planning for timely adjustment of the process and forming a base for tracking changes and trends.

In the course of the work, quantitative and qualitative research methods were used. The latter was used to collect secondary data, which was carried out by desk research, during which, with the active use of the Internet, various literary sources of a practical and theoretical nature were analyzed.

Concerning primary data, in the context of a sufficiently high level of novelty of the question, the most appropriate data collection method was to conduct an expert survey.

The target group was the enterprises' heads and specialized subdivisions of industrial enterprises working in the MIC. The survey was conducted in the form of a questionnaire. The indicators of the average value and median were used to process the results.

3 Results

To understand the stages of digitalization of intra-company planning, let us turn to the concept of "Industry 4.0", according to which the process of digitalization of the enterprise should include:

- computerization of all workplaces, composition of a reliable system of transmission, automated processing, and visualization of the collected data [10];
- equipping all used machinery with measuring devices connected to the data exchange network [11];
- composition of a reliable system of transmission, automated processing, and visualization of the collected data;
- consolidation of the equipment fleet and workstations into a unified information network [12];
- integration into the enterprise management system of intelligent subsystems for automatic interpretation of information received, solution implementation [13].

Based on the above, it is clear that the automation of all tools for intra-company planning is a prerequisite for effective digitalization.

To assess the current level of digitalization of intra-company planning, we took the basic and characteristic functions of the planning and economic department as the "process master", given in the study of MIC, and offered the experts to assess them on the following scale:

- no automation;
- individual actions are automated;
- most of the actions are automated;
- all actions are automated; a single information space has been created.

According to the result processing, let us present the maximum and minimum values of indicators of automation of individual operations of internal corporate planning in Tables 1 and 2, respectively.

Table 1. Maximum values of automation indicators for individual intra-company planning operations. Source: compiled by the authors.

Indicators	Enterprise		
	JSC "VCDB "Polyus""	JSC "Electro Signal"	JSC "Ryazan Radio Plant"
Ensuring consistency and mutual alignment of enterprise plans	1.87	1.8	1.87
Formation and analysis of the company's budget execution	1.73	1.73	1.9

Table 2. Minimum values of automation indicators for individual intra-company planning operations. Source: compiled by the authors.

Indicators	Enterprise		
	JSC "VCDB "Polyus""	JSC "Electro Signal"	JSC "Ryazan Radio Plant"
Calculation and approval of planning and economic standards for the enterprise before the military representation	1.83	1.6	1.4
Control of budget limits	1.3	1.3	1.45

The values presented in Table 1 and Table 2 indicate the automation of individual actions for these intra-company planning operations at the studied enterprises.

Table 3. Average and median values of the evaluation of automation indicators of intra-company planning tools. Source: compiled by the authors.

Indicators	JSC "VCDB "Polyus""	JSC "Electro Signal"	JSC "Ryazan Radio Plant"
The average value of the evaluation of automation indicators for an enterprise	1.66	1.64	1.70
Median value for enterprises	1.71	1.7	1.73

The average and median values of the evaluation of automation indicators of intra-company planning tools are shown in Table 3.

According to the study of the level of automation of intra-company planning tools, given in Table 3, we can conclude that all companies are in the same quality category, but the leading position in it consistently takes JSC "Ryazan Radio Plant".

4 Discussion

A study of various literature sources showed that both foreign and Russian companies are engaged in assessing the level of digitalization of enterprises in general. Foreign ones include MIT Center for Digital Business and Capgemini Consulting [14], Deloitte (Digital Maturity Model) [15], analytical agency Arthur D. Little (Digital Transformation Index) [16], KPMG (Digital Business Aptitude — DBA) [17], the Global Center for Digital Business Transformation (Digitization Piano) [18], Ionology [19] and Acatech [20].

SAP and Deloitte experts have developed a scale of digital maturity of the enterprise [16]. The results of a study by SAP and Deloitte, presented at SAP Forum 2019, showed that, compared with foreign companies, Russian companies' digitalization index is 1.84, foreign companies – 2.08, which corresponds to the stage of fragmented automation correlates with the results of our study. Only 9% of Russian companies correspond to the stage of mature automation [16]. The level of digital maturity of enterprises varies by industry.

As for Russian companies, for example, the Center for Advanced Management has proposed a methodology for assessing the level of digital maturity according to 7 blocks: infrastructure and tools, processes, products, data, models, personnel, culture [21].

Note that at present, there is virtually no data on the evaluation of digitalization of individual processes within the enterprise, including intra-company planning. An exception is the research of the STC "Informtehnika" – a branch of the FSUE NIISU, which includes such assessment sections as design preparation of production, technological preparation of production, automation of production processes, planning and resource management of the enterprise, marketing, and sales management, service and warranty management, management of state contracts (275-FZ), electronic document management, management of inter-plant cooperation, information security, hardware, and software evaluation. A total of 38 evaluation forms were used, on which the digitalization level matrix is built [22]. The results of this study in the open press are given in fragments. The enterprise planning and resource management section is completely absent, limiting the possibility of comparing them with the author's research data.

5 Conclusion

The research direction presented in the article will become increasingly relevant, as in general, in all areas of digitalization of production systems, there is a trend towards the active implementation of modern technology and the growing share of companies that use them in their work. The study materials can be the basis for further analysis and improvement of tools for intra-company planning to improve the efficiency of enterprise production systems management in adapting to the digital economy in the direction of cost optimization, increasing the flexibility and speed of management decisions.

References

1. Sunigovets, O.: Enterprise competitiveness in the digital economy. SHS Web Conf. **67**, 04012 (2019). https://doi.org/10.1051/shsconf/20196704012ISBN:2261-2424
2. Zhu, X.: Introduction: from the industrial economy to the digital economy: a giant leap—research on the "1 + 10" framework of the digital economy. In: Emerging Champions in the Digital Economy. MP, pp. 1–65. Springer, Singapore (2019). https://doi.org/10.1007/978-981-13-2628-8_1
3. Novikov, S.: Transformation of industrial enterprises in the digital economy. Res. World Econ. **11**(5), 90 (2020). https://doi.org/10.5430/rwe.v11n5p90
4. Androshina, I.S., Larina, I.Y.: Company planning and growth in the digital economy. Bull. Russ. Acad. Nat. Sci. **21**(2), 62–67 (2021). https://doi.org/10.52531/1682-1696-2021-21-2-62-67
5. The digital economy in Russia can grow three times by 2025. https://d-russia.ru/tsifrovaya-ekonomika-v-rossii-mozhet-vyrasti-vtroe-k-2025-godu-issledovanie.html
6. Passport of the national program "Digital Economy of the Russian Federation" (approved by the Presidium of the Presidential Council for Strategic Development and National Projects on December 24, 2018 No. 16). http://base.garant.ru/72190282/
7. Atkinson, R.: Driving Toward the Digital Driving Toward the Digital Economy and Digital Economy and Digital Manufacturing Manufacturing (2019). https://www.researchgate.net/publication/265323372_Driving_Toward_the_Digital_Driving_Toward_the_Digital_Economy_and_Digital_Economy_and_Digital_Manufacturing_Manufacturing
8. Jinfang, T., Yiran, L.: Research on total factor productivity measurement and influencing factors of digital economy enterprises. Procedia Comput. Sci. **187**(01), 390–395 (2021). https://doi.org/10.1016/j.procs.2021.04.077
9. Tolstykh, T., Kretova, N., Logun, K., Popikov, A., Kuznetsov, V.: Directions and resource support of information and communication policy of hi-tech enterprises in the conditions of digitalization of the economy. In: Popkova, E. (ed.) Growth Poles of the Global Economy: Emergence, Changes and Future Perspectives, pp. 1275–1285. Springer, Berlin (2020)
10. Ganieva, U.A.: The digital economy and its benefits. TRANS Asian J. Mark. Manag. Res. **10**(4), 29–34 (2021). https://doi.org/10.5958/2279-0667.2021.00011.0

11. Ryan, P.: Distinguishing digital economies from non-digital economies. In: Trust and Distrust in Digital Economies, pp. 26–32 (2019). https://doi.org/10.4324/9781351104845-4.ISBN: 9781351104845
12. Gileva, T.A., Galimova, M.P., Babkin, A.V., Gorshenina, M.E.: Strategic management of industrial enterprise digital maturity in a global economic space of the ecosystem economy. IOP Conf. Ser. Earth Environ. Sci. **816**(1), 012022 (2021). https://doi.org/10.1088/1755-1315/816/1/012022
13. Bazzoun, M.: The digital economy. Int. J. Soc. Sci. Econ. Invent. **5**(09) (2019). https://doi.org/10.23958/ijssei/vol05-i09/157
14. Westerman, G., Bonnet, D., McAfee, A.: The Nine Elements of Digital Transformation. https://sloanreview.mit.edu/article/the-nineelements-of-digital-transformation
15. Digital Maturity Model. Achieving Digital Maturity to Drive Grow. https://www2.deloitte.com/content/dam/Deloitte/global/Documents/Technology-MediaTelecommunications/deloitte-digital-maturitymodel.pdf
16. Digital Transformation – How to Become Digital Leader. Study 2015 Results. http://www.adlittle.com/sites/default/files/viewpoints/ADL_HowtoBecomeDigitalLeader_02.pdf
17. Are You Ready for Digital Transformation? Measuring Your Digital Business Aptitude. https://assets.kpmg/content/dam/kpmg/pdf/2016/04/measuring-digital-businessaptitude.pdf
18. Digital Business Transformation. A Conceptual Framework. https://ru.scribd.com/document/372049639/DigitalBusiness-Transformation-Framework-pdf
19. Digital Transformation of HR. The Experience of Russian Companies. http://obzory.hr-media.ru/cifrovaya_transformaciya_hr_russia
20. The Industry 4.0 Maturity Index. Managing the Digital Transformation of Companies. https://www.acatech.de/wp-content/uploads/2018/03/acatech_STUDIE_rus_Maturity_Index_WEB.pdf
21. KMDA Company Analytical Report "Digital transformation in Russia – 2020: An Overview and Recipes for Success" (2020). https://drive.google.com/file/d/1xVK4lSanDZSCN6kGAHXikrGoKgpVlcwN/view
22. Methodological recommendations on digital production organization at MIC as a tool for monitoring digitalization processes. https://xn--h1aelen.xn--p1ai/wp-content/uploads/2018/05/Karachyov-20.04.18.pdf

Management of the Export of Educational Services in a Modern University

Vladimir V. Zotov[1], Alina A. Pozdnyakova[1](✉), Larisa Yu. Morozova[2], Tatyana S. Grafkova[2], and Marina E. Kalita[2]

[1] The Kosygin State University of Russia, Moscow, Russia
pozdnyakova-aa@rguk.ru
[2] Moscow Pedagogical State University, Moscow, Russia

Abstract. The export of educational services today is the most important indicator of the competitiveness of the university, the degree of its integration into the global educational space. The management of this area is largely focused on the search for new forms and types of interuniversity interaction, which makes it possible to increase the efficiency of educational activities. The main forms of the export of educational services include pre-university training of foreign citizens training in basic educational programs, postgraduate education programs, licensed advanced training and professional retraining programs, short-term educational programs. The last form of the export of educational services includes various types of internships, the potential of which is not sufficiently revealed and used in modern education. Our article is devoted to this problem. For a foreign student coming to Russia, the purpose of the internship is to improve the knowledge of the Russian language in a specific subject area. This determines the modular nature of the formation of internships, on the one hand, and their practical focus, on the other. The principle of modularity is optimal for the implementation of this form of exported educational services, since it allows to develop an internship program of any degree of complexity in accordance with the requests of the "customer" – through the use of ready-made teaching units from the bank of modular materials. To be successful, the marketing strategies of educational organizations must be adapted to the specific needs of the partner countries and be actively promoted in the international educational market.

Keywords: Education management · Educational service · Modular internship · Marketing strategies · Higher education system

1 Introduction

Any modern Russian university teaching foreign citizens participates in the export of educational services provided by the domestic system of higher education. Currently, the process of including Russian universities in the global export of educational service (EES) is gaining momentum in connection with the processes of globalization and internationalization of the entire field of education.

An educational service is an activity aimed at providing the consumer with opportunities for mastering a particular educational program on a paid basis in educational

P. V. Trifonov and M. V. Charaeva (Eds.): DITEM 2021, LNNS 380, pp. 49–54, 2022.
https://doi.org/10.1007/978-3-030-94245-8_7

institutions. EES is part of the process of internationalization of education, since it is oriented towards an international consumer. Professional management of the export of educational services requires understanding of the nature of the educational service being a cultural value itself, on the one hand, and a sort of intangible asset, on the other hand.

The main forms of EES are:

1. pre-university training of foreign citizens at preparatory departments/faculties;
2. testing of foreign citizens in Russian as a foreign language;
3. training of foreign citizens in basic educational programs (including projects of double degrees, "2 + 2" with the issuance of only a Russian diploma, etc.);
4. training of foreign citizens in postgraduate education programs;
5. training of foreign citizens in licensed advanced training and professional retraining programs;
6. training of foreign citizens in unlicensed short-term educational programs.

The last form of EES includes various types of internships, the potential of which is not sufficiently revealed and used in modern education.

2 Materials and Methods

To understand the peculiarities of the functioning of the educational services market, we carried out a comprehensive analysis of domestic and foreign literature on the indicated problem. We would like to emphasize the works concerning: 1) the formation of the strategies of the EES [1–4]; 2) creating content for educational programs [5, 6]; 3) activities of educational institutions in the market [7–10].

The information base of the study was data received from open sources. The sources include: 1) analytical reviews on the problems of economic and social development of regions [11, 12]; 2) summarized statistics related to the study of the teaching staff on teaching and learning [13–17]; 3) materials of websites of foreign universities.

To identify the educational preferences of foreign partners and substantiate the effectiveness of the implemented internship programs, a survey was conducted of foreign interns who in different years studied at the departments of Russian as a foreign language of the Moscow State Pedagogical University and The Kosygin State University of Russia, as well as teachers organizing and managing internship programs.

3 Results

Based on the research of Russian and foreign scientists [1, 2, 4, 6–8, 18], we identified the criteria for the success of EES in six areas: 1) political and geographical, 2) organizational and administrative, 3) organizational and methodological, 4) material and financial, 5) information, 6) scientific and intellectual. The degree of the significance of the criteria in each group was determined by the analysis of the results of a survey of foreign students who completed ten-month and six-month internship programs at The Moscow State Pedagogical University and The Kosygin State University of Russia.

In the first group important (noted by absolutely all participants in the experiment) were: 1) the geographical position of the university within the country, developed urban infrastructure; 2) security in the city and at the university; 3) the convenient location of the university. Moreover, the safety criterion was put in the first place by 80% of foreigners, the development of urban infrastructure was considered important by 82% of the respondents.

The results of our survey turned out to be comparable with statistical data: among students from the countries of the former USSR, the most popular for training are Moscow (25%), St. Petersburg (9%), Omsk region (6%) and Tomsk region (5%). Moscow (26%), St. Petersburg (12%), Kursk region (4%), the Republic of Tatarstan (3%) are more attractive for students from other countries [19].

The following factors became relevant in the second group: 1) highly-qualified staff of the international department; 2) quick feedback (the ability to receive quick answers to emerging questions); 3) quick visa problems solution; 4) tutor teachers who know how to work with foreign students. More than 90% of the respondents prioritized the importance of trouble-free communication with university representatives.

In organizational and methodological sphere foreigners identify such criteria as: 1) a variety of educational programs at the university that are in demand in international markets; 2) agreements between universities on the recognition of internship courses at national universities; 3) double and joint degree programs at the host university; 4) getting an international diploma (Diploma Supplement); 5) the system of credits at the university of training. The first criterion – "a variety of programs" – was chosen by 79% of the respondents. Interest in double degrees and credit system turned out to be lower – at the level of 57% 59%. This fact is explained by the specifics of the surveyed contingent (a significant number of students were from Asia and the former CIS).

From material and financial factors we identified: 1) the financial position of the university; 2) the affordability of prices for educational services; 3) the availability of good classrooms and laboratories at the university; 4) the accessibility of the high-speed Internet and modern computer equipment at the university; 5) comfortable accommodation. In this group, "affordability" and "comfortable accommodation" were selected by 95% of students as the leading criteria. The "high-speed Internet" was chosen by 92% of the respondents.

We identified the following information factors: 1) the availability of information about the university on the Internet; 2) the popularity of the university in Russia and abroad; 3) acceptance of a diploma of a given university abroad. "University status" and recognition of the diploma were the most significant criteria for 92% of the respondents. It is interesting that when choosing a university for an internship, no more than 20% of respondents had complete and reliable information about the universities.

The following factors associated with the quality of education with its focus on globalization and internationalization were identified: 1) scientific schools at the university that are recognized in the world; 2) professionalism of teachers involved in teaching foreigners; 3) teachers' experience of work in foreign universities; 4) disciplines and courses taught in English. In this category, 80% of the respondents gave the first positions to the criteria of "professionalism of teachers" (preference for teachers with PhD and Doctorate degrees) and "disciplines read in English" (less than half of the foreign students surveyed said they know English at B1 Level).

4 Discussion

An internship at a higher school is a form of additional professional education, the purpose of which is to consolidate and improve the knowledge and skills acquired by a student during the period of study. In many cases, internships take place at pilot sites and help trainees gain experience in a specific field [9, 10]. For a foreign student arriving in Russia, the purpose of the internship is to improve the knowledge of the Russian language in a specific subject area. This determines the feasibility of forming internships on a modular basis and their practice-oriented nature.

The principle of modularity is optimal for the implementation of this form of EES, since it makes it possible to form a program of any complexity quickly in accordance with the requests of the "customer" – by using ready-made methodological units from the bank of modular materials [20, 21]. The existence of such a "bank" does not justify the absence of a strategy for preparing internship programs and the chaotic nature of their implementation, observed in individual universities. Without a clear understanding of the goals and objectives of training foreign students, without focusing on the interests of consumers of educational services and on the needs of the labor market, without understanding the prospects for implementing additional education programs, it is difficult for a university to become competitive in the international market. To promote its services, a university needs to have "forward-looking information", that is, data on the development of the market for the coming years.

A university involved in the EES should strive to highlight its market segment in which it will have sustainable competitive advantages. Today, the distribution of the total number of foreign students in Russian universities is extremely uneven: 90% of universities account for only 20% of foreign students, and this trend has persisted for quite a long time [18, 19]. According to analytical reviews, the largest number of foreigners studies in Russian universities teaching art and culture, specialties related to health care, medical and natural sciences, mathematics [19]. This does not mean that universities in other areas cannot offer internship programs that are interesting for foreign consumers [22, 23].

The choice of a market segment can be associated with a specific country and the specialization of the university. So, internships offered at The Kosygin State University of Russia are multifunctional in their focus. Their range is quite wide and is due to the synergy of the educational structure, which gives a "summarizing effect" [24] due to the diversity of educational directions and programs presented at the university. We will name some modular internship programs for foreigners: "The Russian language in the field of culture and art", "The Russian language in the field of design and technology", "The Russian language in the field of Arts", "The Russian language in the field of economics and business", "The Russian language in the field of information technology", "The Russian language in the field of Sciences".

Being short-term forms of training, modular internships have significant mobilization potential and can be used to attract a foreign contingent to a university, thereby increasing the status of this university as an exporter of educational services.

5 Conclusion

Thus, the export of educational services is a promising area of activity for Russian universities, but at present, not all of its forms are sufficiently developed and presented abroad. Expanding and strengthening the system of modular internships, Russian universities are actively involved in the global processes of higher education development. The adaptation of internship programs to the specific needs of the world labor market, the targeted promotion of information on educational websites and exhibitions contributes to the emergence of new Russian players in the export market of higher education.

References

1. Borshchevskaya, E.O.: State and Public Administration of Education. Management of Higher Education. Publishing House of the Polytechnic University, St. Petersburg (2006)
2. Titova, N.L.: The Path of Success and Failure. Strategic Development of Russian Universities. National Research University Higher School of Economics, Moscow (2008)
3. Busemeyer, M.R., Vossiek, J.: Global convergence or path dependency? In: Mundy, K., Green, A., Lingard, B., Verger, A. (eds). The Handbook of Global Education Policy, pp. 145–161. Wiley-Blackwell, Chichester (2016). https://doi.org/10.1002/9781118468005.ch8
4. Frommberger, D., Baumann, F.-A.: Internationalisierung der Berufsbildung. In: Arnold, R., Lipsmeier, A., Rohs, M. (eds.) Handbuch Berufsbildung, pp. 713–724. Springer, Wiesbaden (2020). https://doi.org/10.1007/978-3-658-19312-6_54
5. Chadee, D., Naidoo, V.: Higher educational services exports: sources of growth of asian students in US and UK. Serv. Bus. Int. J. **3**(2), 173–187 (2008). https://doi.org/10.1007/s11628-008-0041-7
6. Graf, L.: Work-based higher education programmes in Germany and the US: comparing multi-actor corporatist governance in higher education. Pol. Soc. **36**(1), 89–108 (2017). https://doi.org/10.1080/14494035.2017.1278872
7. Pelikhov, N.V., Reutov, Yu.I.: Development of an Innovative Scientific and Educational Environment of the Khanty-Mansiysk Autonomous Okrug, Polygraphist, Ugra. Khanty-Mansiysk (2008)
8. Fortwengel, J., Jackson, G.: Legitimizing the apprenticeship practice in a distant environment: institutional entrepreneurship through inter-organizational networks. J. World Bus. **51**(6), 895–909 (2016). https://doi.org/10.1016/j.jwb.2016.05.002
9. Phusavat, K., Setiono, K., Sikora, U.S., Hidayanto, A.N., Delahunty, D.: Adapting social enterprise to improve education quality: key findings at one Bangkok Metropolitan Administration school. Int. J. Manag. Educ. **10**(1), 96 (2016). https://doi.org/10.1504/IJMIE.2016.073366
10. Phusavat, K., Hidayanto, A.N., Kess, P., Kantola, J.: Integrating Design Thinking into peer-learning community: impacts on professional development and learning. J. Workpl. Learn. **31**(1), 59–74 (2019). https://doi.org/10.1108/JWL-03-2018-0055
11. World Bank Group. Europe and Central Asia Economic Update, Spring 2021: Data, Digitalization, and Governance. Washington, D.C. (2021). https://doi.org/10.1596/978-1-4648-1698-7
12. Ranking of countries in the world by education level (2021). https://gtmarket.ru/ratings/education-index
13. OECD: TALIS 2018 Results (Volume I): Teachers and School Leaders as Lifelong Learners. OECD Publishing, Paris (2019)

14. OECD: TALIS 2018 Results (Volume II): Teachers and School Leaders as Valued Professionals. OECD Publishing, Paris (2020)
15. OECD: Positive, High-achieving Students?: What Schools and Teachers Can Do. OECD Publishing, Paris (2021)
16. OECD. Education at a Glance 2021. OECD Indicators (2021). https://doi.org/10.1787/b35a14e5-en
17. Bondarenko, N.V., et al.: Education Indicators: 2020: Statistical Collection. National Research University Higher School of Economics, Moscow (2020)
18. Arefiev, A.L.: Export of Russian Educational Services: Statistical Collection 9. Center for Sociological Research, Moscow (2019)
19. Academic Mobility of Foreign Students in Russia. Education Facts 7. Institute of Education, National Research University Higher School of Economics (2016)
20. Pozdnyakova, A.A., Chepkova, T.P.: Metasubject approach in the creation of training modules in the language of the specialty. Teacher XXI Cent. **4**(1), 297–315 (2016)
21. Pozdnyakova, A.A., Chepkova, T.P.: Module “Phraseology & Language of Russian Media" in the system of teaching Russian language to foreign trainees-philologists. Teacher XXI Cent. **4**(1), 195–206 (2017)
22. Galyamova, E.F., Kudimana, G.O.: The attractiveness of Russian universities in the international market of educational services. Bul. Udmurt. Univ. Ser. “Econ. Law” **27**(3):21–26 (2017)
23. Makhotaeva, M.Yu., Bakumenko, O.A., Varlamov, G.V.: Export of Educational Services vs Internationalization of the University. Publishing House of the Saint Petersburg University of Management Technologies and Economics, St. Petersburg (2019)
24. Berseneva, T.P.: Synergy: essential characteristics and forms of manifestation. Hist. Philos. Pol. Legal Sci. Cult. Stud. Art. Hist. Quest. Theory Pract. **2**(64), 48–52 (2016)

Convergence Risks of Operational and Project Management for SME Development

Viktor V. Ivanov[1,3](✉), Gabdelakhat R. Latfullin[2], and Anastasia V. Ivanova[1]

[1] Institute of Social Development, Moscow, Russia
[2] State University of Management, Moscow, Russia
[3] HSE University, Moscow, Russia

Abstract. This paper examines the impact of the "innovation line" effect on changes in project management format and their impact on the further development of small businesses (SMEs). Based on the Social Development Theory (Ivanov V. V. SDT), including the Theory of Large Innovation Cycles, the Theory of Change Of Social Organizational Formations, the Theory of Atomization of Jobs and Work Functions, as well as several studies 2018–2021 of SMEs, development of state and municipal coworking spaces to support SME activities in regions, the nature and characteristics of patterns of innovation development were identified. It was determined that the transition to a continuous process of project management in organizations, in conjunction with the SMEs' uberization, will lead to SME degradation due to the systemic loss of their entrepreneurial function. The problem can be solved by transforming the working environment proposed by the authors based on unified digital platforms. It will allow moving from the concept of "innovation line" within the organization to the work of SMEs in a "flexible platform environment of innovation", preserving the entrepreneurial initiative as a mass function.

Keywords: Municipal management · Coworking · Network organization · Innovation line · Social Development Theory

1 Introduction

The widespread acceleration of innovation processes turns the creation of innovation in the conveyor line, which from an advantage in the market for the leaders, gradually becomes a necessary factor in maintaining its presence for each participant. However, large and small enterprises have fundamentally different opportunities in this regard.

The SME innovation environment's current success is that focusing on breakthrough technology is easier for a compact group of inventors unencumbered by operational processes and established relationships.

The modern IT infrastructure that has emerged allows the same groups of developers to increase the innovation flow to the line, but only within the framework of narrow or closely related issues.

At the same time, creating an entire line of innovations across the entire spectrum of SME areas by providing them with small teams is a task that remains unfeasible

P. V. Trifonov and M. V. Charaeva (Eds.): DITEM 2021, LNNS 380, pp. 55–60, 2022.
https://doi.org/10.1007/978-3-030-94245-8_8

for mass application. Thus, the continuity of innovation forms a significant functional limitation of small enterprises, depriving them of competitiveness in the general business environment and putting them on the verge of survival as a class.

An illustration of this problem is the uberization of passenger transportation. This model has consolidated virtually the entire community of small entrepreneurs engaged in taxi transportation in just a few years. It became impossible to find free private carriers. At the same time, the exploitation, risks, and problems of entrepreneurs have increased, and they have lost their main income and autonomy.

Reaching most small entrepreneurs with such models with appropriate results is only a matter of the next 2–5 years.

According to the authors, the changes considered and the problems arising from them are fundamental. To understand them and find the right solutions to overcome the effect of the "innovation line" in SMEs, it is necessary to analyze the nature of innovation and the specifics of the mechanism of their development.

2 Methods

A significant part of the work on developing innovations is devoted to solving practical tasks of a tactical nature. The issues of assessing the fundamental characteristics of the process are most clearly seen in the development of the concept of innovation cycles, which are reflected in many works from the founders [1, 2] to modern Russian [3, 4] and foreign scientists [5, 6].

One of the most recent works [3] on technological modes, unlike the previous ones, allows to significantly expand the possibilities of analyzing the fundamental features of innovation development. However, they are not enough to solve the questions that have arisen within the framework of this work.

According to the authors, the Social Development Theory allows solving these problems (Ivanov V.V. SDT). The choice of this theory is since within its framework, the nature of innovation waves, the development of innovation management, including project management, are considered in detail. At the same time, the Theory has a whole set of the original specialized toolkit, built on its own mathematical apparatus and tested in the process of research.

Within the framework of the theory, its cycle of innovative development is built [7], which claims to be a universal explanation of development throughout the history of building society. Its results obtained independently correlate with the results of another study [8] on the identification and periodization of innovation also within the entire historical period of social development.

According to SDT, all processes of society, including economic, develop in an integrated way as systems. As part of this development, a system of cyclical processes of different intensities but of the same nature, forming a complex structure of cycles (waves), is noted. Changes in each cycle at a certain moment affect the corresponding set of characteristics, including the considered phenomena. The reference development points of all cycles have a small confidence interval and high accuracy in calculating time values.

The basic cycles (waves) of development [9] (Ivanov, 2020), on which the basic innovation cycle with its standardized complex sequence is also based, are the waves defined by SDT as I-Waves (the dates of their changes within the period of social development are 2026; 1937; 1793; 1560; 1183; 573; 414 BC; 2011 BC; 4595 BC; 8776 BC).

Also, according to SDT, each of the above periods (organizational social formation) corresponds to its managerial paradigm, as at each stage, there is a complication of the social structure, which requires an adequate change in its management.

The above cyclic series is not the only one; in addition, within the cycles (waves), there is a fairly detailed division into phases and periods of transition crises between phases of development. All of them also have the specifics of the development of innovative tools, their characteristics and management models, and their corresponding economic production models [7].

3 Results

Innovation, which we perceive only as a modern phenomenon, according to SDT, is a universal tool for the development of society throughout its entire course. At the same time, at each stage of development, the format of innovations changes noticeably, reflecting the peculiarities of stages (formations), including the increase in the speed of appearance of changes [7].

In turn, the mechanism of innovation development affects the transformation of organizations conducting them. The most fundamental influence changing the potential and prospects of organizational formats is observed in the period of change of formations, including the current one.

Thus, based on the SDT provisions, we can conclude that until the 19th-century cycle of innovation, taking into account the latent part, had a period of 610 years. Under these conditions, even its individual phases were measured in decades, and the time of change of innovation in the economy was comparable to a person's life, which is why he did not notice this process as a phenomenon.

Since the 19th century, the phase of the innovation cycle became commensurate with the active phase of human life and even the change of one generation. A new generation of goods often appeared already as an innovation, which forms innovations as a market phenomenon with its own innovation and patent environment for full-fledged turnover of inventions and project management for their implementation.

Since the 1930s, in the current formation, the product life cycle begins to include consecutively several phases of the innovation cycle and require the creation of changes already in the course of its life. Design work has been moving from the regular creation of productions to their regular modernization since 1937. Further, along the cycle reference points (I-wave), there is a formation and modification of both management and the very concept of innovation.

At the moment, we are approaching another fundamental transformation in which the process of innovation will become an "innovation line", merging in continuity with operational management.

In contrast to operational activity, invention, especially in the given parameters, can only be put on the line [10, 11] technologically by forming the inventor line.

Is it possible to make serial inventors and innovators a mass phenomenon, or does the "inventor line" have no alternatives? The future and format of SMEs largely depend on the answer to this question.

4 Discussion

In our opinion, we should not expect the mass emergence of serial inventors in the coming years as part of SME work. Perhaps, within 20–30 years, technologies like TIPS (Inventive Problem Solving Theory) will form such a generation or AI will massively replace humans in this part. However, by this point, the loss of the entrepreneurial spirit will be inevitable and may have the most negative impact on the entire process of society's development.

Under these circumstances, to save SMEs as a class, attention should be paid to the possibility of applying the "inventor line" effect not only to large businesses but also to small businesses.

The answer lies in new formats for organizing the working environment [12, 13]. In order for small entrepreneurs to quickly and easily find new ideas, form new temporary teams without losing momentum and current organization, it is necessary to unite a very large number of interested entrepreneurs in constant interaction.

This was one of the reasons for the emergence and then active development of such a form of a workspace as coworking and accelerators built on similar platforms [14–19].

SMEs are actively developing this format. Municipalities and the state help SMEs by opening coworking spaces at business assistance centers [20].

However, the rapid development of the industry is such that its implementation today answers only yesterday's questions. For the above-mentioned task, it is already necessary to form a unified state network of coworking spaces and their service according to unified standards.

These standards require office space for the entrepreneur and a single digital platform for business management and services, which thus connects the workplaces of other entrepreneurs and their potential staff.

This environment will allow, with sufficient build-up of its capacity and regional coverage, elaborated software to move from "innovation line" and "inventor line" to the new phenomenon of continuously developing "flexible platform environment of innovations", formed by both the demanded innovation service of individual teams, and flexible constant transformation of established teams for new ideas, but already at the small enterprise level.

5 Conclusions

The world community is entering a period of change in the world order, the various characteristics of which, including economic ones, will change significantly. As part of these changes, there is a threat of the loss (degradation) of small entrepreneurs as a class. One of the key reasons for this loss is the emergence of the "innovation line" effect in the productive and social spheres due to the accelerated need for change. In turn, this will lead to an actual convergence of operational and project management, which small entrepreneurs will not fully implement.

The way out of this situation may be in the proper set up of a new format of working environment, which is currently being formed based on the development of coworking spaces but requires a fundamental strengthening in the digital environment with the formation of a business management platform, as well as a certain standardization and transformation of physical workspaces into a single technological complex.

The authors are implementing such a program in a number of projects and are based on merging and developing networks of state and municipal coworking spaces, expanding their software and network functionality.

References

1. Schumpeter, J.: Business Cycles: A Theoretical. Historical and Statistical Analysis of the Capitalist Process, New York (1939)
2. Kondratiev, N.D.: Large Cycles of Conjuncture and the Theory of Foresight. Ekonomika, Moscow (2002)
3. Glazyev, S.Yu.: A leap into the future. Russia in the new technological and world economic structures. Knizhny Mir, Moscow (2018)
4. Yakovets, Y.: Epoch-Making Innovations of the XXI Century. Ekonomika, Moscow (2004)
5. Arrighi, G.: The long twentieth century. Money, Power, and the Origins of Our Times. Verso, London. New York (1994)
6. Meissner, D.: J. Innov. Entrep. **5**, 1–18 (2016)
7. Ivanov, V.V.: Innovation financing cycles on the example of digitalization processes. Step into the future: artificial intelligence and the digital economy. Revolution in management: a new digital economy or a new world of machines. In: Proceedings of the II International Scientific Forum, 3, pp. 321–327. State University of Management (2018)
8. Šmihula, D.: Studia politica Slovaca **2**, 50–68 (2011)
9. Ivanov, V.V.: The global world has fallen. Long live the global world! About the new global network world order. Possible scenarios for the future of Russia and the World: an interdisciplinary discourse. In: Collection of Scientific Papers of the Participants of the International Scientific Conference – XXVIII Kondraev Readings, May 20–22, 2020, Moscow State University (2020)
10. Andreev, V.G.: Energy: Econ. Tech. Ecol. **3**, 67–73 (2012)
11. Takmasheva, I.V., Tyaglov, S.G.: Mod. Europe **4**, 60–72 (2019)
12. Selezneva, I.E., Klochkov, V.V.: Drucker's Bul. **5**, 5–12 (2020). https://doi.org/10.17213/2312-6469-2020-5-5-12
13. Zokirova, N.K., Abdurakhmanova, G., Sagidullin, F.R.: Int. Sci. Rev. **1**, 24–28 (2020)
14. Krause, I.: Foresight **13**(2), 52–60 (2019). https://doi.org/10.17323/2500-2597.2019.2.52.60
15. Sekerin, V.D., Zaitsev, I.A., Gorokhova, A.E.: Drucker's Bul **4**, 5–12 (2020). https://doi.org/10.17213/2312-6469-2020-4-5-12
16. Makarov, V.M., Skvortsova, I.V., Torkkeli, M.: Sci. Tech. Bul. SPb State Polyt. Univ. Econ. Sci. **12**(6), 116–127 (2019)
17. Kadol, N.F.: Drucker's Bul. **2**, 187–200 (2020). https://doi.org/10.17213/2312-6469-2020-2-187-200
18. Samovoleva, S.A.: Drucker's Bul. **5**, 91–103 (2020). https://doi.org/10.17213/2312-6469-2020-5-91-103

19. Grankina, S.: Models of coworking development in regions of Russia under conditions of globalization. In: Proceedings of the II International Scientific Conference GCPMED 2019, "Global Challenges and Prospects of the Modern Economic Development", pp 1312–1318 (2020)
20. Analysis of the market of flexible working spaces in Moscow. Demand and supply. Report of the State Budgetary Institution "Agency of Innovations of Moscow". Moscow (2019). https://innoagency.ru/analiz-coworkings/files/coworkings.pdf

Analysis of the Structural Balance in the Salary of the Firm's Personnel

Vladimir P. Fomin and Natalya A. Igoshina(✉)

Samara State University of Economics, Samara, Russia

Abstract. In modern approaches to the formation of integrated reporting, special attention is paid to social aspects. The basis of social stability of business, first of all, is the level of wages of employees of the firm, as well as compliance with the fair structure of employees based on the level of remuneration. The growth of social tension, as a rule, occurs imperceptibly. The economic analysis of the dynamics of the wage fund due to its structural changes allows us to timely see the dangerous trends of stratification of employees of the company with different incomes. The structural factor is indirectly determined when the results of factor analysis are compared based on the generalized data of the company as a whole and the total results of the analysis in the context of individual groups of employees of the company. However, a direct determination of the effect of the structural factor is also possible. This slightly increases the complexity of the analysis. Social stability management requires appropriate tools for quantitative analysis of the wage fund. These tools are discussed in this article.

Keywords: Sustainable growth · Average salary · Firm's employees

1 Introduction

A high degree of uncertainty for the estimation of the occurrence of specific events in the economic component of human society is combined with the predictability of the direction vector of these events. This vector reflects the general trends of the universe, manifested in practice of human civilization. Trends manifest themselves as a recurring sequence of cause-and-effect phenomena and events. Understanding and interpreting these trends in the ability to follow their requirements increase the chances of success. In particular, this applies to modern scientific ideas about the factors and conditions for achieving the desired high efficiency of the use of resources, including labor resources [1]. These ideas are based largely on the historically confirmed relationship of a high level of business activity results with the proper motivation of individuals involved in this activity [2]. For individual commercial organizations at the micro-level of the economy, this relationship is expressed in the ratio of results and wages. Effective human resources management tools should include an understanding of the logic of direct and inverse cause-and-effect relationships between results, wages and innovations [3]. Therefore, the results of work should be taken into account in all relevant indicators that depend on the labor efforts of individuals and affect the achievement of the desired business results.

P. V. Trifonov and M. V. Charaeva (Eds.): DITEM 2021, LNNS 380, pp. 61–73, 2022.
https://doi.org/10.1007/978-3-030-94245-8_9

For example, in the trading business, the polite behavior of the staff is an important factor of future profit, therefore it should relate to the paid results of work [4]. Justice in determining the income level of each employee of the company is based, first of all, on the general trends (laws) of the universe. Humanity as a whole, as well as every individual employee of the company, is a part of the universe. Therefore, each component or particle of the universe rightly claims to contribute to the formation and transformation of the total mass of the universe that surrounds this particle.

2 Materials and Methods

In domestic science and practice, considerable attention is paid to the problems of rational correlation of various types of production costs, including labor costs [5]. There are also studies abroad related to indicators characterizing the use of labor resources [6]. The methodological basis of these case studies is general and particular methods, for example, analysis and synthesis, deduction and induction.

The effectiveness of remuneration as an active motivational lever is determined by a variety of circumstances. First of all, the objective characteristics of the magnitude of labor results that are limited by physical capabilities of individual individuals are taken into account [7]. The hypothetical upper limit of personnel salary costs can be determined as a percentage of the physically possible maximum value of sales under favorable market conditions and the best combination of extensive and intensive factors for the use of all resources [8]. The lower limit-assumes forced labor for the minimum costs of maintaining the physical condition of employees.

Different levels of qualification, responsibility, different working conditions and the like inevitably cause an objective difference in the level of remuneration. However, the administrative and political authorities can make a subjective moment in this difference. Violation of justice and the general trends of the universe in the distribution of income (ultimately, the material benefits of society) leads to social tension and instability from the micro-level of the economy to civilization as a whole. The lower limit-assumes the minimum costs of maintaining the physical condition of forced labor for employees [9].

Different levels of qualification, responsibility, different working conditions and the like inevitably cause an objective difference in the level of remuneration [10]. However, the administrative and political authorities can make a subjective moment in this difference. Violation of justice and the general trends of the universe in the distribution of income (ultimately, the material benefits of society) leads to social tension and instability from the micro-level of the economy to civilization as a whole.

3 Results

The structural characteristics of products in terms of the labor intensity of specific types of products directly affect the amount of wages of piecework workers. Therefore, at the first stage of evaluating the effectiveness of expenses for the remuneration of personnel of the organization, a factor analysis of changes in the wage fund of workers is carried out. The initial data for such an analysis are presented in Table 1.

Table 1. Primary information for the analysis of the dynamics of the wage fund of workers by factors of production volume.

Types of products	Previous period				Current period			
	Volume of output ($)	Specific weight of types of products (%)	Wage fund ($)	Salary intensity of products ($)	Volume of output ($)	Specific weight of types of products (%)	Wage fund ($)	Salary intensity of products ($)
1	2	3	4	5	6	7	8	9
A	49000	4.12	28000	0.57143	56000	4.49	32090	0.57304
B	800000	67.28	412000	0.51500	794450	63.63	405170	0.51000
C	340000	28.60	105180	0.30935	398000	31.88	114860	0.28859
Total	1189000	100	545180	0.45852	1248450	100	552120	0.44224

For convenience, we introduce the following notations:
Ps – types of products;
V – total volume of output;
Vi – the volume of output of the i-th type of products;
di -specific weight of types of products, (%);
di = Vi/$\sum$Vi = Vi/V;
W – wage fund;
Wi = the wage fund of the i-th type of products;
Wip – wage intensity of products;
Wip = W/V;
Wipi – the wage intensity of the i-th type of products;
Wipi = Vi/Wi;
T(0), T(1) – previous period and Current period accordingly.

Let the previous or basic values of the indicator values be indicated by the symbol (0), and the changed, actual or current values of the indicator values are indicated by the symbol (1). At the same time, a change in a specific indicator is indicated by the symbol "Δ", which is placed to the left of the indicator designation. The impact of individual factors in the projection of each level of detail in the analysis is presented as follows:

The wage fund has changed by: +6940 = 552120 – 545180 in the current period compared to the previous time period. In general or in a formalized expression:

$$\Delta W = W(1) - W(0) \tag{1}$$

This change is primarily due to changes in the volume of products produced (ΔWv) and due to changes in the wage intensity of products (ΔWi). For the enlarged factor analysis of this change, the working formula of the first level of detail is used:

$$W = V^{*}Wip = \sum Vi * Wip = \sum (Vi * Wipi) = \sum Wi \tag{2}$$

The calculation of the influence of factors at the first level of analysis detail takes into account only two factors:

$$\Delta \mathrm{Wv} = \Delta \mathrm{V}^{*}\mathrm{Wip}(0) \tag{3}$$

$$\Delta \mathrm{Wi} = \mathrm{V}(1) * \Delta \mathrm{Wip} \tag{4}$$

The calculated values of these factors are determined:

$$\Delta \mathrm{Wv} = \Delta \mathrm{V} * \mathrm{Wip}(0) = (1248450 - 1189000) * 0,45852 = +27259$$
$$\Delta \mathrm{Wi} = \mathrm{V}(1) * \Delta \mathrm{Wip} = (0.44224 - 0.45852) * 1248450 = -20319$$
$$\Delta \mathrm{W} = \Delta \mathrm{Wv} + \Delta \mathrm{Wi} = 27259 - 20319 = +6940$$

The results of the analysis show that the increase in the wage fund occurred only by the factor of the volume of output, and the factor of wage intensity acted in the direction of a decrease in the wage fund. In general, the calculation gives positive outcomes about the changes of the studied indicators, since the increase of the business scale justifies the increase in its resource base.

Such an analysis of the first level of detail with the identification of the influence of two factors (the volume of products produced and its wage intensity) can be carried out not only on the general indicators of the firm as a whole, but also in the context of individual items of products. The results of this analysis are presented in Table 2 and Table 3.

Table 2. Initial data for factor analysis of the wage fund in the context of individual types of products.

Types of products	Previous period			Current period			Total change
	Volume of output ($)	Salary intensity of products ($)	Wage fund ($)	Volume of output ($)	Salary intensity of products ($)	Wage fund ($)	
1	2	3	4	5	6	7	8
A	49000	0.57143	28000	56000	0.57304	32090	4090.00
B	800000	0.51500	412000	794450	0.51000	405170	–6830.0
C	340000	0.30935	105180	398000	0.28859	114860	9680.0
Total	1189000	0.45852	545180	1248450	0.44224	552120	6940.0

Table 3. Factor analysis of the wage fund in the context of individual types of products.

Types of products	Wage fund ($)		Absolute change in the total wage fund ($)		
	Previous period	Current period	Total	Including due to:	
				Volume of output	Salary intensity of products
1	2	3	4	5	6
A	28000	32090	4090.00	4000.00	90,00
B	412000	405170	−6830.0	−2858.25	−3971.75
C	105180	114860	9680.0	17942.47	−8262.47
Total	545180	552120	6940.0	27259.00	−20319.0
Total amount by factors for individual types of products			6940.0	19084.22	−12144.2
The difference between the total factor change and the sum of the product factor changes				8174.78	−8174.78

It can be noted that for the line "total" in Table 3, the absolute change in the context of individual factors (27259.00 and −20319.00) does not coincide with the sum of these factors in the context of individual types of products (19084.22 and −12144.22). The paradox is that the values of the factors under consideration were calculated using the same formulas (3 and 4). The difference between the values of the factors obtained by direct calculation and by summing up in the context of individual types of products is the same value, but with different signs (+8174.78). Thus, the influence of structural changes in the volume of output is manifested.

In formula (2), the total value of the wage intensity of all manufactured products (Wip) is a complex structure that includes the influence of two factors: the direct wage intensity of individual types of products and the significance of each type of product in the total volume of its output. In other words, the wage intensity of products as a whole for the firm is defined as a weighted average:

$$\mathrm{Wip} = \sum (\mathrm{di} * \mathrm{Wipi}) \tag{5}$$

Formula (5) reveals additional opportunities for a detailed analysis of the firm's personnel remuneration fund and allows you to ascertain the appropriate working formula for factor analysis:

$$\mathrm{W} = \mathrm{V}^* \sum (\mathrm{di}^*\mathrm{Wipi}) \tag{6}$$

Based on the formula (6), it becomes possible to directly calculate the influence of all three factors directly, that is, to directly determine the influence of the factor of structural changes in the total volume of production on the change in the firm's personnel remuneration fund (ΔWd):

$$\Delta \mathrm{Wv} = \Delta \mathrm{V} * \mathrm{Wip}(0) = (1248450 - 1189000) * 0.45852 = +27259$$

$$\Delta Wd = V(1) * \sum [\Delta di{*}Wipi(0)] = 1248450 * [(0.0449 - 0.0412) * 0.57143 + (0.6363 - 0.6728) * 0.515 + (0.3288 - 0.286) * 0.30935] = -8175$$

$$\Delta\, Wi = V(1) * \sum [di(1) * \Delta\, Wipi] = 1248450 * [0.0449 * (0.57304 - 0.57143) + 0.6363 * (0.51{-}0.515) + 0.3288 * (0.28859{-}0.30935)] = -12144$$

$$\Delta W = \Delta WV + \Delta Wd + \Delta Wi = 27259 - 8175 - 12144 = +6940$$

The results of the analysis at the second level of detail demonstrate an increase in the wage fund due to the factor of the volume of products produced, which is a positive moment in the peculiarity of the firm's work, as it indicates the increase of the business scale. A decrease in the amount of the wage fund due to the factor of structural changes in the volume of output can also be justified if these changes are caused by market conditions. The factor of wage intensity of products had a great impact on reducing business costs in terms of wages for workers (−12144). The influence of this factor is due to many reasons, often associated with a decrease in the level of remuneration of individual employees. This includes a reduction in payments for forced downtime, a reduction in overtime increased payments. This factor is also associated with an increase in labor productivity, when the growth of output by one worker outstrips the growth of his salary.

The results of the analysis using the three-factor model differ from the two-factor model in that the factor of the wage intensity of the entire volume of products in the whole enterprise: $\Delta W = -20319$ is divided into two component factors: $\Delta Wd = -8175$ and $\Delta Wi = -12144$.

The influence of the volume factor has not changed: $\Delta Wv = +27259$.

An alternative approach to the implementation of three-factor analysis is based on a direct interpretation of the economic content of individual factors. The volume factor assumes a directly proportional change in wage costs with a change in the volume of output:

$$\Delta JV = JV - 1 = (1248450/1189000) - 1 = 1.05 - 1 = 0.05$$

where JV and ΔJV are the growth rate and the growth rate of the output volume, respectively.

Hence $\Delta\, Wv = \Delta JV * W\,(0) = 545180 * 0.05 = 27259$

The conditionally calculated value of W(1) (as the sum of the products of the actual volumes of individual types of products on their basic wage capacity) differs from the basic value of W(0) due to the influence of two factors at once:

$$W(1) - W(0) = \Delta Wv + \Delta Wd, \text{ therefore } \Delta Wd = [W(1) - W(0)] - \Delta Wv$$

Table 4 shows a step-by-step scheme for calculating the factors considered.

The facts in Table 4 show that the most considerable increase in the wage fund by 27259 due to the volume factor is objective and justified by production tasks. A decrease in the share of relatively labor-intensive types of products in its total volume reduced the wage fund by −8175. A significant factor in reducing the size of the wage fund was a decrease in the unit cost of wages per unit of certain types of products: −12144, which indicates a more rational expenditure of wage funds in the current period compared to the previous one (unproductive payments decreased, labor productivity increased…).

Table 4. An integrated analysis of the dynamics of the wage fund by factors of volume, structure and wage intensity of products (subject to the growth rate of the volume of products JQ = 1,05 = 1248450/1189000).

Wage fund of the previous period	Wage fund for current period			Absolute change in the total wage fund			
	In the norms and structure (conditions) of the previous period	In the norms of the previous period and structure of the current period	Current period as is	Total (c.4-1)	Including due to:		
					Volume of output (c.2-1)	Product structure (c.3-2)	Salary intensity of products (c.4-3)
1	2	3	4	5	6	7	8
545180	572439	564264	552120	6940	27259	−8175	−12144

Of great practical importance is the factor analysis of changes in the wage fund in the context of factors of the number and average salary of personnel, both in the whole firm as a whole, and for individual categories of employees. It is also of great importance that the change in the structure of the firm's employees with different levels of remuneration has an impact on the total amount of the wage fund. The enlarged factor analysis of changes in the wage fund, in this case, is based on the working formula:

$$W = S^{*}Aw = \sum (Sj^{*}Awj) = \sum Wj \quad (7)$$

where S, Sj – the total quantity of employees of the firm and the number of the j-th category (group) of employees who have similar production functions and similar values of wage levels, respectively;

Aw, Awj – the average salary, respectively, for the whole firm and for the j-th category (group) of employees.

W, Wj – the value of the wage fund, respectively, for the firm as a whole and for the j-th category (group) of employees.

The initial data for the factor analysis of the first level of detail of changes in the wage fund as a whole for the firm under the influence of the number, average salary and structure of employees are presented in Table 5.

Table 5. Initial data for the analysis of changes in the wage fund.

Categories of employees	Average number of employees			Average salary		
	Previous period	Current period	Change	Previous period	Current period	Change
1	2	3	4	5	6	7
1. Total employees:						
1.1. Number	272.00	265.00	−7.00	2330.62	2422.64	92.02
1.2. Specific gravity (%)	100.00	100.00				
2. Workers:						
2.1. Number	230.00	224.00	−6.00	2370.35	2464.82	94.47
2.2. Specific gravity (%)	84.56	84.53	−0.03			
3. Specialists:						
3.1. Number	30.00	31.00	1.00	1994.30	2128.39	134.09
3.2. Specific gravity (%)	11.03	11.70	0.67			
4. Other categories:						
4.1. Number	12.00	10.00	−2.00	2410.00	2390.00	−20.00
4.2. Specific gravity (%)	4.41	3.77	−0.64			

Formula (7) has the potential to identify the factors discussed above [the average salary of employees (ΔWa) and the factor of their number (ΔWs)] as a whole for the firm simultaneously with the factors in the context of individual j-x groups of employees [the factor of the quantity of employees (ΔWsj) and the factor of their average salary (ΔWaj)]:

$$\Delta \mathrm{Ws} = \left\{[\mathrm{S}(1) - \mathrm{S}(0)]^{*}\mathrm{Aw}(0)\right\} \tag{8}$$

$$\Delta \mathrm{Wsj} = \left\{[\mathrm{Sj}(1) - \mathrm{Sj}(0)]^{*}\mathrm{Awj}(0)\right\} \tag{9}$$

$$\Delta \mathrm{Wa} = \left\{[\mathrm{Aw}(1) - \mathrm{Aw}(0)]^{*}\mathrm{S}(1)\right\} \tag{10}$$

$$\Delta \mathrm{Waj} = \{[\mathrm{Awj}(1) - \mathrm{Awj}(0)] * \mathrm{Sj}(1)\} \tag{11}$$

The factor's influence calculation of the number of employees (ΔWs) and the factor of their average salary (ΔWa) is presented in Table 5. In this table, the value of the conditional calculated values of the indicator of the wage fund is used both for the firm as a whole and for individual categories of employees. This conditionally calculated value is defined as the product (multiplication) of the actual value of the quantitative indicator (in this case, the number of employees) by the basic value of the qualitative indicator (in this case, it is the average salary). The influence of the quantitative factor in Table 6 is defined as the difference between the conditional calculated value and the basic value of the indicator. The influence of a qualitative factor is defined as the difference between the actual and conditionally calculated values of the indicator.

Table 6. Indirect identification of the factor of structural changes in the quantity of employees along with the factors of their number and average salary.

Categories of employees	The wage fund			The absolute change in the wage fund		
	Previous period	Conditional settlement	Current period	Total	Including due to the	
					Number	Average salary (c.4-3)
1	2	3	4	5	6	7
1. Total employees including by category	633929	617614.65	642000	8071	−16314.35	24385,35
2. Workers	545180	530957.91	552120	6940	−14222.09	21162,09
3. Specialists	59829	61823.30	65980	6151	1994.30	4156,70
4. Other categories	28920	24100.00	23900	−5020	−4820.00	−200,00
5. Total amount by factors for individual categories				8071	−17047.79	25118,79
6. The difference between the total factor change in the whole enterprise and the sum of factor changes by category (p.1–p.5)					733.44	−733.44

4 Discussion

Quantitative assessments of social stability at the micro-level of the economy in every economic entity play a weighty role in ensuring stable economic development [11]. An important component of this assessment is the characteristic of changes in the salary fund of the organization's personnel due to structural changes in the number of employees [12]. To obtain this kind of characteristic, the analysis of changes in the wage fund requires taking into account the influence of the following factors. Technical and economic factors: the volume of products, the structure of the labor intensity of these products, the level of remuneration for individual units of production. Economic factors: the total quantity of employees of the organization, the average salary for certain categories of employees, changes in the structure of employees. It is appropriate to consider technical and economic factors, first of all, in the projection of workers, since they directly carry out the production process [5, 13].

Of great importance are the theoretical prerequisites for assessing the possibilities of sustainable economic growth from the standpoint of neutralizing threats from the development of social tension among the company's personnel. The main point, in this case, is the aggregate assessment of all employees of the company of the fairness of remuneration for their work. This assessment is based on several main points. First of all, a lot depends on understanding the availability of professional excellence. Also important is the demand for this accessibility, which depends on the level of culture and upbringing of the company's employees. Therefore, the company's management should focus its efforts not only on the organization of professional development of its employees, but also on the formation of their aspirations for self-development and personal improvement [14].

An important role in establishing the social stability of the company is also played by the awareness of a fair difference in payment for work of different quantity and complexity. The relationship between remuneration and labor results is evaluated primarily in two main directions. In the macroeconomic aspect, the satisfaction of the interests of the majority of stakeholders is assessed by taking into account all socially significant aspects of the concept of labor results. This includes the demanded quality of labor results, minimizing the costs of achieving them, and the like. A private assessment of the fair wage difference is based on compliance with the labor market situation that develops at each time.

It seems that the well-being of a particular business depends on a rational combination of institutional and private criteria for a fair wage difference. Therefore, the improvement of integrated reporting should be directed towards taking into account the institutional justification of the difference in results and wages for the motivation of the company's employees [15].

At the level of an individual firm, it is of great practical importance to establish the dependence of the amount of wages on labor productivity in the context of individual factors. The main task is to identify the constituent elements of this ratio and establish their quantitative relationship with various production indicators. It is also of great importance to establish feedbacks that reflect the active role of wages in achieving the main business results. It should be borne in mind that all areas of evaluation of the ratio of pay and labor results should be checked for financial opportunities that correspond to the achieved results of the business as a whole.

The relative savings of the wage fund due to the outstripping growth of labor productivity over wage growth is the basis for organizing material incentives for further business development. The development of technological progress reduces the share of wages in the production costs of the business. This reduces the value of outstripping labor productivity growth, which provides savings in wages.

In the conditions of the modern increase in the share of material costs in the cost of production, the role of such a result of labor as the careful use of material resources increases. Therefore, the analysis of the ratio of the amount of remuneration and labor contribution to the achieved business results should take into account the relationship of the expenditure of salary funds with all relevant indicators of the achieved business results. This makes it necessary to take an integrated approach to the analysis of the ratio of results and remuneration, that is, to organically combine this analysis with a complete economic analysis of all aspects of the business.

The integrated approach is conditioned by the practical need to compare individual business results with the target costs for the remuneration of the company's personnel. The analysis of the ratio of wages and labor results requires taking into account a large number of indicators, since it depends on a large number of factors. Therefore, the practical significance of the use of correlation and regression analysis based on the processing of empirical data increases. This approach allows us to formalize and identify the influence of the selected set of main factors on the ratio of remuneration and labor results. However, the dependencies identified in this way will be characteristic only for specific business conditions.

5 Conclusion

The data from Table 6 indicate that the wage fund for the firm as a whole increased by +8071, including due to an increase in the average salary by +24385.35. It was also revealed that the wage fund decreased by −16314.35 due to a reduction in the number of employees. The total results of such an analysis of factor changes in the wage funds for individual groups of employees do not coincide with the overall results of the analysis as a whole for the firm. According to the number factor, the discrepancy is +733.44, and according to the average salary factor, the discrepancy is −733.44. Thus, we observe an indirect manifestation of the factor of structural changes in the number of employees of the firm. The direct calculation of the factor of structural changes in the number of employees involves the transition from the first level of detail of the analysis to the second level of detail.

Formula (7) provides limited opportunities for analyzing wages, without identifying the influence of the factor of structural changes in the number of employees. This factor of structural shifts requires a more complex representation of the calculation of the average salary than a simple division of the total amount of the salary fund by the total quantity of employees of the firm. In this case, the average salary in the whole firm should be determined according to the rules of the weighted average:

$$\mathrm{Aw} = \sum(\mathrm{dj} * \mathrm{Awj}), \tag{12}$$

where: dj is the specific weight of the j-th category (group) of employees of the firm in the total number of all employees.

The second level of detail of the analysis of changes in the wage fund as a whole for the firm takes into account three factors. Here, along with the factors ΔWs and ΔWa, which are determined by formulas 8 and 10, the factor of structural changes in the number of employees (ΔWd) is additionally calculated. The working formula for this level of detail is presented as:

$$\mathrm{W} = \mathrm{S} * \sum(\mathrm{dj} * \mathrm{Awj}) \tag{13}$$

Individual factors are calculated using the following formulas:

$$\Delta\mathrm{Ws} = [\mathrm{S}(1) - \mathrm{S}(0)] * \left\{\sum[\mathrm{dj}(0) * \mathrm{Awj}(0)]\right\} \tag{14}$$

$$\Delta Wd = S(1) * \left\{\sum\left[dj(1) - dj(0)\right] * Awj(0)\right\} \quad (15)$$

$$\Delta Wa = S(1) * \left\{\sum dj(1) * \left[Awj(0) - Awj(0)\right]\right\} \quad (16)$$

Formulas 14, 15 and 16 are used to determine the quantitative value of three factors (the number of employees, structural shifts in their composition and the average salary for certain categories):

$$\Delta Ws = (265 - 272) * 2330.62 = -16314.35.$$

$$\Delta Wd = 265 * \{(0.8453 - 0.8456) * 2370.35 + (0.1170 - 0.1103) * 1194.30$$
$$+(0.0377 - 0.0441) * 2410.00\} = -733.44$$

$$\Delta Wa = 265 * \{(2464.82 - 2370.35) * 0.8453 + (2128.39 - 1194.3) * 0.1170$$
$$+(2390.0 - 2410.0) * 0.0377\} = +25118.79$$

A direct calculation of all three factors, including structural changes in the composition of the number of employees (the number of all employees, structural shifts in their composition, the average salary of individual categories of employees) requires a transition from the first level of detail to the second level using a three-factor working formula that includes the specific weights of individual categories of employees.

References

1. Berdnikova, L.F., Mikhalenok, N.O., Frolova, V.A., Sukhacheva, V.V., Krivtsov, A.I.: Human resource management system development at smart university. In: Uskov, V.L., Howlett, R.J., Jain, L.C. (eds.) Smart Education and e-Learning 2020. SIST, vol. 188, pp. 327–337. Springer, Singapore (2020). https://doi.org/10.1007/978-981-15-5584-8_28
2. Fomin, V.P., Igoshina, N.A.: Assessment of the intercompany relationship between business interests and human capital. In: Ashmarina, S.I., Horák, J., Vrbka, J., Šuleř, P. (eds.) Economic Systems in the New Era: Stable Systems in an Unstable World, pp. 390–397. Springer, Cham (2021). https://doi.org/10.1007/978-3-030-60929-0_50
3. Fomin, V.P., Igoshina, N.A., Manyaeva, V.A.: Evaluation of the financial security of the competitive environment of the firm. In: Ashmarina, S., Mantulenko, V. (eds.), Proceedings of the 2nd International Scientific Conference GCPMED 2019 – Global Challenges and Prospects of the Modern Economic Development. European Proceedings of Social and Behavioural Sciences, vol .79, pp. 1462–1467 (2019). https://doi.org/10.15405/epsbs.2020.03.210
4. Bibarsov, K.R., Khokholova, G.I., Okladnikova, D.R.: Conceptual basics and mechanism of innovation project management. Eur. Res. Stud. J. **20**(2B), 224–235 (2017)
5. Haber, J., Schryver, C.: How to Create Key Performance Indicators (2020). https://www.cpajournal.com/2020/04/30/icymi-how-to-create-key-performance-indicators/
6. Fomin, V.P., Potokina, E.S.: Trends in optimizing the formation of consolidated reporting in holding companies in the context of global digitization. In: Ashmarina, S.I., Vochozka, M., Mantulenko, V.V. (eds.) ISCDTE 2019. LNNS, vol. 84, pp. 233–242. Springer, Cham (2020). https://doi.org/10.1007/978-3-030-27015-5_29
7. Melikhov, Y., Maluev, P.A.: Personnel Management. Portfolio of Reliable Technologies. Dashkov and Co., Moscow (2018)

8. Shatalov, M.A., Mychka, S.: Modern paradigm of the organization of the mechanism of management of the financial security of the enterprise. Bus. Strat. **8**(40), 13–21 (2017). https://doi.org/10.17747/2311-7184-2017-8-13-21
9. Dudin, M.N., Frolova, E.E., Lubenets, N.A., Sekerin, V.D., Bank, S.V., Gorohova, A.E.: Methodology of analysis and assessment of risks of the operation and development of industrial enterprises. Qual. Access Success **17**(153), 53–59 (2016)
10. Bazarov, T.: Personnel Management. Academy, Moscow (2017)
11. Rasskazov, S., Rasskazova, A., Deryugin, P.: Corporate Governance. Infra-M, Moscow (2020)
12. Speringer, M., Schnelzer, J.: Differentiation of Industry 4.0 models. In: The 4th Industrial Revolution from Different Regional Perspectives in the Global North and Global South (2019). https://doi.org/10.13140/RG.2.2.35510.55363
13. Rouse, M.: Digital Economy. Techtarget, Newton (2016)
14. Naumova, O.A., Tyugin, M.A.: The technique of monitoring the financial security of the economic entity based on the assessment of financial threatening risk. Sci. Vect. Togliatti State Univ. Ser. Econ. Manag. **2**(33), 34–41 (2018)
15. Bruskin, S.N., et al.: Business performance management models based on the digital corporation's paradigm. Eur. Res. Stud. J. **20**(4A), 264–274 (2017). https://doi.org/10.35808/ersj/833

Management of Operations and Business Processes: Problems of Digitalization and Development of Production Enterprises in Modern Russia

Viktor A. Aksenov[1](✉), Igor B. Khmelev[2], and Alla M. Yakusheva[3]

[1] Lobachevsky Nizhny Novgorod State University, Nizhny Novgorod, Russia
[2] Plekhanov Russian University of Economics, Moscow, Russia
[3] Nizhny Novgorod Institute of Management Russian Academy of National Economy and Public Administration, Nizhny Novgorod, Russia

Abstract. The authors consider the reasons for the recent interest in new approaches related to operational management in the scientific and practical literature. On the basis of the conducted research, the dynamics of business interaction in the organizations of commercial enterprises of Russia is analyzed. They study the dynamics of the use of information technologies and the digitalization of business processes, and try to identify the relationship between their changes and operational management. The authors conclude that it is not easy to assess these changes unambiguously. It is noticeable that the search for existing concepts and approaches to adapting management to new conditions gives very ambiguous, contradictory results. In the presence of positive trends in the small manufacturing business, significantly related to its digitalization, the overall state of large manufacturing enterprises is far from the world standards of efficiency. The results here are only somewhat related to production management. The article puts forward a hypothesis about the impossibility of a rapid change in the situation within the existing system of relations. The use of modern concepts of "Lean Production", Kanban, Kaizen, Just-In-Time, and digitalization runs into the disintegration of the status-role structure and organizational culture that developed during the Soviet period and, despite the shortcomings, the distortions more closely corresponded to the requirements of the fourth and fifth technological orders. The authors offer their own vision of management policy.

Keywords: Technological operations · Business processes · Operational management · Innovation · Information technology · Automation · Digitalization · Organizational culture · Status and role structure · Restructuring · Outsourcing

1 Introduction

Modern trends in the development of management practice are accompanied by the appearance in the scientific and practical literature of an increasing number of concepts

P. V. Trifonov and M. V. Charaeva (Eds.): DITEM 2021, LNNS 380, pp. 74–82, 2022.
https://doi.org/10.1007/978-3-030-94245-8_10

and approaches associated with the management of production processes at enterprises. These are "Lean manufacturing" (Lean manufacturing, Lean production), and Kanban, Kaizen systems, and TPM (Total Productive Maintenance), and JIT (Just-In-Time) [1]. In our opinion, this reflects some objective and subjective processes in the development of practice.

2 Materials and Methods

For a number of years (from 1995 to 2019), we conducted a study of intra-organizational processes at commercial enterprises in a number of regions of Russia (Moscow, Nizhny Novgorod, Penza, Yaroslavl, Kostroma, Vladimir regions, the Komi Republic, Bashkortostan, Udmurtia and a number of others) [2]. A model of the status-role structure of the organization was developed [3].

The methodology was developed within the framework of the Nizhny Novgorod Branch of the Association for Management Development (Chairman V.A. Aksenov) in cooperation with the Department of Production Management of the Faculty of Economics of the Lomonosov Moscow State University (consultant prof. Marshev V.I.) at the Department of Management of the Nizhny Novgorod Commercial Institute (Head of the Department Aksenov V.A.). The methodology is described in more detail in our article in the journal "Financial Markets and Banks" [2].

The results of the research were partially presented in the collection of articles of a scientific and practical nature "Entrepreneurial Management in Russia" edited by V.A. Aksenov, in the materials of the international conference of the Russian-American Center at Krasnoyarsk State University with the participation of the University of Washington, the US Federal Agency for International Development [4], international conferences at Lomonosov Moscow State University on the history of managerial thought and business [5], at the Financial University under the Government of the Russian Federation on CSR and business ethics [6]. Some of the materials were obtained by us in collaboration with Professor Gary D. Kruger (now a professor at the University of Wisconsin) when implementing the grant he received to study the process of transition of state-owned enterprises of Russia to the market [7].

Today, summing up the results of many years of work, we are processing the received data. More than 3,500 people were interviewed; the materials of more than 300 objects (firms, enterprises, divisions, departments, services) were analyzed. The research periods are allocated by us in accordance with the known key events of history. This article is intended to familiarize colleagues with the preliminary results. In our opinion, they are of some interest.

3 Results

Since the beginning of the reforms of the 90s, most of the production enterprises were relatively independent objects in structure, mainly related to technological processes (see Fig. 1).

They were usually unprofitable. The standard products produced were not focused on the needs of a specific customer. There was no marketing. Supply and sales, allocated

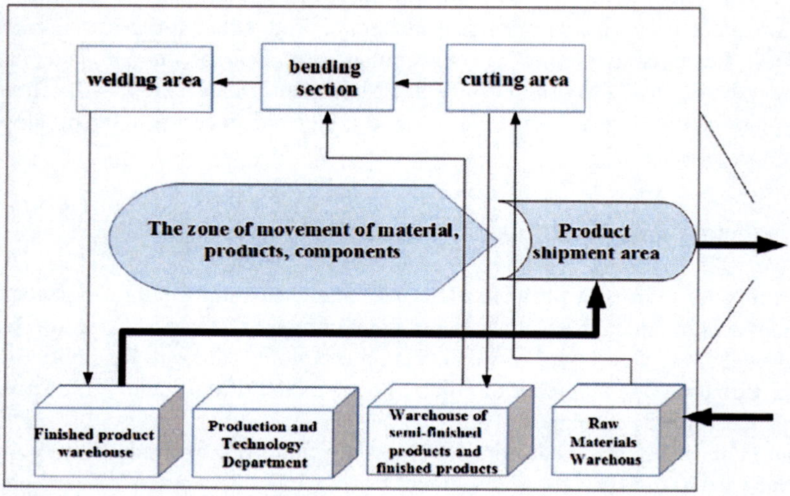

Fig. 1. The structure of the production enterprise. *Source:* printed from the enterprise database by the authors.

as independent firms, also focused on the necessary technological operations, trying to sell goods at the expense of organizational efforts.

Management, which appeared during the formation of the classical concepts of Taylor and Fayol, implied the management of technological operations as the organizational and technological lowest level of the "pyramid" of organization management. At the same time, the management decision was considered as the main tool. The manager of the lower level, in accordance with the tasks that he received "from above", was obliged to coordinate the technological process, not allowing deviations from the specified parameters, solving the problems that arose. Innovations in these conditions belong to the competence of a higher level of management. As a rule, this is the purchase of new equipment, the production of new products. Entering new markets does not affect

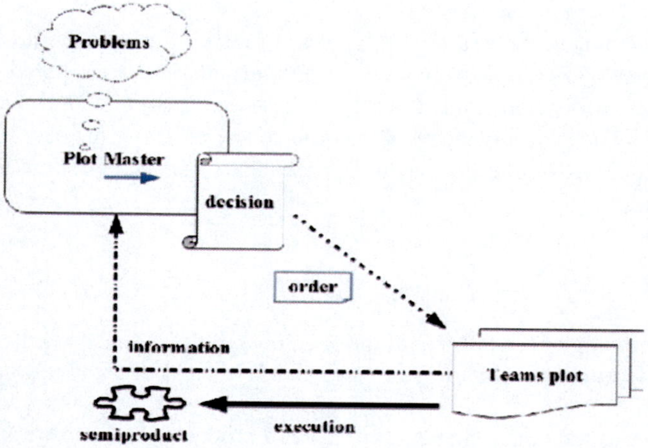

Fig. 2. Operation management based on management decisions. *Source:* compiled by the authors.

this level of management. In general, the "deviation control" represented the following scheme (see Fig. 2).

To achieve the main task of operational management (maintaining the normal course of the technological process and reducing costs), we tried to use traditional methods (see: Fig. 3) on the basis of rational organization of workplaces (Passport, Instructions), rationing (Photo of the working day, Schedule).

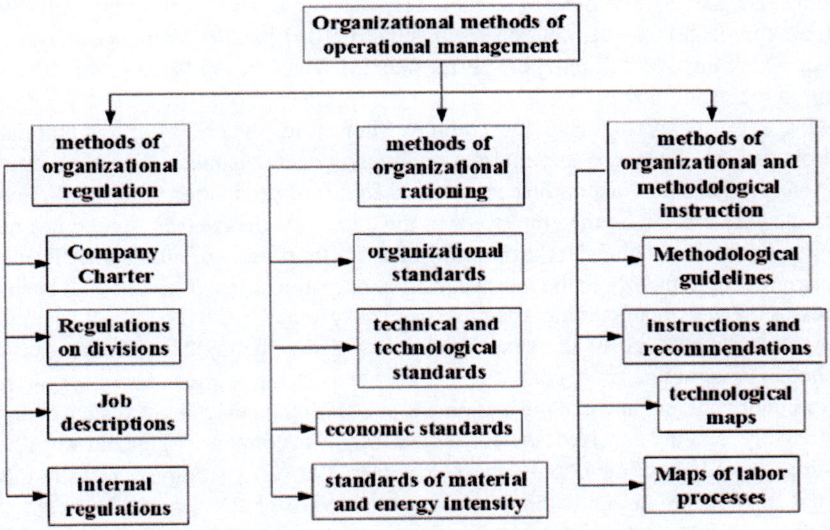

Fig. 3. Methods of operational management. *Source:* [8].

Information technologies (hereinafter referred to as IT) were used mainly in the automation of accounting, less often, personnel, warehouse, and commodity. Everyone knows the 1C, PARUS programs. In the Nizhny Novgorod region, a company associated with us has installed its DIANA programs to more than 2000 customers. Integrated control systems (ACS), for example, GALAKTIKA, were rare and were installed mainly by large companies such as SBERBANK, GAZPROM, LUKOIL (NORSI, TSNILHI), NIZHNOVENERGO (HPP). But they also had problems with the integration of regional divisions and branches. Only the transition to fiber-optic communications, broadband access and digitalization has made IT a mass phenomenon in operational management. Today, this is available even to small manufacturing enterprises.

With the development of the Russian market in the 2000s, the penetration of marketing into the practice of managing production enterprises, the restructuring of business based on "spin-off" technologies [9], the creation of vertically and horizontally integrated companies [10], business process management is becoming increasingly relevant [11]. Top-level managers are beginning to be interested in the implementation of CRM systems, Kanban, Kaizen systems, and TPM (Total Productive Maintenance), and JIT (Just-In-Time) [12].

Moving away from hierarchical structures to flexible, matrix, project, network structures leads to the focus of attention not on objects, sites, workshops, but on business processes [13]. End-to-end networks connecting all objects with the process of satisfying customers, consumers, lead to an expansion of the sphere of attention of the lower-level manager. In addition to coordinating technological operations, he is becoming increasingly concerned about economic issues (cost, calculation of time and resources, etc.). Operational management invades a higher sphere (level) tactics and even strategies. The reason for this trend, in our opinion, is not only the acceleration of STP, the development of innovative processes, but also the growth of education of specialists, qualifications, and their psychological readiness for such work. That is, not only objectively such an approach is needed, but also subjectively it is possible. Complex innovations appear which involve technological performers [5]. The ideas of "lean production" are becoming relevant [15].

IT also creates prerequisites for complex innovations, where integrated automated control systems of enterprises acquire a qualitatively new character. There are custom software products that take into account the specifics of production enterprises. Sometimes they are created by the employees of the enterprise themselves, sometimes managers are limited to technical tasks or setting tasks for third-party programmers. There are very original options, where the interweaving of operational management with business processes creates organic, rather than mechanical systems.

As an example, one of the small enterprises of the Nizhny Novgorod region can be cited. The structure of Argon AE LLC is shown in Fig. 1, however, after the new owner bought out and updated the equipment in 2019, the management system changed qualitatively. Having received several regular large customers, Argon did not refuse one-time orders of an individual nature. A potential buyer, visiting the website today, can not only make an application for the manufacture of the product he needs, but also place a specification with exact drawings of the product in the "Order" form. The order goes to the technologist, who uses CAD to bring the specification in line with the requirements of the CNC machines (see Fig. 4), determining the necessary tool and product specifications.

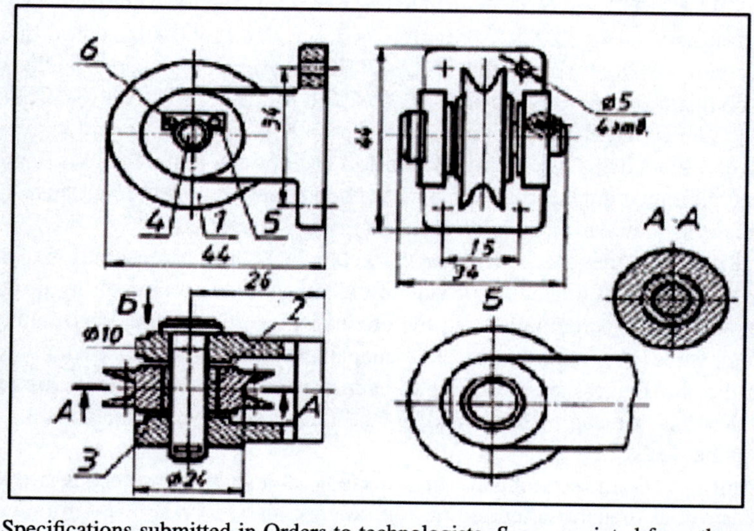

Fig. 4. Specifications submitted in Orders to technologists. *Source:* printed from the enterprise database by the authors.

The order is automatically assigned to the corresponding manager, who makes a calculation and forms a Contract and a Work Order with a schedule of work execution for individual stages (operations, semi-finished products, etc.) on the basis of the Order. The work Order with a Schedule is brought to each production, site, reflected on the contractor's internal screen with an indication of the deadline for execution in real time, as well as the employee's earnings for completed work or semi-finished products. In order to clarify the details of the Order, the manager can organize a video conference of a potential buyer with a technologist directly on the website. Here, the technologist can help the client make an Order-specification or submit drawings of the product needed by the buyer, made on the basis of the Application and even a verbal description of the product. Videos with negotiations, specifications and drawings are stored in the cloud in the Archive of the project, to which the client receives the right of access. If the customer agrees, a Contract with an electronic digital signature (hereinafter referred to as an EDS) of the manager and a Work Order with a work schedule appear on the site. They are sent to the customer in a convenient form (e-mail and messenger are connected to the site). The customer sends the signed Contract in the absence of an EDS in a convenient way for him. Together with the Contract (sometimes instead of it), an Invoice is formed for payment with an indication of the form and condition ("prepayment, letter of credit, etc. An advance in the amount of_ _ _ has been transferred, the amount of___has been paid"). The settlement account of Argon AE LLC becomes visible to the customer with the signing of the Contract, as well as the Schedule of Order execution in real time. The update is being updated every minute. Upon request, the Customer can see the status of Order execution at any time and find out the details from the manager in the video conference mode. The video cameras record the process of picking, packing, shipping the product, checking its quantity and quality. In the case of laboratory analysis, the data from the equipment is transferred to the site and attached to the Act of Acceptance and Transfer, which is formed on the site and signed by the project manager and an authorized person of the customer using EDS or otherwise. All materials are stored in the cloud in the project Archive until it is closed (payment of the Invoice, writing of the Act by the customer).

This type of work, when the site managers, foremen are responsible for operational management, and the project manager is responsible for managing the order (project), organizing and controlling his business process, allows the enterprise to work quite steadily for several years.

During 2008–2019, the number of small manufacturing enterprises enjoying the benefits of digitalization slowly grew. However, at large industrial enterprises, the expansion of the concept of operational management encounters the problem of the readiness of teams to actively participate in innovations [16]. This is manifested not only in the rejection of the proposed system of labor incentives, motivation, but also in the existing organizational culture, the status and role structure of organizations (see Table 1).

Our analysis of these data in other articles [3] allows us to draw a conclusion about the preservation of the patriarchal nature of the first and the disintegration, degradation of the second, their inconsistency not only with the fifth, but even with the fourth technological order. Failure to understand this by managers leads to the development of destructive

Table 1. Rating of the importance of employees depending on the roles in the organization. 2008 and 2018 Survey of ordinary employees. *Source:* compiled by the authors.

Name of formal roles	The degree of significance of the roles in 2008 and 2018		Naming of informal roles
Head-manager	5/5	5/3	Leader
Problem-solving analyst	3/4	4/3	Idea generator
Task manager	4/4	4/3	Teacher-educator
Allocator resource	5/5	4/4	Emotional stabilizer
Coordinator of works	4/4	4/4	Resonator
The controller-the accountant	3/4	1/2	Critic
Executor	4/4	5/4	Executor

interpersonal conflicts [17] and stresses of various typologies in the conditions of ordinary critical situations (see Table 2). According to research data, the socio-psychological situation at the lowest level of management in large industrial enterprises is often close to critical, often causes a state of conflict, sometimes stress and frustration [18].

Table 2. Types of stressful critical situations and their consequences in the period 2019–2021. *Source:* compiled by the authors.

Types of critical situations	Factors	The effects of stress
Failure of group goals Dissatisfaction with over-expectations Submission of super-requirements An attempt to set unrealistic goals	Intensity Duration	Apathy, decreased volitional efforts, loss of interest in work, dissatisfaction with membership in this group, violation of role expectations, group norms
Unexpectedly tough competition The occurrence of super-difficult tasks in the course of activity Loss of orientation in the course of solving the problem Lack of time Discomfort from the conditions of activity (cold, lack of sleep, monotony, oversaturation of interpersonal contacts)	Behavior of the manager Group behavior	Disorientation, loss of group initiative, confidence, depression, aggressiveness, suppression of criticism and the appearance of fear of force, restriction of personal freedoms, conflict, the appearance of closed groups, their competition with formal structures, the survival of objectionable, open sabotage of the decisions of the head, the collapse of informal and formal working structures

The classification of critical situations that we observed in the area of 2008, 2014, 2020 (2008, 2014, 2020 are indicated conditionally, the material covers all years in this range) allows us to speak about their similar nature and approximately the same consequences of group stress in cases of unfavorable development of the situation. Only the factors that affect the result change. During the crisis of 2008, the intensity and duration of negative events prevailed, as well as the behavior of specific groups in the organization. But after 2014, and especially during the epidemic of 2020, the behavior of the manager comes to the fore. We hope to devote a special article to this topic, which we are preparing on the basis of the collected materials.

4 Discussion

The world and Russian society are going through a rather difficult period in its history today. Someone connects this with the competition between the great powers [9], someone talks about the global change of technological structures [19], digitalization [6] and artificial intelligence [16], the change of generations of the managerial elite [14] and the political restructuring of the world community. In our opinion, the main contradiction lies in the discrepancy between the ongoing modernization of the economy in the country and the culture of organizational relations that have developed at the largest industrial enterprises, the collapse of the traditional ethics of labor behavior and its motives, and not only among the new, but also among the older generation.

5 Conclusion

In our opinion, the solution of this large-scale complex of problems is impossible on the basis of digitalization. Here it is necessary to restructure the business on the basis of outsourcing, change the system of personnel training at enterprises with the creation of corporate universities, colleges, technical schools, basic departments of universities directly in production, in workshops and departments with the development of diploma projects, graduation papers commissioned by prospective employers. In a large number of cases, it is necessary to completely replace the inefficient top and middle level of production management of companies or remove managers using the practice of anti-crisis management.

References

1. Anisimov, V.V.: Mod. Probl. Theory Mach. **4**(1), 168–171 (2016)
2. Aksenov, V.A., Berezin, S.V., Yakusheva, A.M.: Fin Markets Banks **3**, 8–13 (2020)
3. Aksenov, V.A.: The role of management in the operational management of organizations. In: Marshev, V.I. (ed.) History of Managerial Thought and Business: National Models of Management Personnel Training. XX International Conference. The History of Managerial Thought and Business. Management and the Roles of Managers: Yesterday, Today, Tomorrow: Materials of the Conference on June 28–30, 2019, pp 15–18. Lomonosov Moscow State University (2019)

4. Aksenov, V.A., Sumin, Yu.A.: Critical situations in the life of a commercial enterprise and the behavior of a manager: the problem of training. In: Abstracts of the International Conference "Problems Of Management Consulting in Russia". June 26–27, 1997, pp 5–11. Editorial and Publishing Center of Krasnoyarsk: Krasnoyarsk State University (1997)
5. Aksenov, V.A.: On the innovative activity of Russian enterprises. In: Marshev, V.I. (ed.) History of Managerial Thought and Business: National Management Models: Yesterday-Today-Tomorrow. X International Conference; Moscow, Lomonosov Moscow State University. Faculty of Economics, June 29–30, 2009, Conference Materials pp 18–23 (2009)
6. Belyaeva, I.Y., et al.: Corporate strategies and technologies in the context of global challenges. In: Belyaeva, I.Yu., Danilova, O.V. (eds.). KnoRus Publishing House, Moscow (2021)
7. Kruger, G.D.: Strategies of transition to the market of former state-owned enterprises of Russia. In: Aksenov, V.A., Panov, A.I. (eds.) Entrepreneurial Management in Russia. Collection of Articles, pp 176–203, Nizhny Novgorod (1997)
8. Bukhalkov, M.I., Kuzmin, M.A.: Ration Remun. Labor Ind. **1**, 15–25 (2019)
9. Bondarenko, T.G., et al.: Espacios **39**(36) (2018)
10. Khmelev, I.B., Kruglov, S.V.: Econ. Entrep. **4**(1), 549–555 (2014)
11. Holloman, D.: Corporate Spin-offs: Examining the Effects on the Entrepreneurial Orientation Extraction-Financial Performance Relationship. University of New Orleans (2008)
12. Sedelnikova, I., Fedorov, O., Yakusheva, A.: Int. J. Econ. Fin. Manag. Sci. **7**(6), 210–214 (2019). https://doi.org/10.11648/j.ijefm.20190706.15
13. Davenport, T.: J. Bus Analitics **1**(2), 73–80 (2018). https://doi.org/10.1080/2573234X.2018.1543535
14. Ayoub, K., Payne, K.: J. Strateg. Stud. **6**(5–6), 793–819 (2016)
15. Shibanov, K.S.: Econ. Entrep. **1**(102), 1045–1049 (2019)
16. Bebris, A.O., Khmelev, I.B.: Econ. Entrep. **12–1**(41), 712–717 (2013)
17. Kondratiev, E.A.: Probl. Theory Pract. Manag. **3**, 109–118 (2015)
18. Yadov, V.A., et al.: The state of labor relations at the enterprises of the Russian industry. In: Yadov, V.A. (ed.) Labor Relations: The State and Trends of Development in Russia. Collection of Scientific Articles. Samara University, Samara (2013)
19. Schwab, K.: The Fourth Industrial Revolution: Transl. from English. Moscow, Publishing house "Eksmo" (2020)

Transformation of Hospitality Industry Under the Covid-19 Crisis

Elena S. Yakushova(✉), Natalya A. Cherepovskaya, Irina Y. Litvin, Olga N. Chemodanova, and Tatiana M. Gainochenko

Financial University under the Government of the Russian Federation, Moscow, Russia
esyakushova@fa.ru

Abstract. The pandemic crisis, which has had a great impact on all aspects of human life, has revealed the need to create new models of organization management and develop quality management standards. Universal quality management in the modern vision of the European EFQM model and the updated ISO 9000 standards is focused on digitalizing the business environment, stimulating innovation and developing human capital. The year 2020 has become a serious test for all sectors of the economy, but especially for the hospitality industry. The purpose of this study is to identify the features of the development of the quality management system on the example of collective accommodation facilities in the Nizhny Novgorod region and to develop recommendations for improving the sustainability of the hotel business in the conditions of COVID-19. Within the framework of this study, an analysis of the financial condition of temporary accommodation facilities in the Nizhny Novgorod region was carried out and the main trends in the development of the hotel services market in this region in the pre-crisis period were identified. The study is based on the methods of variance, correlation and factor analysis, calculations were carried out using a specialized program SPSS Statistics. The practical significance of the study is to develop recommendations for improving the competitiveness of the hotel business in Nizhny Novgorod on the eve of large-scale anniversary events.

Keywords: Quality management system · Tourism · Sustainability · COVID-19

1 Introduction

Currently, tourism is one of the largest industries in the world, which affects the well-being and quality of life of people, the environment and the sustainable development of territories. In the pre-crisis period, the share of tourist services in GDP averaged 10.4% worldwide (in 2019 in Russia-4.8%). In 2020, due to the pandemic and quarantine restrictions, this figure fell to 5.5% [1]. According to the UN World Tourism Organization (UNWTO), in 2020, the international tourist flow decreased by 74% compared to 2019.

P. V. Trifonov and M. V. Charaeva (Eds.): DITEM 2021, LNNS 380, pp. 83–89, 2022.
https://doi.org/10.1007/978-3-030-94245-8_11

In order to survive in the conditions of the pandemic crisis, the hotel business needs to improve its business processes by introducing advanced quality standards into its practice. The priority of the new model of excellence of the European Foundation for Quality Management (EFQM) (adopted in 2019) is customer-oriented and meeting the needs of all stakeholders to achieve long-term success [2]. European Foundation for Quality Management and a series of standards ISO 9000 are integrated and complement each other. ISO 9000 quality standards have been developed for 16 subsectors of tourism. ISO/PAS 5643: 2021 "Tourism and related services" establishes requirements and recommendations for hotels in the tourism sector to prevent the spread of coronavirus and provide safe travel services.

In the world scientific community, there are different approaches to assessing the effectiveness of the hotel business and the quality of hotel services. Most researchers evaluate the quality based on consumer reviews [3–6], using special questionnaires or analyzing posts on major online booking platforms [7].

The quality of Internal services (ISQ) is determined by three main categories of organizational and personal practice: management systems, social systems and personal characteristics [8, 9], measures the sustainability of an organization using an integrated approach: sustainability as a process, sustainability as a result [10], as well as the subsequent impact on the effectiveness of the organization [11], the performance of the hotel.

It is important for hoteliers to take into account changes in consumer demand after the COVID-19 pandemic and restore the confidence of travelers [12, 13]. Study [14] it has shown that women are more demanding than men regarding the security measures that should be taken in hotels. In the study [15], it is noted that robotics and the introduction of artificial intelligence (AI), cleanliness and hygiene, health and healthcare can become promising areas for the development of quality management in hotels [16]; it is noted that the COVID-19 pandemic is an external shock that can change not only the actions of customers during the aggravation of the pandemic situation, but also affect their long-term behavior [17].

According to [18], hotels with a high level of innovation in service receive a higher online quality rating from customers. Thus, based on the knowledge gained with the help of big data, hotel companies can increase the innovation of their services [19].

In the Russian hotel industry, the implementation of certification according to international quality standards is slow. This is due to a number of reasons: a high level of depreciation of the hotel fund, a low level of qualification of administrative, managerial and production personnel, an insufficient level of demand for domestic tourism.

The purpose of this study is to identify the features of the development of the quality management system on the example of collective accommodation facilities in the Nizhny Novgorod region and to develop recommendations for improving the sustainability of the hotel business in the conditions of COVID-19. The choice of the hotel industry of the Nizhny Novgorod region as an object of research is due to the fact that in August 2021 Nizhny Novgorod will widely celebrate its 800-th anniversary.

2 Methodology

According to official statistics, in 2020, 526 hospitality facilities (hotels, guesthouses, hostels, recreation centers and others) were registered in the Nizhny Novgorod Region, 130 of them (24.7%) are located directly in Nizhny Novgorod city. From 2005 to 2019, the number of places in hospitality facilities in this region increased by 2 times, the number of rooms increased in +76.0%.

Over the past fifteen years, the tourism flow to the Nizhny Novgorod region has increased more than 2.5 times: from 505.7 to 1269.3 thousand people. In general, thanks to the implementation of regional tourism development programs, there is a positive dynamics of growth in the number of tourists placed in hospitality facilities in relation to 2005 in absolute terms, the average growth rate for the study period was +6.8%. But at the same time, the linear trend based on annual growth rates is downward.

At the first phase of the study, we used the SPARK-Interfax database to form the sample, which included operating enterprises corresponding to such parameters as:

1) region of activity: Nizhny Novgorod region;
2) the main type of activity (according to OKVED – All-Russian Classifier of Types of Economic Activity): the activity of providing places for temporary residence (code 55.1 and 55.2);
3) form of ownership: private.

In the second phase, we analyzed the financial and economic activities of the sample objects and grouped them by the size of assets, which allowed us to identify the main trends in the development of the hotel business in the region.

In the third phase, we developed recommendations for the hotel business on improving the quality management system during the crisis period, taking into account the original classification of accommodation facilities, which was invented by us at the second phase.

3 Results

The actual sample size was 97 enterprises (out of 132 possible). More than half of the surveyed sample (58.7%) are small private hotels and guesthouses that employ less than 10 staff members. The amount of revenue of these accommodation facilities ranges from 1 to 5 million rubles (Table 1).

Table 1. The distribution of accommodation facilities by the number of personnel and the volume of revenue for 2019.

Stuff number, people	Revenue per year, million rubles								Total
	0,1–1	1–5	5–10	10–20	20–50	50–100	More than 100	No data	
less than 6	8	15	5	3		3	1	5	40
6–10		5	4	6	2				17
11–15			3	2	5				10
16–50		1		2	8	6	2		19
51–100				1	2	1	2		6
101–200							4		4
501–1000							1		1
Total by column	8	21	12	14	17	10	10	5	97
Percentage, %	8.2	21.6	12.4	14.4	17.5	10.3	10.3	5.2	100

Source: Compiled by the authors using database "SPARK-Interfax"

At the next step, the sample population was grouped by the amount of assets. For grouping, we used unequal intervals, which were selected in such a way that the coefficient of variation in them did not exceed 45%. Thus, all the objects of the study were divided into 7 categories (Table 2).

Table 2. Distribution of hospitality facilities by assets, thousand rubles.

Indicator	Less than 300	300–1500	1500–000	6000–30000	30000–80000	80000–300000	More than 300000
Frequency	10	18	20	21	10	10	7
Average	170.30	724.17	2912.95	15982.10	41176.10	140547.90	980712.00
Standard deviation	74.13	296.58	963.36	6201.53	9272.95	45712.29	439052.05
Coefficient of variation, %	43.5	41.0	33.1	38.8	22.5	32.5	44.8
Category	7	6	5	4	3	2	1

Source: Compiled by the authors

Analysis of the financial stability of hotel business in the Nizhny Novgorod region showed that 44.3% of companies have a capitalization coefficient of less than 1.0. Negative values of this coefficient were obtained due to negative values in the balance line

"Capital and reserves" for 35% of hotel enterprises. Most of these companies were in categories 5 and 6, whose assets range from 0.3 to 6.0 million rubles.

4 Discussion

Based on the results of the study, a project was developed to develop a quality management system in temporary accommodation facilities in an unfavorable economic and sanitary-epidemiological situation, which takes into account the requirements of state bodies and new requests of Russian tourists. Below is a brief description of the main sections of the project.

Section 1. Sanitary safety of guests and employees. In this situation, the key condition for the functioning of a hotel of any category is strict compliance with the regulations and methodological recommendations of Rospotrebnadzor. This section provides for measures related to the purchase of special equipment (thermometers, recirculators, etc.), the creation of a stock of disinfectants and antiseptics, the organization of regular treatment of premises, redevelopment of public spaces, etc. A separate item is devoted to the process of admission of hotel staff to work – these are the regulations for conducting current examinations, laboratory examinations for a new coronavirus infection and vaccination.

Section 2. Digitalization of business processes (in order to minimize staff costs and minimize contacts between hotel employees and guests). In particular, small private hotels (categories 6–7) are recommended to introduce such digital services as online registration, communication with staff in the "live chat" mode and a mobile concierge. As for medium-sized hotel enterprises (categories 3–5), it is proposed to improve their quality system through two-way integration of PMS and CRM systems. This will allow the accommodation facility to expand the opportunities for working with the loyalty of guests and stimulating sales. Large hotel complexes in Nizhny Novgorod (category 1–2) have a high level of digitalization. But there are often personnel problems here: a shortage of employees with developed IT competencies and a long period of adaptation to organizational innovations due to a large number of employees and a complex organizational structure. In such cases, the management needs to focus on the development of a corporate culture that will allow retaining valuable personnel and ensuring more rapid implementation of various kinds of innovations in the company's activities.

Section 3. Diversification of the number of rooms. One of the promising areas of development of the hotel business is the redevelopment of vacant premises into coworking offices. In this case, the hotel can increase its income by providing additional services to corporate clients (hot meals, dry cleaning, a fitness room, a swimming pool, etc.). A more radical option for asset diversification is the organization of lofts or office apartments in the hotel fund. As a rule, such accommodation facilities are rented for a longer period than ordinary hotel rooms.

5 Conclusion

The COVID-19 crisis has become a real test for the tourism and hospitality industry around the world. The closure of borders and the introduction of strict restrictions had a negative impact on the activities of hotel enterprises and jeopardized their continued existence. In such conditions, only a strategically oriented enterprise can survive, which improves its quality management system taking into account sanitary and epidemiological rules and quickly adapts to new customer requests.

References

1. World Data Atlas: world and regional statistics, country data, maps and ratings. https://knoema.ru/atlas/topics/Tourism/Total-contribution-of-tourism-to-GDP/Total-contribution-to-GDP-share-percent
2. Fonseca, L., Amaral, A., Oliveira, J.: Quality 4.0: The EFQM 2020 model and industry 4.0 relationships and implications. Sustainability **13**(6), 3107 (2021). https://doi.org/10.3390/su13063107
3. Zobnina, M., Rozhkov, A.: Listening to the voice of the customer in the hospitality industry: Kano model application. Worldwide Hosp. Tour. Themes **10**(4), 43–448 (2018). https://doi.org/10.1108/WHATT-03-2018-0020
4. Xu, X., Li, Y.: The antecedents of customer satisfaction and dissatisfaction toward various types of hotels: a text mining approach. Int. J. Hosp. Manag. **55**, 57–69 (2016). https://doi.org/10.1016/j.ijhm.2016.03.003
5. Hsiao, Y.-H., Hsiao, Y.-T.: Online review analytics for hotel quality at macro and micro levels. Ind. Manag. Data Syst. **121**(2), 268–289 (2021). https://doi.org/10.1108/IMDS-04-2020-0214
6. Lau, A.: New technologies used in COVID-19 for business survival: insights from the hotel sector in China. Inf. Technol. Tour. **22**(4), 497–504 (2020). https://doi.org/10.1007/s40558-020-00193-z
7. Chan, I.C.C., Ma, J., Ye, H., Law, R.: A comparison of hotel guest experience before and during pandemic: evidence from online reviews. In: Wörndl, W., Koo, C., Stienmetz, J.L. (eds.) Information and Communication Technologies in Tourism 2021, pp. 549–556. Springer, Cham (2021). https://doi.org/10.1007/978-3-030-65785-7_52
8. Wu, X., Wang, J., Ling, Q.: Managing internal service quality in hotels: determinants and implications. Tour. Manage. **86**, e104329 (2021). https://doi.org/10.1016/j.tourman.2021.104329
9. Melián-Alzola, L., Fernández-Monroy, M., Hidalgo-Peñate, M.: Hotels in contexts of uncertainty: Measuring organisational resilience. Tour. Manag. Perspect. **36**, e100747 (2020). https://doi.org/10.1016/j.tmp.2020.100747
10. Trachuk, A.V., Linder, N.V.: The impact of financial capital on innovative behavior of industrial companies finance. Theory Pract. **25**(1), 51–69 (2021)
11. Ksenofontov, A.A., Sh Kamaletdinov, A., Trifonov, I.V., Trifonov, P.V., Cherepovskaya, N.A.: Using Russian Federation taxes information and analytical system to monitor economic condition of the manufacturing industry. IOP Conf. Ser. Mater. Sci. Eng. **828**(1), 012014 (2020). https://doi.org/10.1088/1757-899X/828/1/012014
12. Baum, T., Hai, N.T.T.: Hospitality, tourism, human rights and the impact of COVID-19. Int. J. Contemp. Hosp. Manag. **32**(7), 2397–2407 (2020). https://doi.org/10.1108/TR-05-2019-0188
13. Nhamo, G., Dube, K., Chikodzi, D.: Impacts and implications of COVID-19 on the global hotel industry and Airbnb. In: Counting the Cost of COVID-19 on the Global Tourism Industry. Springer, Cham, pp 183–204 (2020). https://doi.org/10.1007/978-3-030-56231-1_8

14. López, R.R., López-Felipe, T., Navajas-Romero, V., Menor-Campos, A.: Lessons from the first wave of COVID-19. What security measures do women and men require from the hotel industry to protect against the pandemic? Int. J. Environ. Res. Pub. Health **18**(5), 2232 (2021). https://doi.org/10.3390/ijerph18052232
15. Jiang, Y., Wen, J.: Effects of COVID-19 on hotel marketing and management: a perspective article. Int. J. Contemp. Hosp. Manag. **32**(8), 2563–2573 (2020). https://doi.org/10.1108/IJCHM-03-2020-0237
16. Hu, F., Teichert, T., Deng, S., Liu, Y., Zhou, G.: Dealing with pandemics: an investigation of the effects of COVID-19 on customers' evaluations of hospitality services. Tour. Manag. **85**, 1–14 (2021). https://doi.org/10.1016/j.tourman.2021.104320
17. Gössling, S., Scott, D., Michael Hall, C.: Pandemics, tourism and global change: a rapid assessment of COVID-19. J. Sustain. Tour. **29**(1), 1–20 (2021). https://doi.org/10.1080/09669582.2020.1758708
18. Shamim, S., Yang, Y., Zia, N.U., Shah, M.H.: Big data management capabilities in the hospitality sector: Service innovation and customer generated online quality ratings. Comput. Hum. Behav. 121, 1–14 (2021). https://doi.org/10.1016/j.chb.2021.106777
19. Prandi, C., Nisi, V., Ribeiro, M., Nunes, N.: Sensing and making sense of tourism flows and urban data to foster sustainability awareness: a real-world experience. J. Big Data **8**(1), 1–25 (2021). https://doi.org/10.1186/s40537-021-00442-w

Choosing an Organizational Change Management Strategy for the Younger Generation on the Base of EDAS Methods Analysis

Elena Y. Ganshina[1], Irina L. Smirnova[1](✉), and Svetlana P. Ivanova[2]

[1] Financial University under the Government of the Russian Federation, Moscow, Russia
ilsmirnova@fa.ru

[2] Moscow State University of Psychology and Education, Moscow, Russia

Abstract. Strategic management of organizations is becoming more complex and multi-tasking because of transformation of traditional management systems in the face of growing instability. The effectiveness of the organization's adaptation to the conditions of the new reality largely depends on the chosen strategies for dealing with change, as well as the ways to overcome resistance to change. This article examines the question of what strategies for overcoming resistance to organizational change are to be considered the most effective when interacting with representatives of the younger generation. The obtained data enable us to judge not only the current views of young people on management approaches to the implementation of innovations, but also to predict which strategies will bring maximum results in the future. This article focuses on the analysis of the best strategic alternatives using the EDAS (the Evaluation based on Distance from Average Solution) methods. The multi-criteria decision-making model proposed in the article shows that the strategy of introducing change can, in certain cases, be fair as a statistical generalization of the target group of respondents' opinions, but at the same time more accurate adjustment is possible only when taking into account, which reasons of resistance to change prevail in organization. Analysis of the research results might help managers to organize the process of upcoming change within an organization, predict a reaction of young workers to it, and develop a plan of measures to prevent negative opposition.

Keywords: Reasons for resistance to changes in organization · Strategy for overcoming resistance to management change · EDAS in decision sciences · Multi-criteria decision-making methods

1 Introduction

In the context of active transformation and instability of the external and internal environment of organization, the problem of analyzing and choosing strategic alternatives in the frame of organizational change and overcoming the resistance of employees to these

P. V. Trifonov and M. V. Charaeva (Eds.): DITEM 2021, LNNS 380, pp. 90–97, 2022.
https://doi.org/10.1007/978-3-030-94245-8_12

changes is becoming more and more urgent. According to different studies, approximately 25% of projects related to organizational change are successful in the long term [1], that is, in ¾ of cases, the stated goals and benefits of the implemented change strategy are not achieved. To adapt organization to changing conditions effectively, the organization's management must be open to interaction and ready for changes, such as to revise their powers, qualification requirements, transformation of the management hierarchy, etc.

One of the promising areas for overcoming resistance to new management approaches may be the "youthification" of the company's staff. Management of companies believes rightly that the younger generation is more inclined to change than the adult generation. Therefore, the main research issue of this article is the contextual relationship between tasks of modern organizations to change their management systems and the perceptions of employees aged 18–22 about how the management of companies should inform them about the upcoming changes and convince them of the need for these changes.

We set a goal using methods of multi-criteria decision-making to check the results within data on which methods of working with changes can be considered the most effective. This article addresses the following tasks:

1. To conduct a survey of future young professionals to identify their basic ideas about how the company's management should interact with employees, depending on various reasons for resistance.
2. Using the method of multi-criteria decision-making, EDAS (the Evaluation based on Distance from Average Solution [2]), to offer a tool for choosing the best strategy to overcome the younger employees' resistance to organizational change.

The rest parts of the article are organized as follows: Sect. 2 summarizes the relevant literature review. The research methodology is detailed in Sect. 3. Section 4 discusses the main results of processing survey results using methods of multi-criteria decision-making. Finally, the research is completed in Sect. 5 by presenting the main findings.

A significant amount of works has been published on the problems of organizational change [3, 4], but a clear-cut opinion on the impact of the strategic choice of tools for overcoming resistance to organizational change and sources of resistance in specific age groups of employees of the organization has not been formed yet.

The most common definition of change management in Western practice is "the application of a structured process and set of tools for managing change by people to achieve the desired business result" [5].

In the framework of organizational transformation, changes affect both the organization itself and the teams and individuals within the organization. Change management is the development of a strategy for a structural transition from the current state to the desired state in a certain way in order to achieve maximum acceptance and support for changes at all levels of the organization. Therefore, at the stage of implementing changes, a correct understanding of the specific basic ideas of the target audience plays a key role. If the audience supports some basic belief, then behavior based on other beliefs will seem incomprehensible and alarming to the audience [6].

The concept of the scientific study of management change was formulated by John Kotter with four main reasons why people resist change: a) Parochial self-interest, b) Misunderstanding and lack of trust, c) Different assessments, and d) Low tolerance for change [7]. Subsequently, these four main blocks of reasons were detailed, not contradicting, but rather complementing the original Kotter's classification.

Higgs and Rowland analyzed whether management behavior is related to the underlying assumptions of different approaches to change, concluding that the root cause behind many change problems is lack of employee trust in leadership [8]. Other researchers have got more attention to the entrenched standards of attitudes and behavior of employees, such as peer pressure, fatigue with changes, inertia of complex organizational systems [9], various fears associated with the likelihood of losing the job, habitual connections and contacts, fear of uncertainty and the unknown [10].

At the same time, researchers show that resistance to change is not always directly related to the inadequacy of personnel qualifications for the introduction of certain innovations. For example, employees are ready to apply technical innovations for their personal goals, but energetically resist the introduction of the same technologies into professional activities which are perceived as a set of routine practices [11].

Initially Kotter and Schlesinger proposed five strategies for different management situations when dealing with change: Education and communication, Participation and involvement, Facilitation and support, Negotiation and agreement and Manipulation and co-optation [7]. Following the logic of Kotter and later authors [12, 13], the authors of this article also group methods of work with changes into five main blocks, each suggesting a certain strategy for interacting with young audience in organization (Table 1).

It should be noted that a number of ways to work with staff, such as corporate information sources, teambuilding, directives are obvious and have been practiced widely for a long time, while others, for example, game practice (gamification), appeared just recently. According to a number of authors, a serious gaming movement is developing so actively so that "we can respond to the challenges of the present moment" and will allow us to work with changes with the least resistance [14].

From a theoretical point of view, any task, process or context can be gamified, because the main goal of this form of interaction is to "increase user engagement" [15]. User engagement, communication, interaction, recognition and promotion are the keywords in relation to the impact of gamification in the business environment [16]. Therefore, complementing the expansion of Kotter's classification by strategies for dealing with changes, we include gaming technologies as part of Kotter's "Participation and involvement" strategy.

2 Materials and Methods

Comparing the sources of resistance to change with strategies for working with changes enables to draw up a matrix questionnaire and conduct a survey which reveals which strategies for working with changes are considered by the younger generation to be effective and which, on the contrary, are ineffective. The results of the survey make it possible to carry out a further analysis within the EDAS method.

In the first phase of this research, respondents were surveyed where each participant assessed the effectiveness of all five strategies in dealing with each of the above concerns, based on their personal opinion. Consequently, the overall result of ranking strategies by each respondent was deduced where the highest score corresponded to the most effective method when working with all concerns, the lowest corresponded to the least effective one. The research involved young people born in 2000–2003, getting higher education in Moscow universities. A total of 331 people took part in the survey.

Having a certain ranking of strategies for managing resistance to organizational change, we proceeded to analyze the data using the EDAS method in order to select the best potential management decision. Since the EDAS technique is described in sufficient details in the scientific literature, we will outline the main stages of the study briefly. At the first stage, it is necessary to form an initial decision-making matrix for a set of alternatives, choosing the most important criteria for evaluating alternative solutions, in our case, these are sources of resistance to organizational change. It is also important to indicate the alternative options, in our case, these are management strategies.

Table 1. EDAS decision-making matrix.

Strategies/sources of resistance	Parochial self-interest	Misunderstanding and lack of trust	Different assessments	Low tolerance for change
Indicator weight	0.25	0.25	0.25	0.25
Facilitation and support	2.53	2.51	2.61	2.8
Education and communication	3.41	3.6	3.5	3.26
Negotiation and agreement	3.39	3.8	3.49	3.46
Manipulation and co-optation	3.11	2.17	2.54	2.45
Participation and involvement	2.7	3.03	2.96	2.61
avg (average column)	3.028	3.022	3.02	2.916

Source: Compiled by the authors.

Further, the EDAS method includes the calculation of positive and negative distances from the matrix means in accordance with the objectives of the criteria. The next step is to calculate the values of the weighted sums of positive distances from the mean (SPi) and negative distances from the mean (SNi) for all alternatives. Then the SPi and SNi values are normalized for all alternatives. This methodology is completed by the calculation of the score (ASi) for all alternatives and the ranking of the obtained results.

3 Results

The results of the research have confirmed the generalized data on the distribution of the effectiveness of resistance management strategies in the opinion of the younger generation. Thus, the analysis shows their applicability. According to the conducted analysis, the proposed method is efficient and the results are valid. It is important to note that the weights of all 4 types of sources of resistance were initially set the same (0.25 for each one) for the purity of the experiment. It is assumed that none of the types of sources of resistance to change prevails over the rest ones in the surveyed group of respondents (Table 2).

Table 2. Results of summary calculations of coefficients and ranking of alternatives by the EDAS method (option 1).

Strategies/odds	SPi	SNi	NSPi	NSNi	ASi	Rank
Facilitation and support	0.000	0.127	0.000	0.152	0.076	4
Education and communication	0.149	0.000	0.826	1.000	0.913	2
Negotiation and agreement	0.180	0.000	1.000	1.000	1.000	1
Manipulation and co-optation	0.007	0.150	0.038	0.000	0.019	5
Participation and involvement	0.001	0.058	0.004	0.612	0.308	3

Source: Compiled by the authors.

The results illustrate that the rating of the effectiveness of the strategies both in the sociological survey and as a result of recalculation according to the EDAS method is as follows:

- first come organizational methods of working with changes, involving the exchange of rational arguments of persuasion in favor of changes,
- second, corporate events, teamwork, trainings, that is, interaction primarily aimed at team building, and not encouraging individual initiative, and
- and only third are gaming technologies that focus on personal contribution and individual achievements.

The honorable fourth place is occupied by information methods of interaction associated with the broadest possible informing of employees about the essence of the changes being made. And, quite expectedly, the top-down approach completes the top five with directive approaches, that is, making changes from top to bottom through strict instructions and control.

Since the average solution is determined by arithmetic mean in this method, the EDAS method can be efficient for solving stochastic problems. However, the variety of management situations is disproportionately wider than the generalized average. In real-world problems, we are probably faced with some alternatives that need to be judged against many conflicting criteria. Making decisions on multiple criteria (where the weight of each criterion is entered) refers to making decisions in such situation. The example of the obtained data of the sociological survey shows the procedure of the proposed method and its applicability. However, the most interesting results we get in the sensitivity analysis, using simulated weights for the criteria to check the stability and reliability of the results of the proposed method.

As we indicated earlier, in the initial calculation of decision-making, the weight of all criteria was the same (0.25 each). That is, we assume that all four sources of resistance are equally present among the group of respondents. If one of the sources of resistance prevails, then the priority of strategies can also be revised. For example, narrow ownership interest prevails in a group of respondents due to specific organizational circumstances. In this case, as seen in Table 3, the indicators of the effectiveness of strategies are changing and the most effective strategy for introducing changes will no longer be "Negotiation and agreement" strategy, but "Education and communication" one. At the same time, the effectiveness of directive methods also increases significantly. Thus, the change in indicators depending on the weight of the studied criteria allows, in general, to illustrate the effectiveness of the application of EDAS methods precisely in the multi-criteria assessment of alternative solutions.

Table 3. Results of summary calculations of coefficients and ranking of alternatives by the EDAS method (option 2).

Strategies/odds	SPi	SNi	NSPi	NSNi	ASi	Rank
Facilitation and support	0.000	0.164	0.000	0.000	0.000	5
Education and communication	0.126	0.000	1.000	1.000	1.000	1
Negotiation and agreement	0.120	0.000	0.948	1.000	0.974	2
Manipulation and co-optation	0.027	0.000	0.215	1.000	0.607	3
Participation and involvement	0.000	0.108	0.000	0.341	0.171	4

Source: Compiled by the authors.

4 Discussion

A list of the most common sources of resistance to change among young employees when carrying out changes in the organization, as well as the most common practices when working with these resistance sources enable us to conduct a survey in order to identify which strategies for dealing with changes are considered as the most effective by respondents. The survey carried out in the course of this research and the obtained results show that representatives of the younger generation as a whole gravitate towards rational methods of interaction.

The younger generation is mostly focused on rationality in management and teamwork. They are convinced that precisely rational influence during meetings, discussions, strategic sessions, where reasoned opinions for and against changes are presented, is the most effective method of interaction. It is important to note that strategies based on individual initiative, direct competition, importance of personal achievements, that is, everything that distinguishes gaming techniques, does not evoke obvious, explicit support from the respondents, as might have been expected before.

At the next stage, the generalized data we initially obtained were verified using the EDAS method, one of the multi-criteria decision-making methods. The ranking of alternative strategies for managing resistance to organizational change according to this method fully confirmed the results of the respondents survey, taking into account the fact that the weight of each group of reasons for resistance was equal.

At the same time, if organization focuses on the most accurate adjustment of management interaction making changes for a young audience of employees, it is advisable to use the extended EDAS methodology. It allows taking into account the dominance of one or more reasons for resistance to organizational change, changing the weight of this reason in the initial decision-making matrix, which means it is more adaptable to the whole variety of managerial situations. Thus, we are intending to continue this research modifying the weight of a particular group of reasons for resistance and analyzing the correlation with the ranking of strategies.

5 Conclusion

In conclusion, it should be noted, that understanding the preferences of the young audience using the EDAS method enables companies to develop a clearer strategy for increasing controllability within introducing management change, both in production and management.

References

1. Lipman, V.: New Study Explores Why Change Management Fails – And How to (Perhaps) Succeed (2013). https://www.forbes.com/sites/victorlipman/2013/09/04/new-study-explores-why-change-management-fails-and-how-to-perhaps-succeed/?sh=797675e07137
2. Keshavarz Ghorabaee, M., Zavadskas, E.K., Olfat, L.: Multi-criteria inventory classification using a new method of evaluation based on distance from average solution (EDAS). Informatica **26**(3), 435–451 (2015). https://doi.org/10.15388/Informatica.2015.57
3. Adizes, I.: Mastering Change – Introduction to Organizational Therapy. The Adizes Institute, Santa Barbara (2016)
4. Beck, D., Cowan, C.: Spiral Dynamics: Mastering Values. Best Business Books, Russia, Leadership and Change (2010)
5. An Introduction to Change Management. https://www.prosci.com/resources/articles
6. Hofstede, G.: Uncommon Sence About Organizations: Cases. Publications Ltd, London, Studies and Field Observations. Sage (1994)
7. Kotter, J.P., Schlesinger, L.A.: Choosing strategies for change. Harv. Bus. Rev. **4**, 130–139 (2008)
8. Higgs, M., Rowland, D.: All changes great and small: exploring approaches to change and its leadership. J. Change Manag. Refram. Leader. Organ. Pract. **5**(2), 121–151 (2005)
9. Litchenstein, B.M.: Evolution or transformation: a critique and alternative to punctuated equilibrium. In: Moore, D. (ed.) Academy of Management Best Paper Proceedings, pp. 291–295. Academy of Management, Vancouver (1996)
10. Athota, V.S.: Lessons from failures. In: Mind over Matter and Artificial Intelligence. Palgrave Macmillan, Singapore, pp 49–56. (2021). https://doi.org/10.1007/978-981-16-0482-9_5
11. Kolygina, D.I., Kapuza, A.V.: Primary school teachers' resistance to change as a factor behind their use of ICT. Monit. Pub. Opin. Econ. Soc. Chang. **4**, 424–444 (2020). https://doi.org/10.14515/monitoring.2020.4.810

12. Burnes, B.: Why does change fail, and what can we do about it? J. Change Manag. Refram. Leader. Organ. Pract. **11**(4), 445–450 (2011)
13. Sammut-Bonnici, R., Wensley, R.: Darwinism, probability and complexity: market-based organisational transformation and change explained through the theories of evolution. Int. J. Manag. Rev. **4**(3), 291–315 (2002)
14. Tettegah, Sh.Y., Noble, S.U. (eds.): Emotions, technology, and design. In: Emotions and Technology. Academic Press, Cambridge, Massachusetts (2016). https://doi.org/10.1016/C2014-0-00787-3
15. Muntean, M., Cabău, L.: Business Intelligence Approach in a Business Performance Context. Austrian Computer Society. Band 280 (2011)
16. Banou, C.: Re-Inventing the Book: Challenges from the Past for the Publishing Industry. Chandos Publishing, Kingston-upon-Hull (2017). https://doi.org/10.1016/C2015-0-04339-8

The Use of Digital Technologies in Management Processes

Olga E. Ustinova[1(✉)], Tatiana A. Ason[1], Tatiana B. Izzuka[1], Elena A. Ermakova[2], and Svetlana N. Milovidova[1]

[1] Financial University under the Government of the Russian Federation, Moscow, Russia
oeustinova@fa.ru

[2] Yuri Gagarin State Technical University of Saratov, Saratov, Russia

Abstract. The development and widespread dissemination of digital technologies in many sectors of the economy has led to radical changes in the management of companies, which in turn creates new problems in organizing their effective functioning. This necessitates the search for new approaches in management. Existing structures of Business Process Management are faced by socio-technical changes such as digitalization, that are fundamentally changing the ways business units cooperate and companies interact. The status of digitalization in companies has changed from a desirable to a fundamental requirement. In modern conditions, the use of digital technologies is becoming one of the key factors in ensuring the competitiveness and investment attractiveness of companies. This paper presents the results of a study of the level of digitalization of management processes in Russian companies. Based on the results of the study, a gradation of companies using neuro- and digital technologies in making management decisions is proposed.

Keywords: Digitalization · Business processes · Management processes · Digital technologies

1 Introduction

The technological revolution and the globalization of business have created a complex, competitive and ever-changing business environment. Scientists and practitioners set themselves ambitious tasks that require solutions in the field of management. It is worth noting that sustainable competitive advantage no longer arises solely from position, scale, and ability to manufacture or sell. Modern management technologies are needed for companies to cope with the changing business environment of tomorrow. To achieve a lofty goal, it may also require a restructuring of the philosophical foundations, the elimination of pathologies of the formal hierarchy, etc. Digitalization of business processes allows for the automation of unstructured tasks [1, 2], and the Internet of Things and blockchain enable decentralized processes [3–5].

Process improvement has long been a top priority for decision makers. Given the diversity of emerging digital technologies, companies face a high degree of uncertainty about their use. They are changing the way companies operate, and this trend will

P. V. Trifonov and M. V. Charaeva (Eds.): DITEM 2021, LNNS 380, pp. 98–104, 2022.
https://doi.org/10.1007/978-3-030-94245-8_13

continue in the future, providing organizations with the opportunity and sometimes forcing them to change their business processes [6]. The problem is that many companies still lack knowledge of what technologies they should use to improve their business processes. For example, a McKinsey study found that only 7% of 850 senior executives believe their organization understands the value of digitalization [7].

The processes are subdivided into main and auxiliary and managerial ones; however, they can be classified according to the recurrence of the intensity of knowledge, interdependence and variability [8, 9]. The concept of business process management (BPM) is based on the practice of monitoring work performance in order to ensure stable results and exploit opportunities for improvement [10]. From a lifecycle perspective, BPM includes identification, definition, modeling, implementation and execution, monitoring and control, and process improvement [11]. Combining knowledge of information technology and management sciences [12], BPM is a prerequisite for efficient and effective work [13]. For a long time, approaches to process improvement have been criticized for the lack of guidance on how to implement process improvement in practice [14, 15]. Several authors [17] have proposed a structure that allows them to independently generate ideas for process improvement. The value of these achievements is undeniable, as there is no single approach that can derive and prioritize ideas for improving processes based on digital technologies.

However, given the benefits of digitalization, most companies are not prepared to harness its full potential. One of the reasons is the use of digitalization in solving individual problems, and not the business processes of the company as a whole. Due to the fact that individual tasks have their own characteristics, it becomes more difficult to scale the created technology. Taking into account the above, we have designated the goal of studying the level of use of digital technologies in the management of Russian companies in order to further, taking into account the identified problems, develop optimal directions for the successful implementation of digital technologies in management processes.

2 Materials and Methods

As part of the construction of the trajectory of the program part of the study, including the study of the level of digital and neuro-technologies used, it was decided to conduct a quantitative study in the form of a questionnaire survey of company executives. This approach will help to optimally implement the assigned tasks, as well as identify the features of digitalization of management in companies. The calculation of the sample size for a survey with a known size of the general population is made according to the following formula 1:

$$n = (t^2 \cdot \sigma^2 \cdot N)/(N \cdot \Delta^2 + t^2 \cdot \sigma^2) \quad (1)$$

where n is the volume of the sample, Δ is the value of the permissible error in shares, N is the volume of the general population, t is the confidence coefficient (reliability criterion), σ is the root of the variance or the measure of dispersion of the studied attribute characterizing the magnitude of the deviation from the mean values in the general population... According to Rosstat [17], the number of active enterprises in 2019 amounted to 3,121,865 organizations.

Thus, the determination of the required sample size with a total target audience of 3,121.8 thousand rubles. companies is:

$$n = (22 * 0.25 * 3121.8)/(3121.8 * 0.052 + 22 * 0.25) = 355 \text{ respondents.}$$

In fact, 557 executives, managers and owners of companies took part in the survey.

3 Results

Digital transformation processes in companies can proceed in different ways. Some companies, adhering to the corporate structure, try to carry out digitalization in stages, sequentially, while others, as a rule, active and innovative organizations are rapidly updating their in-house infrastructure. Heads of companies and managers of structural units in modern conditions are aware of the importance and inevitability of digital transformations of business processes. So, answering a question regarding the business processes of internal divisions subject to digital transformation, the leaders and managers of the company noted the following:

- planning systems, including production and investment;
- management and accounting;
- logistics and communications;
- office work, document flow and reporting;
- interaction with customers and target audience;
- search and selection of personnel, personnel control system;
- information protection and control systems;
- processes of interaction between departments;
- risk analysis;
- product quality control and others.

Many organizations are actively introducing digitalization into their business, celebrating its effectiveness. The effect of the introduction of digital transformation systems allows managers to make more accurate forecasts, optimize material and labor costs, ensure the proper quality of goods and services provided to consumers, etc. Answering the question about indicators that most fully and accurately reflect the positive impact of modern digital technologies on management processes, company managers noted their importance on a 5-point scale (Fig. 1).

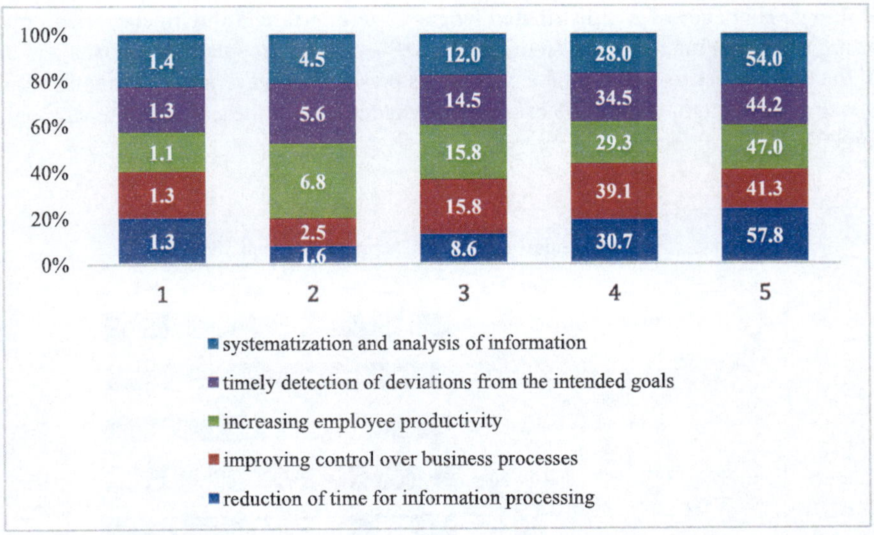

Fig. 1. The impact of digital technologies on management processes.

The following indicators were marked with the greatest positive influence of digital technologies in making managerial decisions: reduction of time for information processing (57.8%) and systematization and analysis of information (54.0%).

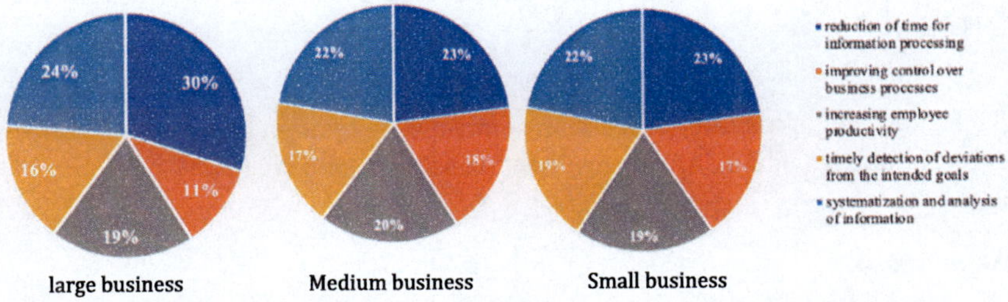

Fig. 2. The importance of using digital technologies in management processes in the context of small, medium and large companies.

Analyzing the results obtained, based on the highest score (Fig. 2), it is worth noting similar responses for all the criteria noted by representatives of small and medium-sized businesses. At the same time, managers of large companies were given special importance to systematize and analyze information (30%). At the same time, the criterion "improving control over business processes" for these companies is not so important (11%) than for medium and small businesses.

Based on the results of the study, a gradation of companies using neuro- and digital technologies in making management decisions is proposed, depending on the frequency of their use, "often-rarely-irregularly-never". So, on an ongoing basis, companies (more than a quarter of companies 26.3%) use wireless technologies, big data, and the Internet of things. He never uses a distributed ledger system, artificial intelligence, neuromarketing when making decisions (more than 26.9%). On an irregular basis, managers use all the types of neuro- and digital technologies presented, on average 27% of all companies. Rarely enough, about 20% of company executives use these types of technologies (Fig. 3).

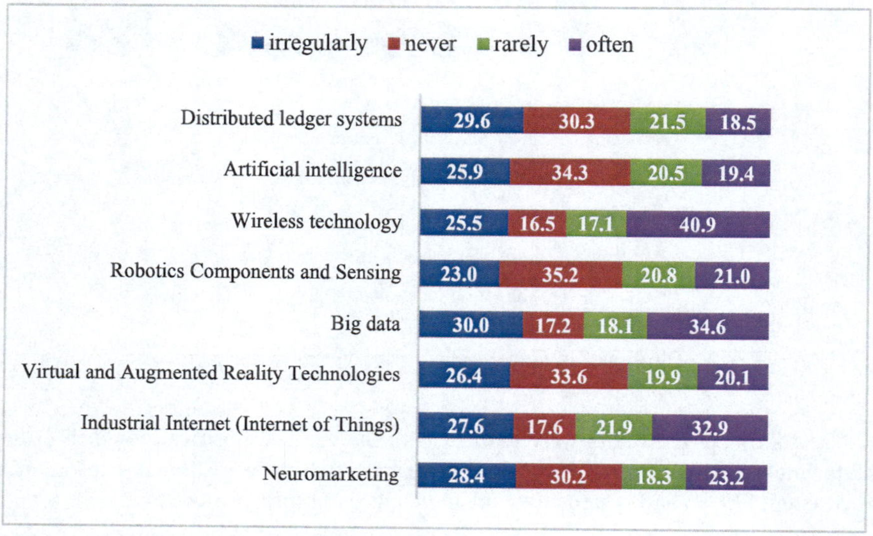

Fig. 3. The use of neuro- and digital technologies in making management decisions by Russian companies.

Thus, the level of maturity of Russian companies in relation to the use of neuro- and digital technologies in making management decisions can be defined as average.

4 Discussion

It is known that an organization consistently goes through different levels of maturity. For any company there comes a time when it is necessary to harmonize business processes and adapt them to the existing realities on the market. Otherwise, they may lose their competitive positions.

The results of the study show the level of maturity of Russian companies regarding the digitalization of management processes. For more mature Russian companies, factors such as a clear business rationale, capabilities and access to technology are not a major concern as they have the appropriate expertise in this area. In turn, the deficit of these factors is characteristic of other, so-called "passive" companies. Most likely, this can be justified by the lack of practical experience. For seasoned companies implementing new projects and investing heavily in innovative technologies, it is important to have access to talented people who provide leading positions in the industry. In addition, an important factor is a competition with other projects, as well as safety.

Therefore, we can conclude that organizations achieve the greatest success when they fundamentally rethink their business processes, including management ones, towards their digitalization. It is important to emphasize that each new development must build on the foundation of the previous one, whereby a natural cycle of changes starts. Business leaders should identify the areas of the business that are most profitable to transform with digital technologies, and then focus them on completely reconfiguring one or two selected areas of the business. Their implementation will also require reconfiguring operational processes, changing collaboration methods, and possibly completely rethinking the business model.

5 Conclusion

Digital transformation of business processes provides companies with the opportunity to increase productivity, reduce costs, improve the quality of work planning and business management, create a positive company image, improve customer service processes and achieve market leadership positions. Within the framework of the study, the features of the intentions of managers of different types of companies regarding the introduction of digitalization into management processes are highlighted. It must be emphasized that different areas of activity require different volumes of digital transformation. At the same time, this criterion may change over time. Nevertheless, the digital transformation of business processes is necessary for companies that are in tough competition, for which a delay in the start of digitalization or even abandonment of it may negatively affect its market positions.

References

1. Van der Aalst, W.M.P., Bichler, M., Heinzl, A.: Robotic process automation. Bus. Inf. Syst. Eng. **60**, 269–272 (2018). https://ezpro.fa.ru:2696/10.1007/s12599-018-0542-4
2. Willcocks, L., Lacity, M.: Conflict resolution in business services outsourcing relationships. J. Strateg. Inf. **26**, 82–100 (2017). http://eprints.lse.ac.uk/69810/
3. Oberländer, A.M., Röglinger, M., Rosemann, M., Kees, A.: Conceptualizing business-to-thing interactions—a sociomaterial perspective on the Internet of Things. Eur. J. Inf. Syst. **27**, 486–502 (2017). https://ezpro.fa.ru:2696/10.1080/0960085x.2017.1387714
4. Zarkadakis, G., Jesuthasan, R., Malcolm, T.: The 3 ways work can be automated (2016). https://hbr.org/2016/10/the-3-ways-work-can-be-automated
5. Viryasitavat, W., Da Xu, L., Bi, Z., Sapsomboon, A.: Blockchain-based business process management (BPM) framework for service composition in industry 4.0. J. Intell. Manuf. **12**, 133 (2018). https://doi.org/10.1007/s10845-018-1422-y
6. Matt, C., Hess, T., Benlian, A.: Digital transformation strategies. Bus. Inf. Syst. Eng. **57**(5), 339–343 (2015). https://doi.org/10.1007/s12599-015-0401-5
7. Gottlieb, J., Willmott, P.: The digital tipping point (2014). http://www.mckinsey.com/business-functions/business-technology/our-insights/the-digital-tipping-point-mckinsey-global-survey-results
8. Becker, J., vom Brocke, J., Heddier, M., Seidel, S.: In search of information systems (grand) challenges. Bus. Inf. Syst. Eng. **57**, 377–390 (2015). https://ezpro.fa.ru:2696/10.1007/s12599-015-0394-0

9. Zelt, S., Schmiedel, T., vom Brocke, J.: Understanding the nature of processes: an information-processing perspective. Bus. Process. Manage. J. **24**, 67–88 (2018). https://ezpro.fa.ru:2696/10.1108/bpmj-05-2016-0102
10. Dumas, M., La Rosa, M., Mendling, J.: Fundamentals of business process management. Springer-Verlag, Berlin Heidelberg. (2013). https://doi.org/10.1007/978-3-642-33143-5
11. Recker, J., Mendling, J.: The state of the art of business process management research as published in the BPM conference. Bus. Inf. Syst. Eng. **58**(1), 55–72 (2016). https://aisel.aisnet.org/bise/vol58/iss1/6
12. van der Aalst, W.P.M.: Business process management: a comprehensive survey. International Scholarly Research Notices, pp. 1–37 (2013). http://www.vdaalst.com
13. De Bruin, T., Rosemann, M.: Towards a business process management maturity model. In: Bartmann, D., Rajola, F., Kallinikos, J., Avison, D., Winter, R., Ein-Dor, P. (eds.). Proceedings of the 13th European Conference on Information Systems, Regensburg (2005). https://aisel.aisnet.org/ecis2005/37
14. Adesola, S., Baines, T.: Developing and evaluating a methodology for business process improvement. Bus. Process Manage. J. **11**(1), 37–46 (2005)
15. Allen, S.: Harvard business review (2015). https://hbr.org/resources/pdfs/comm/verizon/HBR_Report_Verizon_Workforce.pdf
16. Vanwersch, R., et al.: A critical evaluation and framework of business process improvement methods. Bus. Inf. Syst. Eng. **58**(1), 43–53 (2016). https://doi.org/10.1258/jicp.2012.012m16
17. Rosstat. Demographic indicators of enterprises by type of economic activity in 2019. https://www.gks.ru/free_doc/new_site/population/trud/tab_trud1.htm/

Management Based on the "Lean Production" Concept in the Industry of the Tula Region: Experience, Problems and Prospects

Tatyana A. Fedorova[1,3](✉), Elena A. Fedorova[2,3], and Nadezhda O. Kozlova[2,3]

[1] Tula State Lev Tolstoy Pedagogical University, Tula, Russia
[2] Institute of Consulting LLC, Tula, Russia
[3] Financial University under the Government of the Russian Federation, Tula Branch, Tula, Russia

Abstract. The purpose of the research was to get an insight into the experience in implementing "lean manufacturing" methods by industrial enterprises of the Tula region and develop recommendations for achieving sustainable success in that area. Reviewing the "lean production" implementation results showed that enterprises have sufficient reserves to reduce losses and increase labour productivity. To achieve these goals, no additional investment is required, so many enterprises have joined the National Project "Labour Productivity and Employment Support" in order to obtain the necessary competencies and attract investments at preferential rates for manufacturing improvements. Already in the first year, the introduction of "lean manufacturing" methods allows enterprises to significantly reduce the time of the production process, reduce unproductive losses, increase output, product quality and labour productivity. Analysis of various factors affecting the "lean production" implementation effectiveness discovered the need for changes in the development strategy, management system and organisation of production and auxiliary processes based on the implementation of accepted standards and the use of the intellectual potential of employees. Taking into account the traditions in Russia in the field of standardisation and personnel training in various qualifications and levels, the developed recommendations will make reducing losses and increasing labour productivity sustainable even in the absence of any state support.

Keywords: Production management techniques · Resource saving · Productivity

1 Introduction

The history of "lean production" as a business management concept was initiated at Toyota. The astounding success achieved in that globally famous company has pushed the management of organisations and enterprises in industries and countries to implement "lean manufacturing". In Russia, at the moment, the implementation of "lean manufacturing" in industry is going through a new stage, when the implementation is taking place

P. V. Trifonov and M. V. Charaeva (Eds.): DITEM 2021, LNNS 380, pp. 105–110, 2022.
https://doi.org/10.1007/978-3-030-94245-8_14

in the context of a sharp reduction in the number of employees, economic sanctions and overcoming the global COVID-19 epidemic.

Studies conducted in Russia in 2019–2020 reflect aspects of the positive impact of the introduction of innovative manufacturing process management methods. It is noted that those methods influence the improvement of the financial condition of enterprises [1], the development of an entity in the digital economy [2], and in the case of large-scale implementations, the socio-economic development of the region as a whole [3].

The main principle of Toyota's "lean manufacturing" is uniform work based on the use of the intellectual potential of employees (their proposals for improving production processes), which ensures the high quality of the cars produced regardless of the country where the production is located [4]. It can be assumed that it is precisely the "combination of incompatible", namely standardisation and employee initiative that gives the outstanding result of the efficient production of high-quality equipment recorded in Toyota all over the world.

The "lean manufacturing" methodology assumes the reduction of all types of losses and the conservation of all types of resources, such as materials and components, time and labour costs. As a result of the multiplier effect, both labour productivity and production profitability increase.

The introduction of "lean" principles of organising production in Russian companies faces difficulties noted by researchers. One of the main among them is resistance to change, mistrust of employees in the management of enterprises, lack of respect for employees on the part of managers, low labour motivation [5].

It should be added that the effective implementation of new methods for organising production is hindered by the multilevel management system at Russian enterprises and organisations, as well as by the high wage gap between employees and management personnel. In this regard, it is necessary to address the problem of the systemic nature of changes at enterprises during the "lean production" implementation: what changes in management systems should it be accompanied by, what indicators of changes should be used for correction and subsequent planning of enterprise activities.

This research summarises the experience in implementing "lean manufacturing" at enterprises of the Tula region. Approaches to shaping a strategy for the introduction of new methods in manufacturing companies are proposed, including changes in the planning and management system, which will consistently ensure the economic effect of the changes.

2 Materials and Methods

In the proposed research, the goal was to study the impact of the introduction of "lean manufacturing" methods on the performance of industrial enterprises in the Tula region and develop recommendations for achieving sustainable success in that area. To achieve this goal, such challenges as studying the experience in introducing new approaches to organising production at industrial enterprises of the Tula region were solved; addressing issues arising during the implementation; developing recommendations to ensure sustainable results.

To achieve the set goal of the research, the methodological apparatus of system analysis, which allows us to consider the introduced "lean production" as part of a higher-order management system – the management system of an enterprise as a whole, was used.

For the research, the data presented on the Tula region's portal of national projects were used [6]. It should be noted that the portal contains the results of only organisations participating in the National Project "Labour Productivity and Employment Support", which receive some support in the form of concessional lending and training at the expense of federal and regional budgets. In entities implementing new approaches to organising production processes on their own, the results may be different.

3 Results

In the Tula region, the introduction of "lean production" at industrial enterprises has entered an active phase as part of the implementation of the Federal Programme (2018), and then the national project "Labour Productivity and Employment Support" (2019–2021).

Currently, 60 enterprises are participating in the national project in the Tula region. In 2020, Process Factories were opened at Aerosol Novomoskovsk LLC and in the Octava cluster. During the project implementation, more than 4,000 employees of enterprises, including more than 700 members of working groups engaged in changes in processes at their enterprises, were trained. Training programmes are designed for specialists and workers of various profiles and include issues not only directly related to lean production methods but also quality management, personnel management, labour rationing etc. [6]. All enterprises involved in the project note positive results. So, in 2020, the introduction of "lean production" at "TKF" Yasnaya Polyana" OJSC provided a reduction in the cooling time of the filling by 24 h, which made it possible to increase the production of gingerbread by 13%. At the Shchekinskiy Linoleum enterprise, equipment downtime was reduced by 30%, and at the Tula macaroni factory, due to a 69% reduction in changeover time, the total output was increased by 12%. At Present Packaging LLC, by means of work standardisation and changes in the production process, the process development was increased by 44% [6].

It should be noted that the labour productivity indicator calculated as the ratio of revenue to the average headcount depends on not only the organisation of the production process of an enterprise. In the event of a decrease in demand for the production of basic products, the company's revenue may decrease. With the release of reserves, it becomes possible to additionally make other products with the same capacities, which allows maintaining the achieved level of labour productivity and jobs, including skilled workers, which is quite essential in modern conditions.

In this regard, the problem of how to most economically integrate the "lean manufacturing" methodology into the development strategy of an enterprise arises. Given the Toyota experience, the solution to this problem is the introduction of lean manufacturing methods into enterprise standards for production, support and management processes. However, it was noted at none of the enterprises considered, whose experience was studied during the research, that the introduction of new approaches influenced management processes and led to a reduction in management levels and management personnel. This also indicates that, with new manufacturing process management methods, they do not use or insufficiently use the second component of "lean production", namely the intellectual potential of workers.

The research into the experience gained allows us to offer for enterprises, that implement "lean manufacturing" methods, recommendations concerning:

1) Enterprise development strategies;
2) Enterprise management systems;
3) Organisation of main and auxiliary production processes.

The development strategy, in addition to the innovation and investment section, should include a section dedicated to the principles of standardisation of all processes. As these principles, it is recommended to use as follows:

- An approved standard is mandatory;
- Every employee has the right to propose a change in an operation standard.

The company's development strategy highlights new investment cycles associated with upgrading production facilities or opening new production facilities. Within each investment cycle, a project should also be evaluated according to the enterprise standard, during which, labour productivity indicators should be used along with economic efficiency indicators.

The management system of an enterprise should reflect the new role of personnel as a carrier of intellectual potential. It is necessary to track the dynamics of indicators of the people who have completed training and advanced training and the number of proposals submitted for improving processes, including those who have completed training. To involve staff in process improvement activities, personal interaction between managers and employees is required, which increases trust in managers.

Reducing losses and non-productive costs as a result of improved processes should become a source of financing for the growth of personnel remuneration. Remunerative incentives should be understandable for the staff.

It is necessary to set the goal of achieving the minimum required number of levels in the management system hierarchy (depending on the complexity of production).

In the production and auxiliary process standards, indicators to assess the performance of processes should be used: the time of a production process, the volume of work in progress, the time to change the process, etc. Deviations of the actual values of indicators in any direction should be the subject of discussion of all stakeholders in the process. Staff should have the right to propose new process indicators.

4 Discussion

The positive aspects of the "lean manufacturing" implementation identified as a result of the research are also noted by other authors, for example, in reducing the cost of production [7], increasing the economic efficiency of the enterprise [5], increasing the competitiveness of the organisation [8], improving the efficiency of enterprise management [9] and designing organisational changes [10].

To consolidate organisations' positive results obtained at the first stages of the introduction of new processes, it is very important that this area of activity becomes permanent without turning out to be some kind of campaign temporarily carried out while the national project is in effect. This will depend on the factors presented in this research as recommendations. Some of them correspond to the results of other research activities, in particular, the involvement of the management of enterprises in the new organisation of production, which face the challenge of overcoming resistance to changes on the part of the team [11], continuous personnel training and professional personnel development [12], the need to change the wage level [13].

5 Conclusion

The research into the experience of introducing "lean production" at industrial enterprises of the Tula region showed that the enterprises received the first positive results. However, there is still a risk that the existing problems may subsequently level out all the results achieved at the first stage of implementation, if the new production organisation methods fail to become a common thing.

The proposed recommendations for changes in the enterprise development strategy, management system, organisation of production and auxiliary processes, which should accompany the introduction of a new labour organisation system, allow ensuring the stability of the results achieved at the enterprises of the Tula region.

It can be assumed that enterprises, that have fully mastered the new approach to organising production, will have a lower level of necessary costs, which will enhance their investment attractiveness and will allow attracting additional investment resources from outside. In the case of using a systematic approach to changing processes based on "lean production", not only the economic growth of individual enterprises in the region will be ensured but also an increase in the socio-economic development of the region as a whole.

References

1. Slabinskaya, I.A., Bradarskaya, A.A.: Lean manufacturing as a way to improve the financial condition of an enterprise. Belgorod Econ. Newslett. **2**, 163–168 (2019)
2. Papanova, S.Yu., Kravchenko, E.Yu., Bolotova, I.S.: Lean manufacturing: a necessary factor in the development of a modern organisation in the digital economy. Newslett. Belgorod Univ. Coop. Econ. Law **2**, 111–121 (2020)
3. Popov, D.A., Tinyakova, V.I.: "Lean production" as a driver of the socio-economic development of a region. Sci. Result. Econ. Res. **5**(1), 52–59 (2019). https://doi.org/10.18413/2409-1634-2019-5-1-0-6
4. Talman, E.: Toyota: TPS in Russia (2021). https://kachestvo.pro/kachestvo-upravleniya/berezhlivoe-proizvodstvo/toyota-tps-v-rossii/
5. Magomedova, N.I.: "Lean production" as a factor in increasing the economic efficiency of enterprises. MEPS: management. Econ. Politics Sociol. **2**, 34–39 (2019). https://doi.org/10.24411/2412-2025-2019-00017

6. National projects: a new quality of life in the Tula region (2020). News from the national project "Labour Productivity and Employment Support". https://нацпроекты71.рф/news/video-tulskaya-oblast-stala-luchshey-v-strane-po-r/, https://нацпроекты71.рф/news/video-berezhlivye-tekhnologii-uskoryayut-proizvods/, https://нацпроекты71.рф/news/video-kak-tulskie-predpriyatiya-povyshayut-proizvo/, https://нацпроекты71.рф/news/48-predpriyatiy-regiona-vnedryayut-metody-berezhli
7. Pavlova, E.A.: The lean manufacturing concept as a tool for reducing the cost of production: mechanisms for integration into the management system of russian industrial enterprises. Econ. Soc. **11**, 865–868 (2019)
8. Armashova, G.S.: Lean production as a factor in increasing the competitiveness of a business entity. Newsl. Voronezh State Univ. Eng. Technol. **81**(2), 336–340(2019). https://doi.org/10.20914/2310-1202-2019-2-336-340
9. Malakhova, A.A., Starova, O.V., Arefiev, V.A.: Lean manufacturing as a factor in improving management efficiency. Econ. Entrepreneurship Law **10**(3), 615–634 (2020). https://doi.org/10.18334/epp.10.3.100428
10. Nikitina, N.O.: Management of the design of organizational changes according to the lean manufacturing methodology. Colloquium-J. **1–8**, 69–72 (2020). https://doi.org/10.24411/2520-6990-2019-11200
11. Adler, Y.: Lean manufacturing with a human face. Qual. Manage. Methods **2**, 10–13 (2017)
12. Goffe, N., Monusova, G.: Labor productivity: socio-economic prerequisites for growth. World Econ. Int. Relat. **61**(4), 37–49 (2017). https://doi.org/10.20542/0131-2227-2017-61-4-37-49
13. Fedorova, T.A., Lomovtsev, D.A., Potvorov, A.I.: Analysis of the effectiveness of the implementation of the national project "labour productivity and employment support". Econ. Anal. Theory Pract. **20**(3), 396–415 (2021). https://doi.org/10.24891/ea.20.3.396

State, Power, Management and Law: History and Modernity

Lela V. Chkhutiashvili[1(✉)], Nana V. Chkhutiashvili[1], Aleksander M. Gubin[1], and Galina F. Golubeva[2]

[1] Kutafin Moscow State Law University (MSAL), Moscow, Russia
lvachhutiashvili@msal.ru
[2] Plekhanov Russian University of Economics, Moscow, Russia

Abstract. The development of the digital economy should allow the Russian economy to integrate into the global context. The program "Digital Economy of the Russian Federation" dictates new requirements for the system of public administration. But despite the fact that the implementation of the Program requires serious simplification in the interaction of market operators and the state, interdepartmental interaction, it does not fully respond to the challenges of digital transformation of the public administration system. It should lead to a radical change in the approaches to the organization of the work of public authorities through the digitalization, reducing the "bureaucratic burden" when issuing electronic documents. It deals with the formation of conditions ensuring the development of information technologies and effective interaction of a state and business, which covers legal regulation, information infrastructure, personnel and information security, etc. To effectively achieve the main goal of improving the quality of public services, reducing the costs of public authorities, increasing the trust of citizens and businesses in the authorities, as well as supporting the decisions they make, it is important to create ecosystems of public and private information systems and digital platforms for interaction (transmission, data exchange).

Keywords: Remote provision of state and municipal services · Give rise to power · State · Protection of citizens' interests · Historical development · Model of public administration · The rules of law · Of the main types of models of public administration

1 Introduction

In modern conditions of development, the question of the relationship between the state and law, as well as power and management, plays an important role in the life of society.

In different models of government, the state and law, expressed in power, interact in different ways. What is the most acceptable model of public administration in the modern world? To do this, we will give a brief description of the main types of models of public administration and give examples of their practical implementation.

P. V. Trifonov and M. V. Charaeva (Eds.): DITEM 2021, LNNS 380, pp. 111–116, 2022.
https://doi.org/10.1007/978-3-030-94245-8_15

2 Materials and Methods

The solution to the task was carried out through comparative, logical and statistical analysis, and structure analysis, mutual respect, and the dynamics of the phenomena in the documentation and information areas on the basis of information from books, newspapers, legal acts and the Internet. The descriptive method, including the techniques of observation, interpretation, comparison, and generalization, is fundamental for the undertaken research.

3 Results

To date, the concept of e-government in Russia is practically implemented. According to statistics, 10 million people visit the reference and information payment Internet portal per month. However, the portal of public services is not used by all citizens. It is easier for people to get this or that service in a proven way. The process of informatization of the state system is slow, but it is still happening. Fortunately, the e-government method will never be able to replace the existing system of state power, moreover, but the correct introduction of new technologies in order to improve the quality of work of the bodies providing public administration will give a positive result.

We consider it appropriate to mention some of the problems of domestic e-government. First, stable broadband Internet is not widespread throughout our country. Some citizens do not have access to it, which is why they simply cannot interact with e-government. Naturally, this factor significantly slows down the development of digital public services. Secondly, it is the uneven development of e-government in different parts of the country [1]. Due to the fact that the information and communication infrastructure in some regions is rather poorly developed, e-government in these places cannot reach its full potential. It is simply inconvenient for people living in such regions to receive certain services in electronic format, so they prefer traditional methods. In order to solve this problem, it is necessary to provide significant support to regions with an undeveloped information environment. Finally, and thirdly, the incompetence of some civil servants significantly hinders the development of Russian e-government. It often happens that people who provide public services do not understand how to work with information technologies and, having no desire to understand this, continue to provide services in traditional ways. Therefore, it is necessary to expand courses to improve information literacy among employees of the state apparatus.

E-government already plays a huge role for Russian society. Every day it becomes more functional, more convenient, more practical. Problems related to the implementation of its activities, of course, exist, but they are quickly resolved. One thing is for sure – the course of development is set correctly. The main thing is not to deviate from it and actively join the process of informatization. Today, the role of e-government is huge. It contributes to the integration of our country with information technologies, which in the future can play a crucial role for our development.

Separately, I would like to say about the role of e-government during the COVID-19 pandemic. Due to the fact that the system was sufficiently developed at that time, it helped to prevent the deterioration of the epidemiological situation. E-government made it possible to eliminate unnecessary contacts between people. Even during the outbreak of the epidemic, when almost all organizations were closed, the system continued to work, albeit with some restrictions. The pandemic has given invaluable experience to our e-government, and now an increasing number of services are being provided completely contactless.

4 Discussion

First, we will highlight the classical management model based on the concept of Max Weber. Its concept is based on the principles of rational bureaucracy: minimal state presence in the system of economic and social regulation, standardization of management procedures, strict executive discipline, personal responsibility for the quality of decisions made [2]. For example, France during the restoration of the monarchy (1814–1830).

In this model, the state is a clearly formalized structure, and the law is the standard for regulating public relations. As a result, power is distributed depending on the position held and the scope of the assigned powers.

The next model is the socialist model of public administration. At the forefront is a common goal-building a society of universal equality and social justice [3]. This model of governance is characterized by democratic centralism, the domination of state ownership and Marxist-Leninist ideology, and the prohibition of political pluralism. That is, in fact, the power in the state belongs to the state bodies and the party apparatus. States with such a model of governance include the USSR, certain countries of Eastern Europe and Central Asia after the Second World War.

A fairly common model is the so-called "good government" [4]. It is characterized by: a powerful control center; the presence of a strategy that has a scientifically based character; a rational division of powers, and so on. Within the framework of this model, law, as a public regulator, is of paramount importance. In addition, this concept is based on a dialogue with the population. Consequently, power belongs to the population of the state, and the state is based on a social contract. This model of public administration is most clearly represented in Japan.

In contrast to the previously mentioned one, there is a "network model" [4]. Here there is a close interaction between state and non-state structures in order to reach a common agreement, since the government does not implement the so-called function of serving the population. Power also belongs to the population. An example is modern France, where public investigations are often conducted with the participation of non-State entities to protect the rights and interests of citizens.

The law embodied in the State legislation is most clearly expressed in the "institutional model" [5]. The main goal is to implement and protect the interests of citizens from abuse by the state (state bodies and other structures). I will not give a practical example here, since this model is not actually found in its pure form.

There is also a "corporatist model" [6] of public administration: power belongs to several large political groups that have representation in parliament and the Government. The form of the state vaguely resembles an aristocracy, and the rules of law are created in the interests of the dominant political groups. For example, Italy after the end of the First World War.

The most common is the "state-supermarket model" [7]. The state provides public services to the population, and the population, in turn, is a customer and a consumer. Accordingly, the law is designed to regulate the processes of providing services. Power is the prerogative of citizens who dictate the need for certain public services. An example is the United Kingdom ("Citizens' Charter" [8]).

The classical and socialist models were quite common in most countries of the world. The first appeared approximately from the beginning of the 19th century; the second-in the first half of the 20th century. After the end of the Second World War, some states adopted the socialist model of government, while others tried to find their way under capitalist rule.

The latest model that will be considered by us is the most relevant and new one, referred to as the "e-government model", which is currently under development and is characterized by the informatization of state authorities capable of supporting interaction with the population and economic structures of the state. The advantages of the presented new model are: simplification of bureaucratic procedures, introduction of a unified electronic accounting system, reduction of document processing time, reduction of corruption and increase of trust in the institutions of power on the part of citizens, and this increases the level of political culture and political consciousness. The openness and transparency of government activities achieved through the use of the latest information and communication technologies facilitates the interaction of the population with state authorities, opening full access to the necessary state information for all citizens.

To implement this model, it is necessary to attract as many citizens as possible to the information society. Perhaps this is where the greatest difficulty arises, since a large percentage of the elderly will not only not see the need for informatization, but also will not be able to adapt. The current problem can be solved, since it is possible to create a certain organization that will help those people who need help to get used to it.

The concept of "e-government" implies more efficient and less costly administration, improving the mechanism of interaction between the people and the government. The program is being developed on the basis of several regulatory acts: the Decree of the Government of the Russian Federation "On approval of the State Program of the Russian Federation "Information Society (2011–2020)", the Federal Laws "On Electronic Signature", "On the organization of the provision of State and Municipal services", the Decree of the Government of the Russian Federation "On the Federal State information system providing the process of pre-trial (out-of-court) appeal of decisions and actions (Inaction) committed in the provision of State and Municipal services", etc.

The Unified Portal of Public Services (Gosuslugi), created in 2009, allows citizens to get information about the state and municipal services provided, which ensures the openness of state bodies. The key role of Public Services is that on this portal you can submit a request for the provision of a particular service, pay for it and get the result as soon as possible.

5 Conclusion

Based on all of the above, it is safe to say that there are a large number of different types of models of public administration. There cannot and will not be an ideal model of public administration that would satisfy absolutely all segments of society for all States. Therefore, the improvement of existing models, provided the integration of government Internet resources, will help the state to reach a new, successful level of public administration.

The variety of existing models of public administration directly depends on the historical development of a particular country. Within this development, the relationship between such categories as law, power, and the state changes. First of all, the state and law interact, which subsequently give rise to power. This is due to the peculiarities of the current system and management style, as well as the mentality of the people. The business community will be stronger, in our opinion, the more closely it will be connected with civil society, as well as with non-governmental organizations that defend the interests of the population, and all Russian citizens [9].

There is a strong influence on the Russian public administration system of such factors as the global coronavirus pandemic COVID-19 and measures to overcome it, sanctions pressure from other states on the Russian economy, and the development of information technologies in the world. This, in turn, contributes to improving the system of information support for the activities of public authorities at all levels, introducing the practice of remote provision of state and municipal services, respectively, reducing the cost of providing these services and reducing their cost to citizens and organizations.

To effectively achieve the main goal of improving the quality of public services, reducing the costs of public authorities, increasing the trust of citizens and businesses in the authorities, as well as supporting the decisions they make, it is important to create ecosystems of public and private information systems and digital platforms for interaction (transmission, data exchange).

References

1. Chkhutiashvili, L.V., Chkhutiashvili, N.V.: Digitalization of public administration: myths and reality. In: Molodchik, A.V., Valko, D.V. (eds.), Management in Modern Systems. Proceedings of the X All-Russian (National) Scientific and Practical Conference of Scientific, Scientific and Pedagogical Workers and Postgraduates, pp. 259–263 (2020)
2. Maslovsky, M.V.: Max Weber's Theory of Bureaucracy and Modern Political Sociology. Publishing House of the National University of Economics, N. Novgorod (1997)
3. Speransky, A.V.: The Soviet Political System in the History of Russia: Features of Geopolitical and Regional Development. SV-96, Yekaterinburg (2011)
4. Okhotsky, E.V.: State management: on the way to a modern model of state management. Bull. MGIMO (Univ.) **3**, 115–127 (2014)
5. Duran, T.V., Kostin, V.A.: Institutional management in the system of types of public administration. Soc. Power **2**(30), 10–14 (2011)
6. Shapkin, I.N.: Corporatism or pluralism? Institutional models of interaction between the modern state and society. Age Globaliz. **1**(25), 24–39 (2018)
7. Abramova, O.G., Mukhaev, R.T.: Modern state management in search of a new paradigm. In: Problems and Mechanisms of Evaluating the Effectiveness of the System of State and Municipal Management: Theory and Practice (Collection of Abstracts of Reports and Articles of the International Scientific and Practical Conference of Russian and Foreign Universities and Plekhanov Russian University of Economics with the Participation of Representatives of State and Municipal Authorities). Tretyakov Publishing House, Moscow, pp. 212–224 (2016)

8. Kail, Y.: Foreign experience in implementing standardization of state and municipal services. Natl. Interests Prior. Secur. **21**(78), 14–23 (2010)
9. Chkhutiashvili, L.V., Chkhutiashvili, N.V.: Economic security of a person and society in the digital economy. In: Sokolov, A.V., Frolov, A.A. (eds.), Opportunities and Threats of the Digital Society. Materials of the Conference of the All-Russian Scientific and Practical Conference, pp. 232–236 (2020)

Sustainable Marketing: The Importance of Consumer Competence to Modelling the Rational Consumption

Olga V. Firsanova(✉), Veniamin G. Baikov, Alla M. Alekseeva, and Nadezhda V. Naumenko

Saint Petersburg State University of Economics, Saint Petersburg, Russia

Abstract. Research background: a contemporary consumer is becoming demanding of the goods production and promotion methods. Sustainable marketing is used by many companies that "rely" on the principles of sustainable development in their activities, strive to implement rational production and model rational consumption among their consumers. Other companies mimic the former with greenwashing. As a result, consumers lose confidence in companies, their declared values of sustainability and the principles of sustainable development in general. Therefore, responsible companies need to not only promote new products to market through sustainable marketing but also educate consumers to recognise unfair greenwashing practices by other manufacturers. By increasing the competence of consumers, companies will be able not only to reduce the negative effect of unfair product promotion practices but also help build confidence among consumers by modelling rational consumption. Purpose of the research: to conceptually substantiate modelling the rational consumption based on taking into account the levels of consumer competencies and their impact on the conscious choice of the "green product" by contributing to the creation of a culture of sustainable consumption about society. Methods: theoretical analysis, scientific synthesis, scientific adaptation, discourse analysis, scientific generalisation. Results and novelty: a conceptual justification of a sustainable consumption model has been developed by demonstrating the need and sufficiency of the relationships with consumers with various competencies for their education based on the information content about the sustainability of companies and their products, which helps to strengthen the values of sustainable development and reduce the negative effects of greenwashing.

Keywords: Green marketing · Sustainable development · Green economy · Greenwashing

1 Introduction

Today, everyone recognises the active and destructive, in the overwhelming majority, influence of mankind on nature. Goal 12 "Ensuring the transition to rational consumption and production patterns" emphasises the need for the continuation of life on the planet, the creation of rational consumption models in society, and the rational production models

P. V. Trifonov and M. V. Charaeva (Eds.): DITEM 2021, LNNS 380, pp. 117–124, 2022.
https://doi.org/10.1007/978-3-030-94245-8_16

in business. On the one hand, companies should commit themselves to channelling part of their profits towards the benefit of universal values, the needs of future generations and the prosperity of the planet, and, on the other hand, consumers should be responsible and, therefore, educated to make a conscious choice of those products from those companies that actually implement principles of sustainability.

The implementation of Goal 12 involves sustainable marketing activities which are designed to "link" responsible consumption as a model of consumer behaviour and goods produced in the corresponding production of companies.

It should be stated that marketing has not succeeded at all in this challenge. In extremely negatively assessing the role of marketing as a "driver" of sustainable consumption, some scientists call it unsuccessful [1], and even the opposite of sustainability [2]. At the same time, some of them highlight the fact that there is significantly greater market success than expected for those companies that make additional commitments to model sustainable consumption [1]. At the same time, almost all scientists declare the relevance, imperative and even necessity of marketing for sustainable development [3].

While declaring their commitment to sustainable development, many companies in reality only mimic with the so-called greenwashing [4] instead of sustainable marketing. Such activities undermine the confidence of consumers and society in general in companies, responsible business and sustainable development in general.

In striving for sustainable consumption and consciously choosing a product, consumers are able to significantly influence the results of market activities of such irresponsible companies by giving their financial resources to those who really implement responsible production and true sustainable marketing. Consequently, the preparedness of consumers, their competence in the field of sustainable consumption and sustainable development, in general, are the most important area of research, that is a new idea.

It is in this context that the article will present a conceptual rationale for the importance of consumer competencies in modelling sustainable consumption through responsible marketing activities. It is this message that is the purpose of the article.

The problems are solved as follows:

- based on temporal and substantive criteria, a theoretical basis for the study of sustainable marketing, greenwashing, and responsible consumer behaviour; the buying process, consumer competencies and other aspects of sustainability has been created;
- with the discursive and other types, an analysis of the theoretical base with the definition of the "foundation" for the study, concepts, their relationship in processes etc. was carried out;
- with scientific methods, the obtained results are generalised and the conceptual significance of consumer competencies in modelling sustainable consumption is substantiated.

In the article, the concepts of "sustainable", "responsible" and/or "rational" are understood as synonyms; the concepts "ecological" and/or "green" are also understood as synonyms.

2 Overview of the Theoretical Basis of the Research

Many consumers understand and realise the importance of responsible consumption [5] (Nielsen 2015), but following this behaviour model is associated with significant cognitive and temporary consumption efforts and tangible costs; as a result, consumers often refuse to choose "green goods" [6] which does not contribute to the growth of consumer satisfaction with ecological products. Consequently, consumer distrust of both companies and sustainable development, in general, is growing [7].

Many managers view responsible production and consumption activities as meaningless [8]. This position of management at some companies reinforces the already serious gap between sustainable production with consumption goals and the financial performance targets of firms [6]. Therefore, many companies take the path of dishonesty and deliberately use greenwashing instead of sustainable marketing, effectively deceiving consumers by presenting their products as those that are produced in a sustainable manner and are reasonably part of sustainable consumption [9].

Greenwashing is a promotion through content (various kinds of advertising, signs and information on packages, slogans, wording in the product's unique selling proposition, ideas in positioning and promotion etc.), which contains false statements about the company's measures in the field of sustainability without specifying specific information about that [10]. The risk of this practice is the threat of loss of confidence in sustainable consumption on the part of consumers [11]. Companies mimicking those, that adhere to the principles of sustainability, deliberately hide any information that may be negatively perceived by consumers in leaving in the content they produce only that which promotes a positive image among customers [12]. Such companies are more often trying to allocate more funds for the greenwashing content itself, rather than really improve their product to "pull" its properties and attributes to meet the requirements for environmentally friendly products.

However, consumers are becoming more informed and competent; it is becoming increasingly difficult for companies to hide information [13]. Consumers with a high level of competence can broadcast information content to consumers with a low level of competence [14] that reveals the deception of mimicking companies by reducing the effectiveness of their greenwashing practice. Consequently, there is a need for a conceptual substantiation of the role of consumer competence in the creation of a sustainable consumption model.

3 Methods

The research used the methods of theoretical analysis, scientific synthesis, scientific adaptation, discourse analysis, scientific generalization, etc. based on scientific literature and aimed at shaping a conceptual justification for the role of consumer competencies in modelling sustainable consumption by contributing to the implementation of Sustainable Development Goal 12. Through the application of these methods, the consumers' competence is "included" as an integral part of making a purchase of an eco-friendly product [4]. Manvi and his colleagues in their work proved [7] that the development of sustainable marketing activities taking into account the competencies of the consumer leads to a decrease in purchases under the influence of greenwashing.

Thus, these methods made it possible to determine the competencies of consumers (a necessary condition) along with sustainable marketing measures (a sufficient condition) and conceptually substantiate modelling the sustainable consumption with the additional effect of reducing consumer susceptibility to the greenwashing phenomenon.

Sources of scientific literature were selected based on necessity and sufficiency to achieve the research goal and fulfil the assigned tasks. The main pool of the reference sources has been published in leading scientific journals over the past five years. Sources were selected based on an assessment of the fundamental nature of the authors' conclusions and their determining significance for the conceptual justification.

4 Results

The choice of a "green product" is a challenging task for a responsible consumer, therefore, it also simplifies the consumers' competence.

The greenwashing practice is diverse. Consumers say they do not trust organic food [7], i.e. distrust company claims, packaging designs and other methods of promoting sustainable products. Many note that companies are more likely to use "fluffy language" [7]. Consequently, under the influence of greenwashing, consumers, and also those in charge, develop suspicions of dishonesty including those of sustainable companies.

The consumers have developed a critical attitude towards the environmental friendliness of products also because not all of them understand how to check the information about those products. It is no coincidence that the second most common difficulty in assessing the environmental friendliness of a product is the use of scientific words by consumers [7]. Consumers may be reluctant to buy eco-friendly goods because they are more expensive than conventional goods, or by assuming that the company making such goods is seeking to increase brand credibility. Thus, even if a company really cares about the environment, the potential audience can only perceive it as another way of promoting.

Companies, that are truly committed to improving the environment, are constantly confronted with this distrust from consumers, and moreover, the number of such consumers is growing. At the same time, such consumers will listen to other highly competent consumers as sustainable consumption experts. There is a need for the competent application of sustainable marketing tools to promote "green goods".

One of the options for solving the problem is to actively deal with consumers. At the same time, it is not enough to just ensure a constant exchange of information, especially if consumers have reduced confidence in the content. It is relevant and necessary to implement latent education for recognising the signs of greenwashing and eliminating false content from real actions in order to improve the environment. Consumer education will not only improve their competence but also contribute to promoting the loyalty to a company that adheres to sustainable development.

Interacting and educating consumers is a challenging task for many companies. For work, specialists from among the opinion leaders, who "own" large audiences, are usually involved; but they can also create mistrust (consumers understand how the opinion leader earns). One of the solutions to the problem is to place the relevant content on a platform most suitable for promoting it. For example, the social network Instagram is the most popular among young users, where the information and educational content related to sustainable development can be presented in a format suitable for the network users.

The information provided by a company may also include content created by visitors ("User-generated Content"), since it also contributes to the promotion of customer loyalty and is in demand by them according to the research. It is not enough to just post press releases or the outcomes of events. The increasingly popular content format, that demonstrates internally organising and conducting sustainable consumption events, is also in demand among network visitors; it can be presented in the image format for the Instagram social network, video reports for the YouTube social network, or both for the TikTok network".

By conducting activities aimed at raising consumer awareness and educating them in an understandable and demanded form, sustainable marketing contributes to the promotion of a sustainable consumption culture thereby reducing the risk of consumer distrust in the future content of a sustainable company and encouraging them to cooperate. Thus, by educating users and increasing their competence, a company can develop high competencies by letting them distinguish greenwashing practice from sustainable marketing activities and make an informed choice in favour of "green products" without any significant efforts.

Figure 1 shows the algorithm for the behaviour of consumers with high and low competencies, shows the key stages of making a purchase decision, the effect of greenwashing on behaviour and what a conscious choice of a competent user, who adheres to sustainable consumption, can be.

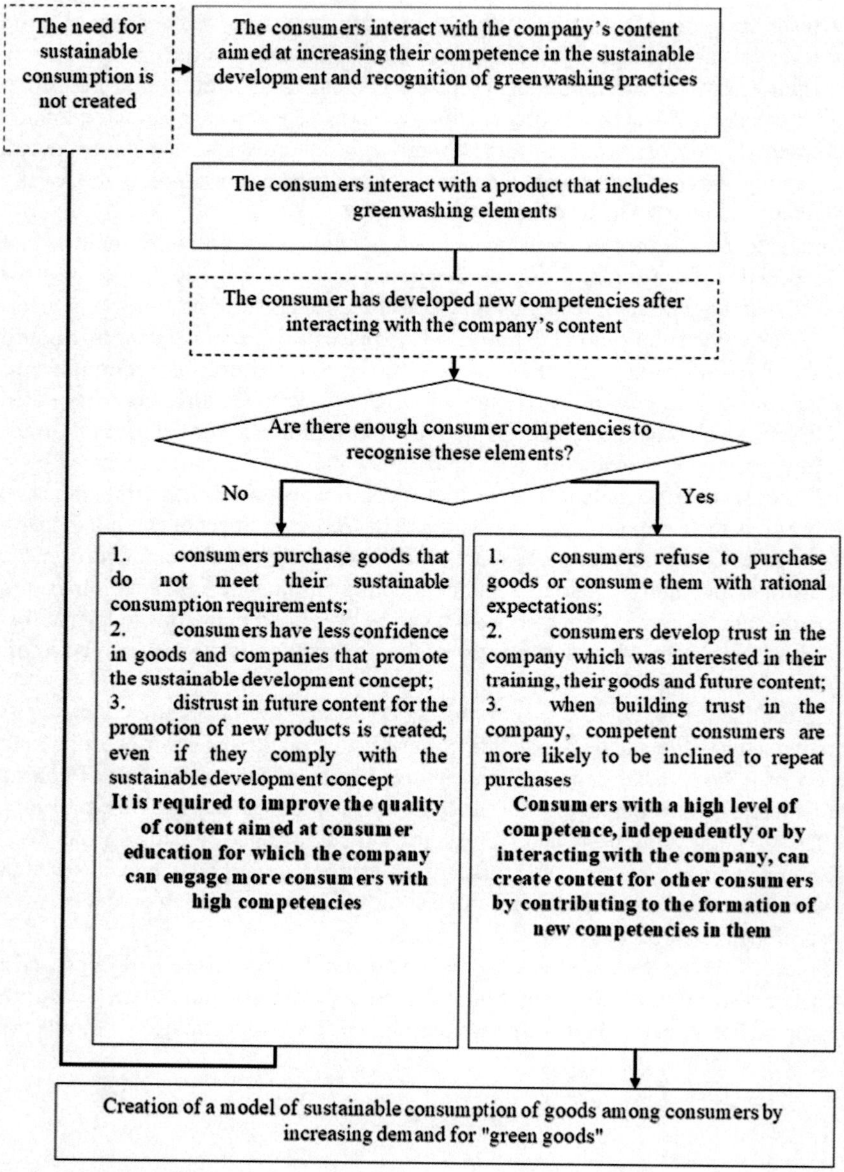

Fig. 1. Modelling the sustainable consumption of "green goods" taking into account the role of consumer competencies.

5 Discussion

The obtained results of the conceptual substantiation of the importance of consumer competencies for the implementation of Goal No. 12 are new. They are most applicable to the B2C sector, especially the FMCG market, which somewhat limits the application of the results in market practices.

The overwhelming majority of studies on sustainable consumption and the role of marketing in it recognise the leading role of marketing in influencing consumer beliefs and attitudes towards companies, their products and sustainable development in general. The studies emphasise that responsible consumers inclined to buy "green goods" during such purchases are faced with restrictions caused by such factors as price, brand and availability. Therefore, it is important to understand how consumers "come" to a purchase and how they make their choice. Therefore, the conceptual substantiation of the role of consumer competencies in the development of a sustainable consumption model presented in the article becomes even more important.

In the scientific literature, the topic of consumer competencies in the presented context is explored superficially, in a number of sources there are only references to this [15, 16]. Nevertheless, a priori, among the sustainability advocates exploring the problem area of responsible marketing, the key role of consumers as decision-makers is recognised in moving towards sustainability, informed choice of products, supporting environmental initiatives, environmental activities, healthy lifestyles, etc.

Other studies highlight the complexity of consumers' switching to sustainable consumption, which lies in the fact that the choice of "green goods" involves mindfulness, constant internal analysis of the factors of choice, understanding and rejection of greenwashing, which stimulates the wrong choice of mimicking products. This choice has not yet become a routine for the vast majority of users. The growth of competence in sustainable consumption significantly reduces this tension, and with sustainable marketing, it is possible to create cognitive-value connections between the responsible consumer and the "green good" by simulating sustainable consumption.

6 Conclusion

In sustainable marketing research, focusing on the permanent change in purchasing behaviour is critical to shaping sustainable consumption patterns. Many users understand and recognise the importance of responsible consumption, but the costs of switching to truly green products are sometimes so high that they represent a significant barrier to changing consumer behaviour towards sustainability. Greenwashing also plays a role in this by significantly weakening the market position of sustainable companies, and promoting mimicking green products to target audiences. By virtue of the above, the characteristics of consumer behaviour "competence" highlighted in the course of the research being an integral part of the conscious choice of a "green product" by a responsible consumer should become the goal of sustainable marketing activities for companies interested in effective algorithms for consumer relationships when modelling sustainable consumption.

Studying the conscious choice of responsible consumers with a high level of competence can provide important information for the subsequent formation of sustainable consumption patterns and the creation of metrics for assessing the introduction of environmental, ethical norms and values, a sustainable consumption culture in general into the everyday life of society members.

The research results are mainly applicable for the B2C market, especially for the goods of the FMCG group; for the B2B market a separate study is required.

References

1. White, K., Hardisty, D.J., Habib, R.: The elusive green consumer. Harvard Bus. Rev. 124–133 (2019). https://hbr.org/2019/07/the-elusive-green-consumer
2. Trudel, R.: Sustainable consumer behavior. Cons. Psychol. Rev. **2**(1), 85–96 (2019). https://doi.org/10.1002/arcp.1045
3. Hunt, S.D.: Strategic marketing, sustainability, the triple bottom line, and resource-advantage (R-A) theory: securing the foundations of strategic marketing theory and research. AMS Rev. **7**(1–2), 52–66 (2017). https://doi.org/10.1007/s13162-017-0090-y
4. Cifuentes, J.: Environmental Management. Circular Economy, -Green Wash-Waste (2018). https://www.researchgate.net/publication/327464080_Environmental_Management_Circular_Economy-Green_wash-Waste
5. Nielsen, N.V.: The sustainability imperative: new insights into consumer expectations. The Nielsen Company (2015). https://www.nielsen.com/wp-content/uploads/sites/3/2019/04/Global20Sustainability20Report_October202015.pdf
6. Jakubczak, A., Gotowska, M.: Green consumerism vs. greenwashing. Eur. Res. Stud. J. **XXIII**, 167–181 (2020). https://www.ersj.eu/journal/1816
7. Manvi, K., Sharma, A., Jain, V.: GREENWASHING: a study on the effects of greenwashing on consumer perception and trust build-up. Res. Rev. Int. J. Multidiscip. **04**(01), 607–612 (2019). https://zenodo.org/record/2552401#.YTcoho4zZhE
8. Galpin, T., Whittington, J.L., Bell, G.: Is your sustainability strategy sustainable? Creating a culture of sustainability. Corp. Gov. **15**(1), 1–17 (2015). https://doi.org/10.1108/CG-01-2013-0004
9. Wu, Y., Zhang, K., Xie, J.: Bad greenwashing, good greenwashing: corporate social responsibility and information transparency. Manage. Sci. **66**(7), 3095–3112 (2020). https://doi.org/10.1287/mnsc.2019.3340
10. Bhatnagar, A., Verma, S.: Celebrity footprint in greenwashing. In: 4th National Conference on Recent Trends in Humanities, Technology, Management & Social Development, vol. 9, pp. 22–37 (2019). https://www.researchgate.net/publication/332859085_Celebrity_Footprint_in_Greenwashing
11. Gatti, L., Pizzetti, M., Seele, P.: Green lies and their effect on intention to invest. J. Bus. Res. **127**, 228–240 (2021). https://doi.org/10.1016/j.jbusres.2021.01.028
12. Zengrui, X., Wang, Y., Ji, X., Cai, L.: Greenwash, moral decoupling, and brand loyalty. Soc. Behav. Personal. Int. J. **49**, 1–8 (2021). https://doi.org/10.2224/sbp.10038
13. Firsanova, O.V., Baikov, V.G.: Crowdfunding as a new form of consumer integration into the process of developing innovative products. In: Yuldasheva, O.U. (eds.), Innovations in Marketing Models of Russian Companies in the Era of Digital Transformation: Collective Monograph. Publishing House of the St. Petersburg State University of Economics, St. Petersburg, pp. 98–114 (2018). https://www.elibrary.ru/download/elibrary_35464758_82164742.pdf
14. Baikov, V.G.: Marketing of online consumer involvement in developing innovative products in the early stages of the life cycle. Thesis of Ph.D. in Economics. St. Petersburg State University of Economics, St. Petersburg (2019). https://unecon.ru/sites/default/files/dissbaykovvg.pdf
15. Miller, T., Maxwell, R.: Greenwashing Culture. Routledge, London (2017). https://doi.org/10.4324/9781315659251
16. Lukinovic, M., Jovanovic, L.: Greenwashing – fake green/environmental marketing. Fundam. Appl. Res. Pract. Lead. Sci. Schools **33**, 15–17 (2019). https://doi.org/10.33531/farplss.2019.3.04

Managing the Labor Adaptation Process of Foreign Graduates of Russian Universities

Alina A. Pozdnyakova[1(✉)], Galina V. Kuznetsova[1], Elena B. Nikitaeva[1], Vladimir V. Zotov[1], and Larisa Yu. Morozova[2]

[1] The Kosygin State University of Russia, Moscow, Russia
apozdnyakova@live.ru
[2] Moscow Pedagogical State University (MPGU), Moscow, Russia

Abstract. The number of foreigners studying at Russian universities is quite significant and, despite the difficult circumstances associated with the pandemic, has not decreased in recent years. Many foreigners arriving to study in Russia set the goal to get a job in a Russian company. The motives are different, but the result that graduates want to see is the same: a well-paid job in a Russian-speaking collective, the possibility of professional growth and adaptation in a Russian-speaking environment. To achieve this result, a foreign graduate must not only be well prepared professionally, but also acquire certain socio-cultural skills, communication skills in a foreign cultural environment, in a multicultural work collective. All this allows us to speak about the importance of labor adaptation of foreigners and the need to consider this problem in the framework of scientific research. The purpose of this article is to identify the factors that hinder the successful labor adaptation of foreigners in the Russian-speaking work collective. The objectives of the study are: 1) analysis of ethnic (socio-ethnic) characteristics that hinder the successful adaptation of foreign employees; 2) identification of the ratio of ethnic and personal components in the structure of the adaptation mechanism; 3) determination of ways and means of successful adaptation of foreign employees in Russian companies.

Keywords: Foreign graduates · Employers · Employment · Migration policy · Socio-cultural adaptation · Professional adaptation

1 Introduction

The federal project "Export of Education" of the national project "Education" provides not only an increase in the number of foreign students by 2024, but also the possibility of employing the best of these students: at least 5% of foreign graduates must be employed in Russian companies. And, although this act tells about "the most demanded areas of training", many foreign graduates are striving to get a job in Russian companies, especially from the CIS countries (Turkmenistan, Tajikistan, Uzbekistan) and countries with an unfavorable political situation (Afghanistan, Iraq, Syria). Some students start working in Russian companies already at the stage of their studies at the university, but

P. V. Trifonov and M. V. Charaeva (Eds.): DITEM 2021, LNNS 380, pp. 125–131, 2022.
https://doi.org/10.1007/978-3-030-94245-8_17

this experience is not always positive. The reason for this is the lack of linguistic, legal, financial awareness, on the one hand, and inability to work in the Russian-speaking environment, on the other. The purpose of this article is to identify the factors that hinder the successful labor adaptation of foreigners in the Russian-speaking work collective. The objectives of the study are: 1) analysis of ethnic (socio-ethnic) characteristics that hinder the successful adaptation of foreign employees; 2) identification of the ratio of ethnic and personal components in the structure of the adaptation mechanism; 3) determination of ways and means of successful adaptation of foreign employees in Russian companies.

2 Methods

Since this study is at the junction of several sciences (economics, sociology, psychology, cultural studies, social pedagogy), the study involved data from different areas of knowledge. We would like to emphasize the works concerning:

1. Peculiarities of the ethnic behavior of foreign employees in a foreign professional environment [1, 2];
2. Factors and conditions of adaptation of foreign employees to the professional environment [3–7];
3. Managing the process of preparing foreign students for work in a multicultural work collective [8–10].

In this regard, it became important to study the regulatory framework for the employment of foreign students and graduates in Russia and the subsequent analysis of proposals for improving Russian legislation taking into account the best foreign practices.

The information base of the study was data from open sources: 1) analytical reviews on the problems of economic and social development of countries and regions; 2) generalized statistical data on the dynamics of labor migration; 3) materials of websites of foreign universities.

Main research methods used: survey, poll, interview.

3 Results

Management of the process of labor adaptation is personalized, since the labor process, regardless of its orientation, is socially and ethnically determined. In this case, we mean such components of labor culture as value orientations, norms and patterns of interaction in a new collective, social and psychological characteristics of people's consciousness and behavior [11, 12]. Experts in the field of labor relations note that representatives of different ethnic groups behave differently in the same work situations and when performing separate (similar, identical) work operations [13]. This largely explains the emergence of interethnic conflicts in the workplace and the dissatisfaction of workers with the labor process. The proposition that work in a foreign cultural (or multicultural) work collective should be specially trained seems to be obvious, but is not reflected either in the theory of learning or in the practice of foreigners' labor activity. Note the works of Shutes [14], Zhuo & Wang [15], Hellgren & Serrano [16], Cordini & Ranci [17], Wang [13], which analyze the mistakes made by migrant workers in the course of communication with clients of various service sectors – medical, educational, consumer. Authors of different fields of knowledge actually come to a common conclusion: it is difficult for a person to adapt to working conditions in a new collective, because of the burden of his own "ethnic attitudes and prejudices" associated with belonging to the national corporate culture [18–20].

The survey among students and graduates from Russian schools has confirmed that the vast majority of respondents face issues to adopt while working at Russian speaking teams.

The most striking fact is that those issues were seen in engagement with leadership (17% of participants with C1 level Russian skills said that they were unprepared to interact with a Russian speaking leader), in dealing with public opinion (37% of participants feared communicating with future colleagues because of lack of knowledge of working traditions), in having informal relations at work (42% of participants were afraid of mobbing).

Some researchers believe that successful adaptation requires the abandonment of elements of the native culture in favor of elements of a new one and this can be acquired in the process of communication [19, 21, 22]. However, as practice shows, not all "new elements" have a bad effect on social well-being and professional opportunities: in a situation with labor adaptation, they play in favor of a foreign graduate.

It is obvious that the revealed socio-ethnic characteristics of corporate cultures interfere with the successful adaptation of foreign employees in Russian companies, so it is necessary to talk about the correct assessment of the ratio of ethnic and personal components in the structure of labor relationships, as well as search for ways and means of labor adaptation of graduates.

4 Discussion

The study revealed insufficient attention of scientists to the problem of labor adaptation of foreign graduates. We have singled out a number of works devoted directly to this problem, but most of them are not fresh enough and therefore do not cover the entire spectrum of real-life problems. Meanwhile, the difficulties faced by foreign university graduates in different countries are not static, but tend to change over time. They largely depend on factors such as the specificity of the culture and traditions of the country where a graduate will work, the openness and tolerance of the host country. Individual personal factors (motivation, disposition to adapt in a multinational work collective, willingness to work on oneself, a sufficient level of self-esteem, emotional stability and maturity) also have an impact.

Several extensive studies [15, 23–25] confirm the dual nature of adaptation problems: the existence in the host country of stereotypes and prejudices towards specific groups of foreigners, on the one hand, and the fact that these foreigners have their own "ideas" about the host country, on the other hand, seem to be the most serious factors in the unsuccessful adaptation of foreign students. Hence is the importance of social contacts with "volunteers" – compatriots who arrived in the country earlier, or with citizens of the host country, focused on helping foreigners. In the case of competent volunteer assistance, foreign employees have a better understanding of the culture of the host country, and the level of social and cultural stress decreases [14, 21, 26, 27].

The management of labor adaptation of foreign students provides for the development of a certain system of measures already in the course of the educational process. This system is to a certain extent individual for each ethnic group and can be oriented towards different time periods. There are several stages in the structure of labor adaptation.

4.1 A Preliminary Stage

At this stage, the student's interest in mastering the future specialty in Russian is manifested. This is expressed in the desire to obtain information about universities that train specialists in this specialty, the validity of the diplomas of these universities, Russian scientific schools and scientific areas, the prospects for employment in their country and abroad.

Note that labor adaptation is not a problem exclusively for foreign students and university graduates. Our compatriots also face problems entering a new academic environment or getting into a new work collective. But for international students and graduates, new customs, forms and codes of social life and behavior are superimposed on standard problems [28]. The readiness of foreigners to adapt in this case is a decisive factor in the success of its implementation.

4.2 Active Educational Stage

This stage includes the direct adaptation of a student within the educational process at a Russian university. Here, students not only master disciplines in Russian, but also receive specific guidelines for organizing their own educational and professional activities in a foreign country.

The need for knowledge of the language goes beyond the tasks of labor adaptation (it is assumed that the graduate already speaks Russian at a sufficient level), but ultimately it is of decisive importance for establishing social contacts and involvement in productive social, cultural and work activities.

4.3 Active Psychological Stage

Social and psychological adaptation of foreign students occurs in parallel with the educational one. However, because the implementation of the main functions within the educational process is accompanied by formal control by the teacher, the psychological adaptation of a foreign student can be significantly delayed in comparison with the educational one.

Psychologists say that anxiety and stress are factors that all students who enter new educational institutions, all employees who come to a new place of work encounter, and that high levels of stress are harmful not only to health, but also to the effectiveness of the educational or work process [28]. For foreigners, stress levels increase due to the need to adapt to life in a foreign country [21]. Therefore, it is advisable to talk about psychological support for foreign graduates in the process of their labor adaptation.

Although Furnham and Bochner [29] argue that psychological adjustment is a kind of "cultural chauvinism" because it implies the "destruction" (or humiliation) of elements of the native culture. This seems to us only partly correct: instead of partially lost, learners acquire new social and cultural skills that matter in the new workforce and therefore are critical to the professional development of the employee [30].

4.4 Stage of Stable Functioning

At this stage, the integration of the foreigner in the collective takes place within the framework of the educational environment, and we can already talk about the degree of their adaptation in society. The educational and sociocultural adaptation of a foreign student ends when he begins to realize himself as a member of the study group, when he feels comfortable communicating with native speakers of the Russian language. Labor adaptation is the next stage of adaptation that cannot be completed if the previous ones have not been passed. Incomplete adaptation of a student often leads to problems, including those of an interethnic nature.

The main ways of labor adaptation of future foreign graduates can be: 1) non-formalized support (individual guidance of a student by a teacher or tutor), 2) participation in professional events (industrial exhibitions, professional skills competitions, etc.), 3) personal growth trainings (business games, professional quests, etc.), 4) professional immersion (practice, work in an international research group, etc.).

This study confirms Wachtal's assertion that the success of labor adaptation is largely associated with motivation and is determined by it [31]. The desire to adapt to the work environment is a more important determinant of adaptation than the response to the environment in which the adaptation takes place. The results of the survey show that although some respondents believed that their native cultures were misunderstood by the host society, the acceptance of new values was assessed by them as something positive.

If the educational institution has a competently organized system of adaptation of foreign students and new students in a short time become motivated for success and good studies, then such students are promising specialists [32, 33].

In other words, the success of adaptation of a foreign graduate directly depends on a well-built educational process management system, as well as on the professional and personal qualities of the teaching staff.

5 Conclusion

The proposals expressed are based on the belief that many difficulties faced by foreign graduates can be eliminated or mitigated through thoughtful, systematic work carried out by the educational organization in cooperation with enterprises and companies. Although the implementation of these proposals will require some additional efforts on the part of the university, these measures seem to us to be justified, since they are aimed at facilitating the labor adaptation of foreign employees in Russian companies.

References

1. Firfiray, S., Cruz, C., Neacsu, I., Gomez-Mejia, L.R.: Is nepotism so bad for family firms? a socioemotional wealth approach. Human Res. Manage. Rev. **28**(1), 83–97 (2018). https://doi.org/10.1016/j.hrmr.2017.05.008
2. Hudson, S., González-Gómez, H.V., Claasen, C.: Legitimacy, particularism and employee commitment and justice. J. Bus. Ethics **157**(3), 589–603 (2017). https://doi.org/10.1007/s10551-017-3685-1
3. Li, J., Nuno Guimarães Costa, P.N.: Chinese expatriates' adjustment process in Portugal: on the road to guanxi replication. Manage. Res. J. Iberoamer Acad. Manage. **14**(2), 166–187 (2016)
4. Chen, J.: Internationalization of Chinese firms: what role does guanxi play for overcoming their liability of outsidership in developed markets? Thunderbird Int. Bus Rev. **59**(3), 367–383 (2017). https://doi.org/10.1002/tie.21877
5. Ganzaroli, A., De Noni, I.: The rise of the Chinese regional cluster specializing in fashion in Lombardy: an evolutionary analysis. J. Enterpr. Commun. People Places Glob. Econ. **11**(4), 491–513 (2017). https://doi.org/10.1108/JEC-11-2015-0052
6. Lee, S.W.: Circulation East to East: understanding the push-pull factors of Chinese students studying in Korea. J. Stud. Int. Educ. **21**(2), 170–190 (2017). https://doi.org/10.1177/1028315317697540
7. Ferris, G.R., Ellen, B.P., McAllister, C.P., Maher, L.P.: Reorganizing organizational politics research: a review of the literature and identification of future research directions. Ann. Rev. Org. Psyc. Org. Behav. **6**(1), 299–323 (2019). https://doi.org/10.1146/annurev-orgpsych-012218-015221
8. Calder, M.J., et al.: International Students attending Canadian universities: their experiences with housing, finances, and other issues. Canad. J. High. Educ. **46**(2), 92–110 (2016). https://doi.org/10.47678/CJHE.V46I2.184585
9. McManus, R.: A methodology to understand student choice of higher education institutions: the case of the United Kingdom. J. High. Educ. Pol. Manage. **39**(4), 390–405 (2017). https://doi.org/10.1080/1360080X.2017.1330806
10. Gbadamosi, A.: The anatomy of international students' acculturation in UK universities. Ind. High. Educ. **32**(2), 129–138 (2018). https://doi.org/10.1177/0950422218760335
11. Sabetova, T.V.: Corporate culture as a tool of a company's competability improvement within labour market. RJOAS **7**(55), 47–53 (2016). https://doi.org/10.18551/rjoas.2016-07.07
12. Lebedeva, T.V.: On the statutory regulation of adaptation of educational migrants. Migr. Law **3**, 24–29 (2019)
13. Wang, B.X.: Developing Guanxi in the West: Chinese expatriates' adjustment in Europe. In: Guanxi in the Western Context, pp. 67–101. Springer, Cham (2019). https://doi.org/10.1007/978-3-030-24001-1_4
14. Shutes, I.: The employment of migrant workers in long-term care: dynamics of choice and control. J. Soc. Pol. **41**(1), 43–59 (2012). https://doi.org/10.1017/S0047279411000596
15. Zhuo, R., Wang, C.: Working pressure of expatriate managers in China international engineering. In: Zhang, Z., Shen, Z.M., Zhang, J., Zhang, R. (eds.) LISS 2014, pp. 747–751. Springer, Heidelberg (2015). https://doi.org/10.1007/978-3-662-43871-8_107
16. Hellgren, Z., Serrano, I.: Transnationalism and financial crisis: the hampered migration projects of female domestic workers in Spain. Soc. Sci. **6**(1), 8 (2017). https://doi.org/10.3390/socsci6010008
17. Cordini, M., Ranci, C.: Legitimising the care market: the social recognition of migrant care workers in Italy. J. Soc. Pol. **46**(1), 91–108 (2017). https://doi.org/10.1017/S0047279416000398

18. Tsui, A.S., Nifadkar, S.S., Amy, Y.O.: Cross-national, cross-cultural organizational behavior research: advances, gaps, and recommendations. J. Manage. **33**(3), 426–478 (2007). https://doi.org/10.1177/0149206307300818
19. Lin, Z., Zhao, Z., Lin, Z.: Culture, expatriation and performance: case of Chinese multinational enterprises. Chin. Manage. Stud. **10**(2), 346–364 (2016)
20. Shneikat, B., Abubakar, A., Ilkan, M.: Impact of favoritism/nepotism on emotional exhaustion and education sabotage: as moderated by gender. In: Fourth 21st Century Academic Forum Conference, Harvard University, Boston (2016)
21. Kim, Y.Y.: Communication and Cross-cultural Adaptation: An Integrative Theory. Multilingual Matters, Clevedon (1988)
22. Herberg, E.N.: Ethnic groups in Canada: adaptations and transitions. Am. J. Soc. **95**(2), 22–329 (1989). https://doi.org/10.1086/229289
23. Schuetze, H.G.: Social and cultural adjustment of foreign students: Experience from Western Europe. In: Ebuchi, K. (ed.) Foreign Students and the Internationalization of Higher Education: Proceedings of OECD/Japan Seminar on Higher Education and the Flow of Foreign Students, pp. 144–159. Research Institute for Higher Education, Hiroshima University, Hiroshima, Japan (1989)
24. Rohrlich, B.F., Martin, J.N.: Host country and reentry adjustment of student sojourners. Int. J. Intercultural Relat. **15**(2), 163–182 (1991)
25. Finsterbusch, K. (ed.): Sociology 92/93 (21st, Annual Duschkin Publishing Group, Guilford (1992)
26. Gudykunst, W.B., Hammer, M.R.: Strangers and hosts: an uncertainty reduction based theory of intercultural adaptation. In: Kim, Y.Y., Gudykunst, W.B. (eds.) Cross-cultural adaptation: Current Approaches, pp. 106–139. Sage, Newbury Park (1988)
27. Kaikai, S.M.: Accommodating diversity. Coll. Teach **37**(4), 123–125 (1989)
28. Perrucci, R., Hu, H.: Satisfaction with social and education experiences among international graduate students. Res. High. Educ. **36**(4), 491–508 (1995)
29. Furnham, A., Bochner, S.: Social difficulty in a foreign culture: an empirical analysis of culture shock. In: Bochner, S. (ed.) Culture in contact: Studies in cross-cultural interaction, pp. 161–198. Pergamon, Oxford (1982)
30. Disman, M.: The process of adaptation of immigrants to Canadian culture and communities. In: Shankar, Y. (ed.) Proceedings of the settlement and Integration of New Immigrants to Canada Conference, pp. 45–63. Fatuity of Social Work & Centre for Social. Welfare Studies, Wilfrid Laurier University, Waterloo (1990)
31. Wachtel, P.I.: Cognitive style and style of adaptation. Percept. Motor Skills **35**, 779–785 (1972)
32. Skorobogatova, V.I.: Legal regulation of foreign students and graduates employment in Russia and abroad. Bul. Saratov State Law Acad. **1**(138), 246–255 (2021). https://doi.org/10.24412/2227-7315-2021-1-246-255
33. Swanson, M.K., Swanson, B.A.: The role of motivation and adjustment in selection of expatriate academics in China: an intensive study of a Chinese university with one of the largest foreign faculty staffs. Int. J. Qual. Res. Serv. **2**(4), 251–261 (2017). https://doi.org/10.1504/IJQRS.2017.088110

Corporate Culture Management in a Multinational Organization

Sergey N. Nikitaev[1], Tatiana Yu. Nechaeva[1], Elena R. Laskareva[2](✉), Elena N. Tarasova[3], and Irina S. Solunova[3]

[1] The Kosygin State University of Russia, Moscow, Russia
[2] St. Petersburg State University, St. Petersburg, Russia
laskarevaelena@gmail.com
[3] Russian Technological University (MIREA), Moscow, Russia

Abstract. Corporate culture acts as a part of the national culture and is largely determined by its characteristics. However, building a competent system of relationships between employees and the formation of a positive corporate culture is impossible in case of refusal to recognize the importance of the national component, as well as without taking into account the national-ethnic characteristics of employees. The corporate culture of companies located on the territory of Russia has special characteristics and greater closeness, but at the same time it is a kind of "conglomerate of ethnotypes" that, to a greater or lesser extent, influence the formation of the "management policy" of the organization. If we consider corporate culture through the prism of national cultures, then we can find many interesting features, such as prolonged socialization of employees, a plurality of views, assessments, behavior patterns, etc. The description of corporate culture as a social phenomenon from the standpoint of ethnoeconomics and ethnopedagogy seems extremely relevant, and the development of approaches to the analysis of technologies for managing the corporate culture of a multinational structure is the purpose of our article. Revealing the social and ethnic characteristics of corporate cultures, their analysis in terms of the consolidation of cultural and intellectual resources contribute to the successful adaptation of employees of multinational organizations and the formation of a strong corporate culture.

Keywords: Corporate culture · Ethnic culture · Socio-cultural adaptation · Professional adaptation · Change management

1 Introduction

At present time the study of corporate culture is a kind of scientific trend. Corporate culture is studied by many sciences such as economics, sociology, psychology and pedagogy. And this interest is not accidental; it is due to the complexity of the phenomenon itself, on the one hand, and its insufficient theoretical coverage, on the other.

In fact, most Russian organizations are now multinational. The specificity of Russia as a country tolerant towards all peoples and cultures, as well as the non-monocultural nature of the population, the socialist past common to many close countries, allow us

P. V. Trifonov and M. V. Charaeva (Eds.): DITEM 2021, LNNS 380, pp. 132–137, 2022.
https://doi.org/10.1007/978-3-030-94245-8_18

to talk about the special mission of Russia and in terms of the formation of corporate culture as a special social phenomenon.

The corporate culture of companies located on the territory of Russia has special characteristics and has greater closeness, in which they see the influence of the Russian mentality, but at the same time it is a kind of "conglomerate of ethnotypes" which, to a greater or lesser extent, influence the formation of the "management policy" of the organization. Indeed, if we consider corporate culture through the prism of national cultures, then we will be able to discover many interesting features, such as prolonged socialization of employees, a plurality of views, assessments, behavior patterns, etc. The description of corporate culture as a social phenomenon from the standpoint of ethnoeconomics and ethnopedagogy seems extremely relevant and is the purpose of our article.

2 Methods

The analysis of scientific literature in various fields of knowledge has become the most important research method for us. The analysis covered the works of Russian and foreign authors. We will name those of them who had a significant impact on the conclusions we draw in our study. These are the works that consider the issues of corporate culture management in a multinational organization [1–5], as well as analyze the national characteristics of the behavior of foreign employees in a foreign professional environment [6–8].

In the course of the work, it was revealed that the amount of research in the field of human resource management practices and, especially, in terms of innovative practices in the education sector, is rather small, and this explains the importance and relevance of this study [9].

To obtain experimental data, a questionnaire was conducted among university staff and international companies providing educational services. The purpose of the survey was to identify the socio-ethnic characteristics of corporate cultures that hinder the successful adaptation and integration of employees of different ethnic groups.

3 Results

The results of the study should be assessed from two perspectives – theoretical and practical. Identification of "white spots" in the theory of corporate culture helps to develop recommendations for the implementation of innovations in the practical sphere. In our opinion, in relation to the management of the collective of an international organization, the following theoretically undeveloped questions are relevant:

- To what extent are the personal and ethnic components correlated in the structure of corporate culture?
- To what extent is it possible to mitigate negative influence caused by specific social phenomena common in some ethnic cultures?
- What measures of influence (administrative, psychological, and pedagogical) should be taken in the event of acute interethnic conflicts?

The analysis of the survey results helped to form a package of assessment materials for managers, including diagnostic tests to identify ethnic stereotypes that influence the formation of corporate culture in the following areas: 1) "individualism – collectivism"; 2) "manager – subordinate"; 3) "formal – informal"; 4) "tradition – innovation". This approach made it possible to identify not only the attitudes of employees with a clear ethnic position, but also to capture the moods of that part of the multinational collective that does not have stable socio-ethnic preferences.

To assess the practical significance of the proposed materials, the technique of Bordovskaya [10], was used which highlights:

- effectiveness in terms of time of recording (final and current results);
- effectiveness depending on scientific interests and goals of the subject of scientific and pedagogical activity (direct and indirect);
- effectiveness in terms of significance (significant and insignificant results).

Control tests carried out at different stages of the assessment activity made it possible to conclude the validity of the proposed test materials.

4 Discussion

Corporate culture is often defined as the sphere of the spiritual and material life of the collective, the dominant moral norms and values in it, the adopted code of conduct, long-established rituals and traditions [2]. In our opinion, the concept of "corporate culture" is much broader, since the "norms and values" of a particular organization do not arise by themselves, but are a reflection of the mental attitudes of a particular society, a particular ethnoculture [1].

Corporate culture determines a set of labor behavior patterns that are acquired by an organization in the process of adaptation to the external environment and are processed by it through internal resources (intellectual, mental, material), but within the specified sociocultural, ethnic attitudes [4, 11, 12]. Thus, corporate culture acts as a part of the national culture and is largely determined by its characteristics. According to the researchers, it is practically impossible to build a competent system of relationships between employees and form a "positive corporate culture without taking into account the national characteristics of employees, for example, only by borrowing ideas from companies from other countries or on a completely synthetic basis" [13]. It is supposed that any head of the organization should be aware of it, but it is not observed in most structures (both Russian and foreign). Meanwhile, any national culture has qualities that can be productively used in the formation of a corporate culture aimed at effectively achieving the company's goals [2, 5].

In our opinion, the corporate culture of a multinational organization has a complex character and is formed by disclosing the "ethnotypes" or "ethno-complexes" that are present in the collective. The components of the complex can be presented in the mind of the bearer of the culture or remain unconscious, but are manifested in specific verbal and non-verbal images. In this sense, we agree with Hofstede's views expressed in a number of his works [14, 15]. The results of Hofstede [15] show that even in a large multinational company with a strong corporate culture, national differences are of paramount importance. An approach that focuses only on corporate culture, at best, leads to the development of theories of "nationality". In the worst case, the study of corporate culture outside the context of national culture leads to the opposition "own culture – foreign culture" and puts a person in a situation of constant choice.

Effective management of the process of forming the corporate culture of a multinational organization should take into account its specificity as a group of people united by a common collective goal, on the one hand, as well as a "conglomerate of ethnotypes", on the other [16, 17]. It can be assumed that it is the social technology of corporate culture management that helps to take them into account and make the most of its multicultural status. In practical terms, this allows:

- to focus on the ethnic uniqueness of the people who form the collective;
- to identify and use hidden opportunities (knowledge and professional skills) of employees belonging to different cultures;
- to integrate separate knowledge into the general system of "shared" knowledge of the organization;
- to integrate different cultures within the organization and build on their basis the metaculture and meta-knowledge of the multinational organization [18].

Some researchers believe that any consolidation of cultures within one common culture leads to a departure from one's own ethnicity, abandoning elements of the native culture in favor of elements of a new one, which can be acquired in the process of communication [16, 19]. However, this is only partially true: with competent management, corporate culture is the zone where national elements merge into a single complex and become an instrument for solving collective problems. This becomes possible if there are special training programs for a multinational company.

Summarizing the considered provisions, we emphasize that any activity on the formation of corporate culture involves a conscious impact on employees (educational, developmental). This gives rise to a number of interrelated issues and urgent tasks in the field of human resource management, and the need to integrate various scientific approaches to the study and research of knowledge management of multinational personnel, the study of the characteristics of motivation of employees of different ethnic groups become obvious [20, 21].

In this regard, it is correct to talk about ethno-deterministic principles of management of the formation of corporate culture in a multinational collective. These include:

- the principle of an individualized approach;
- the principle of taking into account a specific combination of cultures;
- the principle of priority of the rules of the host country.

It is obvious that the identification of the socio-ethnic characteristics of corporate cultures, their analysis in the aspect of the consolidation of cultural and intellectual resources contribute to the successful adaptation of employees of multinational organizations and the formation of a strong corporate culture.

5 Conclusion

The topic under study is complex and voluminous enough to be covered in detail within the framework of one article, so we dwelt only on some of its aspects. We considered the corporate culture of a multinational organization as a "conglomerate of ethnotypes", identified the features that characterize corporate culture as a social phenomenon, outlined approaches to the analysis of technologies for managing the corporate culture of a multinational organization.

References

1. Karahcarovsky, V.V., Shkaratan, O.I., Yastrebov, G.A.: The Russian Labour Culture and Foreign Influence. Strana Oz Publ, Moscow (2015)
2. Sabetova, T.V.: Corporate culture as a tool of a company's competability improvement within labour market. RJOAS **7**(55), 47–53 (2016). https://doi.org/10.18551/rjoas.2016-07.07
3. Ya, D.: Corporate culture change management. Econ. Alt. **2**, 296–312 (2019)
4. Mohsen, A., Neyazi, N., Ebtekar, S.: The impact of organizational culture on employees performance: an overview. Int. J. Manag. **11**(8), 879–888 (2020). https://doi.org/10.34218/IJM.11.8.2020.079
5. Jabid, A.W., Buamonabot, I., Fahri, J., Arilaha, M.A.: Organizational politics and job satisfaction: mediation and moderation of political skills. Binus Bus. Rev. **12**(1), 1–9 (2021). https://doi.org/10.21512/bbr.v12i1.6226
6. Hudson, S., Claasen, C.: Nepotism and cronyism as a cultural phenomenon? In: Aßländer, M.S., Hudson, S. (eds.) The Handbook of Business and Corruption: Cross-Sectoral Experiences, pp. 95–118. Emerald Publishing Limited (2017). https://doi.org/10.1108/978-1-78635-445-720161006
7. Firfiray, S., Cruz, C., Neacsu, I., Gomez-Mejia, L.R.: Is nepotism so bad for family firms? a socioemotional wealth approach. Human Res. Manag. Rev. **28**(1), 83–97 (2018). https://doi.org/10.1016/j.hrmr.2017.05.008
8. Hudson, S., González-Gómez, H.V., Claasen, C.: Legitimacy, particularism and employee commitment and justice. J. Bus. Ethics **157**(3), 589–603 (2017). https://doi.org/10.1007/s10551-017-3685-1
9. Okhotnikov, O.V., Kazakova, Yu.E.: Employment of graduates as a problem of the russian education system. Bul Ural Fed Univ. Ser. Econ. Manag. **18**(3), 431–449 (2019). https://doi.org/10.15826/vestnik.2019.18.3.022
10. Bordovskaya, N.V.: Dialectics of Pedagogical Research: Logical and Methodological Problems. Russian Christian State Institute publishing house, St. Petersburg (2001)
11. Safina, D.: Favouritism and nepotism in an organization: causes and effects. Proc. Econ. Fin. **23**, 630–634 (2015). https://doi.org/10.1016/S2212-5671(15)00416-5
12. Ali, Q., Brandl, J.: HRM research in Pakistan: existing approaches and future directions. J. Manag. Sci. **4**(2), 170–192 (2017). https://doi.org/10.20547/jms.2014.1704203
13. Odegov, Y.: Corporate culture as a factor increasing productivity of employees. Curr. Trends. Stand. Payroll Indus. **2**, 40–45 (2015)
14. Hofstede, G.: The cultural relativity of the quality of life concept. Acad. Manag. Rev. **9**(3), 389–398 (1984). https://doi.org/10.5465/AMR.1984.4279653
15. Hofstede, G.: Culture's Consequences: Comparing Values, Behaviors, Institutions, and Organizations Across Nations, 2nd edn. Sage, Thousand Oaks (2001)

16. Gudykunst, W.B., Hammer, M.R.: Strangers and hosts: an uncertainty reduction based theory of intercultural adaptation. In: Kim, Y.Y., Gudykunst, W.B. (eds.) Cross-cultural adaptation: Current Approaches, pp. 106–139. Sage, Newbury Park (1988)
17. Bain, P., Park, J., Kwok, C., Haslam, N.: Attributing human uniqueness and human nature to cultural groups: distinct forms of subtle dehumanization. Group Proces. Intergr. Rel. **12**(6), 788–805 (2009). https://doi.org/10.1177/1368430209340415
18. Gorokhova, E.: Social Technology of Knowledge Management in Multinational Organizations. Editus Publ, Moscow (2013)
19. Herberg, E.N.: Ethnic groups in Canada: adaptations and transitions. Amer. J. Soc. **95**(2), xxii–329 (1989). doi:https://doi.org/10.1086/229289
20. Yao, C., Arrowsmith, J., Thorn, K.: Exploring motivations in Chinese corporate expatriation through the lens of confucianism. Asia Pacific J. Human Res. **54**(3), 312–331 (2016)
21. Tone, K.: Examining the moderating effect of work motivation on the lecturer performance: a contribution to organizational commitment and competence. Res. Bus. Manag. **5**(2), 1–16 (2018). https://doi.org/10.5296/rbm.v5i2.12773

Modern Opportunities for Optimizing Business Processes to Achieve Sustainable Development

Lyudmila N. Usenko, Victoria A. Guzey(✉), and Alina S. Bijieva

Rostov State University of Economics, Rostov-on-Don, Russia

Abstract. The study revealed the opinion of the scientific community about the existence of prerequisites for the emergence of the fourth industrial revolution, which are based on the development of technologies that have a significant impact on the main production aspects. The study of the impact of Industry 4.0 technologies on business processes in order to achieve sustainable development was carried out. The significance of the concept of sustainable development for business activities is investigated. The authors propose a classification of sustainable development goals in accordance with the components of sustainable development. The study examines the essential content and features of business process modeling based on process standards (notations), as well as the need to optimize them to achieve sustainable development of economic entities. The article formulates a sequence of works on optimization of business processes, adjusted for the specific conditions of functioning of business entities. The analysis of the competitive advantages of business participants, potential for implementation when participating in the implementation of the UN Sustainable Development Goals, is carried out. It is revealed that the achievement of sustainable development is not possible in the absence of the introduction of new technologies, expressed in the implementation of process innovations.

Keywords: Business process · Industry 4.0 · Sustainable development goals · Agenda 2030 · Sustainable development · Competitive advantages

1 Introduction

Currently, there is an increasing interest in research related to the transition to sustainable development of economic entities. By sustainable development, the scientific community means a process of development in which the satisfaction of existing urgent needs does not adversely affect the ability of generations to meet their needs in the future [1], which is a set of transformations carried out in order to increase competitive advantages, in terms of the formation and implementation of production processes that meet not only environmental requirements, in terms of the safety of products, works, services provided, as well as public needs.

In 2015, the 2030 Agenda for Sustainable Development was adopted [2], within the framework of which the UN member states established an updated list of challenges for humanity, formulated in the form of seventeen Sustainable Development Goals [3].

P. V. Trifonov and M. V. Charaeva (Eds.): DITEM 2021, LNNS 380, pp. 138–150, 2022.
https://doi.org/10.1007/978-3-030-94245-8_19

In accordance with the provisions of the 2030 Agenda, it is essential to increase the involvement of business representatives in the process of implementing the set of goals and objectives of sustainable development.

At the same time, it is necessary to point out the fact that it is impossible to transition to sustainable development in the absence of significant changes in business processes of a production nature. Such radical transformations are currently being implemented and are due to the rapid development and introduction of Industry 4.0 technologies. The process of industrialization of public life has led to the transformation of industrial development paradigms, through the emergence of technological breakthroughs, later called "industrial revolutions". In total, there are three industrial revolutions associated with:

- with a significant increase in the level of mechanization of production processes;
- with the introduction of electricity in production processes;
- digitalization of production processes.

Currently, the scientific community believes that in the context of the industrial industry, the prerequisites for the emergence of the next (fourth) industrial revolution have already been formed, which is expressed in ensuring the integration of the Internet and a set of "smart" technologies into the digital processes of industrial enterprises [4–7]. The fourth industrial revolution will thus be closely related to the term "Industry 4.0". Within the framework of Industry 4.0, it is expected to form qualitatively new business models and value propositions that can solve socially important issues by forming a set of links between endogenous and exogenous factors of industrial production [8].

The implementation of Industry 4.0 technologies is ensured by a fundamental transformation of the fundamental production aspects. According to experts, the fourth industrial revolution will occur due to the development of technologies of autonomous robotics, cybersecurity, big data, augmented reality, simulation, horizontal and vertical integration of systems, the Internet of Things, cloud technologies, and additive manufacturing [9].

The fundamental positive results from the implementation of these technologies will be the following:

- reduce production costs, management costs, and logistics costs;
- reduction of the length of the period of entering the market of new products;
- improving the efficiency of working with the customer base;
- increased savings in the course of production activities due to the increase in production scale, together with an increase in the efficiency of using the company's resources (material, labor, financial) [10, 11].

This article examines the fundamental aspects of ensuring the transformation of models and business processes of business participants in the transition to Industry 4.0 technologies, in order to achieve the goals and objectives of sustainable development.

2 Materials and Methods

Since 1992, in accordance with the Strategy of the World Business Council for Sustainable Development, the so-called "environmental efficiency" was developed, which meant a hypothetically ideal economic business model, in which an increase in the level of production would naturally be combined with a reduction in the level of negative impact on the environment, together with an increase in the efficiency of resource use.

This strategy of environmental efficiency has received the status of a modern vector of business development, within the framework of the "Earth Summit" held in Rio de Janeiro. Implementation of the environmental efficiency strategy, according to the participants of the Agenda for the 21st century [12], is possible through the involvement of private business structures. In the future, the essential role of business in achieving "socially sustainable" globalization was repeatedly emphasized. In particular, the World Economic Forum 1999 was an international platform for consolidating this thesis. In this regard, the United Nations Global Compact initiative was implemented [13], in the context of which the fundamental aspects of popularizing the concept of social responsibility of business and consolidating the practice of forming and implementing reporting that includes aspects of social responsibility were formulated.

In the future, the sustainability of business structures, in terms of economic, environmental and social components, was developed within the framework of the developed Family of Dow Jones Sustainability Indices (DJSI), a world-wide rating that contributes to the analysis of the activities of economic entities. The dissemination of the concept of sustainable development in the practical activities of private business, contributed to the formation and development of the system of non-financial reporting of business. This contributed to the emergence of the "GRI's History" [14], which was developed by the Global Reporting Initiative (GRI), an international organization in the field of standardization. The guidelines set out the principles and indicators for reporting, in the context of economic, environmental and social components. The use of these indicators was addressed not only to multinational corporations, but also to representatives of medium and small businesses. The importance of the GRI Guide to Sustainability Reporting as a standard for non-financial reporting is difficult to overestimate, since its introduction has helped streamline reporting, in terms of eliminating the practice of generating episodic environmental reports or including sections on the environmental component in the annual report. Thanks to the GRI Guidelines on Sustainable Development Reporting, the practice of creating integrated documents on sustainable development has spread.

In 2015 The United Nations has adopted seventeen Sustainable Development Goals, which have contributed to a paradigm shift in global development. Thanks to the formation of the Sustainable Development Goals, there was an understanding of the problems that cannot be ignored due to their importance, which is associated, among other things, with the processes of globalization and the increasing level of interaction between the countries of the world. Of particular concern is the growing global problem of overpopulation of the Earth, combined with the progressive shortage of available resources, which has contributed to the need for the formation and implementation of sustainable production and consumption models. The development of such models implies the use of innovative solutions, through the most effective use of technological, financial and expert resources by business structures. Thus, business structures are a key factor in the implementation of the Sustainable Development Goals, contributing to the attraction of the necessary investments for the purpose of research and development.

To ensure sustainable development, in the context of the formulated seventeen goals, as well as 169 tasks that contribute to the achievement of these goals, a comprehensive set of indicators is currently being developed, eventually numbering several hundreds of them.

In addition to these developments, there are already separate projects, thanks to which companies will be able to achieve sustainable development and evaluate their own contribution to this process. Thus, at the initiative of the Global Partnership for Effective Development Cooperation, ten indicators were formulated to help determine progress in achieving the Sustainable Development Goals. In addition, in 2015, in accordance with the UN Global Compact, the SDG Compass project was developed [15], which promotes the formation of business strategies by companies based on the Sustainable Development Goals in order to form an idea of the contribution made by companies to their achievement. In addition to this project, the Business and Industry Advisory Committee to the OECD (BIAC) has issued a document to help explain the Sustainable Development Goals to private businesses.

The study of issues related to the essential content and features of business process modeling has attracted the attention of the scientific community, starting with the works of M. Porter, according to whom business processes were the most important component of the strategic development of economic entities. Over time, the description of the essence of business processes was deepened and detailed. Scientists M. Hammer and D. Champi formulated the essence of a business process as "an activity that uses one or more resources at the input and creates a result that is of value to the client" [16]. A slightly different view of the concept of business process was formed by the author A.-V. Scheer, who interpreted it as "a sequence of works characterized by a single natural or cost criterion of the result". In this context, the fundamental point is the set of consecutive works, that is, the content of the business process. According to the point of view proposed by the International Organization for Standardization, a process is understood as "a set of interrelated and/or interacting activities that use inputs to achieve the intended result" [17].

It should be noted that some business processes may not generate value for consumers. Business processes that are at the bottom of the hierarchy structure can be unprofitable in some cases. The consumer value is formed by the main processes, the implementation of which is not possible in the absence of control and auxiliary processes. Focusing only on the business processes that form the consumer value and ignoring the rest of the business processes is erroneous, since only a comprehensive study of all the business processes of the organization, without exception, will help determine the level of maturity of the processes of the economic entity under study.

In the practice of the functioning of companies, different ways of classifying business processes are distinguished, depending on the criteria included in the study. Some companies prefer to highlight the main business processes in order to focus on them. The scientist M. Porter carried out the classification of business processes based on the value chain, highlighting:

- the main (primary) processes that cover all areas of the company's activities, starting with suppliers and ending with consumers, forming consumer value during the formation and sale of products;
- auxiliary (supporting processes), which by themselves do not create consumer value, and their main purpose is to provide the necessary level of support to the company, in the context of providing information flow management, financial management, etc.;
- development processes, which form a new value chain in the context of the main and auxiliary processes, in terms of a new level of indicators (the development of external links, the growth of the level of qualification of employees, etc.).

The implementation of business process optimization should be carried out in a complex, thereby ensuring the manifestation of the synergy effect, contributing to the growth of the effectiveness of the functioning of the economic entity. The development of the company's business processes should be carried out through a detailed analysis of the sequence of their implementation, as well as their internal content and interaction with related processes. This study is carried out using the business process modeling method, which is a set of measures that allow the analyst to formulate an idea of the business process that exists or is expected to be implemented. The modeling method is applicable to the main, auxiliary and developing processes.

The detail of the business process model depends on the purpose of the study and can range from the most detailed level to the most abstract. After selecting the level of detail, the business processes are described graphically, using process notation. The most commonly used process standards (notations) are the following:

- VAD notation, which makes it possible to simulate the highest-level processes that ensure the creation of consumer value;
- EPC notation is used for complex business processes, allowing them to be modeled, analyzed, and redesigned (if necessary);
- BPMN 2.0 notation has the function of supporting simulation modeling and is designed to optimize business processes that have an end-to-end nature;
- IDEF notation is used in automated production systems;
- UML notation is used to provide software design, job descriptions, and to implement IT system specifications.

The above notations are tools that must meet the requirements of Industry 4.0, within the framework of modeling and optimizing business processes. The most preferred standard currently is the BPMN 2.0 notation. When comparing this standard with the IDEF notation, it turned out that they both allow for multi-level modeling and are suitable for automating business processes; however, the IDEF process notation is somewhat inferior to the BPMN 2.0 standard, in terms of graphical visualization capabilities and functionality.

3 Results

In business practice, there are different classifications of sustainable development goals, from the point of view of assessing the existing opportunities and "bottlenecks" for achieving sustainable development by private business structures. The traditional triune approach to sustainable development, which includes economic, environmental and social components, characterizes ("socially inclusive and environmentally sustainable economic growth") [18]. When dividing the Sustainable Development Goals into economic, environmental and social goals, the difficulties in differentiating social and environmental goals should be taken into account. In addition, the triune approach is now increasingly complemented by two dimensions – managerial [18], which are interpreted differently as institutional or political. And, in addition, some researchers highlight the information component, in connection with which, in our opinion, it is advisable to correlate the sustainable Development Goals in accordance with the components (Table 1).

Table 1. The proposed classification of the Sustainable Development Goals in accordance with the components.

Components of sustainable development	Sustainable development goals
Economic component	Goal 1: (No poverty) Eliminate poverty in all its forms everywhere Goal 7: (Affordable and clean energy) Achieve universal access to affordable, reliable, sustainable and modern energy for all Goal 8: (Decent work and economic growth) Promote sustained, inclusive and sustainable economic growth, full and productive employment and decent work for all Goal 9: (Industry, innovation and infrastructure) Build strong infrastructure, promote inclusive and sustainable industrialization, and promote innovation Goal 10: (Reduced inequalities) Reducing inequality within and between countries Goal 12: (Responsible consumption and production) Ensuring the transition to rational consumption and production models
Environmental component	Goal 6: (Clean water and sanitation) Ensure the availability and sustainable use of water and sanitation for all Goal 7: (Affordable and clean energy) Achieve universal access to affordable, reliable, sustainable and modern energy for all Goal 11: (Sustainable cities and communities) Ensuring the openness, security, resilience and environmental sustainability of cities and human settlements Goal 13: (Climate action) Taking urgent action to combat climate change and its consequences Goal 14: (Live below water) Conservation and sustainable use of oceans, seas and marine resources for sustainable development Goal 15: (Live on land) Protect and restore terrestrial ecosystems and promote their sustainable use, manage forests, combat desertification, halt and reverse land degradation and halt biodiversity loss

(continued)

Table 1. (*continued*)

Components of sustainable development	Sustainable development goals
Social component	Goal 1: (No poverty) Eliminate poverty in all its forms everywhere Goal 2: (Zero hunger) Eliminate hunger, ensure food security and improve nutrition, and promote sustainable agriculture Goal 3: (Good health and well-being) Promote healthy lifestyles and promote well-being for all at all ages Goal 4: (Quality education) Ensure inclusive and equitable quality education and promote lifelong learning opportunities for all Goal 5: (Gender equality) Achieving gender equality and empowering all women and girls Goal 6: (Clean water and sanitation) Ensure the availability and sustainable use of water and sanitation for all Goal 8: (Decent work and economic growth) Promote sustained, inclusive and sustainable economic growth, full and productive employment and decent work for all Goal 10: (Reduced inequalities) Reducing inequality within and between countries Goal 11: (Sustainable cities and communities) Ensuring the openness, security, resilience and environmental sustainability of cities and human settlements Goal 16: (Peace, justice and strong institutions) Promote a peaceful and inclusive society for sustainable development, ensure access to justice for all, and build effective, accountable and inclusive institutions at all levels
Institutional component	Goal 11: (Sustainable cities and communities) Ensuring the openness, security, resilience and environmental sustainability of cities and human settlements Goal 16: (Peace, justice and strong institutions) Promote a peaceful and inclusive society for sustainable development, ensure access to justice for all, and build effective, accountable and inclusive institutions at all levels Goal 17: (Partnerships for the goals) Strengthening the means of implementation and strengthening the work of the Global Partnership for Sustainable Development
Information component	Goal 4: (Quality education) Ensure inclusive and equitable quality education and promote lifelong learning opportunities for all Goal 9: (Industry, innovation and infrastructure) Build strong infrastructure, promote inclusive and sustainable industrialization, and promote innovation Goal 16: (Peace, justice and strong institutions) Promote a peaceful and inclusive society for sustainable development, ensure access to justice for all, and build effective, accountable and inclusive institutions at all levels Goal 17: (Partnerships for the goals) Strengthening the means of implementation and strengthening the work of the Global Partnership for Sustainable Development

Source: Compiled by the authors based on information from the Division for Sustainable Development Goals Department of Economic and Social Affairs United Nations (https://sdgs.un.org/goals)

Of particular importance for enhancing the involvement of private business in the process of achieving the Sustainable Development Goals is the formation of a set of long-term positive effects for business participants implementing the Sustainable Development Goals. The specified complex is presented in the table below (Table 2).

Table 2. Long-term positive effects for business participants implementing the Sustainable Development Goals.

Sustainable development goals	Positive effects for business participants
Goal 1: (No poverty) Eliminate poverty in all its forms everywhere	1. Increasing the level of consumer demand 2. Expanding the scope of the presence markets 3. Formation of new sales markets, due to the emergence of "socially inclusive" categories of goods, products, works, services 4. Employment growth
Goal 2: (Zero hunger) Eliminate hunger, ensure food security and improve nutrition, and promote sustainable agriculture	1. Increasing the level of consumer demand, in connection with the satisfaction of primary needs 2. Formation of new sales markets, due to the emergence of "socially inclusive" categories of goods, products, works, services 3. Ensuring the transfer of innovative technologies in the agro-industrial complex 4. Employment growth
Goal 3. (Good health and well-being) Promote healthy lifestyles and promote well-being for all at all ages	1. Increasing the level of consumer demand 2. Employment growth 3. Increasing opportunities for the formation and implementation of projects in the health sector, including projects with the equal participation of public and private partnerships
Goal 4. (Quality education) Ensure inclusive and equitable quality education and promote lifelong learning opportunities for all	1. Increasing the skill level of the workforce 2. Expanding the number of innovative proposals in the field of education and culture
Goal 5. (Gender equality) Achieving gender equality and empowering all women and girls	1. Employment growth 2. Increase in the level of qualification of the labor force, due to increased competition in the labor market 3. Increasing the level of consumer demand, in connection with the overcoming of gender inequality and the expansion of demand from the female population
Goal 6. (Clean water and sanitation) Ensure the availability and sustainable use of water and sanitation for all	1. Development of innovative technologies in the energy sector 2. Increase in the level of supply of equipment in the field of energy-saving technologies 3. Cost reduction due to localization of production processes 4. Creating conditions for receiving subsidies and tax benefits from the state
Goal 7: (Affordable and clean energy) Achieve universal access to affordable, reliable, sustainable and modern energy for all	1. Ensuring the transfer of innovative technologies in the energy sector 2. Increase in the level of supply of equipment in the field of energy-saving technologies 3. Cost reduction due to localization of production processes 4. Creating conditions for receiving subsidies and tax benefits from the state

(*continued*)

Table 2. (*continued*)

Sustainable development goals	Positive effects for business participants
Goal 8. (Decent work and economic growth) Promote sustained, inclusive and sustainable economic growth, full and productive employment and decent work for all	1. Reducing the number of currency, socio-economic and political risks 2. Stabilization of the level of consumer demand
Goal 9. (Industry, innovation and infrastructure) Build resilient infrastructure, promote inclusive and sustainable industrialization and innovation	1. Development of opportunities for localization of production processes 2. Formation of new sales markets, as well as consumer markets 3. Increase in the level of productivity of existing markets, in the context of various sectors of the economy 4. Ensuring the possibility of bringing products, goods, works, and services of domestic economic entities to foreign markets
Goal 10. (Reduced inequalities) Reducing inequality within and between countries	1. Growth in the level of effective demand 2. Attracting broad segments of the population to work 4. Reduction of social tension in the region of presence
Goal 11. (Sustainable cities and communities) Ensuring the openness, security, resilience and environmental sustainability of cities and human settlements	1. Formation of prerequisites for improving the sustainability of cities ("Internet of things", "smart cities") 2. Reducing the number of environmental and social risks 3. Increasing opportunities for the formation and implementation of projects with the equal participation of public and private partnerships
Goal 12. (Responsible consumption and production) Ensuring the transition to rational consumption and production models Goal 13. (Climate action) Taking urgent action to combat climate change and its consequences Goal 14. (Live below water) Conservation and sustainable use of oceans, seas and marine resources for sustainable development Goal 15. (Live on land) Protect and restore terrestrial ecosystems and promote their sustainable use, manage forests, combat desertification, halt and reverse land degradation and halt biodiversity loss	1. Ensuring the availability of resources and natural capital in the long term 2. Creating conditions for the formation of new markets and product differentiation, in order to increase the level of production of ecological, innovative, certified types of products, as well as solutions, in order to increase competitiveness 4. Creating conditions to meet demand and achieve loyalty from environmentally and socially responsible consumers 5. Providing a set of measures to expand existing opportunities for participation in the system of public "sustainable" procurement
Goal 16. (Peace, justice and strong institutions) Promote a peaceful and inclusive society for sustainable development, ensure access to justice for all, and build effective, accountable and inclusive institutions at all levels	1. Reducing the number of political (institutional) and social risks

(*continued*)

Table 2. (*continued*)

Sustainable development goals	Positive effects for business participants
Goal 17. (Partnerships for the goals) Strengthening the means of implementation and strengthening the work of the Global Partnership for Sustainable Development	1. Creating conditions for establishing relationships with stakeholders) 2. Formulation of possible mutually beneficial projects, including participation in joint public and private partnerships

Source: Compiled by the authors based on information from the Division for Sustainable Development Goals Department of Economic and Social Affairs United Nations (https://sdgs.un.org/goals)

The role of private business structures in ensuring the achievement of the sustainable Development Goals, within the framework of the allocated institutional component, is realized through participation in joint public and private partnership projects.

The implementation of business process optimization is necessary to achieve sustainable development of economic entities, which will ensure:

- identification of "bottlenecks" in the functioning of economic entities;
- increase the transparency of the company's functioning for stakeholders, and therefore to increase their loyalty level;
- conducting a correct assessment of the impact of the processes carried out on the final results of the functioning of the economic entity;
- identification and automation of the most significant business processes;
- improving the quality characteristics of products, goods, works, and services.

4 Discussion

The implementation of business process optimization involves, first of all, the use of a systematic approach. In the course of the study, we studied various points of view on the stages of implementing business process optimization identified by researchers. Taking into account the specific conditions of activity of economic entities, we have determined the following sequence of work to optimize business processes:

- implementation of planning, through the selection of the business process to be optimized, with the allocation of goals, objectives, as well as the formulation of the scope of the required changes. At this stage, a detailed analysis of the process to be transformed is carried out;
- transformation, which consists in formulating the necessary changes, taking into account the tasks set and forming an ideal model of the transformed business process;
- providing financial, labor and material resources for the project to optimize the business process to be optimized;
- conducting test events to identify the results of the redesign, especially if the optimized business process involves several structural divisions of the company;

- implementation of an optimized business process, with the provision of corrective changes in the functioning of the company, in connection with the transformations carried out;
- evaluate the effectiveness and adjust the optimized business process with a specified frequency.

Passing all stages of business process optimization without exception is possible and necessary in the presence of significant changes to the company's business processes, including if the specified business process has a significant impact on the activities of the economic entity.

One of the necessary and required improvements in the activity of an economic entity is the provision of conditions for its automation, which is of particular importance in the context of the introduction of Industry 4.0 technologies. Automation of activities acts as an actual improvement, implemented using stages similar to the sequence of optimization of the company's business processes. The implementation of automation as a possible result of improvements can transform the sequence of changes by:

- changing the number of steps in the sequence;
- reduce unnecessary steps in the process;
- reducing the number of matching points;
- transformations of the sequence of stages and approval cycles;
- running some processes simultaneously;
- adjustment of "bottlenecks";
- eliminate or provide granularity.

5 Conclusion

Based on the results of the study, it is possible to characterize the set of fundamental trends that determine the participation of business in the implementation of the Sustainable Development Goals.

As a key trend, we should note the desire of companies to implement the sustainable Development Goals in the developed business strategies, taking into account the industry specifics that determine their functioning. Only in some exceptional cases, the social initiatives implemented in the business strategy are not related to the specifics of the company's activities.

The contribution of private business to the achievement of individual sustainable development Goals in some cases contributes to the manifestation of synergy effects, involving the implementation of interrelated sustainable development goals. In the course of achieving the Sustainable Development Goals, business participants are consolidated, as part of a joint response to emerging global challenges, through the creation of partnerships that are multilateral in nature. The development of mutually beneficial partnerships by business participants helps to reduce the possibility of various risks, as well as reduce the costs of implementing social projects. Strengthening cooperation between business participants, as well as with the state and non-profit organizations, allows you to fully realize the existing opportunities to achieve the Sustainable Development Goals, representing an effective tool that contributes to the realization of existing advantages and the achievement of goals.

In addition to the above-mentioned results of the study, it is necessary to highlight the importance of approving the sustainable Development Goals, which over time have transformed into certain control points, the achievement of which will allow us to fully assess the effectiveness of all business participants. The need to provide answers to emerging global challenges has contributed to the emergence of the 2030 Agenda, which in turn has become a kind of motivator for business participants to develop and implement completely new solutions and models. Competition in the markets of goods and services contributes to the formation of companies' desire to meet the necessary standards, showing interest in achieving the sustainable Development Goals, in the context of their field of activity. In addition, business participants integrate the sustainable Development Goals into their own corporate strategies, including those that are "charged" for other areas of operation, as well as those related to addressing global macroeconomic challenges. In general, the contribution of business participants to achieving sustainable development is realized through the implementation of their activities as such. This determines the vector of development of companies that adequately assess the impact they have on the environment, which makes it possible to formulate priority directions for the development of business structures in the long term, reflecting the existing achievements, within the framework of the generated reports. Achieving sustainable development of business entities is not possible without ensuring the optimization of business processes. The introduction of Industry 4.0 technologies requires a thorough study of business processes for their optimization. The decision to optimize business processes is made on the basis of the priority principle. The fundamental criteria for selecting a business process to be optimized are the following:

- the importance of the business process to be transformed, determined on the basis of an assessment of the effect received by the company from its implementation, in the course of achieving the set goals and objectives. In this case, it is essential to have the potential advantage of optimizing the highest priority business process, which can bring the greatest positive effect from the redesign;
- the existing gap between the actual and required state of the business process;
- the degree of achievability of changes in the business process, in the context of achieving the optimal ratio between the costs necessary for its transformation and the potential benefits after optimization.

The implementation of Industry 4.0 technologies is necessary and relevant to ensure the optimization of the most important business processes of the company, taking into account the ratio of costs for optimizing business processes and the effect obtained from the transformations carried out. Any unsuccessful adjustments will lead to inefficient labor costs of a financial and material nature.

References

1. Kemp, R., Loorbach, D.: Governance for sustainability through transition management. In: EAEPE 2003 Conference, Montreal, 16–19 October (2003)
2. Transforming our World. The 2030 Agenda for Sustainable Development. https://documentsddsny.un.org
3. Ban Ki-moon Urges Business Leaders to Embrace New Sustainable Development Goals. https://sustainablebrands.com/read/leadership/un-secretary-general-calls-on-businesses-to-help-finance-sustainable-development
4. Kondratyuk, T.V.: Fourth industrial revolution: what competencies are necessary for employees? Strat. Dec. Risk Manag. **3**, 58–69 (2018)
5. Nalbandian, G.G., Khovalova, T.V.: The concept of the internet of energy in Russia: drivers and prospects. Strat. Dec. Risk Manag. **3**, 60–65 (2018)
6. Tarasov, I.V.: Technologies of industry 4.0: influence on increasing the productivity of industrial companies. Strat. Dec. Risk Manag. **2**(107), 62–69 (2018)
7. Tarasov, I.V., Popov, N.A.: Industry 4.0: transformation of production factories. Strat. Dec. Risk Manag. **3**, 38–53 (2018)
8. Pfeiffer, S.: The vision of "Industrie 4.0" in the making—a case of future told, tamed, and traded. NanoEthics **11**(1), 107–121 (2017)
9. Embracing Industry 4.0 – and Rediscovering Growth. https://www.bcg.com / capabilities / operations / embracing-industry_4.0_rediscovering-growth.aspx
10. Rojko, A.: Industry 4.0 concept: background and overview. Int. J. Interact. Mob. Tech. **11**(5), 77–90 (2017)
11. Trachuk, A.V., Linder, N.V.: Technologies of the fourth industrial Revolution: a study of technological adoption by industrial companies using the method of nonlinear principal components (NLPCA). Manag. Sci. Mod. World **1**(1), 4–14 (2018)
12. Agenda for the Twenty-First Century. https://www.un.org/en/chronicle/article/global-health-priority-agenda-21st-century
13. GRI's History. https://www.globalreporting.org
14. United Nations Global Compact. https://www.unglobalcompact.org
15. Hammer, M., Champi, D.: Reengineering of the Corporation: A Manifesto of the Revolution in Business. St. Petersburg University Press, St. Petersburg (1997)
16. SDG Compass. https://media.rspp.ru/document/1/0/b/0b6a1b82138f3b7d363eb80a291a5ae3.pdf
17. ISO 9000. Quality Management Systems. Basic Provisions and Dictionary (Russian Version) (2015). ISO 9001:2015 The Process Approach https://www.iso.org/files/live/sites/isoorg/files/archive/pdf/en/iso9001-2015-process-appr.pdf
18. Sachs, J.: The Age of Sustainable Development. Columbia University Press, New York (2015)

Institutional Support for the Growth of Manufactural Industries in the Russian Federation

Pavel V. Trifonov(✉), Andrei A. Ksenofontov, Aleksandr S. Ksenofontov, Anvar Sh. Kamaletdinov, and Roman V. Seryshev

Financial University under the Government of the Russian Federation, Moscow, Russia
{PVTrifonov,AAKsenofontov}@fa.ru

Abstract. In Russia, ensuring a high level of competitiveness of the national economy is possible only with the realization of a non-resource model. We need to make advanced development of the high-tech and innovative sector in industry. The reasons for the obstacles to the development of industrial sectors are found. These include lack of a financial base for investing in production, low workload, and low profitability of industrial capacities. The necessity of using the principle of complexity in eliminating the identified problems has been substantiated. Currently, studies in the field of economic development of the Russian Federation have become widely known. To this day, the basis of the revenue side of the federal budget of our country is the revenues from the import of hydrocarbon fuel. This circumstance makes the Russian economy extremely unstable in relation to external influences. In fact, the economic well-being of the country is associated with fluctuations in world prices for hydrocarbon fuel. Therefore, it is necessary to develop a new model of sustainable economic development based on industrialization with the use of technical and technological innovations.

Keywords: Industrial policy · Legislative system · State program · Federal target program · Megaproject

1 Introduction

The development of the modern economy is associated with the use of innovations in production and the transition to the sixth technological order [1]. Many countries of the world will actively apply industrial policy tools for a more successful transition to a new level of economic development. State stimulation of innovation activity is often used [2]. One of the reorganization trends is the use of new forms of functioning of the public private partnership mechanism. This mechanism can make it possible to use national and corporate intellectual more effectively, material, and financial resources to transform national industry based on breakthrough technologies [1]. This will allow forming competitive advantages using the capabilities of the developed innovative sector of the economy [3, 4].

P. V. Trifonov and M. V. Charaeva (Eds.): DITEM 2021, LNNS 380, pp. 151–156, 2022.
https://doi.org/10.1007/978-3-030-94245-8_20

Increasing the competitiveness of our country economy is possible only with the use of a non-resource model [5]. The focal course of growth should be the high-tech sector of the economy [6]. The current situation requires the adoption of procedures in the area of legal regulation of the introduction of innovative technologies in industries. One of the levers may be the mechanism of public private partnership. The performed economic analysis showed that this mechanism will be mainly interesting for representatives of large business.

An analysis of the costs of technological innovations showed that their structure has not changed for 16 years. The main source of financing is the company's own funds. Low utilization and profitability of manufacturing capacities reduce the formation of a financial base for innovative changes. It is necessary to develop and implement mechanisms for the return of Russian capital from offshore zones and incentives for financial institutions.

2 Materials and Methods

The article is aimed at studying the institutional support for the development of industries in the territory of the Russian Federation. Companies engaged in industries are the subject of research. The object of research is their economic activity in the field of introduction of innovative industrial technologies. General scientific methods of cognition were used in the research: deduction and induction, synthesis, and analysis, etc. Statistical methods were used as special methods. Data for the research were obtained from the portal of the Federal State statistics service.

The analysis of the data showed that in industry the maximum level of various forms of cooperation is observed in the "Production and distribution of electricity, gas and water". The value of the indicator was 85%. The minimum level of cooperation is observed in the manufacturing industry. The indicator was 52%. The lowest level of cooperation is observed in the high-tech sector of industry. It should be noted that this sector of production is the economy 5 and 6 technological modes development driver.

Therefore, in developed countries, the government, together with business, attaches great importance to the development of cooperation institutions. In Russia, state support for industrial development should use not only financial mechanisms. Organizational mechanisms aimed at improving and creating new forms of public private partnership should also be applied.

3 Results

The first results of the reform of development institutions in the Russian Federation were the concentration of investment functionality in the hands of a single center and the optimization of the existing structure of operators of the institutional structure. The most pressing issue from the recent reform is the task of linking the existing coordination center of the investment entity and the center for the implementation of innovations in Russia. It is also necessary to select investment points and objects for the development of innovations in industrial sectors.

A very important observation in this article is to identify the optimal proportion of levels of technological development of sectors in the structure of the manufacturing industry.

Table 1 shows the optimal structure of manufacturing sectors in terms of manufacturability for the period from 2015 to 2019 in the countries of Southeast Asia in comparison with the same in Russia.

Table 1. Structure of manufacturing industry in Russia [7].

Sector	2015	2016	2017	2018	2019
High-tech	6.5	6.9	6.6	7.1	3.8
Medium-tech high level	23.3	21.9	19.5	19.5	22.9
Medium-tech low level	45.8	45.5	50.1	47.7	54.3
Low-tech	21.5	22.9	20.7	22.1	19

Based on the above table, there is a gap in the shares of high-tech and medium-high-tech sectors in Russia and the countries of Southeast Asia. This means the current technological lag for our country as critical for the formation of competitiveness [8].

The competitiveness of manufacturing industries in the foreign market is characterized by economic effects arising from the interaction of these industries in foreign markets. An important characteristic of competitiveness is the international trade balance of manufacturing sectors (see Table 2). The value of exports of manufactured goods for the period 2012–2019 increased from 199 to 209 billion dollars.

It should be noted that a number of industries have positive dynamics of export potential growth – food industry (100%), wood processing (25%), mechanical engineering (10%). At the same time, industries that have a significant contribution to the export structure, such as the production of coke and petroleum products, the metallurgical and chemical industries have negative dynamics, respectively by 18% and 17%. This is due to the regression of the world market of petroleum products and the decline in the market of raw materials.

In such segments of the manufacturing industry as chemical production, mechanical engineering and the food industry, there is a negative balance between exports and imports. This indicates the unprofitability of these industries in foreign trade. At the same time, the share of export products in mechanical engineering is increasing. However, as the results of the above analysis show, this increase does not solve the main problem – the transition to profit in foreign trade. In general, it should be noted that there is no significant strengthening of the positions of manufacturing industries in the foreign market.

Table 2. Balance of international trade in manufacturing sectors [3, 5].

Sub-sectors	Operations	2012	2015	2018	2019
Manufacturing	Exports	199.9	238.3	193.8	209.8
	Imports	261.3	265.2	243.2	221.8
	Balance	−61.4	−26.9	−49.4	−12
Food industry	Exports	4.389	6.27	6.831	8.085
	Imports	42.5	40	28.9	29.6
	Balance	−38.111	−33.73	−22.069	−21.515
Textile industry	Exports	1.3	1.5	1.4	1.5
	Imports	18.3	17.6	14.7	16.1
	Balance	−17	−16.1	−13.3	−14.6
Chemical industry	Exports	32.6	29.2	24	27.4
	Imports	46	46.5	40.3	43.6
	Balance	−13.4	−17.3	−16.3	−16.2
Mechanical engineering	Exports	26	26.5	28.3	29.1
	Imports	148	137	110	113
	Balance	−122	−110.5	−81.7	−83.9
Coke and petroleum products	Exports	95.96	115.67	58.02	78.15
	Imports	9.9	7.3	4.5	5
	Balance	86.06	108.36	53.52	73.15
Metallurgical production	Exports	58.7	52.3	48.1	54.6
	Imports	23	20.6	16.3	17.8
	Balance	35.7	31.7	31.8	36.8
Other sub-sectors	Exports	5.8	7	7.3	7.1
	Imports	11.2	12.7	8.7	9.5
	Balance	−5.4	−5.7	−1.4	−2.4

4 Discussion

This area of research is very important for the formation of sustainable development of the Russian economy. Digitalization of the economy is a lever for its development. In the article [9], a simulation of the macroeconomic production function, which is an analogue of the Koba-Douglas production function [10], is made. It is noted that the lag in economic development is associated with the lack of modern industrial technologies.

The analysis carried out in the work showed:

1. The volume of the domestic consumption market for industrial goods in Russia does not fully correspond to the strengthening of Russia's geopolitical positions, sustainable socio-economic development that ensures the quality of life of the country's population. The specific level of consumption of industrial goods in Russia is more than 3 times less than in Germany, South Korea and Italy, countries that occupy close positions in the ranking of countries in terms of GDP [6].
2. A significant part of the products in the domestic consumption market is provided by imports: in the whole market by 30% and in such market segments as chemical and engineering products more than 50%.
3. In 2011–2019, the domestic consumption market grew slightly by only 12%, and in a number of segments, the volume of product sales fell.
4. Most of Russia's demand for high-tech products is provided by imports. The share of civilian electronic products produced by Russian industry organizations in the total volume of the domestic electronics market (by revenue) was 31%. Moreover, in 2011–2019, the dependence in these products is only increasing.

5 Conclusion

The article analyzes the high-tech sector of the manufacturing industry and its role in the development of domestic products competitiveness. This circumstance is possible only with the implementation of a non-resource model for the development of the Russian economy. It is revealed that an important condition for the transition to a neo-industrial model of economic development is the implementation of the principle that the technological levels of sectors within the industrial cluster of economic sectors are balanced.

The factors of a negative nature and their correlation according to the degree of influence on the development of the manufacturing industry are determined. These include: a weak financial base for investing in the production capacities of enterprises, the existing low efficiency of production capacity management, the orientation of state support to short-term projects to support medium-tech and low-tech industries, a low degree of implementation of domestic leading industrial enterprises in global value chains.

The necessity of using the principle of complementarity in minimizing risks and eliminating the identified problems is substantiated.

References

1. Cao, Y.H., You, J.X.: The contribution of environmental regulation to technological innovation and quality competitiveness an empirical study based on Chinese manufacturing enterprises. Chin. Manag. Stud. **11**(1), 162–172 (2017). https://doi.org/10.1108/CMS-12-2016-0252
2. Bleys, B., Whitby, A.: Barriers and options for alternative measures of economic welfare. Ecol. Econ. **117**, 162–172 (2015). https://doi.org/10.1016/j.ecolecon.2015.06.021
3. Sazanova, S.L., Valenczyk, R., Yerznkyan, B.A., Ryazanova, G.N.: Relationship of economic communications and productive consumption: theoretical aspect. In: Popkova, E.G., Sergi, B.S. (eds.) ISC 2019. LNNS, vol. 129, pp. 357–364. Springer, Cham (2020). https://doi.org/10.1007/978-3-030-47945-9_40

4. Rogulenko, T.M., Lesnikh, Y.G., Zakutnev, S.E., Zhurakhovskiy, A.S., Kandybko, N.V.: Innovation and financial performance of industrial companies doing business in Russia. In: Bogoviz, A.V., Ragulina, J.V. (eds.) ISCI 2019. LNNS, vol. 280, pp. 279–288. Springer, Cham (2021). https://doi.org/10.1007/978-3-030-80485-5_34
5. Prodchenko, I.A., Gulkova, E.L., Karp, M.V., Tipalina, M.V., Rykova, I.N.: Effective innovation policy in the industrial sector: the case of Russia. In: Bogoviz, A.V., Ragulina, J.V. (eds.) ISCI 2019. LNNS, vol. 280, pp. 269–278. Springer, Cham (2021). https://doi.org/10.1007/978-3-030-80485-5_33
6. Trachuk, A.V., Linder, N.V.: The impact of financial capital on innovative behavior of industrial companies. Fin. Theory Prac. **25**(1), 51–69 (2021). https://doi.org/10.17747/2078-8886-2016-1-80-89
7. World Bank Open Data. https://data.worldbank.org/
8. Bogachev, Y., Trifonov, P.V., et al.: Manufacturing as a driver of technological development of the Russian economy. Int. J. Eng. Adv. Techn. **9**(1), 3221–3224 (2018). https://doi.org/10.35940/ijeat.F9260.109119
9. Kamaletdinov, A.Sh., Ksenofontov, A.A.: Modeling of income of socio-economic systems on the basis of the production function. Fin. Theory Prac. **22**(1), 118–127 (2018). https://doi.org/10.26794/2587-5671-2018-22-1-118-127
10. Daly, H., Cobb, J.: For the Common Good. Beacon Press, Boston (1989)

Imperatives and Driving Forces of the Prospective Economy

Development of Social Economy and Its Impact on Society in the Period of Covid-19

Peter Plavčan(✉)

Danunius Academy, Sládkovičovo, Slovakia
peter.plavcan@vsdanubius.sk

Abstract. This article sums up knowledge on social economy development seen from the point of view of international and Slovak documents from its establishment until now. A part of it is also a historical and structural definition of the organizational structure of the social economy. The social economy is that part of national economies that constantly provides new concrete and clear possibilities of organization an implementation of innovations in the economy. At the beginning of the 21st century, one could have witnessed that there are big marginalized groups of citizens present in the worlds' society that require specific social and economic measures to improve their status using social economy. The study furthermore presents the social economy current state during COVID-19 period, unclear state economy policies of the individual states and international organizations and proposes innovations to support the improvement of the state of the entrepreneurial environment in the Slovak republic, using social economy. It also lists proposals on the legal regulation measures. The research results can be used as an argument for implementing of social economy in national economies around the world.

Keywords: Social economy · Social enterprise · Disadvantaged person · Vulnerable person · COVID-19 crisis

1 Introduction

The economic model of the state with the traditional private sector and state public sector of the second half of the 20th century has proven as insufficiently functional when solving social problems of the society. The pattern of mixed economy is a basis of a social state that focuses on the notorious market failures and on the implementation of economic measures that are applicable for market failures, being the distribution of the profit and costs and the anti-cyclic measures. At the beginning of the 21st century, however, one could have witnessed that there still are big marginalized groups of citizens that require specific social and economic measures to improve their status. These measures are jointly called social economy.

The social economy is defined as the sum of production, distribution and consumption activities done in public interest, by rendering socially useful service for the society as such or for the unlimited group of physical persons, disadvantaged persons or vulnerable persons. That is why we call it solidarity economy.

P. V. Trifonov and M. V. Charaeva (Eds.): DITEM 2021, LNNS 380, pp. 159–165, 2022.
https://doi.org/10.1007/978-3-030-94245-8_21

The 20-ties of the 21st century are marked by a period characterized by the presence of COVID-19. The fall of the world economy, frustrations of the citizens being concerned about their health and unclear state policies throughout continents require effective measures. The knowledge on new technologies' development dynamics of internet platforms and innovative internal markets proves their advantages in comparison to conventional markets in this segment of economy. This knowledge creates a theoretical basis for the preparation of the proposal on possibilities of usage of the internet platforms in context of the international and Slovak law in order to disseminate knowledge on social economy and social businesses [1].

2 Materials and Methods

The goal of this study is to present knowledge on the development of the social economy and the possibilities of its implementation in the pandemic and post pandemic period of COVID-19. A further goal of this study is to propose measures to support the improvement of the unfavourable state of the economy of small and middle-sized companies. The proposition of the measures is focused on the creation of the functional system of further education including consultancy to physical and legal persons active in small and middle-sized companies in order to integrate their business activities into social economy.

We rely on the thesis that there is not enough information on social economy system in Slovak society as well as on possibilities that are available not only for disadvantaged and vulnerable physical persons, but also for small and middle-sized businesses whose entrepreneurial activity is being taken apart. As anti-thesis, we provide the opinion that in Slovak economy there is enough information on creation, performance and propagation of social economy, while this is indirectly supported by medial inactivity of competent state organizations for labour, social affairs and family. The research problem of this study is the search for the working system of measures to support small and middle-sized companies towards their transformation into social businesses or businesses with social impact. The main research methods in this paper are special methods used in the context of social sciences. The historical-logical analysis of the social economy, the method of analogy, the transfer of the basic principles of the social economy to the Slovak and world economy are used.

3 Results

During the last two centuries, the Czech economists came up with socio-economic approaches where profit was not a priority. Based on this tradition that reaches up to the 19th century, during the period of the first Czech-Slovak republic (1918 until 1938), many non-profit organizations associating inhabitants were founded.

The mixed economy favoured great portfolio of companies and organizations such as cooperatives, joint ventures and associations that helped to address socially important questions of general interest concerning cyclic unemployment, misbalance in regional development and relationships between retailers and consumers.

Social economy in the literature used to be connected with the third sector countries [2] as well as with thoughts on its development from the legal, economic standpoint and the point of view of the historic development of the social economy organizations. Modern view on social economy is connected with its new role as a solver to current social problems and needs of citizens, when the markets and state were failing.

The term "social economy" is not terminologically stable in foreign English written literature. To describe social economy, they often use solidarity-based economy, work integration social economy, market-oriented social economy [3], or even alternative economy, plural economy and intermediate economy. In foreign and French literature it is e.g. économie solidaire, économie plurielle, l´autre économie [4, 5].

The National Liaison Committee for Mutual, Cooperative and Associative Activities (CNLAMCA) prepared and published in 1980 the Social Economy Charter that introduces the definition of social economy as a group of organizations that do not belong to public sector, that democratically interact with members having the same rights and duties and applying specific mode of ownership, distribution of profit, whereas the surpluses have been used to expand the organization and to improve its services for its members and for the company [6].

From the institutional point of view, the "social economy" includes the main cooperatives, joint ventures and associations, as well as foundations. The cooperatives, joint ventures, associations, foundations have rendered it inevitable to lead a stable dialogue on European policies of common interest. In November 2000, the European Standing Conference of Cooperatives, Mutual societies, Association and Foundations (CEP-CMAF) was founded. The currently widely accepted definition is to be found in the Social Economy charter, as follows [7]:

1) Primacy of people and of the social objective over capital
2) Voluntary and open membership
3) Democratic control by the membership (does not apply to foundations, as these have no members)
4) The combination of the interests of members, users and society general interest
5) The defence and application of the principles of solidarity and responsibility
6) Autonomous management and independence from public authorities
7) Reinvestment of the essential surplus to carry out sustainable development objectives, services of interest to members or of general interest.

The European Standing Conference of Cooperatives, Mutual societies, Associations and Foundations was founded in 2000 and changed its name to the Social Economy Europe in 2008. Social Economy Europe works in fields of the general interest of its members and acts in compliance with the rule of subsidiarity, while securing added value by means of organizations´ measures. It pursues the following values: primacy of people and of the social objective over capital, democratic control by the membership, the combination of the interest of members, users and society general interest, the defence and application of the principles of solidarity and responsibility, reinvestment of the essential surplus to carry out sustainable development objectives, services of interest to members or of general interest, voluntary and open membership and autonomous management and independence from public authorities [8].

The significance of Social Economy is also being confirmed by the European Union states, the Legal Orders of which have Social Economy acts incorporated, e.g.: Belgium, Spain, Greece, Portugal, France, Romania, and also Slovak Republic.

In 1989, the European Commission published a Communication titled: Businesses in the "Economie Sociale" sector: Europe´s frontier-free market, where it has introduced the social and economic significance of Social Economy [9].

In 1996, a European research network became an important bearer of research activities on social economy bearing the name of the first research project The Emergence of the social businesses in Europe in French (L'Emergence de l'Entreprise Sociale en Europe – EMES), which deals with following main topics: social businesses, social economy, solidarity economy and social entrepreneurship. It associates 14 research institutions, 330 individual members and the network of partners who support EMES mission all over the world.

Furthermore, the European Council approved the Resolution on the Promotion of Social Economy as a Key Driver of Economic and Social Development in Europe in 2006 [10].

There is a parliamentary group called "The Parliamentary Social Economy Intergroup of European Parliament" active in the European Parliament since 1990. This Parliamentary Intergroup has also asked the European Commission to implement the Action plan for Social Economy in 2017.

In 2013, the European parliament adopted the Report on the contribution of cooperatives to overcome the crisis. The document was created by Patrizia Toia, it recognized the Social Economy as a social partner and as a key subject for the fulfilment of goals of Lisbon strategy and the contribution of cooperatives towards overcoming of the financial and economic crisis [11].

The European Economic and Social Committee, the Consulting Committee of European Union has been publishing numerous publications, reports and opinions on the contribution of social economy to the achievement of various public policy goals. The European Commission has recently adopted two important initiatives concerning social businesses, the group of companies that are an inseparable part of social economy: the Social Business Initiative (SBI) of the European Commission and the Regulation on European social entrepreneurship funds.

Social economy innovations during COVID-19 period.

The Social Economy in the European Union states includes ca. 2,8 million social economy subjects. The Social Economy enterprises in the European Union are the associations, cooperatives, foundations and mutual societies, as well as social businesses that not only create 6,3% of the employment in the European Union states, but also exert high integration influence in the economic field during the current COVID-19 period, and also in the social sphere focused on social, educational or environmental goals that are important for citizens. This integration influence has a special significance in rural, municipal, regional and other social communities of interest, to the time of decreased social contact among people.

According to the Social economy and COVID-19 crisis document: the current and future roles, the recommended political actions of national and regional governments have defined a shared vision about the future to "build back better" and ensure that social economy organizations actively contribute to this vision [12].

In order to support the improvement of the unfavourable state in the economy sector, where 99,7% of business subjects in the economy of Slovak republic are active, the following is being proposed:

1) Conduction of analysis of the impact of the COVID-19 period on community, regional and national level of economy management, in order to propose measures to improve the status on all levels of economy management,
2) Conduction of analysis of the impact of the COVID-19 period on the endangered groups of citizens, not only on disadvantaged persons and vulnerable persons in social economy, but also on physical and legal persons owning small or middle-sized businesses that are currently losing their legal and material securities in their business,
3) Conduction of the tendency prediction of the lasting impacts of COVID-19 period and prediction of potential legal amendments on social economy during the so called post-COVID period after 2020,
4) Preparation of new legal version of the Social Economy act as well as on social businesses according to identified risks of the current COVID-19 period, applicable for selected groups of physical and legal persons.
5) Proposal of measures that would increase awareness of social economy
6) Proposal of measures aiming to improve further education and consultancy to endangered groups of citizens, mainly to physical and legal persons active in small and middle-sized businesses in order to achieve the integration of their business into social economy,
7) Proposal of measures aiming to support the directed and systematic research of social, educational, legal, economic, social, psychological, medical and further impacts of the COVID-19 period,
8) Proposal of measures to be implemented in order to increase financial support for social economy subjects in broader social economy sector.

Simplification of the rules for establishment and economy of the generally beneficiary businesses providing services focused on social, educational, cultural and health care field in given economic environment, whereas in the legal form it is necessary to respect the community, regional and national opinions;

Simple and transparent information system about social economy subjects, especially about current support of social economy by financial tools [13].

4 Discussion

In Slovak society, there is not enough information on social economy system and on opportunities this system provides not only to disadvantaged and vulnerable physical persons, but also to small and middle-sized entrepreneurs. This thesis has been confirmed. Small and middle-sized businesses are currently suffering because of the world,

European and national economy crisis, as well as because of incorrect economic measures that were implemented by the governing agencies, while receiving only small state support in order to mitigate part of the economic impacts.

The media campaign and the usage of dynamics of the new technology development in the field of internet platforms and innovation internet markets might directly provide enough information requested by business sphere [14]. These information on creation, implementation and practicing of social economy when establishing new social companies or when transforming existing small and middle-sized businesses into social economy businesses is inevitable to support the small and middle-sized businesses in the world economy.

5 Conclusion

The social economy is that part of national economies that constantly provides new possibilities of organization an implementation of innovations in economy. These innovations in economy are positively accepted by citizens because they present a new fair view on selected parts of business activities in economy. These are the so-called fair trade, ecologic groceries and ethical financing. These social innovations change the thinking of people about economy, society and contribute to bigger all-society solidarity in the current COVID-19 period and in the period to follow. The social economy businesses have in the current period of more or less closed economy also significant integration function, where they are in a position to impact marginalized groups of citizens by their activities, even more so now than any time before. However, the future of the social economy also appears to be in higher spending on education, which increases educational outcomes and reduces the social impact on the population [15].

References

1. Plavčan, P., Funta, R.: Justičná Rev. **10**, 934–957 (2019)
2. Shragge, E., Fontan, J.M.: Social Economy. Black Roses Books, London (2000)
3. Defourny, J., Favreau, L, Laville, J.L.: Introduction to an international evaluation. In: Spear, R., Defourny, J., Favreau, L., Laville, J.L. (eds.) Tackling Social Exclusion in Europe: The Contribution of the Social Economy. Ashgate, Aldershot (2002)
4. Laville, J.L., Cattani, D.: Dictionnaire de l´autre économie. Gallimard, Paris (2006)
5. Laville, J.L.: L´économie solidaire. Une perspective Internationale. Hachette Littératures, Paris (2007)
6. Monzon, J.L.: CIRIEC – España **10**, 19–29 (1987)
7. The Social Economy Charter. https://www.socialeconomy.eu.org/wp-content/uploads/2020/04/2019-updated-Social-Economy-Charter.pdf
8. Social Economy Europe. https://www.socioeco.org/bdf_organisme-478_en.html
9. The social economy in the European Union. Summary of the Report drawn up for the European Economic and Social Committee by the International Centre of Research and Information on the Public, Social and Cooperative Economy (CIRIEC). https://op.europa.eu/en/publication-detail/-/publication/486d651a-a3a2-4902-b5a8-8228f48a3b24/language-en?_publicationDetails_PublicationDetailsPortlet_source=search

10. Council of the European Union. The promotion of the social economy as a key driver of economic and social development in Europe (2015). https://www.eesc.europa.eu/en/documents/promotion-social-economy-key-driver-economic-and-social-development-europe-council-conclusions
11. European Parliament (2009). Report on the contribution of cooperatives to overcoming the crisis. https://www.europarl.europa.eu/doceo/document/A-7-2013-0222_EN.html
12. Social economy and the COVID-19 crisis: current and future roles. oecd.org/coronavirus. https://www.oecd.org/coronavirus/policy-responses/social-economy-and-the-covid-19-crisis-current-and-future-roles-f904b89f/
13. Plavčan, P.: Sociálna a vedomostná ekonomika. MSD, Brno (2020)
14. Plavčan, P., Funta, R.: Some economic characteristics of internet platforms. Danube (2020)
15. Plavčan, P.: The comparison of PISA educational results with gross domestic product in member states of European Union. In: International Conference on Innovations and Science education, Prague, 20–22, March 2019

Transformation of Educational Space in the System of Digitalization of the Economy

Elena V. Altukhova[1,2](✉), Veronika V. Nikeryasova[1], Yana Y. Ivanova[1], Maxim A. Markov[1], and Olga V. Romanchenko[1]

[1] Plekhanov Russian University of Economics, Moscow, Russia
[2] Russian Presidential Academy of National Economy and Public Administration, Moscow, Russia

Abstract. The current state of the economy and the globalization processes affecting it require significant changes in all spheres of society, including the system of highly qualified personnel training. A list of professions has been identified by now as well as the priorities for their development for the next five years. The work aims to study new trends in education development based on the analysis of specific methods and tools that allow not only to change significantly the situation, but also to introduce new levers, tools, approaches and methods, to form a new paradigm for solving certain problems. The implementation of new tools and methods requires a significant change in the educational space itself. A new challenge today is to adapt to the digital environment that has become part of the education system. Based on a comparison of various aspects of the system of offline and online learning, conclusions about the formation of an independent direction of online education with specific content and specific environment were made herein. The views of different researchers regarding the impact of digitalization on educational content, which has become more structured, have been studied. It is concluded that it is important to understand the fundamentals that form the existing dependencies, the need to use creative approaches is considered. As a novelty, it is proposed to consider the use of new digital educational technologies as an additional impetus for the transformation of the educational space, allowing to meet the interests of students, business and the labor market.

Keywords: Digital educational technologies · Information security · Digital marketing · Digital economy

1 Introduction

An important condition for the development of the digital economy is not only legal regulation and information security, but also the availability of qualified and adapted to the new conditions personnel who will ensure the sustainable development of the entire system. During the pandemic, the field of education received a special impetus for the development of new independent areas.

The list of professions and priorities of their development for the next years has been revealed by now. The directions to be considered in programs of development

P. V. Trifonov and M. V. Charaeva (Eds.): DITEM 2021, LNNS 380, pp. 166–171, 2022.
https://doi.org/10.1007/978-3-030-94245-8_22

of the educational organizations have been defined and the results to be achieved for maintenance of target indicators have been formulated.

The development of the technology of achieving results remains the important task, which will allow the introduction of new tools, methods and technological approaches, using a new methodology for solving certain problems.

According to the McKinsey Global Institute, 2 to 50% of work expressed in man-hours can be automated by 2036, and this share can reach from 46 to 99% by 2066, which will contribute to the demand for programmers, engineers, designers, advertisers, marketers. These areas are associated with a creative approach and a sufficiently high level of computer literacy.

2 Methods

A comprehensive approach based on the comparison and analogy of the analyzed data, which determine the priority areas of education in the digital transformation system, is used herein.

A modern specialist must have not only universal skills, but also a unique specialization that can ensure its demand in the labor market. A practice-oriented approach that ensures the success and demand for training of future professionals, which requires the direct participation of practitioners in the process of training graduates is of particular importance for the construction of modern competencies [1].

The indicator "The best universities in terms of demand for graduates by employers" used in the university rating system is one of the tools to incite the interest of universities in the training of in-demand personnel. Ensuring competitive innovation development in a difficult demographic situation today requires additional efforts from universities.

Given that modern knowledge is born at the intersection of different sciences, an interdisciplinary approach becomes especially important. A qualified economist with basic mathematical training or an IT specialist who understands the peculiarities of business processes will have a better chance of success [2].

In this regard, industry universities today have taken a course on multidisciplinary, and began to develop related areas. In order to maintain their position in the global education market, many universities consider it necessary to enter the international scientific and educational space. The degree of globalization of the university in terms of the share of foreign students and teachers can be seen in the QS World University Ranking or THE (Times Higher Education) ranking.

The rapid introduction of online learning, which stimulates the development of blended learning (blended learning) and online courses MOOC (Massive on-line open course), is among the most interesting digital innovations. The explosive growth of online learning could be observed throughout 2020 (Fig. 1).

Market growth outstripped the forecasted values by 16%. Today, the digitalized part of the market accounts for up to 4% of its total volume.

The use of e-learning and the introduction of distance learning technologies can significantly reduce the classroom load within the educational program by increasing the proportion of independent work and contact extracurricular work between the student and the teacher.

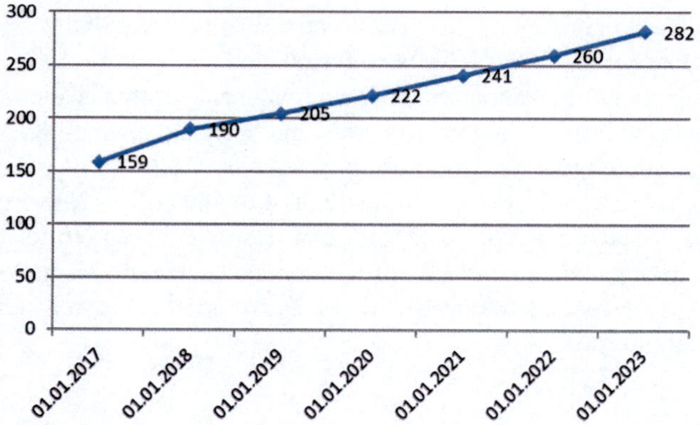

Fig. 1. Forecast growth in the volume of the global online education market, billion US dollars. *Source:* Global Market Insights, EdTechXGlobal, Education International.

The increase in the number of offered courses and students requires a serious modernization of the convenience of the services themselves, functionality, as well as the options for presenting the material. It is interesting to use smart-didactics aimed at the formation of personal and professional success of the learner based on open dialogical interaction with a teacher or artificial intelligence [3].

An important area of using digital technologies in education is the development of digital libraries and digital campuses of universities, as well as the introduction of Bloomberg and Thomson Reuters functionality for monitoring and analyzing financial information into the educational process. In this system, it is important to maintain a balance between what is actually needed and what is necessary [4].

The most resource-intensive for a university in the digital transformation mode will be digital marketing, an integral part of which will be a modeling method that allows predicting further trends in order to maintain feedback with all stakeholders.

3 Results

A serious effect of the pandemic was added to the new requirements of the labor market and modern digital realities, which forced the system to develop in an accelerated format, forming new directions of development, tools and methods for implementing educational programs [5]. Considering the current trends in the development of the online segment, some key features can be identified (Table 1).

The presented forms of training characterize the significant differences between the new services.

It should be noted when comparing the offline and online approaches (Table 2), that online services not only cause the automation of processes, but also stimulate the development of new approaches and teaching methods [6].

This fact determines the need for a mixed form, which on the basis of the offline segment will include the highest quality online services, which will seriously improve both the educational technologies themselves and the form of presentation of educational content [5].

Table 1. Modern forms of online learning.

Form of education	Description	Advantages
Microlearning	The program is divided into short frames (3–5 min)	The speed of assimilation
Artificial Intelligence	Individual training	Personal educational content
Mobile Learning	Content adapted to mobile devices	Ease of use
Gamification	Study of the material on a specific situation	Works well in children's education
Project training	Working on a specific project	In addition to basic knowledge, additional skills are acquired

Table 2. Comparative characteristics of the forms of education.

Form of study	Advantages	Disadvantages
Online form	– speed of mastering the material – the versatility of the platforms used – high speed of processing results – availability – continuity of education forms the ability to self-education	– insufficient digital user skills – low level of security of services – technical problems with the settings of teaching devices – insufficient degree of reflection on the information received by the listener
Traditional form	– development of creative and communication skills – individual approach – quality control of results – live communication	– fixed time frames for classes – long period of results monitoring – subjective assessment of results

It can be assumed that competition will single out the best programs and educational institutions capable of training a high-level specialist [7], will take into account the interests and capabilities of various groups of the population, and will ensure the continuity of the educational process.

All this will increase the listener's ability to master the material in the most acceptable form, expanding the range of sources of reliable information.

4 Discussion

Modern system of digitalization of education is a rather costly model. The introduction of tests significantly reduces the costs of the educational institution. Formally, the computer checks the test results much faster than the teacher does. However, the machine will not be able to identify the strengths and weaknesses of the listener in creative assignments [8].

In the studies of Kozlova, Popova, Terelyansky digital transformation of education is presented as a solution to the problems of bridging the digital divide [9]. Thus, educational technologies are tuned to personalize the educational process through the use of convenient digital technologies.

The first demonstrative lesson of the pandemic showed that it was impossible to supersede the teacher-lecturer as an essential unit that ensures the formation and development of creative abilities. At the moment, research continues on the technique of presenting material to the student audience to make the courses taught more accessible. This aspect is discussed in the works of Julian Kirchherr based on teaching principles for the circular economy [10].

The improvement of teaching technologies in the digital economy is presented in the study by Adel Ben Youssef [11], which connects the entrepreneurial skills of students with the digitalization of the economy, which today is reflected in the creation of start-ups on the basis of universities [12] and also reflects a personal approach in the system of organizing project training education.

When developing a model of professional competencies, it is necessary to take into account such basic criteria as structuredness, comprehensibility, simplicity and adaptation to market requirements [13].

5 Conclusion

The labor market and the personnel training system should become a single tool that solves the tasks set for the economy, ensuring the creation of a motivation system for mastering the necessary competencies for the development of digital transformation. The education system should train personnel for the "knowledge" economy, while taking into account the needs of business. In turn, the business must provide an environment for the growth and self-realization of the employee. The alignment of these interests will provide additional impetus for the formation of a technological breakthrough [14].

This study is limited to considering the features that are typical mainly for the Russian education system. In the near future, globalization processes will force Russian universities to become part of the international market for educational services, which will be actively promoted by a mixed education model based on online services. In the course of adapting to the new reality, many universities are trying to find their place in the system of the world scientific and educational environment, while maintaining their uniqueness and competitive advantages [15].

Digital transformation will make it possible to change the corporate culture of the university and optimize many internal processes, which will increase the efficiency of interaction between the departments of the educational institution during the transition to a new educational paradigm.

The use of digital technologies will make it possible to focus the educational process not only on fulfilling the requirements of the educational standard, but also will tune in to the formation of the professional culture of the future specialist, forming a desire for constant self-improvement using information services and technologies, which will contribute to acceleration of the learning processes automation, and incite the formation of educational process consistency.

References

1. Kupriyanovskiy, V.P., et al.: Skills in the digital economy and challenges of the education system. Int. J. Open Inf. Technol. **5**(1), 19–24 (2017). https://cyberleninka.ru/article/n/navyki-v-tsifrovoy-ekonomike-i-vyzovy-sistemy-obrazovaniya/viewer
2. Sizova, I., Khusiyanov, T.: Labor and employment in the digital economy: problems of the Russian labor market. Bull. St. Petersburg University **4**, 378 (2017)
3. Sharonin, Y.V.: Digital technologies in higher and professional education: from personality-oriented smart-didactics to blockchain in targeted training of specialists. Mod. Prob. Sci. Educ. **1** (2019). https://science-education.ru/ru/article/view?id=28507
4. Zhao, Y., MaríaPinto, L.A., Sánchez, G.M.C.: Digital competence in higher education research: a systematic literature review. Comput. Educ. **168**, 104212 (2021). https://doi.org/10.1016/j.compedu.2021.104212
5. Altukhova, E.: Using the results of intellectual activity in the educational process. Bull. PRUE G. V. Plekhanov **4**, 39–45 (2018). https://doi.org/10.21686/2413-2829-2018-4-39-45
6. Zain, S.: 13c – digital transformation trends in education. In: Future Directions in Digital Information Predictions, Practice, Participation Chandos Digital Information Review, pp. 223–234 (2021). https://doi.org/10.1016/B978-0-12-822144-0.00036-7
7. Bondarenko, T., Kolmakov, V.: Distance learning as an active educational technology: assessment of the feasibility of implementation. Azimuth Sci. Res. Pedagogy Psychol. **7**(3), 53–57 (2018). https://cyberleninka.ru/article/n/distantsionnoe-obuchenie-kak-aktivnaya-obrazovatelnaya-tehnologiya-otsenka-tselesoobraznosti-vnedreniya/viewer
8. Altukhova, E.: Competence-based approach in the context of digitalization of the economy. Bull. PRUE G.V Plekhanov **2**, 66–73 (2020). https://doi.org/10.21686/2413-2829-2020-2-66-73
9. Kozlova, N.: Digital technologies in education. Bull. Maikop State Technol. Univ. **1**, 83–91 (2019). https://doi.org/10.24411/2078-1024-2019-11008
10. Kirchherr, J., Piscicelli, L.: Towards an education for the circular economy (ECE): five teaching principles and a case study. Resour. Conserv. Recycl. **150**, 104406 (2019). https://doi.org/10.1016/j.resconrec.2019.104406
11. Youssef, A.B., Boubaker, S., Dedaj, B., Carabregu-Vokshi, M.: Digitalization of the economy and entrepreneurship intention. Technol. Forecast. Soc. Chang. **164**, 120043 (2021). https://doi.org/10.1016/j.techfore.2020.120043
12. Barinova, N., Nazarova, T.: Theoretical approaches to the commercialization of innovations. Stage Econ. Theory Anal. Pract. **4**, 49–59 (2018). https://doi.org/10.24411/2071-6435-2018-10036
13. Dneprovskaya, N.: Assessment of the readiness of the Russian higher education for the digital economy. Econ. Stat. **4**, 16–28 (2018). https://doi.org/10.21686/2500-3925-2018-4-16-28
14. Tereliansky, P., Kuznetsov, N., Ekimova, K., Lukyanov, S.: Transformation of education in the digital age. Univ. Manag. Pract. Anal. **22**(6), 36–43 (2018). https://doi.org/10.15826/umpa.2018.06.056
15. Popova, O.: Transformation of higher education in the conditions of the digital economy. Manag. Issues, 158–160 (2018). https://cyberleninka.ru/article/n/transformatsiya-vysshego-obrazovaniya-v-usloviyah-tsifrovoy-ekonomiki/viewer

Formation of a Model of Functional Working Ages

Irina B. Durakova, Marina G. Kholyavka, Tatyana I. Rakhmanova, Ekaterina V. Mayer, and Ivan V. Grigorov(✉)

Voronezh State University, Voronezh, Russian Federation

Abstract. The article attempts to substantiate the possibility of increasing the success of senior workers through the implementation of the synthesis of biomedical and behavioral drivers. To carry out this interdisciplinary study, it was determined that it is advisable to refer those who have reached an average of 20–25 years of age to the adult category of the working-age population, including on the basis of the stasis of tubular bones growth. Among the stages of the life cycle, there are 3 working periods based on the chronological age: 25–44, young age; 45–59, middle age; 60+, old age. Biomarkers of aging were selected for analysis, reflecting the rate of development of age-associated multifactorial pathologies. Using statistical methods, a set sample of the population was determined, represented by different age groups of respondents (patients) suffering from certain diseases or relatively healthy. Scientific images about life cycle phases were exposed. Measures were identified to preserve the potential for success of elderly employees in accordance with life cycle phases, intended for personnel services as well as directly for employees. Research methods: statistical groupings, modeling, comparative analysis, calculation of forecast indicators. A new element of research that distinguishes it from similar ones is, firstly, the employment of biomedical and interdisciplinary approaches. Secondly, the distribution of practical measures by the phases of the employee's life cycle, which makes it possible to diagnose regional thresholds of personnel aging and to carry out preventive and therapeutic measures, as well as those aimed at improving working conditions.

Keywords: Functional age · Life cycle phases · Successful longevity

1 Introduction

People are more important than profit. This is a statement underlying the modern management paradigm and expressed by the formula "person – organization – victory". Winning, synonymous with explosion, push or success, is the only contemporary opportunity to meet the new demands of the consumer market while taking into account the Olympic motto: "Faster, higher, stronger". The idea of sporting in management fell within the period of the silver economy, justifying the need for the formation of competitive advantages in organizations, taking into account a different demographic situation, a new layout of jobs, a shortage of specialists and managers on the labor market, generations unpredictably facing different value and target settings [1–3]. Unlike Japan,

P. V. Trifonov and M. V. Charaeva (Eds.): DITEM 2021, LNNS 380, pp. 172–176, 2022.
https://doi.org/10.1007/978-3-030-94245-8_23

the Netherlands, Germany and a number of other countries classified in the age table as demographically old, Russia is still on the verge of problems that those states have already faced [4, 5].

The transition to the paradigm of successful old age in recent decades is based on the understanding of wide variability of the forms of aging, the presence of special mechanisms (for example, selection, optimization and compensation that contribute to the replacement of "losses" in old age), and the accumulated base of evidence concerning the fallacy of age-related stereotypes widespread in society [6].

2 Methods

The article proposes the implementation of one of the approaches, which consists in substantiating the assumption of regionally heterogeneous aging of employees, presenting functional working ages based on the employee's life cycle phases, which increases the likelihood of replenishing age-related "losses" and extending labor longevity in a sporting-oriented paradigm.

3 Results

It was revealed that physiological indicators reach a peak from the age of 20 to 30 years old, and then decline by about 1 percent per year. The boundaries of aging are nationally heterogeneous, scientific thought testifies to the shifting representation of the worker's life cycle stages.

It is advisable to refer those who have reached an average of 20–25 years of age to the adult category of the working-age population, including on the basis of the stasis of tubular bones growth. Therefore, among the life cycle stages, we distinguish 3 working periods based on the chronological age: 25–44, young age; 45–59, middle age; 60+, old age. But according to one of the general laws of gerontology, everyone and everything inside everyone is aging at a different rate. This is confirmed by the data, when at the same chronological age the degree of aging of individuals differs [6]. Therefore, we substantiate the need to introduce two concepts: biological age which would reflect the level of wear of the functional structure or the organism as a whole, and functional age reflecting the age-related dynamics of physiological functions and functional reserves, the ability of a person to function. Consequently, performance will directly depend on functional age, which, in turn, will be determined by human health.

The main causes of premature mortality of the able-bodied population as the object of research were determined: 1) diseases of the circulatory system, 2) neoplasms, 3) accidents, poisonings, injuries. At the bottom of the list are infectious and parasitic diseases. It was found that in the 35–45 age group the mortality rate from diseases of the circulatory system is slightly lower, but higher from accidents, poisonings and injuries. The highest mortality rate from neoplasms was found in the group of 56–60 years old, and from cardiovascular diseases, in the group of 46–55 years old, which, according to literature sources, indicates an early manifestation of atherosclerosis [7].

The study showed an increase in the number of people with diseases of the blood, blood-forming organs and certain disorders involving the immune mechanism, including those diagnosed with anemia. The cohort of people suffering from high blood pressure has dramatically increased.

Based on the results of clinical examination of able-bodied persons, firstly, a positive (by 25%) dynamics of those suffering from diabetes mellitus was determined starting from the age of 46, the consequences of the disease can lead to a decrease in the body's functional capabilities and disability. It is known that diabetes mellitus increases mortality 2–3 times, the risk of coronary heart disease and myocardial infarction, 2 times, kidney pathology, 17 times, gangrene of the lower extremities, 20 times, arterial hypertension, more than 3 times, and rates first among the causes of blindness [8].

Secondly, it was found that, starting from the age of 35, with half of the examined patients the level of total cholesterol (CS) exceeds the acceptable level, which is a risk factor for the development of atherosclerosis. With age, the level of cholesterol tends to gradually increase, since by the age of 65, the rate of absorption of cholesterol in the intestine increases from 50% to 80% [9]. Thus, the rate of atherosclerotic vascular lesions will only increase with age. Therefore, it is important to monitor the diet and lifestyle, which can mitigate the effect of aging on the metabolism of cholesterol.

The causes of the main non-infectious diseases morbidity of the working-age population (hypertension, tumor diseases) were revealed: acceleration of the pace of life, technical progress, the use of new synthetic materials in industry, construction, everyday life, deterioration of the environmental conditions of human life, changes in the quality of nutrition, which is consistent with the literature data [10].

It follows from our forecast based on statistical data that the number of patients with diseases of the circulatory system, including high blood pressure, will increase by ~30% in the next five years. This may lead to a further increase in the proportion of persons with disabilities and an increase in premature mortality from these diseases, which significantly reduces the level of the population's working capacity.

It was defined that the task of diagnosing functional age comes down to assessing the following, practically important indicators:

- indicators of the actual "biological age", which are difficult, and often impossible to affect (pulse wave velocity, range of near vision, hearing threshold), but giving an idea of the individual aging rate;
- labile indicators that relate to aging, but can be changed by physical training, diets, therapeutic prophylactic impacts: determination of physiological age, psychological age, characteristics of metabolic indicators, etc., which also gives a wide scope for replacement and corrective therapy;
- factors of the functions reserve, risk factors and statistical indicators of mortality and life expectancy.

According to the predicted values, in the population of the object under study in the period from 2021 to 2025, the number of groups of 65–69 years old and from 70 years and older is very likely to escalate due to an increase in life expectancy. In accordance with this, the traditional view of the content of life cycle phases is changing.

While the phase of young age is characterized by "expansion", that is, readiness to make plans, breakthroughs, learn new things, as well as to overcome the first professional difficulties and achieve the first success, still the "middle life" phase is characterized by the presence of sustainable experience, the ability to work independently and acquire expert competence. The 60+ phase, which was traditionally seen as a time of slow working decline, putting things in order, finding new goals, often out-of-work, yet now it is acquiring new content. In accordance with the aging of the region (more than 7% of the population aged 65 and over), the need for employment increases in this category, together with the preservation of its ability to work and achieve professional success.

Reducing the discrepancy between chronological and functional age is possible with the involvement of three parties: health care, employer and employees.

First, it is appropriate to determine the reasons for the decline in the functional and cognitive abilities of workers. The study revealed that the share of malignant neoplasms, mental disorders and behavioral disorders is increasing in the structure of regional disability. Efforts in early diagnosis, prevention and treatment of psychosis and dementia is a factor in reducing the severity of the problem.

Secondly, a long-term universal program is proposed to maintain the functional capabilities of the body, including the prevention and treatment of age-related diseases.

The program involves the preliminary stage, its goal being to bring to health and foster a healthy lifestyle, including a course of preliminary body cleaning, normalization of body weight and body shaping, correction of physiological age by means of physiotherapy exercises, sleeping schedule, autopsychotherapeutic treatment, correction of individual nutrition pattern, etc.

The goal of the main stage is to constantly maintain the minimum aging mode. In the structure of activities – the ecology of water (filters), food, air and home (air conditioners, ionizers, humidifiers and air purifiers) Constant internal body cleaning, detoxification, reduction of glucose and free blood ferrum levels. Control of the endoecology of the body, a special iodine-replacement diet, psychoemotional correction, growth and development correction, external maintenance.

4 Discussion

Research of functional working ages has traditionally been based on three life cycle phases [11]. At the same time, there appeared justifications for dividing phase 25–44 and including the fourth phase of 35–44, which makes it possible to obtain a detailed idea of the working line of life, since it was during this period that half of the examined patients demonstrated an increased risk of developing atherosclerotic vascular lesions. This period is very important, since at the initial stage it is easier to influence the speed of the pathological process and slow down the clinical manifestations of the disease, which, as a rule, lead to a decrease or loss of working capacity. The vector of modern developments is aimed, in addition to the tasks set in the article, at the problems of working longevity, depending on the involvement and emotional burnout of workers [12]. It was revealed that with great experience and competence, young and old workers alike experience low stress. When competencies are deficient, stress with the elderly is more pronounced than with the young [13]. The connection between age and work involvement is positive if the significance of the task, interaction outside the organization, or both of those performance characteristics are high [14]. The interaction between the individual characteristics of the employee and the requirements of the employer enhances psychological adaptation and well-being with age [15].

5 Conclusion

The results of the study confirmed the necessity and possibility of preserving the potential of older workers during the transition of organizations to the sporting-oriented management paradigm. For organizations, long-term programs are advisable to diagnose and slow down pathological accelerated aging of employees. The main limitations for improving the dynamics are ageism and untimely changes in working conditions on the part of employers, neglect of employees' health and their lack of initial literacy to prolong their own active longevity.

Acknowledgments. The reported study was funded by RFBR, project number 20–010-00263.

References

1. Iparraguirre, J.L.: Economics and Ageing. Springer, Cham (2019). https://doi.org/10.1007/978-3-030-29013-9
2. Evangelista, R.: The Business of winning. CEP Press, Atlanta (2001)
3. Rogelj, V., Bogataj, D.: IFAC-PapersOnLine **52**(13), 2680–2685 (2019). https://doi.org/10.1016/j.ifacol.2019.11.612
4. Safarova, G.L., Kalmykova, N.M., Safarova, A.A.: Adv. Geront. **31**(2), 162–169 (2018)
5. Timonin, S., Danilova, I., Andreev, E., Shkolnikov, V.M.: Recent mortality trend reversal in Russia: are regions following the same tempo? Eur. J. Popul. **33**(5), 733–763 (2017). https://doi.org/10.1007/s10680-017-9451-3
6. Lowsky, D.J., Olshansky, S.J., Bhattacharya, J., Goldman, D.P.: J. Geront. Ser. A Bio. Sci. Med. Sci. **69**(6), 640–649 (2014). https://doi.org/10.1093/gerona/glt162
7. Kotsis, V., Antza, C., Doundoulakis, I., Stabouli, S.: Curr. Pharm. Des. **23**(999), 3200–3204 (2017). https://doi.org/10.2174/1381612823666170328142433
8. Harding, J.L., Pavkov, M.E., Magliano, D.J., Shaw, J.E., Gregg, E.W.: Global trends in diabetes complications: a review of current evidence. Diabetologia **62**(1), 3–16 (2018). https://doi.org/10.1007/s00125-018-4711-2
9. Mc Auley, M.T., et al.: BMC Sys. Bio. **6**(130), 1–21 (2012). https://doi.org/10.1186/1752-0509-6-130
10. Peters, R., et al.: Therap. Adv. Chronic Disease **10**, 2040622319880392 (2019). https://doi.org/10.1177/2040622319880392
11. Herrmann, N.: Erfolgspotential ältere Mitarbeiter. Den denografischen Wandel souverän meistern. Carl Hanser Verlag (2008)
12. Salmela-Aro, K., Upadyaya, K.: J. Voc. Behav. **108**, 190–200 (2018). https://doi.org/10.1016/j.jvb.2018.08.002
13. Yaldiz, L.M., Truxillo, D.M., Bodner, T., Hammer, L.B.: J. Voc. Behav. **107**, 182–194 (2018). https://doi.org/10.1016/j.jvb.2018.04.005
14. Gostautaite, B., Buciuniene, I.: J. Voc. Behav. **89**, 109–119 (2015). https://doi.org/10.1016/j.jvb.2015.05.001
15. Rodrigues, F.R., et al.: J. Voc. Behav. **118**, 103400 (2020). https://doi.org/10.1016/j.jvb.2020.103400

Influence of External Factors on the Creative Agricultural Economy

Yulia S. Zatrova(✉)

Moscow Timiryazev Agricultural Academy, Russian State Agrarian University, Moscow, Russia

Abstract. A fairly new direction of the creative agrarian economy is considered and a brief description of this concept is given in the article. Subsequently, a study is carried out based on statistical methods on the dependence of the creative agricultural economy and external factors, which include the infrastructure with one of the significant indicators, such as digitalization. The prerequisites for this study were some concepts that replace the meaning of the creative agrarian economy. It is often confused with the digital economy, with an innovative economy, therefore, the main goal of this work is to approve more accurately and in more detail the very concept of a creative agricultural economy and to determine the factors that affect or do not affect its development. As a result of the work, it was established that all factors except tourism do not significantly affect the development of a creative agricultural economy. And tourism, or rather agritourism, is the most promising and favorable direction for the development of new unique creative products. Consequently, the development of a creative agricultural economy in this direction has significant prospects. The article examines in which countries and why these prospects exist. At the same time, agritourism has its own nuances, which are considered herein in more detail and reveal the essence of a creative agricultural economy.

Keywords: Creative agricultural economy · Agritourism

1 Introduction

The authors believe, that a creative agrarian economy is not the improvement of the old product, but the creation of a radically new product or service in the agricultural sector. This economy also includes new farming methods that are fundamentally changing the traditional components of agriculture. In this case, human capital is the key factor. In other words, an individual is a subject of the world economy. Being one of the ways to obtain new information by connecting two or more pieces of data, the idea is of great importance [1].

The creative agrarian economy is based on knowledge, creativity and humanity which are equally important. Knowledge includes the level of education, upbringing, personal health. Creativity includes a set of personal qualities and personality traits to find new things, make decisions and act outside the box. Humanity is an essential element. This concept includes honesty, justice, loyalty, hard work, patriotism and the whole set of

P. V. Trifonov and M. V. Charaeva (Eds.): DITEM 2021, LNNS 380, pp. 177–184, 2022.
https://doi.org/10.1007/978-3-030-94245-8_24

qualities characteristic of the dignity of a Human. Creativity seems to be something that an individual can relatively easily achieve or create [2].

The purpose of this work is to study this issue, using the methods of statistical analysis.

2 Methods

Is there an interconnection between digitalization, globalization, agritourism, infrastructure and a creative agrarian economy, what is the structure of the interconnection and how to measure its strength – these are the research problems set by the authors. These problems can be solved using the correlation analysis. The tightness of the correlation interconnection between the variables can be qualitatively characterized using the data presented in Table 1.

Table 1. The tightness of the linear interconnection.

Correlation coefficient value	The tightness of the linear interconnection
$0.8 \leq \lvert r \rvert \leq 1$	Strong interconnection close to functional
$0.5 \leq \lvert r \rvert < 0.8$	Medium strength interconnection
$0.3 \leq \lvert r \rvert < 0.5$	Moderate interconnection
$0.2 \leq \lvert r \rvert < 0.3$	Weak interconnection
$0 \leq \lvert r \rvert < 0.2$	Very weak interconnection

Source: According to Mkhitaryan [3].

Based on a specially developed methodology, I CAE (index of creative agricultural economy) was calculated for 20 leading countries according to The Global Competitiveness Report 2019 and for Russia. The calculation of this index was carried out on the basis of normalized indicators as follows: three indices were calculated, namely the knowledge index, the health index and the morality index, each of which consisted of several indicators. Ultimately, for each country, the first CAE was calculated using the arithmetic mean. The summary values for the studied countries are shown in Fig. 1.

The standardization procedure is as follows:

$$x_i = \frac{x_i - \underline{x}}{S}, \quad (1)$$

where i = 1, 2, …. n,

The obtained data satisfy the condition $\underline{x} = 0$ и $S_x^2 = 1$. As follows:

$$\begin{aligned} \underline{x} &= \tfrac{1}{n}\sum_{i=1}^{n} x_i = \tfrac{1}{n}\sum_{i=1}^{n} \tfrac{x_i - \underline{x}}{S} = \tfrac{1}{S} * \tfrac{1}{n}\sum_{i=1}^{n} (x_i - \underline{x}) \\ &= \tfrac{1}{S} * \tfrac{1}{n}\left(\textstyle\sum_{i=1}^{n} x_i - \tfrac{1}{n}\sum_{i=1}^{n} \underline{x}\right) = 0 \end{aligned} \quad (2)$$

$$S_x^2 = \frac{1}{n}\sum\nolimits_{i=1}^{n} (x_i - \underline{x})^2 = \frac{1}{n}\sum\nolimits_{i=1}^{n} \left(\frac{x_i - \underline{x}}{S}\right)^2 = \frac{1}{S^2} * \frac{1}{n}\sum\nolimits_{i=1}^{n} (x_i - \underline{x})^2 = 1. \quad (3)$$

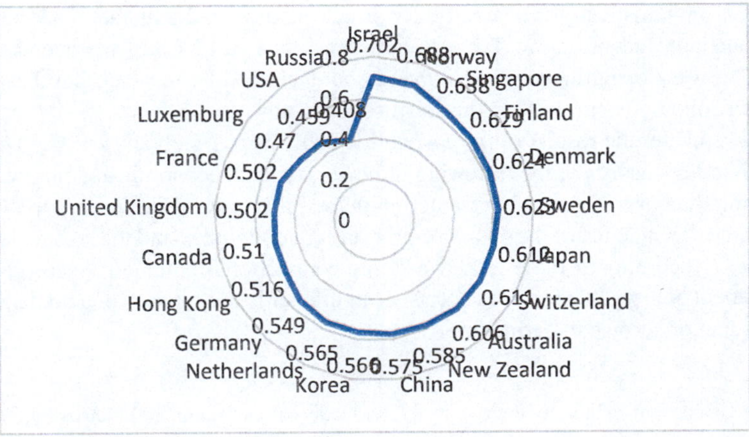

Fig. 1. CAE Index for 2019.

In some cases, data require unification. In this case, it is necessary to use the principles of calculating the variables of the second type, which are calculated by the following formula:

$$x_i = \frac{x_{max} - x_i}{x_{max} - x_{min}}. \tag{4}$$

Accordingly, if $x_i = x_{min}$, then $x_i = 1$, and if $x_i = x_{max}$, then $x_i = 0$.

The calculation of individual indices of the external environment is carried out in a similar way. First, the most significant indicators for each are selected, the data are normalized and unified, and then the index is calculated for each country. The year under study is 2019. The calculations used statistical data from well-known international organizations and institutions.

3 Results

Figure 2 shows the distribution of the countries under consideration depending on the value of I env. Germany is among the top three and actually ranks second, since Switzerland and Great Britain have equal values (0.687), while Germany has 0.684.

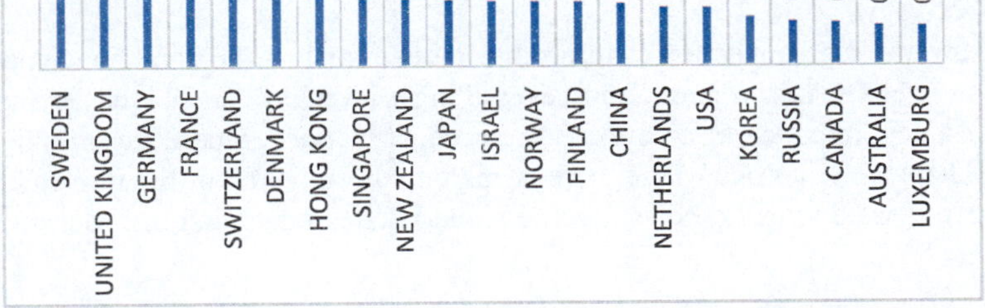

Fig. 2. Countries depending on the value of the environmental index.

Based on statistically processed data, it can be assumed that there is a moderate interconnection (according to Table 1) between I env. and I CAE, the actual value is 0.334. The very insignificant indicators can be explained by the fact that the creative agrarian economy is currently at the stage of formation.

Now consider the results of the correlation analysis of I glob and I CAE. The globalization index consists of the following indicators: social, economic and political globalization indicators. The value presented in pivot Table 2 is 0.176, in turn, the value of 0.206 is the dependence of the creative agrarian economy index and the social globalization index. According to Table 1, the weak interconnection is obtained. Accordingly, the development of globalization does not affect significantly the creation of a fundamentally new product or service in agriculture.

Table 2. Linear correlation coefficient I CAE and external environmental factors by country in 2019.

Index	Designation	Value
Environmental index	*I env.*	0.334
Infrastructure Index	*I* inf	−0.141
Globalization index	*I* glob	0.176
Tourism index	*I* tuor	−0.530

Infrastructure factors were analyzed in the next phase of the research. This index includes the following indicators: digitalization indicators, value added in the agricultural sector, sales of new commercial vehicles and agricultural machinery and the quality of roads, railway infrastructure, air transport infrastructure.

As a result, the actual value of I inf and I CAE is −0.141. This indicates that there is a very weak (according to Table 1) inverse interconnection.

Thus, it turns out that the creative agrarian economy is developing regardless of digitalization. Nevertheless, in this case, digitalization can be a tool for the development of a new creative product in the agro-industrial complex [4].

The most significant interconnection with I tour and I CAE is noted, the value is −0.530, according to Table 1, the tightness of the linear interconnection of the average force is inverse. It can be concluded that there is an interconnection between a tourist destination and a creative agricultural economy. Tourism events or tourism specialized events in rural areas or in agricultural enterprises are the most striking example of creativity [5]. Today tourism is the most favorable direction for the development of this type of economy.

The inverse interconnection confirms that it is extremely difficult (although possible) to create something new where "almost everything is invented", and it is easier and faster to create a new direction or a new creative product (service) where "everything is the same". There are countries where there is an "unplowed field" for the implementation of creative ideas. Figure 3 shows that the United States ranks first in I tour, the value

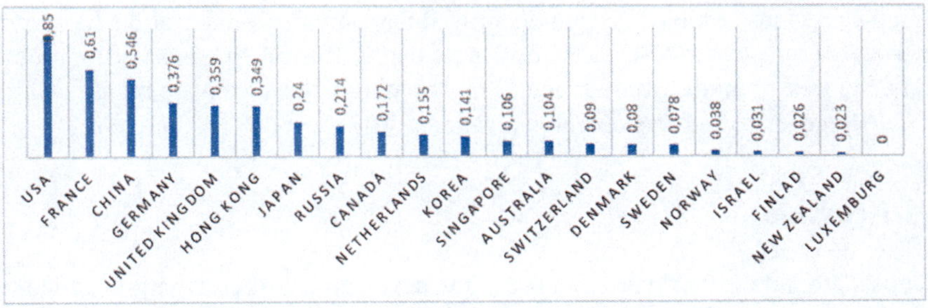

Fig. 3. Distribution of countries depending on the I tour.

of which is 0.850, it is followed by France with the value 0.610. In these countries, a certain maximum of development of the tourism industry has been reached.

Figure 4 shows that France is the most visited country in the world in 2019, all directions are developed and it is not so easy to create something new.

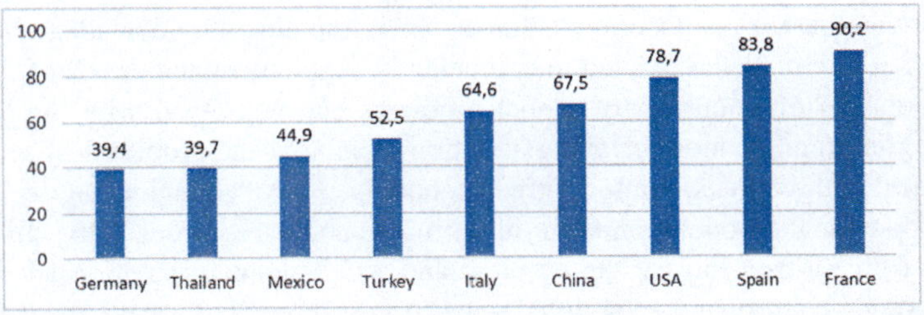

Fig. 4. The most visited countries in the world in 2019, million tourists. *Source:* Compiled by the authors based on the data from the World Tourism Organization (UNWTO).

In developing countries, a good result can be achieved even by using a ready-made creative solution. This is evidenced by developing economies, for example, Myanmar, where, according to Fig. 5, the growth amounted to 40.2% in 2019.

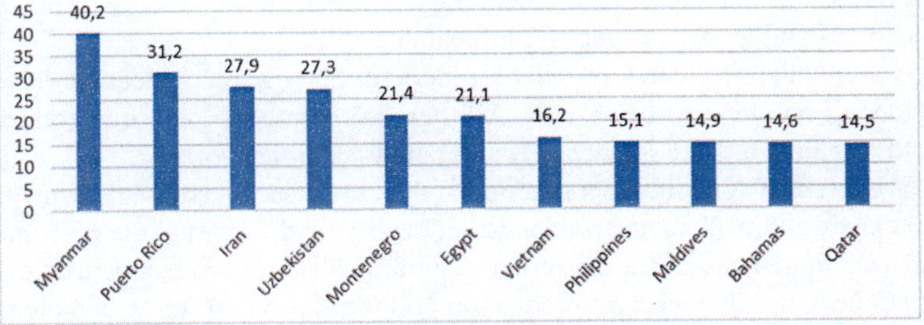

Fig. 5. Tourist destinations of the world at the end of 2019, an increase in%. *Source:* Compiled by the author based on the data from the World Tourism Organization (UNWTO).

In the previous sections, the authors proved that human capital is at the heart of the creative agrarian economy. The scientists note that in Russia, "the most acute problem will be the lack of innovative individuals in agriculture, those who are highly qualified and able to generate and implement innovations" [6].

4 Discussion

Scientists around the world are looking at the development of creativity in agriculture from different angles. Chinese scientists believe that creative agriculture is the integration of modern agricultural technology and local culture. Being a new developing agricultural industry, it is seen as an effective solution to the problems of environmental pollution and the development of the rural economy of China [7]. The development of agritourism requires careful planning. This is due not only to the economic effect that it can make, but also to long-term social and cultural sustainability [8]. However, it should be noted that there are a number of weaknesses and threats [9]. Moreover, the emergence of a new stream of activity entails social changes, which can be latent. The third point is clearly seen here, at the heart of which a creative agrarian economy is being built. It is very important that there is no "ax effect", that is, when implementing new ideas, building a new one, not to violate the old, but to competently supplement and develop it.

The solution of a number of global problems can be solved with the help of a creative agricultural economy, paying attention to solving the problems of sustainable development and the transition to a "green economy" [10]. "Greening" agriculture will provide food for the population with no harm to natural resources" [11]. In order to implement this, a new look at the problem and a non-standard solution are required, both in a single country and on a global scale. An example, in this case, can be vertical farms of various modifications, which may in the future be able to feed large megacities [12]. Besides, new directions of farming are based on ecological farming.

This type of alternative direction fully corresponds to the three aspects of creative agricultural economy: first, it combines experienced knowledge and scientific discoveries; secondly, the health (of the creator), an example can be in this case Albert Howard, who is considered the "father" of modern organic agriculture; and thirdly, the moral component, which consists in the rejection of chemical means of protecting animals and plants and keeping animals in accordance with their specific features, taking into account their need for movement. According to Albert Howard: "the health of soil, plants, animals and humans is one and inseparable" [13].

Food production is closely related to a moral imperative, and adequate food is the cornerstone of human well-being [14]. All these aspects are closely interconnected in the world economy and are important for all countries of the world.

In the context of agricultural post-productivity, rural spaces acquire new functions or enhance existing ones. Thus, the production of quality food, as part of the agro-industrial complex, and tourism are common activities in rural development strategies [15]. Some scholars believe that knowledge, production and innovation are three complementary phenomena that interact with each other and represent the most important sources of change and the basis of the modern agricultural system [16].

Our work is based on other three aspects: knowledge, health and morality (the basis of a creative agricultural economy). It is on these three aspects that the prudent creative development of agriculture depends.

5 Conclusion

Based on the results obtained, it can be concluded that the availability of creativity does not depend on infrastructure, digitalization, and globalization. If we develop these areas in the hope that there will be the development of creative products, then this is not the case. On the contrary, creativity creates an infrastructure, and in the future, there is an evolutionary development, that is, a natural development and improvement of quality. "Infrastructure is akin to a living system that brings more and more people together in increasingly complex economic and social relationships" [17].

Globalization is the habitat of a creative product. Moreover, all these components are necessary for the development of a creative agricultural economy.

References

1. Paap, J.: Mapping the technological landscape to accelerate innovation. Foresight STI Governance **14**(3), 41–54 (2020). https://doi.org/10.17323/2500-2597.2020.3.41.54
2. Stephensen, J.L.: Talking the creative economy into being. In: Thorup, M. (ed.) Intellectual History of Economic Normativities, pp. 201–215. Palgrave Macmillan US, New York (2016). https://doi.org/10.1057/978-1-137-59416-7_13
3. Mkhitaryan, V.S. (ed.): Data analysis: Textbook for Higher Education Institutioins. Yurayt Publishing House, Moscow (2020)
4. Zatrova, Y.S.: Digitalization as a tool of a creative agrarian economy. Digit. Finan., 14–24 (2020)
5. The Agricultural Creative Economy: Needs, Opportunities and Market Analysis. A Report to the Second Regular Session of the 123rd Legislature. http://www.maine.gov/agriculture/mpd/information/agcreative.pdf
6. Paptsov, A.G., et al.: Forecast of scientific and technological development of the plant growing industry, including seed production and organic farming in Russia, in the period up to 2030. Novosibirsk State Agrarian University, Novosibirsk (2019)
7. Lin, B.K., Lv, Q.H.: Research on the motivations and paths of collaborative technology innovation for Chinese creative agriculture. J. Appl. Sci. **13**(10), 1774–1780 (2013)
8. Fitriana, W.: The role of triple helix actors for agro-tourism development in west Sumatera. MIMBAR Jurnal Sosial dan Pembangunan **33**(2), 217–225 (2017)
9. Fatimah, S., Wazni, Q.K.: Development of tourism village based on agro-creative and disaster consideration, case of Cikondang village, Cingambul district, Majalengka regency. IOP Conf. Ser. Earth Environ. Sci. **466**(1), 012022 (2020)
10. Bulatov (eds): Prospects for economic globalization. KNORUS, Moscow (2019)
11. Paptsov, A.G., Shelamova, N.A.: Green economy is a new direction of sustainable development. Agric. Econ. Russia **11**, 67–73 (2014)
12. Zatrova, Y.: Creative agrarian economy in crop production. Theor. Appl. Econ. **4**, 118–128 (2020). https://doi.org/10.25136/2409-8647.2020.4.34118
13. Howard, A.: Medical "Testament" on nutrition (correspondence concerning the Cheshire Medical Testament on Nutrition). BMJ **1**(4090), 1106 (1939)

14. Kornilova, E.N., Kornilov, N.I.: Interactive methods of social and ecological teaching of students in the course "Ecology." Sci. Bull. Southern Institute Manag. **3**, 81–84 (2014)
15. Pizarro Gómez, A., Šadeikaitė, G., García Delgado, F.J.: The world of Iberian Ham and its tourist potential in the Sierra De Huelva (Andalusia, Spain). Eur. Countryside **12**(3), 333–365 (2020). https://doi.org/10.2478/euco-2020-0019
16. Cukanovic-Karavidic, M., Karavidic, S., Pejanovic, R.: Agro-economic development based on the economy of knowledge, business and social skills. Ekonomika Poljoprivrede **65**(3), 1241–1257 (2018)
17. Rifkin, D.: The Third Industrial Revolution: How Horizontal Interactions Change Energy, Economics and the World as a Whole [Tret'ya promyshlennaya revolyuciya: kak gorizontal'nye vzaimodejstviya menyayut ehnergetiku, ehkonomiku i mir v celom]. 2-e izd. Moscow, Al'pina non-fikshn (2015)

Military Research and Development as a Driving Force for the Economy of the Future

Olga N. Gilkova(✉)

RUDN University, Moscow, Russia

Abstract. Military research and development in the modern world is not only one of the main factors in building up the military power of a state, strengthening its strategic independence and ensuring significant technological advantages over a potential adversary, but also a major factor providing a positive impact both on the development of scientific and technological progress in general, and on the development of the state's economy. Military research and development can contribute to the creation of a powerful national innovation system and the shaping of the knowledge-based economy of the future. The study aims to try to prove that, despite the fact that military research and development requires the use of significant material, financial and intellectual resources, its indirect contribution to civilian spheres can be very significant and this type of research and development can rightfully be considered one of the driving forces of the economy of the future. The methodological basis of this study was analysis and synthesis, logical and comprehensive approaches to assessing the role of military research and development in the development of the economy of the future. The study shows that the states leading in terms of spending on military research and development not only dominate in the creation and development of weapons, military equipment and ammunition, but are also today the most developed countries in the world in the field of robotics, information and other types of civilian technologies. The study also provides particular cases of peaceful use of military research and development results.

Keywords: Innovation · Scientific and technological progress · Military and technological superiority · Spin-off effect

1 Introduction

As is known in the modern world, innovation is the central driving force behind economic growth and development of states, increasing labor productivity and economic efficiency, improving the daily life of people [1]. Science today has become an instrument of cognition and transformation of the surrounding reality in which people are living. In this regard, the need for state funding for research and development (R&D) and stimulation of innovation processes is beyond doubt.

However, there is a special kind of R&D – military R&D, which is the development, testing and evaluation of weapons, military equipment and ammunition. Military R&D is a relatively new area of activity and interests of industrial corporations, taking into

P. V. Trifonov and M. V. Charaeva (Eds.): DITEM 2021, LNNS 380, pp. 185–189, 2022.
https://doi.org/10.1007/978-3-030-94245-8_25

account its modern scale, and most importantly, the impact it has on the level of technical and economic development of the state.

The accelerated development of military R&D began in the 1940s. Before the outbreak of the World War II, the main stream of innovations in the field of weapons and military equipment came, as a rule, from individual amateur inventors [2]. Scientific research was carried out on a very modest scale, and their financial support for military research projects was extremely small [3].

Over time, military R&D began to turn into the main factor in building up military power and transforming the armed forces. At the same time, this type of R&D began to acquire special significance not only for the military-political leadership of countries, but also for industrial corporations. Since military R&D is almost entirely funded by the state and only a small part of it is funded by private business (and, as a rule, this part is subsequently reimbursed by including of the resulting weapon system in the selling price), military R&D is very attractive to industrial firms, all the more so, by completing them, these firms receive a number of additional benefits, including the right to patents obtained as a result of military R&D [4].

In recent decades, in many countries of the world, a sharp increase in the share of military R&D compared to the share of civilian R&D has become a characteristic trend. And if the answer to the question about the need for state funding of military R&D in order to build up the military power of the state, strengthen its strategic independence and ensure significant technological advantages over a potential adversary is obvious, then the answer to the question about the impact of this type of R&D both on the development of scientific and technological progress as a whole and the development of the national economy can be considered controversial.

The purpose of this study is to try to prove that, despite the fact that military R&D requires the use of significant material, financial and intellectual resources, its indirect contribution to civilian spheres can be very significant and this type of R&D can rightfully be considered one of the driving forces of the economy of the future. The objectives of this study are to identify the leading countries in terms of spending on military R&D and to analyze the level of their technological and innovative development in the civilian sphere.

2 Methods

The methodological basis of the study was the analysis and synthesis of official statistics from the UN and the OECD on military R&D expenditures, as well as a logical and comprehensive approach to assessing the role of military R&D in the development of the economy of the future.

3 Results

An analysis of the official statistics of the UN and OECD shows that the concentration of military R&D in the world is even higher than the concentration of military production in general. Only a small number of organizations in a limited range of countries are capable of creating advanced weapons or related technology. Military R&D activities are concentrated in a small number of industrialized countries. In these leading countries, government funding of military R&D accounts for a fairly high share of the total budgetary expenditures on R&D for civil and military purposes.

For example, in 2019, the United States allocated 43.5% of all government R&D spending to defense. Turkey ranked second with 17.3%. The UK ranked third with 15.2%, the fourth place was occupied by South Korea with 15.0%, the sixth – by France with 7.5%, the eighth – by Germany with 3.8%, the ninth – by Japan with 3.5% [5].

In countries leading in terms of military R&D expenditures, such expenditures also account for a large share of total government support for new equipment and technology.

For example, in 2019, the United States spent 0.285% of its GDP on military R&D. South Korea, with a share of 0.169%, ranked second in this indicator. Compared to other leading nations, the US has spent a much larger share of GDP on military R&D: more than 12 times than Japan, more than 8 times than Germany (0.035%), more than 3 times than the UK (0.080%), and more than 5 times than France (0.048%) [6].

Thus, as can be seen, the list of leaders in military R&D spending includes the most industrially and technologically advanced countries today, while being not the most militarized ones, including Oman, Saudi Arabia and a number of others [7].

The greatest attention to military R&D is shown by the United States, whose scientific and technological potential is currently the highest in the world. Moreover, the United States, which today, as the analysis shows, accounts for about 75% of all world expenditures on military R&D, can be considered the undisputed leader in military development [8].

In recent decades, US federal spending on military R&D has grown rapidly [9]. In the future, the priority of military R&D in US innovation and industrial policy is likely to remain.

Undoubtedly, the desire of the military-political leadership of the United States to maintain military-technological superiority over other states is the main reason for such a significant attention of this state to military R&D. However, there is another, non-obvious reason – the spin-off effect, which means the use of military technologies for civilian purposes.

The fact is that military R&D is often accompanied by scientific and technical discoveries that can later be used for peaceful purposes.

As history shows, the civilian impact of military development can be very high. At one time, military R&D contributed to the creation of a jet engine, electronic computers, microelectronic devices, which were then widely used in the civilian sphere.

In general, it can be stated that at the turn of the 20th – 21st centuries, military R&D turned out to be a factor that markedly influenced many aspects of social development in the United States. Military R&D, thanks to the spin-off effect, not only stimulated a sharp leap in the development of science in the United States, but also the creation of a powerful national innovation system and the formation of a knowledge-based economy.

In the last century, many inventions that were originally created for military purposes had a significant impact on the development of the US industry [10].

For example, the development of a control system for an unmanned aircraft by Honeywell in the 1950s later served as the basis for the creation of equipment for the blind landing of civil aviation aircraft [11].

Widespread in the United States is the use of gas turbine engines in the electric power sector to cover peak loads in power systems, which were originally created for combat aircraft [12].

That is, military R&D, in addition to its main purpose, is capable of accelerating the general scientific and technological progress and thus stimulating the development of the economy.

The main forms of influence of military R&D results on the civilian sphere of the economy include: the use of certain types of products created as a result of military R&D for civil purposes; the spread in the civil economy of new planning and management methods developed for military programs (for example, the method of systems analysis), etc.

4 Discussion

There are various approaches to assessing the impact of military R&D on the development of the national economy and scientific and technological progress in general. For decades, military R&D has been the subject of analysis by researchers who have tried to develop theories about the impact that R&D has on the economy. In general terms, there is a point of view that military R&D affects any type of activity to varying degrees.

Some of the researchers argue that military R&D is inherently a destabilizing force for the national economy, which inevitably leads to an increase in military purchases and an increase in military spending by the state, and, consequently, to an increase in government spending in general, which entails, firstly, the increased tax burden, which falls on taxpayers, and secondly, the strengthening of inflationary trends in the economy and an increase in the cost of living.

For example, this argument has been aggressively pursued in Acland-Hood, "Military research and development expenditure," SIPRI Yearbook 2015, p. 287: "In the presence of large and growing expenditures [on R&D], one can expect… the creation of pressure in favor of increasing military spending in the long term, regardless of what political relations will prevail in the future" [13]. The author revised his thesis somewhat in SIPRI Yearbook 2016. Nevertheless, such arguments are quite widespread.

Also, Thee puts on p. 118: "Each new stage of R&D efforts becomes a starting point for a transition to a new spiral of the arms race" [14]. But practice shows that states can allocate significant funds for military R&D without creating an incentive for the production and procurement of the results of this work.

In addition, economists have repeatedly attempted to assess the burden on a national economy by diverting resources, in particular capital and highly skilled labor, to military R&D instead of channeling it to other purposes, and the level of return on the output of products obtained upon the transfer of the results of military developments into the civilian sphere, however, these attempts were not very successful [15].

5 Conclusion

Thus, due to the rational use of the spin-off effect, the civilian payoff of military developments is very high, as demonstrated by the example of the United States. In addition to directly stimulating the development of science, military R&D can help create a powerful national innovation system and shape the knowledge economy of the future.

References

1. Romanenko, I.I., Romanenko, M.I.: Eng. Bul. Don. 4 (2018)
2. Faramazian, R.A., Borisov, V.V.: Transformation of Defense Economy. The 20th – Beginning of the 21st Centuries, Nauka, Moscow (2006)
3. Gilkova, O.N.: Mod. Sci.: Act. Prob. Theory Prac. Ser: Econ. Law **1**, 78–81 (2018)
4. Stockholm International Peace Research Institute. https://sipri.org/
5. Organization for Economic Cooperation and Development. A Database of Research and Development Statistics. https://stats.oecd.org/Index.aspx?DataSetCode_GBARD_NABS2007
6. United Nations. https://www.un.org/
7. Center for Analysis of the World Arms Trade. https://armstrade.org/
8. UNESCO Science Report. http://unesdoc.unesco.org/images/0023/002354/235407e.pdf
9. Gilkova, O.N.: About military and civil R & D in the USA. In: Materials of the International Scientific and Practical Conference Scientific Center "Disput", pp 64–66. "Marker", Vologda (2018)
10. Military Enterprise and Technological Change: Perspectives on the American Experience. Smith, M.R. (ed). The MIT Press (1985)
11. Gromov, L.M., Faramazian, R.A.: The Military Economy of Modern Capitalism. Voenizdat, Moscow (1975)
12. Sword and Business Alliance: (Military-Industrial Complexes of the Imperialist States). In: Bykov, O.N., Bugrov, E.V. (eds.), IMEMO AS of the USSR. Mysl, Moscow (1988)
13. Acland-Hood, M.: Military research and development expenditure. In: SIPRI, World Armaments and Disarmament: SIPRI Yearbook. Oxford University Press, Oxford (2015)
14. Thee, M.: Science – based military technology as a driving force behind the arms race. In: Gleditsch, N.P., Njolstad, O. (eds), Arms Races: Technological and Political Dynamics. Sage Publications, London (2018)
15. Ledermann, L.L.: Introduction and General Summary in a Review of the Relationship Between Research and Development and Economic Growth Productivity, Washington (2011)

Ecosystem Strategies of Retail Banks: A View of Russian Innovators

Igor B. Turuev(✉) and Evgeniya O. Shashkina

Moscow State Institute of International Relations, Moscow, Russia
i.turuev@inno.mgimo.ru

Abstract. The article raises the problem of filling the ecosystems of non-financial services created by the largest Russian banks. The relevance is determined by the controversial nature of the issue of combining banking and non-financial activities by credit institutions both at the level of the scientific community and the mega-regulator of the Russian financial market. The study was aimed to determine the factors stimulating the creation of ecosystems of non-financial services by banks, as well as the importance of multifunctional platforms for a single bank and the entire economy. The methods of theoretical systematisation and comparative analysis of Russian banking ecosystems are used. Based on the symbiosis of the economic, sociological and psychological approach, the main consumer trends, the reasons for their formation, the industries that have formed to meet the needs of users, as well as the opportunities for the banking sector have been formulated. The following phenomena of consumer behaviour have been highlighted: sharing economy, club consumption, status consumption, other consumption via digital channels. The main risks for banks are in the slowdown in the growth rate of demand for the most profitable products – mortgage and car loans, as well as in the rate of deposit inflow. Opening opportunities lie in the growth of the number of clients of banking Internet acquiring, POS credit financing, more extensive data collection for personalisation of offers, etc. The response to new competing traditional banks is the creation of investment (for example, Sberbank) or partner (for example, Tinkoff) ecosystems of non-financial services.

Keywords: Digital economy · Banking sector · Ecosystems of non-financial services · Synergies

1 Introduction

The approach to organising banking is changing: today, the focus is not on the sale of financial services, but on the satisfaction of the client's needs. Increasingly, the development strategies of the largest banks began to use not only the words like *digitalisation*, but also *ecosystem*, which means not just transferring part of the sales of their products to digital channels, but also attempts to build a system of interaction with complementary non-financial industries, which causes concern of the mega-regulator in connection with the possible occurrence of systemic risks.

P. V. Trifonov and M. V. Charaeva (Eds.): DITEM 2021, LNNS 380, pp. 190–197, 2022.
https://doi.org/10.1007/978-3-030-94245-8_26

Research in the field of digital platforms in the banking sector focuses mainly on the relationship of fintech companies, including challenger banks and incumbent banks. So, the works analyze the value of technological innovations for the banking sector [1], ways of interaction between fintech startups and banks [2], the competitive advantages of each of the participants, i.e. banks and IT companies, in the fight for the client [3].

Since the development strategy of banks in the field of service delivery channels is increasingly becoming customer-focused, and banking services are not just remote, but semantic [4], and embedded [5], digital non-banking ecosystems, that is, a symbiosis of financial and non-financial services, will be of primary scientific interest for us. At the same time, the literature places dividing lines between the ecosystems of financial services created by banks and ecosystems of non-financial services [6, 7]. Most often, articles provide a case-study analysis of existing ecosystems that are built both around banks and around IT companies, telecom operators, etc. [8–10].

In our opinion, there is a gap in the theoretical base of research on the problem of building an ecosystem business, namely, the most promising digital markets, as well as the reasons for their emerging, are not studied. In the course of research, we must test the following hypothesis: "banks are modifying the customer service model not only under the influence of competition from fintech companies, but also companies from other sectors of the digital economy, which appear and develop in response to changing consumer habits".

2 Materials and Methods

The purpose of this article is to identify the cause and effect of the construction of ecosystems of non-financial services by banks. In the process of research, we must solve the following tasks: to classify consumer trends/phenomena of consumer behaviour; to determine the factors of formation of each trend; to systematise the risks and opportunities for banks from each trend; to formulate the synergistic effect of ecosystem development.

The authors propose an interdisciplinary study based on general scientific methods: analysis, systematisation, classification, comparison, and extrapolation. The information base for the comparative analysis was the press releases of Russian banks, the creators of ecosystems.

Since the focus of the study is the client – an individual and his needs, the analysis of global consumer trends of 2021 prepared by Euromonitor International was used to classify new industries and markets, and the most relevant, in our opinion, trends were highlighted, namely, sustainable development, instant access to goods and services, combination of physical and digital reality (phygital reality), personalised approach, reduction of discretionary spending [11]. In addition, older trends, such as the promotion of social networks and instant messengers and conspicuous consumption were proposed. For each group of trends or phenomena, an analysis of the relevant literature (economic, social and psychological) was carried out to identify the causes of its occurrence, as well as the industries that emerged as a result of changes in consumer habits. Further, the authors will offer an analysis of the risks and opportunities of new markets for banks.

3 Results

The points of contact between ecosystem creators and service providers are through the lens of the fastest growing industries that appear not only because it has become technically possible but due to economic and socio-psychological changes in consumer behaviour. Table 1 presents the main trends in the modification of consumer behaviour, a list of new industries or new channels for the provision of traditional services, as well as the strategies of banks' response wishing to strengthen their positions in the retail segment to structural shifts in the economy.

Table 1. Socioeconomic shifts in digital society consumer habits and banking strategies.

Phenomenon	Economic reasons for the appearance of the phenomenon	Non-economic (psychological, social) causes of the phenomenon	Industries that have appeared or have been significantly modified as a result of the implementation of the phenomenon	Banking sector risks	Opportunities for the banking sector
Sharing economy/collaborative consumption	1. Decrease in purchasing power and creditworthiness of the population 2. Increase in the cost of borrowed resources	1. The tendency of the population to frequently change their place of residence; 2. The spread of post-materialistic values	1. Carpooling, car sharing, bike sharing; 2. Rent of residential real estate; 3. Services for finding accommodation while travelling 4. Crowd-funding	Decrease in demand for mortgage and car loans	1. Internet acquiring; 2. Consumer loans; 3. Escrow accounts; 4. Insurance policies for rented objects; 5. Bank crowdlending platforms
Club consumption	*Supply side:* Transition from a supply economy to a demand economy *Demand:* Rationalisation of consumption, reduction of transaction costs of information retrieval	*Supply side:* It has become technically possible to receive instant feedback from the consumer *Demand:* 1. It has become technically possible to quickly receive information about all offers on the market; 2. Introversive consumer confident	1. Streaming platforms 2. Social networks, messengers	–	1. Internet acquiring; 2. Marketing activities

(*continued*)

Table 1. *(continued)*

Phenomenon	Economic reasons for the appearance of the phenomenon	Non-economic (psychological, social) causes of the phenomenon	Industries that have appeared or have been significantly modified as a result of the implementation of the phenomenon	Banking sector risks	Opportunities for the banking sector
Status consumption	Expansion of the range, quantity and quality of status services	1. Strengthening the influence of materialistic values, a shift towards "materialism of services"; 2. Social networks as a channel for promoting conspicuous consumption	1. Online purchase of package tours; 2. Foodtech	–	1. Internet acquiring; 2. POS lending/consumer loans/credit cards; 3. Currency exchange operations; 4. Insurance for those travelling abroad
Other consumption via digital channels	1. Decrease in real disposable income; 2. Increasing financial literacy of the population, increasing risk appetite	1. The value of time; 2. The need to always be in trend	1. Electronic commerce; 2. Housing ecosystems; 3. Online education and job search; 4. Medtech; 5. Remote state services; 6. Instant transfers, personal finance management	Decline in the market share of banks in deposits, payments and transfers, investment brokerage. Decrease in turnover for stationary acquiring	1. Internet acquiring; 2. POS lending/credit cards

Source: compiled by the author

4 Discussion

4.1 Sharing Economy and Risks for the Banking Sector

The phenomenon of the last decade is the collaborative consumption, or the sharing economy in which not the goods themselves are purchased, but the rights to access and use these goods [12].

The economic reason for the phenomenon is the instability of monetary incomes, high unemployment after the global crisis of 2008–2009, and there is a clear correlation between uncertainty about the future and the shift from investment goods to rental consumption patterns [13].

Psychological reasons: the tendency of millennials to migrate, frequent change of residence, as a result; spreading the idea of sustainable development: overcoming environmental problems, preserving natural resources, overcoming the problem of overproduction and overconsumption [14–16].

In connection with the development of the trend of the sharing economy, banks may face risks of a decrease in demand for mortgage and car loans, that is, for the most profitable products for the bank.

4.2 Channels of Stratification of Society in the 21st Century

Another trend is the integration of consumers into consumption communities (club consumption), which is most often manifested in the field of leisure.

Social networks/messengers which segment consumers by interest groups can also be attributed to the manifestations of the phenomenon. For the millennial generation, chatting is becoming the preferred mode of communication due to the rise in introversion in society [17].

Social networks act as aggregators of opinions and structure users into groups with similar consumer characteristics, which is subsequently used to build an individual consumer demand curve and conduct first-degree discrimination. It is worth noting that chatbots for ecosystem companies are not an independent earning unit, but an element in the value chain.

4.3 Materialism of Services in a Digital Society

Others' opinions have a strong influence on the volume and structure of consumption of the population: empirical studies confirm a positive correlation between the habit of individuals to compare themselves with their neighbours and materialistic inclinations [18].

One of the drivers of the development of status consumption, in our opinion, is the promotion of social networks. Friends and acquaintances on social networks, broadcasting the consumption of prestigious goods and services are becoming a factor in the formation of materialistic aspirations, and, as a result, of consumption for the sake of status. At present, materialism is shifting towards "materialism of services", that is, consumption in excess of the need for premium services, which is an indicator of the status of an individual in society.

4.4 Other Trends in Changing Consumer Habits

Ecosystem banks include companies from other sectors of the digital economy in their project portfolio: e-commerce, medical tech, online education and job search, personal finance management services and instant transfers, etc.

Online trading is considered to be the fastest growing of non-financial services [19]. This industry, like social media, is of great value in terms of accumulating data on customer preferences.

When it comes to financial services, banks are faced with a strong competitive environment emerging from fintech startups. A niche that banks have never focused on is detailing the composition of consumer spending, spending analysis and recommendations on where and how to save money. Meanwhile, consumers are less and less focused on bank deposits as an instrument of savings, one of the reasons for which is an increase in the level of financial literacy and an increase in the risk appetite of the population.

As a result of the analysis of possible directions for combining financial and non-financial services on one site, two areas of synergy can be formulated:

1. Integration of banking services on non-financial platforms.

Banks often partner with non-financial service providers and post banking offers on their sites. This is most typical for industries with a high average purchase order: travelling, online shopping, etc. (Fig. 1).

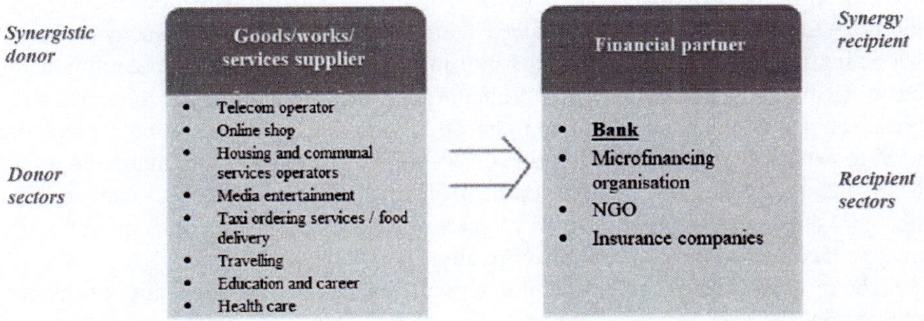

Fig. 1. Synergistic effect of financial services on non-financial platforms. *Source*: compiled by the author.

2. Monetising the audience of banks through cross-selling of non-financial products.

It is implemented by embedding these complementary services via an API into an online or Internet bank. Customer data aggregators monetise their existing resource by placing advertisements on their websites and other marketing activities, for example, special offers from partners or cross-selling (Fig. 2).

With such a synergistic effect, banks as platform owners become digital rentiers and receive additional commission income [20].

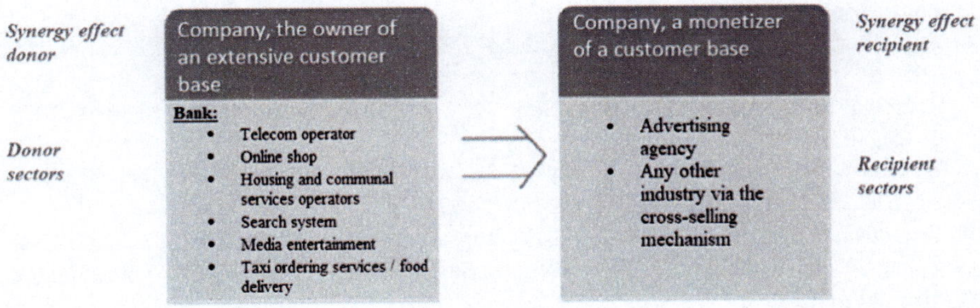

Fig. 2. Synergistic effect of monetisation of communication relations. *Source*: compiled by the author.

Ecosystems can be investment (Sberbank, VTB) and partner (Tinkoff). The main reason for choosing an investment target is the potential profitability of an asset.

5 Conclusion

In general, as a result of theoretical analysis, our hypothesis about the impact of all digital industries was confirmed.

Currently, the pattern of consumption of everyday and investment goods is changing; therefore the role of the bank in the value chain is being modified. Banks are increasingly becoming not transformers of savings into investments (loans), but an institution for ensuring the daily transactions of the client. For banks, in addition to the mandatory presence of a digital channel of communication with the client, these trends mean the need to expand the range of transactional services for the client, reorientation to a fee-based income model and the development of a partner network (ecosystem of non-financial services) to realise a two-way synergistic effect. This will compensate for lost interest income and increase customer loyalty.

The results obtained can be useful to commercial banks from the point of selection of potential partners for integration.

Directions for further research can be: 1) conducting cross-country comparisons on the principles of building banking ecosystems of non-financial services; 2) comparison of banks' strategies in the development of retail and corporate banking ecosystems.

References

1. Chen, M.A., Wu, Q., Yang, B.: Rev. Financ. Stud. **32**(5), 2062–2106 (2019)
2. Hornuf, L., Klus, M.F., Lohwasser, T.S., Schwienbacher, A.: How do banks interact with fintech startups? Small Bus. Econ. **57**(3), 1505–1526 (2020). https://doi.org/10.1007/s11187-020-00359-3
3. Mirchandani, A., Gupta, N., Ndiweni, E.: Int. J. Econ. Financ. Iss. **10**(5), 331–343 (2020)
4. Skinner, K.: Digital human: the forth revolution of humanity includes everyone. Mann, Ivanov and Ferber, Moscow (2019)
5. King, B.: Bank 4.0. Banking Everywhere, Never at a Bank. Olimp – business publishing house, Moscow (2020)
6. Strelnikov, V.V.: Financ. Credit **26**(9), 2054–2068 (2019). https://doi.org/10.24891/fc.25.9.2054
7. Trushina, K.V., Smagin, A.V.: Bank Serv. **12**, 7–11 (2019). https://doi.org/10.36992/2075-1915_2019_12_7
8. Bykanova, N.I., Solovey, J.A., Gordya, D.V., Konshina, L.A.: Econ. Inf. Technol. **47**(1), 91–100 (2020). https://doi.org/10.18413/2687-0932-2020-47-1-91-100
9. Rosanova, N.: Soc. Econ. **2**, 14–29 (2019). https://doi.org/10.31857/S020736760004132-4
10. Zhang-Zhang, Y., Rohlfer, S., Rayasekera, J.: Sustainability **12**, 8907 (2020). https://doi.org/10.3390/su12218907
11. Top 10 global consumer trends 2021. https://www.euromonitor.com/top-10-global-consumer-trends-2021/report
12. Egina, N.: Financ. Credit **25**(9), 1971–1986 (2019). https://doi.org/10.24891/fc.25.9.1971
13. Abraham, R., Harrington, C.: Mod. Econ. **6**, 51–64 (2015). https://doi.org/10.4236/me.2015.61005
14. Cinjarevic, M., Kozo, A., Berbekovic, D.: SE J. Econ. Bus. **14**(1), 49–60 (2019). https://doi.org/10.2478/jeb-2019-0003
15. Ciocoiu, C.N.: Theor. Emp. Res. Urban Manage. **6**(1), 33–43 (2011)

16. Smidt, M.: The sharing economy and its paradoxes: a sociological study of sharing communities in Russia. Rus. World **28**(2), 148–171 (2019). https://doi.org/10.17323/1811-038X-2019-28-2-148-171
17. Kaur, H., Anand, S.: Segmenting generation Y using the big five personality traits: understanding differences in fashion consciousness, status consumption and materialism. Young Consumers **19**(4) 382–401 (2018). https://doi.org/10.1108/YC-03-2018-00788
18. Chan, K., Prendergast, G.: Materialism and social comparison among adolescents. Soc. Behav. Pers. **35**(2), 213–228 (2007). https://doi.org/10.2224/sbp.2007.35.2.213
19. It's time to make consumer became the company's center again. https://www.pwc.ru/ru/retail-consumer/publications/gcis-2019-ru.pdf
20. Stepnov, I., Kovalchuk, Y.: Platform capitalism as the source of digital rentier's superprofit. MGIMO Rev. Int. Rel. **4**(61), 107–124 (2018). https://doi.org/10.24833/2071-8160-2018-4-61-107-124

Some Promising Areas for the Development of Behavioral Economics

Anton Y. Borisov(✉), Elena V. Aleksandrova, Marina I. Skripnikova, Elena A. Rozhanskaya, and Natalia A. Razina

Plekhanov Russian University of Economics, Moscow, Russia

Abstract. The article is devoted to a new direction of modern behavioral economics, which has an interdisciplinary character and integrates psychology and economics. The study's relevance is due to the lack of a systematic approach to the theoretical understanding of this direction in economic science. The report examines four main areas of prospective research in the framework of behavioral economics. The work objective is to substantiate the theoretical essence in the systematic research approach and synthesis of different versions of the interpretation of this theory. Since this is one of the new directions in economics, new knowledge requires deepening, systematization, integration into the existing theory, and practice. The study's objective is to show that the progress of these directions is a logical development of the research program of behavioral economics. The tools of behavioral economic theory reveal the psychological basis and dynamics of decision-making by economic agents. The theoretical and methodological basis of the research was the principles of economic and comparative analysis, deduction, and archival search. An analysis of sources on the development of behavioral economics contributes to a step-by-step exploration of this theory. The novelty of the research lies in the modern approach to achievements in this field, structuring, organizing, and systematizing the available knowledge.

Keywords: Behavioral economics · Economic psychology · Consumer behavior theory · Economic experiment · Choice architecture · Paternalism

1 Introduction

Behavioral economics is inherently interdisciplinary. It emerged at the junction of economic and psychological research (primarily within the framework of cognitive psychology). Over the past 40 years, many deviations from the rational model of behavior (i.e., distortions and heuristics) have been discovered by researchers working in this area. Behavioral economics is widely accepted, but it is too early to say that its research capacity has been exhausted.

The theoretical and methodological basis for writing the article is a systematic approach that thoroughly considers the problem. During the research, general scientific **methods** of analysis, inductive and deductive methods, and the economic method were used, based on which the essential content of the studied issue is determined. The article uses publications of well-known scientists-economists, materials of dissertations, and articles by specialists in economics, psychology, and sociology.

P. V. Trifonov and M. V. Charaeva (Eds.): DITEM 2021, LNNS 380, pp. 198–202, 2022.
https://doi.org/10.1007/978-3-030-94245-8_27

2 Materials and Methods

At the moment, it would be possible to distinguish four main directions of the development of behavioral economics:

Deepening of knowledge about the features of human thinking that influence economic decision-making.

Checking existing economic models for compliance with new data.

Analysis of the possibilities of application of behavioral economics.

Systematization of the main provisions of the theory.

Each point should be discussed in more detail.

In the current stage of the development of behavioral economics, interdisciplinary connections are becoming more and more diverse. For example, the so-called "methods of registering brain activity", including functional magnetic resonance imaging (fMRI), are used today to study human economic behavior. With fMRI, one can see which part of the brain is active at the moment. These studies refer to one of the "types" of behavioral economics, the so-called "neuroeconomics", which lies at the intersection of economics and neurobiology – a combination that would have seemed quite exotic just a short time ago [1].

The prospect of obtaining new data on human behavior today is primarily associated with new, non-standard research methods, including those that use the latest advances in technology [2]. Once it was technological progress that facilitated the emergence of behavioral finance: it was the development of computer technology in the 1980s that made it possible to analyze a large array of data and detect several anomalies in financial markets, many of which were brilliantly explained by Nobel laureate Richard Thaler in his articles [3].

A poorly explored area in both behavioral economics and psychology in general (the question is wider than disciplinary boundaries) remains the influence of various factors on forming the human cognitive system [4]. Applied specifically to behavioral economics, the question can be posed as follows [5]: how fair is the data on heuristics and distortions affecting the economic behavior of a European or an American to a person who was formed in a completely different sociocultural environment?

There are several criteria for comparative analysis in this direction. This may be gender, age, race, profession, nationality, religion, etc. [6]. However, given the current attitude of radical tolerance in Western society today, some of these studies might not seem quite "politically correct". However, they would be interesting and informative.

Also, the mechanism of forming heuristics and distortions is still not completely clear, and, in this regard, two important questions remain open:

Which of them are innate, and which are acquired?

How can I change them? Or how can you influence their formation?

It is relevant to **check the existing economic models for compliance with new data on real economic behavior** [7]. Many models in neoclassical theory are based on the assumption of the rational behavior of an individual [8]. Thaler noted that there was a period when the model in which agents were "more rational" was considered truer [9]. Now, perhaps, some need to be supplemented or adjusted.

3 Results and Discussion

The possibilities of applying behavioral economics remain extremely limited to date. The methods of so-called "libertarian paternalism" or "nudges" resemble marketing on a state scale. The essence of libertarian paternalism is that public and private institutions can "legitimately influence behavior while respecting freedom of choice" [10]. The term was coined by two scholars, Richard Thaler and Cass Sunstein. Thaler is an American economist studying behavioral economics (born September 12, 1945, in New Jersey, USA. Professor Emeritus of Behavioral Sciences and Economics at the University of Chicago School of Business), and Sunstein is a legal scientist (born September 21, 1954, Concord, Massachusetts). These scientists are the winners of the 2017 Alfred Nobel Memorial Prize in Economics for their contributions to behavioral economics. Thaler wrote, "In the process of transforming economics into a more mathematically rigorous science, the economics profession seems to have lost its good intuition for human behavior" [11]. The paper was first published in 2003 in the American Economic Review. The authors presented their ideas in more detail in subsequent articles and books. They suggest that libertarian paternalism is paternalistic in the sense that "it tries to influence choices in ways that make choices better, judged by them" [10]. "The concept of paternalism specifically requires a restriction of choice. It is libertarian in that "people had the right to refuse certain arrangements if they wished to do so" [10]. Rejection of an arrangement "preserves freedom of choice" [10]. Thaler and Sunstein published a book called "Nudging". It fully defends this position. The authors called their concept of coercion ("nudging") to the right choice a seemingly paradoxical term – "libertarian paternalism". "If policymakers want to get citizens to make the economic decision they want without restricting their freedom of choice, we need to nudge them in the right direction through the option that is offered by default. For example, to stimulate pension savings, it is better to transfer employees to such a system automatically, and those who do not agree should refuse in a pronounced way. If you offer people an active choice between two options, they are likely to choose the "leave as is" option, but not because it is better, but because people tend to "cognitive distortion" (bias) in favor of maintaining the status quo" [11].

"Libertarian paternalism is similar to asymmetric paternalism, which refers to policies designed to help people who behave irrationally and therefore do not promote their interests, but only minimally interfere with people who behave rationally. Such a policy is also asymmetric in the sense that it should be acceptable both for those who believe that people behave rationally, and for those who believe that people often behave irrationally" [10]. However, it has already shown its effectiveness in solving a number of issues. The strategy of "libertarian paternalism" is aimed at nudging a person to an optimal decision dictated by reason, and not by feelings and emotions.

By implementing a policy that includes elements of the architecture of choice, the state could try to partially solve the problems associated with the improvement of the population. This issue has long been an economic one since the financing of the medical support system is a very significant expenditure for any developed state. The epidemic of obesity and diabetes, according to some estimates, can cause damage to the economy, measured in trillions (!) of dollars [12]. Undoubtedly, in this area, a "marketing" policy of the state based on the achievements of behavioral economics would be useful. The task could be formulated as follows: to change the structure of consumption and the lifestyle of citizens in favor of a healthier policy.

Paradoxically, the population of developed countries is living longer and longer, but at the same time, the number of people with severe chronic diseases is increasing. Part of this problem is also related to the economics and peculiarities of the food market, which is saturated with "unhealthy" food, the harm of which the buyer is not always able to assess. Probably, the solution to the problem should also go "from above".

We should also look for other ways to use behavioral economics to solve acute economic and social problems. For example, there is a positive experience of adopting the law on the deliberate consent to be an organ donor, which has already saved hundreds of thousands of lives in many European countries [13].

Today, behavioral economics is a very voluminous but not structured set of data on subrational human behavior. Against this background, **structuring, organizing, and systematizing the existing knowledge seems to be an extremely important task**. This is urgently needed, including the wide dissemination of knowledge obtained within the framework of this discipline.

To date, the educational literature on behavioral economics is extremely limited. In a few manuals, the provisions of the theory are most often given as they are discovered in chronological order, which does not reflect the structural and logical relationships necessary for a holistic understanding of the theory. However, attempts to combine the main provisions of behavioral economics into one logically interrelated and consistent scheme, if any, have not been widely publicized. There are, however, some classification schemes for cognitive distortions and heuristics, of which more than several hundred have already been discovered today.

The problem is probably connected with the fact that it is very difficult to unambiguously determine the underlying reasons for deviations from the rational model of behavior in some cases. This question concerns complex aspects of the functioning of the human psyche and lies far beyond the disciplinary framework of economic theory [14].

Today, even a person with a good economic education, who does not have basic knowledge in psychology, often cannot understand the provisions of behavioral economics [15].

4 Conclusion

Behavioral economics is a modern, multifaceted and promising field of research, within the framework of which a number of important discoveries have already been made about the features of human economic behavior. However, discoveries often raise new questions and broaden research horizons: the realization that individuals often behave subrationally presents economists with new problems that must be solved.

New knowledge, first of all, requires deepening, systematization, integration into existing theory, and practice.

Materials of the article are of theoretical and practical importance for employees of research institutes, higher educational institutions, as well as various economic structures interested in theoretical and practical aspects of behavioral economics.

References

1. Borisov, A.Y.: Mod. Sci. Serv. Econ. Law **11**, 3–6 (2017)
2. Borisov, A.Y.: Risk Resour. Inf. Supply Compet. **1**, 71–76 (2017)
3. Thaler, R.: The Making of Behavioral Economics. Eksmo, Moscow (2017)
4. Veblen, T.: The Place of Science in Modern Civilization. New York (1919)
5. Mitchel, W.C.: The Backward Art of Spending Money. New York (1937)
6. Schumpeter, J.A.: The history of economic analysis. In: Avtonomov, V.S. (ed.) vol 1. Economic School, St. Petersburg (2001)
7. Khvinteliani, V.G.: The Evolution of the Theory of Value in the 19th Century: Cost and Marginalist Versions. Nauka, Moscow (2003)
8. Chamberlin, E.: The theory of monopolistic competition: reorientation of the theory of value. In: Leikina, E.G., Rozovsky, L.Y. (eds.) Ekonomika, Moscow (1996)
9. Thaler, R.: Logos **1**(97), 141–154 (2014)
10. Libertarian paternalism. https://ru.qaz.wiki/wiki/Libertarian_paternalism
11. Forced to choose: why the Nobel Prize was given to a "paternalist libertarian". https://www.rbc.ru/economics/09/10/2017/59db3a7f9a7947c5c18d2b7b
12. The World Bank assessed the impact of the level of obesity on the economy. https://tass.ru/ekonomika/7712035
13. A transplant? I am in! The project on transplantation and organ donation [Electronic source]. http://donor4life.ru/
14. Commons, J.R.: Economics of Collective Action. New York (1950)
15. Commons, J.R.: Institutional Economics. New York (1934)

Improving the Methodology for Assessing the Effectiveness of Management Organisations in the Housing and Utility Sector

Yulia V. Larionova[1](✉), Irina V. Alekseeva[2], Elena A. Repina[3], and Ekaterina F. Kuzmenkova[2]

[1] Moscow State University of Civil Engineering, Moscow, Russia
LarionovaYUV@mgsu.ru
[2] Rostov State University of Economics, Rostov-on-Don, Russia
[3] Southern Federal University, Rostov-on-Don, Russia

Abstract. The relevance of the topic of this research is due to the current situation in assessing the performance of management organisations (MOs) in the housing and utility sector in constituent entities of the Russian Federation. At present, each region is developing its own activity performance assessment system; on that basis, they develop a rating of a MO. MOs with a low rating are gradually "dropping out" from the housing and utility market, while MOs with a higher rating, on the contrary, are strengthening their positions by increasing the apartment block area serviced. However, each constituent entity of the Federation develops and approves its own assessment criteria, which often do not reflect an objective assessment of companies' activities. Based on the foregoing, we have finalised the current management organisation performance assessment methodology in the Moscow region, which will allow not only to make a more objective assessment of activities but also transfer that experience to other Russia's regions. The purpose of the article is to improve the management organisation performance assessment methodology for the subsequent creation of the management organisation rating in any constituent entity of the Russian Federation. To achieve the goal of the research, the following tasks were set: 1) to study the approved methods for assessing the management organisation performance; 2) to analyse the approved management organisation performance methodology in the Moscow region; 3) to evaluate the management organisation performance of the activities of DOMZHILSERVICE LLC according to the approved methodology; 4) to consider and propose possible areas for improving the management organisation performance methodology; 5) to assess the economic efficiency of the activities of DOMZHILSERVICE LLC according to the improved assessment methodology; 6) to determine the management organisation rating according to the approved and improved methods.

Keywords: Management organisation · Organisations rating · Apartment blocks · Management organisation performance · Range of activities · Financial stability

P. V. Trifonov and M. V. Charaeva (Eds.): DITEM 2021, LNNS 380, pp. 203–213, 2022.
https://doi.org/10.1007/978-3-030-94245-8_28

1 Introduction

Currently, the management organisation performance in the housing and utility sector is assessed by calculating the indicators of the management organisation performance assessment methodology approved in each constituent entity of the Russian Federation and building an appropriate rating. The indicators in such methods differ, which suggests different approaches and principles incorporated in the assessment and the impossibility of adequately assessing the management organisation performance [1–6].

As part of the research, as a basic assessment method, we used the Moscow region's management organisation performance assessment methodology as the most complete although not devoid of shortcomings. As a significant drawback of this methodology is the absence of a criterion for assessing the financial and economic activities of a management organisation, which is important for both the leaders of the management organisation and the owners of the common property of apartment blocks. During the management organisation activities, financial and economic problems often arise (due to high indebtedness to utility providers, low collectability, lack of direct agreements between the owners of common property and utility providers etc.). Timely and regular assessment of financial and economic indicators will allow to promptly identify these problems and rationally develop a management mechanism without forcing the organisation into bankruptcy. The owners of the common properties of apartment blocks also enjoy advantages: understanding when to think about changing the management organisation because of its unfavourable financial and economic situation [7–9].

An element of the scientific novelty of the study is the improvement of the management organisation assessment methodology for the creation of a rating of organisations that manage apartment buildings based on the criteria approved by the order of the Ministry of Municipal Affairs and Housing of the Moscow Region dated 13 July 2018 No. 113-RV "Approval of the Regulation on Assessing the Performance of Organisations that Manage the Housing Stock in the Moscow Region" through the introduction of the financial and economic module [10].

The practical significance of the research results lies in the possibility of using an improved methodology for assessing the performance of management organisations to create a rating of management organisations in the housing and utility sector throughout the Russian Federation.

2 Methods

During the research, we considered three methods for assessing the performance of management organisations. According to the Methodology for Assessing the Activities and Creating the Rating of Organisations that Manage Apartment Blocks in Moscow, the management organisation activity is assessed at three levels, the level of satisfaction of the population with the management organisation activities; the level of reliability of the management organisation; the level of violations in the work of management organisations [11]. As criteria for assessing the effectiveness of the organisation that manages apartment blocks in the Chuvash Republic, the following have been identified: work to ensure the technical condition of the common property of apartment blocks, to carry out current and major repairs; the quality of the maintenance of the adjacent territory; quality of service provision; training of management organisation employees; management quality [12].

In the Moscow region, the criteria for assessing the performance of the activities of organisations that manage the housing stock are the maintenance of the courtyard areas of the apartment blocks, parking in courtyards on lawns and locking devices installed without authorisation in the courtyard including boom gates; maintenance of apartment blocks including the current maintenance of the entrances of apartment blocks; arrangement for operations control to the population through the unified operations control service portal; ensuring that management organisations uninterruptedly provide public utility services to residents of apartment blocks; payment discipline of the management organisations in settlements with the utility providers; implementation by the management organisation of the plans for the current repair of the entrances of the apartment blocks established by the municipal programmes; serviceability of automated fire protection systems of apartment blocks; the management organisations' cooperating with the district police officers to identify "rubber apartments" and illegal migrants in apartment buildings; participation of management organisations in the regional action "Open Day at a Management Organisation"; assistance to management organisations in the creation of Apartment Block Councils; signing by the management organisation of apartment blocks' datasheets to prove operational readiness in the autumn-winter period with the chairman of the Apartment Block Council; holding a reporting annual general meeting of owners of apartment blocks; assessment of the management organisation activities by the chairman of the Apartment Block Council [10].

As a result of the analysis, it was concluded that the most complete and detailed is the Methodology for Assessing the Effectiveness of Organisations that Manage the Housing Stock of the Moscow Region. Given below are the main provisions of the management organisation performance assessment methodology (by the example of the Moscow region). However, this methodology does not fully reflect the current range as well as financial and economic indicators of management organisations. In this regard, it is advisable to supplement it with two sets of indicators, such as the range of activities and financial stability.

1. **Range of activities:**

It includes the indicators as follows:

- *Change in the gross floor area of apartment blocks, %:*

$$P_{house\,change} = \frac{P_{house} - P^{\mathrm{H}}_{house}}{P_{house}} \times \frac{12}{N} \times 100\%, \tag{1}$$

where $P_{house\,change}$ – change in the gross floor area of houses, %;

P_{house} – gross floor area of apartment blocks under management at the reporting date, thousand m^2;

P^{H}_{house} – gross floor area of apartment blocks under management as of the beginning of the reporting period, m^2;

N – number of months in the reporting period, unit.

10 points: at $P^{H}_{house} < P_{house\ change}$;

10 points: at $P_{house\ change} < P^{H}_{house}$.

- Change in the number of apartment blocks, %:

$$N_{house\ change} = \frac{N_{house} - N^{H}_{house}}{N^{H}_{house}} \times \frac{12}{K} \times 100\%, \quad (2)$$

where $N_{house\ change}$ – change in the number of apartment blocks, %;

N_{house} – number of apartment blocks under management at the reporting date, units;

N^{H}_{house} – number of apartment blocks under management at the beginning of the reporting period, units;

N – number of months in the reporting period, units.

10 points: at $N^{H}_{house} < N_{house\ change}$;

10 points: at $N_{house\ change} < N^{H}_{house}$.

- *Income from management of apartment blocks, thousand roubles:*

$$N_{i.manag.change} = N_{i.manag.} - N^{H}_{i.manag.}, \quad (3)$$

where $N_{i.manag.change}$ – change in management income, thousand roubles;

$N_{i.manag.}$ – management income at the reporting date, thousand roubles;

$N^{H}_{i.manag.}$ – management income at the beginning of the reporting period, thousand roubles;

10 points: at $N^{H}_{i.manag.} < N_{i.change}$;

10 points: at $N_{house\ change} < N^{H}_{i.change}$.

2. **Financial stability:**

- *Profit from apartment block management activities:*

$$Pr_{\%} = \frac{I - E}{I} \times 100\%; \quad (4)$$

$$Pr_{\%\text{change}} = Pr_{\%} - Pr^{H}_{\%}, \quad (5)$$

where $Pr_{\%change}$– change in profit from management activities, %;

$Pr_{\%}$ – profit from the apartment buildings management activities at the end of the reporting period, %;

$Pr^{H}_{\%}$ – profit from the apartment buildings management activities at the beginning of the reporting period, %;

I – income received for the reporting period for the apartment buildings management, thousand roubles;

E – direct expenses for the apartment buildings management during the reporting period, thousand roubles.

50 points: at $Pr_{\%\text{change}} < 0$;

50 points: at $0 < Pr_{\%\text{change}}$.

- *Overdue indebtedness of inhabitants in respect of apartment block management activities* [13]*:*

$$D_{manag.\%} = \frac{D_{manag.} \times K}{I} \times 100\%, \tag{6}$$

$$D_{manag\%change} = D_{manag.\%} - D^{H}_{manag\%}, \tag{7}$$

where $D_{manag.\%}$ – overdue indebtedness of premises owners accumulated for the entire maintenance period as of the reporting date in relation to the average monthly income of management organisations, %;

$D_{manag.}$– overdue indebtedness of premises owners accumulated for the entire maintenance period as of the reporting date, thousand roubles;

I – income from apartment buildings management services received during the reporting period, thousand roubles;

N – number of months in the reporting period, unit.

0 points: at $D_{manag\%change} < 0$;

30 points: at $0 < 3_{manag\%change}$.

- *Overdue indebtedness by management organisations to utility providers:*

$$D_{pco.\%} = \frac{D_{pco} \times N}{D} \times 100\%; \tag{8}$$

$$D_{pco\%change} = D_{pco.\%} - D^{H}_{pco\%}, \tag{9}$$

where $D_{pco.\%}$– overdue indebtedness of management organizations to utility providers in relation to the average monthly income of the management organizations from the provision of municipal services, %;

D_{pco} – overdue indebtedness of management organizations in respect of the provided municipal services accumulated for the entire service period, as of the current date, thousand roubles;

I – income received for the reporting period from the provision of utility resources, thousand roubles;

N – number of months in the reporting period, units.

30 points: at $D^{H}_{pco\%} = 0$;

0 points: at $D^{H}_{pco\%} < D_{pco\%change}$, owners entered into direct contracts with utility providers [14];

30 points – at $D_{\%change} < D^{H}_{\%}$.

The rating position is determined by the sum of the points scored according to the evaluation criteria. Table 1 shows the distribution of points in the approved and improved methodology [10].

Table 1. Procedure for awarding rating stars to management organizations.

Number of stars	Approved methodology, points	Improved procedure, points
5	Rpoint > 350	Rpoint > 460
4	300 < Rpoint < 350	400 < Rpoint < 460
3	200 < Rpoint < 300	300 < Rpoint < 400
2	100 < Rpoint < 200	150 < Rpoint < 300
1	Rpoint < 99	Rpoint < 150

where Rpoint – number of points awarded to management organisations based on the assessment results

3 Results

As a base for the research, the management organisation DOMZHILSERVICE LLC, which runs apartment blocks with a total area 976,718.10 m^2, was used.

As a result of calculating additional sets of indicators of the refined methodology for assessing the performance of the management organisation, the results were obtained as follows:

1. **Range of activities:**

- *Change in the gross floor area of apartment buildings, %:*

$$P_{house.change} = \frac{400973.1 - 384921.6}{384921.6} \times \frac{12}{12} \times 100\% = 4.17\%$$

In 2020, DOMZHILSERVICE LLC increased the serviced area by 4.17% compared to 2019.

Score: 10 points.

- *Change in the number of apartment blocks, %:*

$$N_{house.change} = \frac{37 - 36}{36} \times \frac{12}{12} \times 100\% = 2.78\%$$

In 2020, DOMZHILSERVICE LLC increased the number of apartment buildings by 2.78% compared to 2019.

Score: 10 points.

- *Income from running apartment blocks, roubles:*

$$N_{h.manag.change} = 176727302 - 172429531 = 4297771$$

In 2020, in DOMZHILSERVICE LLC the income from running apartment blocks increased by 4,297,771 roubles compared to 2019.

Score: 10 points.

2. **Financial stability:**

- *Profit from the apartment block management activities, %:*

$$Pr_{\%}^{2020} = \frac{176727302 - 174747160}{176727302} \times 100\% = 1.12\%;$$

$$Pr_{\%}^{2019} = \frac{172429531 - 169143627}{172429531} \times 100\% = 1.9\%;$$

$$Pr_{\%}^{2018} = \frac{169334872 - 164996839}{176727302} \times 100\% = 2.56\%;$$

$$Pr_{\%change} = 1.9 - 1.12 = 0.78\%.$$

In 2020, the profit from the management of apartment buildings decreased by 0.78% compared to 2019. Over time, the profit of DOMZHILSERVICE LLC decreases every year.

Score: 50 points.

- *Overdue indebtedness of residents for apartment blocks management services, %:*

$$D_{manag.\%} = \frac{338628 \times 12}{176727302} \times 100\% = 2.3\%;$$

$$D_{manag\%change} = 2.3 - 2.1 = 0.2.$$

In 2020, at DOMZHILSERVICE LLC, overdue indebtedness associated with apartment blocks management services increased by 0.2%.

Score: −30 points.

- *Overdue indebtedness of management organisations to utility providers:*

The owners of the common property of apartment blocks entered into direct agreements with utility providers.

Score: −0 points.

Table 2 shows the final results of assessing the performance of apartment blocks' activities according to the existing (approved) and improved methods [10].

As can be seen in Table 2, according to the current methodology (which does not take into account the range of activities and financial stability), DOMZHILSERVICE LLC is gaining 330 points, is included in the second group of management organizations, and is receiving 4 stars of the rating. When calculating the rating of management organizations according to the improved methodology, this organization is already gaining 280 points and is included in the fourth group of management organizations, and is receiving only 2 stars.

Table 2. The results of calculating the rating of management organisation DOMZHILSERVICE LLC.

Item	Approved methodology, points	Improved procedure, points
1. Maintenance of courtyards of apartment blocks [15–17]	−15	−15
2. Maintenance of apartment blocks, including current maintenance of entrances to apartment blocks	$N1$: 10 $N2$: 30	$K1$: 10 $K2$: 30
3. Arrangement for control services to inhabitants through the unified control service portal	20	20
4. Ensuring that management organisations uninterruptedly supply utility services to the residents of apartment blocks	15	15
5. Payment discipline of the management organisations in making settlements with utility providers	30	30
6. Implementation by the management organisations of plans for the current repair of apartment block entrances established by municipal programmes	0	0
7. Serviceability of the automated fire protection systems of apartment blocks	20	20
8. Cooperation of management organisations with district police officers to identify "rubber apartments" and migrants illegally residing in apartment buildings	20	20
9. Participation of management organisations in the regional action "Open Day at a Management Organisation"	30	30

(continued)

Table 2. (*continued*)

Item	Approved methodology, points	Improved procedure, points
10. Assistance by management organisations in the creation of Apartment Block Councils	40	40
11. Signing by management organisations of data sheets to prove operational readiness of apartment blocks in the autumn-winter period with the chairman of the Apartment Block Council	30	30
12. Signing by management organisations of data sheets to prove operational readiness of apartment blocks in the autumn-winter period with the chairman of the Apartment Block Council	30	30
13. Evaluation of the management organisation performance by the chairman of the Apartment Block Council	70	70
14. Range of activities		Change in the gross floor area of apartment blocks, %: 10 Change in the number of apartment blocks, %: 10 Income from running apartment blocks, roubles: 10
15. Financial stability		Profits from running apartment blocks, %: −50 Overdue indebtedness by residents for apartment block management services, %: −30 Overdue indebtedness by management organisations to utility providers: 0
Points, total	330	280
Rating, stars	4	2

4 Conclusion

The article presents the results of analysis of the existing management organisation performance assessment methods and the creation of the rating of organisations, an improved methodology for assessing the management organisation performance is proposed, the rating of the management organisation in question is determined.

Having considered the methodology for assessing the management organisation performance in the Moscow region, we concluded that this methodology does not fully feature the current position of the range or financial and economic indicators of the management organisation. An improved assessment methodology taking into account the "Range of Activities" and "Financial Sustainability" modules has been developed. Calculations of the rating of the management organisation DOMZHILSERVICE LLC according to the approved methodology showed that the company has a high rating (4 stars) and is in the leading group of the Moscow region's organisations with 330 points out of 350 possible. This indicator is an excellent result and contributes to the involvement of new apartment blocks in management.

When analysing the effectiveness of the activities of DOMZHILSERVICE LLC according to the improved methodology, it was found that the entity every year increases the area served and the number of apartment buildings (this indicates the growth of the organisation), while the profit decreases every year. According to the new methodology, the company gains 280 points and falls into the fourth group by gaining only 2 stars. This can be explained by a number of internal problems of the organisation, an unoptimised management system, a decrease in the economic efficiency of activities, a staffing table that is not staffed with the required employees. It is difficult to identify all these problems for ordinary owners of the common property of apartment blocks and make the right decision when choosing a management organisation.

Thus, the existing methodology for assessing the management organisation performance for the Moscow region is not complete, nor it reflects the objective position of the management organisation DOMZHILSERVICE LLC. Only when the missing modules are included, the real position of the company in the rating can be seen; it determines the stability of the organisation's position and increases the chances of servicing large volumes of space by attracting new customers. Consequently, the application of the improved methodology for assessing the management organisation performance helped reflect the actual information about the company's position. In this regard, it is advisable to adopt an improved methodology for widespread use in other regions of the Russian Federation.

References

1. Zamotaylova, D.A., Kurnosova, N.S., Reznikov, V.V.: Methodological approaches to assessing the effectiveness of management organisations. Proc. Kuban State Agrar. Univ. **64**, 18–23 (2017). https://doi.org/10.21515/1999-1703-64-18-23
2. Samokhin, A.V.: Consumer assessment of the residential property management efficiency. Newsl. Civil Eng. **4**(45), 118–121 (2014)
3. Larionova, Y.V.: Focal points of the state policy in the housing and utility sector. Real Est: Econ. Manag. **3**, 18–22 (2018)

4. Kolyhaeva, Y.: The development of information technologies in the sphere of housing service and utilities as a factor of national life quality increase. In: International Conference on Research Paradigms Transformation in Social Sciences 2014. Procedia Soc. Behav. Sci., vol. 166, pp. 557–561 (2015)
5. Finnigan, R., Meagher, K.D.: Past due: combinations of utility and housing hardship in the United States. Soc. Persp. **62**(1), 96–119 (2019)
6. Zamiralov, A., Khodorchenko, M., Nasonov, D.: Detection of housing and utility problems in districts through social media texts. In: 9th International Young Scientist Conference on Computational Science (YSC 2020). Procedia Comput. Sci., vol. 178, pp. 213–223 (2012)
7. Kirillova, A.N., Skripnik, O.B.: Methods for assessing the effectiveness of reforming the housing and utility sector. Real Est. Econ. Manag. **1**, 11–17 (2012)
8. Gabovoy, P.G., Kirillova, A.N. (eds.): Economics and Management of the Housing and Utility Sector. Prosvetitel, Moscow (2019)
9. Gabovoy, P.G. (ed.): Urban Management and Modernisation of the Housing and Utility Infrastructure. Prosvetitel, Moscow (2013)
10. Order of the Ministry of Municipal Affairs and Housing of the Moscow Region No. 113-RV dated 13 July 2018 “Approval of the Regulations for Assessing the Performance of Organisations that Manage the Housing Stock in the Moscow Region”. https://mgkh.mosreg.ru/dokumenty/normotvorchestvo/rasporyazheniya/16-08-2018-11-49-21-rasporyazhenie-ot-13-07-2018-113-rv-ob-utverzhdeni
11. Methodology for Assessing Activities and Creating a Rating of Organisations that Manage Apartment Blocks in Moscow: order of the State Housing Inspectorate of Moscow dated 23 November 2015 N 342. https://www.mos.ru/upload/documents/files/1383/prikaz_moszhilinspektsii_ot_23112015_342_ob_utverzhdenii_metodiki_otsenki_deyatelnosti_i_formirovaniya_reytinga_organizatsiy_osushchestvlyayushchikh_upravlenie_mnogokvartirnymi_domami.pdf
12. Recommendations for Assessing the Performance of Apartment Blocks Management Organisations: order of the Ministry of Urban Planning and Development of Public Infrastructure of the Chuvash Republic dated 13 June 2007 No. 04-13/117. https://docs.cntd.ru/document/444772563
13. Prices Rates Hous Utility Sec **4**(78), 6–8 (2018)
14. Larionova, Y.: Basic procedures for switching to direct contracts with utility providers in order to ensure sustainable development of the housing and utility sector. Real Est. Econ. Manag. **3**, 45–48 (2019)
15. Matek, B.: An examination of voluntary green power programs at U.S. utilities using behavioral science principles. Elect. J. **29**, 55–63 (2016)
16. Larionov, A.: Strengths and weaknesses of energy-saving management in housing and public service: Russian experience. J. App. Sci. **14**, 2374–2379 (2014)
17. Larionova, Y., Smirnova, E.: Substantiation of ecological safety criteria in construction industry, and housing and communal services. IOP Conf. Ser. Earth Environ. Sci. **543**, 012002 (2020)

Controlled Chaos as a Tool of Modern Globalization

Igor S. Ivanchenko[1(✉)], Ksenia Y. Boeva[2], Alla V. Temirkanova[2], and Hui Jiang[3]

[1] Rostov State University of Economics RINH (RSUE), Rostov-on-Don, Russia
[2] Southern Federal University, Rostov-on-Don, Russia
[3] Don State Technical University, Rostov-on-Don, Russia

Abstract. The prerequisite for this study was the increased criticality in international economic relations, which was particularly acute after the financial crisis of 2008. Then authoritative international institutions such as the UN, WHO, etc. began to lose leverage in a globalized economy. Leading world powers increasingly began to resort to a policy of brute force and sanctions to achieve their national interests. The study aims to find the cause of regularly occurring hotbeds of tension, social conflicts, and color revolutions in the modern world. The method of historical analysis of the evolution of the theory of economic doctrine, starting from the fifteenth century, was applied. The article notes that the process of globalization, accelerated after the Bretton Woods Conference in 1944, began to falter in the second decade of the twenty-first century. Then the policy of the new neo-mercantilism began to be resorted to not only by developing countries, but also by economically developed countries. The desire to maintain a unipolar world and retain economic dominance has led Western political scientists to create the theory of controlled chaos. The unenviable economic and social prospects for the victim country of the practical application of the theory of controlled chaos on its territory were shown. There are double standards applied by developed countries in economic management, which are based on the postulates of different economic schools. These are the main reason for the numerous bloody conflicts when governments undesirable to Western countries are replaced.

Keywords: Neoclassical economic theory · New neo-mercantilism · Multipolar world · Crisis of international relations · Color revolutions

1 Introduction

In the last two decades, we have witnessed a sharp increase in economic and political contradictions between developed and developing countries. It seems that developed countries have stopped respecting international law when interacting with the rest of the world. The practice of international relations has become based exclusively on brute military force, on unleashing local wars, on imposing economic sanctions, on unilateral breaking of international agreements, on blackmail and violence. Let us try to understand the reasons for this transformation of world politics, and what factors underlie this

P. V. Trifonov and M. V. Charaeva (Eds.): DITEM 2021, LNNS 380, pp. 214–218, 2022.
https://doi.org/10.1007/978-3-030-94245-8_29

negative process. Let us try to understand why the media increasingly began to mention the theory of controlled chaos as a tool for the relationship between developed and developing countries.

The process of development of any complex system, which undoubtedly includes today's globalized society, is not smooth and homogeneous. It represents a transition from a state of order through a phase of transition (destruction of structure) into a state of chaos, and then through a phase of self-organization into a state of ordered development. If we turn to scientific publications, among economists, financiers [1], mathematicians [2], physicists, chemists and biologists [3] studies of chaotic behavior of systems are becoming increasingly popular. For example, in physics, there is the law of increasing entropy, which states that nature itself tends toward disorder, and the degree of chaos in the universe is constantly increasing.

In human society, a chaotic state can occur not due to objective reasons, but when it is artificially influenced by. When a complex social system is in a state of chaos at a bifurcation point, the way out of it is possible in any direction, the prediction of which is impossible [4]. However, if at this point there is a certain purposeful influence on the system from the outside, then its further development can go in the desired direction for the external force [5].

In the "Methods" section, it is noted that the study is based on a historical analysis of the evolution of the theory of economic doctrine. The "Results" section provides an analysis of the concept of controlled chaos. The section "Discussions" substantiates the authors' concept that the main reason for the strategy of controlled chaos is a policy of double standards in the application of economic theory in practice in today's globalized world. The "Conclusion" summarizes the findings of the research.

2 Materials and Methods

It is necessary, in our opinion, to refer to economic theory and apply the method of historical analysis of the evolution of the theory of economic doctrines, because economy always determines politics, but not vice versa. It is essential to find the cause of the analyzed negative phenomenon in the modern globalized world, when developed countries refuse to compete fairly with developing economies, and prefer to switch to terrorist activities to contain competitors, using the theory of controlled chaos.

3 Results

Having analyzed the scientific literature on the practical implementation of the theory of controlled chaos, the authors can conclude that there are almost no publications on this topic in English-language publications. The authors found only one article by an American political scientist [6] in which he accuses the Russian government of applying the theory of controlled chaos in Ukraine. American political technologists know how to keep their secrets! There are not many publications by domestic authors either. Their content is generally duplicated. Therefore, we can conclude that the theory of controlled chaos, applied in modern geopolitics, needs a detailed study so that our country can adequately withstand provocations from foreign secret services. Some authors of articles have argued that since the 1970s American liberals, under the guise of defending democracy and human rights, have actually declared a new type of world war, creating a mess in a country with an unwanted regime [7]. The ideological centers of controlled chaos are such organizations as the Bilderberg Club and the Santa Fe Institute. The methods they devised to wreak havoc on the economy of a developing country and then gain control over its natural resources were concretely embodied in the activities of the IMF, WTO and the World Bank [8].

Steven Mann, a theorist of controlled chaos, an American diplomat and political technologist of color revolutions in the post-Soviet space, has lifted the veil on the secrets of creating hotbeds of tension in different parts of the world (including the USSR and modern Russia) in his book "Chaos Theory in Strategic Thinking" [9]. According to the author of this paper, the theory of controlled chaos is characterized by the following features:

- it can only be applied to complex dynamic systems;
- these systems can be both chaotic and ordered, but not periodic;
- these systems must have a strong dependence on initial conditions: there is a slight change in the initial parameter of the system leads in the future to unpredictable properties and behavior of this system;
- the theory of controlled chaos allows predicting the behavior of at least a weakly chaotic system.

Steven Mann openly urges the United States to use the ideological virus as a weapon of mass destruction to achieve geopolitical goals: "The United States must... infect target populations with ideologies of democratic pluralism and respect for human rights. Thanks to advances in communications and the ease of global travel, the ideological virus will reproduce itself and spread rather chaotically. In this way, our national security will be best ensured" [9]. Steven Mann advanced an interesting idea that has been a leitmotif of U.S. foreign policy for the past several decades. "We must learn to see chaos and permutations as possibilities, rather than seeking stability as an illusory goal" [9]. Therefore, the economy of any victim country was destroyed, turned into chaos. This was the case in Iraq, Yugoslavia, Syria, Libya [10].

4 Discussion

The process of globalization of the world economy accelerated in the 1990s and tied countries and continents together, facilitating the free movement of goods, scientific ideas, capital and labor across national borders. The process of globalization had to connect all scientific, production and financial chains firmly into a single mechanism. The level of modern internationalization of the world economy, the enormous dependence of the production sector and the financial market of any country on the world commodity and capital markets should have led long ago to the peaceful integration of national economies into a single well-oiled production mechanism. It had to work without artificially created failures on the part of the governments of the leading economies. But this has not happened so far, and the reason for this phenomenon, revealed in the study, was very serious.

Having analyzed the history of economic thought since the 15th century, it can be concluded that there are two irreconcilable economic schools in economic theory: classical [11–13] and mercantilism [14–16] (or in later times: neoclassical and neomercantilism). All other schools of economics are either their side branches or a synthesis of neoclassicism and neomercantilism. The economic theory of the dominant scientific school is reflected most directly in the methods of economic management in developed capitalist countries. In the U.S., for example, the economic programs of presidents correlate with the basic tenets of the economic mainstream.

We analyzed the economic policies of American presidents over the past hundred years. It was found that the highest rates of economic growth were observed in the U.S. in the years when the economy was managed according to mercantilist prescriptions. Regular cycles in this country's economic policy, associated with a change of leader in the economic mainstream, the search for answers to international challenges and financial crises, have been identified. The paradox is that developing countries continue to live steadily according to the recipes of mercantilists. Developed countries, following the traditions of the classical school laid down by Adam Smith over the past forty years; consume more material goods than they produce, despite the growth of foreign public debt. Developing countries, on the other hand, tend to export as many goods and services as possible in the bosom of the mercantilist theory, while increasing the volume of foreign exchange reserves [17]. Such an international system of "mutually beneficial" economic cooperation can last long enough. However, practice shows that countries developing according to mercantilist recipes win the economic race against the developed countries. In this regard, the attempt made by Donald Trump to change this situation and make the U.S. a great economic power again has failed. The American financial elite, the main beneficiary of the globalization process, has fiercely resisted the return of American businesses from developing countries to the United States. Therefore, in the next few years, there may be a change of the world economic leader. In order to prevent this, developed countries began to use a new weapon in international relations to fight their competitors in the form of the theory of controlled chaos [18].

5 Conclusion

State-supported economies in developing countries are growing faster than those in developed liberal. To curb this trend, developed countries have advanced and successfully tested in practice the theory of controlled chaos, which, in fact, is a new form of colonial policy. The chaos theory created by mathematicians and physicists has been adapted to socio-political processes. Steven Mann's methodology for influencing complex nonlinear systems allows us to model an unstable state in the economy by controlling political events. Without a declaration of war, without the use of armies, the human and natural resources of the victim country are at the full disposal of the aggressor country. Therefore, developing countries must unite in political and economic alliances to protect their national interests collectively. Examples of new partnerships between developing countries are such associations as BRICS, ASEAN, or the EAEU. The process of transition from a bipolar world in the 1980s (USSR-USA) to a monopolar world in the 1990s (US), and now back to bipolar again (US-China) should not prove disastrous for the rest of the world. Russia cannot allow itself to be drawn into the orbit of imperial unequal relations of the developed countries of the West with the developing states. This can only be guaranteed by a drastic improvement in the quality of education and science, an acceleration of the innovative development of the Russian economy, and a transition from commodity exports to exports of high-tech goods. This will reduce poverty and undermine the social base of protest movements in our country.

References

1. Klioutchnikov, I., Sigova, M., Beizerov, N.: Chaos theory in finance. Procedia Comput. Sci. **119**, 368–375 (2017). https://doi.org/10.1016/j.procs.2017.11.196
2. Xu, F., Lai, Y., Shu, X.: Chaos. Sol. Fract. **117**, 125–136 (2018). https://doi.org/10.1016/j.chaos.2018.10.005
3. Alves, P.R.L.: Chaos. Sol. Fract. **1**, 1–9 (2019). https://doi.org/10.1016/j.csfx.2019.100002
4. Petrov, A.P.: Sci Tech Stat Peter the Great St Petersburg Polytechnic Univ. Human Soc. Sci. **8**(4), 77–87 (2017).https://doi.org/10.18721/JHSS.8408
5. Oxley, L., George, D.: Economics on the edge of chaos: some pitfalls of linearizing complex systems. Environ. Model. Softw. **22**(5), 580–589 (2007). https://doi.org/10.1016/j.envsoft.2005.12.018
6. Cordesman, A.: America's failed approach to chaos theory: the complexity crisis in U.S. strategy. Center for strategic and international studies (2015). https://www.csis.org/analysis/america%E2%80%99s-failed-approach-chaos-theory
7. Suleymanova, L.: The influence of the "Controlled Chaos" technology on the geopolitical situation in the Greater Middle East. Open J. Polit. Sci. **7**, 189–196 (2017). https://doi.org/10.4236/ojps.2017.72015
8. Chen, L., Chen, G.: Controlling chaos in an economic model. Physica A Stat. Mech. Appl. **374**(1), 349–358 (2007). https://doi.org/10.1016/j.physa.2006.07.022
9. Mann, S.R.: Parameters (US Army War College Quarterly), vol. 22, pp. 54–68 (1992)
10. Savchenko, V.A.: Steven Mann's "Chaos Theory" and corporate governance. Upravlenie **1**(15), 72–79 (2017). https://doi.org/10.12737/24704
11. Pirgmaier, E.: The neoclassical Trojan Horse of steady-state economics. Ecol. Econ. **133**, 52–61 (2017). https://doi.org/10.1016/j.ecolecon.2016.11.010
12. Stivachtis, Y.: Liberal democracy, market economy, and international conduct as standards of 'civilization' in contemporary international society: the case of Russia's entry into the 'community of civilized states.' J. Eurasian Stud. **6**(2), 130–142 (2015). https://doi.org/10.1016/j.euras.2015.04.001
13. Strunz, S., Bartkowski, B.: Degrowth, the project of modernity, and liberal democracy. J. Cleaner Prod. **196**, 1158–1168 (2018). https://doi.org/10.1016/j.jclepro.2018.06.148
14. Conti, T.V.: Mercantilism: a materialist approach. Scandinavian Econ. Hist. Rev. **66**(2), 186–200 (2018). https://doi.org/10.1080/03585522.2018.1465847
15. Mahdi, A.S.: Saudi neomercantilism in the oil price war. Rev. Econ. Polit. Sci. **5**(1), 2–20 (2020). https://doi.org/10.1108/REPS-10-2019-0134
16. Nelson, D.R.: Facing up to Trump administration mercantilism: the 2018 WTO trade policy review of the United States. World Econ. **42**(12), 3430–3437 (2019). https://doi.org/10.1111/twec.12875
17. Ivanchenko, I.S.: Optimization of Russia's international reserves structure: theoretical approaches, practical implementation. Probl. Econ. **1**, 64–80 (2017). https://doi.org/10.32609/0042-8736-2017-1-64-80
18. Demidov, A.V.: Geopolitics J. **4**, 86–93 (2014).https://doi.org/10.7256/2305-560X,2014.2.10355

Human Capital and Economic Development

Din Khaomin[1], Yulia A. Dianova[2](✉), Mariya A. Kuznetsova[3], and Irina O. Bogdanova[4]

[1] Southern Federal University, Rostov-on-Don, Russia
[2] Penza State Technological University, Penza, Russia
dianova@penzgtu.ru
[3] Rostov State University of Economics, Rostov-on-Don, Russia
[4] Don State Technical University, Rostov-on-Don, Russia

Abstract. Human capital largely determines the economic development of a country. In this regard, the purpose of the research is to identify prerequisites for improving human capital in the contemporary economic space. The article presents the research findings according to the range of stated problems. The analysis of the national projects "Demography", "Healthcare", "Education" and "Culture" from the point of view of the possibility of their influence on the human capital development is given. Based on the data sheets of national projects included in the "Human Capital" package, official statistics and sociological survey data, a conclusion about the human capital development patterns. The important role of investment in education as the main element of human capital is outlined. The novelty of the research lies in the ideas, approaches and principles of human capital for economic development proposed on the basis of an analysis of the regulatory and scientific literature of both foreign (Austria, USA) and domestic experiences. The discovered relationship between the goals, results and conditions of the functioning of the system under study for the long-term implementation of the goals of enhancing the competitiveness of the Russian national economy in the today's context can be in demand and successfully implemented. Economic development ideas through knowledge acquisition and innovation accumulation, the use of human capital management functions, integrative and technological approaches, as well as the principles of stimulating innovation, management of investments in human capital are proposed as upcoming trends for the enhanced human capital.

Keywords: National project · Human capital development model · Education · Human capital

1 Introduction

Today, human capital is considered as a key indicator of national competitiveness and the level of economic development of countries [1]. In order to successfully meet challenges set, preserving and increasing human capital for the economic development of the country becomes of paramount social importance. This is due to the fact that human capital undoubtedly contributes to an increase in labour productivity and is made up of

P. V. Trifonov and M. V. Charaeva (Eds.): DITEM 2021, LNNS 380, pp. 219–224, 2022.
https://doi.org/10.1007/978-3-030-94245-8_30

the volume of investment. Investments in human capital determine the vector of state development. Now that the digital economy has caught on, the role of human capital increases, and the demand for specialists ready to quickly adapt to technological challenges grows [2]. This certainly requires education upgrading [3].

The focus on tackling this problem is reported in the Innovative Development Strategy of the Russian Federation, the National Security Strategy of the Russian Federation, which identify national interests related to maintaining the quality of intellectual potential, IT penetration in the society, improvement of the people's quality of life, and socio-economic progress of the country [4].

The problem of human capital as a factor of economic development was addressed by both domestic (Posysaev and Kudryashova [5], Sleptsova and Ryndina [6], O.P. Ovchinnikova and N.E. Ovchinnikova [7], Chechina [8] and others) and foreign (Becker [9], Schultz [3] and others) researchers. Meanwhile, the importance of preserving and developing human capital is geared toward further search for ways and mechanisms to solve this problem in practice.

The novelty of the research lies in the proposed meaningful ideas (economic development through the accumulation of knowledge and innovation, the use of human capital management functions), concepts (integrative, technology-based) and principles (fostering innovations, investments and improvements) of human capital as a factor of economic development.

The research is aimed at identifying and analysing prerequisites for the improvement and development of human capital in a rapidly developing post-industrial economic space. The main tasks are as follows: based on the scientific literature analysis, to define the concept of "human capital", consider underlying problems and trends in the development of human capital as a factor of production, identify the key patterns of human capital development in the context of the developed National Projects in the Russian Federation.

2 Methods

To accomplish the tasks set in the introduction, the methods of analysis and synthesis of historical-pedagogical and scientific literature were used, which was selected according to the major challenges of the research. As a source base for the research, regulatory, scientific and methodological documents from the middle of the twentieth century to the present were chosen.

3 Results

In state policy, human capital is considered as a set of qualitative attributes of the country's population, which are an investment target, have a value and can be characterised by yielding a return [5].

Human capital is a driver of economic growth of any entity, enterprise, and the state as a whole. The creative and practical skills of a person, his/her knowledge and skills should have quantitative and qualitative characteristics on the basis of which programmes to improve and promote human capital in a specific area of public life will be developed. It is becoming an increasingly relevant objective to consider this issue by the pivotal role of human capabilities in achieving the goals and objectives on the part of an economic entity [10].

Competition is the driving force behind the innovative development of an economic system [11]. In this regard, the intensity and efficiency of its labour can be considered as essential characteristics of human capital. Labour intensity is most often understood as the stress level for labour in performing a labour function per unit of time. Consequently, working conditions should be suitable for them to unlock their labour potential in full, without restrictions, which is possible only given there are favourable resources. Labour productivity is a quantitative indicator that characterises labour efficiency in assessing the volume of a product or service provided per unit of time. At the same time, the intensity and efficiency of human labour depend on not only the personality itself but also external conditions associated with state conditions, work environment and subsidies.

In Russia, the situation is taking an unfavourable turn for the human capital development; there is a tendency for the best specialists to outflow abroad, a decrease in fertility rates and an increase in mortality among people of working age. These factors negatively affect the qualitative characteristics of human capital, and as a consequence, the economy as a whole [6].

The history of the development of the theory of human capital originates in the works of Nobel laureates Schultz and Becker. Schultz, who proved that the share of human capital in the most advanced countries is more than 80% of the national wealth [3]. According to Becker, human capital is a set of innate abilities and acquired knowledge, skills and motivations, the effective use of which contributes to an increase in income and other benefits [9].

In solving problems related to the preservation and increase of human capital as a factor influencing the economic development of a country, a significant emphasis is put on the implementation of national projects, which are combined into 3 packages, "Human Capital", "Comfortable Environment for Life" and "Economic Growth" [12].

The first area can be characterised by the national projects "Healthcare", "Education", "Demography" and "Culture". The deadline for the project implementation is from 1 January 2019 to 31 December 2024. The table shows the amount of funds for the implementation of the above projects according to the document "Main Lines of the Budgetary, Tax and Customs Tariff Policies for 2020 and the Planning Periods of 2021 and 2022" (approved by the Russian Ministry of Finance) (Table 1).

Table 1. Parameters of financial support of national projects from the federal budget.

	Healthcare	Education	Demography	Culture
Amount of financing, billion roubles	1,371.4	728.2	3,263.2	109.8

The amount of funding indicates the readiness of the state to invest in the education, health care, culture and the development of demography that are human capital components.

The national project "Health" is aimed at addressing the advancement of children's health care, raising the level of professionalism of doctors, preventing and eliminating diseases, and developing methods of providing primary health care. The national Education project is aimed at promoting the global competitiveness of Russian education and the country's entry into the top ten world leaders in terms of this indicator, as well as fostering a socially conscious and moral human being. During the implementation of the national Demography project, it is planned to increase life expectancy, increase the proportion of citizens leading a healthy lifestyle and methodically engage in physical culture and sports, as well as an increase in the total fertility rate. The goal of the Culture National project is to create a cultural environment, support creative people and develop digital culture. Thus, the national projects included in the "Human Capital" package should contribute to upward advancement in economy, raise the level of well-being of the population, as well as create comprehensive opportunities for realisation of personal potential.

We can agree with Becker, Machlup and Drucker that education is a vital investment in human capital. Machlup's idea of the importance of education as a branch of the economy was substantiated by him in the 1970s of the twentieth century. The versatility of education is manifested in close cooperation with science, production, the labour market, and IT penetration. The importance of knowledge for innovation was emphasised by Gretchenko and Kaverina [13]. It has been observed that getting any level of education can increase a person's income. Wide access to education (especially now with the active introduction of online learning), high levels of education lead to an increase in labour productivity and economic growth.

Becker, Schultz and Minser referred to investments in human capital as any investment in a person aimed at improving his/her abilities and increasing his/her economic potential. As the key indicators of human capital, these scientists named expenses associated with medicine (health preservation), advanced training, educational systems of different levels and continuing adult education.

The studied and described human capital development models are worth noticing. The Nobel Prize Laureate in Economics (2018) Romer proposed a model in which labour and "scientific talent" are used at various stages of production. It turns out that human capital is also involved in research and development. From Romer's model it follows that the "scientific talent" reserves in a country are associated with the amount of resources required for research, and therefore affects the economic development.

The integrative approach to human capital management considered in Chechina's study is of particular interest. The researcher proposes a human capital management model as an innovative economic development subject. The peculiarity of the model lies in the use of management functional in the creation, assessment, and use of human capital. This allows you to get the maximum efficiency from the involvement of human capital in the functioning of the studied system, which has a positive effect on increasing the competitiveness of Russia's national economy [8]. Human capital management makes it possible to keep track of costs according to the level of professionalism of the subject [14, 15].

In studying the stages of human capital development in economic history, O.P. Ovchinnikova and N.E. Ovchinnikova distinguish a technological approach (a combination of all technologically related sectors of the economy). Researchers believe that it is due to human capital in terms of quality education and scientific knowledge that a technological breakthrough is possible, which will ensure the innovative development of mankind [7].

The human capital creation is based on principles that reflect the features of the problem under study. Nobel laureate in economics Pissarides emphasises the relevance and importance of incentives to induce a person to apply their knowledge and skills for specific purposes. The principle of motivating innovation will ensure the growth of human capital [16]. In the context of the innovative development of an entity, researcher Krakovskaya, highlights the principle of managing investments in human capital [17]. The said principles are applicable to the formation of human capital in the socio-economic conditions of the development of an innovative economy.

4 Discussion

The analysis of the data obtained indicates that the problem of preserving human capital as a factor influencing the economic development of the country is evidenced in regulatory documents, discussed by both domestic and foreign researchers. A review of contemporary literature demonstrates a high interest in this problem from the point of view of its connection with education, since in the global sense of the word, the human capital preservation largely depends on the quality of education, which is an essential condition for the country's economic development.

5 Conclusion

In summary, the following conclusions can be drawn: the country's economic development is directly dependent on human capital. In shaping human capital, it is considered promising to use economic development ideas through the accumulation of knowledge and innovation, the use of functions of human capital management, integrative and technological approaches, as well as the principles of encouraging innovation, management of investments in human capital.

References

1. Khaitov, Sh.N.: Development of innovative processes and their impact on the labour quality. Scientific results. Econ. Res. **6**(2), 13–18 (2020)
2. Abdurakhmanova, G., Shayusupova, N., Irmatova, A., Rustamov, D.: The role of the digital economy in the development of the human capital market. Int. J. Psychol. Rehabil. **24**(07), 8043–8051 (2020)
3. Schultz, T.W.: Investment in human capital. Am. Econ. Rev. **L1**(1), 1–17 (1961)
4. Kuzminov, Y.I., Ovcharova, L.N., Jacobson (eds.): Human LI capital as a factor in the social and economic development. In: Brief Version of the Report: Speech to the XVII April International Scientific Conference on the Economic and Social Development Problems. Publishing House of the Higher School of Economics (2016)

5. Posysaev, M.A., Kudryashova, O.K.: Modern investments in human capital. In: Problems of the Economic Growth and Sustainable Development of Territories: Files of the III International Research-to-Practice Conference. Internet conferences. Part I. FGBUN VolRC RAS (Vologda Scientific Centre of the Russian Academy of Sciences), Vologda, pp. 303–304 (2019)
6. Sleptsova, E.V., Ryndina, T.I.: State policy for the human capital development in Russia. J. Econ. Bus. **3–1**(61), 180–182 (2020)
7. Ovchinnikova, O.P., Ovchinnikova, N.E.: Stages of human capital development in the economic history. Newsletter of PNRPU (Perm National Research Polytechnic University). Soc. Econ. Sci. **3**, 338–349 (2018)
8. Chechina, O.S.: Human capital management model for the innovative development of sectoral economic systems. Innov. Invest. **3**, 268–271 (2012)
9. Becker, G.S.: Human Capital: A Theoretical and Empirical Analysis, with Special Reference to Education. University of Chicago Press, New York (1975). http://www.nber.org/chapters/c3730.pdf
10. Aleshko, R., Petrova, L., Ivanova, E., Plotnikova, A., Melnikov, M., Antonov, V.: Human capital in the digital economy format. Int. J. Eng. Adv. Technol. **9**(1), 7517–7523 (2019). https://doi.org/10.35940/ijeat.A2201.109119
11. Khudyakova, E.G.: Assessment of human capital and its effectiveness for the innovative development of an enterprise. Abstract of a Ph.D. thesis in Economics, Ufa (2019)
12. Leonidova, G.V., Golovchin, M.A.: National education project and its potential impact on the human capital development. Territory Dev. Proble. **4**(102), 7–25 (2019)
13. Gretchenko, A.I., Kaverina, N.A.: Implementation of the principles of social justice in contemporary social policy (by the example of the national project Human Capital). Vestnik NSUEU **2**, 71–85 (2020)
14. Hilorme, T., Perevozova, I., Shpak, L., Mokhnenko, A., Korovchuk, Y.: Human capital cost accounting in the company management system. Acad. Account. Fin. Stud. J. **23**(2), 1–6 (2019)
15. Sergeeva, S., Vagaeva, O., Dianova, Y., Denisov, M., Voskrekasenko, O.: Professionalism as a personal priority of a higher school teacher and a condition for high-quality training of a graduate. In: The Third Annual International Symposium "Education and City: Education and Quality of Living in the City" (2020)
16. Bugaevsky, K.A., Peshikov, O.V.: History of Nobel Prize winners in economics as reflected in collectibles. Manag. Mod. Syste. **2**(26), 3–13 (2020)
17. Krakovskaya, I.N.: Principles of managing the human capital investments. Scientific and Technical Reports of SPbSPU (Peter the Great St. Petersburg Polytechnic University). Ser. Econ. Sci. **4**, 184–190 (2007)

Development of Emigration Processes in Russia Under Conditions of an Unstable Epidemiological Situation

Lidia N. Roshchina(✉), Aleksander M. Basenko, Svetlana V. Vishnyakova, Svetlana V. Solyonaya, and Konstantin D. Andryushchenko

Rostov State University of Economics, Rostov-on-Don, Russia

Abstract. An increase in the emigration of citizens from Russia due to the spread of the new COVID-19 infection, therefore, the article attempts to consider the problem of the emigration outflow of citizens from Russia in terms of a worsening epidemiological situation. The process of emigration outflow from Russia during the period of an unstable global epidemiological situation was studied. Method of statistical analysis of Rosstat and the Organization for Economic Cooperation and Development databases; a computational method, methods of comparison and description, content analysis method of scientific articles were used. The negative consequences of emigration processes occurring in Russia due to the ongoing COVID-19 pandemic have been identified. An attempt was made to determine the scale of emigration from Russia to the most demanded geographical directions in 2020. An assessment of the real economic losses from the emigration of citizens from Russia was made based on predecessor methods. The article analyzes the actual problem of objective assessment of the scale of emigration from Russia and thereby substantiates the need to develop a fundamentally new approach to the regulation of the emigration processes in Russia in order to prevent uncontrolled escalation of their scale. The scientific novelty of the research is the definition of principles for the development of a new approach to the regulation of emigration processes aimed at improving social, financial and economic instruments through the adoption for the future of the national migration policy in the field of emigration by authorized federal government bodies of Russia.

Keywords: Emigration from Russia · COVID-19 · Scale and consequences · Improvement of Russia's migration policy

1 Introduction

The deterioration of the global epidemiological situation hinders the systematic development of international migration. Thus, the rapid spread of COVID-19 caused a sharp and spontaneous increase in migration flows under the threat of a subsequent radical restriction of the international mobility of citizens.

The relevance of the study lies in the fact that Russia, acting as a recipient and donor of a significant number of international migrants, also faced the unprecedented challenge

P. V. Trifonov and M. V. Charaeva (Eds.): DITEM 2021, LNNS 380, pp. 225–232, 2022.
https://doi.org/10.1007/978-3-030-94245-8_31

of COVID-19, which negatively affected all indicators of socio-economic development, including international migration.

For a more detailed study of the problem, the works of Voronina, Suvorova, Volokh [1], Soldatova, Pivkina [2], Maslennikov, Linnikov, Maslennikov [3], Sturova [4], Vorobyeva, Rybakovsky [5], Docquier, Lodigiani [6] and other researchers addressing various aspects of the impact of COVID-19 on the development of emigration from Russia.

The novelty of the study includes the establishment of the scale of emigration from Russia during the COVID-19 period, the identification of the urgent problems, consequences and real economic losses from emigration, as well as determination of the principles for developing a new approach to regulating emigration from Russia.

Hypothesis. There is an assumption that the COVID-19 pandemic provoked an increase in the emigration of Russian and foreign citizens from Russia, which led to an aggravation of social, financial and economic problems within the country.

The goal of this research is to analyze the scale, consequences and problems of the development of emigration from Russia during the COVID-19 period.

Problems:

1. to conduct a statistical analysis of the scale of emigration from Russia during the COVID-19 period (2020);
2. to identify the volumes of the emigration outflow of net capital from Russia in 2020;
3. to calculate real economic losses as a result of the emigration of citizens from Russia during the period of COVID-19;
4. to determine the most popular emigration destinations in 2020;
5. to identify urgent problems and consequences of emigration and propose possible ways to resolve them.

2 Methods

The following methods were used to attain the objectives: the method of statistical analysis made it possible to determine the scale and the most demanded directions of emigration of citizens from Russia during the COVID-19 period, as well as to identify the volumes of the emigration outflow of net capital from Russia in 2020; the comparison method made it possible to correlate the data obtained in the course of statistical analysis with similar indicators in 2019; using the methods of content analysis of Russian and foreign literary sources and the method of description, urgent problems and consequences of emigration from Russia and suggest possible ways to resolve them were identified; based on the use of a computational (numerical) method, the calculation of real economic losses as a result of the emigration of citizens from Russia during the period of COVID-19 was made.

The selection of literary sources was carried out on the basis of the criteria of relevance, novelty, reliability and accessibility. All sources used were published in peer-reviewed publications (RSCI, SCADT, Web of Science) in the period 2016–2021.

3 Results

According to the Rosstat report "Socio-economic situation in Russia", the number of emigrants from Russia in the period from January to November 2020. increased by 73.2 thousand people, or by 20.0% in comparison with the indicator for the same time period in 2019 (Table 1) [7].

Table 1. The number of citizens who left Russia from January to November 2019 and 2020.

	The number of citizens who left Russia (Jan-Nov 2019)	The number of citizens who left Russia (Jan-Nov 2020)	Growth, %
Total	366789	439995	+20%
To the CIS countries	321687	378531	+18%
To non-CIS countries	45102	61464	+36%

Obviously, a significant part of those who left are labor migrants who previously stayed in Russia and were forced to re-emigrate due to the deterioration of the sanitary and epidemiological situation.

Nevertheless, the thesis is true that emigration from Russia to non-CIS countries in 2020 has acquired completely new dimensions, having reached the maximum indicator in the last 10 years (Fig. 1).

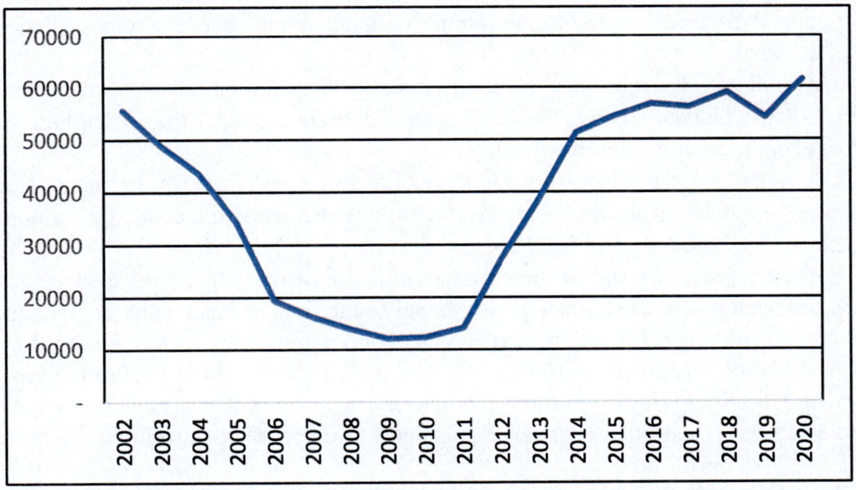

Fig. 1. Dynamics of emigration of citizens from Russia to non-CIS countries (2002–2020). *Source:* authors compilation, based on data from the Federal State Statistics Service [7].

In particular, in 2020, Vietnam, India, China and Syria were the most demanded destinations for emigration from Russia among the non-CIS countries (Fig. 2).

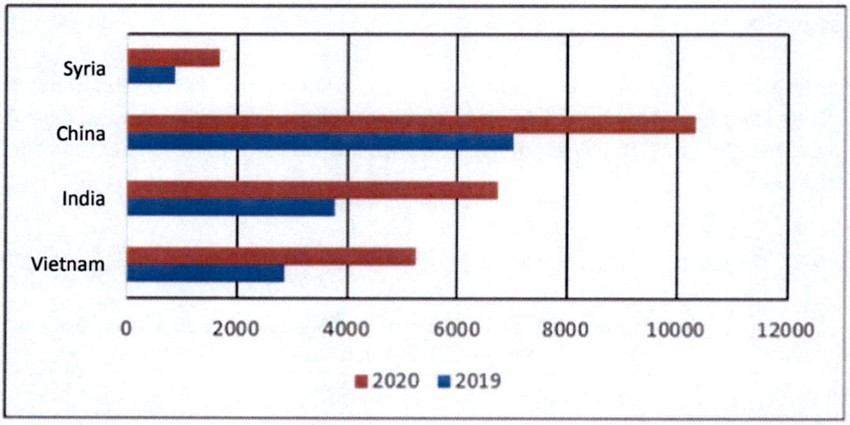

Fig. 2. Emigration from Russia to non-CIS countries in 2020 in comparison with 2019. *Source:* authors compilation, based on data from the Federal State Statistics Service [7].

In accordance with the official data from the United Nations (UN), the demanded directions of emigration from Russia (exclusively for persons with Russian citizenship) in 2020 include Germany, USA, Israel, Spain and Canada [8].

When examining the scale of emigration from Russia during the COVID-19 period, the continuing complexity of their objective assessment should be noted.

This is due to the fact that Rosstat is mainly based on data on the place of registration of citizens when conducting a statistical analysis of emigration. Based on this, citizens registered in Russia who actually live abroad are not identified as emigrants. Consequently, the real scale of emigration to a large extent exceeds the indicators of official statistics.

This indicates the relevance of the increasing emigration outflow of citizens from the country, which entails a number of negative consequences for the social, financial and economic development of Russia.

One of them is the reduction of highly qualified specialists among Russian citizens and representatives of the working class in various sectors of the country's economy against the background of COVID-19.

For example, as a result of the emigration of Russian highly qualified specialists, there has been a serious shortage of personnel in the field of healthcare, engineering, computer technology (IT), business analytics, programming and communication [9].

At the same time, in view of the massive outflow from Russia of representatives of working specialties (citizens of foreign countries), there was a shortage of employees of food enterprises, seamstresses, service personnel, workers in the agro-industrial sector [10].

A further escalation of the outflow of able-bodied citizens from Russia may soon lead to a prolonged recession in the development of national production, and, accordingly, to a dramatic increase in consumer prices for goods and services [1].

An equally destructive consequence of emigration is the reduction of national capital due to its export abroad, including by the private sector (individuals) [11].

Capital outflow can take various forms, including concealment of profits legally obtained in the domestic market and the export of funds abroad, remittances made by foreign migrants working in Russia, etc.

Thus, according to TASS, "net capital outflow from Russia by the private sector for the six months of 2020 increased by 24% compared to last year and amounted to $28.9 billion against $23.3 billion (at the rate of 65 rubles/dollar ≈ 1.514 trillion rubles) for the first half of 2019" (Table 2) [12].

Table 2. Comparison of the volume of net capital outflow by the private sector from Russia in the period from January to July 2019 and 2020.

Net capital outflow by the private sector in 2019, billions of dollars	Net capital outflow by the private sector in 2020, billions of dollars	Growth (%)
23.3	28.9	+24%

According to the Central Bank of the Russian Federation (July 2020), the volume of remittances abroad made by labor migrants working in Russia amounted to $837 million [13].

Correlating this indicator with the data in Table 2, we get that the share of remittances made by migrants in the period from January to July 2020 in the total volume of net capital outflows by the private sector is approximately 2.9%, while the share of other forms of capital outflows accounts for almost 87%.

Actual capital outflows can be large because large amounts of capital are exported illegally. The approximate estimate of the illegally exported capital is $105 billion [14].

The most obvious reason for the emigration capital outflow is the emergence of citizens of the need to invest in a more stable economy of other countries in order to avoid the complete loss of their savings against the background of COVID-19 [6].

Referring to the methodology of [3], estimate the real economic losses from emigration from Russia during the COVID-19 period, based on the economic indicators of 2020 and according to the following formula:

$$\text{CLE} = \text{PCGDPCP} * (\text{ARA} - \text{AAE}) * \text{Crhdi} \tag{1}$$

where *CLE* – the average cost of the "life" of one emigrant (economic losses from the emigration of one citizen);

PCGDPCP – per capita GDP in constant prices (728 thousand rubles in 2020);

ARA – average retirement age (≈62.5 years);

AAE – the average age of an emigrant (≈35 years);

Crhdi – coefficient of the ratio of human development indices (human development index of Russia in 2020 – 0.824).

Substitute the data into the formula presented earlier and make the calculation.

$$\text{CLE} = 728000 * (62.5 - 35) = 20.02 \text{ million rubles.}$$

Consequently, the average "cost" of economic losses as a result of the emigration of one person in 2020 amounted to approximately 20.02 million rubles.

The index of Ukraine (0.779) was chosen as a correlated human development index, since in 2020 the largest number of Russian citizens emigrated there.

Having made the calculation, the authors found that the average "cost" of economic losses from the emigration of one person to Ukraine during the pandemic is 15.56 million rubles.

These estimates are not unambiguously accurate, since they do not take into account the individual characteristics of each emigrant, including age, education and qualifications. Nevertheless, the application of this methodology makes it possible to approximately determine the average values of economic losses from the emigration of citizens from Russia.

According to the analysis, the development of the processes of emigration from Russia is accompanied by a number of social, financial and economic problems.

4 Discussion

The problems identified as a result of the analysis can be considered as grounds for taking measures to improve the modern migration policy of Russia, which does not provide for tools to curb the emigration outflow of citizens, in particular, in an unstable epidemiological situation.

Such instruments could be state preventive measures to curb labor emigration in the event of an acute shortage of labor within the country. For example, one of the preventive measures could be the introduction of the prohibition on the emigration of representatives of certain specialties, the acute shortage of which is experienced by certain sectors of the economy. Assuming that the introduction of such a measure will largely contribute to the resolution of a number of national economic problems, there is an understanding of the negative consequences that will entail the limitation of the international mobility of Russian citizens.

Thus, there is a likelihood of the emergence of illegal channels of international migration [15], which will increase the risk of the spread of dangerous diseases and pandemics that endanger the life and health of citizens around the world [5], as evidenced by the consequences of COVID-19.

Given the complex nature of the problem, there is a need to develop a fundamentally new approach to its solution [4]. It should be based on the following principles: legality, flexibility, implementation stages, moderate loyalty.

Legality is understood as the development and adoption of measures of emigration regulation through their mandatory inclusion in regulatory sources.

The principle of flexibility means the possibility of making adjustments to the structure of the applied regulatory measures in connection with the occurrence of unforeseen circumstances (for example, a worsening of the epidemiological situation).

Due to the fact that the problem of emigration from Russia is comprehensive, it is important to observe the principle of phasing, conditionally highlighting specific levels of this problem, at each of which the development and testing of regulatory measures will be carried out for both Russian and foreign citizens. This principle will allow to check the effectiveness of the measures being developed and make adjustments to their content at the initial levels.

The principle of loyalty is reduced to state containment of emigration by providing preferential conditions for citizens operating in the sectors of the national economy.

At first, the practical application of the listed principles can be expressed in the provision of privileges for specialists, the lack of which is experienced by strategically important sectors of the Russian economy (nuclear industry, mechanical engineering, high technology, chemical complex). As the economic effect from the application of the proposed principles increases, there will be a possibility of a systematic transition to other industries, covering a large number of specialists and workers.

Ultimately, the application of such principles will make it possible to form a powerful regulatory instrument with unlimited potential for managing emigration, and can also stimulate the development of the processes of re-emigration (return) of citizens to Russia. This will create an additional effect for the favorable development of the country, and most importantly, will allow solving a number of complex problems.

One of the most significant problems in the period of COVID-19 was the acute shortage of specialists in the field of virology in a number of regions of Russia [16]. Meanwhile, according to the Organization for Economic Cooperation and Development for 2020, 950 medical specialists carry out their professional activities in Canada, and in the USA the number of practicing doctors from Russia is 6291 [17].

Partial or complete re-emigration of doctors to Russia should be viewed exclusively in a positive way, not only as a source of development of domestic medicine, but also as a factor ensuring the provision of high-quality medical support to the population in the context of a worsening epidemiological situation.

5 Conclusion

The COVID-19 pandemic has largely contributed to the exacerbation of the problem of emigration from Russia. This was expressed in an increase in the number of emigrants, a sharp increase in the volume of exported capital, as well as in the aggravation of a number of national problems. On the other hand, COVID-19 can be perceived as a challenge for government measures to improve Russia's migration policy in the field of emigration [2].

As a result of the study, the following conclusions can be formulated:

1. The emigration outflow of citizens from Russia during the COVID-19 period (2020) increased by 20% compared to 2019.
2. There is a significant increase in the volume of net outflow of Russian capital by 24%.
3. The average "cost" of economic losses as a result of the emigration of one person during the COVID-19 period was approximately 20.02 million rubles.
4. There is a sharp increase in emigration to non-CIS countries by 36%.
5. The problem of an objective assessment of the scale of emigration from Russia has been substantiated, and the most pressing problems and consequences of emigration have been identified. The necessity of developing a new approach to the regulation of emigration processes in Russia has been substantiated. Principles for such an approach have been proposed and characterized.

Solving the existing problems and developing a new approach to migration regulation will help curb the emigration outflow of citizens from Russia and, thus, create conditions for the planned socio-economic development of the country in the near future [18].

References

1. Voronina, N.A., Suvorova, V.A., Volokh, V.A.: Power **4**, 26–32 (2020). https://doi.org/10.31171/vlast.v28i4.7422
2. Soldatova, S.S., Pivkina, K.R.: Student **2**, 260–265 (2020)
3. Maslennikov, V.V., Linnikov, A.S., Maslennikov, O.V.: Fin.: Theory Pract. **22**(2), 54–65 (2018). https://doi.org/10.26794/2587-5671-2018-22-2-54-65
4. Razin, M.V., Karavaev, E.S., Sturova, N.A.: Human Soc.-Econ. Soc. Sci. **11**(1), 92–95 (2020). https://doi.org/10.23672/q9915-2893-8497-e
5. Vorobieva, O.D., Rybakovsky, L.L., Savinkov, V.I.: Population **4**, 4–13 (2016)
6. Docquier, F., Lodigiani, E., Rapoport, H., Schiff, M.: J. Dev. Econ. **120**, 209–223 (2016)
7. Federal State Statistics Service (2021). https://rosstat.gov.ru/storage/mediabank/6WHhWc1N/oper-12-2020.pdf
8. United Nations: Population Division. https://www.un.org/en/development/desa/population/migration/data/estimates2/countryprofiles.asp
9. Izvestia: personnel request – there is a shortage of consultants and engineers in Russia (2020). https://iz.ru/977457/anna-ivushkina/kadrovyi-zapros-v-rossii-voznik-defitcit-konsultantov-i-inzhenerov
10. Rambler/finance: there is a shortage of employees in working specialties in the labor market. https://finance.rambler.ru/realty/43861261-na-rynke-truda-obrazovalsya-defitsit-sotrudnikov-rabochih-spetsialnostey/
11. Safonova, O.V., Medvedeva, M.V.: Mod. High Tech. Reg. Appl. **4**(56), 147–157 (2018)
12. Net capital outflow from Russia in January-June 2020 increased by 24%. https://tass.ru/ekonomika/8937285
13. Migrants are withdrawing money from Russia in record volumes. https://www.mk.ru/economics/2020/09/18/cb-migranty-vyvodyat-dengi-iz-rossii-v-rekordnykh-obemakh.html
14. About $105 billion is illegally withdrawn from Russia annually on average. https://www.rbc.ru/finances/09/12/2015/566824a89a7947a41a89d73b
15. Mangin, S., Zenou, Y.: Econ. Lett. **148**, 83–86 (2016)
16. A shortage of doctors has been diagnosed in the regions. https://www.rbc.ru/newspaper/2020/02/06/5e39418c9a794713cb0a26bd
17. Stat: Immigrants by detailed occupation (OECD International Migration Database). https://stats.oecd.org/Index.aspx?DataSetCode=MIG
18. Coulibaly, D., Gnimassoun, B., Mignon, V.: J. Macroecon. **66**, 32–41 (2020)

Industrial Engineering Methods for Reforming Manufacturing Subsystems of Enterprises

Anna S. Vikhrova(✉)

MIREA – Russian Technological University, Moscow, Russia
vihrova@mirea.ru

Abstract. As the tasks of the state armaments program were fulfilled in connection with the successful solution of its goal of ensuring the share of modern weapons in the Russian army on average to 70%, the volume of the state defense order after 2020 began to decline. To compensate for the load on the production capacities made available, the President of the Russian Federation set the task of bringing the share of civil and dual-use products at the enterprises of the military-industrial complex of Russia to 30% by 2025, and to 50% by 2030. This task is being successfully fulfilled: if the share of such products in 2018 was 20.9%, then in 2020 it has increased to 25.6%. At the same time, diversification takes place with different intensity in various branches of the defense industry. For example, in aircraft construction, the share of civilian products in 2019 was 34.1%, in the radio-electronic industry only 14.6%, in shipbuilding 19.1%. To resolve the current contradiction between the objective need to mobilize resources to preserve and develop the potential of the defense industry complex by accelerating the diversification and the state of production activities' organization, the article proposes to use a set of methods for reforming production subsystems of enterprises of the military-industrial complex. They are based on industrial engineering methods. Some of them are widely known and described in detail in the literature, others require additional explanation. In particular, proposed procedure establishes rational boundaries for defense enterprises' specialization.

Keywords: Engineering · Production · Subsystem · Defense · Enterprises · Outsourcing

1 Introduction

At the moment, there is a contradiction between the objective need to mobilize resources to preserve and develop the potential of the defense industry complex through accelerated diversification of military-industrial complex enterprises, increasing the competitiveness of products and the state of the organization of production activities which hinders their implementation.

The purpose of the study is to identify engineering methods that can be used when reforming the production subsystems of enterprises of the military-industrial complex.

P. V. Trifonov and M. V. Charaeva (Eds.): DITEM 2021, LNNS 380, pp. 233–238, 2022.
https://doi.org/10.1007/978-3-030-94245-8_32

2 Materials and Methods

The research materials are based on the study of scientific literature and the practice of engineering companies, which made it possible to highlight the principles of building their business model, which are useful in developing research recommendations regarding the organization of production activities at defense enterprises.

The main source of information on the activities of the military-industrial complex was the data of the Chuvash defense enterprises whose production subsystems implement the full production cycle, in addition, they represent various branches of the defense industry, integrators, and suppliers of various levels.

3 Results

Engineering as a special type of business uses specific methods used in the course of work on projects for the construction of facilities or the creation/reforming of new industries. Among them are methods of: market analysis and applicable technologies, development of economic and technological feasibility studies, conceptual design and modeling; simulation, engineering calculations and prototyping; organization of project management; reengineering of business processes; development of design and estimate documentation; setting requirements for time standards and qualifications of employees; organization of contract tendering, etc. Let us consider what methods are advisable to use to solve the problems of reforming the production subsystems of defense enterprises. An enlarged list is shown in Fig. 1.

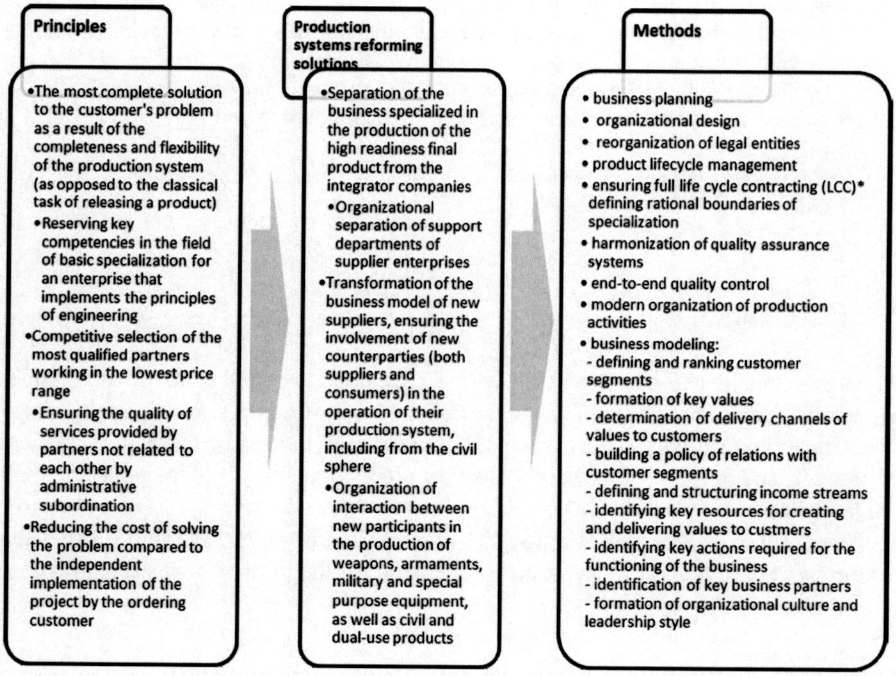

Fig. 1. Interrelation between the main categories of the research. *Source:* compiled by the author.

Some of the methods listed in the figure are well tried and tested in practice, their content and features are discussed in a large number of scientific publications. For example, on business planning, one can use the works of Petrov [1] and Ernst & Young [2]. Organizational design methods are detailed in the works of Barinov [3], Daft, Murphy, and Wilmott [4]. The works of Womack and Jones [5], Liker [6] and Kazintseva [7] are devoted to the presentation of the essence of modern concepts of organizing production activities. The regulations for the reorganization of legal entities have been thoroughly worked out.

At the same time, some questions of the implementation of the principles of engineering in reforming the production activities of the defense industry enterprises remain open. In particular, the methods for determining the rational boundaries of specialization of enterprises in their new configuration require clarification. As a rule, making a decision on the specialization of production and its reverse side, cooperation, is preceded by the calculation of the economic effect (E) from the development of specialization and cooperation according to the known formula:

$$E = [(C_1 - C_2) - (S_{tr1} - S_{tr2})]V_2 - Ef_n \Delta Ca + \Delta Pr \quad (1)$$

where C_1 and C_2 – unit cost before and after specialization;

S_{tr1} and S_{tr2} – transportation spendings per unit of production before and after specialization;

V_2 – the volume of production after specialization;

Ef_n – standard (industry-specific or corporate) capital investment efficiency ratio;

ΔCa – additional capital investments required for the implementation of production specialization;

ΔPr – additional profit obtained by improving product quality due to specialization of production [8].

At the same time, the procedure for establishing rational boundaries of specialization should include the following main stages:

1. Defining a simple or multiple product or product line that is in demand in the market within one or more industries. The complexity of the product may vary. The works of this stage are implemented using industrial marketing methods [9];
2. Calculating the required production volumes to achieve break-even operation of the allocated industries, in accordance with the methodology for calculating the multi-product break-even point [9];
3. Development of a plan of technical measures for the division of production, including the separation of production assets, power supply, the creation of additional auxiliary divisions, the solution logistics issues, regime, etc.;
4. Resolving issues of resource support for the activities of the new enterprise, including material and technical supply and personnel;
5. Determination of the composition, volumes and sources of costs required for the separation of a subsidiary (affiliate) enterprise and the organization of the initial stage of its work;
6. Carrying out an inventory of property, registering the procedure for dividing assets in accounting, drawing up a separation balance sheet;

7. Calculating the economic efficiency of a specialization decision based on the results of the previous stages;
8. Formalization of the results in the form of a economic and technological feasibility study (PFS) of the expediency of specialization (business plan) and presentation to the owners of the parent company for making a decision on the rational boundaries of specialization and further steps to create a subsidiary.

Taking into account the increase in the number of interacting economic structures in connection with the reform of the production subsystems of defense enterprises, special attention should be paid to maintaining the integrity of the design concept of the final product being created in the manufacture of its parts, assemblies and modules, as well as their quality parameters. When manufacturing a product within a single legal entity, this problem is solved by the methods provided for by the corporate quality management system. In the case of developed cooperation, the situation becomes more complicated and requires additional actions to harmonize the quality management systems of the enterprises involved in production. A combination of the following three approaches is not the only, but the most optimal way to solve this problem: CALS technologies in product design for the generation of a "digital twin" of the product and accurate reproduction of its specific elements by various participants in cooperation; methods for harmonizing quality management systems of cooperation participants and tools for organizing the process of creating, replicating, servicing and disposing of a product on the basis of life cycle contracts.

The origins of the digital twin concept date back to the beginning of the current century and reflect the evolution of the CALS and PLM methodology [10]. Digital twin technologies are just beginning to spread across the defense industry. The first successful examples of their use are available at the enterprises of JSC Russian Helicopters, JSC Concern VKO Almaz-Antey [11], JSC TVEL (State Corporation Rosatom) [12], JSC "USC" [13].

When solving the problem of harmonizing the quality management systems of the customer and suppliers, as noted in the monograph by Bobryshev and Ivanov [14], it is advisable to use three groups of methods. The first one is based on the use of corporate building technologies and it involves:

1. unification of the customer and suppliers into a vertically integrated holding;
2. acquisition of a large block of shares in supplier enterprises to indirectly influence their activities to ensure one's own interests;
3. cross exchange of manufacturer and supplier shares.

The second group of methods for ensuring product quality in the context of an expanding number of independent suppliers is based on technologies of training, coordination and control. The first is verification (audit) of the supplier's quality systems by the customer. This practice has been most developed in science-intensive and diversified industries. For example, "Airbus" prepares its suppliers to undergo an audit according to its own method, IPCA (Industrial Process Control Audit) in a special training center "Air Business Academy" [15].

The third group of quality assurance methods along the entire value chain is based on capital concentration technologies. There are first successful examples of the implementation of this idea. As noted at the 5th Forum "Information Technologies for the Defense Industrial Complex – 2016" held in Chelyabinsk, "recently a consortium was created in the Urals, a fairly new and completely different form of interaction between the members of the association. The consortium "Digital Enterprise. Typical information system of defense industry" sets the goal of creating a high-quality IT platform independent of imports for the automation of enterprises of the military-industrial complex. The consortium includes the government of the Chelyabinsk region, FSUE RFNC-VNIIEF, academician E.I. Zababakhin RFNC-VNIITF, PA "MAYAK" and the Ural system integrator "LANIT-Ural"" [16].

More specific results are demonstrated by another consortium "TransITion", which is a collective supplier of high-quality domestic "PLM-complexes of a heavy class for defense enterprises outside any industry specialization, taking into account the requirements of high-tech industries of aircraft construction, engine building, shipbuilding. In 2020, new versions of the products were released that make up the "TransITion" software package for design and product lifecycle management" [17].

A variety of the organizational and economic structures under consideration are dual-use scientific and industrial clusters, "consortia of organizations with the participation of defense industry enterprises united by common value chains, localized in a specific territory and having certain characteristic aspects" [18]. A number of scientists see the creation of such clusters as an opportunity to increase the level of development of an innovative economy in the Russian Federation [19].

4 Discussion

In contrast to the established practice of concentration of technological redistributions of a complete production cycle within one legal entity, the article proposes a different mechanism for organizing the production subsystem of a defense enterprise. It is based on the outsourcing of auxiliary functions and the organization of the work of the parent company on the principles of engineering, involving the organization of network interaction between related companies connected by the common production process. Such a formulation of the question provides for the separation of ineffective links from the operating enterprise, which, on the one hand, increases the efficiency of its main production chain, and, on the other hand, creates the basis for the transformation of the selected industries into a self-sufficient business.

5 Conclusion

When reforming the production subsystems of defense enterprises using the principles of engineering, it is necessary to use a set of specific methods that are largely characteristic of this type of business. These include such methods as: business planning, organizational design, reorganization of legal entities; product lifecycle management; ensuring full life cycle contracting; determination of the rational boundaries of specialization; comprehensive control and harmonization of quality assurance systems; modern organization of production activities and business modeling.

References

1. Petrov, K.N.: How to develop a business plan. 69 ready-made business plans. Williams, Moscow (2018)
2. Pruitt, P., Bornstein, J., Ford, B.R.: The Ernst & Young Business Plan Guide. Alpina Publisher, Moscow (2018)
3. Barinov, V.A.: Organizational Design. INFRA-M, Moscow (2019)
4. Daft, R., Murphy, D., Wilmott, H.: Organization Theory and Design. Piter, Saint Petersburg (2013)
5. Womack, J., Jones, D.: Lean Thinking. Banish Waste and Create Wealth in Your Corporation. Alpina Business Books, Moscow (2013)
6. Liker, J.: Dao Toyota. 14 Management Principles of the World's Leading Company. Alpina Business Books, Moscow (2005)
7. Kazintsev, A.V.: Six Sigma in Russia. Methodology for Reducing Losses, Defects, Costs. Printing House "News", Six Sigma Association, Moscow (2009)
8. Sergeev, I.V., Veretennikova, I.I.: Economics of the Organization (Enterprise). Publishing House Urait, Moscow (2016)
9. Lyukmanov, V.B., Mandych, I.A., Bykov, V.M., Vikhrova, A.S.: Financ. Manage. **2**, 23–34 (2016)
10. Vichugova, A.: Digitalization of production and digital twins: combining PLM, IoT and Big Data (2020). https://www.bigdataschool.ru/blog/digital-twin-plm-big-data.html
11. Balaki, D.A., Kersky, E.V.: J. "Almaz – Antey" **1**, 10–18 (2020)
12. Borovkov, A.I.: Strategies **1**, 34–38 (2020)
13. Borovkov, A.I., Ryabov, Y.A., Ageev, A.B.: Information technology in shipbuilding-2019. In: Materials of the 20th International Scientific-Practical Conference "Morintech-Practice". Development and Application of Digitalis Twins in Nautical Architecture and Shipbuilding, Saint Petersburg, pp. 9–14 (2019)
14. Bobryshev, A.D., Ivanov, I.A.: Organizational Technologies for Involving Supplier Counterparties in the Work of the Company's Production System to Ensure the Quality and Competitiveness of Products. MALMIT, Moscow (2015)
15. The difference between QMS audit for compliance with customer requirements (2020). https://aviationunion.ru/news_second.php?new=17759
16. "Digital Enterprise" Consortium: implementation of the state's strategy for import substitution in defense industry. https://xn--h1aelen.xn--p1ai/wp-content/uploads/2017/05/SC_2016-14-15.pdf
17. "TransITion" Consortium at the 9th Forum on digitalization of the military-industrial complex "ITMIC-2020". https://ascon.ru/news_and_events/news/3144
18. Antipina, E.A., Zhurenkov, D.A., Sheloumov, M.A.: Gov. Order: Manag. Plac. Prov. **56**, 126–131 (2019)
19. Mandych, I.A., Bykova, A.V.: Russ. Tech. J. **9**(2), 88–95 (2021). https://doi.org/10.32362/2500-316X-2021-9-2-88-95

Assessment of Economic Advantages of Implementation of Technology for Production of Manufactured Articles from Biodegradable Materials

Svetlana V. Deminova(✉), Elena A. Kirpichenko, Liudmila A. Altynnikova, Mariia E. Khanenko, and Oksana A. Stepanova

Orel State University of Economics and Trade, Orel, Russia

Abstract. The article is concerned with the study of the situation on disposable tableware market in terms of modern environmental problems and economic realities. The process of formation of this industry on the territory of Russia and modern problems of its functioning are considered. Much attention is paid to considering the specifics of making a business in this area. An assessment of the economic benefits of introducing a technology for the production of manufactured articles from biodegradable materials based on a reduction in the production of disposable plastic tableware with a subsequent transition to the use of biodegradable samples has been made. Under the influence of various factors, the disintegration of tableware made of biodegradable materials occurs in a month. The main components from the point of view of raw materials are the following: corn starch, polypropylene, food additives. The basis is corn starch, its share is at least 72%. This type of raw material has all the necessary qualities and the most important criterion is absolute safety for the consumer. The content of polypropylene in this type of production is insignificant and, in combination with other materials, it becomes biodegradable. Strength and durability can also be noted as the advantages of this tableware.

Keywords: Investment project · Economic efficiency · Discounted approach · Project payback · Disposable tableware · Plastic tableware · Biodegradable materials

1 Introduction

The modern economic realities of the activities of manufacturing enterprises are associated with cardinal changes in a number of factors, in particular, the introduction of various kinds of innovations, new technologies, the growth of competition, etc. [1].

The historical aspect of the development of the Russian market testifies to the fact that up to 1998 it functioned due to imports. A change in the situation was facilitated by the introduction of a 70% duty on the import of the above category of goods. This fact contributed to the development of domestic production.

P. V. Trifonov and M. V. Charaeva (Eds.): DITEM 2021, LNNS 380, pp. 239–245, 2022.
https://doi.org/10.1007/978-3-030-94245-8_33

Official research on the topic "Russian market of disposable tableware: results of 2019, forecast until 2022" carried out by NeoAnalytics suggests that in 2019 the volume of disposable tableware market amounted to 38 billion. rub. and, in general, domestic producers dominated the market as their share was 86.3%.

This situation also developed due to the growth of domestic producers in the market and the increased production volume, as well as a 50% duty on the value of products from foreign manufacturers. In general, 2019 is characterized by a 1% increase in the production of disposable tableware, namely, 14.5 million pieces were produced. Based on the situation, plastic disposable tableware prevails in the domestic market (80% of the market volume) – the tableware in which the main material is polystyrene and polypropylene [2].

The Russian market is also characterized by the monopolization of suppliers of raw materials for the production of disposable tableware. Polypropylene is produced by only three plants in Russia.

Belokurova and Solokhin note in their research: "many plastic products can contain harmful stabilizers, heavy metals salts and other toxic substances, and all this when heated, especially when reused, can enter the body, namely, therefore, disposable tableware cannot be reused" [3].

At the same time, the development of a new niche in the market for this type of product can be noted – the production of tableware using biodegradable materials. This production is characterized by a high degree of environmental friendliness, ease of the recycling process.

Thus, according to a number of authors, in particular, Ivanov and Nikitin, "based on this, biodegradable tableware is a great solution for personal use by people and in business. It can be made from biodegradable polymer materials obtained from lignin-containing biomass (straw, sunflower peelings, flax shive, etc.) with the addition of a small amount of synthetic polymer materials such as polyethylene and its copolymers" [4]. However, this type of production is more expensive.

Therefore, as Ivanova notes in her work: "it is advisable for the manufacturers of plastic tableware to provide state support for their development of the production of substitute goods (paper and biodegradable tableware), which should include similar measures, and it is advisable to use an environmental tax to incite such a transition" [5].

Summing up the above it can be stated that it is advisable to develop the production of disposable tableware, including those made from biodegradable materials. Moreover, various authors, e.g. Nechaeva, note in their studies that "an acute problem in Russia is the disposal of waste, including disposable tableware (regardless of the material used)" [6].

2 Materials and Methods

Consider the implementation process of a project for disposable tableware production from biodegradable materials using the example of Introplastica LLC [7]. The IP GROUP enterprise is one of the leaders in the Russian Federation among companies producing the disposable tableware, paper cups and products for baking. The manufactured products are represented by a wide assortment in the middle and low price segments of the Retail and HoReCa markets.

When starting a new project, it is required to perform a number of activities aimed at:

- conducting market analysis;
- assessment of the competitive situation in the region;
- development of a business plan detailing the necessary financial calculations;
- purchase of the necessary production equipment, raw materials;
- attraction of personnel with the required qualifications;
- development of measures for the further promotion of products, etc.

Investment project development as the main form of investment activity is the main point that ensures the further development of the enterprise.

In general, as Pereverzeva notes in her works, "an investment project is a justification of the economic feasibility, volume and timing of capital investments, including the necessary project documentation developed in accordance with the legislation of the Russian Federation, as well as a description of practical actions for implementation of investments (business plan)" [8].

Thus, the basis for making investment decisions, including for the enterprises engaged in the production of disposable tableware, is an assessment of the economic efficiency of investments, that is, the ratio of the results obtained to the costs incurred, resources, taking into account the cost of the capital used to implement the project. An investment project can be considered effective if it is profitable [9–11].

The economic efficiency of investments is a complex indicator, since for its calculation the influence of many factors, changes in the economic situation, including in the taxation system should be considered. The complexity is also due to the fact that the calculation takes into account the information corresponding to a long period [12–14].

In general, the sequence of projects efficiency assessment should be presented in the following sequence:

1. Forecasting cash flows based on the collection of the necessary information.
2. Calculation of indicators and their subsequent analysis.

It should be noted that when comparing income and payments over the investment project life, attention should be paid to the comparison of the rate, changes in inflation and fluctuations in loan rates. The main concept that determines the efficiency of investments is the indicator of the net present value.

3 Results

Consider the prospects for the implementation of an investment project for the manufacture of disposable tableware from biodegradable materials on the basis of practical data provided by the enterprise.

As shown above, products made from biodegradable materials are of particular importance. Following the latest trends, "IntroPlastika" LLC is constantly developing in this direction. For example, a project was proposed for the production of biodegradable tableware with the following parameters:

- main material – corn starch;
- length – 165 mm;
- fork width – 25 mm, spoon width – 38 mm, knife width – 20 mm.

In addition to corn starch (72%), the material and raw material base includes polypropylene (8%) and food additives (20%). Various equipment is involved in the production process, namely:

- extruder (rolls out the mass mixed from the components of raw materials);
- thermoforming machine (forms fabricated parts);
- compressor;
- molds.

Even given the obviousness of the prospects of this project, its effectiveness should be evaluated using software tools. First of all, the current costs of the project were analyzed.

The data from the analysis show that the estimated time frame for the project is 4 years. It is assumed that the total expenses for the production and implementation of the project will amount to 12,879 thousand rubles, while their value will be practically the same during the period under review.

Further, the effectiveness of the project was assessed based on the inflow and outflow of funds within its life cycle.

Analyzing the cash flow, it should be kept in mind that its growth rate turns into stagnation after the second year of the estimated plans for the production of biodegradable tableware. This fact may indicate the risks associated with the implementation of the project. Therefore, it is advisable to analyze the forecast data of the assets and liabilities of the organization, as well as the financial results of its activities.

For example, according to the results of the last year of the project, the total assets will grow from 399 thousand rubles to 6542 thousand rubles and the company's own funds will rise from a negative mark to 5748 thousand rubles. The accounts payable will be four times lower than the accounts receivable, and their constant dynamic level will remain. All of these points are characterized exclusively from the positive side.

A graphical representation of the project income and expenses dynamics is presented in Fig. 1.

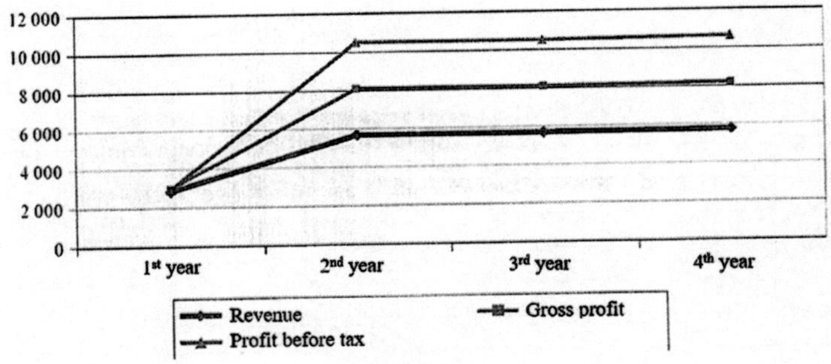

Fig. 1. Project revenues and costs analysis.

The obtained tabular and graphic data confirm the profitability of the proposed project. It is assumed that during the second year of implementation, it will recoup the costs and bring a net profit in the amount of 1,929 thousand rubles. Furthermore, there will be positive dynamics in the increment of the financial results of the research object.

In addition to making a profit when making investments in the project, the goal is to achieve and maintain a sustainable financial position of the organization, which is assessed by calculating the system of financial state indicators.

The return on assets ratio reflects that the level of return on total investment in the project for a specified period of time will be 34.8%. The return on equity shows each one ruble of the organization's own funds will bring 41 kopecks of net profit during the implementation of the project. The return on sales shows that the share of profit in the total revenue of the research subject will be 33.6%.

In addition to the system of financial and economic indicators for analyzing the project efficiency, it is advisable to use a discounted approach, the essence of which is to assess the ratio of the results obtained with the value of the costs incurred in terms of the interests of various groups of project participants. In this case, the criterion for the efficiency of investment in a project is the net present value (NPV), which is the amount of discounted net income over its entire life.

The data obtained show that the payback period will be 1.27 years. For the specified period, the amount of net profit and depreciation will compensate for the investment. At the end of the fourth year net cash flow taking into account inflationary processes will amount to 3124 thousand rubles, in turn, according to forecasts, the net present value (NPV) will be 4254 thousand rubles. The positive value of this indicator indicates the feasibility of investing money in a project for the production and sale of biodegradable tableware.

In the course of the study, an assessment of the degree of the project's resistance to various influences was carried out, as well as the analysis of its sensitivity by constructing break-even boundaries, determining the maximum levels of production volumes.

This process included the selection of a key indicator of investment efficiency, the selection of factors whose magnitude is in doubt, the establishment of their upper and lower boundaries, the calculation of the key indicator, as well as the construction of a project sensitivity graph (Fig. 2).

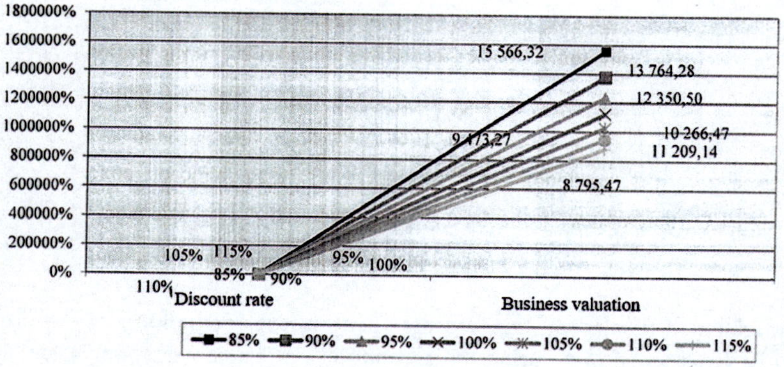

Fig. 2. Graph of the sensitivity of the project.

Thus, the financial and economic assessment showed that almost all indicators have a positive trend and high values, investment in the production and sale of biodegradable tableware will pay off before the end of the project.

For each business entity, the implementation of any project is associated with the involvement of additional material and financial resources. Undoubtedly, the goal of this process is to maximize profit with minimal cost, that is, in addition to the implementation of the project as such, its effective solution is the main priority [15, 16]. In turn, the system of economic efficiency indicators characterizes the results of the organization's functioning in absolute and relative terms, thereby making it possible to assess the potential for the implementation of projects of various orientations.

References

1. Alekhina, L.L., Burtseva, M.N., Savina, A.G., Ilminskaya, S.A., Skobliakova, I.V.: Directions of perspective development of innovational infrastructure of Russian regions. In: Popkova, E.G. (ed.) ISC 2018. LNNS, vol. 57, pp. 219–228. Springer, Cham (2019). https://doi.org/10.1007/978-3-030-00102-5_23
2. Irinina, O.I., Sukhanova, K.A.: Eco-friendly dishes and packaging for food products: realities and prospects. Serv. PLUS **3**, 65–75 (2020). https://doi.org/10.24411/2413-693X-2020-10308
3. Belokurova, E.V., Solokhin, S.A.: Disposable plastic tableware – a danger to the environment and human health. Mod. Technol. Civ. Defense Emerg. Response **1**(5), 24–27 (2014). https://www.elibrary.ru/item.asp?id=25655094
4. Ivanova, E.S., Nikitin, D.V., Chirkin, A.M., Khairov, R.F., Kuzmenko, S.L.: The choice of equipment for the production of disposable tableware. Dostizheniya Nauki i Obrazovaniya **12**(34), 7–8 (2018). https://www.elibrary.ru/item.asp?id=35646931
5. Ivanova, O.A.: Economic measures to stimulate the production of biodegradable disposable tableware. Entrepreneur's Guide **13**, 157–163 (2020). https://doi.org/10.24182/2073-9885-2020-13-1-157-163

6. Nechaeva, E.O.: The impact of disposable tableware on the environment. Vestnik Sovremennykh Issledovaniy **1.12**(28), 17–19 (2019). https://www.elibrary.ru/item.asp?id=36944978
7. LLC: IntroPlastika. https://introplastik.ru/
8. Pereverzeva, V.V.: The mechanism of project financing in the implementation of investment projects. Vestnik Rossiyskogo Ekonomicheskogo Univ. G. V. Plekhanova **3**(99), 24–31 (2019). https://www.elibrary.ru/item.asp?id=35095327
9. Popkova, E.G. (ed.): Growth Poles of the Global Economy: Emergence, Changes and Future Perspectives. LNNS, vol. 73. Springer, Cham (2020). https://doi.org/10.1007/978-3-030-15160-7
10. Soboleva, Y.P., Parshutina, I.G.: Management of investment attractiveness of the region by improving company strategic planning. Indian J. Sci. Technol. **9**(14), 1–11 (2016). https://doi.org/10.17485/rjst/2016/v9i14/91522/
11. Tsoloev, M.: Regulation of insurance activities in Russia. Entrepreneur's Guide **13**(2), 104–110 (2020). https://doi.org/10.24182/2073-9885-2020-13-2-104-110
12. Kovalevskaya, N.Y., Molokova, D.N.: Budget efficiency as the main criterion for choosing projects for integrated development of the territory. In: Proceedings of the 7th Online International Scientifically-Practical Conference: Problems of the Economy and Construction Management in Conditions of Ecologically Oriented Development, pp 253–259 (2020). http://lib-catalog.bgu.ru
13. Kovalevskaya, N.Y., Vasilenko, O.A.: Methodological aspects of investment efficiency assessment: project approach. In: Proceedings of the 7th Online International Scientifically-Practical Conference: Problems of the Economy and Construction Management in Conditions of Ecologically Oriented Development, pp 242–252 (2020). http://lib-catalog.bgu.ru
14. Malyavkina, L.I., Ilyukhina, I.B., Marchenkova, L.M., Samorodova, E.M., Starikova, L.I.: Modeling well-balanced development of innovational and investment activities of economic systems. In: Popkova, E.G. (ed.) ISC 2018. LNNS, vol. 57, pp. 883–895. Springer, Cham (2019). https://doi.org/10.1007/978-3-030-00102-5_95
15. Golaydo, I., Parshutina, I,. Gudimenko, G., Lazarenko, A., Shelepina, N.: Evaluation, forecasting and management of the investment potential of the territory. J. Appl. Econ. Sci. **2**, 618–635 (2017). https://www.elibrary.ru/item.asp?id=41773416
16. Lazarenko, A.L., Orlova, S.A., Rykova, I.A., Golaydo, I.M., Uvarova, E.E.: Increase of effectiveness of public welfare through improvement of the system of investment management. In: Popkova, E.G. (ed.) HOSMC 2017. AISC, vol. 622, pp. 16–29. Springer, Cham (2018). https://doi.org/10.1007/978-3-319-75383-6_2

Modern Strategies of Socio-economic Development

Breakthrough Technologies for Socio-economic Development of Russia Based on Personalized Pension Accounts

Evgeny V. Kostyrin(✉) and Evgeny V. Sokolov

Bauman Moscow State Technical University (National Research University), Moscow, Russia
kostyrinev@bmstu.ru

Abstract. The economic effect of the accumulation of the financial resources for the provision of pensions to the citizens of the Russian Federation for 1992 to 2020 in case the system of pension provision of citizens based on personified pension accounts is used in practice instead of the existing system of pension provision is calculated in the article. All constituent elements of the economic effect of the use of personified pension accounts are described in detail and illustrated by calculations: the accumulated funds of working citizens for pension, the economic effect due to the exit of working citizens from the "shadow". The economic effect from the use of personified pension accounts from 1992 to 2020, instead of the existing system of pension for the citizens of Russia, is estimated at 992.9 trillion rubles, while the main contribution (76.05% of the total amount) accounts for the exit of working citizens from the "shadow". The introduction of personified pension accounts, along with the economic effect, will allow solving many important social tasks: to defeat poverty and sharply increase incomes of all categories of citizens of the Russian Federation by increasing labor productivity and the number of working-age citizens of the Russian Federation making contributions to the Pension Fund of the Russian Federation; to increase labor productivity in the Russian Federation; to bring working-age citizens out of the "shadow"; to accumulate funds in personified pension accounts of all categories of Russian citizens.

Keywords: Social contributions · Budgets · Investment part · Insurance pension · Economy of Russia

1 Introduction

The current pension system in the Russian Federation does not encourage working-age citizens to increase productivity and efficiency of labor since there is no accurate and transparent mechanism for calculating pensions, it is not clear how the pension ratio is formed, why, in particular, it depends on the length of service and does not depend on the number of contributions to the Pension Fund of Russia, how it changes in the case of continued employment after retirement age, how to dispose of income from investment of pension funds in the Russian economy. Other issues complicate the understanding of the future size of pension, but most importantly, there are no clear rules, factors, and tools

P. V. Trifonov and M. V. Charaeva (Eds.): DITEM 2021, LNNS 380, pp. 249–255, 2022.
https://doi.org/10.1007/978-3-030-94245-8_34

with which you can increase the size of the pension [1]. All of the above requires new approaches to pension composition, such as the use of personalized pension accounts.

Analysis of scientific works of domestic and foreign researchers [2–12] devoted to solving the problem of pension provision for working citizens showed that most researchers pay too much attention to the legal aspects of this problem, leaving without deep analysis the economic and mathematical apparatus of formation of optimal parameters and values of contributions to pension and the factors affecting them. Thus, Starodubtseva [5] carries out a comparative legal analysis of the pension system of Russia and China, identifies the main differences, features, and current trends in reforming the pension system of both states, ignoring the factors affecting the size of the old-age retirement pension, and does not offer new mechanisms for financing the pension provision of Russian citizens. Barkinkhoeva [2] compares the modern pension system of the USA and Russia, as well as the stages of the formation of the pension system in Europe. In the work of Gao, Zhang, Yang & Li [9], the influence of health and education benefits on the estimated size, structure, and redistribution of the Chinese social security system is modeled, but economic and mathematical models, mathematical and instrumental methods for determining the optimal and scientifically justified amounts of deductions for health, medical care and pension provision of citizens are not considered. The article Kostyrin, Sokolov & Zozulya [13] proves the advantage of the methodology for financing an old-age retirement pension based on deductions to personalized pension accounts, compared with the existing pension provision in the Russian Federation.

The closest point of view to the authors on the methodology of pension provision for citizens is a study conducted by Chia & Tsui [14], in which the authors modeled the minimum size of the Central Reserve Fund (CRF) necessary to meet the pension needs of older people in Singapore.

Thus, a literature review showed the lack of works aimed at the development and implementation of a new promising technology of financing pension for citizens of the Russian Federation based on personified pension accounts, as well as mechanisms for implementing effective foreign experience in managing pension accounts and pension provision financing in domestic practice based on mathematical and economic models that most accurately describe the movement of cash flows and scientifically substantiate the prospects and effectiveness of the implemented breakthrough technologies.

A unified theoretical and methodological approach to creating mathematical models of management processes of pension provision of working citizens of the Russian Federation has not been developed. The task of their scientifically accurate mathematical modeling is not finally solved, ***which determines the relevance of the research topic***. Mathematical modeling of pension financing technologies is also relevant and represents a scientific and practical problem.

The objective of this study advocates the development of breakthrough technology for socio-economic development of Russia based on personal pension accounts, which stimulates the growth of labor productivity and exit from the "shadow" of working-age citizens, as well as the evaluation of the economic effect from the use of personal pension accounts from 1992 to 2020 instead of the current system of pension provision in Russia.

2 Materials and Methods

Table 1 presents the initial data for modeling the economic effect of using the pension system based on personified pension accounts for 1992 to 2020 instead of the existing financing system of pension provision for Russian citizens. All the data presented in Table 1 are taken from open sources or calculated by the authors. Column 1 of Table 1 shows the number of the modeling row, and column 2 shows the number of the modeling year in order. The first year of modeling in Table 1 is 1992, since the regulation on the Pension Fund of Russia, which contains the procedure for paying insurance premiums by employers and citizens to the Pension Fund of Russia, was approved by Resolution No. 2122–1 of the Supreme Council of the Russian Federation of December 27, 1991 [15].

Table 1. Initial data for modeling.

1*	2	3	4	5	6	7	8	9	10	11
1	1992	6	31.6	22.7	1.3044	1.233	148.3	82.2	31.6	34.5
2	1993	58	28.0	197.2	10.664	10.212	148.5	82.7	31.4	34.4
3	1994	220	28.0	740.5	38.315	36.124	148.4	83.4	37.7	27.2
4	1995	472.4	28.0	1587.3	88.941	74.366	148.5	84.3	37.3	26.8
5	1996	790.2	28.0	2655.1	133.489	115.108	148.3	85.01	36.8	26.4
…………………………………………										
28	2019	47.9	22.0	126.4	8.693	5.394	146.8	86.2	32.0	28.6
29	2020	49.4	22.0	130.4	8.937	5.655	146.7	86.1	32.5	28.2

Notes: 1 – Item No.; 2 – Year; 3 – Average salary, RUB, in thousands/month; 4 – Contributions to the Pension Fund, %; 5 – Contributions to the Pension Fund per year, RUB, in thousands; 6 – The budget of the Pension Fund, RUB, in trillions; 7 – Contributions to mandatory pension provision in the Russian Federation, RUB, in trillions; 8 – Population, mln.; 9 – The number of the working-age population, mln.; 10 – The population is younger than the working-age, mln.; 11 – Population over the working age, mln. *Source:* Compiled by the authors.

Column 3 of Table 1 shows the size of the average monthly salary according to the official data of Rosstat [16]. Column 4 shows the percentage of deductions from the Payroll budget for working citizens of the Russian Federation to the Pension Fund.

Column 5 of Table 1 shows the number of annual contributions in rubles to the Pension Fund. The values of the budget of the Pension Fund of Russia for 1992–2020, presented in column 6 of Table 1, are taken from the relevant federal laws [17]. Column 7 of Table 1 shows the amounts of contributions to mandatory pension provision in the Russian Federation by year according to the data provided in the reports on the execution of the budget of the Pension Fund of Russia in the corresponding period [18, 19]. The population of the Russian Federation by various age categories (columns 8–11, respectively) is taken from the data of the Internet resource [20].

3 Results

The simulation results are given in Table 2. Column 3 shows the number of the working-age population making contributions to the Pension Fund of the Russian Federation, taking into account the expected exit from the "shadow" of working-age citizens of the Russian Federation for five years when introducing personalized pension accounts into the pension system of Russian citizens [13]. Column 4 of Table 2 shows the actual number of the working-age population making contributions to the Pension Fund of Russia. The data in column 4 is obtained by dividing the number of insurance premiums of enterprises, institutions, organizations, and other economic entities by the number of annual contributions to the Pension Fund of the Russian Federation. The number of contributions of working citizens transferring funds to the Pension Fund of Russia per year (column 6 of Table 2) is obtained by multiplying the number of the working-age population making contributions to the Pension Fund when switching to personalized pension accounts (column 3 of Table 2) the number of contributions to the Pension Fund per year (column 5 of Table 1). Average transfers to the Pension Fund of Russia per pensioner per month (column 8 of Table 2) obtained by dividing the number of contributions of all working citizens making contributions to the Pension Fund of Russia per year (column 6 of Table 2) by the number of pensioners (column 7 of Table 2).

Table 2. Simulation results.

1*	2	3	4	5	6	7	8	9	10	11	12
1	1992	62.6	54.2	6.0	1.43	34.5	3.443	2.4	12.51	12.76	0.178
2	1993	68.3	51.8	58.7	13.46	34.4	32.585	23.48	121.773	124.46	4.339
3	1994	73.9	48.8	220.4	54.70	27.2	167.631	88.16	1075.4	1099.68	29.675
4	1995	79.5	46.9	472.4	126.16	26.8	391.996	188.96	3511.9	3606.83	102.374
5	1996	85.1	43.4	790.2	225.94	26.4	713.130	316.08	8276.5	8538.86	260.703
...											
28	2019	86.2	42.7	47.8	10.89	28.6	31.749	19.15	2154.5	2546.94	744.910
29	2020	86.1	43.4	49.4	11.23	28.2	33.191	19.76	2315.7	2762.29	755.043

Notes: 1 – Line number; 2 – Year; 3 – The number of the working-age population making contributions to the Pension Fund when switching to personalized pension accounts, mln.; 4 – The actual number of working-age citizens making contributions to the Pension Fund, mln.; 5 – Aver-age salary, RUB, in thousands/month; 6 – The number of deductions of all working citizens transferring funds to the Pension Fund of Russia, per year, RUB, in trillions; 7 – Number of pensioners, mln.; 8 – Average transfers to the Pension Fund of Russia per pensioner per month, RUB, in millions; 9 – Average insurance pension per month (40% of salary), RUB, in thousands; 10 – The investment part of the pension per employee at the end of the year, RUB, in thousands; 11 – The accumulated part of the pension per employee, taking into account the capitalized interest on the deposit at the end of the year, RUB, in thousands; 12 – The cumulative effect of working citizens of the Russian Federation coming out of the "shadow", RUB, in trillions. *Source:* compiled by the authors.

The average insurance pension per month (column 9 of Table 2) is determined based on the need to provide a pension of no less than 40% of wages, i.e., the data presented in column 9 of Table 2 is equal to 40% of the value presented in the corresponding row in column 5 of Table 2.

The investment part of the pension per working citizen of Russia at the end of the year, presented in column 10, is the difference between the average transfers to the Pension Fund of Russia per pensioner per month (column 8) and the average insurance pension per month (column 9) multiplied by 12 (number of months per year), taking into account the balance of the investment part of the pension per working citizen of Russia at the end of the previous year. The funds of the funded part of the pension are stored on bank deposits at 2% per annum. Therefore, the amount of funds, taking into account the capitalization per employee, shown in column 11, is the investment part of the pension per employee at the end of the year with interest accrued on this balance. As can be seen from the last line of Table 2, the investment part of the pension of one working citizen for 1992 to 2020 is equal to 2,76 RUB million; thus, the accumulated amount for the pension of all working citizens will amount to 237.8 RUB trillion.

In addition to the funds accumulated for the pension of working citizens, there is a huge effect due to the exit of working citizens from the "shadow", which is shown in column 12. The effect of the exit of working citizens of the Russian Federation from the "shadow" will amount to an increasing total of 755.0 RUB trillion (last row, column 12 of Table 2).

4 Discussion

Table 3 shows all the components of the economic effect of the use of personal pension accounts during 1992 to 2020 instead of the existing RF pension system, which amounts to 992.9 RUB trillion, while the main contribution accounts for the exit of working citizens from the "shadow", the value of which is 755.0 RUB trillion, or 76.05% of the total amount.

Table 3. Structure of the economic effect.

Name of the economic effect source	Accumulated funds of working citizens for pension provision, RUB trillion	As a result of the exit of working citizens from the "shadow", RUB trillion	TOTAL, RUB trillion
Economic effect value	237.8	755.0	992.9
The share of the economic effect source in the total amount	23.95%	76.05%	100.00%

5 Conclusion

The introduction of personalized pension accounts, along with the economic effect, will allow solving a number of important social tasks:

1) defeat poverty and sharply increase the incomes of all categories of the RF citizens by increasing labor productivity and increasing the number of working-age citizens making contributions to the Pension Fund of the Russian Federation;
2) increase labor productivity in the Russian Federation;
3) remove working-age citizens from the "shadow";
4) accumulate funds on personalized pension accounts of all categories of citizens of the Russian Federation, which can be used as cheap domestic credit resources for the development of the Russian economy;
5) sharply reduce the corruption capacity of pension provision for Russian citizens, as social contributions will reach working citizens of the Russian Federation.

References

1. Sokolov, E.V., Kostyrin, E.V.: Breakthrough Technologies for Financing Healthcare, Pension Provision, and the Russian Economy. Nauchnaya biblioteka, Moscow (2018)
2. Barkinkhoeva, A.A.: Econ. Manage. Prob. Sol. **11**(1), 9–11 (2015)
3. Kuznetsova, O.N., Shloma, A.V., Kovaleva, N.N., Ermakova, L.V.: Bul. Bryansk State Univ. **3**(37), 197–203 (2018)
4. Nechaev, A.V.: Soc. Phen. **2**(9), 6–14 (2018)
5. Starodubtseva, K.A.: Pension provision in Russia and China: comparative analysis. In: Russia and China: Problems of Strategic Cooperation: A Collection of the Eastern Center, vol. 17, pp. 46–50 (2016)
6. Shirokikh, S.V.: Soc. Econ. Human J. Krasnoyarsk State Agrarian Univ. **1**(11), 133–143 (2019)
7. Albeda, W.: Europ. J. Law Econ. **2**(10), 121–124 (2000)
8. Bouchet, M., Marchiori, L., Pierrard, O.: Pension reform in a worst case scenario: public finance versus political feasibility. J. Pension Econ. Financ. **2**(16), 173–204 (2017). https://doi.org/10.1017/S1474747215000451
9. Gao, Q., Zhang, Y., Yang, S., Li, S.: The divided Chinese welfare system: do health and education change the picture? Soc Pol Soc **2**(17), 227–244 (2018). https://doi.org/10.1017/S1474746417000100
10. Garbat, M.: Paradigmata poznani **2**, 278–283 (2014)
11. Low, L.: Soc. Pol. Soc. **3**(3), 301–310 (2004). https://doi.org/10.1017/S14747464040
12. Turner, J., Guenther, R.: A comparison of early retirement pensions in the United States and Russia: the pensions of musicians. J Aging Soc Pol **4**(17), 61–74 (2005). https://doi.org/10.1300/J031v17n04_04
13. Kostyrin, E.V., Sokolov, E.V., Zozulya, V.V.: Int. J. Supply Chain Manage. **8**(3), 321–333 (2019)
14. Chia, N.C., Tsui, A.K.C.: Life annuities of compulsory savings and income adequacy of the elderly in Singapore. J Pension Econ Fin **1**(2), 41–65 (2003). https://doi.org/10.1017/S1474747203001239

15. The Supreme Council of the Russian Federation: Resolution of the Supreme Council of the Russian Federation No. 2122-December 1 27, 1991 (as amended on 05.08.2000) "Issues of the Pension Fund of the Russian Federation (Russia)" (together with the "Regulations on the Pension Fund of the Russian Federation (Russia)", "The procedure for paying insurance premiums by employers and citizens to the Pension Fund of the Russian Federation (Russia)") (1991).
16. Federal State Statistics Service. http://rosstat.gov.ru
17. Law of the Russian Federation No. 2296-February 1 6, 1992 "On a budget of the Pension Fund of the Russian Federation for Q1, 1992"
18. The Supreme Council of the Russian Federation: Resolution of the Supreme Council of the Russian Federation No. 5705-I of 10.09.1993 "On the execution of the budget of the Pension Fund of the Russian Federation for 1992" (1993)
19. Federal Law No. 21-FZ of 09.08.1994 "On the execution of the budget of the Pension Fund of the Russian Federation for 1993"
20. Population size. https://xn----8sbncanfseqabeh4bfdo0g7e4c.xn--p1ai/

New Economic Realia of the State Agricultural Policy

Nataliya A. Alekseeva(✉), Oksana G. Dolgovykh, Elena V. Aleksandrova, Larisa A. Istomina, and Zinaida A. Mironova

Izhevsk State Agricultural Academy, Izhevsk, Russia

Abstract. The government measures taken in the field of agriculture over the past two or three years have not fully met the requirements of food security and require adjustment. The purpose of the study is to substantiate the directions for improvement of the state support for agriculture. Research methods: system approach, dialectical, monographic, economic and statistical methods. As a result of the study, various scientific points of view on the role and importance of the state in the regulation of markets, individual sectors and sub-sectors of agriculture, the categories of agricultural producers, the main directions of development of agricultural economies, and the forms of innovative cooperation have been identified. The role of farming in the economies of different countries, the need for cooperation of producers has been emphasized. The main adjustments in the state program for the development of agriculture in Russia until 2025 have been considered. The official methodology for evaluation of the effectiveness of budget funds has been critically assessed based on a plan-fact analysis. The scientific novelty of the research lies in the development of recommendations for improvement, directions, forms, methods of state support provision: a shift in emphasis in supporting producers to supporting infrastructure projects, taking into account the level of demand for state support and the efficiency of budget funds spending by commodity producers and others.

Keywords: State support · Subsidies · Agricultural production cooperative · Efficiency assessment

1 Introduction

Agriculture has always been a priority and strategic sector of the economy in terms of ensuring food security of the country, growth of the export potential of the economy, a point of economic growth during the period of economic recession and state support obtaining. Agricultural commodity producers use capital at a reduced weighted average price due to subsidies [1, 2]. The priority of state financing of the industry creates the preconditions for its outstripping growth and development [3]. However, the volume, the directions of state support for agriculture, as well as the official methodology for efficiency assessment of state transfers are not adequate to its contribution to economic development. The question of the adequacy of structural and dynamic changes in agriculture to the volume, forms and directions of its state support is becoming relevant.

P. V. Trifonov and M. V. Charaeva (Eds.): DITEM 2021, LNNS 380, pp. 256–260, 2022.
https://doi.org/10.1007/978-3-030-94245-8_35

The theoretical basis of the study was the scientific works of foreign and domestic scientists on the issues of state agricultural policy and its efficiency. The scientific novelty of the research lies in the assessment of changes in the state agricultural policy in Russia and in the regions in connection with the main world challenges. The purpose of the study is to substantiate the directions of improving state support for agriculture. The objectives of the research are to study the main provisions of the concepts of agricultural regulation in the national economies of the world over the past two or three years and to develop directions for improving state support for agriculture.

2 Materials and Methods

The review of foreign and domestic scientific research in economically leading and developing countries such as the USA, China, Russia, the EU countries, India, Japan, Israel was carried out. The study involved government decrees, the state program for the development of agriculture in Russia. The empirical basis of the study is the annual reports of agricultural production cooperatives for 2014–2019, operating in the zone of risky farming in the Udmurt Republic. Special economic incentives for efficient land management are created in cooperatives. Moreover, in the advanced Western countries, the agrarian economies of which are based on individual farming, the number of people supporting the conclusion on the necessity of cooperation of the farmers grows increasingly.

3 Results

The researchers evaluated the efficiency of joint innovations of the government, research institutes, universities, colleges, enterprises, which contribute to the accelerated transformation of scientific and technological advances into production, the coordinated development of science, technology and economics, the creation of more mature technologies [4]. A number of countries have begun to adhere to a more flexible and "humanized" policy of "integral communication interventions" in the interests of innovation at the expense of aggressive and ill-considered policies of imposing technological innovation on farmers [5]. It was defined that those farmers who participate at the beginning of the supply chain as suppliers of products to the main production have higher income and productivity due to government subsidies. However, excessive subsidies reduce the profit of the farmers [6]. New principles of subsidies have been developed: subsidies should not encourage farmers to take more risks than they could take without subsidies; the insurance subsidy should not compete with other forms of direct support [7].

On the European continent, the main problem is whether a common agricultural policy in the EU increases the sustainability of national agricultural systems [8]. It was found out that the tools to support farmers in the Netherlands that were in force until 2020 contributed more to the reliability of systems than to adaptation and transformation. The instruments currently being developed in the EU should be more in line with the specifics of the policy of European states. To implement transformations in innovation systems in agriculture, missions and submissions or their combinations are required concerning agroecology, digital transformation, bioeconomy [9]. The subsidization of the use of digital technologies in agriculture is possible [10]. Digital technologies and audit of technological processes by unmanned aerial vehicles can significantly reduce the specific energy consumption per unit of production, its cost, the need for subsidies [11, 12]. A more modern way of promoting digital technologies is the trust of farmers in more accurate information than their empirical knowledge [13].

It was figured out in China, that there is a direct relationship between the volume of research on the creation of biotechnological products by private companies and an increase in demand for subsidized products [14]. The area of cultivated land near urbanized territories in China is decreasing in the long-term horizon, however, there has been an upward trend in the area used for aquaculture production [15]. In China, an attempt has been made to transform smallholders into sustainable small businesses by examining the relationships between plot size, amount of fertilizer applied by type, yield and profit. Strong producers, main crops to cultivate, cost saving and efficiency tools have been identified [16]. In a number of developing countries, they are moving from the production of grain crops to the production of high-value products in clusters in which small farmers have the opportunity to learn modern agricultural technologies [17]. The coronavirus pandemic has revealed a number of significant imbalances in the agricultural economy of India [18]: shortage and surplus of migrant labor, critical crop losses, closure of rural markets, reduced purchasing options, limited access to food, shortage in storage facilities for long-term storage of crops; price increases due to lack of products.

The mechanisms for structural transformation of agriculture have been developed in Russia: clustering of products producers and processors; improvement of taxes, depreciation and the use of depreciation charges; ranking of investment activity subjects; taking antimonopoly measures and others [19, 20]. The state program for the development of agriculture and regulation of markets for agricultural products, raw materials and food until 2025 expanded the forms of small business support through grants: the Agrostartup grant can be used to reimburse the costs of cooperatives, as well as procurement of machinery, transport, equipment from producers and its installation and delivery, purchase of cattle, purchase of agricultural products from the members of the cooperative, the project documentation development etc.; the Agroprogress grant can be used to ensure an increase in production in priority sub-sectors of the agro-industrial complex.

The rates for the subsidies distribution for grants are increased if the producers have planned certain types of work: phosphorization, gypsum plastering of sown areas, crop insurance or fulfill certain conditions. For all cattle breeding subsectors, the interest rate is reimbursed for the purchase of agricultural machinery, machinery, equipment, commercial replacement young cattle, construction, reconstruction and modernization of farms, breeding and genetic centers. A new area of support is the partial reimbursement of the direct costs incurred for the creation and modernization of a number of agricultural facilities. During the course of the program, there should be a reorientation from compensatory forms of state support to incentive subsidies. As for the incentive subsidy, milk production remains the most in demand – 63 regions chose this direction in 2021.

Nevertheless, it is believed that state support funds provide only simple reproduction [21–23]. According to academician Altukhov [24], only 75% of the amount established by the WTO rules is spent on support. In connection with the COVID-19 pandemic in 2021, support for agriculture is expected to decrease by 5.6% (17 billion rubles) in Russia. At the same time, a 1.5% increase in production is forecasted.

In Russia, a share of state support for scientific and technical development programs in the field of agriculture is low [25]. The existing mechanism for the allocation of budgetary funds is low effective when there is no ranking of organizations for the provision of targeted state support [26]. At the end of 2017, 86% of the profits of the agro-industrial complex companies were formed at the expense of subsidies received from the state.

4 Discussion

The state agriculture development program is internally contradictory in terms of goals and resources. Underdeveloped infrastructure is unlikely to expect an increase in the export potential of the industry with a reduction in support for agriculture. The federal methodology for subsidies efficiency assessment is not associated with benefits and costs [22, 27] and is less progressive than methods applied in the regions [2].

5 Conclusion

In general, the Russian practice of agricultural regulation will have to turn to greater support for the infrastructure projects and consumers, rather than producers, to incentive subsidies. As for the distribution of subsidies, it is necessary to consider the target orientation of costs, the beneficial use of natural and climatic conditions, the differentiation of recipients of subsidies according to the demand for state support and the efficiency of budget funds. The development of inter-farm cooperation should receive its own targeted support. All these are the directions for future research.

References

1. Alekseeva, N.A.: Udmurt. Sci. **2**(92), 12–14 (2020)
2. Alekseeva, N.A., Sokolov, V.A.: Manag. Acc. **2**, 4–11 (2021)
3. Theory and practice of the development of the agri-food system in the Udmurt Republic. Izhevsk (2020)
4. FuJun, W., Zhou, Y., JieLi, D.: "Government-industry-university-research- promotion" Collaborative innovation mechanism construction to promote the development of agricultural machinery technology. IFAC-PapersOnLine **51**(17), 552–559 (2018)
5. Brian, R., Cook Satizábal, P., Curnow, J.: Humanising agricultural extension: a review. World Dev. **140**, 105337 (2021). https://doi.org/10.1016/j.worlddev.2020.105337
6. Kang, K., Wang, M., Luan, X.: Comp. Indust. Eng. **152**, 107058 (2021). https://doi.org/10.1016/j.cie.2020.107058
7. Hazell, P., Varangis, P.: Glob. Food Sec. **25**, 100326 (2020). https://doi.org/10.1016/j.gfs.2019.100326
8. Buitenhuis, Y., Candel, J.L., Feindt, P.H.: Does the common agricultural policy enhance farming systems' resilience? Applying the Resilience Assessment Tool (ResAT) to a farming system case study in the Netherlands. J. Rural Stud. **80**, 314–327 (2020). https://doi.org/10.1016/j.jrurstud.2020.10.004

9. Klerkx, L., Begemann, S.: Agricult. Syst. **184**, 102901 (2020). https://doi.org/10.1016/j.agsy.2020.102901
10. Ehlers, M., Huber, R., Finger, R.: Food Policy **100**, 102019 (2021). https://doi.org/10.1016/j.foodpol.2020.102019
11. Vorotnikov, I.L.: Izvestiya Orenburg State Agricult. Univ. **4**, 35 (2019)
12. Dragaytsev, V.I.: Mach. Equip. Vil. **7**, 30–34 (2013)
13. Gardezi, M., Stock, R.: Growing algorithmic governmentality: interrogating the social construction of trust in precision agriculture. J. Rural Stud. **84**, 1–11 (2021). https://doi.org/10.1016/j.jrurstud.2021.03.004
14. Deng, H., Hu, R., Jin, Y.: Impact of government policies on private R&D investment in agricultural biotechnology: evidence from chemical and pesticide firms in China. Tech. Forec. Soc. Change **147**, 208–215 (2019). https://doi.org/10.1016/j.techfore.2019.07.011
15. Xia, M., Zhang, Y., Zou, W.: Land Use Policy **90**, 104266 (2020). https://doi.org/10.1016/j.landusepol.2019.104266
16. Zhang, Q., Chu, Y., Cui, Z.: Glob. Food Sec. **26**, 100444 (2020). https://doi.org/10.1016/j.gfs.2020.100444
17. Otsuka, K., Mubarik, A.: World Dev. Persp. **20**, 100257 (2020). https://doi.org/10.1016/j.wdp.2020.100257
18. Kumar, P., Singh, S.S., Drews, M.: Agricult. Syst. **187**, 103027 (2021). https://doi.org/10.1016/j.agsy.2020.103027
19. Kholmuminov, S.: Econ. Soc. **11**(66), 715–721 (2019)
20. Khulkhachieva, G.D.: NGIEI Bull. **3**(70), 134–142 (2017)
21. Rodionova, O.A., Pertsev, A.A.: Sci. Pract. J. **2**, 41–48 (2019)
22. Kholodova, M.A.: Econ. Ecol. Territ. Form. **4**(1), 14–24 (2020)
23. Eskiev, M.A.: Sci. Notes Crimean. Eng. Pedag. Univ. **4**(66), 288–294 (2019)
24. Altukhov, A.I.: Solving the current challenges of agriculture development, do not forget its problems. Sci. Works Free Econ. Soc. Russ. **223**(3), 488–495 (2020)
25. Zatsepina, G.N., Polyakov, S.N.: The role and importance of innovative development of the crop growing industry in agriculture. In: Karamnova, N.V. (ed.), Sustainable Development of the Region's Economy (II Shalyapin Readings): Materials of the All-Russian Scientific-Practical Conference, pp. 126–130 (2019)
26. Semin, A., Maltsev, N., Loginov, A.: Econ. Agricult. Russ. **9**, 2–11 (2019)
27. Esembekova, A.U.: The effectiveness of the implementation of state support for the agro-industrial complex of the Kurgan region. In: Sukhanova, S.F. (ed.), Scientific and Technical Support of the Agro-Industrial Complex in the Implementation of the State Program for the Development of Agriculture Until 2020: A Collection of Articles Based on the Materials of the International Scientific and Practical Conference Dedicated to the 75th Anniversary of the T.S. Maltsev Kurgan State Agricultural Academy, pp. 46–51 (2019)

Income Growth or Redistribution: What Reduces Poverty?

Lyudmila I. Nivorozhkina(✉), Aleksandra A. Tregubova, Tatiana V. Toropova, and Galina V. Pavlenko

Rostov State University of Economics, Rostov-on-Don, Russian Federation

Abstract. The trend of decline in the poverty level of Russians, which began in 2000, is the result of the governmental social policy to support vulnerable groups of people. In this context, it is of scientific interest to find out what determines these trends: an increase in incomes of all population groups or a redistribution of funds to the poorest segments? At the micro level, the answer to this question can be obtained by analyzing the response of households to these processes and thus clarify and compare the effects of income growth and their redistribution on poverty reduction. In this study, based on the Ravallion-Dutt decomposition method, the contributions of two factors to the change in the level of relative poverty and inequality between 2000 and 2017 were assessed: changes in average income and redistribution of income between subgroups of people, including taking into account the availability of hidden income. It is shown that the decrease in the level of poverty in this period was mainly ensured by the general growth of income with a weak intensity of the processes of redistribution of income from high-income groups of households to low-income groups. The impact of hidden incomes on the studied dynamics of poverty reduction manifested itself during the 2008 crisis, when significant measures were taken to support socially vulnerable groups, and the subsequent valorization of pensions in 2010. In addition, there was a tendency towards the localization of hidden incomes mainly among the poorest households.

Keywords: Poverty · Income · Redistribution · Decomposition

1 Introduction

One of the key directions of the governmental social policy is to reduce poverty and income inequality. Recent publications touch on the impact of high and rapidly growing levels of inequality in many countries of the world on the rate of economic growth and the dynamics of recovery from the economic recession caused by the COVID-19 pandemic [1]. In this context, it is of particular relevance to find out what effect the income redistribution policy has on the rate of economic growth in the short and long term [2–5].

The main mechanism for the redistribution of households' incomes is state regulation of this process: the aggregate reduction of poverty and inequality is caused by direct taxes

P. V. Trifonov and M. V. Charaeva (Eds.): DITEM 2021, LNNS 380, pp. 261–267, 2022.
https://doi.org/10.1007/978-3-030-94245-8_36

and transfers, including pensions, which account for most of government spending. So, in 2014, 16.3% of all federal budget expenditures were directed to the Pension Fund, and in 2019 – already 18.2% (as estimated according to the Ministry of Finance [6, 7]). In 2021, the Ministry of Finance plans to send almost 21% of all federal budget expenditures to the Pension Fund [8].

Since 2000, indicators of poverty and inequality (after 2007) have shown a steady decline trend (Fig. 1).

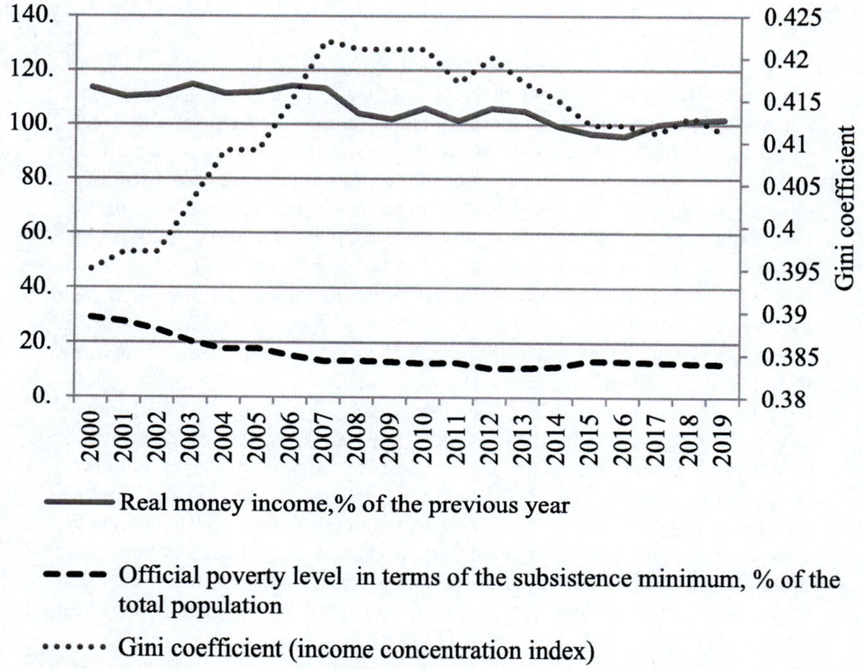

Fig. 1. Real money income, official poverty level in terms of the subsistence level, inequality in the distribution of money income, 2000–2019 (Rosstat official data [9]).

What determines these trends: a proportional increase in the income of all groups of the people, or is there a mechanism for redistributing funds to the poorest strata? The answer to these questions can be obtained by identifying how households respond directly to this process and thus clarify and compare the effects of income growth and income redistribution on poverty reduction.

It is known that in the official reporting of organizations, enterprises, tax authorities, a certain amount of cash in the hands of the population may not be reflected, i.e. hidden income. Whereas it is tax revenues that provide the state with the ability to carry out a redistributive policy. In this context, it is of scientific interest to identify how the presence of hidden incomes in the budget of Russian households affects poverty reduction, and what is the mechanism of this impact [10, 11].

The purpose of this paper is to clarify the impact of income growth effects and their redistribution on reducing the level of relative poverty, taking into account the hidden component in the income structure.

2 Methods

The most famous poverty characteristic with a number of useful analytical properties is the FGT family of indices FGT [12]:

$$FGT = \frac{1}{N}\sum_{i=1}^{n}\left[\frac{Z - Y_i}{Z}\right]^{\alpha}, \tag{1}$$

where Y_i is per capita income, N is the population size, n is the number of the poor, Z is the subsistence minimum (poverty line). For $\alpha = 0$, the index gives an estimate of the proportion of the population with incomes below the subsistence minimum; for $\alpha = 1$ – an estimate (in % of the poverty line) of the missing income (poverty gap); for $\alpha \geq 2$ – the significance of large values of the missing income increases, therefore this indicator characterizes the severity of poverty.

According to Datt and Ravallion [13], the decomposition of the change (for the period between t1 and t2) of the poverty estimate into the sum of the increment component, the redistributive component and the remainder is given by the formula:

$$P_2 - P_1 = \underbrace{[P(\mu_{t_2}, \pi_{t_1}) - P(\mu_{t_1}, \pi_{t_1})]}_{C_1} + \underbrace{[P(\mu_{t_1}, \pi_{t_2}) - P(\mu_{t_1}, \pi_{t_1})]}_{C_2} + R, \text{ for } t_1, \tag{2}$$

$$P_2 - P_1 = \underbrace{[P(\mu_{t_2}, \pi_{t_2}) - P(\mu_{t_1}, \pi_{t_2})]}_{C_1} + \underbrace{[P(\mu_{t_2}, \pi_{t_2}) - P(\mu_{t_2}, \pi_{t_1})]}_{C_2} + R, \text{ for } t_2, \tag{3}$$

where (P2-P1) is the difference in poverty between periods t_1 and t_2, C_1 is the growth effect, C_2 – is the redistribution effect, R is the indecomposable remainder, $P(\mu_{t_2}, \pi_{t_1})$ is the FGT index for the first period if the total incomes Y_i^{t1} grow μ_{t_2}/μ_{t_1} times, $P(\mu_{t_1}, \pi_{t_2})$ is the FGT index for the second period if the total incomes Y_i^{t2} grow μ_{t_1}/μ_{t_2} times.

According to the Shapley's approach [14], the exact decomposition of the FGT index into the effect of growth and distribution is given by the formula:

$$P_2 - P_1 = \underbrace{\frac{1}{2}\left([P(\mu_{t_2}, \pi_{t_1}) - P(\mu_{t_1}, \pi_{t_1})] + [P(\mu_{t_2}, \pi_{t_2}) - P(\mu_{t_1}, \pi_{t_2})]\right)}_{C_1} + \underbrace{\frac{1}{2}\left([P(\mu_{t_1}, \pi_{t_2}) - P(\mu_{t_1}, \pi_{t_1})] + [P(\mu_{t_2}, \pi_{t_2}) - P(\mu_{t_2}, \pi_{t_1})]\right)}_{C_2}. \tag{4}$$

Initial Data. The largest source of data on the living standards of the population at the micro level is the project "The Russia Longitudinal Monitoring Survey – Higher School of Economics" (RLMS-HSE) [15], which contains representative information on socio-demographic characteristics, income and expenditures of households and individuals. For the purposes of this study, an indicator of monthly disposable resources of a household (hereinafter referred to as income) was additionally formed. An income equal to 50% of the median of household income distribution was taken as the threshold (line) of relative poverty. For each decomposition, the unit of analysis is the individual, although income is measured at the household level, which is equivalent to using households as the observation unit but weighted by household size.

The initial data from 2000 to 2017 inclusive were converted into panels for three consecutive years; all subsequent calculations were carried out according to the data of the panel's final year. Households with hidden income were considered to be those in which, for two consecutive periods, current expenditures exceeded available resources by at least 10%. The methodology and empirical assessment of the size of the hidden income component of Russians, based on the Pissarides-Weber model [10], is presented in [11, 16].

3 Results and Discussion

Results reflecting the effect of growth and redistribution of household incomes are presented in Fig. 2 for all three measures of poverty. Decomposition was performed using the Shapley's approach to obtain an accurate decomposition without remainder. It should be noted that the growth effect reflects changes in nominal income values and therefore includes an increase in inflation. To eliminate the impact of inflation, in the graphs below, the effects of growth and redistribution are presented as shares of the difference in poverty indices between periods t_1 and t_2.

The effect of redistribution is of primary interest for analysis; therefore, the transfer-sensitive measure FGT for $\alpha = 2$ seems to be the most suitable for analysis. However, for comparative purposes, the graphs show the effects of growth and redistribution on poverty change using all three measures. The decomposition of poverty indices into increments due to the effect of income growth and due to the redistribution effect indicated that for all periods, both the growth component and the redistribution component had a negative sign, which indicates that they contributed to poverty reduction. However, the effect of income growth had a more significant effect on poverty reduction, and redistributive processes during the study period played a relatively small role in changing the poverty level (Fig. 2).

The FGT decomposition graph at $\alpha = 0$ indicated that from 2000 to 2008 the share of the effect of income growth on poverty reduction gradually increased, and then sharply decreased and remained almost unchanged until 2011–2014. The increase in the share of the redistributive effect during this period can be explained by the active social policy of the government, designed to prevent a drop in the income of people during the crisis of 2008, which was then followed by the valorization of pension payments. Since 2014, the decline in poverty has been driven mainly by an increase in average income rather than a redistribution of income in favor of the poor.

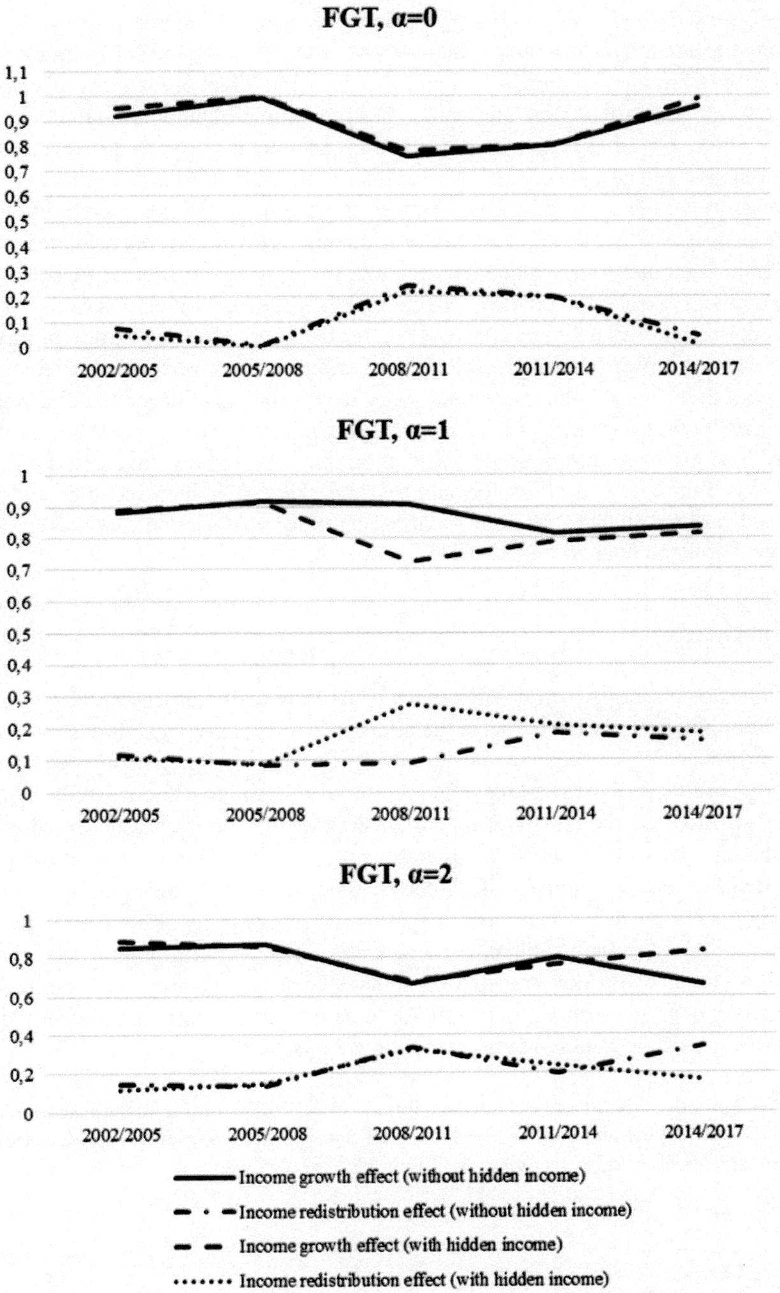

Fig. 2. Decomposition of FGT index on the income growth and redistribution effects.

The following FGT plot with $\alpha = 1$ shows the effect of income growth and redistribution on the amount of missing income until the poverty line. Since 2008 to 2011, the presence of hidden incomes led to an increase in the effect of income redistribution. That is, it can be assumed that hidden incomes predominantly increased the incomes of the poor and moved these households with hidden incomes closer to the poverty line.

The shape of the graph for FGT at $\alpha = 2$ also confirms that the effect of redistribution was weaker than the effect of growth. However, after 2011, the presence or absence of hidden incomes led to the opposite trend of income growth and redistribution effects. When hidden incomes were taken into account in disposable resources of households, income growth accelerated poverty reduction, and while latent incomes were not taken into account, it slowed down this process. As for the effect of redistribution of incomes, hidden incomes slowed down the effect of redistribution on poverty, while their absence accelerated them. As already mentioned, when $\alpha = 2$, the significance of large values of the missing income increases and the FGT index in this case characterizes the severity of poverty. The presence of hidden incomes among poor households reduced the severity of poverty. Taking into account the fact that the share of hidden incomes constantly decreased during the study period, it can be assumed that hidden incomes gradually shifted towards the poorest households.

4 Conclusion

An analysis of the impact on the dynamics of poverty of the processes of growth and redistribution of household income at the micro level revealed a weak intensity of redistribution processes. The decrease in the level of relative poverty in the period under study was largely due to an increase in income of all groups of households. This fact casts some doubt on the effectiveness of social policy measures aimed at maintaining the standard of living of especially vulnerable groups, such as large families and single-parent families, etc. The impact of hidden incomes on the dynamics under study was generally insignificant, but the sensitivity of the effects of redistribution and growth to the values of the FGT index with $\alpha = 1$ and $\alpha = 2$ allowed confirming the effectiveness of measures taken by the government to eliminate the possible consequences of the 2008 crisis and regulate pension payments in 2010, as well as the fact that hidden incomes tended to concentrate mainly among the poorest households.

Acknowledgments. The reported study was funded by RFBR according to the research project No. 19-010-00009 "Inequality and Poverty in Post-Soviet Russia: Dynamics and Determinants with Adjustments for Hidden Incomes of Households".

References

1. Poverty and Shared Prosperity, 2020: Reversals of Fortune. https://doi.org/10.1596/978-1-4648-1602-4
2. Cingano, F.: OECD Soc, Empl Migration Work Papers 163 (2014). https://doi.org/10.1787/5jxrjncwxv6j-en

3. Berg, A., Ostry, J.: Inequality and unsustainable growth: two sides of the same coin? (2011). https://www.imf.org/external/pubs/ft/sdn/2011/sdn1108.pdf
4. Castelló-Climent, A.: J. Econ. Inequal. **8**(3), 293–321 (2010). https://doi.org/10.1007/s10888-010-9133-4
5. Persson, T., Tabellini, G.: NBER Work Paper w3599 (1991). https://www.nber.org/system/files/working_papers/w3599/w3599.pdf
6. Reports on the execution of the federal budget. https://minfin.gov.ru/ru/perfomance/budget/federal_budget/budgeti/
7. Laws on the Pension Fund budget. https://minfin.gov.ru/ru/statistics/docs/law_pension/
8. Anti-crisis measures. https://www.rbc.ru/economics/25/06/2020/5ef1c34b9a79471d3a3b1697
9. The Unified Interdepartmental Statistical Information System (UISIS). https://www.fedstat.ru/
10. Pissarides, C., Weber, G.: J. Pub. Econ. **39**, 17–32 (1989). https://doi.org/10.1016/0047-2727(89)90052-2
11. Nivorozhkina, L., Arzhenovskiy, S., Galazova, S.: Int. J. Econ. Bus. Admin. VII **I**(4), 635–646 (2020). https://doi.org/10.35808/ijeba/614
12. Foster, J.E., Greer, J., Thorbecke, E.: Econometrica **52**, 761–766 (1984). https://doi.org/10.2307/1913475
13. Datt, G., Ravallion, M.: J. Dev. Econ. **38**, 75–295 (1992). https://doi.org/10.1016/0304-3878(92)90001-P
14. Shapley, L.: A value for n-person games. In: Kuhn, H.W., Tucker, A.W. (eds.) Contributions to the Theory of Games, vol. 2, pp. 303–317. Princeton University Press, Princeton (1953). https://doi.org/10.1515/9781400881970-018
15. RLMS-HSE. http://www.cpc.unc.edu/projects/rlms, http://www.hse.ru/rlms
16. Nivorozhkina, L.I., Arzhenovskiy, S.V.: Fin. Res. **3**(68), 27–39 (2020)

Ecologically Oriented Marketing in Digital Economy: Substance, Formation Management, Social Significance

Natalia P. Ketova(✉) and Viktor N. Ovchinnikov

Southern Federal University, Rostov-on-Don, Russia
nketova@sfedu.ru

Abstract. The prerequisites of the study of potential for solving the problems of formation and realisation of modern tools, principles and strategies of ecologically oriented marketing emerged in the context of the growing importance of innovative "green" technologies in all sectors of economy. The study aims to develop and verify the algorithm for formation of ecologically oriented marketing in digital environment, expounding the main tools for integrated policy of its realisation and development of ecologically oriented marketing policy tools. Methods of statement and testing of hypotheses, factor and comparative analysis, benchmarking, justification of algorithms, tabular interpretation of empirical and factual information, etc. were used. The article presents the main tools, formulated by the authors, of integrated policy of ecologically oriented marketing formation focused on solving the problems of justification and successful realisation of its model. It is emphasised in the conclusion of the article that ecologically oriented marketing is one of the socially significant components in the paradigm of perception of ideas, requirements and resources of "green economy" by business structures and reasonable consumers; it is expected to become quite popular and highly-demanded in the near future.

Keywords: Ecologically oriented marketing · "Green economy" · Ecological marketing tools · Digital environment

1 Introduction

The prerequisites for research in the field of ecologically oriented marketing, with its growing importance, are predetermined by the increasing awareness by modern business and the society in general of the need to achieve common well-being through efficient use of natural resources and simultaneous return of waste products of everyday life and production into a new production cycle. The realisation of ecologically oriented marketing strategies accounts for the relevance of this problem area; it is recognised as an important factor in meeting the needs of a community oriented towards thrifty use of natural resources subject to depletion, rational consumption of unexpendable resources

P. V. Trifonov and M. V. Charaeva (Eds.): DITEM 2021, LNNS 380, pp. 268–273, 2022.
https://doi.org/10.1007/978-3-030-94245-8_37

and simultaneous creation of reasonable demand for environmentally friendly goods and services. This actualises the stimulation of joint well-coordinated actions of public authorities, business structures, civil institutions and the population aimed to develop and widely introduce innovative "green" technologies in all sectors of the Russian economy, to increase the funding of research and innovation centres and complexes engaged in development of resource-saving technologies and production of equipment for waste management, sewage water and emissions treatment; to improve environmental safety.

The theoretical basis of the research with the results presented in the article is formed on the basis of fundamental provisions of modern marketing, theories of sustainable economic growth and "green economy", reproduction of socio-natural management systems, innovative factors of ecological marketing development in digital economy.

The novelty of the approach to the problem of development and promotion of eco-friendly, or "green", marketing in the market environment lies in the original statement and conceptual development of the algorithm for formation of this particular type of marketing in the digital environment from the position of the need to develop its model, with a view to ensure sustainable and dynamic mode of development and inculcation of environmentally friendly marketing strategies and technologies stimulating consumers to buy, and manufacturers – to produce and sell environmentally friendly organic foodstuffs and other goods. It is their production and sale that requires innovative technologies, ecologically safe raw materials, special equipment, etc.

The hypothesis of the research, with the results forming the basis of this article, represents the assumption that ecologically oriented marketing is one of the socially significant components in the paradigm of perception of ideas and requirements of "green economy" by certain business structures and reasonable consumers, focused on thrifty use and rational consumption of natural resources, demand for environmentally friendly goods and services. In the context of digitalisation, as a modern trend of social development, consumers should have an opportunity to organise personal ecological safety with the purpose to achieve positive changes in their life-sustaining environment through development of appropriate digital platforms. The above hypothesis predetermined the purpose and objectives of the research. The purpose of the research, with the results presented in the article, is the original development of the algorithm for formation of ecologically oriented marketing in digital environment, expounding the main tools of integrated policy of its realisation – as a quite promising and socially significant tool in terms of treatment of its model in the Russian economy. The research objectives are formulated as follows: to characterise the economic nature of ecologically oriented marketing, to show the importance of its development as a trend in modern economy, to consider the possibility to "fit" it into the green economy, to assess the prospects of digitalisation in this sphere.

2 Methods

The methods used for implementation of objectives outlined in the introduction represent a complex of tools: generation and verification of hypotheses, factor and comparative analysis, benchmarking, algorithmisation, etc.

3 Results

The range of problems considered in the article assumes revealing the substance, factors of development and structure of the marketing mechanism for the formation of environmentally friendly economy in the conditions of its digitalisation. It is obvious that the issues of environmental consciousness should be addressed not only at the level of the global community, a nation state or enterprises, but also at the level of individual consumers – economic actors [1].

The ecological crisis is a consequence of many factors, including the crisis of ethical and moral standards of the consumer civilisation. Ecological problems of varying level are regularly covered by the mass media; thousands of blogs, non-profit websites and forums are devoted to them in the Internet. This shows that the statement of the given problem by the world community in the mid-20th century has shifted to a qualitatively new level – involvement of all social groups in the solution of environmental problems [2].

The authors believe that in order to solve the above targets, certain pricing and non-price mechanisms of influencing the consumer market should be realised. Their absence may cause artificial price increase which leads to decreased demand and, consequently, drop in production [3]. An example of such a situation is the increase in prices for tobacco products in the present conditions.

The second group of measures aimed at stabilising the ecological condition represents non-price mechanisms. Their substance is reassessing marketing as such and raising it to the level of human aspirations, values and spirit (Marketing 3.0 concept). Marketing 3.0 supplements emotive marketing with the resource of human soul marketing [4], including in the sphere of ecology where the man and his needs are in the focus. The realisation of such targets – to make the world a better place – is promoted by PR actions in a number of companies (for instance, replacing plastic ware and packaging bags with biodegradable ones in Zelenyi Perekrestok stores, introduction of uniforms from recycled bottles for employees of several Moscow stores, broadcasting of eco-advice and recommendations for invigoration of children, etc. in advanced baby stuff departments).

The table below lists a number of tools providing the solution to the objectives in the sphere of production of environmentally friendly commodities, stimulation of their competent promotion to the market, etc. (Table 1).

Thus, the extension of the scope of "green" marketing is indicative of the growing market for environmentally friendly products, their being in demand, on the one hand, and the awareness of the need to realise the policy for this type of marketing, on the other hand.

Table 1. Main tools of integrated policy for ecologically oriented marketing formation from the position of managerial objectives for realisation of its model. Compiled by the authors from [4–11].

No.	Policy tools of ecologically oriented marketing formation from the position of management and solution of problems in the sphere of production of environmentally friendly items and their promotion to the market	Policy tools for ecologically oriented marketing with regard for the potential for solving managerial objectives in the sphere of consumption of environmentally safe products by the population
1.	Requirements and restrictions of legislative nature: increased accountability of manufacturers for production of environmentally safe foodstuffs, high-quality raw materials for their fabrication, animal feed, quality of drinking water, etc.	Correct information for consumers: ecologically transparent labelling providing reliable product information (in particular on food, baby stuff), ecological certification, complete information for consumers on Internet portals, etc.
2.	Mandatory environmental certification and labelling for producers, stricter requirements for suppliers of food, drinking water, etc.	Raising consumer awareness, increased environmental literacy in the field of marketing – as concerns both shopping centres staff and the general public
3.	Competitions, grants and subsidies for development of environmentally safe foodstuffs, new recipes, creation of environmentally friendly technologies in food industry	Extension of eco-friendly advertising in all media, including in the Internet space, in particular, in social media; explanatory information on prevention of environmental violations
4.	Introduction, where appropriate and necessary, of environmental management and quality management systems at production facilities; tightening of environmental standard systems in the area of environmentally friendly production and its promotion to the market	Stimulation of establishment and active operation of consumer protection organisations, youth unions providing environmental education of consumers, realisation of eco-friendly projects for children and young people, summer eco-schools, etc.
5.	Introduction of eco-labels for producers of foodstuffs, cosmetics, etc. with information on environmental impact on the consumers' organism throughout the life cycle, on possible negative side effects of consumption of contaminated products, on eco-safety	Toughened environmental safety requirements in respect of food, personal care products, detergents, etc. – for pregnant women and nursing mothers in the context of requirements for improvement of ecological environment, with a view to protect mothers and children

4 Discussion

The emergence of the environmental marketing concept is accounted for by the awareness of actual relevant threats by the world community in the second half of the 20th century, including those voiced at the meetings of the Club of Rome: overestimating the natural resource potential of different regions and the planet; underestimating the economic damage from environmental pollution; ignoring the "excessive consumption" of natural environment and ecological risks potential [12].

The commencement of research in the sphere of environmental marketing dates back to 1975, when the American Marketing Association (AMA) held its first seminar on "ecological marketing". Following its discussions, a book was published defining it as "green" or "ecological" marketing which involves "all activities designed to generate and facilitate any exchange intended to satisfy human needs or wants such that satisfying of these needs and wants occur with minimal detrimental input on the national environment" [12].

According to Peattie [13], the evolution of green marketing takes place in three stages: the first stage is "environmentally sensitive" marketing, 1970s; the second stage is "ecological" (environmental) marketing, 1980s; the third stage is "sustainable" green marketing. It gained prominence in the late 1990s and early 2000s, but is still quite modern even today. In the current conditions, "green marketing" is interpreted as a holistic and responsible strategic management process that identifies, anticipates and satisfies stakeholders' needs for a reasonable remuneration that has no negative impact on human or natural ecological well-being [14]. "Sustainable green marketing" assumes companies' efforts at designing, promoting, pricing and distributing products that will not harm the environment [5–10]. At the same time, this type of marketing is characterised, quite rightly, by the dominating socially-significant accent [11].

However, the authors of this article consider it appropriate to introduce the term ecologically oriented marketing into scholarly discourse, which will accentuate not only the fact that this kind of marketing is ecological, but also that it is oriented at achieving specific goals, such as environmental friendliness at all stages of production, maximum preservation of natural resources in terms of delivery of goods (including biodegradable packaging), responsible storage and organisation of sale of "green" goods, waste disposal using the methods that are safe for the biosphere. The expediency of introduction of this term is confirmed by a number of fixed phrases in relation to this type of marketing: customer-oriented, value-oriented, socially-oriented [15].

5 Conclusion

The research results make it possible to draw the following summary. First, ecologically oriented marketing as a term seems quite timely for its introduction into scholarly discourse, in order not only to focus on reorientation, updating and improvement of marketing principles, but also to ensure its perception as having significant prospects for building an innovative marketing environment matching the environmental realities of the future. Second, the realisation of its concept in the digital economy should be based on a set of actions including components such as: support for ecological innovations, digitalisation of ecosystem services, maximum loyalty to eco-friendly companies, use of new eco-technologies [5–10]. Digital eco-PR, well-developed eco-communications system, "green advertising", etc. are also important, being expedient for use already in the near term.

References

1. Zintsova, M.V., Vorontsova, M.A., Zintsov, K.S.: Russ. Foreign Econ. J. **8**, 94–104 (2019)
2. Green marketing. https://www.marketch.ru/marketing_dictionary/marketing_terms_z/green-marketing/
3. Counteracting marketing: examples, tasks and methods. https://promdevelop.ru/business/zadachi-i-metody-protivodejstvuyushhego-marketinga/
4. Kotler, P., Kartajaya, H., Setiawan, I.: Marketing 3.0: from Products to Consumers to Humankind Issues. Eksmo, Moscow (2011)
5. Environmental marketing as a tool for implementing an enterprise's environmental strategy. https://studwood.ru/995637/ekologiya/ekologicheskiy_marketing_instrument_realizatsii_ekologicheskoy_strategii_predpriyatiya
6. Zaitseva, D.S., Krakovetskaya, I.V.: Bul Kemerovo State Univ. Ser: Pol. Soc. Econ. Sci. **2**(2) (2016)
7. Environmental marketing. https://studme.org/192990/ekonomika/ekologicheskiy_marketing+&cd=4&hl=en&ct=clnk&gl=en
8. Club of Rome. https://ru.wikipedia.org/wiki/%D0%A0%D0%B8%D0%BC%D1%81%D0%BA%D0%B8%D0%B9_%D0%BA%D0%BB%D1%83%D0%B1
9. Dovgotko, N.A.: Marketing in the field of environmental management as a modern tool for managing socio-natural-economic systems. In: Actual Problems of Agribusiness Development in the Context of Modernizing the Economy: Collection of Scientific Papers Based on the Materials of the International Scientific and Practical Conference of StSAU (Stavropol, December 2012), pp. 89–90. AGRUS Stavropol State Agrarian University, Stavropol (2012)
10. Ecotechnologies. https://ecoportal.info/category/ekotexnologii/
11. Yashalova, N.N., Krylova, N.P., Levashov, E.N.: Sci. J. ITMO Ser. Econ. Ecol. Manag. **3**, 123–131 (2020)
12. Henion, K., Kinnear, T.: The Ecological Marketing. American Marketing Association, Columbus (1976)
13. Peattie, K.: Mark. Rev. **2**, 129–146 (2001)
14. Environmental marketing and its basics. https://www.marketing.spb.ru/lib-mm/strategy/eco_marketing.htm
15. Vitalieva, E.M., Kurbatova, Y.A., Yatsechko, S.S.: News of higher. Edu. Inst. Ser. Econ. Fin. Prod. Manag. **01**(27), 97–101 (2016)

Priority Strategic Guidelines for the Building Complex Companies of the Regions of Russia

Dmitry A. Makarov(✉), Nadezhda A. Polovnikova, Aleksander A. Leontiev, Svetlana A. Nikolikhina, and Grigorii A. Uraev

Emperor Alexander I St. Petersburg State Transport University, St. Petersburg, Russia

Abstract. Premises of research comprises the relevance of finding a solution to the problem of forming and implementing priority strategic guidelines for the development of construction enterprises and organizations, mainly by increasing technological and innovative potentials that form their economic potential; that potential contributes to the growth and development of the economy of the construction industry and the region. The purpose of the study is to identify priority strategic guidelines for the development of the building complex companies of Russian regions that meet the targets of the fourth scientific and technological revolution Industry 4.0 and achieve national goals and strategic objectives for the development of the Russian economy and its regions in the long term. In the course of the research, the generally accepted methods of scientific research were used, including the method of the general theory of systems development, the method of systemic and comparative analysis, synthesis, analogy, generalization, methods of economic analysis. The key features of development inherent in the nature of the development of objects and economic entities of the economy have been identified. This made it possible to propose a systematization and classification of the formation of priority strategic guidelines for the economic development of construction companies in the medium and long term, reflecting the relationship with the target guidelines for achieving national goals and strategic objectives for the development of the Russian economy and its regions and meeting the targets of the fourth scientific and technological revolution Industry 4.0.

Keywords: Construction companies · Signs of economic development · Aims for development · Change management

1 Introduction

The current trends in globalization and the development of the world economy are changing significantly under the influence of the factors of the fourth scientific and technological revolution Industry 4.0, the impact of the coronavirus infection (2019-nCoV) and the imposition of trade barriers aimed at achieving the goal of reducing global hydrocarbon emissions. It creates new trends for the development of the Russian economy operating under conditions of economic sanctions and investment restrictions, the impact of man-made, environmental, political and socio-economic risks. In these conditions, the solution to the problem of ensuring sustainable economic growth and

P. V. Trifonov and M. V. Charaeva (Eds.): DITEM 2021, LNNS 380, pp. 274–281, 2022.
https://doi.org/10.1007/978-3-030-94245-8_38

socio-economic development of the national economy and its economic entities in the medium and long term is an urgent problem.

The identification of priority strategic guidelines for the companies of the building complex of the regions of Russia focused on achieving national goals and strategic objectives of the Russian economy and its regions in the medium and long term has determined the purpose of the study. In accordance with the purpose of the study, the following tasks have been identified, providing for the need:

- generalization of the results of research work, integration and systematization of theoretical and practical knowledge in the research area;
- clarification of the characteristics of the philosophical and economic category "development" and identification of key features of economic development of economic entities, including construction companies (hereinafter referred to as CC);
- systematization and classification of priority strategic directions of economic development of construction companies.

2 Methods

A productive search for a solution to the problem under study involves the use of generally accepted methods of scientific research, one of which is system analysis. System analysis makes it possible to assess the change in the development of the activities of industrial enterprises and organizations and the change in the management system of the process of their development.

Following a systematic approach made it necessary to clarify the methodological prerequisites for substantiating the priority strategic guidelines for the economic development of companies in the building complex. This has prompted to disclose of the characteristics of the concept of "development" as a philosophical and economic category. This concept is not well-established until recently, which complicates the possibilities of identification, systematization and classification of signs of economic development.

From the standpoint of the concept of the general theory of systems development [1], the characteristic of the studied phenomenon "development" is a categorical concept that reflects the patterns of systemic transformations and changes in the form of stability and instability, forms of movement and change, any manifestations of quantitative and qualitative changes in the transition from the old to the new type movement, which is characterized by regular, directed, irreversible and qualitative changes in an object, its connections and processes.

It is reasonable to consider such an interpretation of the phenomenon of "development" inherent in a material object as a methodological basis for the formation of key features of economic development of an economic entity, positioned by purposeful and qualitative systemic changes in the connections and processes of a functioning (analyzed) entity. However, this composition of the key features of "development" is limited for the purposes of systematizing and classifying the economic development of economic entities.

The need to expand opportunities for identifying key signs of development prompts us to use the method of comparative analysis of research results. The founder of the theory of innovation Joseph Schumpeter [2] characterizes the economic growth by quantitative changes in the scale of increase in production and consumption of the same goods over time; the economic development is characterized by positive qualitative changes in the form of innovation in production, in products and services, in management and other types of economic activities. In the works of A. Lewis [3] and W. Rostov [4], economic development and economic growth are considered as structural transformations of production activities, as a result of which less productive types are replaced by more productive ones. Russell L. Ackoff defines development as "the process in which individuals increase their abilities and desires to satisfy their own needs and legitimate desires, and those of others" [5]. The existence of opportunity and ability of its implementation is legitimately recognized as one of the key features of the economic development of economic entities. Following the logic of this interpretation, the economic development of a construction company implies the ability and opportunity of its supervisors (owners, top managers) to make qualitative changes in updating construction equipment, introducing new technologies in production and management, increasing the volume of innovative products and the competitiveness of the organization as well as its products (and in other areas reflecting qualitative changes).

3 Results

The identified key features of the economic development of economic entities were a methodological prerequisite for the systematization and classification of priority strategic guidelines of the construction companies of Russia in the medium and long term (Table 1).

Table 1. Systematization and classification of priority strategic guidelines of construction companies. *Source*: compiled by the authors.

Classification attribute	Key feature of the development vector	Systemic characteristics of the economic development vector
Qualitative changes in connections in the relationship of construction participants	New quality of business partnerships between construction participants	Qualitative changes in business partnerships between participants in the construction business accelerating the return on investment and the commissioning of production facilities and installations
	New quality of connections in the organization and leadership of the team	Qualitative changes in the team work of the CC team, which increase the efficiency of managing the economic development of the organization

(*continued*)

Table 1. (*continued*)

Classification attribute	Key feature of the development vector	Systemic characteristics of the economic development vector
Qualitative changes in the methods of sustainability, growth and development of the CC economy	New quality in the development and implementation of CC business strategies	Qualitative changes in the methods of justifying business strategies and in ensuring their effective implementation
	New quality in ensuring sustainability, growth and development of the CC economy	Qualitative changes in the organization's sustainability
		Qualitative changes in ensuring the growth potential of the CC economy
		Qualitative changes in the management of the CC economic development based on ensuring balanced economic, information and environmental security
Qualitative change in the impact of innovations on the economic development of the construction company	Application of new technologies and production methods Innovations in the development of production of construction products Application of new methods in the organization of production	Mastering new technologies for the production of construction products and new methods of organizing construction production and labor of workers
	Development of new (innovative) construction products	Mastering the production of new (innovative) construction products
	Innovations in the management of the construction organization	Qualitative changes in the application of new technologies in management
		Qualitative changes in the application of new methods in management
New quality of structural transformation	Qualitative structural changes in ensuring the economic development of the construction company	Structural changes associated with the formation of an innovative type of growth in the volume of work, the development of the domestic material and technological base of the construction company, the reorientation of investment in technological production innovations

(*continued*)

Table 1. (*continued*)

Classification attribute	Key feature of the development vector	Systemic characteristics of the economic development vector
		Qualitative changes in the sphere of structural transformation
The abilities and opportunities of qualitative changes in ensuring the potential of economic development of the construction company	Abilities and opportunities to successfully compete in the foreign construction market	Ability and opportunity to form and develop stable competitive positions in the foreign construction market
	The ability and opportunities to increase the competitive potential in the regional construction market	Ability and opportunity to form and develop established market positions
	Abilities and opportunities to implement qualitative change in capacity building of the comparative advantages of the construction organization	Ability and opportunity to qualitatively measure the potential of economic development, to assess, analyze, forecast and monitor economic development
		Ability and opportunity to exercise comparative advantages in the development and implementation of targeted programs, strategies and mechanisms of economic development with a focus on building the competitive potential of organizations, taking into account the acceleration of return on investment, increasing the efficiency of resources used, increasing production intensification and reducing production costs
		Ability and opportunity to shift comparative advantages into a competitive advantage

(*continued*)

Table 1. (*continued*)

Classification attribute	Key feature of the development vector	Systemic characteristics of the economic development vector
		Ability and opportunity to be a leader of positive qualitative changes in ensuring balanced economic, socio-economic and environmental-socio-economic development of the construction organization
		Ability and opportunity to form an innovative economy and use its advantages in strengthening the innovative potential of a construction organization and reducing their dependence on foreign technology
		Ability and opportunity to qualitatively manage economic development, applying new methods of employee motivation in production management, in the capital and profit of a construction organization
		Ability and opportunity to successfully manage sustainable growth and development of a construction organization as a result of balanced management of economic and environmental risks in reducing damage to the environment

4 Discussion

The proposed systematization and classification of priority strategic guidelines of construction organizations was a methodological prerequisite for substantiation and selection of priority strategic guidelines for the economic development of the building complex companies of the country's regions, as it reflects the potential economic development of organizations based on their ability to carry out constant changes focused on the introduction of new technologies in the field of construction production and management, the use of new construction machines, mechanisms and innovative construction material.

The need to adapt construction companies to constant changes in their activities gives rise to the need for the formation and development of a new approach to management – change management [6–8]. This calls for the formation of an adequate strategy for managing changes in the system of strategic management of the economic development of construction companies.

At the same time, the identification of priority directions for the formation of strategic guidelines for the economic development of organizations of the building complex of the regions of Russia necessitates their harmonization with national goals and strategic objectives of the growth and development of the Russian economy and its regions in a comprehensive relationship with the targets of the fourth scientific and technological revolution Industry 4.0. The implemented concepts of economic development of countries with advanced technologies reflect, in their essence, the targets of their development, adequate to the stages of industrial development known as industrial (scientific and technical) revolutions [9–12].

Today, it is necessary to develop digital software products for progressive economic development based on the use of innovative technologies in the production sector. This was a prerequisite for the transition to the fourth industrial revolution (the beginning of the 21st century to date), presented in the conceptual model of neo-industrial development, which was developed by K. Schwab [13]. This model reflects the need not only to expand the already known and used digital technologies in the productive sector, but also to use them in the service sector, politics, social, environmental and other areas of related activities, which provides for the removal of a number of systemic restrictions [14].

Strengthening the impact of scientific and technological progress on economic development predetermined the prerequisites for the formation of priority targets for long-term socio-economic development, goals and strategic objectives for the development of the Russian Federation and the strategy of its economic security [15–19]. In accordance with the decree of the President of the Russian Federation of May 7, 2018 No. 204 "On national goals and strategic objectives for the development of the Russian Federation for the period up to 2024" [17], nine national development goals of the Russian Federation are identified, among which the goals of technological development of the Russian economy in a complex relationship with the introduction of digital technologies in the economy and social sector. At the same time, achieving the set goals of technological development is problematic without structural changes associated with the formation of an innovative type of economic growth, the development of the own material and technical resources, and the reorientation of investments in technological innovation.

In this regard, it is important to properly measure, analyze, evaluate and build up the capacity for economic development of construction companies in an integrated relationship with their technological and innovative development potential.

5 Conclusion

Determination of priority strategic guidelines for the development of construction companies required finding a solution to a number of theoretical, methodological and applied problematic issues that require concretization of the concept of "development" in relation to economic entities, identification of key features of economic development, their systematization and classification, substantiation of criteria and methods for managing economic development.

The priority directions of the formation of strategic guidelines for the development of the building complex companies of the regions of Russia, identified in the course of the study, reflect the relationship with the targets for achieving national goals and strategic objectives for the development of the Russian economy and its regions in the medium and long term and meet the targets of the fourth scientific and technological revolution Industry 4.0.

At the same time, priority strategically significant guidelines should be considered those that can also minimize the damage caused to humans and the environment. In this regard, the justification of the parameters, criteria and methods for managing economic development, the development of methods for assessing its impact on the quality of economic growth of construction companies, its efficiency and competitiveness in a complex relationship with the search for methods for effective management of economic and environmental safety should be considered an urgent direction for further research.

References

1. Urmantsev, Y.: Evolutionics or General Theory of the Development of Systems of Nature, Society and Thinking. LENAND, Moscow (2020)
2. Shumpeter, Y.: Theory of economic development. Progress, Moscow (1982)
3. Lewis, A.: The theory of economic growth. Richard D. Irwin, Homewood (1955)
4. Rostov, W.W.: Econ. Hist. Rev. **12**(1), 1–16 (1959)
5. Akoff, R.: Akoff on management. Peter, St. Petersburg (2002)
6. Graetz, F., Smith, A.C.: J. Change Manag. **10**(2), 135–154 (2010)
7. Harrington, J.: Excellence in change management. Standards and Quality, Moscow (2008)
8. Shironina, E.M.: Univ. Bul. **7**, 26–31 (2019). https://doi.org/10.26425/1816-4277-2019-7-26-31
9. De Vries, J.: The Industrious Revolution: Consumer Behavior and the Household Economy, 1650 to the Present. Delo, Moscow (2008)
10. Jevons, H.S.: Econ. J. **41**(161), 1–18 (1931). https://doi.org/10.2307/2224131
11. Earth Policy Institute. http://www.earth-policy.org
12. Rifkin, J.: The Third Industrial Revolution: How Lateral Power is Transforming Energy, the Economy, and the World. Palgrave Macmillan, New York (2011)
13. Schwab, K.: The Fourth Industrial Revolution. Eksmo, Moscow (2017)
14. Gubanov, S.: Economist **4**, 3–10 (2017)
15. Zamulin, O.A., Sonin, K.I.: Econ. Issues **1**, 11–36 (2019). https://doi.org/10.32609/0042-8736-2019-1-11-36
16. The concept of long-term socio-economic development of the Russian Federation for the period up to 2020. http://www.consultant.ru/document/cons_doc_LAW_82134/28c7f9e359e8af09d7244d8033c66928fa27e527
17. On national goals fnd strategic tasks of the development of the Russian Federation for the period up to 2024. http://kremlin.ru/events/president/news/57425
18. On the Strategy of Economic Security of the Russian Federation for the Period up to 2030. https://www.garant.ru/products/ipo/prime/doc/71572608
19. Ardzinov, V.D., Chepachenko, N.V., Leontev, A.A.: Econ. Manag. **26**(11), 1227–1235 (2020). https://doi.org/10.35854/1998-1627-2020-11-1227-1235

Comprehensive Assessment of Socio-economic Differentiation of Municipal Districts in the Old-Cultivated Region (Voronezh Region)

Natalia V. Yakovenko[1](✉), Roman V. Ten[1], and Evgeniya M. Shirokova[2]

[1] Voronezh State University of Forestry and Technologies named after G.F. Morozov, Voronezh, Russia
yakovenkonv@vgltu.ru
[2] Voronezh State University, Voronezh, Russia

Abstract. The current period of development of the Russian economy, characterized by its instability and the need to improve the management system at the country's regional level, requires an objective assessment of the uneven development of spatial socio-economic systems. In recent decades, Russia has developed a stable trend of polarizing the level of socio-economic development, sharp property stratification and differentiation of living standards by region. The scientific search for modern economic geographers is determined by the trend direction towards regionalization and municipalization of research objects. In this regard, the study and structuring, addition and creation of a new theoretical-methodological basis for assessing the socio-economic differentiation of municipal development constitute the main task of most regional studies and socio-economic geographical disciplines. The study developed and practically tested a methodology for a comprehensive assessment of socio-economic development of municipalities in the region using new mathematical tools and methodological approaches. Computer technologies, general and special-purpose tools were used for data processing, including statistical ones – Novo Forecast, PPS Statistica, Stadia, as well as QGIS technologies, MapInfo, etc. The practical significance of research results is determined by the applied nature of a number of conclusions and recommendations, in particular, the development of tools and the author's methodology for assessing the territorial differentiation of socio-economic development of municipalities in the region.

Keywords: Socio-economic differentiation · Voronezh Region

1 Introduction

The modern development of social relations in most states is characterized by the localization of economic and social processes that occur at the regional and intraregional levels and are caused by the prevailing factors of population settlement and the territorial organization of production activities. All these conditions and factors lead to increased territorial heterogeneity. Differentiation makes it difficult to pursue a unified

P. V. Trifonov and M. V. Charaeva (Eds.): DITEM 2021, LNNS 380, pp. 282–288, 2022.
https://doi.org/10.1007/978-3-030-94245-8_39

policy of socio-economic transformations and the national market formation, increases the danger of regional crises and interregional conflicts. Improvement of the mechanisms for managerial decision-making at the regional and municipal levels should be carried out by optimizing methodological approaches to a complete systematic high-quality diagnostics of the regional socio-economic development, meeting the needs of all interested users. The strengthening of heterogeneity makes it difficult to pursue a unified policy of socio-economic transformations and the formation of the regional and national markets, increases the risk of regional crises, weakening the integrity of the region and the state as a whole [1–7].

The creation of a scientific basis for managerial decision-making in the field of the regional, i.e. territorially differentiated socio-economic policy, requires an objective assessment of unequal conditions of municipalities.

2 Materials and Methods

The theoretical and methodological research basis was formed by the works of leading domestic and foreign experts addressing the issues of spatial inequality and socio-economic differentiation of development. The following methods were used in this study: economic-statistical analysis, abstract-logical method, econometric modeling. To process static data, the following software products were used: Novo Forecast, PPS Statistica, Stadia, as well as QGIS technologies, MapInfo, etc.

3 Results

When studying the regional differentiation of socio-economic development of municipal districts, it is proposed to use a methodology based on an integral indicator, which is a characteristic of the socio-economic development of municipalities and is carried out from the standpoint of combining the socio-economic development level. When forming a system of indicators of the municipal socio-economic development level, the main requirements for their selection are taken into account: materiality (indicators should be useful in making adequate managerial decisions); compatibility (a set of indicators should fit into the existing system of information flows); reliability (the ability to objectively measure the values of indicators); differentiation (indicators should objectively reflect the differences between municipalities); completeness (a set of indicators should take into account the achieved development level of various spheres of activity of a municipality); simplicity (indicators should be understood by a wide range of analysts); adequacy (a set of indicators should take into account the specifics of activities of each municipality).

At the first stage, the time range for studying the territory is chosen. At the second stage, the estimated indicators of sustainable development of the socio-economic system of municipalities are selected depending on research purposes and objectives. Since the socio-economic system consists of two subsystems – the social and economic ones, the selected assessment indicators make up two groups of indicators – social and economic ones (Table 1) [8, 9].

Table 1. The system of indicators of a comprehensive assessment of municipalities by the socio-economic development level.

	Name of indicator	Calculation method
Economic indicators	Volume of fee-based services	Volume of fee-based services per capita
	Commissioning of residential buildings	Commissioning of residential buildings per thousand
	Total retail and catering turnover	The ratio of the sum of retail trade and catering volume per capita
	Volume of investments in fixed assets	Volume of investments in fixed assets per capita
	Average salary of corporate employees	Amount of average salary paid to corporate employees
Social indicators	Fertility rate	Average number of births during the year per thousand in the middle of the current year
	Mortality rate	Number of deaths during the year per thousand in the middle of the current year
	Crime rate	The ratio of the number of registered crimes per 100,000 population
	Road density	The length of paved roads per 10,000 square kilometers of territory
	Hospital bed capacity	Number of hospital beds per thousand

Today, the problem of assessing the differentiation of socio-economic development lies in the lack of uniform approaches and standards of assessment indicators and their quantity, as well as their distribution by assessment group [10–13].

At the first stage, the statistical indicators presented in Table 1 are taken as initial indicators of socio-economic development of the territory. After determining the estimated indicators, it is necessary to calculate the average sustainability coefficient (Y_{avg}), the average coefficient of development dynamics (D_{avg}) and the final integral coefficient of socio-economic development (I):

$$I = \frac{Y_{avg} + D_{avg}}{2}. \quad (1)$$

When determining Y_{avg}, it is necessary to lend weight to each initial social and economic indicator based on the reference value. If the indicator stimulates development, i.e. its growth has a positive effect on the region:

$$K_i = \frac{x_i}{max_i}. \quad (2)$$

If the indicator has a negative effect, the following formula is used:

$$K_i = \frac{min_x}{x_i}, \tag{3}$$

where K_i – level of development of a region on the i-th indicator;
x_i – indicator value;
$\max x_i$, $\min x_i$ – reference value, which can be taken as threshold values of indicators.

Then the integral indicators of the level of economic and social sustainability are sequentially calculated (4). To get rid of negative values, it is necessary to square the indicators, find their arithmetic mean and define the square root:

$$Y_j = \sqrt{\frac{\sum_{i=1}^{n} k_i^2}{n}} \tag{4}$$

where Y_j – an indicator of sustainability of each component;
n – number of indicators.

Then the sustainability coefficient (geometric mean value) is determined.

The sustainability coefficient for (i) year is calculated by the formula:

$$Y_i = \sqrt[2]{Y_{econ} \times Y_{social}}, \tag{5}$$

where Y_{econ} – economic sustainability,
Y_{social} – social sustainability.

Finally, Y_{avg} is calculated by the following formula:

$$Y_{avg} = \frac{Y_{i1} + Y_{i2} \cdots + Y_{in}}{n}, \tag{6}$$

where n – number of indicators.

When calculating the average coefficient of development dynamics D_{avg}, x_i in formulas (2) and (3) is found as follows:

$$x_t = \frac{x_n}{x_{n-1}} \times 100, (\%) \tag{7}$$

where x_t – indicator of relative dynamics as of the previous year in (%).

$$x_i = \frac{x_{t1} + x_{t2} \cdots + x_{tn}}{n}, \tag{8}$$

where n – number of indicators.

Then the calculation by formulas (2–6) is repeated to find D_{avg}.

The final integral coefficient of socio-economic development is calculated by using formula (1).

In this case, it is proposed to distinguish five (5) clusters by level of socio-economic differentiation: with a high level of development, an above-average level of development, an average level of development, a below-average level of development and an extremely low level of development (depressed level). When determining a group interval, it should be borne in mind that the value of a characteristic of many socio-economic phenomena and consequently of the final integral indicator vary unevenly and on a significant scale, therefore, this study uses an unequal interval that is progressively increasing in arithmetic progression:

$$h_{i+1} = h_i + a, \tag{9}$$

where h_i – the interval value of the first group (depressed municipalities);

α – constant number (a constant), positive number – when intervals increase progressively, and negative number – when intervals decrease progressively.

In this case, having estimated the range of variation of the final integral indicator, the authors set the values $h_i = 0.010$ and $\alpha = 0.005$, respectively.

The quintessence of the entire scientific research carried out by the authors is the derivation of the final integral indicator of sustainable socio-economic development of municipalities in the Voronezh Region and consideration of its territorial aspect (Fig. 1).

Vorobyevsky, Nizhnedevitsky and Ternovsky districts (<0.577) are among the territories with a depressed level of development. During the period under study, they were within the lagging group both in terms of absolute indicators and the dynamics of changes in these indicators. In terms of geographical location, it is necessary to highlight their edging or border position in the Voronezh Region, consequently – an unfavorable economic and geographical position. It should be recognized that the development of these territories and the way out of the crisis are impossible without the support of regional and federal authorities in the form of financial support instruments.

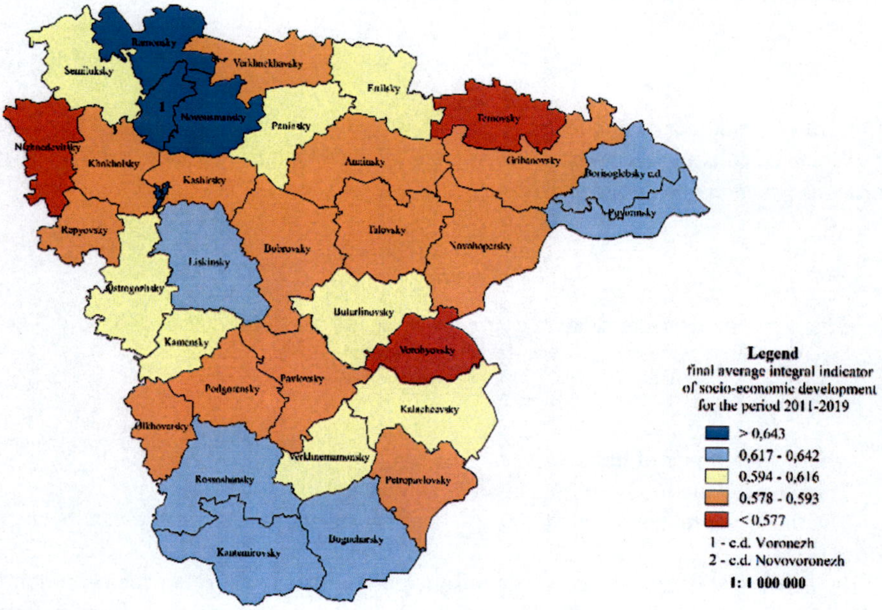

Fig. 1. The final average integral indicator of sustainable socio-economic development for the period of 2011–2019. *Source:* compiled by the authors.

A cluster with a below-average level of development (0.578–0.593) is formed by thirteen (13) districts, Pavlovsky, Bobrovsky, Anninsky districts can be distinguished among them. According to the new regional development strategy until 2035, the specified districts should become the growth drivers for the entire territory of the Voronezh Region as a whole. These territories have good starting indicators, but they are slightly behind the regional average ones in terms of development dynamics.

Eight (8) districts form a cluster with an average level of development (Kamensky, Semiluksky, Kalacheevsky, Paninsky, Ertilsky, Verkhnemamonsky, Ostrogozhsky, Buturlinovsky districts), the territories of which are directly adjacent to the districts of a higher level of development. It should be noted that such districts as Paninsky and Ertilsky ones were included in this group due to the good growth dynamics of socio-economic development indicators, although initially they are outsiders in absolute indicators.

Clusters with an above-average level of development are located in the south of the region (Rossoshansky, Kantemirovsky, Bogucharsky districts), in the east (Borisoglebsk urban district, Povorinsky district), in the center – Liskinsky district. These territories have a favorable economic and geographical position with access to the main highways and railways. It should be noted that in this cluster, according to most indicators of socio-economic development and sustainability, neighboring districts were included in the same groups by development level of the territory. In the south, these are Kantemirovsky and Bogucharsky districts, in the east – Povorinsky district and Borisoglebsky urban district. This confirms their close economic and social ties and makes it possible to consider the proposals for their unification into one municipality in the south and east of the region, respectively, to be considered justified.

A cluster with a high level of development forms the administrative center of the Voronezh Region – Voronezh urban district and the adjacent districts – Ramonsky and Novousmansky ones, as well as Novovoronezh urban district. These territories are characterized by the best indicators of socio-economic development and in some cases exceed the indicators of other districts in the region by tens and hundreds of times. Therefore, there is an inverse proportionality, and according to the relative indicators of dynamics of the same indicators, these constituent territories were included in the lagging or depressed clusters.

4 Discussion

The calculations clearly demonstrated that during the period under study, there is a decrease in differentiation in terms of economic indicators and indicators of municipal administration activities, while social differentiation during this period is growing.

Thus, the differentiation of municipalities as a whole on the integral indicator slightly decreases. The data obtained through the analysis of differentiation of socio-economic development should become an information basis for managerial decision-making in the regional policy implementation and make it possible to determine priority areas for regulating the socio-economic development of municipalities in the Voronezh Region. Based on the research results, it can be argued about the leading role of economic indicators. They set the main parameters of spatial differences between municipalities. Consequently, economic indicators should become the main benchmarks in developing measures of regional differentiated socio-economic policy.

5 Conclusion

Thus, the sustainable socio-economic development of the Voronezh Region implies the development of all its territorial units. The research results can be used in plans and strategies for the development of the region as a whole and of each municipality in particular, for the medium and long term. It should be borne in mind that the territory of the region is a priori heterogeneous in terms of socio-economic development, and it is not possible to achieve "ideal" homogeneity. Nevertheless, the maximum possible smoothing of territorial disparities in these conditions is one of the fundamental tasks of regional policy and local governance.

The regional policy to reduce territorial differentiation should be aimed at ensuring sustainable, dynamic socio-economic development of "problematic" municipalities with a gradual equalization of the levels of their socio-economic development in comparison with the most developed municipalities in the long term.

Acknowledgments. The reported study was funded by RFBR, project number 19-35-90021. The reported study was funded by RFBR, project number 19-29-07400-mk.

References

1. Akhmeduev, A.: Reg. Probl. Econ. Transf. **6**, 37–51 (2017)
2. Baranov, S., Skufyina, T.: Econ. Issues **8**, 54–75 (2005)
3. Zencheva, N.V., Gerasimov, B.I.: Economic Analysis of Differentiation of Socio-Economic Development of Russian Regions. Publishing House of Tambov State Technical University, Tambov (2005)
4. Lapin, A.V.: Socio-economic differentiation of territories: theoretical and methodological aspects, assessment and analysis. Perm (2018). https://elis.psu.ru/node/502245
5. Bulochnikov, P.A., Smirnov, K.B.: SPB Econ. J. **4**, 68–75 (2019)
6. Pobedin, A.A.: Reg. Econ.: Theory Pract. **2**, 92–100 (2009)
7. Polynev, A.O.: Interregional economic differentiation: methodology of analysis and government regulation. Editorial URSS, Moscow (2011)
8. Yakovenko, N.V., Komov, I.V., Ten, R.V.: Adv. Econ. Bus. Manag. Res. **79**, 201–203 (2019)
9. Yakovenko, N.V., Porosenkov, Y.V.: Bul. Voronezh State Univ. Ser.: Geogr. Geo. **2**, 10–21 (2013). https://elis.psu.ru/node/502245
10. Amirova, E.A.: Probl. Econ. Transf. **1**, 65–70 (2020). https://doi.org/10.26726/1812-7096-2020-01-65-70
11. Jovovic, R., Draskovic, M., Delibasic, M., Jovovic, M.: J. Int. Stud. **10**(1), 255–266 (2017). https://doi.org/10.14254/2071-8330.2017/10-1/18
12. Kazakov, M.Y., Mitrofanova, I.V.: J. Volgograd State Univ. Econ. **22**(1), 44–54 (2020). https://doi.org/10.15688/ek.jvolsu.2020.1.5
13. Litwiński, M.: Econ. Law **16**(4), 449–458 (2017). https://doi.org/10.12775/EiP.2017.031

Conceptual Model of the Stakeholders of the Modern Regional Socio-Ecological-Economic System

Gor A. Abramyan(✉), Dmitry A. Shevchenko, Oxana L. Savenko, and Alexey N. Yeletsky

Southern Federal University, Rostov-on-Don, Russia
gorabramyan@sfedu.ru

Abstract. The issues of regional development, caused by the significant volatility of the socio-economic situation at the macro level, form new requirements for the executive authorities responsible for economic growth and improvement of the quality of life in the region. The purpose of the study is to describe the conceptual model of the stakeholders of the socio-ecological-economic system at the meso-level, which would meet the modern challenges of regional development. The article presents the author's model of the stakeholders of the modern regional socio-ecological-economic system, lists the internal and external stakeholders of this system, describes their impact on the competitiveness of the territory and reflects the key directions of development of this system. The internal stakeholders of the regional socio-ecological-economic system as a quasi-corporation included those operating at the regional level: government bodies, development institutions, financial organizations, scientific organizations, business operating already in the region, and thus the most investing in it, the local population, as well as public organizations. External stakeholders included those functioning at the federal and international level: national and foreign competing territories and partner territories, authorities, development institutions, financial organizations, scientific organizations, potential investors, tourists and migrants, mass media, as well as public organizations dealing with environmental issues, energy efficiency and energy conservation. Within the framework of the study, an analogy was drawn between efficiency indicators of corporate activities and the indicators characterizing the development of the regional socio-ecological-economic system. A number of relevant methods of economic research were also used during the research.

Keywords: Stakeholder · Socio-ecological-economic system · Competitiveness · Investment attractiveness

1 Introduction

Due to economic turbulence and extremely high competition for limited resources both in external and internal markets, the dynamic development of territories, in particular regions characterized by a multi-vector trajectory, is becoming more and more difficult. It

P. V. Trifonov and M. V. Charaeva (Eds.): DITEM 2021, LNNS 380, pp. 289–295, 2022.
https://doi.org/10.1007/978-3-030-94245-8_40

should be noted that competition tends to intensify not only for investment resources, but also for productive forces in general, which form the final territorial product of any region. In order to determine the key directions of the development of the region in the current conditions, it is necessary to take into account the theoretical and methodological basis of its competitiveness, and the competitiveness of the region, in turn, should be based on a comprehensive understanding of the interaction of various factors and economic agents that form it. These tendencies in particular form the prerequisites for the study and determine its relevance.

The concept of a region as a socio-ecological-economic system should be considered as a theoretical and methodological basis for the competitiveness of a region.

The conducted scientific and literary review of the theoretical basis of the research can be represented by the following results.

The broadest interpretation of the concept of a socio-ecological-economic system means by this system a part of the territory, where the intensity of connections between elements of nature and economy significantly exceeds the set of elements directed from outside or outside the system [1].

At the same time, a number of scientists have studied the topics of such systems in the context of various aspects.

Some of the authors in their works paid special attention on the theoretical and methodological foundations for assessing the socio-economic systems of the region [2].

Moreover, considerable attention is paid preeminently to the region as a sophisticated, multi-element, socio-cultural complex, which is an object of management [3].

A number of studies are concerned with the analysis of factors that limit the development of regions due to the complexity and multidimensionality of regional socio-ecological-economic systems operating in the context of globalization and increasing competition [4].

Other works are concerned with the ecological and economic content of the structure of regional systems as the basis for the formation of the implementation of policies aimed at achieving the goals of sustainable development of the region [5].

Menshutkin and Filatov proposed four cognitive models of the ecological-socio-economic system of the region, aimed at assessing the possibilities of improving the living standards of the population, rational use and protection of the environment, the development of certain sectors of the economy and social sphere of the region [6].

Despite the importance of the studies under consideration, the description of socio-ecological-economic systems at the meso-level from the point of view of the concept of the region as a quasi-corporation is practically absent or is of a fragmentary nature.

The novelty of the research is due to the author's definition of the modern socio-ecological-economic system at the regional level, as well as its presentation as a quasi-corporation with accompanying stakeholders and assets.

The author hypothesizes that modern regions are increasingly becoming independent economic agents, the competitiveness and investment attractiveness of which depends on the degree of satisfaction of the needs of stakeholders.

The purpose of the study is to describe the conceptual model of the stakeholders of the socio-ecological-economic system at the meso-level.

The research tasks include:

- the overview of the theoretical basis of the study;
- the adaptation of the stakeholders theory at the level of the regional economy;
- the determination of the structure and content of the region's assets as a quasi-corporation;
- the presentation of the stakeholders model of the socio-ecological-economic system in terms of the formation of investment attractiveness and competitiveness of the region.

2 Results

The authors believe that the regional socio-ecological-economic system (RSEES) itself is an open, dynamic, indeterminate and self-reproducing system that develops under the influence of factors of the internal and external environment and is characterized by functional and physical connections between the social, ecological and economic spheres of the region's life.

At the same time, today the region as a socio-ecological-economic system has many characteristics of a corporation, which allows us to consider this system and therefore the region itself as a quasi-corporation functioning in a competitive environment, in which various subjects interact.

It is advisable to consider the motives of the economic behavior of these subjects through the prism of the theory of corporate stakeholders, which became widespread in the second half of the 20th century [7].

The structure of stakeholders of any modern corporation can be divided into internal and external stakeholders. Internal and external stakeholders of the corporation are shown in Fig. 1.

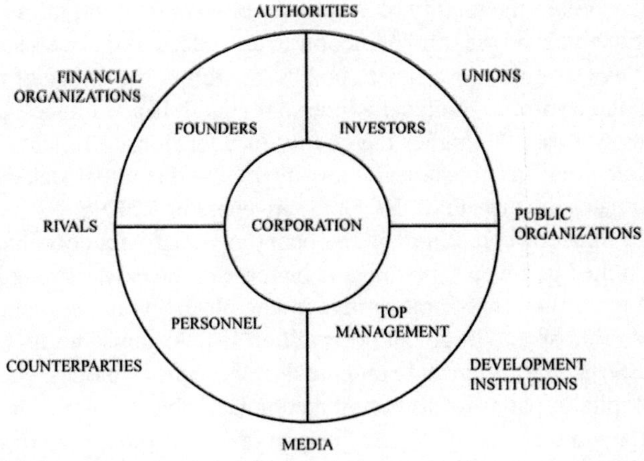

Fig. 1. Internal and external stakeholders of the corporation. *Source:* compiled by the authors based on research materials.

Preserving the essence and semantic content, but taking into account the specifics and characteristics of administrative-territorial units, the theory of stakeholders can be projected at the level of the region or RSEES as a quasi-corporation, which also has its own internal and external stakeholders and functions within the framework of their interactional relationships (Fig. 2).

Fig. 2. Internal and external stakeholders of RSEES. *Source:* compiled by the authors based on research materials.

The consumers of the territorial product, the peculiarity of which lies in their dualism – both internal and external stakeholders of the territory can act in their role at the same time, should be singled out as a separate stakeholder. The territorial product, in turn, should be understood as the totality of a complex of resources and infrastructure of the territory, material goods (in the form of income and goods) and services, socio-economic policy pursued by territorial government bodies, as well as the image of the region [8]. In other words, the territorial product includes both tangible and intangible assets of the region. It is obvious that the higher the quality of a territorial product, the higher the degree of satisfaction of its consumers – key internal and external stakeholders, which ultimately stimulates the growth of the competitiveness of RSEES.

Thus, in the modern conditions of the economic space functioning, the RSEES, which are not limited to the preservation and increase of the existing tangible assets, but are focused on the formation of competitive intangible assets in the first instance, have an advantage and prospects in global competition [9]. At the same time, the tangible assets of RSEES include its current strategic development resources, such as: natural-geographical capital, spatial-infrastructural complex, objects of the socio-economic sphere. The intangible assets of RSEES include human capital, organizational capital and the image of the territory [10]. In order to withstand the competition for resources, the region must create a favorable environment for development, conducive to the effective implementation of existing and attracted resources both in the short term and in the medium and long term – in order to solve this problem, special attention should be paid to the development of intangible assets. Moreover, the RSEES has at its disposal the region's financial assets – budgetary and private investments – for the development of such an environment.

Consideration of the RSEES competitiveness from the point of view of the structure of its assets confirms that today the region is increasingly acquiring the characteristics and targets of an independent economic agent, thereby turning into a quasi-corporation that has its own unique instruments of economic policy and is a full participant in economic relations between economic entities.

The assets of RSEES as a quasi-corporation consist of tangible assets, intangible assets and financial assets – these assets form the basis of regional competitiveness (Fig. 3).

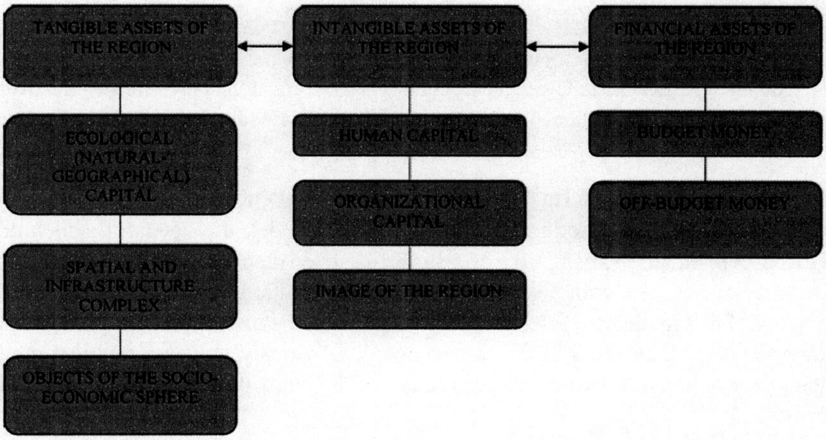

Fig. 3. Structure and content of RSEES assets as a quasi-corporation. *Source:* compiled by the authors based on research materials.

Intangible assets of the region are of particular importance in modern conditions. Obviously, within the framework of the concept of a region as a quasi-corporation in regional management, the tasks similar to the tasks of corporate image and reputation forming, as well as increasing the investment attractiveness of a brand arise. At the same time, the processes of goal-setting and assessment of the effectiveness of the region become largely identical to corporate management and business processes.

Accordingly, it is also possible to draw an analogy between the resulting RSEES indicators and corporate performance indicators:

- SP – sales proceeds;
- SP energy intensity – capital productivity;
- the number of labor resources – the average number of employees;
- budgetary efficiency (tax incentives, subsidies, etc.);
- profitability;
- actual disposable cash income – actual accrued wages, etc.

Thus, having considered the structure and content of the region's assets and the resulting indicators of its development as a quasi-corporation, it is possible to represent schematically the model of the stakeholders of the modern RSEES (Fig. 4):

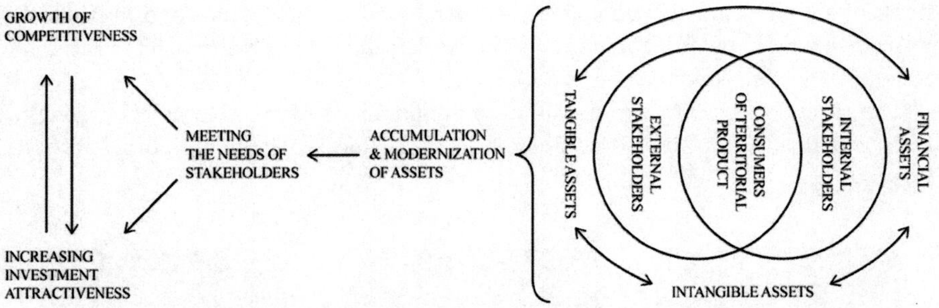

Fig. 4. Model of RSEES stakeholders in the context of the formation of investment attractiveness and competitiveness of the region. *Source:* compiled by the authors based on research materials.

The subjects of this model are internal and external stakeholders who form consumers of the RSEES territorial product. As objects of the model, the needs of stakeholders should be defined, the satisfaction of which is carried out through the accumulation and modernization of a complex of assets – tangible, intangible, financial. Within the framework of this model, the interdependence between the investment attractiveness and the competitiveness of the RSEES can be noted. Obviously, this is a multicomponent model, the elements of which are characterized by both multidirectional and cyclic connections.

3 Methods

Special mention should go to a systematic approach, used by the authors along with other methods in the framework of this study, based on which the RSEES is considered with its interconnections and relations within the elements. The study also used the method of analogy between the indicators of corporate performance and indicators characterizing the development of the regional socio-ecological-economic system. It is also worth noting that the following methods of economic research were used in the research process: dialectical method, abstract-logical method, including analysis, synthesis, analogy and formalization, as well as comparison.

4 Conclusion

The fact that the representation of the region as a socio-ecological-economic system with its stakeholders is the most important element of regional policy can be singled out as the main conclusion of the study. Disregard to the corporate aspects of regional management can only hinder the development of the territory. Within the framework of the proposed RSEES model, both internal and external stakeholders were listed, the region's assets were presented, and interactions and links between the model elements were indicated, which have a significant impact on the competitiveness of RSEES.

The need for an empirical analysis of the degree of influence of one or another stakeholder on the level of competitiveness of RSEES through statistical analysis and construction of a multivariate econometric model using the example of one of the regions of the Russian Federation should be noted as further research prospect.

References

1. Vershinina, L.P.: Instruments and methods of complex socio-ecological-economic assessment of the state of territories on the example of the Rostov region. Regionalnaya Ekonomika i Upravleniye: Elektronnyy Nauchnyy Zhurnal **4**(32) (2012). https://eee-region.ru/article/3212/
2. Bessonova, E.A., Kelesh, Y.: Methodological approaches to assessing the socio-economic system of the region. Probl. Reg. Econ. **2**(35), 18–24 (2018)
3. Nasaraia, Z.: The region as a socio-economic system and object of management. Eur. J. Humanit. Soc. Sci. **6**, 68–71 (2020). https://doi.org/10.29013/EJHSS-20-6-68-71
4. Dzagoyeva, I., Temmoyeva, S., Litovka, N., Shokumova, R., Musayeva, B.: Development of the system of criteria and indicators of improvement of organizational conditions of the balanced development of the region. Revista San Gregorio **32** (2020). https://doi.org/10.36097/rsan.v1i32.1072
5. Mishenin, Y., Koblianska, I., Medvid, V., Maistrenko, Y.: Sustainable regional development policy formation: role of industrial ecology and logistics. Entrepreneurship Sustain. Issues **6**(1), 329–341 (2018). https://doi.org/10.9770/jesi.2018.6.1(20)
6. Menshutkin, V.V., Filatov, N.N.: Experience in assessing the state and forecasting of the socio-ecological-economic system: on the issue of sustainable development of the region. InterCarto InterGIS **1**(26), 30–44 (2020). https://doi.org/10.35595/2414-9179-2020-1-26-30-44
7. Freeman, R.E.: Strategic Management: A Stakeholder Approach. Pitman, Boston (1984)
8. Sachuk, T.V.: On the issue of territory branding, or who needs a territory brand and why. Munitsipalnoye Imushchestvo: Ekonomika, Pravo, Upravleniye **1**, 30–36 (2016)
9. Vazhenina, I.S., Vazhenin, S.G.: Horizons of territorial competition in the modern economic space. Obshchestvo i Ekonomika **3**, 67–85 (2011)
10. Larionova, N.A.: Mechanisms of management of intangible assets of the region. Vestnik SevKavGTU **2**, 109–112 (2006)

Paradigm of Modeling a Business Development in Agro-Industrial Region

Igor V. Mandritsa[1(✉)], Olga V. Mandritsa[2], Irina V. Solovieva[3], and Anna A. Ter-Grigoryants[1]

[1] North Caucasus Federal University, Stavropol, Russia
imandritsa@ncfu.ru
[2] Branch of Russian Technological University (MIREA), Stavropol, Russia
[3] Nevinnomyssk State Humanitarian and Technical Institute, Nevinnomyssk, Russia

Abstract. The specificity of agro-industrial regions in depression is manifested in lagging behind the average growth rates of the economy, the quality of life and employment of population, subsidized dependence, insufficient use of production factors, technology, innovation, low potential of small, medium and large businesses. The development of state and regional entrepreneurship support programs does not ensure the complete activation of business, which increases the relevance of the search for paradigms and the modeling strategies for getting out of the situation. Systems theory and systematic analysis, matrix modeling of industrial and entrepreneurial activities, systemic dynamic modeling, statistical analysis, theories of uneven economic development and spatial business projections, conceptual methodological approaches to the formation of an entrepreneurship development strategy were used in the study. Formation of the paradigm of modeling a business strategy in an agro-industrial region is based on the study of theories, methods and analysis of the role and trends of changes in the position of small and medium-sized enterprises ("SMEs") in the economic system of a region. As a strategic priority for overcoming the depressive phase by regions of the North Caucasus Federal District, researchers call an integrated strategic business management system based on the combination of institutional factors: regional authority and entrepreneurship. The use of a matrix modeling model in the SME development planning strategy allows removing the imbalance between production and consumption and improving the results of functioning of SMEs as part of the economic system of the region.

Keywords: Small-sized business · Matrix modeling · Development strategy

1 Introduction

The current stage of economic development is characterized by transformation processes towards high-quality renewal at the national level, with a simultaneous unevenness of their course throughout the country. Transformations and their results are most clearly visible mainly in regions with propulsive industries and businesses. These regions are characterized by rapid growth in entrepreneurship, employment and living standards.

P. V. Trifonov and M. V. Charaeva (Eds.): DITEM 2021, LNNS 380, pp. 296–303, 2022.
https://doi.org/10.1007/978-3-030-94245-8_41

They are in sharp contrast to meso-formations of traditional spheres of activity related to traditional agricultural and agro-industrial production.

The research purpose is to rethink existing approaches and develop a paradigm for modeling an entrepreneurship development strategy based on a mathematical description of tasks of the main market relations agents in the agro-industrial territory.

2 Methods

The study of new properties of economic and mathematical models and their applicability to various technologies and technological processes used by entrepreneurs is relevant for creating an entrepreneurship development strategy (as part of state policy at the regional level). A number of works consider the aspect of modeling economic processes in the agro-industrial complex [1–3]. The approach of systemic dynamic modeling based on works by Dimov, Ilyasov, Makarova, et al. [4], who revealed the essence of integrity of SME functioning as part of the economic system of a region as a research object, was chosen as the fundamental one. A fundamentally new situational-strategic planning system is presented in the works by Merkulova [5], where long-term strategic planning and short-term situational planning are combined into a single system.

When describing the integrity of functioning of SMEs as part of the economic system of a region, one should apply a matrix modeling method, which is described in the works on planning the economic systems of a territory. The input-output model (Leontyev) makes it possible to form an optimal relationship between production sectors of individual regions, described in the studies by Kustov within a dual-sector model [6].

The experience of studying the processes of lagging behind and outstripping growth of some or other meso- and macro-formations is presented in theories of exogenous and endogenous pressure on the economy, developed in the works by foreign scientists Solow – Swann (exogenous growth theory), Lucas – Romer, Myrdal, Friedman, Perroux, Richardson, Hagerstrandt, Hirsch and their followers (endogenous growth theory based on the assessment of internal factors of the regional development) [7].

Methods and tools of the modern theory of uneven economic development and spatial business projections, business development in the country's depressed territories are presented in publications of the authors (Anchishkin, Asaul, Barkan, Goncharov, Bogdanov, Zainullin, Kleiner, Konoplyanik, Kravtsova, Meshkov and others).

The analysis of entrepreneurial activity at the regional level is widely presented in the works by Russian and foreign researchers, which emphasize the unevenness of regional development and the importance of small and medium-sized businesses as an institutional factor in the development of the economic system [8].

The presented conceptual methodological approaches to the formation of an entrepreneurship development strategy as the main potential for the territorial development form the basis for developing a paradigm of modeling a strategy for small and medium-sized business development in agro-industrial regions of the South of Russia.

3 Results

The high level of differentiation of the national economic space and the uneven economic development of Russian territories are emphasized in works [9–11], the authors of which differentiated the regions by such criteria as innovation-investment activity, economic growth rate, the index of conditions for small business development. The ranking results showed that 67 regions (79% of their total number in the country) have lagged behind the other regions for over 10 years [7], including the agro-industrial regions of constituent entities of the North Caucasus Federal District, experiencing similar problems of depressed development.

The formation of the paradigm of modeling a business strategy in an agro-industrial region is based on the study of theories, methods and analysis of the role and trends in the position of SMEs in the economic system of a region on the example of the Stavropol Territory. According to official statistics [11] from 2015 to 2019, the number of SMEs decreased by 17% and amounted to 23,253 enterprises, the number of employees decreased by 22.8% and amounted to 99.6 thousand people, which is 8% (66.9 thousand people) in 2019 of the total employed population of the region (1,315.7 thousand people), while in 2015 it amounted to 10% (129.1 thousand people), i.e. decreased by 2% (29.5 thousand people). The ability to bring economic benefits to the regional economy in terms of added value as part of the gross regional product ("GRP"), as follows from the dynamics of the share of GRP in the revenue (turnover) of SMEs, decreased from 38% to 30%, which is characterized by a decrease in business activity and the scale of positive participation of SMEs in the regional economy. In 2019, in the revenue of all SMEs in the region by type of activity, the largest share (54%) falls on trade and services, the second sphere in terms of revenue (19%) is agriculture, and the third one (14%) is construction. In the Stavropol Territory, it mainly prevails in trade in comparison with those spheres where added value is created (agriculture, processing industries, construction), which is so necessary for the region's economy. SMEs reduce the value of their business, do not create or apply technology with a greater share of added costs (jobs, fixed assets, new technology and other resources).

It appears necessary to create an integrated SME strategic management system based on the combination of institutional factors: regional authority and entrepreneurship.

The main difficulties in regulating markets at the regional level in the strategic planning of entrepreneurship development are associated precisely with the balance of production and consumption. There is either "surplus" or "deficit" of some types of products and goods. Overproduction of products, works or services or excessive pricing for a type of activity leads to an imbalance between production and consumption. There is a need to regulate the imbalance based on the matrix model of the SME industry as the sum of production capacities of SME agents in the territory with its dynamic characteristics (Table 1).

The matrix modeling (Table 1) revealed a paradigm of strategic SME development planning, which indicates two target functions for two agents of the economy, namely, an increase in the gross regional product for regional authorities and an increase in the SME production capacity for entrepreneurs. It should be noted that these target indicators may not coincide in their direction vectors, i.e. to fill the domestic market of goods (services, works) the regional government will manage the balance of consumption for the benefit of external resource suppliers $V_{m+l1} V_{m+l2} \ldots V_{m+1m}$ and producers $Y_{m+1(1)} + Y_{m+1(2)} + \ldots = Y_{m+1}$, which will also lead to an increase in the gross regional product, but will

Table 1. Matrix model of implementing the paradigm of SME development strategy in an agro-industrial region.

	Own suppliers of SMEs in the region	Types (directions) of production of SMEs in the region m_i	Production results (SME products) Domestic market of the territory V_{mi}	Third-party consumers External market outside the territory V_{mi}	
Produced goods and services (types, industries)		$X_{11}\, X_{12} \ldots = X_{1m}$	$Y_{1(1)} + Y_{1(2)} + \ldots = Y_1$	$V_{1l}\, V_{1\,l1} \ldots V_{1L}$	M2
		$X_{21}\, X_{22} \ldots = X_{2m}$	$Y_{2(1)} + Y_{2(2)} + \ldots = Y_2$	$V_{2l}\, V_{2\,l1} \ldots V_{2L}$	M1
		…	…	…	
		Inter-production (intersectoral) supply of resources within the territory for production (SME assets)	Final consumed product (import) (SME revenue)	Export supplies (export) (SME revenue)	Production capacities of SMEs in the region Mmi_i
			$\Sigma Y_{mi}(V_{mi})$	$\Sigma Y_{mi}(V_{mi})$	
		$\Sigma X_{mi}(V_{mi})$	$Y_{m(1)} + Y_{m(2)} + \ldots = Y_m$	$V_{ml}\, V_{m\,l1} \ldots V_{mL}$	
Similar supplied (imported) goods and services	Third-party suppliers	$X_{m1}\, X_{m2} \ldots = X_{nm}$			ΣM_{mi}
	$V_{m+l1} \ldots V_{m+1L}$	$V_{m+l1} V_{m+l2} \ldots V_{m+1m}$	$Y_{m+1(1)} + Y_{m+1(2)} + . = Y_{m+1}$	Region's losses in value added (losses of regional product)	
	…	…	…		
	Supply sources (import)	Supply of resources from outside (import) for production consumption	Supply of similar products from outside (import) for final consumption	$\Delta RP = (-\Sigma X_{mi}(V_{mi}) - -\Sigma Y_{mi}(V_{mi}))$	
	$V_{nl} \ldots V_{nL}$	$-\Sigma X_{mi}(V_{mi})$	$-\Sigma Y_{mi}(V_{mi})$		
		$V_{n1}\, V_{n2} \ldots V_{nm}$	$Y_{n(1)} + Y_{n(2)} + \ldots = Y_n$		
Depreciation, wages, profits and sales tax		$Z_{A1}\, Z_{A2} \ldots Z_{Am}$	$W_{A(1)}\, W_{A(2)}$		
		…			
		$Z_{V1}\, Z_{V2} \ldots Z_{Vm}$	$W_{V(1)}\, W_{V(2)}$ Redistribution		
		…	$W_{M(1)}\, W_{M(2)}$		
		$Z_{M1}\, Z_{M2} \ldots Z_{Mm}$			
		$T_1\, T_2 \ldots T_m$ Labour and workforce	$T_{(1)}\, T_{(2)}$	Region's losses on underemployed capacity of SMEs (loss of jobs and resources)	
		$F_1\, F_2 \ldots F_m$ Fixed and circulating assets	$F_{(1)}\, F_{(2)}$		
		$R_1\, R_2 \ldots R_m$ Production resources	$R_{(1)}\, R_{(2)}$	$\Delta M = (-\Sigma T_{mi} - \Sigma F_{mi} - \Sigma R_{mi})$	

negatively affect the capacity of the region's SMEs on these types of goods (works, services) $\sum M_{mi}$.

The paradigm of strategic SME development management will consist in point regulation and control of deviations (deltas) of these factors of the matrix model as target functions, namely, their minimization:

$$\begin{cases} f(\Delta\text{RP}) = min\left(-\sum X_{mi}(V_{mi}) - \sum Y_{mi}(V_{mi})\right) \\ (\Delta\text{M}) = min\left(-\sum \text{T}_{mi} - \sum \Phi_{mi} - \sum \text{P}_{mi}\right) \end{cases}. \quad (1)$$

A delta (difference) in deviations by indicators of this matrix is only dynamic, temporary deviations in the process of production, economic and trading activities of all matrix participants (factors).

4 Discussion

According to the authors, modeling of a SME development strategy in an agro-industrial region is based on a paradigm, which implies the use of an institutional approach of interaction between the main agents of the regional market, on the one hand, represented by regional (territorial) executive authorities (its ministries, curators of the SME industry) and all active participants in SMEs subjectively united into a single "industry".

The success of implementation of this paradigm will be based on combining the efforts of regional authorities and entrepreneurship, and the authors propose to implement its verbal mathematical description by developing a matrix model of production and consumption, reflecting the structure of costs for the production and consumption (distribution) of SME products, as well as the newly created value for the local economy. SMEs are used for various types of entrepreneurial activities in the territory, which allows them to be combined into a certain common sphere of the economy – small and medium-sized business of a region.

The implementation of the matrix modeling model is provided by regulatory measures and the interaction between the main agents. Table 2 presents 18 key indicators of an integrated system of strategic business management in an agro-industrial region in the form of a matrix. Jointly regulating and exercising control over these indicators, the market agents, regional authorities and their departments will ensure the possibility of system implementation and achievement of the set goals, namely, an increase in the gross regional product, reduction of unemployment for government authorities and an increase in the SME production capacity and additional jobs for entrepreneurs. The dual task of strategy agents should not lead to "damage" incurred by any of the participants.

In recent years, Russian and foreign scientists have conducted theoretical and practical research on a territorial development strategy, in particular – entrepreneurship as the main potential of the market economy. Kvint formed the theory of strategizing [12], which was further studied in Pavlova's works on the development of regional strategies [13], presented in the legislative documents "Strategic priorities of the Far Eastern and Volga Federal Districts" and others [14]. The growth pole theory by foreign scientists Perroux, Budville, Potier, Lasuen, Torre [15] and the theory of nationally oriented clusters in the works by Sherin [16] are used in the development of regional strategies. Examples of successful development of Russian territorial entities on the basis of adopted strategies are characterized by their belonging to propulsive industries, the presence of other factors of outstripping growth.

Table 2. Integrated system of strategic entrepreneurship management in an agro-industrial region.

	Input	SME value added research				
Input	Analysis according to costs – output scheme, SME production cost	Analysis of consumer demand for CURRENT SME products in accordance with the SME development strategy	Models of creation of SME capacity, stocks and growth of fixed assets; investment policy for SMEs in the Territory	Analysis of demand for products, goods and services produced by SMEs		Output
	Production functions – planning and control of the SME business process: the introduction of new management technologies	Income distribution and redistribution; regional tax policy for SMEs: protectionism or protective measures	Analysis of depreciation of SME fixed assets Implementation of new technologies for CONTROL of SME activities by the Ministries and Agencies of the Territory	Research on SME income from investment in new technologies and production		
Research on the value of SME assets less liabilities	Analysis of consumer demand for NEW SME products in accordance with the SME development strategy	Analysis of behavior of credit institutions for SMEs as part of the strategy – incentives, subsidies	Monetary policy by the Ministries and Agencies of the Territory for SMEs	SME financing and liquidity problems within the Territory Analysis of SME turnover	Research on SME profits and losses on the capital account	Research on the value of SME assets less liabilities (study of losses or growth)
	Analysis of the demand for imported goods and services in the Territory similar to SME agents in the Territory	Policy on short-term assistance to SMEs under the chosen strategy, points of SME growth	Financing and liquidity problems, government support, loans to SMEs	Analysis of deviations in SME strategy indicators		
		Output	SME forecast by key indicators Research on SME added value (regional product)			

5 Conclusion

Thus, the recommended paradigm of an integrated system of strategic business management in an agro-industrial region will make it possible to bring it into a strategic factor of the economy for overcoming the stage of depression by agro-industrial regions. In such a system, the integration of institutional factors is ensured: regional authority and entrepreneurship. The government becomes responsible for the development of tools for regulating regional markets, conditions for promoting business, coordinates interactions with SMEs and monitors the fulfillment of all conditions, prepares an action program to get out of the depression to the stage of sustainable development. Entrepreneurs in the form of SMEs play the role of market agents with the growth of production-economic initiatives, replenish budgets, create profitable jobs and domestic demand, create the prerequisites for economic growth and a way out of the depressive state of the economy of an agro-industrial region.

It should be pointed out that under "any" strategy for the development of the territory's economy at the current stage, the following additional mechanisms should be developed that create favorable conditions for SMEs in the region, namely:

1. Modernization of infrastructure of an economic development strategy object, i.e. the environment of factors influencing SMEs in terms of its economic functioning (methods of planning, compliance with technologies and regulatory requirements, as well as operational accounting and control of selected indicators) by subjects of control;
2. Corruption control by all indirect agents, participants in the SME development strategy, including credit institutions to control speculative transactions, the cost of a loan;
3. Control of monopolies on the territory of the studied region in terms of prices and freedom of access to SME product markets;
4. Control opposite to control of monopolies – reasonable protectionism to improve the investment climate and support SMEs for the purpose of creating a successful image.

References

1. Akindinov, V.V., Kuryanov, A.V.: Modeling of economic processes in the agro-industrial complex. In: Gulyaeva, G.Y. (ed.) Innovative Approaches in Science and Education: Theory, Methodology, Practice. Science and Education, Penza (2017)
2. Sokolova, I.V.: Mathematical models in the economy of the agro-industrial complex. Kuban State Agrarian University named after I.T. Trubilin, Krasnodar (2020)
3. Statistical and econometric research on socio-economic systems of an agrarian-oriented region: collection of scientific papers based on the materials of the VIII International Scientific and Practical Conference, Stavropol, May 28. Publishing House "AGRUS", Stavropol (2018)
4. Dimov, E.M., et al.: Infocom Technol. **16**(1), 81–96 (2018). https://doi.org/10.18469/ikt.2018.16.1.09

5. Merkulova, Y.V.: Guidelines for the Implementation in the Economy of a Situational Strategic Planning System. Publishing House "Academy of Natural Sciences", Moscow (2017). https://doi.org/10.17513/np.230
6. Kustov, E.F., Dli, M.I., Kuzavko, A.S.: Bul. Moscow Power Eng. Inst. **2**, 161–165 (2012)
7. Kuznetsov, Y.V., Kaisarova, V.P.: Univ. Bul. **8**, 104–110 (2019). https://doi.org/10.26425/1816-4277-2019-8-104-110
8. Karatysh, V.E.: Econ. Sustain. Dev. **1**(37), 150–155 (2019)
9. Valiakhmetov, I.R.: Assessment and development of the innovation-investment potential of mesoeconomic systems. Ph.D. thesis in Economics, Kazan (2019)
10. The National Investment Climate Ranking in Constituent Entities of the Russian Federation. https://asi.ru/investclimate/rating/
11. Statistical Yearbook. Stavropol Territory. 2020: Statistical Book. C78 North Caucasus Statistic Service, Stavropol (2020)
12. Kvint, V.L.: The concept of strategizing, v 1. NWMI RANEPA, St. Petersburg (2019)
13. Pavlova, N., Fomin, A., Ismailov, T.: RISC **3**, 365–372 (2011)
14. Kozyrev, A.A.: Econ. Ind. **13**(4), 434–447 (2020). https://doi.org/10.17073/2072-1634-2020-4-434-447
15. Torre, A.: Revue d'économie régionale et urbaine **3**, 409–437 (2010). https://doi.org/10.3917/reru.103.0409
16. Sherin, V.A.: Rus. Econ. Int. J. **1**, 34 (2016)

Efficiency of Tax Instruments in Stimulation of Investment Activity

Elvina A. Omshanova(✉) and Alla Yu. Chalova

Plekhanov Russian University of Economics, Moscow, Russia
Omshanova.EA@rea.ru

Abstract. Various preferential taxation schemes designed to stimulate investment activity have gained new momentum as part of the implementation of policy documents on realisation of national projects. The analysis of the prescriptive legal regulation has shown that preferential taxation schemes, depending on the sphere, can be as follows: national and supranational; territory-specific: special economic zones, advanced development territories, free ports, innovation centres, special administrative districts; depending on the form of investment and innovation activities: special investment contracts, regional investment projects, agreements on protection and promotion of capital investments, investment tax credit. Addressing the ways to increase the efficiency of tax instruments intended to stimulate investment activity is the key objective of fiscal, tax and custom-tariff policy in modern conditions. The purpose of the research is the formulation of proposals to improve the evaluation of efficiency of tax instruments used to stimulate investment activity within the framework of special investment projects. The basic research methods involved system analysis, comparative analysis of the existing methods for evaluation of efficiency of tax instruments aimed to stimulate investment activity. The research result lies in the development of approaches to assessing the efficiency of special investment contracts.

Keywords: Fiscal performance · Economic efficiency · Special investment contracts · Investments · Tax incentives · Tax expenditure

1 Introduction

The expansion of investment activity of economic agents is of great importance in the conditions of adaptation to new economic realities after the fall in oil prices and quarantine restrictions. Investments are able to ensure due modernisation of production, increased labour productivity and, therefore, improved competitiveness of output. In accordance with the established national objectives for the development of the Russian economy pursuant to the Decree of the Russian Federation President No. 474 as of July 21, 2020, it is necessary to provide the growth of investments in fixed capital by 70% as compared to the figure of 2020. In 2011–2019 the share of investments in Russia's GDP averaged 20.8%, with the average annual investment growth of 1.9%. In order to restore and secure sustainable dynamics of the Russian economy, the annual growth of

P. V. Trifonov and M. V. Charaeva (Eds.): DITEM 2021, LNNS 380, pp. 304–311, 2022.
https://doi.org/10.1007/978-3-030-94245-8_42

investments by 10–15% is required with the share of investments in GDP amounting to 20–22%. Among the main sources of investment funding are legal entities' own funds (in 2020 – 57%, an increase by 15 p.p. as compared with 2011). A whole range of tax instruments is used for stimulation of investment activity, including: tax incentives for capital expenses, research and development, accelerated depreciation, investment tax credit, investment tax deduction, as well as tax incentives granted to investors within the framework of preferential taxation schemes: regional investment projects, special investment contracts, priority development areas, free economic zones, special administrative districts. Given the need to stimulate investment and innovation activities, as well as to attract investment in high-tech sectors of the economy, the introduction of a special investment contract mechanism in 2015 is a promising tool for stimulation of investment activities. In this regard, the efficiency of tax instruments within the framework of SPIC (special investment contracts) becomes particularly important, with regard for the relevant budget expenditure and the investor's fulfilment of its obligations under contracts, as well as ensuring the balance of fiscal powers of the Russian.

2 Materials and Methods

The assessment of tax incentives for investment activity is included in the fiscal process within the framework of the developing institute of tax expenditure, in order to ensure the balance of budgets in the Russian fiscal system and in view of the need to meet the objectives of the national projects. Tax expenditure is represented by tax incentives or preferences the objectives of which can be achieved through public funding in the form of subsidies or other direct costs of state budgets. The Russian Ministry of Finance uses the foregone revenue method in estimating the tax expenditure. The assessment of tax expenditure is based on unified methodological approaches that include the following elements:

1. systematisation of tax expenditure into a single register;
2. appointing a tax expense supervisor;
3. development of tax expenditure evaluation methodology based on assessing the feasibility and efficiency of tax expenditure, which supposes evaluation of effectiveness – fiscal, economic, social, formation of statutory documents for tax incentives. Economic efficiency takes into account the growth rate of investment in fixed capital and of certain indicators of taxpayer's financial and economic activity: output of goods (scope of work, services), profit, share of innovative goods, etc. The analysis of tax instruments efficiency within the framework of SPIC is carried out with regard for related tax expenditure and the investor's achievement of obligations under the contract, the system of accounting and monitoring of the total scope of governmental support. The tax expenditure includes as well tax incentives. In addition, the analysis was also carried out on the basis of estimating the number of SPIC taxpayers, total number of SPICs, the volume of actual investment and the scope of factual tax payments [1–4].

3 Results

Between 2016 and 2019, 45 SPICs to the total amount of RUB 805.5 billion were concluded at the federal level according to the SPIC registry, with the largest number of contracts awarded in the automotive industry (15 SPICs), while the greatest amount of investment commitments were registered in the chemical industry (RUB 432.7 billion, representing 54% in the total SPIC investment commitments) (Fig. 1).

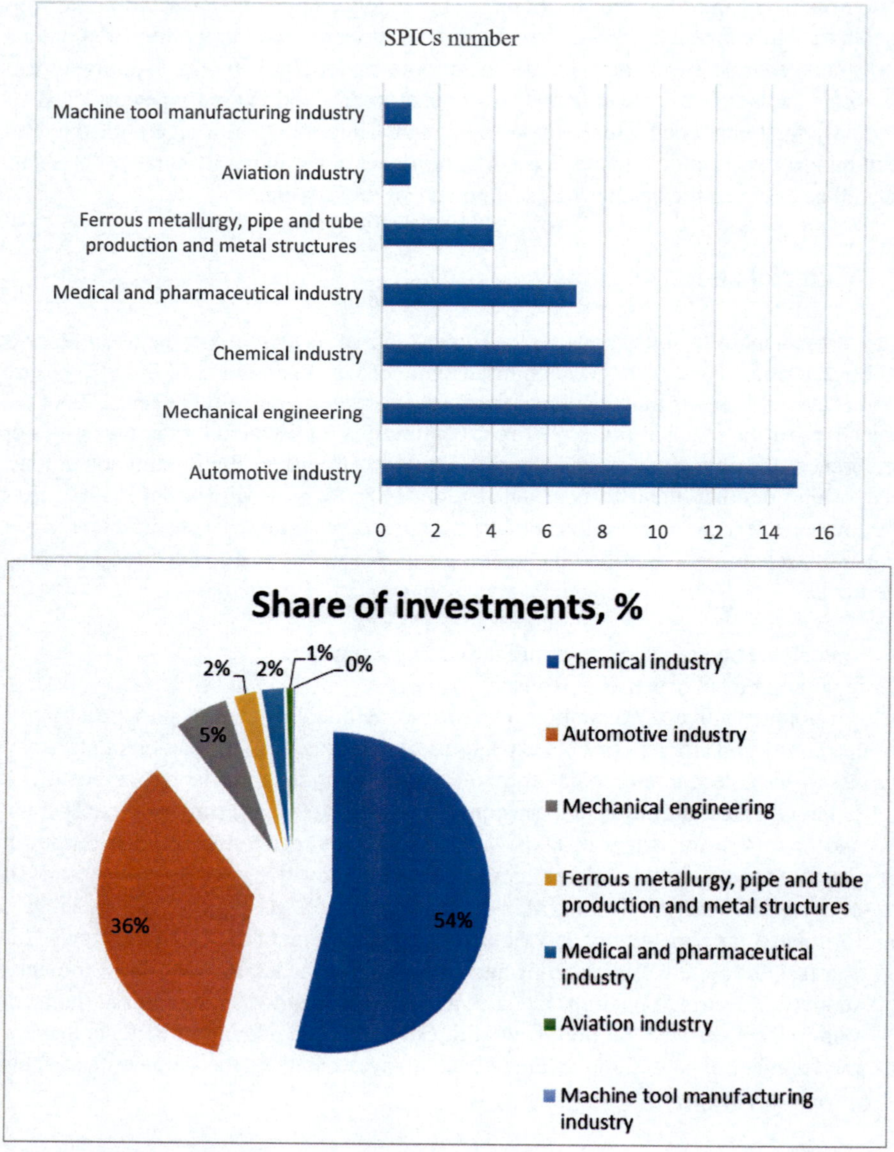

Fig. 1. Number of SPICs and investment structure in terms of industries (%). *Source:* Compiled by the authors [5].

The reconfiguration of the tax mechanism for investment incentives in terms of SPIC (Table 1) was introduced by amendments to the federal law "On Industrial Policy in the Russian Federation", in the Tax and Budget Codes, starting from 2019 [1, 2, 6].

Table 1. Specific features of a special investment contract.

Criteria	SPIC 2.0
Investment project	Inculcation or development and introduction of new technologies in the mass fabrication of industrial products
Minimum investment amount	No requirements established
Tax incentives	Corporate income tax, property tax, transport tax, land tax
Guarantees	Invariability of business conditions for the duration of a SPIC
Participants	Mandatory participation of the Russian Federation, a constituent entity of the Russian Federation, a municipality
Conclusion procedure	Tendering procedures
Period of validity	Depending on the scope of investment

Source: Compiled by the authors.

Under the adopted changes, the SPIC mechanism shifted towards introduction or development and realisation of new technologies in serial manufacture of industrial products. The analysis of the modified conditions of SPIC made it possible to reveal several specific features, among them [3, 7–10]:

1. three-tier architecture of contract participants represented by executive authorities and management bodies: federal, regional executive authorities and local self-government bodies.
2. tendering procedure for conclusion of SPICs, which, on the one hand, makes this tool rather cumbersome, while on the other hand, increases the quality of investment projects;
3. removal of restrictions on the volume of investment, which extends the range of potential investors, including small and medium business, participants of territorial clusters.
4. setting the duration of SPIC 2.0. depending on the volume of investment became a compensation for withdrawal of restrictions on the scope of investment: 15 years for SPIC with the volume of investment less than 50 billion rubles and 20 years for those over 50 billion rubles.
5. optionality of tax incentives for corporate income tax, i.e. the investor has the right to choose: to apply reduced corporate income tax rates relative to the entire taxation base or to the tax base under SPIC activities;
6. separate accounting for SPIC activities;
7. setting the conditions for the forfeit of tax incentives. The loss of tax incentives is connected with the scope of budgetary support to a participant of SPIC 2.0 in the form of subsidies or tax expenditure, which will lead to the need to track achievement of the 50% threshold of budgetary support and may cause some practical difficulties.

When assessing the efficiency of SPIC, it is necessary to consider their attractiveness for the state, which includes: development of significant industries, introduction of new technologies in industrial production, perfection of high-tech production processes, improvement of investment attractiveness of a particular territory. According to the reports of the Federal Tax Service of Russia, the tax expenditure of the state budget system increased 5.4 times from 2017 to October 2020: from 1,209.4 billion rubles to 6,539.5 billion rubles; the income tax burden decreased by 0.9 p.p. – from 3.8% in 2017 to 2.9% in 2020; however, the number of taxpayers using SPIC did not change according to the tax reporting data (Tables 2 and 3).

Table 2. Tax expenditure for SPIC participants in 2017–2018.

Parameter	2017		2018	
	Legal entities – regional investment project participants (RIPs)		Legal entities – regional investment project participants (RIPs)	
	Included in the register of RIP participants and special investment projects	Those who applied for a benefit	Included in the register of RIP participants and special investment projects	Those who applied for a benefit
Corporate income – total	57,182,759	39,774,048	147,063,079	81,718,307
Corporate expenses – total	44,086,751	24,606,874	183,017,028	59,903,927
Taxation base for tax calculation	7,472,006	14,798,534	10,146,921	21,230,932
Amount of calculated income tax transferrable to the budgets of Russian Federation entities	284,979	2,268,158	161,305	3,141,543
Amount of tax shortfall due to application of zero and reduced tax rates	1,209,417	691,549	1,801,487	1,101,870
Including to the federal budget	224,153	443,956	275,855	636,513

(continued)

Table 2. (*continued*)

Parameter	2017		2018	
	Legal entities – regional investment project participants (RIPs)		Legal entities – regional investment project participants (RIPs)	
	Included in the register of RIP participants and special investment projects	Those who applied for a benefit	Included in the register of RIP participants and special investment projects	Those who applied for a benefit
Amount of corporate loss calculated upon taxation	4,428,163	0	50,348,806	0
Number of taxpayers	12	7	26	12
Including loss-making organisations (units)	4	0	15	0

Source: Compiled by the authors.

Table 3. Tax expenditure for SPIC participants in 2019–2020.

Parameter	2019	As of 01.10.2020
	Legal entities – participants of special investment contracts	Legal entities – participants of special investment contracts
Corporate income – total	61,120,868	464,233,542
Corporate expenses – total	56,229,340	435,680,233
Amount of losses of preceding tax periods that reduced the taxation base	131,739	146,219
Taxation base for tax calculation	11,201,022	38,212,160
Amount of calculated income tax	12,793	1,102,924
Including to the federal budget	X	X

(*continued*)

Table 3. (*continued*)

Parameter	2019	As of 01.10.2020
	Legal entities – participants of special investment contracts	Legal entities – participants of special investment contracts
To the budgets of the constituent entities of the Russian Federation	12,793	1,102,924
Amount of tax shortfall due to application of zero and reduced tax rates	2,225,527	6,539,507
Including to the federal budget	333,973	1,146,366
Amount of corporate loss calculated upon taxation	6,441,234	9,788,307
Number of taxpayers	10	12
Including loss-making organisations	5	4

Source: Compiled by the authors [5, 11].

4 Discussion

Tax instruments aimed to stimulate investment activity include the mechanism of tax incentives at the federal and local level [4, 12]. The existing procedure for tax investment benefits and their scope leads to imbalance of federal and regional tax powers in the sphere of stimulation of economic agents' investment activity. In this regard, further improvement of the institute of assessment of tax expenditure in the investment sphere is required in order to provide a due balance of fiscal powers of the Russian Federation and its constituent entities. The evaluation parameters of tax incentives instruments should include standard assessment units, including the efficiency of SPIC investors, the indices of SPIC impact on industry and regional development as well as the integral index of efficiency of taxation instruments of investment activity stimulation.

5 Conclusion

The core results of the research are represented by the identified problems of assessing the efficiency of tax instruments for stimulation of investment activity: poor linkage of efficiency criteria with the objectives and mechanism of tax incentives. Taking into account the existing SPIC practice and the SPIC-specific preferences, it is possible to talk of further expansion and advancement of this instrument. The accent in the updated SPIC 2.0 is shifted to encouragement of long-term investment in the scientific and technological sphere along with the creation of innovation clusters; however, to achieve efficient use of this tool, it is necessary to adopt a number of regulatory acts.

References

1. Evstafieva, Y.: Bul. Inst. Econ. Rus. Acad. Sci. **3**, 152–167 (2019). https://doi.org/10.24411/2073-6487-2019-10038
2. Evstafieva, Y.V.: Econ. Strat. **22**(1(167)), 64–71 (2020)
3. Andrievsky, K.V.: J. Rus. Law **1**, 147–158 (2020). https://doi.org/10.12737/jrl.2020.011
4. Goncharenko, L.I., et al.: Assessment of Tax Incentives' Efficiency. Ru-Science, Moscow (2020)
5. SPIC 1.0 Registry. https://gisp.gov.ru/spic2/pub/spic/search/
6. Evstigneev, E.N., Viktorova, N.G.: Tax Management and Tax Planning in Russia. INFRA-M, Moscow (2020). https://doi.org/10.12737/606
7. Goncharenko, L.I., Vishnevskaya, N.G.: Econ. Taxes. Law **12**(4), 121–131 (2019). https://doi.org/10.26794/1999-849X-2019-12-4-121-131
8. Serikova, M., et al.: Entrep. Sustain. **7**(3), 2419–2435 (2020). https://doi.org/10.9770/jesi.2020.7.3(64)
9. Hans, G., Ulrich, S., Maik, T.S.: Econ. J. **129**(620), 1731–1781 (2019). https://doi.org/10.1111/ecoj.12588
10. Sembiyeva, L.M., Serikova, M.A., Zhagiparova, A.O., Korzeb, Z.: Bul. Natl. Acad. Sci. Republic Kazakhstan **5**, 132–142 (2019). https://doi.org/10.32014/2019.2518-1467.132
11. Tax expenditures of the Russian Federation 2015–2021. https://minfin.gov.ru/ru/document/?id_4=124754-prilozhenie_1
12. Serikova, M., et al.: Entrep. Sustain. **7**(4), 2764–2783 (2020). https://doi.org/10.9770/jesi.2020.7.4(13)

The Problems of Digitalisation in the Eurasian Economic Union

Irina N. Platonova(✉) and Maria A. Maksakova

MGIMO University, Moscow, Russia
platonova@mgimo.ru, m.maksakova@inno.mgimo.ru

Abstract. The transition to digital transformation of economy is viewed by the EAEU countries as an opportunity to increase the competitiveness and sustainability of the national economy by joining efforts within the framework of their integration. It is within the framework of integration-based associations that the barriers to increasing the scope of digital transactions effectuated in the conditions of free competition may be eliminated. The purpose of the research is to verify the potential of digitalisation in terms of influencing the development of integration in a number of countries, in particular the EAEU, and thus increasing their international competitiveness. The article presents the approaches of various authors to defining the concept of "digital economy" and highlights the international component in its socio-economic substance. It substantiates the need to coordinate the actions of EAEU countries upon the implementation of the Main Directions for the Realisation of the EAEU Digital Agenda until 2025. It is shown that the realisation of the national digital transformation programme in concordance with the supranational level makes it possible to improve the competitive position of a country by the example of Russia which has significantly improved its positions in the Global Innovation Index. The need to strengthen the coordination of the EAEU countries in the field of digitalisation has been substantiated: new barriers separating the countries by the technological principle may arise without it, and this will hinder the realisation of the principle of "four freedoms" necessary for enhancement of the integration.

Keywords: Digital transformation · Eurasian economic union · Innovations · Digitalisation · Integration

1 Introduction

From the standpoint of achieving the economic growth against the backdrop of the slow recovery of the global economy, the digital transformation of the entire reproduction process is the basis for the realisation of the innovative development strategy. Artificial intelligence and new technologies occupy an important place in this strategy owing to the digitalisation, allowing to achieve maximum possible effect from the use of investment resources and human capital.

Developing the Solow Growth Model as an important source of long-term economic growth, D. Romer considers generation of new knowledge and innovations as a condition for strengthening the international division of labour and due development of the

P. V. Trifonov and M. V. Charaeva (Eds.): DITEM 2021, LNNS 380, pp. 312–317, 2022.
https://doi.org/10.1007/978-3-030-94245-8_43

international commodity turnover in a competitive environment [1]. In these terms, the use of digital technologies may be theoretically considered as an important growth factor in transition of economy to the Fourth Industrial Revolution, when the introduction of innovations plays a key role for sustainable development of the world economy. However, according to the estimates of Klaus Schwab, head of the World Economic Forum, about 17% of the world population have no possibility to use electricity, and therefore – to benefit from the results of (yet) the second industrial revolution, so the digitalisation of their economies is so far delayed in time [2].

Currently, the scale of high-tech production sector in the developed countries and the efficiency of use of high technologies (commercialisation opportunities, integration of innovations into production activities) provide for due scientific, technical, innovative and economic potential of a country and suppose transition to digital economy [3]. Therefore, a number of economists single out, for good reason, digital economy from the great number of innovations required for transition to the new technological order [4].

In defining the backbone of "digital economy" (DE), it is necessary to point to the dual approach because of the rapid growth and diversity of digital activities in the last decade. First, "it is the economy based on digital technologies such as telemedicine, distance learning, sale of media content (cinema, TV, books, etc.)" [5]. Second, "digital economy is economic reproduction with the use of most complex digital technologies, including cloud storage and introduction of robotic cloud platforms" [6, 7]. However, these definitions of DE as such do not take into account the impact of digitalisation on the deepening international interaction of countries, on the one hand, as well as the impact of the countries' integration based on the common digital agenda on the issues of strengthening of their economies in the international competition, on the other hand.

Given the multifaceted nature of digital technologies, the purpose of the research is to assess the potential of digitalisation in terms of influencing the integration of a number of countries, in particular within the EAEU, and enhancing their international competitiveness.

The analysis of the international experience shows that the leading countries have formed "common digital agendas" coordinating and regulating the digitalisation mechanisms at the interstate level. An example is the Europe 2020 Strategy and the Digital Agenda for Europe which envisage "the creation of a single European digital space aimed to ensure sustainable and inclusive growth of the EU economy" [8].

2 Materials and Methods

To assess the possibility of digital transformation of economy with a view to promote integration of a number of countries, in particular within the EAEU, and to strengthen their international competitive position, a number of objectives were identified that might give an answer to this question.

These objectives include:

- specification of the substance of DE from the standpoint of its impact on strengthening the inter-country integration, in particular in the EAEU;

- the use of comparative analysis in respect to the content of the national digital transformation programmes for the economies of the EAEU countries in accordance with the digital agenda of the EAEU;
- identification of further development areas of digital transformation regulation in the process of strengthening the Eurasian integration in the EAEU.

The criterion analysis of the points of view defining the nature of digital economy made it possible to clarify its essential characteristics arising from the international interaction and the integration of the EAEU countries.

To identify the conditions under which the digitalisation of economy will contribute to the integration of the EAEU countries, the qualitative analysis of the national digital programmes of the EAEU countries was used in terms of their compliance with the document of supranational regulation – "Main Directions for the Realisation of the EAEU Digital Agenda until 2025" which was adopted and approved by the results of the meeting of the Supreme Eurasian Economic Council in October 2017. This made it possible to conclude relatively successful realisation of this document, however, not sufficient to deepen the countries' integration-based cooperation.

3 Results

The analysis of the national and foreign researchers' approaches to the definition of the "digital economy" concept allowed the authors to identify several aspects:

1. Broad and narrow definition of DE, depending on how its coverage of the reproductive process is taken into account. For instance, the broad definition of "the digital economy makes it possible to consider the totality of industries connected with the emergence of new technologies using digital platforms, new technologies, robotics, smart technologies, etc. The narrow definition reduces digitalisation to the use of the Internet. The digital economy in the strict sense accounts for about 5% of the global GDP (gross domestic product), and DE in the broad sense – 22% of the global GDP" [9].
2. When defining the substance of DE, many sources focus on "technologies and related changes in the ways of interaction between economic agents and economic processes" [10]. According to a number of American analysts, one of the reasons hindering the development of due definition of DE is rapid replacement of technologies by new, more progressive ones [11, 12].

Following the study of various approaches to specification of the "digital economy" concept, it was found out that the penetration of digital activity into various spheres of economy got a foothold at the beginning of the 21st century, at the period of strengthened economic globalisation and expansion of integration processes that make it possible to remove restrictions and collectively use the digital technologies to an increasing extent, for instance, digital platforms, thus having a positive impact on competitive positions.

The comparative analysis of the national digitalisation programmes of the EAEU countries' economy in terms of their compliance with the requirements of supranational regulation makes it possible to conclude that the realisation of the EAEU collective document proceeds at a relatively slow pace.

The generalisation of measures for digital transformation of economy in the EAEU countries envisaged in their national projects makes it possible to highlight their main characteristics.

1. In Armenia and Kyrgyzstan, the national programmes for digitalising the economy aim to solve the most common problems ("setting up a digital government, development of digital skills, infrastructure, cybersecurity and institutional framework" [13, 14].
2. The programme of the Republic of Belarus focuses on creating conditions for participation of world IT companies in the Belarusian economy, development of digital environment and use of cryptocurrency, training of IT personnel, introduction of latest financial technologies and tools, development of e-government [15].
3. A more comprehensive digital transformation is envisaged by the Digital Kazakhstan programme which involves the implementation of the smart city concept, the development of digital infrastructure, telecommunications, information security for the realisation of the Digital Silk Road project [16].
4. The international cooperation aspect is reflected most extensively in the programme "Digital Economy of the Russian Federation" that provides for the formation of a system of legal regulation of digital economy, introduction of civil-law transactions based on digital technologies and creation of a competitive environment for transmission, processing and storage of data [17]. The realisation of the digital agenda within the EAEU is one of the goals of Russia's national development until 2024, being enshrined in the federal project "Digital public administration" which provides for coordination in the sphere of digital transformation with all EAEU countries.

The success in this direction is evidenced by the fact that whereas in 2010 Russia ranked 64th in the Global Innovation Index it moved up to the 46th position in 2019, which can be recognised as a positive result of the realisation of the programme "Digital Economy of the Russian Federation", which is slightly better than the rating of EAEU countries that in the aggregate are in 50th position.

At the same time, Russia's position could be even better since the development of digital technologies is not proceeding as quickly as the changes taking place in the global economy and within the framework of regional integration associations. For instance, in the European Union, the policy of four freedoms (goods, services, capital, people) is supplemented by the fifth component within the framework of the Digital Single Market Strategy adopted in 2015 – free exchange of data owing to the elimination of all national barriers in the digital environment between the member states of the European Union for the purpose of facilitation of the digital transactions development practiced in the conditions of free competition.

4 Discussion

In the process of development of the "Eurasian Economic Union Digital Agenda until 2025" [18], free exchange of data and elimination of all national barriers to digital environment between the EAEU member countries has become one of the problems of scientific discourse, considering not only the unequal level of digital transformation of national economies, but also the existing discrepancies in understanding the concept of "digital economy". Given that digitalisation has extensively penetrated the service sector (for instance, accounting and financial services), the problem of ensuring security in the use of digital technologies (Internet of things) is promoted to the forefront [19].

With the advent of a more complex use of the digital factor, DE is already considered as a segment of IT industry, being auxiliary for the products and services of the IT sector and based on the development of uniform protocols for their use [20]. This approach presupposes coordination of actions of the EAEU countries and free exchange of digital data which is currently at an early stage.

5 Conclusion

The comparative analysis of the national programmes for digital transformation of economy in the EAEU countries in accordance with the document "Main Directions for the Realisation of the EAEU Digital Agenda until 2025" showed that the national digitalisation programmes basically coincide, which objectively ensues from the modern global challenges. But at the same time, considering the peculiarities of the economy of each country, as concerns organisation and regulation of the transition to the digital environment, some significant differences can be traced.

Each EAEU member country shall have to ensure improvements in the quality of digital infrastructure, as well as the introduction of broadband communication, formation of electronic government, enhancing the level of digital literacy; however, the measures taken in this direction at the moment are more characteristic of the stage of informatisation and automation of economic processes, while the digital transformation of economies is at its early stage, excluding Russia and Kazakhstan.

Nevertheless, the EAEU countries are working towards the realisation of the digital agenda of the Eurasian Economic Union until 2025, since the digitalisation of national economies can promote integration into the EAEU as a strategic direction of interaction between the countries.

The adopted national documents in the field of digital transformation of the economy are not always consistent at the supranational level, at the same time the conceptual apparatus contains differences in comprehension of particular characteristics. This concerns, in the first place, the core of the DE concept. The elimination of such discrepancies in the national digital transformation programmes in the EAEU countries is of practical importance since if they persist, then some barriers separating the countries according to the technological principle may arise, and this will hinder the realisation of the principle of "four freedoms" that is necessary for strengthening the integration and achieving free exchange of data in the digital environment between the EAEU member states.

References

1. Romer, D.: The solow growth model. Adv. Macroec. McGraw-Hill, New York (2011)

2. Nankervis, A., Connell, J., Montague, A., Burgess, J. (eds.): The Fourth Industrial Revolution. Springer, Singapore (2021). https://doi.org/10.1007/978-981-16-1614-3
3. Proskurin, S.D.: Fund. Res. **9**(2), 404–410 (2016)
4. Yudina, T.N.: Theor. Econ. **3**, 14 (2016)
5. Kling, R., Lamb, R.: Comp. Soc. **29**(3), 17–25 (1999). https://doi.org/10.1145/572183.572189
6. Chueshev, A., Melekhova, O., Meshcheryakov, R.: Cloud Robotic Platform on Basis of Fog Computing Approach. In: Ronzhin, A., Rigoll, G., Meshcheryakov, R. (eds.) ICR 2018. LNCS (LNAI), vol. 11097, pp. 34–43. Springer, Cham (2018). https://doi.org/10.1007/978-3-319-99582-3_4
7. Evsyutin, O.O., Mescheryakov, R.V., Melman, (Kokurina) A.S.: Signal Proc. **179**, 107811 (2021). https://doi.org/10.1016/j.sigpro.2020.107811
8. European Commission (2020). Europe 2020 Strategy. https://ec.europa.eu/info/business-economy-euro-0/economic-and-fiscal-policy-coordination/eu-economic-governance-monitoring-prevention-correction/european-semester/framework/europe-2020-strategy_en
9. Starodubtseva, E.B., Markova, O.V.: Bul Astrakhan State Tech Univ. Ser. Econ. 2, 7–15 (2018). https://doi.org/10.24143/2073-5537-2018-2-7-15
10. Tultaev, T.A., Tultaeva, I.A.: Fund Res **9**(2), 408–413 (2014)
11. Barefoot, K., et al.: Defining and measuring the digital economy. Bureau Econ. Anal. (2018). https://www.bea.gov/system/files/papers/WP2018-4.pdf
12. Bukht, R., Heeks, R.: Int. Org. Res. J. **13**(2), 143–172 (2018). https://doi.org/10.17323/1996-7845-2018-02-07
13. Armenia's Digital Transformation Agenda until 2030 as of (2017). https://d-russia.ru/tsifrovaya-povestka-respubliki-armeniya.html
14. Programme of digital transformation of Kyrgyzstan "Taza Koom" as of (2017). https://d-russia.ru/tsifrovaya-povestka-kyrgyzskoj-respubliki.html
15. State programme for development of digital economy and information society in the Republic of Belarus for the years 2016–2020 as of (2016). https://nces.by/wp-content/uploads/psm235.pdf
16. State programme "Digital Kazakhstan" for the years 2017–2020. Strategic development plan of the Republic of Kazakhstan until 2020 (2017). https://adilet.zan.kz/rus/docs/U100000922_#z0
17. Government of the Russian Federation (2017). Programme "Digital Economy of the Russian Federation" as of (2017). http://government.ru/docs/28653/
18. Legal web portal of the EAEU. On the Main Directions for the Realisation of the Digital Agenda of the Eurasian Economic Union until 2025 (2017). https://docs.eaeunion.org/pd/ru-ru/0101963/pd_28072017
19. Evsyutin, O.O., Kokurina, A.S., Mescheryakov, R.V.: Comp. Optics **43**(1), 137–154 (2019). https://doi.org/10.18287/2412-6179-2019-43-1-137-154
20. Mescheryakov, R.V., Iskhakov, A.Y., Evsyutin, O.O.: SPIIRAS Proc. **19**(5), 1089–1122 (2020). https://doi.org/10.15622/ia.2020.19.5.7

Development of Approaches to Assess the Effectiveness of the National Projects Implementation

Alla Yu. Chalova(✉) and Elvina A. Omshanova

Plekhanov Russian University of Economics, Moscow, Russia
tchalova.ay@rea.ru

Abstract. This article is aimed to develop and justify the approaches to the development of methods for evaluating the effectiveness of national projects in Russia in order to increase the effectiveness of their financing and strengthen the impact on economic growth. The integration of project expenditures of the federal budget of Russia into the structure of program expenditures determined the use of the existing methodology for assessing the effectiveness of the implementation of state programs in the Russian Federation. In order to unify the approaches while assessing the effectiveness of budget expenditures for the implementation of national projects and the comparability of their results with the results of the implementation of state programs, the authors propose a methodology for evaluating the effectiveness of national projects in Russia. The proposed methodology takes into account not only the results of the execution of expenditures for the implementation of national projects from various sources of financing (i.e. resource component), but also such important elements of efficiency assessment as the fulfillment degree of tasks aimed at achieving the goal of the federal project. The authors took into account the significance of each task in achieving the project goal, the degree of achievement of the federal project indicators, as well as the significance of each indicator. This allows, in turn, to consider the specifics and priorities of various federal projects that are part of a specific national project; to more objectively assess the effectiveness of project financing implementation on the basis of a comprehensive integrated assessment.

Keywords: Federal project · State program · Budget expenditures · Integral evaluation indicator

1 Introduction

The need to resume sustainable progressive economic growth in Russia has determined the consolidation of priority areas of its strategic development in the format of national projects. The integration of national projects into the system of program budgeting in Russia was carried out in order to achieve the unity of the system of state strategic planning and the implementation of a unified state policy [1, p. 9].

P. V. Trifonov and M. V. Charaeva (Eds.): DITEM 2021, LNNS 380, pp. 318–324, 2022.
https://doi.org/10.1007/978-3-030-94245-8_44

There was the integration of federal projects into state programs in 2019 and their details at the level of the subjects of the Russian Federation in the format of regional projects. Therefore, the task of assessing the overall effectiveness of the implementation of the national project as a multi-component tool for achieving quantitative and qualitative indicators identified in the national development goals of the Russian Federation has become urgent [2]. This requires a certain specification of the methodology for calculating the effectiveness of project financing, matching the volume of approved budget allocations with the degree of achievement of the indicators provided for by the projects. Thus, it implies the need for scientific research in the absence of a single comprehensive methodology for evaluating the effectiveness of national projects.

2 Materials and Methods

The complex design of the national project with numerous structural elements determines the multiplicity of indicators by which the achieved results of the project and the national goals assigned to it should be evaluated [3, p. 37]. In these conditions, the question of the criteria aspects of evaluating the effectiveness of the implementation of national projects arises with all urgency: it should be carried out only on the basis of achieving the target indicators planned by them. On the other hand it is necessary to take into account the following: the financing development timeliness and completeness allocated for the implementation of the project, the timeliness of the implemented measures, the significance of their implementation in achieving the goal, ensuring the quality of the implementation of the event when achieving quantitative indicators, and others [3, p. 42].

These components of the assessment are partly reflected in the existing methodology for evaluating the effectiveness of state programs in the Russian Federation [4]. However, the evaluation of the effectiveness of the implementation of state programs according to this method is associated with a number of its disadvantages [5, p. 65]: the use of the level of execution of cash expenditures under the subprogram while assessing the degree of achievement of the target indicators of state programs [6, p. 84], the same approach to all indicative indicators of the state program, regardless of the significance of their achievement on the results of the implementation of the goals of the subprogram (state program) [7, p. 76].

Taking into account the above-mentioned shortcomings of the methodology for evaluating the effectiveness of state programs in the Russian Federation, it requires significant refinement and adaptation in order to use it in order to assess the effectiveness of the implementation of national projects and federal projects.

3 Results

Currently, the Accounts Chamber of the Russian Federation evaluates the effectiveness of the implementation of national projects only according to the degree of cash execution of federal budget expenditures for their implementation. The assessment of the cash execution of national projects for 2019–2020 showed that in 2020 the level of their cash execution amounted to 97.4% of the approved volume in the consolidated budget list, which is 5.9% points higher than the level of their execution in 2019 (91.5%) [8,

9]. Despite the high level of cash execution of expenditures for the implementation of national projects, it was not possible to fully fulfill budget assignments for them again [10].

The increase in the level of cash execution of expenses for the implementation of federal projects does not reflect the increase in the effectiveness of their implementation. It is obvious that the assessment of the effectiveness of the implementation of national and federal projects should be comprehensive, taking into account not only the financial aspects in terms of the use of budget funds, but also the significance of each event and its corresponding indicators in achieving the goals and indicators of national projects.

According to the authors, taking into account the multicomponent structure of national projects, as well as the presence of various criteria and principles of their formation, one of the most balanced, comprehensive and objective methods for evaluating the effectiveness of their implementation is an integrated efficiency assessment.

The use of an integral evaluation indicator allows combining indicative criteria of various content and meaning (quantitative, cost, qualitative), which increases the level of evaluation of the effectiveness of state programs and national projects [11, p. 391–392].

Basing on the key features of the formation of national projects [3, p. 37], as well as the best foreign practice of budgeting state programs and projects [12–14], the authors have developed a comprehensive methodology for evaluating the effectiveness of national projects. The algorithm of the proposed method is presented below.

Stage 1. The Effectiveness of the Tasks.

At this stage, the degree of fulfillment of the tasks aimed at achieving the goal of each federal project that is part of the national project is assessed. The authors propose to use the following approach when calculating the significance of the implemented tasks of the federal project:

- if a project task is defined as a key one, then it is necessary to give it a significance coefficient equal to 1, since this task directly affects the achievement of the project goal;
- if the project task is defined as not a key one, then it is necessary to give it a significance coefficient equal to 0.5, since the result of this task indirectly affects the achievement of the project goal, but the task remains necessary for its implementation.

The evaluation of the effectiveness of the implementation of the federal project in the context of the fulfillment of its tasks using the significance coefficients proposed above is carried out according to the following formula:

$$\mathrm{DFt} = \sum \mathrm{AT} * \mathrm{SC} / \sum \mathrm{TP} * \mathrm{SC} \quad (1)$$

where:

DFt – the degree of fulfillment of the tasks of the federal project;
AT – the actual number of completed tasks;
TP – the planned number of tasks;
SC – coefficient of significance of a specific task.

Stage 2. The Degree of Compliance of Cash Expenditures of Budgets for the Implementation of the Federal Project with their approved volume:

$$DCe = (CEfb + CEcb + CEebf + ACebf) / (AEfb + AEcb + AEebf + PCebf) \quad (2)$$

where:

DCe – the degree of compliance of actual expenses with the planned level of project expenses;

CEfb and AEfb – respectively cash and approved expenses for the implementation of the project at the expense of the federal budget;

CEcb and AEcb – respectively cash and approved expenses for the implementation of the project at the expense of the consolidated budgets of the subjects of the Russian Federation;

CEebf and AEebf – respectively cash and approved expenses for the implementation of the project at the expense of the budgets of state extra-budgetary funds;

ACebf and PCebf – respectively, the actual and planned costs for the implementation of the project at the expense of extra-budgetary funds.

The result obtained will be used to evaluate the effectiveness of the use of funds in the project at the third stage.

Stage 3. The Efficiency of using the Budget Funds of the Budgetary System of the Russian Federation and Extra-Budgetary Funds Aimed at the Implementation of the Federal Project:

$$Euf = DFt / DCe \quad (3)$$

where:

Euf – efficiency of using funds in a federal project;

DFt – the degree of fulfillment of the tasks of the federal project;

DCe – the degree of compliance of actual expenses with the planned cost level.

The significance coefficients in this formula are already included in the indicator of the degree of fulfillment of the tasks of the federal project.

Stage 4. The Degree of Achievement of the Results (Indicators) of The Federal Project based on an Integrated Assessment.

It is advisable to calculate the effectiveness of the project by indicators, taking into account the significance coefficients for each indicator and its contribution to the achievement of the final goal of the project, which is determined by experts. At the same time, such an assessment should be carried out in two intermediate stages according to the formulas:

$$DAIi = Iai * ACi / Ipi * ACi \quad (4)$$

where:

DAIi – the degree of achievement of a specific indicator of the federal project;

Iai – the actual value of a specific indicator of the federal project;

Ipi – the planned value of a specific indicator of the federal project;

ACi – the coefficient of significance of a specific indicator;
или по формуле:

$$\text{DAIi} = \text{Ipi} * \text{ACi}/\text{Iai} * \text{ACi} \tag{5}$$

if the achievement of the expected result of the project is associated with a decrease in its value relative to previous periods (for example, the level of mortality, morbidity, accidents, etc.).

$$\text{DARfp} = \sum\nolimits_1^n \text{SAi}/N \tag{6}$$

where:
DARfp – the degree of achievement of the results of the federal project;
$\sum$ SA – the sum of the results of achieving all the indicators of the federal project;
N – the number of indicators.

Stage 5. Evaluation of the Effectiveness of the Federal Project.

The overall assessment of the effectiveness of the implementation of the federal project is calculated using the following formula:

$$\text{Eifp} = \text{Euf} * \text{ADRfp} \tag{7}$$

where:
Eifp – efficiency of the implementation of the federal project;
Euf – efficiency of using funds in a federal project;
ADRfp – the degree of achievement of the results of the federal project.

Next, it is important to set the boundaries of the degree of effectiveness of the implementation of the federal project. Taking into account the strategic importance of national projects for achieving the national development goals of the Russian Federation, their effectiveness should be evaluated with minimal deviations of all parameters from their established planned values. The authors suggest using the following ranges of the boundaries of the degrees of efficiency of the federal project implementation (EIFP) while conducting the final assessment:

EIFP $\geq$ 99.1% – high efficiency degree;
EIFP $\geq$ 97.5%–99% – efficiency above average;
EIFP $\geq$ 95.1%–97.4% – average efficiency degree;
EIFP $\geq$ 90.1%–95% – satisfactory efficiency degree;
EIFP < 90% – unsatisfactory efficiency degree.

Stage 6. Integrated Assessment of the Effectiveness of the National Project.
The final efficiency indicator can be calculated by using the following formula:

$$\text{EInp} = \left(\sum \text{EIfp}\right)/\text{N} \tag{8}$$

where:
EInp – the effectiveness of the implementation of the national project;
EIfp – efficiency of the implementation of the federal project;
N – the number of federal projects that are part of the national project.

4 Discussion

The developed methodology for evaluating the effectiveness of federal projects takes into account not only the financial component of the project implementation for all sources of project financing, but also the degree of fulfillment of all its tasks, as well as their significance in achieving the project goal, the degree of achievement of the project indicators, their significance in achieving the target result. The universality of the proposed methodology allows to consider the specifics and priorities of specific federal projects by adjusting the significance coefficients of the evaluated measures and indicators. The authors do not exclude the need for a more in-depth analysis of the reasons for the deviation of the actual and achieved values of the project indicators from their planned levels during the evaluation. In order to do this, it is necessary to introduce a procedure for multi-factor analysis of project indicators and take into account the profile indicators that characterize the effectiveness of the work of the department responsible for the implementation of the national project [15, p. 26].

5 Conclusion

The algorithm for evaluating the effectiveness of national projects through federal projects proposed by the authors does not exclude expert work at each stage. At the same time, it will allow identifying defective indicators, linking the project's target indicators with the actual results of its activities and the overall goal of the project. This will require clarifying the system of target indicators of the federal project in the context of all its constituent activities.

In order to improve the effectiveness of the implementation of federal project activities, it is also necessary to bring their resource provision into line with the actual results achieved in the reporting period, establishing the relationship "the degree of implementation of the event and the amount of funding". Following the recommendations proposed above will ensure an increase in the efficiency of project financing in Russia, and will allow achieving the planned results.

References

1. Buchwald, E.M. (eds.): National projects in the system of strategic planning in the Russian Federation. INFRA-M, Moscow (2019). https://doi.org/10.24158/tipor.2019.2.8
2. Lysenko, A.A., Shaibakova, L.F.: Problems in the implementation of national projects of the Russian Federation. Scientific electronic library "CyberLeninka" (2021). https://cyberleninka.ru/article/n/problemy-pri-realizatsii-natsionalnyh-proektov-rossiyskoy-federatsii
3. Kudelich, M.I.: The system of regulatory requirements for evaluating the effectiveness of national projects: actual problems. Financial J. **4**, 36–49 (2019). https://doi.org/10.31107/2075-1990-2019-4-36-49
4. The Government of the Russian Federation. "On approval of the Procedure for the development, implementation and evaluation of the effectiveness of state programs of the Russian Federation" (No. 588). https://base.garant.ru/198991/
5. Margolin, A.M.: Ways to improve the methods of evaluating the effectiveness of state programs. Econ. Policy **11**, 54–81 (2018). https://doi.org/10.18288/1994-5124-2018-6-54-81

6. Yuzhakov, V.N., Dobrolyubova, E.I., Alexandrov, O.A.: How to evaluate the effectiveness of the implementation of state programs: questions of methodology. Econ. Policy **6**(10), 79–98 (2018). https://doi.org/10.18288/1994-5124-2015-6-04
7. Rykova, I.N., Gubanov, R.S.: Criteria for evaluating the effectiveness of the implementation of investment projects in the field of territorial infrastructure development. Econ. Educ. **2**(105), 73–81 (2018). https://www.elibrary.ru/item.asp?id=32717665
8. The Ministry of Finance of the Russian Federation. The report "Execution of the federal budget and budgets of the budgetary system of the Russian Federation for 2019" (2019). https://minfin.gov.ru/common/upload/library/2020/09/main/Ispolnenie_federalnogo_budzheta_2019_god.pdf
9. The Ministry of Finance of the Russian Federation. The report "Execution of the federal budget and budgets of the budgetary system of the Russian Federation for 2020 (preliminaryresults)" (2020). https://minfin.gov.ru/common/upload/library/2021/03/main/Ispolnenie_2020_god.pdf
10. National projects: targets and key results. Official website of the Russian Government. http://static.government.ru/media/files/p7nn2CS0pVhvQ98OOwAt2dzCIAietQih.pdf
11. Jasińska-Biliczak, A., Krzysztof, M.: Measuring the integrated effectiveness of regional development: directions for regional government. European Res. Stud. J. **XXII**(1), 389–403 (2020)
12. Benito, B., Solana, J., Moreno, M.-R.: Explaining efficiency in municipal services providers. J. Prod. Anal. **42**(3), 225–239 (2013). https://doi.org/10.1007/s11123-013-0358-7
13. Kim, J.: From line-item to Program Budgeting. Korea Institute of Public Finance, Seoul (2010)
14. Miller, G.J., Hildreth, W.B., Rabin, J.: Performance-based budgeting. Westview Press, Colorado (2010). https://asu.pure.elsevier.com/en/publications/performance-based-budgeting-an-aspa-classic
15. Buryk, Z.M.: Types and criteria for evaluating the effectiveness of public administration. Int. Sci. J. 10–1(20), 23–28 (2016). https://www.elibrary.ru/item.asp?id=27319591

Modern Personnel Under the Conditions of Rotational Shiftwork

Vladimir V. Elgin(✉), Nadezhda P. Sheveleva, and Natalia G. Savkina

Tyumen Industrial University, Tyumen, Russia

Abstract. The purpose of this study is to identify the degree of influence of narcissistic accentuation of personality traits, as a base for the formation of neuroses, on the effectiveness of the implementation of rotational shiftwork; determination of directions for improvement of the situation. The results of a 13-year study of the dynamics of the presence of narcissistic accentuation of personality traits among students of technical specialties of a correspondence university using graphic tests, the Belov test and the Beck depression scale, showed that this accentuation is observed in 3.8–14.7% (varying by years) of cases, n = 480 people (2% of the sample), the links between narcissistic accentuation of personality traits and the level of anxiety ($K_{acc} = 0.77$) and self-esteem ($K_{acc} = 0.86$) were identified. The opinion of employers (representatives of the companies Gazprom, SIBBUR, Transneft, Rosneft, etc.) was studied, it was found that the presence of a narcissistic accentuation of personality traits affects the quality of work, productivity and the ability to adapt, in particular, under the conditions of a rotational shiftwork. The recommendations for management of work with the neurotized contingent are formulated. Thus, in certain cases, a significant proportion of employees of oil and gas companies, primarily young people, systematically need psychological assistance and support to increase and maintain efficiency and productivity, reduce the number of production losses and emergency situations, improve the quality of life and stability of personnel at the enterprise, especially in terms of rotational shiftwork.

Keywords: Narcissistic accentuation of personality traits · University students · Graphic tests · Socio-economic conditions · Shift work method · Psychological support

1 Introduction

In recent years, many large companies have been implementing programs for the adaptation of young specialists. However, these programs are limited to young people gaining knowledge about the specifics of the enterprise, its production process, creating conditions for their participation in scientific and technical activities. At the same time, the coverage of adaptation measures of an educational and training nature is insufficient. Thus, participation in the Gazprom PAO Young Specialists Training School is 29% across the entire surveyed sample of Gazprom Dobycha Nadym LCC.

P. V. Trifonov and M. V. Charaeva (Eds.): DITEM 2021, LNNS 380, pp. 325–330, 2022.
https://doi.org/10.1007/978-3-030-94245-8_45

With regard to the health of the employed young specialists, the enterprises usually limit themselves to a general assessment of generally accepted indicators of physical condition for the employees of the enterprise. The state of mental health and psychological state of young professionals remains outside the attention of corporate management. At the same time, the level of neurotization among young people is growing, which reduces production productivity [1–5].

Meanwhile, oil and gas production facilities of oil and gas companies, which form the basis of their production process, belong to the highest hazard class of production facilities. Therefore, personnel errors in performing their duties or non-compliance with established requirements can cause economic damage to the enterprise, lead to environmental emergencies, and even human casualties (2013 explosion at the Bovanenkovskoye field).

Another circumstance that cannot be ignored when assessing the mental health and psychological state of a person is the rotational method of work, which is widely used at the enterprises of the oil and gas industry. Usually, the duration of intraregional shifts does not exceed 2 weeks, while the interregional shifts last 1–1.5 months. During the pandemic in the spring of 2020, the duration of shifts at many enterprises was increased by 2–3 times.

About 38% of the total number of personnel of Gazprom Dobycha Nadym LCC, PAO Gazprom, are covered by the rotational organization of work, including (as of 01.07.2017):

- Interregional form – 6% (522 people),
- Intraregional form – 28% (3686 people).

"The main object of Gazprom Dobycha Nadym LCC is the large oil and gas condensate field Bovanenkovskoye, which is located in the Arctic zone of Russia. An analysis of the economic and psychological characteristics of employees in the gas fields of the enterprise showed that with a positive attitude towards their work (satisfaction at the level of 70% of the surveyed employees), 9–14% of employees plan to change their place of work in the near future. The problems of managing rotational personnel are mainly associated with insufficient knowledge of the medical, social and psychological problems of using this method of work organization" [6]. However, nowadays at the enterprises of the oil and gas complex, the number of specialists providing psychological support to personnel remain clearly at an insufficient level. For example, Gazprom Dobycha Nadym LCC has one psychologist per about 5 thousand employees.

In this regard, the purpose of this study is to identify the degree of influence of the narcissistic accentuation of personality traits as a basis for the formation of neuroses on the effectiveness of the implementation of rotational shiftwork; to identify the areas for improvement of the situation [7–10].

2 Methods

The authors have been conducting a survey of students of technical specialties of the university by means of graphic tests for thirteen years (M. Dukarevich test "Non-existent animal" and K. Koch test "House-tree-man" [11, 12]. The temperament test (Belova) and the Beck Depression Scale were used in parallel. Moreover, the questioning and interviewing of students and employers was carried out. The results were subjected to analysis of variance and correlation (association and contingency coefficients were calculated).

3 Results

The research covered the correspondence students of technical specialties of the university, employed as workers and managers of the lower and middle level of management, engineering and technical personnel by such oil and gas companies as Gazprom, Rosneft, Transneft, SIBBUR and a number of others, 100% of the respondents were male, sample size was 408 people (2% sample), age – from 21 to 58. According to the graphic tests described in the article, the percentage of pronounced neurotic personalities among the interviewees varied in the range of 3.8–14.7% in the student group. The association coefficients in the analysis of the dependence of the joint manifestation of narcissistic accentuation of personality traits (according to the proportion of subjects in the total sample) and the increased level of anxiety, as well as low self-esteem, were 0.77 and 0.86, respectively, which confirms their mutual dependence (the coefficient of association was more than 0.5).

The fact that in the sample under consideration 77% were persons with a strong nervous organization – phlegmatic and sanguine, 15% melancholic and 8% choleric (according to A. Belov test [13]) is of great interest. However, the overwhelming majority expressed the opinion that the optimal duration of the shift, which does not require long-term recovery in the physical and psychological sense is 2 weeks, the shifts with duration of 1–15 months are more convenient organizationally, if the object of work is very remote from the permanent place of residence, the employees, in this case, require a certain psychophysiological adaptation upon returning home, moreover, some discontent and difficulties in family are observed. A shift of 3–4 months or more is a serious psychological test, followed by emotional burnout, loss of interest in life, depression of varying severity, unwillingness to do anything and maintain relationships with loved ones, often in a protracted form, decreases control over aggression.

Moreover, on the basis of long-term observations (from 2007 to 2019), the features of the behavior and work of neurotically organized individuals in the conditions of the shift method were identified:

1. Significant tension (difficulty in falling asleep in the presence of strangers, etc.). Inability to organize the repletion of urgent wants in conditions of reduced comfort, etc.
2. Increased propensity towards conflict, resentment, the requirement for a special attitude, selfishness, and desire to resolve the conflict on the principle of "win-lose".
3. Inability to work in a team.
4. Emotional exhaustion can occur by the end of the second or even the first week. The frequency of contact with loved ones and labor productivity decrease, and irritation increases. A shift longer than 2 months can trigger irreversible personal changes.

For comparison, in individuals without pronounced signs of neurotization: increased fatigue occurs in 2–2.5 weeks, in 1–1.5 months it increases, by the end of 3–4 months of the shift it can develop into aggression with a sharp decrease in labor productivity.
5. Family relations are either not established or are quickly destroyed [often with the departure for the shift], perhaps, the existence of several families – at the main place of residence and at the place of shift work, abandonment of long-term personal plans.
6. There are two possible types of behavior: on the one hand, workaholism, contributing to accelerated emotional burnout, an increase in the number of mistakes; on the other hand, the fulfillment of only the minimum requirements to avoid serious complaints.
7. Frequent job changes.
8. In the event of emergency situations, the following is possible:

    a. retraumatization, leading to irreversible consequences for the psyche, the development of a full-fledged neurosis up to disability, a long period of adaptation;
    b. a deliberate desire to get into the epicenter of events, an unjustified risk, which generates losses in productivity.

9. Labor productivity is 1.4–1.7 times lower (in comparison with workers without pronounced signs of neurotization).

4 Discussion

Definitely, all of the above requires psychological support. However, within the framework of psychological work, one should adhere to certain characteristics and practice the use of certain tools that help a person to realize and work out his psychological problems, to become more emotionally stable.

First of all, it is very important to prevent the transmission of the manager's anxiety to subordinates. A modern manager has an important function of a buffer, extinguishing negative information that is unnecessary for the work process, and creating a positive, safe atmosphere [14, 15]. The anxiety of a manager, perceived by subordinates with a narcissistic accentuation of personality traits negatively affects their psyche as a parental figure in the overwhelming majority of cases.

In the case of working with neurotically organized individuals, it is recommended to use tools to relieve anxiety in the process of work, such as:

- Preliminary detailed articulation of the forthcoming work, its algorithm, control measures, consideration of possible scenarios for the development of events, possible strategies of behavior, in case of implementation of one or another option;
- Stability and transparency of requirements for the quality of work, discipline, communication conditions; stability of the emotional background, bringing communication to the "adult-adult" level;
- Avoidance of cross-transactions "parent-child" (the boss is right and knowledgeable, the subordinate is initially wrong, not aware and guilty) in communication; appeal to the empirical experience of the individual;
- Avoidance of emotional pressure, non-constructive criticism;

- Strict adherence to the boundaries of the personality of the manager and the employee, demonstration of adequate ways to protect their own boundaries without aggression, so that individuals experiencing difficulties with this can learn from positive experience. Therefore, a particularly strict selection of candidates for leadership positions based on psychological status is recommended.
- More reasonable and limited use of various educational and motivational methods and technologies, fraught with repeated psychotrauma (provocative; methods based on fierce competition, subjective assessments, etc.).

Taking into account the above, it would be advisable to take the following measures at the level of enterprises and organizations of the oil and gas industry:

1. To develop and implement special adaptation programs for young specialists in terms of the psychological and neurological state in the first three years of their work at the enterprise.
2. To increase the number of specialists providing psychological support to personnel, including psychological testing, counseling, and the development of individual recommendations for correcting the psychological state of young people.
3. To improve the system of selection of management personnel, including at the initial stages of selection testing and conversations with a psychologist, in order to avoid broadcasting the manager's anxiety to subordinates and, as a consequence, a decrease in staff productivity.

5 Conclusion

Thus, there is a noticeable level of neurotization of modern personnel (3.8–14.7%) and it systematically needs psychological assistance and support to increase and maintain efficiency and productivity, reduce the number of production losses and emergency situations, and the stability of personnel at the enterprise. When working with narcissistically organized workers the efforts should be aimed at increasing their level of self-esteem, reducing the level of anxiety. The companies-employers should develop effective organizational and management decisions.

References

1. Yakovleva, L.S.: The selfie phenomenon as a form of transformation of personality Narcissism in the information society. Phil. Human Inf. Soc. **1**(27), 70–79 (2020)
2. Chikalina, V.A.: Analysis of the influence of social networks on the growth of narcissism. Quest. Stud. Sci. **8**(36), 6–9 (2019)
3. Giniyatova, E.V., et al.: Narcissism in modern visual culture: the selfie phenomenon. Bul Tomsk State Univ. Cult. Art Hist. **35**, 5–13 (2019)
4. Babikova, M.R., Tagiltseva, Y.: Communication strategies and tactics as one of the factors in the formation of vulnerable behavior of young people in the network. Polit. Ling. **5**(77), 171–176 (2019)
5. Gurina, E.S.: Clinical picture of love relationships with a partner of the narcissistic type. SMALTA **4**, 10–14 (2019)

6. Elgin, V.V., Pereladov, A.M.: Improving the motivational mechanism of the shift personnel of gas production enterprises. Econ. Siberia Context Global Challenges XXI Century Collection Articles **1**, 50–64 (2018)
7. Cherednichenko, G.A.: Situation on the labor market of graduates of the system of higher and secondary vocational education. Educ. Iss. **1**, 256–282 (2020)
8. Fennel, M.: How to improve self-esteem. Astrel, Moscow (2007)
9. Posada-Arias, O., Avendaño-Ramírez, J., Arias-Pérez, J.: Intellectual capital management and university research group performance: mediating effects of machiavellianism and organizational narcissism. J. Inf. Knowl. Manag. **4**, 1850047 (2018)
10. Campbell, K.W., McCain, J.: Theoretical perspectives on narcissism and social media: the big [and beautiful] picture. In: Handbook of Trait Narcissism: Key Advances, Research Methods, and Controversies, pp. 443–453 (2018)
11. Sheveleva, N.P., Kuznetsov, I.S.: Assessment of the level of socio-psychological adaptation of students. Sci. Rev. Human Res. **4**, 12–18 (2013)
12. Sheveleva, N.P., Savkina, N.G.: Narcissistic accentuation of the character of modern students, factors, consequences and prevention. Soc. Sociol. Psych. Pedag. 5, 126–130 (2020)
13. Darvish, O.B.: Age-related psychology. VLADOS-PRESS, Moscow (2003)
14. Rasskazova, E.I., Ivanova, T.: Psychological self-regulation and subjective well-being in professional activity. Psych. J. Higher Sch. Econ. **4**, 626–636 (2019). https://doi.org/10.17323/1813-8918-2019-4-626-636
15. Mikhailova, O.R.: When is an insult perceived as a joke? Personal and situational factors of disabling the moral responsibility of a cyberbullying witness. J. Sociol. Soc. Anthrop. **2**, 55–92 (2019)

Improving Pricing as a Factor of Russia's Economic Growth

Vladimir E. Novikov[1](✉) and Maria V. Dolgova[2]

[1] Research Institute of Russian Customs Academy, Lyubertsy, Russia
[2] Plekhanov Russian University of Economics, Moscow, Russia

Abstract. The prevailing view, in the course of economic reforms, was represented by the position that price and foreign trade liberalization in the near term is supposed to ensure progressive economic development. However, the subsequent years showed that the market was not able to solve all of the emerging problems. In this context, the study aims to develop due directions for optimization of pricing in the conditions where market mechanisms cannot provide the necessary level of competition and solve the social problems. The research methods include analysis and synthesis, induction and deduction, logical and systemic analysis. The paper reveals the dominant role of prices in regulating the economic processes of production, sale and consumption of domestic and foreign goods; cost optimization; creation of added value. It has been proved that the reduction of customs-duty rates does not always contribute to the improvement of competitiveness of the national goods at the foreign and domestic markets, to reduction of prices and costs. The article substantiates the need for strengthening the state regulation of prices in Russia in the conditions of inability of free markets to provide competitive pricing, cost reduction and better satisfaction of public needs. The authors propose specific measures for the development of state regulation of prices aimed at stimulating the growth of the national economy in modern conditions. The efficiency of the stated measures requires their comprehensive and coordinated realisation.

Keywords: Prices · State regulation · Monopoly pricing · Competitiveness

1 Introduction

Continuous improvement of competitiveness is an indispensable factor of economic development. It is achieved by building up and efficient use of the intellectual potential, which allows satisfaction of social needs and increasing the country's competitiveness to the fullest extent.

One of the major tools to stimulate economic growth is the price. According to the value law, the price is formed in a competitive environment and thus stimulates reduction of production costs and improvement of consumer properties of goods, which provides for labour savings.

Therefore, the goal of the society and its managing systems in the economy is to fully maintain the competitive environment, and in its absence – to take measures aimed to introduce temporary or permanent price regulation. In particular, price regulation is

P. V. Trifonov and M. V. Charaeva (Eds.): DITEM 2021, LNNS 380, pp. 331–337, 2022.
https://doi.org/10.1007/978-3-030-94245-8_46

necessary in the natural-monopoly sectors. As to the other sectors, it is necessary to continually monitor the level of market competitiveness, to create due preconditions for the growth of competition, and in the absence of such an opportunity – to introduce price regulation, to improve the methods and forms of regulation. J. Stiglitz writes that as the structure of the economy changed – the transition from agrarian to industrial and informational economy – the importance of the constraints of the market mechanism increased [1]. K. Polanyi notes that markets have always been mediated by the state; state intervention limiting the market scope resumes as soon as the market ceases to meet its social obligations [2].

Despite the price regulation in the natural monopoly sectors and a number of other spheres, prices in Russia continue to rise steadily, which is indicative of the weak competition in most of the commodity markets and of the weakened stimulative role of prices in cost reduction and improvement of product quality. However, competition works efficiently only where the legal framework supports the optimal functioning of the market. For instance, many countries regulate medication prices to make medical drugs more affordable. To achieve due efficiency of the pricing policy, it is necessary to promote the maximum possible level of competition and to ensure compliance of prices with the quality of products [3] as well as the competent use of digital technologies in pricing [4].

The policy to combat inflation through monetary measures pursued in Russia [5] as well as the policy to restrain income growth leads to decreased consumer demand, disruption of normal proportions of social production and consumption, slowdown of economic growth rates. The economic growth is possible only in the conditions of acceleration of scientific and technological progress, increased labour productivity, which requires increased funding of education and science, development of new products and technologies, demand for their production. But this problem can be solved only by increased rate of consumption by the population, guaranteed equal consumption of national and imported goods in the Russian markets [6], by restraining the outflow of capital abroad, strengthening the role of the state in the investment policy.

2 Methods

The purpose of this paper is to develop ways to optimize price formation in the conditions where the market mechanisms do not provide the necessary level of competition and cannot solve the social problems and stimulate the economic growth. The methodology of research is based on the principles of scientific objectivity and systematic approach, relies on general scientific methods including analysis and synthesis, induction and deduction, systemic and logical analysis. The used reference and analytical materials aimed to identify the current problems and trends in pricing. The credibility of the scholarly provisions contained in the research is conditioned by the use of statistical data officially presented by the state and international institutional structures.

3 Results

The steady increase in world prices for foodstuffs, metals, oil and cotton since mid-2020 (Fig. 1) stimulates inflation in all national markets. However, the growth of prices in

Russia in 2020 significantly exceeded the inflation rates in the EU countries, Australia, Canada, the USA, Switzerland, Japan and China.

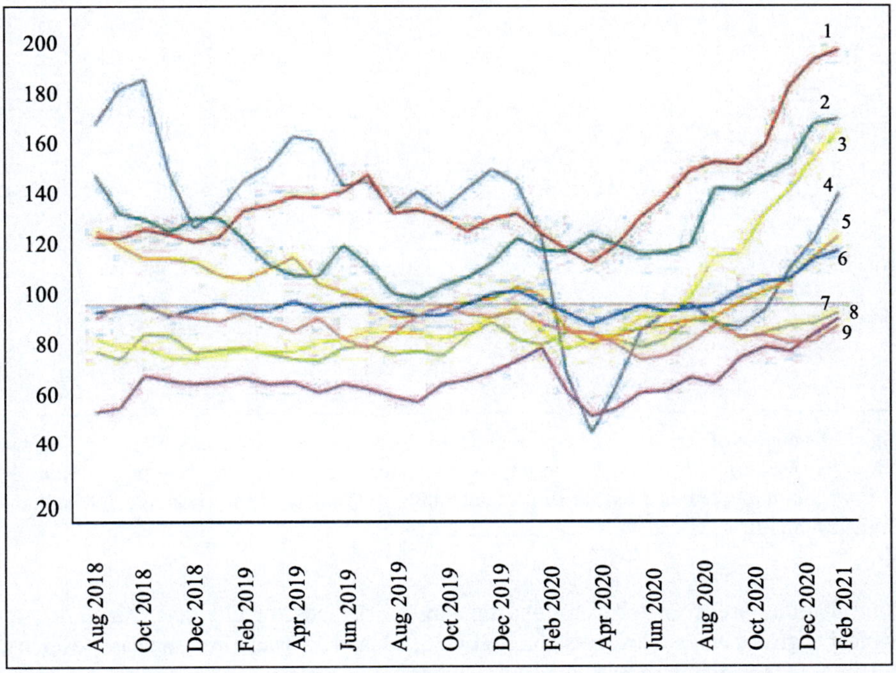

Fig. 1. World price indices for major commodity groups: 1 – Metal index, Index; 2 – Wheat, Index; 3 – Sunflower Oil, Index; 4 – Brent Crude, Index; 5 – Cotton, Index; 6 – Food index, Index; 7 – Coffee index, Index; 8 – Sugar index, Index; 9 – Tea, Kenyan, Index. *Source:*Compiled by the authors according to the data of the International Monetary Fund [7].

High world prices are moneymaking for exporters who gain additional profit, by providing the opportunity to compensate for losses caused by the pandemic. The decline in the national wholesale wheat prices resulting from export restrictions is clearly illustrated in Fig. 2.

Despite the importance of maintaining competitive pricing, Russia still does not have a Law on prices and pricing. Therefore, in many cases, there is no competition in the country in the domain of cost reduction and improvement of products' consumer properties at all stages of the stock movement – from the producer to the end consumer.

In terms of price growth, Russia has been ahead of many countries over the last decade, except for Belarus, Kazakhstan and Ukraine [9]. This is largely due to the fact that our country mainly invests financial resources in extraction of raw materials, while the developed countries – in technologies. The Russian billionaires invest mainly in production of export commodities, while U.S. billionaires – in development of new technologies [10].

The importance of these issues and the need to address them in terms of perfection of the entire pricing chain has been clearly demonstrated in the situation of coronavirus

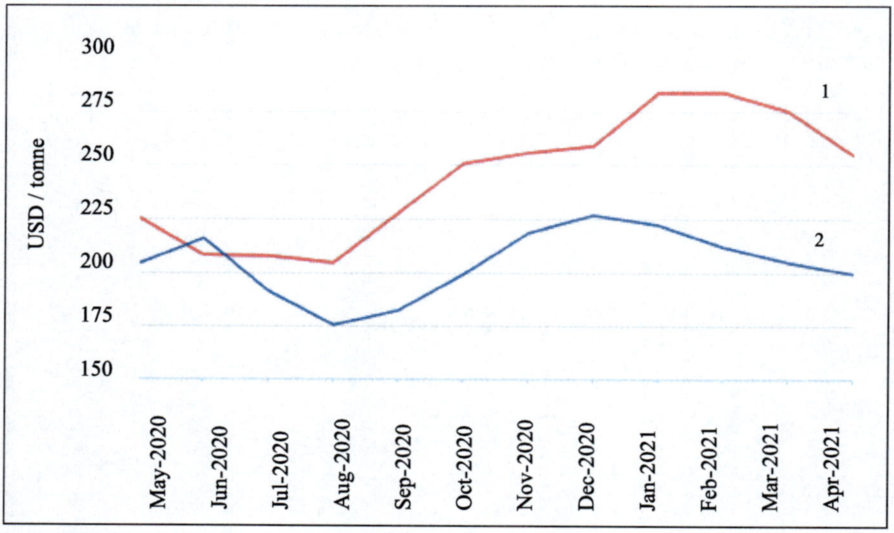

Fig. 2. Dynamics of export prices and manufacturers' wholesale wheat prices in Russia: 1 – export prices for wheat (f. o. b, Black Sea ports); 2 – manufacturers' wholesale wheat prices (grade 3, e. x. w). *Source:* Compiled by the authors according to the data of the Food and Agricultural Organization [8].

control. The absence of proper state regulation of prices for pharmaceuticals led to explosive growth of medical mask prices by 25, 30, and 50% and in some cases even by 4.0 times [11]. In many cases, the increase in retail prices was not related to the growth of wholesale factory prices [12]. Intermediaries also profited from the speculative demand for personal protective gear. For instance, a wholesale company purchased medical masks from the manufacturer at a selling price of 7.29 rubles per unit and sold them to the pharmacy chains at a markup of 957% [13]. Thus, instead of stimulating the growth of production and improvement of masks quality, the current pricing procedure resulted in enrichment of intermediaries and resellers.

In April 2020, the Government introduced maximum retail markups for a number of medical products, including masks, not exceeding 10 kopecks per unit, and marginal wholesale markups of 10% of the purchase price [14]. This pricing procedure was proposed earlier, but was not accepted as contrary to the freedom of the market.

The pricing process in the course of promotion of a number of food products to retail chains is not adequately regulated; these chains set retail prices on their own and thus manipulate them to maximize profits. They drive up retail prices for goods with higher consumer properties and lower down the producers' purchasing price. For instance, proceeding from the possible selling price of organic agricultural products at 80 rubles/kg and possible retail price of 160 rubles/kg, the retail networks offer a purchase price of 25 rubles/kg [15]. Thus, it is intermediaries, not the market, who control the interaction between the producers and the consumers.

Free pricing in this sector leads to accumulation of goods with an expired shelf life; meanwhile they would have been sold under regulated pricing.

Thus, due to the inflated retail prices of food products, retail chains sell most of the newly created value, with significant share of profits passed to other countries since the chains are usually owned by foreign companies.

The continuous price increase hampers the economic development of Russia. In 1992, the time of transition to market relations, the prices increased 26 times within 1 year, which resulted in the decrease of GDP by 19%, of the industrial output – by 18.8%, of the retail turnover – by 39% [16]. The contraction of GDP, in this case, was 19/2600 = 0.007% as calculated per 1% of the increase in consumer prices.

4 Discussion

Under the conditions of insufficient state regulation of prices, a certain interdependence of world and national prices for exported goods is formed [17]. The internal prices for consumers go up, both in case of increase or decrease in world prices. If the world prices rise the producers seek to bring the national prices to the level of world prices and maximize their profits. The foreign-exchange factor plays a significant role [18]. If the world prices drop, then the national producers increase the domestic prices to compensate for the loss of incomes from export, through supplies to the internal market. This process was not in fact regulated by the state for a long time, although some signs of monopolistic behaviour and monopolistic pricing on the part of the producers could be traced. At the same time, taxes and customs duties may have a significant impact on the level and dynamics of the domestic prices [19].

A one-off factor accounting for the increase of the national prices was the abolition of crude oil export duties. The reserves replacement tax in respect of mining resources included in the cost of oil was increased by the amount of abolished export duties, which contributes to the surge of the national oil prices and has an impact on the level of costs and prices for industrial goods, the products of agro-industrial complex and other industries, as well as the level of transport tariffs and those of the public utilities sector. Undoubtedly, the large-scale increase in costs leads as well to the economic slump.

Another factor constraining the economic growth is the non-competitiveness of the goods produced by the domestic manufacturing sector in comparison with the imported ones, as presented in the national market. The reduction of import duties upon joining the World Trade Organization in the conditions when the Russian manufacturing industry was already uncompetitive aggravated the situation of the national producers significantly.

The almost twofold reduction of the weighted average import tariff to 4.8% proved to be the payment for membership in the WTO. In early 2018, this figure amounted to 3.4% in the USA and 8.5% in China [20]. In 2019, the weighted average tariff in the USA surged up to 13.78% [21].

The liberalization of foreign trade quite often negatively affects the national market actors [22]. The Russian manufacturers found themselves unprepared for the competition in the national market due to technological backwardness and obsolete equipment, being unable to upgrade it because of the lack of financial resources and modern home-grown technologies.

The economic growth is negatively affected by the devaluation of the ruble in the conditions of the drop in world oil prices and the growth of world prices for other goods exported from Russia. The depreciation of the national currency leads to the increase in national prices and, as a consequence, the drop in real incomes and a decline in solvent demand. At the same time, the imports prices go up as a consequence of the ruble depreciation, which restrains the scope of imports and increases the competitiveness of the national production. However, this factor does not work in the situation of a decline in effective demand.

5 Conclusion

Thus, one can assert that the pricing policy has "pinched" the national producers, reducing their competitiveness by increasing the national prices and by the imposition of reduced customs import duties.

The prompt measures to regulate the prices for basic commodities, taken by the government, have restrained their growth, but not inflation as a whole.

The following seems to be expedient to ensure the economic growth in the current environment:

- Development of a Law on prices and pricing, that would promote fair competition in the national market;
- Efficient mechanisms to reduce the impact of world prices on the domestic rates on an ongoing basis;
- Development of competition at all stages of production and promotion of goods;
- Increasing the competitiveness of the national products;
- Development of mechanical engineering, electronics, biotechnologies, computer science along with gradual reduction of export of raw materials, fuel and material-intensive products.

References

1. Stiglitz, J.: The Roaring Nineties: Seeds of Destruction. Modern Economics and Law, Moscow (2005)
2. Wright, M., et al.: J. World Bus **56**, 101160 (2021). https://doi.org/10.1016/j.jwb.2020.101160
3. Holtorf, A.-P., Gialama, F., Wijaya, K.E., Kalo, Z.: Value Health Reg. Iss. **19**, 122–131 (2019). https://doi.org/10.1016/j.vhri.2019.04.003
4. Voronkova, E.K., Dolgova, M.V.: Auditors' J. **4**, 124–130 (2019)
5. Aliev, A.T., Vesnin, V.R., Slepov, V.A.: Theory, methodology and practice of pricing in industry. Publishing and Trading Corporation "Dashkov and Co", Moscow (2020)
6. Novikov, V.E.: Bul Rus. Customs Acad. **1**, 110–119 (2018)
7. International Monetary Fund. https://data.imf.org/?sk=471DDDF8-D8A7-499A-81BA-5B332C01F8B9&sId=1547558078595
8. Food and Agricultural Organisation. https://fpma.apps.fao.org/giews/food-prices/tool/public/#/dataset/international
9. The World Bank. https://databank.worldbank.org/reports.aspx?source=2&series=FP.CPI.TOTL&country=#
10. Rakhmanov, A.B.: Vopr. Ekon. **12**, 141–154 (2019)

11. Results of all-Russian monitoring of medical mask price dynamics. https://fas.gov.ru/pages/monitoring-cen-na-maski/10-fevralia-2020-goda
12. Medical mask price growth registered in the vast majority of regions. https://fas.gov.ru/news/29309
13. Federal Anti-Monopoly Service has revealed a nearly tenfold inflation of medical-mask prices by a wholesale firm. https://www.rbc.ru/rbcfreenews/5eac1fc59a79475cc2485c30
14. The Government has introduced state regulation of prices for medical masks. https://www.vedomosti.ru/business/articles/2020/04/06/827240-pravitelstvo-vvelo-gosregulirovanie-tsen
15. Makeeva, Y.: Ladybugs instead of chemicals. https://rg.ru/2020/02/03/reg-ufo/ivan-novichihin-u-rossiian-horoshie-shansy-zavoevat-mirovoj-rynok-organiki.html
16. On development of economic reforms in the Russian Federation in 1992. State Statistics Committee of the Russian Federation. Moscow (1993)
17. Slepov, V., Sakharov, D., Sarukhanyan, Y.: Fin Bus **4**(201), 40–45 (2019)
18. Slepov, V.A., Sarukhanyan, Y.A.: Fin Credit **23**(29), 1737–1749 (2017). https://doi.org/10.24891/fc.23.29.1737
19. Ben-Xi, L., Zhang, Y.Y.: J. Integ. Agr. **19**(12), 2937–2945 (2020). https://doi.org/10.1016/S2095-3119(20)63430-X
20. Belousov, A.: WTO, time has come to make a point. https://customsexpert.ru/articles/vto-prishla-pora-obyasnits.htm
21. The World Bank. https://databank.worldbank.org/reports.aspx?source=2&series=TM.TAX.MRCH.WM.AR.ZS&country=CHN,RUS,USA,WLD
22. Hsieh, C.-T., Li, N., Ossa, R., Yang, M.-J.: J. Int. Econ. **127**, 103370 (2020). https://doi.org/10.1016/j.jinteco.2020.103370

Transformation of the Information and Analytical System for Making Managerial Decisions in Economic Sectors

Ma Xiangzhu[1], Ekaterina V. Parada[2](✉), Olga V. Dymchenko[3], and Yulia A. Kruchanova[2]

[1] Southern Federal University, Rostov-on-Don, Russia
[2] Rostov State University of Economics, Rostov-on-Don, Russia
[3] Don State Technical University, Rostov-on-Don, Russia

Abstract. Financial reporting acts as the main means of communication and the most important element of the information base for making managerial decisions. Therefore, it is a kind of "framework" on which the foundation of strategic and tactical planning and enterprise management is built. In this regard, each enterprise strives to find a management solution that will allow it to gain leadership positions in the competitive struggle and get high financial results. Therefore, the company's economists need a thorough analysis of the market situation, independent planning of their activities in the field of searching for suppliers and buyers, pricing, etc.

Keywords: Information and analytical systems · Financial reporting · Management decision system

Information and analytical systems (hereinafter – IAS) are a necessary part of any business. The value of such systems is determined by the availability of reliable and up-to-date information, as well as the maximum consideration of reasonable user requests. The specifics of building an information and analytical system will be determined by the specifics of the industry in which the enterprise operates. This issue is particularly relevant for multi-profile holdings with widely diversified activities, including manufacturing enterprises, trading and financial structures.

In this case, a single ERP system is being implemented in business units, which solves the problem of forming three classical reporting forms based on primary financial accounting data. In addition, as part of the implementation of this form of IAS, additional tasks are being solved related to information support for monthly income and expense planning for each area of activity and regular analysis of the implementation of plans.

The transformation of the information and analytical system of management decision-making is due to the changed rules of presence in industry markets and the emergence of new technologies (Table 1).

P. V. Trifonov and M. V. Charaeva (Eds.): DITEM 2021, LNNS 380, pp. 338–343, 2022.
https://doi.org/10.1007/978-3-030-94245-8_47

Table 1. The impact of new information technologies on the procedures for collecting and processing information.

Former rule	New rule	Technology
The information appears at the same time and in the same place	Information can be obtained on request anywhere and at any time	Distributed databases and data warehouses, search engines, technologies for searching for specified information
Analytical work can be carried out only by a specially trained employee	Analytical work can be performed by a general specialist	Expert systems
The information is either centralized or decentralized	At the same time, both methods of management can be used, which makes it possible to take advantage of each of them	Distributed group work, telecommunications and networks
Decisions are made centrally – at the highest level of management – by directors or top management	Each employee becomes responsible for making decisions on their area of work	Decision support tools, access to knowledge bases and repositories, knowledge systems
Specialized premises and/or equipment are needed to receive and process information	Everyone can use a variety of information systems from their personal digital device	Receiving and sending information from a place where Internet users are located-technologies, fiber-optic and satellite communication systems, mobile systems
Personal contact with the buyer prevails while getting acquainted with the product or during purchase and sale transactions	Every year more and more sales are carried out through online stores and websites	Initially, a virtual contact with a potential buyer is an opportunity to study the client's characteristics in advance
In order to find the right thing/information, you need to know where it is located	Things/data themselves "tell" the user where they are	Search systems, mobile systems, voice assistants
Approved plans/budgets can be revised only under the influence of extreme circumstances	Plans are reviewed and adjusted promptly, rolling budgeting, as well as beyond budgeting, is becoming popular	Expert systems, flexible planning and risk management systems, high-performance computers

Source: Compiled by the authors

The enterprise has to face competition not only in the consumer market, but also in the capital market, competing with other enterprises for obtaining credit resources or attracting direct investment. The scheme of information formation and its use for the analysis of the company's activities is demonstrated in Fig. 1.

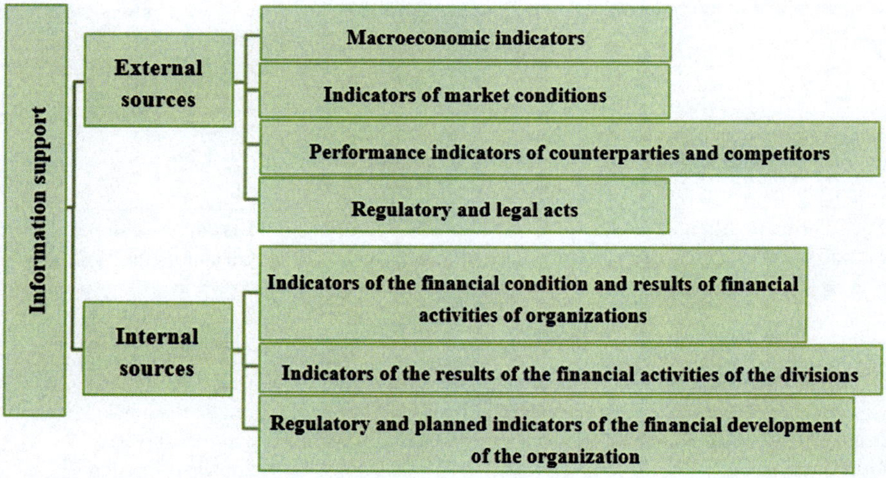

Fig. 1. Information support for economic and managerial analysis of the company's activities. *Source:* Compiled by the authors.

The financial analysis based only on the data of the accounting statements mostly demonstrates the features of an external analysis that is carried out by interested counterparties, shareholders, creditors or government agencies. The analysis is rather limited, since it is based only on the use of accounting statements presented by the forms of the balance sheet, the report on financial results, etc. It does not allow to fully link the financial condition of the enterprise with the ongoing financial and economic processes.

As for small and medium-sized businesses, in order to form an IAS for making current and operational management decisions, it is usually sufficient to use the data of the balance sheet and the report on the financial results of the enterprise, since with the help of analysis of financial reporting data, it becomes possible to identify the main reserves for improving the efficiency of the enterprise. That is the main goal of effective management. If to consider the tasks of the IAS more broadly, for example, to include the preparation of management decisions in them, it will already be necessary to conduct a management analysis, which includes not only an analysis of financial statements, but also an analysis of the industry market and the macroeconomic situation. Such studies are usually carried out by specialized marketing agencies and cost about 50 thousand rubles.

Providing objective information about its economic activities increases the likelihood of attracting external financing for the enterprise, which indicates the special importance of correctly compiled and transparent financial statements. If the economic results of the enterprise are favorable, but the possibility of attracting investment is also high [1].

Financial statements are a system of reporting forms that are compiled on the basis of information from financial accounting, and its purpose is to maximize the information of the entire range of reporting users about the financial results of the company's activities, the dynamics of the financial condition and its prospects. Moreover, such information should be provided in the most accessible and user-friendly form [2].

The principles of forming financial statements according to international standards list different groups of its users: investors, owners, creditors, suppliers, government agencies, employees of the enterprise, etc. [3]. Therefore, Table 2 provides information about the most important groups of users of financial statements, their interests and the sources of information that they use in making their decisions. This grouping is extremely important for the purposes of the initial formation of the IAS, as it gives an understanding to the specialists of the enterprise as to what sources of information they will need in the process of managing the enterprise, as well as what economic indicators are priority for each group of users [2].

Table 2. Users of financial statements and sources of information for management decision-making.

Users	Interests	Information sources
Company managers	- efficient production and financial activities; - up-to-date and reliable information base for making managerial and financial decisions	- internal financial, accounting and production reports of the enterprise; - financial statements of the company
Tax authorities	- payment of relevant taxes to the budgets by an economic entity	- primary and secondary accounting statements; - tax reporting; - data of internal and external audit checks
Shareholders	- monitoring of the financial condition of the enterprise in order to assess the degree of riskiness of the investments made; - analysis of the prospects for increasing the expected amount of dividends;	- financial statements; - annual reporting
Creditors	- analysis of the borrower's solvency; - analysis of the borrower's collateral availability	- financial statements
Suppliers	- determination of the payment security and business activity of the counterparty	

(*continued*)

Table 2. *(continued)*

Users	Interests	Information sources
Buyers	- assessment of the quality of products and the adequacy of the pricing policy	
Employees	- determining the prospects of their employment, salary growth and career growth	
Statistical bodies	- collecting information about the company's activities	- special statistical reporting; - financial statements

The place of the information and analytical system in making managerial decisions is shown in Fig. 2. The external contour of the presented scheme connects together all the functions performed by the IAS of the enterprise: 1 – strategic planning, 2 – tactical planning, 3 – information about material and technical equipment, 4 – assessment of the reliability of the decisions taken, 5 – analysis of approaches to making managerial decisions, 6 – final justification of managerial decisions consistent with the strategy of the enterprise development.

As a rule, three Microsoft software products act as a software platform for creating IAS at enterprises of various sectors of the economy: DBMS for storing and processing information (SQL SERVER); as a database client — Excel spreadsheets; as a means of visualizing reporting-the company's website.

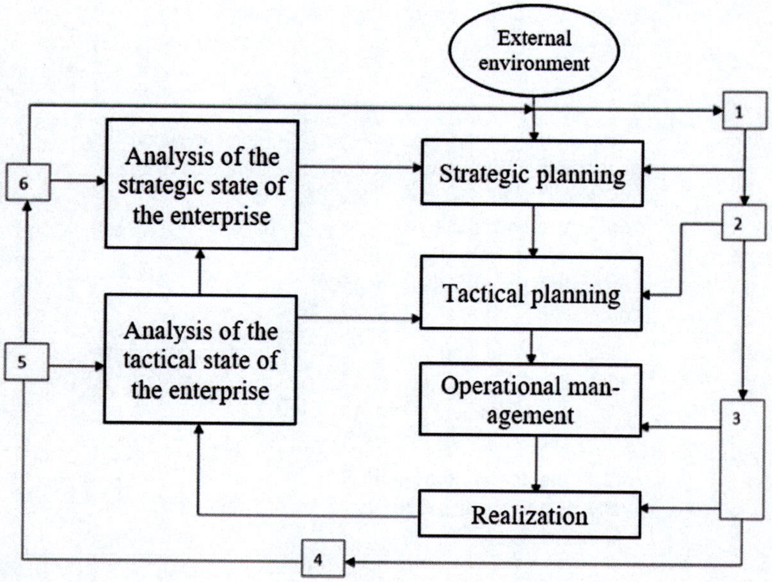

Fig. 2. Information and analytical system in the generalized enterprise management scheme [4].

It is necessary to build an analytical system so that it could have full access to the information of the accounting system in the reading mode by accessing directly all the DBMS tables [5]. The effective operation of the company's IAS implies that information from the ERP system makes up approximately 90% of the volume of all primary information [6]. In addition to this data, plans are uploaded to the system from Excel spreadsheets sent by departments and a small amount from report files of other programs that have historically worked in some businesses.

As input information, transactions on accounting accounts are used directly, colored by analysts with the categories "income/expense", "department", "object". For each industry direction, already in the analytical system, their own groups of income and expense items are formed with their own hierarchy. Similarly, divisions are grouped within business units. All this data is stored in the database of the analytical system, does not affect the accounting program, and at the same time, it remains possible to quickly rearrange the data in accordance with the user request. At the end of the reporting period — a month — financial reports are calculated for all business units, which are stored in the database of the analytical system.

As the experience of using IAS by Russian companies proves, the main groups of reports in the analytical system are represented by a group of management financial reports — balance sheet, OPU, DDS with detailed indicators; a group of different analytical production reports for various departments (reports on flights, transactions, warehouses, customers, etc.); plan/fact analysis for all the departments. In the analytical system of large companies (holdings), the consolidation of reports for various divisions groups is carried out — therefore, an enlarged consolidated report is formed. In addition to the functions of analytical reporting, over time, such a system began to solve tasks that were not quite peculiar to it, complementing in some way the functionality of the accounting system.

References

1. Charaeva, M.V.: Formation of structural and logical model of company's financial standing analysis. In: 2nd International Conference on Social, economic, and academic leadership (ICSEAL), Prague, pp 127–133 (2018). https://apps.webofknowledge.com/full_record.do?product=WOS&search_mode=OneClickSearch&qid=7&SID=D4BW3tdlgkfALz9Qi7w&page=2&doc=19
2. Barmuta, K.A., Akhmetshin, E.M., Andryushchenko, I.Y., Meshkova, G.V., Zekiy, A.O.: Problems of business processes transformation in the context of building digital economy. Entrepreneurship Sustain. Issues **8**(1), 945–959 (2020)
3. Mitrovich, S.: Directions and prospects for the development of information support for economic analysis in Russia. Econ. Anal. Theor. Pract. **9**, 102–106 (2016)
4. Karlova, T.V., Rastorguev, A.V.: Evaluation of the effectiveness of the information and analytical system of the enterprise. Herald Bryansk State Tech. Univ. **4**(65), 82–87 (2018)
5. Evstafyeva, E.M., Charaeva, M.V., Krokhicheva, G.E., Kiryanov, E.O.: Account-analytical ensuring of economic entities for the development of financial control. In: The European Proceedings of Social & Behavioural Sciences (EpSBS), pp 167–172 (2018). https://doi.org/10.15405/epsbs.2019.04.19
6. Uvarova, G.: The analytical system should quickly provide the information necessary for decision-making. Consultant Partner **31**, 10 (2014)

Comparative Analysis of Trends in Housing Development in Russian Regions

Vyacheslav K. Sevek[1(✉)], Kherel M. Naksyl[2], Cheynesh G. Dongak[1], Choduraa S. Manchyk-Sat[1], and Kherel B. Badarchi[3]

[1] Tuvan State University, Kyzyl, Russia
[2] Government of the Republic of Tuva, Kyzyl, Russia
[3] Tuvan Institute for Integrated Development of Natural Resources, SB RAS, Kyzyl, Russia

Abstract. Construction is a key sector of the national economy, one that is inextricably linked to all other sectors. Apparently, it would be difficult to incentivize better education, extend life expectancy, and increase people's income unless housing shortage and poor conditions are addressed. This study sought to compare the status and issues of growth and improvement in housing development in medium term across Russian regions. Statistical methods of economics were used to verify the official statistics available for 1990–2019. Overview of residential housing commissioned over the period identified four stages of the housing industry's development in the country. Thus, it was found out that since 2018, as abolition of the Federal Law No. 93-FZ was impending, and amendments were on the way regarding equity construction with use of escrow accounts, people started to commission more and more detached houses. 56% of all detached houses were commissioned in rural areas. Overall, cities account for 70% of new residential construction, rural areas for 30%. Housing construction varies from region to region depending on how developed the local investment and construction infrastructure is as well as on transport accessibility, income, and purchasing power of the locals. All these factors as well as the persistent sanctions, Russia's actual isolationism in global politics, and the global Covid-19 pandemic have been defined and systematized herein as factors that affect the growth and trends in housing development in Russian regions.

Keywords: Housing development · Growth and trends · Regions of Russia

1 Introduction

Russian state's housing policy focuses on making housing more affordable and available through an increase in construction and decrease in price per square meter. This paper compares the trends in housing development across Russian regions as stipulated in the National Program approved by the Government Decree No. 2227-r dd. November 30, 2012, On Access to Comfortable and Affordable Housing and Public Utilities for Russian Citizens; this Program plans up to 71 million square meters of total floorspace to be commissioned in 2015, up to 76 million in 2017, 83 million in 2018, and 92 million in 2020.

P. V. Trifonov and M. V. Charaeva (Eds.): DITEM 2021, LNNS 380, pp. 344–349, 2022.
https://doi.org/10.1007/978-3-030-94245-8_48

Theory behind the scientific overview of housing development is based on papers by Russian and international researchers [1–5]. The contribution of this paper consists in the novel approach to staging the development of the national construction industry, identifying its growth and development factors with breakdown by regions.

The goal hereof is to compare the status and issues of growth and improvement in housing development in Russian regions.

2 Methods

Methodically, this study relies on literature overview, economic statistical analysis, and expert opinions.

3 Results and Discussion

Analysis of Russia's official housing development statistics for 1990–2019 revealed the following trends in the industry and the effects of general economic processes on it, see Fig. 1.

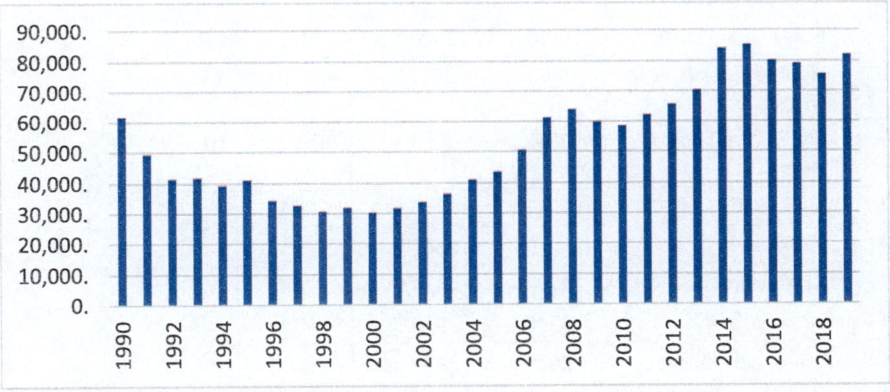

Fig. 1. Commissioning of residential housing in Russia, 1990–2019, million sq. m.[6].

Figure 1 shows the following stages:

- From 1990 to 2000, the industry was in decline after the USSR collapsed.
- From 2000 to 2008, rising oil prices boosted Russia's economy;
- In 2008, the global financial crisis caused a worldwide recession, which affected Russia and its construction industry;
- In 2014, the Crimean crisis was followed by sanctions against certain Russians.

Nevertheless, 2015 saw the peak construction since 1990: 85.3 million sq. m. Since then, the industry has been in decline.

When it comes to breakdown by regions, housing development is largely concentrated in a few larger regions. In million sq. m., Central Federal District had 31.4% of the total commissioned floorspace, and Volga Federal District had 19.7%; together, they had 51%.

In 2019, Moscow Oblast had the highest housing commissioning figures of any region and was ahead of four federal districts. Table 1 shows Rosstat's data.

Table 1. Housing commissioned in federal districts and regions, million sq. m. Table compiled by the authors.

No	Federal districts and regions	2017	2018	2019	% of the national total
1	Russian Federation	79,224	75,658	82,042	100
2	Central Federal District	24,284.9	23,434.5	25,752.3	31.4
3	Volga Federal District	15,640.8	15,222.8	16,190.2	19.7
4	Southern Federal District	9622.1	9067.3	9920.1	12.1
5	Northwestern Federal District	8979.0	9472.7	9448.0	11.5
6	Siberian Federal District	6793.2	6649.0	7437.9	9.1
7	Ural Federal District	6301.7	5964.5	6730.1	8.2
8	North Caucasus Federal District	5081.7	3557.3	4027.5	4.9
9	Far Eastern Federal District	2512.2	2269.4	2513.7	3.1
10	Moscow Oblast	9077.7	8866.6	8613.5	
11	Moscow	3419.0	3541.2	5175.5	
12	Krasnodar Krai	4728.4	4414.9	4532.0	
13	St. Petersburg	3536.1	3950.3	3471.2	
14	Leningrad Oblast	2625.8	2640.6	2930.2	
15	Republic of Tatarstan	2408.1	2409.9	2675.5	
16	Rostov Oblast	2333.9	2347.4	2611.3	
17	Tyumen Oblast	2468.5	2120.4	2558.1	

Housing commissioning peaks in more populated regions where people have greater purchasing power, see Table 1. Data clearly shows which regions develop most dynamically and receive the influx of domestic migrants, including those from the Far North.

Yet in 2018–2019, Sevastopol showed the fastest growth, perhaps due to its unique position as a new region that receives a lot of budget transfers. Interestingly, many of the Top 10 regions had different patterns of growth in 2018 and 2019, as 2018 saw a substantial decline in housing commissioned, see Table 2.

Table 2. Top 10 regions in terms of housing commissioned, 2019, %. Table compiled by the authors.

No	Regions	2017	2018	2019	Increase in 2018, YoY, %	Increase in 2019, YoY, %
1	Sevastopol	174.3	314.3	568.9	180	181
2	Magadan Oblast	6.4	4.3	7.4	67	172
3	Tver Oblast	585.2	424.9	639.7	73	151
4	Krasnoyarsk Krai	1056.5	1148.5	1695.9	109	148
5	Moscow	3419.0	3541.2	5175.5	104	146
6	Lipetsk Oblast	1084.2	903.0	1253.9	83	139
7	Republic of Adygea	244.3	187.0	257.0	77	137
8	Khanty-Mansi Autonomous Okrug, Yugra	812.5	586.0	797.9	72	136
9	Republic of Khakassia	235.9	188.2	247.8	80	132
10	Amur Oblast	181.6	158.7	204.9	87	129

In general, 16 regions had negative growth (less than 100% of the 2018 value) in 2019, whereas in 2018, 49 out of 85 regions (57%) commissioned less housing than in 2017. In this case, 2018 started a cycle of apartment block construction projects that hadn't ended by the end of the year. As new regulations on equity construction went into force in 2019, many companies had registered and started their construction projects in 2018 so that the older regulations would apply.

Regional demographics also determines the structure of newly commissioned housing. Statistics shows that in regions with predominantly rural population, detached houses prevail. Other regions "prefer" apartment blocks.

Breakdown by class has patterns, too: richer regions tend to build more upper-class housing.

Housing availability per person rose by a factor of 1.46 from 1995 through 2019, from 18 m^2 in 1995 to 26.3 m^2 in 2019. Over the past 10 years, i.e., from 2010 through 2019, average housing availability rose by 16.4%.

As for breakdown by districts, Northwesterners lead at 28.12 m^2 per person, followed by the Central (27.53) and Volga (27.31) Federal Districts. North Caucasus Federal Districts close the ranks at 21.82.

Researchers note housing availability in Russia is below the UN standards of 30 m^2 per person [7]. The Housing Development Strategy notes that room availability is also far below that in developed European countries: 0.9 rooms per person against 1.8 in the OECD member states [8].

In a Russian Public Opinion Research Center (WCIOM) survey [9] of 2018, 54% of the respondents reported they needed to improve their housing (39% reported their homes were crowded, and 16% said they lacked housing in ownership).

Demographic trends, mortgage availability, and income are the key drivers of housing demand. Mortgage availability affects demand for apartments in blocks the most.

Bank of Russia reports that despite the absolute historical lows of mortgage rates in 2019, mortgagors were not many [10]. 27%/35% of Russian households would be able to pay their monthly instalments on a new home with 30 m^2/18 m^2 of floorspace per person, respectively.

Rosstat reports a housing affordability rate of 48.7% (percentage of households able to purchase housing, whether through mortgage or not). 53% of the regions had above-average affordability rates. Table 3 below shows top 10 regions in terms of housing affordability in 2019.

Table 3. Housing affordability, %, top 10 regions in 2019.

No	Region	2019, %
1	Yamalo-Nenets Autonomous Okrug	81.6
2	Khanty-Mansi Autonomous Okrug, Yugra	80.2
3	Nenets Autonomous Okrug	78.8
4	Magadan Oblast	76
5	Republic of Adygea	69.9
6	Bryansk Oblast	69.7
7	Kostroma Oblast	68.8
8	Tambov Oblast	68.2
9	Lipetsk Oblast	67.7
10	Vladimir Oblast	66

Apparently, the regions rich in minerals and fossil fuels could boast the best affordability of housing (lines 1 and 2 in the table).

4 Conclusion

Analysis of the housing development industry reveals trends that are important for the growth and change in the sector and concern housing policies and the country's spatial development in general:

- Construction varies from region to region depending on the location (proximity to transport routes), proximity to major logistic centers, suppliers, and manufacturers, which affects the costs of production in the industry;
- Development is mostly concentrated in European Russia, which has a denser population. Central and Volga Federal Districts lead in terms of housing commissioned;

- Some European regions of Russia have higher figures than whole federal districts elsewhere in the country, which indicates a substantial lag in development that some regions and geographical spaces of Russia suffer from (e.g., Siberia and Far East);
- Interestingly, Southern European regions of the country are experiencing a boom in housing development, a trend consistent with domestic migration from the Far North and equivalent territories;
- Income and purchasing power as well as the family structure (number of dependents) directly affect housing demand.

References

1. Asaul, A.N., Asaul, M.A., Lyulin, P.B., Chepachenko, N.V.: Stud. Rus. Econ. Dev. **3**(174), 111–117 (2019)
2. Yuji, C., Niimi, H.Y.: Was the expansion of housing credit in Japan good or bad? Jpn. World Econ. **53**, 100996 (2020). https://doi.org/10.1016/j.japwor.2020.100996
3. Xu, H., Zhou, Y.: Public housing provision and housing vacancies in Japan. J. Jpn. Int. Econ. **53**, 101038 (2019). https://doi.org/10.1016/j.jjie.2019.101038
4. Karji, A., Woldesenbet, A., Khanzadi, M., Tafazzoli, M.: Assessment of social sustainability indicators in mass housing construction: a case study of Mehr housing project. Sust. Cities Soc. **50**, 101697 (2019). https://doi.org/10.1016/j.scs.2019.101697
5. Barkokebas, R.D., Chen, Y., Haitao, Y., Al-Hussein, M.: Achieving housing energy-efficiency requirements: methodologies and impacts on housing construction cost and energy performance. J. Build. Eng. **26**, 100874 (2019). https://doi.org/10.1016/j.jobe.2019.100874
6. Sayenko, I.A.: Prob. Mod. Econ. **1**(57), 137–140 (2016)
7. Kudryavtseva, V.A.: Proc. Univ. Invest. Const. Real Est. **7**(1), 45–52 (2017)
8. Housing Development Strategy (2025). https://minstroyrf.gov.ru/upload/iblock/ec7/Strategiya-zhilishchnoi_-sfery.pdf
9. Analytical Review. https://wciom.ru/analytical-reviews/analiticheskii-obzor/pod-kryshej-doma-svoego
10. Achkasov, Y.K., et al.: Housing development. an analytical memo. https://www.cbr.ru/content/document/file/84168/analytic_note_20191004_ddkp.pdf

Planning in the Process of Implementation of the Strategy of Socio-economic Development of the Region

Oleg V. Boychenko(✉) and Eskender I. Ismailov

Institute of Economics and Management, V.I. Vernadsky Crimean Federal University, Simferopol, Russia

Abstract. The article examines the features of modern approaches to organizing planning in the process of implementing the strategy of socio-economic development of the region, taking into account its main priorities. The relevance of using the strategic approach is determined by the revival of interregional cooperation as a way to identify the competitive advantages of economic agents and the national economy as a whole. It has been established that the legal framework on which regional development strategies were formed is very problematic due to excessive formalization of procedures for the development and approval of regional development strategies. It has been determined that the modern strategy of socio-economic development of regions is aimed at building economically strong and competitive regions that would independently solve current problems and perform their functions without additional attraction of funds from the state budget. It has been determined that an effective solution to the problems of the national economy at the present stage is possible by leveling disproportions in the economic and social development of regions, creating conditions for activating proactive and internal sources of socio-economic growth, and supporting backward regions. The expediency of the symbiosis of the modern strategy of the socio-economic development of the region with the participation of local authorities, various representatives of business and enterprises, public organizations and local governments has been proved.

Keywords: Priorities of contemporary society · Concept of development through strategies · Model of the future

1 Introduction

The development of a program for the integrated development of the region (development concept, development strategy, strategic plan) is a difficult task due to the following reasons. Firstly, strategic planning for the development of the region is an extremely politically loaded process [1]. When developing a strategy for the development of the region, it is necessary to take into account the standpoints and interests of many parties, which may be mutually contradictory and even mutually exclusive. Particularly sharp disagreements may arise at the stage of approval of the final document obtained in the

P. V. Trifonov and M. V. Charaeva (Eds.): DITEM 2021, LNNS 380, pp. 350–354, 2022.
https://doi.org/10.1007/978-3-030-94245-8_49

process of strategic planning, the concept (strategy) of development or the program of socio-economic development.

Secondly, each Russian region is an extremely complex territorial entity with many features of geographic, socio-economic, geopolitical, ethnic and natural character. It is almost impossible to take into account all the distinctive features and factors of the development of the region [2]. In addition, each region is subdivided into municipalities, consequently, it is necessary to consider the socio-economic development of each of them, and this seriously complicates the process of strategic planning. It should also be taken into consideration that the territorial structure of the region is not homogeneous: some municipalities replenish the regional budget, while others are financed by donors. Thus, within the region, socio-economic contradictions are growing between developing and depressed territories [3].

Thirdly, when planning social and economic development, a really huge amount of information needs to be processed. The range and variety of information sources determine the high level of complexity of strategic planning at the regional level. In addition, the reliability and validity of some information sources may raise serious doubts; data from some sources may contradict others. Moreover, as studies have shown, the state statistics bodies cannot provide the entire volume of information required to design a strategy for the development of the region. Strategy developers have to make a lot of effort to select sources of information, analyze and process the data obtained [4].

Fourthly, the elaboration of a regional development strategy requires the involvement of qualified specialists. Strategic planning is both a scientific problem and a management task, and at the same time a line of empirical research. Therefore, various specialists are needed: politicians, scientists, leaders and subject analysts. These specialists are needed not only to write down the outline of the strategy, but also to develop a methodology and create a system of strategic planning at the regional level [5]. It is also necessary to conduct an approval process for the developed strategy, take into account the views of stakeholders, make adjustments, prove the viability of the strategy and present it to the general public. However, external consultants, such as research organizations or Big Four consulting firms, are often hired to develop the strategy. This does not solve the problem of attracting qualified specialists, moreover, it hinders the active participation of interested parties which can influence the effectiveness of the developed strategy.

Fifthly, strategic planning at the regional level is a complex organizational and methodological process. There are no universal methods of management, and there is no universal methodology for strategic planning. Each region should develop an original methodological approach to the organization and implementation of the process of strategic planning of the socio-economic development of the region [6]. To develop such an approach, it is possible to attract consultants, but first of all, this requires trained and experienced specialists [7]. The above factors are enough to devolve the strategic planning of the socio-economic development of the region into an ineffective and pointless process. As a result, the content of the designed regional development strategy is assessed as unsatisfactory, interest groups express dissatisfaction with the strategy and oppose its approval. Ultimately, all this leads to the lack of demand for the designed development strategy, doubts arise about the effectiveness of strategic planning methods [8]. Let us consider from a scientific point of view the central problems and difficulties of strategic planning of the socio-economic development of the region and specify the ways of their solution [9].

One problem is the declarative nature of most strategies, the status of their authenticity compared to the real programs (adopted to implement the strategy) [10]. The next problem is the presence of conflicts between local government structures and groups of influence at the regional level; different visions of regional development priorities, which is formed at different levels of state management of regional development [11]. It should be noted that a significant problem for all regions that have developed strategies for their development is the practical implementation of those strategies [12].

All the above-mentioned problems can be solved on condition of the adoption of a normative legal act on strategic planning and uniform methodological requirements adapted to domestic economic conditions for the development of strategies for the socio-economic development of regions with detailed control (monitoring) of the processes that will occur in the economic and social sphere in the process of implementing the strategy, specifying the basic monitoring indicators for the purpose of consistency between the strategies of various hierarchical levels of management.

The practical implementation of strategies for regional socio-economic development depends on funding [13]. In the context of aggravated competition for the use of resources available at the regional (local) level (including limited financial resources), the problem of determining the priorities for financing regional development projects is of particular importance.

Today, the foremost among such priorities is: provision of financial support to business entities of all forms of incorporation to increase the competitiveness of the region; provision of financial support to investment projects in the field of road transport construction and the creation of transportation infrastructure, the development of local, regional and international transport corridors, telecommunications and energy networks, and the related infrastructure; the financing of the projects which after being implemented will contribute to the development of import-substituting and export-oriented production of goods and services in priority areas defined in the regional development strategy; assistance in the formation at the regional (local) level of infrastructure facilities for investment and innovation development: business centers, business incubators, innovation and consulting centers, venture funds; coordination of their activities in order to implement the priorities defined in the strategy; provision of financial support to both small and medium-sized businesses in order to increase employment and self-employment of the population in the areas identified by the strategy as priorities.

2 Results

Strategy is the result of a strategic planning process. But the stage of implementing the strategy is a logical continuation of work to achieve the goal, for the sake of which the corresponding strategy was developed. Therefore, it is reasonable and logically justified to consider the process of achieving socio-economic development (as a general goal) within the framework of the strategic management process. The strategic management process includes setting strategic goals and objectives, preparing, making and implementing strategic decisions, as well as monitoring their implementation [14].

The fundamental requirements for the system of regional development goals are their attainability (goals as the desired state of the goal-setting object must be realistic), flexibility (as a mechanism for adaptation to the conditions of a certain uncertainty of the future, forecast period), measurability (quantitative characteristics of each goal set), hierarchical structure of the goal system. The stage of strategic choice, according to experts [15], lends itself least to unification.

3 Conclusion

The inclusion of sustainable development principles in long-term socio-economic planning at the regional and municipal levels requires a new methodology and supporting tools. Based on a thorough analysis of the dynamics of sustainability indicators, the main stakeholders and organizations involved in the process, this article explores the best practices of strategic planning for sustainable initiatives. The approaches that support sustainable planning and the corresponding tools were identified using the example of planning in a large industrial and agricultural region, Rostov Oblast. As a result, recommendations were formulated for strategic planning that supports sustainable development at the local and regional levels. These recommendations can be useful for strategic planning in other regions with similar socio-economic characteristics.

In the context of the above, it can be argued that globalization based on the principles of homogenization and unification, instead of consolidating and harmonizing the national interests of countries, causes a disbalance in the development of the world economy. The modification of national strategies for socio-economic development is particularly associated with the expansion and deepening of globalization, the need to "protect" itself from its negative manifestations.

References

1. Federal Law of June 28, 2014 No. 172-FZ: Strategic planning in the Russian federation. https://rg.ru/2014/07/03/strategia-dok.html
2. Egorov, K.Y.: Assessment of the implementation of the strategy of socio-economic development of the region. Samara State Univ. Econ. Bull. **10**(72), 10–16 (2010)
3. World Bank: Russia economic report 39 (2018)
4. Gorlin, Y., Lyashok, V., Maleva, T.: Positive effects and potential risks. Econ. Policy **13**(1), 148–179 (2018)
5. Ilyina, I.N., Kirillova, A.N., Plisetskiy, E.E., Kopychenko, G.S., Rybina, E.G.: Evaluation of the effectiveness of targeted programs (the example of the Tyumen oblast). Reg. Econ. Theor. Pract. **46**(373), 37–55 (2014)
6. Kislitsyna, V.V., Cheglakova, L.S., Karaulov, V.M., Chikisheva, A.N., Kislitsyna, V.V.: An integrated approach to the socio-economic development of the region. Econ. Reg. **13**(2), 369–380 (2017)
7. Kolozhvari, E.S., Glazychev, K.A., Zalomskaya, I.E., Kolozhvari, E.S.: Formation and evaluation of the effectiveness of the implementation of municipal programs in Novosibirsk in the context of the federal law "on strategic planning in the Russian federation." Siberian Financ. Sch. **2**(121), 57–68 (2017)

8. Morozova, N.V., Bondarenko, N.V.: Approaches to assessing the effectiveness of the implementation of regional strategies for socio-economic development. Chuvash Univ. Bull. **3**, 178–182 (2014)
9. Selyukov, M.V., Skachkov, R.A.: Theory and practice of strategic management of the socio-economic development of the region. Fundam. Res. **6**, 516–519 (2012)
10. Tazhitdinov, I.A.: Planning and Audit of the Socio-Economic Development of the Region. Economics, Moscow (2012)
11. Shpakova, R.N., Vasilieva, G.S.: Target-programmed method in regional environmental policy. Internet J. Resour. Conserv. Recycl. **1** (2018). https://doi.org/10.15862/10ECOR118
12. Gurvich, E., Prilepskiy, I.: Impact of financial sanctions on the economy. Econ. Issues **1**, 5–35 (2016)
13. Iskhakova, A.F., Pryadkina, N.N.: Formation of a regional development strategy. Bull. Orenburg State Univ. **8**(144), 110–115 (2012)
14. Gordon, R.J.: Obstacles to innovative development (the example of the United States). Econ. Issues **4**, 49–67 (2013)
15. Dobrolyubova, E.I., Klochkova, E.N., Yuzhakov, V.N.: State Programs in the Regions: Analysis of Practice and Recommendations. RANEPA Publishing House "Delo", Moscow (2016)

Management under Digitalisation: Transformation of Management Paradigm or Traditional Approaches to Management

Relevant Issues of Building a State-of-the-Art Model of Internal Financial Control for Publicly Funded Institutions

Dmitry A. Endovitsky(✉), Larisa S. Korobeynikova, Nadezhda V. Kondrashova, Sergey V. Golovin, and Natalia E. Spiridonova

Voronezh State University, Voronezh, Russia
office@main.vsu.ru

Abstract. As publicly funded institutions become more independent, they face the need to improve their internal financial control. Retrospective study of various economic literature and the authors' own experience point to several issues and flaws of the existing internal financial control systems. There are theoretical, organizational, and legal aspects to the situation. These were where the authors sought for what could be recommended in order to update the internal financial control models. The goal hereof was to present the authors' comprehensive view of how internal financial control systems could be built on a risk-oriented model for use in publicly funded institutions. Methods used herein include cognition methods such as dialectics, analysis, synthesis, and analogies. The study is based on an overview of relevant scientific literature and periodicals on economics. Results and novelty consist in systematizing the issues and flaws of internal financial control. The paper makes proposals on how those could be addressed in order to improve the internal financial control model. It further defines the factors that jeopardize internal control. The authors substantiate why internal financial control systems should be built upon the foundation of procedures practiced by external control agencies. The paper ultimately presents the mechanism of a risk-oriented internal financial control model for publicly funded institutions.

Keywords: Internal financial control · Public institutions · Publicly funded institutions · Risk-oriented model · Risk assessment

1 Introduction

Effective business management requires proper control. The importance of control depends on the status and functional specifics of the organization. State statistics (Table 1) shows that a majority of public institutions work in such key social areas as education, science, health, culture, social security, employment, physical education, and sports. These institutions are mostly publicly funded. Their performance affects the quality of public services rendered to people. Since such institutions are provided with their assets and funding by the state, which acts as the founder, monitoring the status of such assets and the usage of funding becomes crucial.

P. V. Trifonov and M. V. Charaeva (Eds.): DITEM 2021, LNNS 380, pp. 357–362, 2022.
https://doi.org/10.1007/978-3-030-94245-8_50

Table 1. Federal public institutions by activity (as registered in the State Registers) as of Jan 1, 2021.

Activity	Number of institutions
Public institutions, total	13,681
Including	
Education	1,218
Professional, research and engineering activities	998
Health and social services	837
Culture, sports, leisure and entertainment	410

Source: Uniform Interagency Information and Statistics System for State Statistics: www.fedstat.ru

The current trend is to give publicly funded institutions a greater degree of independence, which encourages a move towards better management, including improved internal financial control. This is why building an effective internal financial control model for such institutions is a question that many researchers and practitioners are interested in.

2 Methods

Any effort to improve internal financial control would require thorough analysis and solution of the existing problems. In order to get insight in what today's specialists think of the most relevant issues of internal financial control, the authors hereof collected literature published in 2017–2021 to analyze, group, compare, and generalize its content.

Internal control is the topic of papers by Bahturina, Bulyga, Berdnikova, Vasina, Gvarliani, Demina, Danilov, Ivanova, Kachkova, Kolesnikova, Oborin, Tomskikh, and others. Subject to much discussion is the lackluster legal regulation of internal control, as many more issues arise from it [1–7]. Bulyga notes a lack of a single, integral financial control system in the country, the deficiency of a single methodological platform in standards applicable to different forms of control [1]. Bahturina points out that lack of due legal explanations leads to a broader focus on internal control as control of the facts of economic activity [2]. Kachkova and Demina also state that internal control is often overlooked or only has a formal implementation, which aggravates the lack of a single internal + external control concept [3]. Of interest is the opinion of Kolesnikova, who believes that the uniform, declarative provisions in most legal regulations is what stalls the attempt to make a complete picture of internal control applicably to an organization or its activities / audited entities. The author also notes that the documentation used by ministries and other agencies lacks a practical component, and that definitions of internal control vary in the sectoral legislative framework [4]. In their work, the researchers note that risk-oriented approach to internal control is uncommon in the public sector [1, 5]. A system of internal control effectiveness indicators is missing, which constitutes another major issue [3, 8–11].

Although scientific discussion of internal control has been prolific, many papers offer statements of problem rather than solutions. Analysis of the existing R&D in the area reveals an objective need to develop proposals on how to address the existing shortcomings of internal financial control.

3 Results

A state-of-the-art internal financial control model should first of all be effective, i.e., perform well at a low cost. Addressing the existing issues and flaws will help optimize control and minimize its costs. Other researchers' opinions and the authors' own experience suggest that internal financial control should be analyzed as a multifaceted phenomenon if its issues are to be identified, and solutions are to be proposed (Table 2).

Table 2. Proposals on how to address the issues and flaws of internal financial control systems.

Aspect	Issues and flaws	Proposed action
Theoretical aspect	Lack of a single approach to defining the concepts of internal control and internal financial control	Standardize internal control in public institutions Draft a bill to regulate financial control in the public sector of the economy Organize conferences on internal control to share best practices Publish internal control reporting on the organization's website for public access Ministries and agencies need to cooperate with their subsidiaries to create a single, uniform approach to internal control
	Lack of guidelines	
	Unreliable internal control environment	
Organizational aspect	Internal control systems are critically dependent on the competence of those who develop such systems	Test and certify employees Organize advanced training and retraining
	Internal auditor loses functional independence	Internal auditor should be subordinate to the top manager or to the elected representative body of the organization
	Internal control is formal, i.e., superficial	Consolidate internal control requirements in binding regulations Design disciplinary action for employees whose control activities are only formal; have an effectiveness evaluation system in place
	Arrangement of control activities is not in line with the organization's activities	Develop a control effectiveness assessment system

(*continued*)

Table 2. (*continued*)

Aspect	Issues and flaws	Proposed action
	Internal control model does not cover potential risks	Build a risk-based internal control model
	Lack of internal control effectiveness criteria	Develop a control effectiveness assessment system
Regulatory aspect	Ambiguous interpretation and incomplete regulation of the conceptual framework	Standardize internal control in public institutions
	Lack of a single regulatory framework	Draft a bill to regulate financial control in the public sector of the economy
	Lack of sectoral documentation that could provide internal financial control guidelines for various industries	Public authorities should develop sectoral guidelines for their subordinate agencies and institutions

4 Discussion

As a result, the administration of a publicly funded institution becomes able to set forth its own, local requirements. This freedom leads to a lack of standardized internal control that would feature unambiguous interpretation of terminology, uniform requirements to control procedures, effective cooperation with external control agencies, well-coordinated work of the internal auditor reinforced by effective practices.

Publicly funded institutions operate in the context of a complex, unstable legal system; this complicates control procedures, a factor that compromises their effectiveness. Firstly, during the audit, the person responsible for internal control (i.e., the auditor) should base their conclusions on the legislation and standards that were effective as of the date of this or that economic action. Secondly, an imperfect legal system may facilitate corruption, if it allows ambiguous interpretation. Tsvetkov and Domashuk researched methods for identifying corruption in bidding; they classify the factors above as 'legal sign factors of corruption' [13]. Thirdly, frequent change in law or other regulations increases the risk of violation due to human factor. For instance, many violations in public procurement happen because officers responsible for contracting lack knowledge of procurement law and regulations; their approach to procurement is only formal.

Risk-oriented internal control model is a recognized way to improve control performance. The model prioritizes such entities where violations are most likely. The advantage is that effort concentrated on high-risk entities, which improves detection of potential violators and diverts the auditing effort from low-risk entities. The model helps optimize the material, staff, and financial resources of the auditor. More effective control means better security, less violations, same or even lower costs [14].

Risk-oriented internal control model consists of several steps: risk identification and assessment with respect to the entities that are subject to control, risk response and monitoring.

The accuracy of assessing the risks associated with each supervised entity, and thus the effectiveness of risk minimization effort depend on the quality and underlying principles of the risk assessment system. Depending on the scale of activity or the complexity of its organizational structure, the administration of a publicly funded institution might want to create an in-house risk assessment system by analyzing the following:

- The entity to be assessed, be it a structural unit, an officer, or a particular transaction;
- The detail of the assessment scale (the number of risk levels can vary from 3 to 8; for example, a four-level scale includes low, moderate, medium and high risk);
- Assessment approach: a static approach implies a one-time classification; a dynamic approach allows further reclassification depending on whether violations have occurred;
- Assessment method: numerical method uses numerical valuation of risks; conditional method involves a conditional classification based on features characteristic of this or that class of hazard, e.g., probability of occurrence or consequences thereof; the conditional integral method consists in scoring each parameter on a specific scale and summing them up [14].

Any control activity should focus on detecting and preventing violations and breaches before they happen, making prevention a core function of audit [15]. Long ago, V. I. Lenin wrote in To the Question of Rabkrin Objectives, Their Understanding and Completion that the purpose of control is not merely to "catch" or "expose"; rather, it is to prevent abuse or manifestation of shortcomings. Risk-oriented model for internal control is fully consistent with this idea, as it is based on minimizing or preventing the consequences of a risk rather than reporting the adverse events that have already occurred and auditing on a schedule, as the conventional model would do. An important part of risk-oriented internal control is responding to risk assessment by taking preventive action.

5 Conclusions

Most of the problems highlighted herein are a result of lackluster regulation. Apparently, no guidelines or recommendations regarding standardization of internal control could ever be implemented without involving executive authorities. In the context of today, development of internal financial control should look to the best practices of external control bodies whose legal regulatory framework are undergoing reforms as of writing this paper. Should the control model match, it will greatly simplify the cooperation of external and internal control and help avoid redundant 'double' audits [15].

References

1. Bulyga, R.P.: Account analysis. Audit **5**, 10–17 (2017)
2. YuI, B.: Act is Acc. Manage. Inf. Econ. **2**, 26–31 (2020)

3. Kachkova, O.E., Demina, I.D.: Act Prob. Soc. Econ. Dev. Rus. **3**, 33–39 (2018)
4. Kolesnikova, E.N.: Modern problems of legislative regulation of an internal control. Bul Univ **1**(1), 16–22 (2020). https://doi.org/10.26425/1816-4277-2020-1-16-22
5. Vasina, N.V., Danilov, A.N., Ivanova, N.G.: Human Sci. Humanit. Stud. **1**(31), 160–167 (2018). https://doi.org/10.17238/issn1998-5320.2018.31.160
6. Gvarliani, T.E., Tomskih, S.A., Oborin, M.S.: Int. Acc. **17**, 024–1038 (2017)
7. Berdnikova, L.F., Andrievskaya, E.N.: Karelian Sci. J. **1**(22), 95–97 (2018)
8. Cheng, Q., Goh, B.W., Kim, J.B.: Internal control and operational efficiency. Contemp. Acc. Res. **35**, 1102–1139 (2018). https://doi.org/10.1111/1911-3846.12409
9. Kim, Y., Matkin, D.S.T.: Financial condition and internal control deficiencies: evidence from New York counties. Pub. Budget Financ. **40**, 45–69 (2020). https://doi.org/10.1111/pbaf.12246
10. Domnişoru, S., Ogarcă, R., Dragomir, I.: Audit Financ. **4**(148), 628–643 (2017)
11. Rija, M., Rubino, F.E.: Int. J. Bus. Manage. **5**, 21–36 (2018)
12. Cvetkov, V.Y., Domashuk, P.V.: Distributed intelligent management. State Adviser **2**, 34–37 (2015)
13. Chaplinskij, A.V., Plaksin, S.M.: Prob. Manage. Theory Pract. **2**, 7–29 (2016)
14. Golovin, S.V.: Audit **10**, 30–33 (2020)
15. Endovitsky, D.A., Golovin, S.V., Spiridonova, N.E.: Int. Acc. **11**(473), 1240–1252. https://doi.org/10.24891/ia23.11.1240

Marketing Management in the Creation of Business Models in the Medical Services Market in the Context of Digitalisation

Victoria A. Bondarenko[1(✉)], Aleksander A. Voronov[2], Natalia N. Zubareva[3], and Vyacheslav A. Chisnikov[1]

[1] Rostov State University of Economics, Rostov-on-Don, Russia
[2] Kuban State University, Krasnodar, Russia
[3] Belgorod National Research University, Belgorod, Russia

Abstract. Marketing management seems to be necessary in ensuring that health institutions operate in a consumer-oriented format in the provision of medical services that determine the quality and duration of life of the population. Digital solutions empower customer orientation of the business. The business activities of for-profit health organisations can demonstrate their actual readiness to meet the needs of consumers and the importance of marketing management of business to them. Purpose of research: formulation of the priority format of the marketing business model of a for-profit health organisation relevant for the Russian market in the context of digitalisation. Methods: clarification of the existing business models of commercial medical centres on the Russian market; analysis of empirical research in the industry to identify the current situation, sentiments in the commercial medicine segment, focus on the marketing management implementation and the choice of a priority marketing business model. The introduction of digital technologies in various areas of the healthcare sector plays a significant role in strengthening relationships with consumers for commercial health centres. Digital platforms make it possible to expand the reach of consumers, involve them in permanent communications to diagnose and familiarise themselves with the whole range of possible medical services. It was established that the marketing business model of a multidisciplinary for-profit health organisation is the most acceptable from the point of view of maximising consumer value in the Russian market. The results obtained demonstrate a willingness to implement marketing management and a marketing business model focused on market development.

Keywords: Business model · For-profit health organisations · Marketing · Digitalisation of healthcare

1 Introduction

Marketing management is needed to ensure that institutions operate in a consumer-friendly format. In the modern situation, great importance is attached to the provision of medical services; health, quality and life expectancy of the population depend on them.

P. V. Trifonov and M. V. Charaeva (Eds.): DITEM 2021, LNNS 380, pp. 363–369, 2022.
https://doi.org/10.1007/978-3-030-94245-8_51

Digitalisation contributes to the quality characteristics of services in the form of increasing the availability of consulting services, involving citizens in early diagnosis of health status, and covering the population with remote monitoring within the framework of telemedicine projects. It seems significant to conduct professional medical consultations between specialists of health organisations, for example, to provide services for high-tech medical care (HTMC).

The key to the formation of digital platforms in Russia is laid down in the federal targeted programme Digital Economy of the Russian Federation [1]. With regard to the area we are studying, there is the national Health project, within the framework of which the goal of creating a single digital circuit is being implemented [2]. The importance of creating a digital platform of an ecosystem type for the rationalisation of work and the quality of their interaction with patients is noted [3].

Muslimov and Khabriev and others believe that the digital platform creates additional safeguards to secure the health of citizens [4, 5]. In the United States, the introduction of digital technologies in the form of remote diagnostics of patient health conditions has reduced costs and improved the quality of life of the population [6]. Such examples exist in Japan and several other countries [7].

The role of marketing in digital solutions is growing [8] as digital platforms allow for a convenient format of customer communications. It is possible to focus on consumer preferences with the introduction of marketing management in the segment of commercial medicine, which is created in a client-oriented format by pursuing both social and commercial goals. This keeps up to date the study of business models built by for-profit health centres that orient their activities towards meeting consumer needs.

As a hypothesis, we put forward the thesis that the business activities of for-profit health organisations can demonstrate their relevance to focus on consumer demands and the importance for them of marketing business management in the context of the implementation of digital solutions.

The purpose of the article is to formulate the priority format of the marketing business model of a for-profit health organisation, which is relevant for the Russian market.

As part of the challenges addressed, we study the focus on digital transformation of the industry, clarify the understanding of the marketing business model, analyse the existing business models of for-profit health organisations in the domestic market, analyse the results of empirical research conducted in the segment of commercial medicine and determine the marketing business model of a for-profit health organisation to meet the increased consumer value.

Under a business model, it is customary to understand the activities implemented by the organisation, with the specified format and time parameters of implementation, the resources involved in the process, to create customer value.

Timmers defines a business model as an aggregator of goods, services and information flows to specify all business actors, their interests, sought benefits and areas of profit [9]. It determines the marketing business model, which includes the marketing strategy of an organisation in combination with the implemented business model. A similar point of view is shared by other researchers [10].

Cantrell and Linder define business models as the system that defines the logic of an organisation and the customer value produced by it [11].

Steller emphasises that the business model is designed to streamline the ongoing business processes and simplify the complexity and chaos of efforts for the organization [12].

A number of researchers focus on the study of optimisation of business processes in the interaction of contacting parties [13, 14].

In the current situation, in relation to the construction of marketing business models including business processes aimed at maximum customer-centricity and increase in consumer value, researchers also note the role of digital services and business ecosystems [15, 16].

An equally significant quality analysed by scientists is the ability to fully implement the organisation's marketing strategy through the implementation of a proven business model [17]. We believe that marketing business models are aimed at determining and implementing a business strategy and long-term relationships with consumers [18].

2 Methods

The study of the role of marketing management in the creation of business models in the medical services market in the context of digitalisation suggests the need:

- To clarify the existing business models on the Russian market, implemented by for-profit health centres with a view to meeting consumer needs and making a profit;
- To analyse the results of empirical research in order to identify the current situation, sentiments in the segment of commercial medicine, focus on the implementation of marketing management and the choice of a priority marketing business model.

3 Results

The business models used by health organisations in the Russian market differ depending on the size, potential of the business, goals set, profiling and implemented statements to increase consumer value along with the profit of a health organisation. According to the established practice, it is common to distinguish five main types of business models of for-profit health organisations, i.e. multidisciplinary, multidisciplinary, field-specific, private ambulance and diagnostic laboratory [19, 20].

They depend on the quality of marketing management of a business by orienting its business processes towards increasing customer value. The format of a multidisciplinary business model assumes a branch network, a wide range of services offered to customers and the availability of an actively promoted brand. The multi-profile option does not imply opening branches; patients are offered no more than five options of health care within the framework of the services of narrowly focused specialists. The format of a field-specific centre assumes the depth of the provision of a specific service. When choosing a business model for a diagnostic laboratory, the institution provides services only in the format of collecting materials for conducting the necessary analyses. The private ambulance business model involves the provision of services in the form of emergency medical care, transportation and emergency medical services at the patient's home.

A more sustainable format is a multi-disciplinary business model that will dominate the Russian market. It is a kind of maximum orientation towards emerging consumer needs.

It is possible to measure willingness to implement marketing business models by for-profit health organisations by the emerging sentiment associated with steps to expand customer base and market reach (Fig. 1).

For-profit health organisations are planning to implement solutions within the framework of a marketing business model associated with an increase in the consumer value of the medical services provided. The medical business expresses its readiness to re-equip, launch digital projects, introduce high-tech medical care (HTMC), as well as open new clinics and enter markets in other regions.

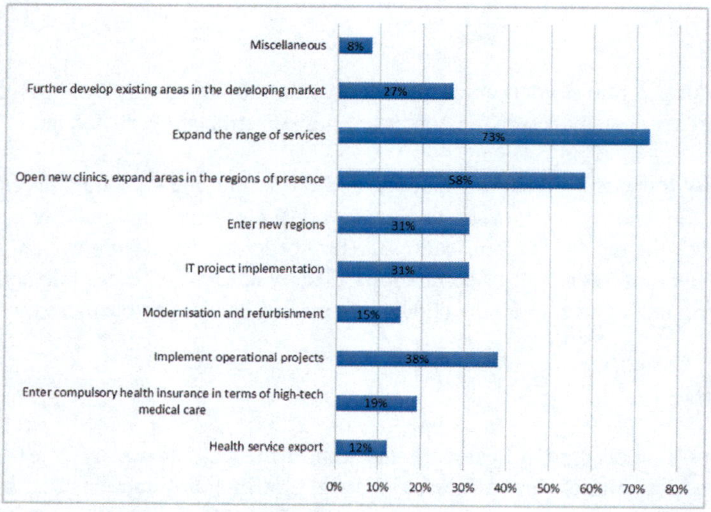

Fig. 1. Projects planned for implementation by for-profit health organisations projects [21].

For-profit health centres are expressing interest in streamlining business processes and improving service quality, which also indicates a commitment to implementing a marketing business model.

The planned efforts to increase customer value for for-profit health organisations are associated with engaging health workers and expanding the range of services provided to customers (Fig. 2).

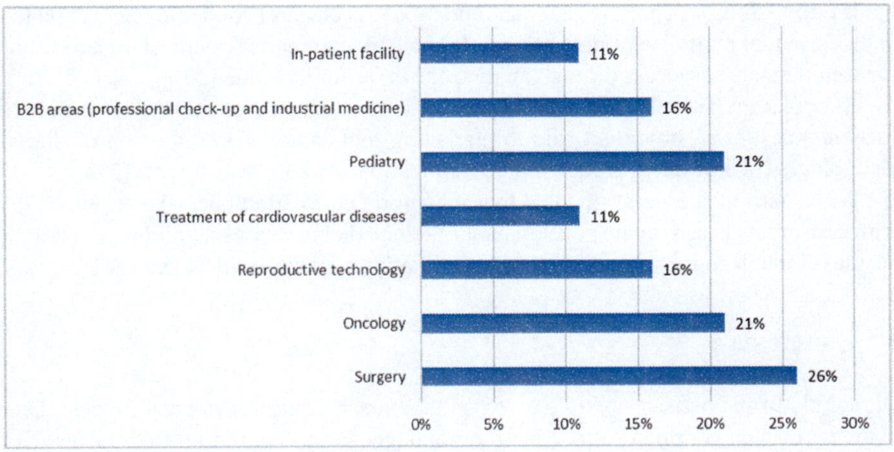

Fig. 2. Planned expansion of the range of medical services provided [21].

The implemented business model is built on the possibility of digital communication with consumers. The planned efforts to automate processes, introduce mobile applications that are convenient for modern patients are obvious (Fig. 3).

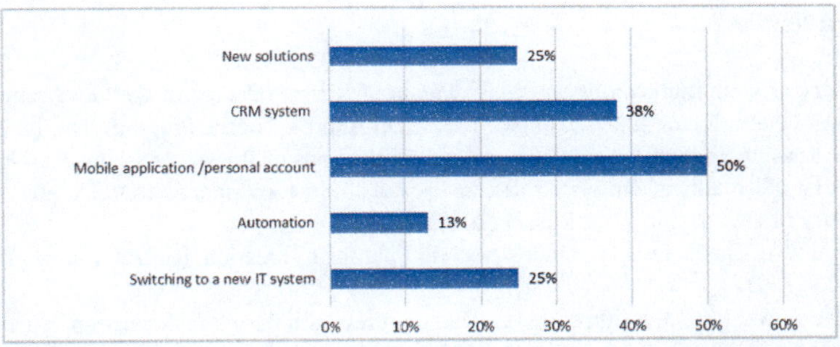

Fig. 3. IT projects planned for implementation and carried out by organisations in the commercial medicine segment [21].

About 38% of industry representatives expressed their readiness to invest more than 500 million roubles, which suggests a focus on the introduction of high-tech medical care, infrastructure development and expanding the range of health services provided [21].

4 Discussion

Domestic researchers, for example, Uvarina, also focus on various business model options that determine the construction of relationships with consumers and differentiation of the service volume [19].

Uvarina, Shushkin believe that marketing tools in business modelling are presented in the format of multidisciplinary and single-industry commercial medical organisations focused on market success through an increase in customer value [20].

Akopova, Przhedetsky, Przhedetskaya, et al. in building relationships with consumers (patients) assign an important role to marketing tools and the introduction of digital technologies in a business model, which implies encouraging investments [22].

We adhere to the point of view that it is possible to strengthen the relationships with consumers based on the development of telemedicine services, building marketing business models in the format of multidisciplinary for-profit health organisations [7].

5 Conclusion

Digital platforms make it possible to expand the reach of consumers and involve them in the permanent communications in order to diagnose and familiarise themselves with the whole range of possible health services.

The business model of a multidisciplinary for-profit health organisation seems to be acceptable from the point of view of maximising customer value. An analysis of the empirical research results in the commercial medicine segment demonstrates the readiness to implement marketing management and implement a marketing business model focused on market development.

References

1. Ministry of Communications and Mass Media of Russia: Decree of the Government of the Russian Federation of 28.07.2017 No. 1632-r On Approval of the Program ‘Digital Economy of the Russian Federation. Ministry of Communications of Russia, Moscow, Russia (2017)
2. Ministry of Health of Russia: National Project “Zdravookhraneniye” (No. 204). Russian Ministry of Health, Moscow, Russia (2018)
3. Gribanov, Y.: Basic models of industry digital platforms creation. Russ. J. Innov. Econ. **8**(2), 223–234 (2018)
4. Muslimov, M.: Digital health care – as a factor of revolutionary transformation in the industry. Mod. Prob. Health Care Med. Stat. **3**, 63–72 (2018)
5. Khabriev, R.U., Yagudina, R.I., Pravdyuk, N.G.: Evaluation of Health Care Technologies. Moscow Information Agency, Moscow (2013)
6. Digital/McKinley: Digital Russia: a new reality. The group’s expert report (2017). https://roscongress.org/materials/tsifrovaya-rossiya-novaya-realnost/
7. Bondarenko, V., Kostoglodov, D., Nekrasova, T.: Telecommunications techniques in the healthcare development: foreign experience and russian realities. In: Galinina, O., Andreev, S., Balandin, S., Koucheryavy, Y. (eds.) Internet of Things, Smart Spaces, and Next Generation Networks and Systems. NEW2AN 2020, ruSMART 2020. Lecture Notes in Computer Science, vol 12526. Springer, Cham (2020). https://doi.org/10.1007/978-3-030-65729-1_27
8. Krasyuk, I., Kirillova, T., Amakhina, S.: Marketing concepts development in the digital economic environment. In: Proceedings from SPBPU IDE 2019: ACM International Conference Proceeding Series, New York, pp 1–6 (2019). https://doi.org/10.1145/3372177.3373304
9. Timmers, P.: Business models for electronic markets. J. Electron. Mark. **8**(2), 3–8 (1998)
10. Ammar, O., Ouakouak, M.L.: The business model as a configuration of value: toward a unified conception. J. Bus. Manage. Sci. **3**(2), 78–84 (2015). https://doi.org/10.12691/jbms-3-2-4

11. Linder, J.C., Cantrell, S.: Changing business models: surveying the landscape, institute for strategic change. Accenture Inst. Strateg. Change **15**, 72–88 (2000)
12. Stähler, P.: Business models as a unit of analysis for strategizing. In: International Workshop on Business Models, pp 4–5 (2002)
13. Boschian, V., Paganelli, P.: Business models for advanced ICT in logistics. In: Lu, M., De Bock, J. (eds.) Sustainable Logistics and Supply Chains. CMS, pp. 15–51. Springer, Cham (2016). https://doi.org/10.1007/978-3-319-17419-8_2
14. Krasyuk, I., Kolgan, M., Medvedeva, Y.: Development of an ecosystem approach and organization of logistics infrastructure. Transp. Res. Procedia **54**, 111–122 (2021). https://doi.org/10.1016/j.trpro.2021.02.054
15. Krasyuk, I., Kirillova, T., Nazarova, E., Dudakov, G., Moshkin, I.: Marketing technologies in the organization of business processes of retail trade. In: IOP Conference Series: Materials Science and Engineering, Krasnoyarsk (2020)
16. Krasyuk, I., Krimov, S., Kolgan, M., Medvedeva, Y., Khukhlaev, D.: Conceptual framework for creating a digital business ecosystem based on marketing. In: IOP Conference Series: Materials Science and Engineering, Krasnoyarsk (2020)
17. Osovtsev, V.A., Przhedetskaya, N.V., Sagidullaeva, M.S.: Conceptual model of adaptive management of strategic marketing: a system approach. Eur. Res. Stud. J. **4**(21), 666–677 (2018)
18. Bondarenko, V.A., Pisareva, E.V.: Justifying companies' marketing models for a winning business strategy. Contemp. Stud. Econ. Financ. Anal. **100**, 41–48 (2018)
19. Uvarina, Y.: Value modeling in the market of private medical services. Sci. Rev. **10**, 527–534 (2014)
20. Uvarina, Y.A., Shushkin, M.A.: Innovative business models of medical centers: marketing toolkit for the analysis of the implementation of business processes. Innovations **1**(207), 99–108 (2016)
21. Commercial medicine market research in Russia 2018–2019. https://assets.ey.com/content/dam/ey-sites/eycom/ru_ru/news/2020/03/ey_healthcare_research_2018-2019_24032020.pdf
22. Akopova, E.S., Przhedetskaya, N.V., Przhedetsky, Y.V., Borzenko, K.V.: Marketing of healthcare organizations. Technologies of public-private partnership. In: Advances in Research on Russian Business and Management. Charlotte, MA (2020)

International Environmental Management Standards as a Tool of Legal Business Regulation

Olga A. Abakumova[1](✉), Andrey V. Lyubanenko[1], and Tatiana M. Vazhenina[2]

[1] Tyumen State University, Tyumen, Russia
o.a.abakumova@utmn.ru
[2] Tyumen Industrial University, Tyumen, Russia

Abstract. The article is devoted to the problem of determining the boundaries of normative regulation of a green economy, sustainable development, balance between the need to preserve nature and economic activity. The research is based on methods: dialectical, economic-statistical, comparative-legal, empirical, expert assessment, monographic description. The role of the system of international standards for achieving sustainable development goals, its importance, and significance for improving the system of relevant international mechanisms and instruments of environmental legal regulation has been substantiated. A review of the mechanisms of environmental standardization is carried out. The authors consider issues of sustainable development; defining the boundaries of regulatory regulation of the green economy; finding a balance between the need to preserve nature, economic activity; integration of sustainable development principles into the policies of nation-states; the importance of the environmental standardization system for business as well. The international standards of the ISO 14000 family and their use in modern corporate practice are analyzed. It is required to integrate the principles of sustainable development into the policies of national states and the introduction of universal environmental standards that combine the best elements of various national norms. All these issues have become more relevant during the COVID-19 coronavirus pandemic, as, due to the ensuing economic crisis, states postponed the introduction of new, more stringent environmental standards, which could require significant financial costs to modernize industrial production.

Keywords: Green economy · Ecology · Environment · Sustainable development · Globalization · Environmental policy · Standardization · Guidelines

1 Introduction

Over the past decades, there has been a significant increase in global awareness of the need for a more global approach to protecting the Earth's environment and natural resources. The most important ecological problem of the 21st century is the need to develop coordinated strategies and mechanisms to combat transboundary environmental threats, from river pollution and acid rain to ozone depletion and climate change in general [1–3].

P. V. Trifonov and M. V. Charaeva (Eds.): DITEM 2021, LNNS 380, pp. 370–381, 2022.
https://doi.org/10.1007/978-3-030-94245-8_52

Conservation of the environment and resources in today's world is actively being thought about, analyzed, and legally regulated on a national, regional, and international scale due to the global importance of these issues. The preservation of the planet's nature requires the consolidation of the countries of the world with different law systems.

At the same time, the specifics of the legal systems of different states, their mentality, the level of legal consciousness, and socio-economic development must be taken into account.

The problem of shaping the environmental responsibility of business and building a "green economy" as a whole has long been widely discussed on the global agenda and is outlined in many policy and legal documents.

Following the environmental policy and the results of environmental monitoring, standards are developed at various levels, which are important tools for sustainability in environmental management.

The application of international environmental standards ensures the necessary level of safety of products, works, and services for the environment, life, and health of people, as well as compliance with environmental requirements, starting from the development of goods (product), its production, sales, and subsequent disposal. In addition, business interest in environmental standardization is also growing due to consumer demand for environmentally friendly products.

2 Methods

The research is based on a dialectical method of knowledge of the institute of global standardization as a socio-legal phenomenon. The specifics of establishing and applying international standards and their use in modern corporate practice were identified. The Comparative (comparative-legal) method was used to consider the international experience of standardization and compare different standards. Using historical and predictive methods, the importance, relevance, and necessity of implementing a system of international standards to achieve sustainable development goals were justified.

The following methods were also applied: economic-statistical, modeling, empirical, expert evaluation, monographic description.

The information background for the study was international normative acts, reports of international organizations, and analyzed literary sources.

3 Results and Discussion

3.1 International Environmental Regulation and National Legislation

Due to globalization and the integration of the modern world, there is a need to unify the regulatory framework of environmental relations at the international level. Such documents generally contain basic principles and values while at the same time being voluntary.

The main global international documents regulating environmental issues and containing the basic requirements for business in the field of human rights include primarily the Rio Declaration on Environment and Development [4]; international standards, including environmental [5, 6]; the UN Global Compact [7]; the Guidelines for

Reporting on Sustainable Development [8], which contains a methodology for reporting, indicators for a comprehensive assessment of the company sustainability, including economic, social and environmental components; UN standards containing human rights requirements for business [8].

The RIO Declaration contains some specific language regarding environmental policies that discourage the export and import of hazardous wastes and trade protectionism achieved through environmental regulation.

Certain documents, in particular, for example – the International Labor Organization Conventions (regarding compliance with environmental standards at work), Recommendations of the Organization for Economic Cooperation and Development (in terms of regulation and development of the green economy), assume their adoption at the national level, becoming, in this case, part of the law system of the acceding state.

Such international level documents express the ideological field of the concept of sustainable development to create and apply a set of legislative and other measures stimulating the development of ecology and environmental protection. First and foremost among such measures are the development and implementation of environmental management standards.

The practical implementation of established requirements (standards) requires their development, adoption, and implementation at the level of economic entities.

Environmental principles were first established by the United Nations Conference on Environment (Stockholm, 1992) [9]. Subsequently, the principles of environmental protection were supplemented by the World Charter for Nature [10].

Further development of environmental standards was carried out under such documents as the Ramsar Convention (1971); the Agreement on the Conservation of Biological Diversity (Rio de Janeiro, 1992); the Helsinki Convention (1974); the UN Convention on Climate Change UNFCCC (New York, 1992); Kyoto Protocol under UNFCCC (Kyoto, 1997); Organization for Economic Cooperation and Development Guidelines for Multinational Enterprises (1999); UN Global Compact (2000); Sullivan Global Principles (1999), etc.

Socio-economic relations in the environmental sector are regulated by normative legal acts at the national level. In essence, state bodies' activity on protection, preservation, and restoration of the environment, rational use and reproduction of natural resources, prevention of the negative impact of economic and other activities (Art. 1 of the Federal Law "On environmental protection").

At the same time, the implementation of domestic legal norms is mandatory, while the documents containing international standards are prescriptive.

Of course, the documents of the international level are universal and cannot fully take into account the national characteristics of each country, but it is possible to adopt basic normative documents at the domestic level based on them.

The participation in the development of such rules and standards by many states and representatives of other interested parties allows the adoption of normative documents considering opposing interests as much as possible.

3.2 International Experience of Standardization in the Field of Environmental Management, Control, and Audit

International Organization for Standardization, ISO was founded in 1946 to promote common international standards that would facilitate the implementation of national technical regulations and requirements imposed on participants in world trade relations.

Currently, 164 states are represented in ISO. The Russian Federation is represented by the Federal Agency for Technical Regulation and Metrology (Rosstandart). Today, developed international standards have become the most important product of the partnership between states and the business community, becoming a tool for forming sustainable development [11].

The unification of standards is based on an in-depth analysis of the requirements in the legal systems of national states. This is based on best practices and wording. As a result, the agreement process formulates uniform minimum technical and environmental requirements acceptable to each of the participating countries. States combine the best elements of different national norms and eliminate the most controversial ones.

As a result, international standards may generally be less stringent than those of individual countries. At the same time, the development of standards does not consider the size or type of organization – the standards are inherently generic (ISO 9000 Quality Standards 1997).

Thus, one of the first standards in this area, which summarizes the best international practices, dedicated to general guidelines and based on the general agreement of experts and representatives from more than 140 countries, was ISO26000:2010 "Guidelines for Social Responsibility", prepared in ISO Technical Council [5].

Moreover, during the development, the opinions of not only the participating countries were taken into account, but also the main categories of stakeholders: government, business, consumers, non-profit organizations, trade union representatives. Establishing best practices in corporate social responsibility, this standard is international in scope, and thousands of business entities worldwide are certified for compliance. At the same time, the standard requirements are not mandatory and are voluntary [12].

3.3 ISO 14000 Series Standards

In general, the activities of environmental organizations and the business community created the basis for the ISO 14000 series of standards (describing the responsibility of business in complying with environmental requirements in production), which have become the most widespread internationally.

The main goal of the standard was to reduce pollution by standardizing procedures and business processes in corporate management [13].

The original development of this standard was based on 10 core principles established by the Boston-based non-profit advocacy organization Ceres:

1. Protecting the biosphere, minimizing harm to it.
2. Economic use of natural resources, including a commitment to use renewable resources.
3. Reduce and dispose of waste, minimization whenever possible.

4. Reduction of energy use, use of renewable resources.
5. Reduction of environmental risks.
6. Production of environmentally safe products and services, including the obligation to inform consumers about the environmental impact of services and products.
7. Compensation for damage, restoration of the environment to the utmost.
8. Disclosure of health and safety incidents to employees.
9. Creation of the position of Director of Environmental Protection.
10. Obligation to conduct and publish an annual audit of the implementation of these principles [14].

The principles mentioned above were defined as extremely stringent and, after their publication, were adopted by only 50 companies.

At the same time, the current economic situation has forced the business community to become more active in developing new environmental ISO standards. The rapid growth of world trade and the emergence of transnational corporations conducting economic activities in dozens of countries led to the difficulty of implementing environmental legislation at the national level, as they had a significant difference, which dramatically complicated production and trade activities.

Therefore, the main goal of developing the new standards was:

- Promoting sustainable development in the field of ecology and environmental protection;
- Harmonization of various existing corporate governance standards – EMAS;
- Creating better international regulation of these processes instead of stricter control by national legal systems.

The direct impetus for the International Environmental Management Standards ISO 14000 came from the UNCED conference in Rio de Janeiro in 1992.

The Conference published Agenda 21, a statement of principles for implementing sustainable development in countries worldwide, including both developing and industrialized nations. One of the Agenda recommendations was that each state should create a national council for sustainable development. This was a major impetus for developing new standards to harmonize environmental management and auditing systems among different countries.

The United Kingdom pioneered environmental management by passing the first Environmental Act (1990). British standards BS 7750 of 1992, the basis for the European Union Guidelines on Eco-Management and Audit Scheme – EMAS 1836/93 became the basic model for the standards series 14000.

EMAS provided a voluntary program for businesses that included standards for environmental auditing, management, and reporting. The existence of such different international standards caused serious concern among members of the business community.

The development of international standards ISO 14000 in environmental management was carried out by the International Organization for Standardization, under which the Technical Committee – Environmental Management (ISO/TC 207) was established in 1993.

The first ISO 14000 standards were prepared by 1996, and there are now more than 20 standards. The uniqueness of the methodology of these standards is determined by its focus not on the quantitative parameters of the interaction with the environment, which implies compliance with the parameters of the concentration of harmful substances or the volume of emissions with the established minimum standards, but on the environmentally oriented management system, integrated into the company environmental management system.

In general, compliance with the EMAS standard is costly, which is a significant barrier to participation among most small and medium-sized enterprises, even in the EU itself. However, its presence is significantly more important to the public in the market in which the company operates.

In the Russian Federation, the ISO 14000 state standards were adopted in 1998. The new standards defined the main stages of the organization's creation, implementation, and adjustment of environmental management.

The ISO 14001 standard has both similarities and differences with the European EMAS system (Table 1).

Table 1. Comparative characteristics of the EMAS system and the ISO 14000 standard.

System name	ISO 14001 (2004)	EMAS III (2009)
Mandatory implementation	Voluntary (advisory)	
Area of action	International standard	European Union countries
Certification frequency	Every three years	Certification is carried out once
Audit frequency	6, 9, 12 months	1, 2, 3 years
Obligation of a business entity to publish documents	Environmental policy	Full environmental reporting on activities

The application of ISO 14000 standards allows business entities to ensure the creation of a system for monitoring their compliance with national legislation in ecology and environmental protection.

According to the documentation, the system of standards provides for the reduction of adverse effects on the natural environment at the following levels:

- Organizational – implementation of environmentally-oriented solutions in the management system of business entities;
- National – adjustments and additions to the national regulatory framework in the field of ecology and environmental protection;
- International – formation of new conditions for the implementation of international trade.

3.4 Challenges of Implementing ISO 14000 at the National Level

Following the requirements of ISO 14000, a business entity must do the following to obtain a certificate of compliance:

11. develop a document defining the organization's environmental policy, with environmental goals and objectives established therein;
12. determine procedures and compliance for significant environmental impacts;
13. establish environmental goals and objectives, which should have a quantitative standard for each area of production activity and level of management;
14. develop an environmental management program, including the designation of responsible persons, means, and deadlines for its implementation;
15. train personnel following the requirements of ISO 14000, including emergency response;
16. create a system for monitoring the parameters of activities negatively affecting the environment;
17. conduct an environmental audit of the management system to determine compliance with the criteria of ISO 14000 (the audit can be carried out by authorized employees of the company and by an external party);
18. carry out periodic adjustments to the goals and objectives of the environmental management program, aimed at improving the environmental performance of the production activities of the business entity.

Any procedures, actions, and results under this standard must be documented in the organization's general document management system. Despite the "voluntary" implementation of ISO 14000 standards, they do not eliminate the obligation to comply with the requirements of national legislation but form additional requirements for the system of monitoring their implementation.

Unfortunately, the ISO 14000 standards and certification procedure have not received a significant spread either in Russian legislation or in the practice of business entities because they have not been implemented in national law and are not binding.

At the same time, a certificate allows demonstrating to customers and the public the existence of a modern management system based on respect for the environment.

For this reason, the first and the only enterprises that have implemented this standard were export-oriented companies for which obtaining a certificate of compliance with ISO 14000 is required to enter foreign markets or organize cooperation with foreign partners.

At the same time, such certification has become a mandatory requirement for the sale of products in certain commodity markets. For example, products can only be supplied to the European Economic Union market with an ISO 14000 certificate.

At the same time, economically weak and developing countries were poorly represented during the development of the ISO 14000 series of standards, and the potential for national application has not been explored. Today, the international expert community is concerned that the standards presented in ISO 14000 may constitute non-tariff barriers to trade against developing countries [13].

Developing countries may also face significant challenges in developing and enforcing environmental regulations according to international standards. For example, the generation of hazardous waste in the production and extraction of minerals is one of the key problems for developing countries involved in rapid industrialization with a high share of exports of their products [15].

Developing countries today are the most environmentally vulnerable, acting as "pollution havens" due to the concentration of industry and the release of toxic substances. As developed countries adopt environmental standards to protect their natural habitats, less developed countries bear the brunt of dirty industries [16].

While certification becomes mandatory for doing business in certain markets, non-certified companies are at a primary disadvantage. Since the cost of ISO 14000 certification can be prohibitive for small and medium-sized companies in developing countries, any legal or de facto requirements a company would have to accept at production would act as a tariff barrier to international trade.

The introduction of this standard provoked heated discussions in the business community, government agencies, and public (environmental) organizations. Thus, in economically developed countries, the standard is too flexible and soft, and their environmental standards may differ significantly in terms of quantitative parameters. Moreover, the lack of quantitative requirements makes critics question whether ISO 14000 can be considered a standard at all.

Equally challenging are the requirements for organizational openness outlined in the standard. The content of the "Environmental Policy" document, as the only one available to the public, is quite general with standard wording. The goals and objectives set in are determined by the enterprise and are usually quite soft. In addition, the standard does not require publication of the environmental audit.

At the same time, since each country creates its accreditation regulations for standard registrars, accreditation procedures and oversight may vary significantly from country to country. In addition, accreditation of a registrar in one country does not guarantee that his work will be recognized in another country.

The standard documents are currently under development, which allows the state and public entities to take a more active part in implementing environmentally significant decisions.

The main reasons why businesses would like to implement an international ISO standard are the image of ISO as a globally authoritative and trustworthy organization for setting international technical standards [17].

In principle, all sides of society are interested in the introduction of and compliance with environmental standards: from governments, which with the help of state green procurement can promote the production of environmentally friendly goods and manufacturers, gaining certain image competitive advantages, consumers interested in buying environmentally friendly goods and services.

The popularity of healthy lifestyles is growing, and consumers are becoming more environmentally conscious. Generally, the number of consumers aware of the importance of eco-products and even willing to pay more for them continues to grow [18].

ISO international standards often become a regulation either through international treaties or by incorporating similar requirements into national legislation.

3.5 Private and Public Environmental Certification Mechanisms

In addition to national and international regulation, in practice, the world's largest corporations adopt their corporate environmental standards. The most common type of cooperation is when various organizations come together to create standards, certificates, and codes of conduct for services and products.

One example of such an agreement is the Global Environmental Sustainable Apparel Coalition – SAC) is a collaboration of organizations, including Nike, Walmart, and Patagonia, that seek to reduce environmental harm from the production of clothing and footwear worldwide.

An equally significant example is a voluntary initiative of the chemical industry, the Responsible Care program, which provides a commitment to environmental protection, health, and safety. The program has now been adopted in more than fifty countries. The Program's Steering Committee developed methodological documents and a self-assessment questionnaire for chemical industry enterprises in industrial and environmental safety.

The program's main goals are to minimize the level of harmful emissions, increase the efficiency of raw materials and energy use, and reduce risks related to environmental protection. Particular attention is paid to informing government agencies and the public about the impact on the environment, personnel safety, and public health.

There are own effective standards of large, mostly transnational companies, such as Nestlé, Shell, etc.

It is also necessary to mention informal, non-statutory documents, such as the Sunshine Standards, the Sullivan Global Principles, and the Good Corporation Standard.

The European Union countries have introduced additional legal requirements for certain enterprises for additional reporting.

Directive 2014/95/EU on the disclosure of non-financial and diversified information by certain large companies and groups entered into force on December 6, 2014, intending to increase the transparency of some large companies in the European Union.

Enterprises of public interest must include in the management report information at least on environmental, social, employment and human rights, anti-corruption, and anti-bribery aspects (Art. 2 of Directive 2014/95/EU).

In addition, a business can use not only national framework approaches but also pan-European approaches (such as EMAS) and international ones based on the Principles of the Global Compact, the Guiding Principles on Business and Human Rights.

The BRICS countries are developing consonant but not similar approaches based on the European experience, taking into account these countries' national objectives, economic specificities, and cultural characteristics.

For example, India, South Africa, and China have already developed national guidelines for corporate social responsibility reporting, disclosing the organization's environmental protection activities.

In Russia and Brazil, businesses' mandatory implementation of corporate social responsibility practices is not legally enshrined. However, it is necessary to note the Social Charter of Russian Business, which was adopted on the initiative of the Russian Union of Industrialists and Entrepreneurs (RSPP) in 2004, to which 270 organizations with a total of over 7 million employees have already joined.

In Russia, under the Federal Law of 27.12.2002 No. 184-FZ "On technical regulation", a system of voluntary environmental certification is in place, and regional environmental standards (from 2021 introduced in the Chelyabinsk region), and private (e.g., Ecocluster, Green Russia) are emerging.

We can note the smooth and continuous growth of reporting, as currently in the www.corporateregister.com database registered 20,405 organizations that publish such reports annually, compared to 19,959 in July 2020, 2018 – 16,418, 2017 – 14,622 (and the register contains information, not about all the reports that are issued in the world – the reports published only in national languages are not included in this database).

Of course, the state of non-financial reporting directly depends on the existence of mandatory requirements for the disclosure of non-financial information, formalized in the form of legislative acts at the domestic level.

3.6 Interaction between Corporate and Government Agencies in the Application of Environmental Standards during the COVID-19 Pandemic

In 2020, according to the approved program of the United Nations, the UN Climate Change Conference COP26 was to be held. It was supposed to agree on stricter environmental standards to reduce CO_2 emissions to curb the temperature rise globally by at least 1.5° [19].

However, due to the pandemic, the event was postponed until 2021. At the same time, the serious economic crisis forced countries to postpone the tightening of environmental standards and loosen them.

At the same time, the main emitters of carbon dioxide worldwide remain the United States and China. These countries emit up to 40% of the world's CO_2 emissions. However, due to the crisis and the sharp decline in demand from the population, there has been a significant reduction in air emissions, leading to an improved environmental situation and the actual implementation of new environmental standards not legally adopted.

The International Monetary Fund (IMF) has lowered its forecast for global GDP to minus 3% in 2020. A significant drop in production is also predicted. However, according to experts, this is a temporary phenomenon. Significant environmental problems continue to dominate, especially in developing countries. For them to have a long-term trend, the assumption of voluntary obligations on corporations is not sufficient. It is necessary to introduce new mandatory requirements at the state level with new environmental standards, which will be extremely difficult in the current economic state.

4 Conclusion

The analysis of the studied topic led to the following conclusions.

The activities of states to protect the environment are important not only for the possibility of their sustainable development but also for the planet as a whole. Therefore, many governments are working to balance economic growth with the preservation (improvement) of the environment.

At globalization of the world economy, it is especially important to unify the requirements for business entities. This reduces administrative costs and allows the flow of monetary capital to different countries, stimulating their economic growth.

Equally important are the environmental requirements for the activities of business entities, which can be unified through the development of international environmental standards.

At the same time, the requirements contained in them can be implemented, or harmonized with national legislation, either through the conclusion of international treaties or by including them in the internal state regulations.

The decision on implementation contributes to the unification and harmonization of national legislation, reduces administrative barriers, and improves the investment climate. This guarantees the attraction of foreign investment, reduces administrative and management costs in production processes.

The introduction of environmental management standards at enterprises creates new management mechanisms focused on more responsible production decisions with a balance of environmental and economic performance. This is important both to improve the enterprise's image and reduce the administrative and legal consequences of possible environmental pollution, increasing the efficiency of the use of material resources.

Russian enterprises, unfortunately, still do not actively apply ISO 14000 certification. It is mostly forced for them when entering foreign markets or when interacting with foreign partners.

Cooperation on sustainable development and the green economy appears to be an important global area of interaction between countries. Such interaction will help better address the economic, environmental, and social problems of member states and make a greater contribution to global sustainable development since there are natural synergies between environmental justice, sustainable development, and corporate responsibility principles and standards.

Unfortunately, sustainable development goals are formulated far more often than they are achieved. The system of international environmental standards can be the necessary tool for Russia to develop and implement high-quality green standards, which will increase the number of environmentally responsible companies producing safe certified products.

References

1. Baer, H.P.: ISO 14000: potential compliance and prevention guidelines for EPA and DOJ. Fordham Environ. Law Rev. **7**(3), 927 (2011)
2. Clapp, J.: The privatization of global environmental governance: ISO 14000 and the developing world. Glob. Gov. **4**(3), 295–316 (1998). https://doi.org/10.1163/19426720-00403004
3. Hemphill, T.: The ISO 26000 guidance on social responsibility international standard: what are the business governance implications? Corp. Gov. **13**(3), 305–317 (2013). https://doi.org/10.1108/CG-08-2011-0062
4. Ecological principles NCO "Ceres". https://ceres.org/
5. Mikulich, C.: The developing world's perspective. Tulane Envirom. Law J. **17**(1), 117–162 (2003). ISO 14000–14001
6. World Charter for Nature. https://www.un.org/ru/documents/decl_conv/conventions/charter_for_nature.shtml
7. Keeton, C.M.: Sharing sustainability: preventing international environmental injustice in an age of regulation. Hous. L. Rev. **48**(5), 1167–1203 (2012)

8. Getman, A.P., Lozo, V.I.: State and legal problems of overcoming the era of globalization ecological crisis. International legal and national regulation of the ecological sphere of society. In: Shemshuchenko, Y.S., Bogolyubov, S.A. (eds.) Collection of articles, Moscow, pp. 154–173 (2011)
9. Declaration on environment and development. http://www.un.org/ru/documents/decl_conv/declarations/riodecl.shtm
10. Bogolyubov, S.A.: Approximation of environmental legislation through their harmonization and development. In: Harmonization and development of environmental legislation of Kazakhstan and Russia, pp 11–29. LLP "Master PO", Astana, Moscow (2013)
11. United nations guiding principles on business and human rights. http://www.ohchr.org/Documents/Publications/GuidingPrinciplesBusinessHR_RU.pdf
12. Esty, D.C.: Stepping up to the global environmental challenge. Fordham Envtl L J. **8**, 103 (1996)
13. Guidance on social responsibility ISO 26000:2010 G. http://spbtpp.ru/wp-content/uploads/2015/11/law_national_standart.pdf
14. Global compact. https://www.unglobalcompact.org/what-is-gc/mission/principles
15. Guseva, Y., Simakova, Y.: Ekologicheskie standarty: nuzhny li oni biznesu? (2018). https://kachestvo.pro/kachestvo-produktsii/standartizatsiya/ekologicheskie-standarty-nuzhny-li-oni-biznesu/
16. Sustainability reporting guidelines G4. https://www.globalreporting.org
17. Scott, I.: Antitrust and socially responsible collaboration: a chilling combination? Am. Bus. Law J. **53**(1), 97–144 (2016). https://doi.org/10.1111/ablj.12073
18. Achieving RMS in the context of the COVID-19 pandemic: the role of the state and supreme audit institutions. https://ach.gov.ru/upload/pdf/Covid-19-SDG.PDF
19. Declaration of the UN Conference on the Human Environment. www.un.org/ru/documents/decl_conv/declarations/declarathenv.shtml

Comprehensive Approach to Assessing the Quality of Trade Services by Its Indicators

Olga N. Gutnikova[1](✉), Liliya E. Pavlunenko[1], and Svetlana Yu. Tsekhla[2]

[1] Institute of Economics and Management (Structural Division at Vernadsky CFU), Simferopol, Republic of Crimea

[2] Tavrichesky Academy (Structural Division at Vernadsky CFU), Simferopol, Republic of Crimea

Abstract. The lack of a unified methodology for assessing the quality of trade enterprises and the inaccuracies identified in a number of regulatory documents were the prerequisites for the development of an adapted approach, the use of which can improve the effectiveness of quality management systems. The work objective is to present and justify the author's approach and tools for assessing the quality of trade services. Various approaches to assessing the quality of services, as well as models developed and recommended in the quality management system, are considered. The service indicators are proposed, the evaluation of which fully characterizes the level of its quality and methods that allow increasing the objectivity of the results obtained. The novelty lies in the developed score-rating of evaluation, which considers the importance of the service indicator, both for its recipient and the expert performing the evaluation. A graphical profile for assessing the level of quality of service using a score-rating model was created.

Keywords: Trading service · Service quality · Service properties · Quality assessment methods · Score-rating scale

1 Introduction

Currently, in the quality management system, which includes a set of processes aimed at ensuring the stability of quality, there are many models for assessing the quality of services offered by domestic and foreign scientists economists, the basis of which is the normative regulation of the variety of indicators, as well as their defined evaluation indicators associated primarily with the sold goods, or the attitude of buyers to the committed operation. However, most of the existing models are designed to assess service enterprises' quality of services, limiting their use in trade.

2 Methods and Models

As approaches to assessing the quality of services, the most interesting models are identified, the essence of which is presented in Table 1.

P. V. Trifonov and M. V. Charaeva (Eds.): DITEM 2021, LNNS 380, pp. 382–396, 2022.
https://doi.org/10.1007/978-3-030-94245-8_53

Table 1. Approaches to assessing the quality of services and the models used.

Approach	Scientists	Model characteristics
From the point of view of the customer's satisfaction with the service provided	V. Zeithaml, A. Parasuraman, L. Berry [1, 2]	"SERVQUAL" method. The model of the method takes into account the gap between the estimates of the "expected" and "minimum acceptable" quality of the service. It shows the range of "tolerance" of customers to the appearance, responsiveness, politeness, professionalism of the staff, the condition of the premises, equipment, compliance with the terms of service provision, the degree of manifestation of individual approaches to service, etc
	G. D. Krylova and M. I. Sokolova [3]	The basis of the model – "SERVQUAL" method. According to the model, the quality of services is evaluated according to ten parameters defined in two directions: technical (technological), service quality, and functionality. The quality of services is considered a measure of how well the level of services provided meets the client's expectations
From the point of view of the customer's perception of the service provided	J. Cronin, S. Taylor [4]	"SERVPERF" method. Unlike the "SERVQUAL" method, this model does not include an assessment of such a parameter as the buyer's expectation. Only the buyer's perception of five dimensions (materials, reliability, response, provision, empathy), including 22 indicators, is subject to evaluation. The model expresses a form of the position referred to as "satisfaction", but is not equivalent to it, but results as the difference between expectation and perception

(*continued*)

Table 1. (*continued*)

Approach	Scientists	Model characteristics
	S. Gounaris [5]	"INDSERV" method. The model implies obtaining a direct assessment of the perceived quality from consumers through a survey. Allows calculating the index of service quality (simple and weighted), provided that the questionnaire includes the need for the consumer to rank indicators according to the degree of importance
From the position of expert assessment of the quality of services	T. A. Ponomareva M. S. Supryagina [6]	A model of sociological method of research (survey, questioning), the peculiarity of which is to obtain information not from the end-user, but the relevant specialist, by systematizing data, expert conclusions obtained by visiting the organization, control by the governing bodies
	A. A. Mishchenko [7]	Mystery shopping is one of the methods of marketing research, a special type of experiment. It involves the use of pre-recruited and trained consumers (usually independent specialists) or professional personnel who have received special training for anonymous business evaluation using a pre-compiled evaluation form. The assessment can be made during the visit, by phone, or via the Internet
From the point of view of consumer requirements and competitors' offers	E. I. Reshetnyak [8]	The "House of Quality" model is a group of interconnected tables that involves evaluating the quality of services from the perspective of customer requirements and competitors' offers, including technical and consumer parameters

(*continued*)

Table 1. (*continued*)

Approach	Scientists	Model characteristics
From the point of view of evaluating individual parameters of product quality	T. N. Nikolaeva, N. R. Egorova [9]	A methodology for evaluating a generalized indicator of the quality of services considers eight evaluated criteria, including characteristics that are normalized for competitive goods: the stability of the assortment, the breadth of the assortment, etc
From the point of view of the employee's professionalism	D. Flanagan [10]	"Critical cases" method. The model of the method consists of a survey of an employee of the enterprise to identify the attitude to critical situations at work and ways to eliminate them. As a rule, based on the results of the survey, a description of the professional situation and the prerequisites for the employee's behavior is compiled, an accurate description of the employee's actions that are considered effective or ineffective for this situation; the consequences of the employee's behavior; the dependence of the results of actions on the employee or external reasons. The model is widely used for personnel certification, development of professional requirements for personnel, and assessment of labor behavior consequences

Most of these methods are sociological. They aim to obtain information from the consumer about his attitude (impression) to the service using various tools (surveys, questionnaires, filling out prepared tables, etc.). These methods are subjective since the estimated parameters, on the one hand, are difficult to compare with the standards; on the other hand, there are no standardized parameters and criteria for their evaluation. The problem, first and foremost, is related to the imperfection of the regulatory framework for assessing the quality of trade services.

3 Results

To develop an adapted approach and tools for assessing the quality of trade services, as well as determining the rational methods for use within the enterprise quality management system, we turned to the current standard – GOST R 55812-2013 [11], which contains the standardized indicators and methods used in the evaluation of service quality. According to the study, the first problem encountered when using the standard is a reference to invalid normative documents: "The All-Russian Classifier of services to the population" [12] and "The All-Russian Classifier of types of economic activities, products and services" [13], which makes it difficult to determine the name of the service for choosing the normalized variety of quality indicators and their evaluation.

The second problem is the clear attribution of services to a certain type of activity. The confirmation is the current Order "On the approval and implementation of the classifier of services in foreign economic activity (KUVED) to the system of state statistics." The classifier states that "…examples of activities classified as services are wholesale and retail trade…". Following the classifier structure, wholesale and retail trade services are classified in the group of "distribution services", which does not fully reflect the essence of trade. In addition, in the structure of the studied group, wholesale trade services are assigned to the code 04200.0.000.00.00.00.000, and retail – to 04300.0.000.00.00.00.000, but the group of retail trade services contains wholesale services, such as services for the sale of cars (broken down by type of vehicles).

Returning to GOST R 55812-2013, it should be noted that difficulties arise with the definition of the type of trading services since their species is considered in another standard – GOST R 51304-2009, according to which services are divided depending on the type of trade carried out by the enterprise (wholesale or retail). The authors found that in the standard, the trade and technological process operations are partially attributed to services, which, in our opinion, contradicts the very definition of the concept of services. According to the definition: "trade service is the result of direct interaction between the seller and the buyer, as well as the seller's activities to meet the needs of the buyer when buying and selling goods." Consequently, attributing warehousing operations (e.g., laying goods in storage) or operations for the selection of goods by customers to services, contrary to the general principles of trade organization, as these actions are mandatory elements of the trade and technological process, which do not always involve the relationship between seller and buyer and involves actions aimed only at the goods (products). The authors of the work, for the choice of tools for evaluating trade services, taking into account the discrepancies identified in the attribution of certain actions to services, developed their classification, clearly differentiating their types by individual parameters, including the role in trade, orientation, purpose, etc. [14].

In our opinion, it is necessary to approach the definition of methods for assessing the quality of services and the rationality of the tools used from three positions: the materiality of the service (material content), its purpose (functions performed), and the role in the trading process (mandatory or additional), which will provide a more accurate choice of normalized indicators. Thus, when assessing the quality of service with material content (e.g., a product packaging service), the normalized indicators can be the appearance of the package, the material of its manufacture, integrity, and other characteristics that form the safety of the product service. Moreover, when evaluating a service that plays an additional role in the organization of trade services, the normalized indicators can be taken into account, for example, the degree of expediency of its provision, the time of provision, or the level of demand by the service consumer [15].

The tools for assessing the quality of services provided by trade enterprises and recommended for optimizing the processes implemented within the framework of the quality management system, in our opinion, should be based on a clear definition, on the one hand, of a group of indicators of the quality of trade service [16], on the other hand, its type, taking into account the characteristics noted above. The basis for choosing a group of indicators is the recommended research method.

The systematized inaccuracies identified by us in GOST R 55812-2013 are shown in Table 2.

Table 2. Analysis of inaccuracies in the list of trade service quality indicators and recommended evaluation methods according to GOST R 55812-2013 [11].

Group of quality indicators	Estimated indicators	Recommended methods	Note
Security of the service	Packaging, goods, equipment, vehicles, personnel, consumer property, fire safety	Expert, organoleptic, instrumental, analytical	- Safety of goods refers to the parameters that characterize the quality of the object of the sale itself and is regulated by other regulatory documents. When assessing the safety of equipment and technical means, there is no instrumental method of assessment, which does not determine the number of technical parameters

(*continued*)

Table 2. (*continued*)

Group of quality indicators	Estimated indicators	Recommended methods	Note
Purpose of the service	Functional and social suitability of the service, staff qualification, information support	Expert, analytical, sociological, organoleptic	The instrumental method of quality assessment is not declared, but it is given among the evaluated parameters the presence of goods of sound quality, which is impossible to determine or confirm in some cases without the use of special technical means
Ergonomics of the service	Hygienic indicators of the room, comfort, and convenience for the buyer	Expert, analytical, instrumental	There is no sociological method among the declared methods, which casts doubt on the very possibility of determining the convenience parameter for the buyer when making a purchase
Ethics	Sales personnel and their characteristics	Sociological	Analysis of personnel behavior only based on sociological research will have a subjective assessment, depending on the psychological state of the buyer
Aesthetics	Design, style solution, composition, building facade, advertising products	Expert	- It is inappropriate to evaluate the design or style decision concerning the service since they do not have a direct impact on the quality of the service; - The assessment cannot be carried out without studying the opinion of buyers, i.e., a sociological method is necessary

(*continued*)

Table 2. (*continued*)

Group of quality indicators	Estimated indicators	Recommended methods	Note
Adaptability	Equipping of retail space, its area, demonstration of goods, work on the preparation of goods, after-sales service	Expert, analytical	It is impossible to evaluate a number of parameters, such as the quality of work on commissioning, assembly, installation, without using the instrumental method with the use of some technical means
Professional level of personnel	The level of professional training and qualification of personnel	Analytical employee testing	To eliminate the subjective approach in the evaluation of each employee, evaluation should be adjusted by an analytical method based on a comparison of actual results with the regulated standards and requirements

Given the inaccuracies in Table 2, which are contained in the standard and complicate the development of an adapted approach and tools for assessing the quality of trade services, we recommend the following indicators of service quality, which should be evaluated taking into account its type (Table 3).

Table 3. Recommended indicators of the quality of trading services and criteria for their evaluation.

Service quality indicators	Applicability	Evaluation criteria
1. Functionality	For all types of services	The conformity of the service is subject to assessment: - Its purpose, targeted orientation, - Declared content and volume, - Availability of personnel for the buyer

(*continued*)

Table 3. (*continued*)

Service quality indicators	Applicability	Evaluation criteria
2. Security	- In conjunction with the evaluation of goods, packaging (if there is a relationship); - When using means that form a certain danger	Subject to evaluation: - Legality of the service provision (licensing, authorization), - Safety of the trade enterprise (compliance with SNIPs and sanitary rules and regulations, fire and electrical safety requirements, accident rate, etc.), - Personnel safety (access to work, hygienic and sanitary condition of clothing, epidemiological safety, etc.), - Safety of personnel actions, - Equipment safety, - Vehicle safety, - Safety of auxiliary materials used in the provision of services (packaging materials, containers)
3. Ergonomics	For evaluating material services	Subject to evaluation: - Terms of service provision, - Compliance with hygienic and sanitary standards necessary for the provision of services - Execution time (duration)
4. Reliability	When providing mandatory services	Subject to evaluation: - Stability (constancy) of receiving, - Timely performance of the service, - The possibility of re-rendering the service (guarantee of reproduction), - Compliance with the approved standards for the provision of services (in the case of its standardization)
5. Adaptability	For all services	Subject to evaluation: - Compliance with the type of trading enterprise, - Availability of the necessary technical means for the provision of the service

(*continued*)

Table 3. (*continued*)

Service quality indicators	Applicability	Evaluation criteria
6. Professionalism of the service performance	For all services	Subject to evaluation: - Personnel qualification - Work experience
7. Aesthetics	For all services	Subject to evaluation: - Ethics of service provision (communication of staff) - Tactfulness of service provision (behavior and politeness of the personnel)
8. Cost-effectiveness	For paid services	The compliance of the price declared for the service with the costs formed during its provision is subject to evaluation

Source: compiled by the authors

Considering the criteria for evaluating indicators that characterize the quality of trade services indicated in Table 3, we have defined methods for evaluating them, focusing on sociological, organoleptic (sensory), and expert analysis methods. The objectivity of the obtained results can be improved by applying instrumental and analytical analysis methods using several modern economic and technical tools reflected in Table 4.

Table 4. Recommended methods and tools for analyzing the quality of trading services.

Service quality indicator	Method	Tools
Functionality	Expert, sociological, instrumental, analytical	Coordinated expert profiling*, questionnaires, and customer surveys. The evaluated parameters are registered by instruments and equipment
Safety	Expert, organoleptic, analytical, instrumental	Conditional expert profiling* Evaluation – by a score-rating assessment** The evaluated parameters are registered with the help of laboratory equipment, comparison with reference samples

(*continued*)

Table 4. (*continued*)

Service quality indicator	Method	Tools
Ergonomics	Organoleptic, sociological, instrumental	Score-rating parameters evaluation characterizing the ergonomics of the service. Evaluation of parameters – by laboratory equipment, surveys, questionnaires
Reliability	Expert, instrumental	Coordinated expert profiling. The evaluated parameters are registered by instruments and equipment
Adaptability	Instrumental, analytical	The parameters are registered with the help of laboratory equipment and their evaluation
Professionalism and ethics	Expert, sociological	Score-rating evaluation The evaluated parameters are registered by surveys and questionnaires
Aesthetics	Expert, sociological	Score-rating evaluation The evaluated parameters are registered by surveys and questionnaires
Cost-effectiveness	Expert	Evaluation – by calculating the cost of the service

* The tools are described in GOST ISO 13299-2015 [17]
** A score-rating evaluation can be used, if necessary when evaluating each parameter of the service quality
Source: compiled by the authors

As follows from Table 4, for many indicators that characterize the quality of the service provided, a score-rating evaluation system is used as a tool, which assumes assigning a certain number of points to the service indicator depending on the level of its current provision or organization, taking into account the significance of this indicator for the consumer (service recipient) [18].

Conducting a quality assessment, in our opinion, requires the participation of the service recipient and awarding points for all these properties. The work authors proposed their assessment system in the absence of a developed assessment scale, presented in Table 5.

Table 5. Score-rating evaluation scale of the quality of provision (organization) of the trading service parameters.

Parameter score*	Characteristics of the estimated quality indicator
5	The assessed indicator is at a high level; there is maximum compliance with the degree and time of its provision or organization expectations of the consumer, and standardized indicators
4	The assessed indicator is at a high level; there is sufficient compliance with the degree and time of its provision or organization to consumer expectations and standardized norms, there are minor deviations from the established values or requirements of the service recipient
3	The assessed indicator is at a low level; there is relative compliance of the degree and time of its provision or organization with expectations of the consumer and standardized norms, there are clearly expressed deviations from the established norms or expectations of the service recipient
2	The assessed indicator is at a low level; for some criteria, there is a discrepancy in the degree and time of its provision or organization to consumer expectations and standardized norms, there are clearly expressed violations of the established norms or expectations of the service recipient
1	The assessed indicator is at a low level; the degree and time of its provision or organization does not fully meet the standardized norms or expectations of the service recipient (in this case, the service is considered to be provided)

* 0 points are not provided by the rating scale
Source: compiled by the authors

Developed score rating of quality parameters of trade services is based, on the one hand, on a comparison of the values of individual and the standardized parameters, on the other – on the level of satisfaction of the service recipient to the extent of its provision [19]. For example, when receiving a consulting service, the buyer was satisfied with the seller's work, while some of the information received, as it turned out, does not correspond to reality (is not reliable), as a result, the purchased goods cannot fully satisfy the consumer. On this basis, the indicator "professionalism of provision and ethics of service" cannot be rated at 5 points, given its apparent inconsistency with the criteria of "reliability" and "sufficiency" of information. Its maximum rating can be 3 points, and in case of recording the negative consequences of providing such a service, the rating can be reduced to one.

It is also necessary to take into account the significance of each indicator for the consumer. For example, the functionality of a service is a priority indicator that most affects its quality. At the same time, the aesthetics of the service may be less important for the consumer, especially in the case of a gratuitous provision.

When evaluating the quality of services, the significance of the evaluated indicators acquires different weights. On the one hand, an in-depth (scientific) approach to evaluation is applied; on the other hand, the degree of consideration of the importance of each evaluated indicator is based on the overall quality level of the enterprise. For this purpose, in our opinion, it is rational to take into account the ratio of its significance when evaluating each indicator.

Based on the different influence of indicators on the overall level of service quality, the authors of the work proposed their ranking taking into account the significance. It seems appropriate to use two assessment scales oriented to the level of training of the evaluation subjects, which can be consumers and specialists (experts).

The scale of the score-rating evaluation system is presented in Table 6.

Table 6. Scale of the score-rating system for evaluating the quality of trading services.

Service property	Maximum number of points	Significance ratio (rating) for the service recipient	Total number of points	The significance ratio when evaluated by an expert (specialist)	Total number of points
Functionality	5	0.25	1.25	0.2	1.0
Safety	5	0.2	1.0	0.2	1.0
Ergonomics	5	0.1	0.5	0.1	0.5
Reliability	5	0.1	0.5	0.2	1.0
Adaptability	5	0.1	0.5	0.1	0.5
Professionalism of service performance	5	0.1	0.5	0.1	0.5
Aesthetics	5	0.1	0.5	0.05	0.25
Cost-effectiveness	5	0.05	0.25	0.05	0.25
Total	40	1.0	5	1.0	5

Source: Compiled by the authors

According to Table 6, the total maximum number of points that a service can receive, taking into account the significant ratio of its properties for both the consumer and the expert, can be 5 points. Based on this, the grading of the evaluation looks like this:

5 points – high-quality service;

4 points – a good quality service;

3 points – a service of satisfactory quality;

2 points – a service of unsatisfactory quality;

1 point – the provision of the service is unacceptable.

When applying a score-rating assessment of the quality of trade services, a conclusion is made about the level of its provision, from quantitative assessment and qualitative description of the result. The quality assessment can be complemented by the results obtained with other analysis tools. In our opinion, it is advisable to conduct a comprehensive assessment. Figure 1 shows a graphical representation of the profile of such an assessment.

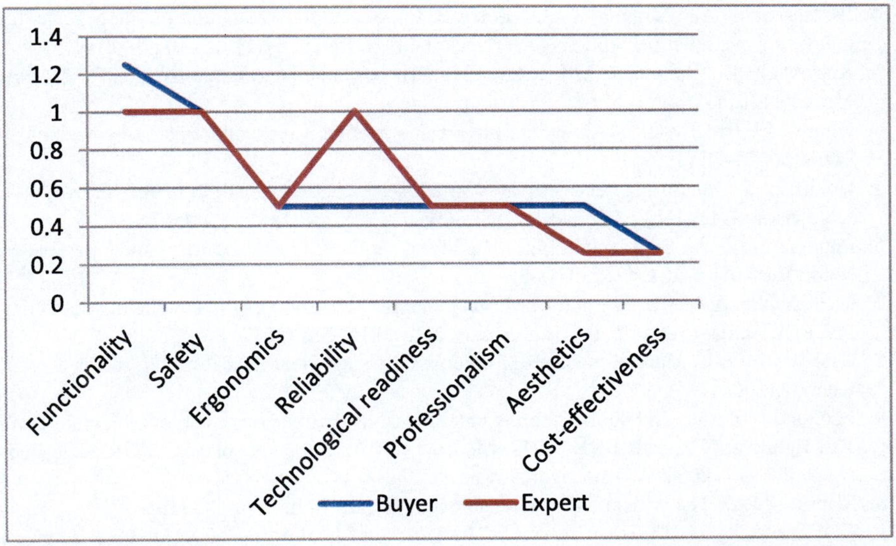

Fig. 1. Profile of a comprehensive assessment of the trade service quality. Compiled by the authors.

The points that are not the points of contact of the estimated parameters from the position of different assessment subjects are subject to the most thorough analysis. In complex analysis, the evaluation of the parameters located at the points of contact can be taken into account as the arithmetic mean of the obtained values. In our opinion, in such an assessment variant, the subjectivity of the service recipient's approach will be leveled by expert opinion.

4 Conclusion

Based on the results of the study of regulatory support for the methodology of quality assessment of trade services, the authors proposed the author's assessment model, which involves an integrated approach to scoring the individual properties of services from the perspective of customer satisfaction and expert vision of the level of assessed characteristics. The use of the proposed tool in practice will allow a more objective assessment of the quality of trade services to identify problems and form a general profile of the level of the enterprise to work out rational management decisions.

References

1. Balvanovich, A.V.: Directions for improving the quality-of-service provision informing interested parties in the field of standardization, certification, metrology, and conformity assessment based on the quality model of Berry, L., Parasuraman, A., Zeithaml, V. Inf. Econ. Aspects Stand. Tech. Regul. Sci. Online J. **1**(17) (2014). http://iea.gostinfo.ru/files/2014_01/2014_01_06

2. Parasuraman, A., Zeithaml, V.A., Berry, L.L.: SERVQUAL: a multiple-item scale for measuring customer perceptions of service quality. J. Retail. **64**(1), 12–40 (1988)
3. Krylova, G.D., Sokolova, M.I.: Marketing: Theory and 86 Situations. UNITY DANA, Moscow (2011)
4. Cronin, J.J., Jr., Taylor, S.A.: Measuring service quality: a reexamination and extension. J. Mark. **56**, 55–68 (1992)
5. Gounaris, S.: Measuring service quality in b2bservices: an evaluation of the SERVQUAL scale vis-à-vis the IND SERV scale. J. Serv. Mark. **19**(6), 421–435 (2005)
6. Ponomareva, T.A., Supryagina, M.S.: Quality of services: qualitative assessment parameter. Mark. Russ. Abroad **1**, 47–53 (2005)
7. Mishchenko, A.A.: The use of the "Mystery shopper" methodology in conducting marketing research. Actual Prob. Aviat. Cosmonautics **3**(13), 216–218 (2017)
8. Reshetnyak, E.I.: Modern methods of enterprise management in the service sector. Factor, Kharkiv (2008)
9. Economics of trade and public catering enterprises: textbook/team of authors. In: Nikolaeva, T.I., Egorova, N.R. (eds.) KNORUS, Moscow (2016). https://docplayer.ru/27995296-Pod-redakciey-t-i-nikolaevoy-i-n-r-egorovoy.html
10. Flanagan, J.C.: The critical incident technique. Psychol. Bull. **51**(4), 33 (1954)
11. GOST R 55812-2013 Trade services: The variety of quality indicators of trade services. Current. https://docs.cntd.ru/document/1200107203
12. The all-Russian classifier of services to the population. Dormant. https://docs.cntd.ru/document/9041894?marker=7D20K3
13. The all-Russian classifier of types of economic activity, products and services. Dormant. https://docs.cntd.ru/document/9041176?marker=64U0IK
14. Sharapov, R.F.: Methodological approaches to the assessment of the quality of retail services. Bull. TSU LBP **4**(48), 80–86 (2011)
15. Gutnikova, O.N., Pavlunenko, L.E.: Classification of services provided by enterprises of the trade sector. Sci. Notes Vernadsky Crimean Fed. Univ. Econ. Manage. **6**(4):38–47 (2020)
16. Gutnikova, O.N.: Problems of regulatory regulation of the assessment of the quality of trade services. Stand. Qual. **3**, 58–62 (2021)
17. GOST ISO 13299-2015 organoleptic analysis: Methodology. Current. https://docs.cntd.ru/document/1200137276
18. Lyubushin, N.P., Bricach, G.E.: The use of the generalized Harrington desirability function in multiparametric economic problems. Econ. Anal. Theor. Pract. **18**(369), 2–10 (2014)
19. Timashev, E.T., Maslova, E.A., Nikishin, A.F.: Service quality as a factor of increase of efficiency of economic activity of trade organizations. Mod. Sci. Res. Innovations **7**, 4 (2015). https://web.snauka.ru/issues/2015/07/56810

The Role of Smart Logistics in the China's Industrial Structure Upgrading

Dmitry A. Shevchenko[1(✉)], Weili Zhao[1,2], Elena V. Fomicheva[1], Weichang Chen[1,2], and Yuchan Wang[1,2]

[1] Southern Federal University, Rostov on Don, Russia
dashevchenko@sfedu.ru
[2] Henan University of Economics and Law, Zhengzhou, China

Abstract. With the coming of the fourth industrial revolution, traditional logistics is subject to digital transformation and enters the stage of smart logistics. Now, the Chinese economy is facing double pressures from developed countries and other developing countries. China's industrial structure is subject to a great pressure of upgrading. In recent decades, China was famous for its infrastructure construction in the world. Smart logistics is the subject of many publications in recent years, but most of these studies focus on the introduction of new technologies, subdomains, and applications. As distinguished from previous studies, this paper is mainly about economic evaluation of smart logistics, its positive effect on the China's industrial structure upgrading. Using historical analysis as a research method, this study adopts the theory of New Structure Economics and explores the relationship between China's industrial structure, logistics development, and infrastructure construction. This study shows that the development of smart logistics is in co-evolution with industrial structure upgrading. With the upgrading of China's industrial structure from labor-intensive and capital-intensive to technology-intensive, logistics has taken the same stages of development which starts from instrumental logistics, interconnected logistics to smart logistics. Smart logistics is endogenous in industrial structure and also can react on it, thus promoting China's economic growth.

Keywords: Digital platform · Logistics development · Industrial structure · Infrastructure construction

1 Introduction

Logistics is an important link between production and consumption and is the basic industry of the national and world economy.

The main trends and conditions of modern world economic development include the following: parallel development of globalization and de-globalization processes, scaling of globalization; multipolarity of development of the world economic system; a change in the ratio of the economies of the leading countries of the world (USA and China, the G7 and the 7 countries with emerging markets, BRICS – RIC and other formal and

P. V. Trifonov and M. V. Charaeva (Eds.): DITEM 2021, LNNS 380, pp. 397–405, 2022.
https://doi.org/10.1007/978-3-030-94245-8_54

informal associations); the revival of protectionism, the unfolding of trade wars (a bright and not the only example is the confrontation between the US and China) and others [1].

The development of logistics industry can deepen the connection between various regions, enhance the integration and penetration between industries, so as to improve national economic competitiveness. Smart logistics is the digital transformation of traditional logistics and is also considered to be an effective measure to solve the current transport efficiency problem. At present, Chinese economy is facing double pressures from developed countries and other developing nations. China's industrial structure is subject to a great pressure of upgrading. China has the most complete industrial value chain and biggest consumer market. Therefore, in order to promote the upgrading of China's industrial structure, it is of great practical significance to analyze the role of smart logistics.

There are many studies on smart logistics in the works of Yassine Issaoui et al. [2], Guimares et al. [3], Jiang et al. [4], Liu et al. [5] and Pan et al. [6], but none of them is about its economic evaluation. As a service industry, logistics is to serve the whole industrial value chain. Smart logistics is digital transformation of traditional logistics and can also indirectly influence the industrial structure. Although Shu et al. [7] have discovered that the dynamic factors of logistics industry are related to industry integration, they do not go far to research on the industrial structure upgrading. Miguel [8] and Zheng et al. [9] research on the implementation of Industry 4.0 and indicate that Industry 4.0 can quicken delivery times, make logistics smarter, but they do not examine the reaction of smart logistics on industry. Douaioui et al. [10] conclude that to accompany the progress of industry 4.0, through the four-industrial revolution, the development of logistics comes up to "smart logistics". Bag et al. [11] find smart logistics and green manufacturing capability have a positive influence with each other. All these papers have already noted that there is kind of relation between logistics and the whole industry, but none of them exactly examine the relation between logistics development and industrial structure.

With the spread of Covid-19 and the worsening global economic situation, Chinese government put forward the policy of accelerating the construction of new technological infrastructure in March 2020. From an economic point of view, strengthening infrastructure construction at a specific time will help to expand demand, stabilize employment in the short term, and promote economic growth, improve people's welfare and enhance national competitiveness in the long run. Moreover, Infrastructure construction is the foundation and precondition of logistics development.

Therefore, this study mainly examines the relation between logistics development, infrastructure construction and China's industrial structure, analyzes the role of smart logistics in the upgrading of China's industrial structure. To achieve the established goal, this study mainly adopts the theory of New Structure Economics and selects historical analysis as a research method. According to the theory of New Structure Economics, the industrial structure upgrading is a dynamic process. The industrial structure of countries in different stages of development are qualitatively different and thus may entail different principles of economic operation [12]. Economic operation is endogenous in industrial structure and can react on it. On the one hand, the industrial structure development will require that the logistics must be as flexible and agile; on the other hand, compared with traditional logistics, smart logistics has comparative advantages due to the application of advanced technologies. Therefore, smart logistics can input new impetus to industrial structure upgrading.

2 Methods

To achieve the established goal, this study mainly adopts historical analysis. Historical analysis is a method of analyzing objective things and social phenomena from the perspective of development and change. As for data collection, because the data for annual logistics cost only can be found in all means from 1991 in China, to keep the data consistent, all the data in this study are collected from 1991. Based on these historical data, this study tries to explore the nature essence of the relation between logistics development, infrastructure construction and industrial structure upgrading. Hence it will give some glimpse of the future.

2.1 Logistics Development

With the rapid development of the national economy, China's logistics industry also maintains a rapid growth rate. The logistics system is constantly improving, and the logistics industry is increasingly mature. The annual logistics amount in China increased from 3 trillion CNY in1991 to 298 trillion CNY in 2019. The logistics amount from January to September 2020 has reached 202.5 trillion CNY, which is expected to continue to grow in the future (Fig. 1). At the same time, the annual logistics revenue in China is also growing, from 0.2 trillion CNY in 1991 to 10.3 trillion CNY in 2019, The logistics revenue from January to September 2020 is 7.4 trillion CNY, which is also expected to maintain growth in the future (Fig. 2). With the increase of the annual logistics amount and the annual logistics revenue, the annual logistics cost in China also maintains a

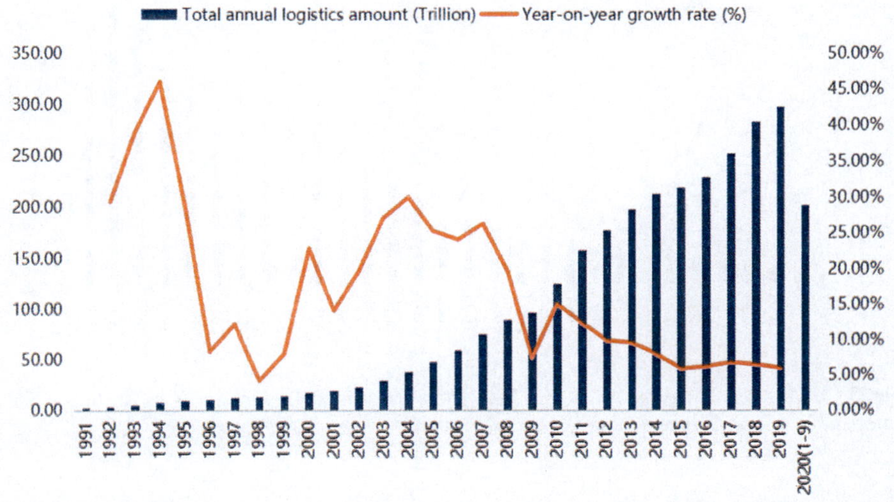

Fig. 1. Total annual logistics amount in China from 1991 to 2020 (Jan.-Sep.), in trillion CNY, comparable prices. Source: China Logistics Yearbook.

growth trend, from 0.5 trillion CNY in 1991 to 14.1 trillion CNY in 2019. The logistics cost from January to September 2020 is 10.4 trillion CNY (Fig. 3). To measure the efficiency of logistics operation, this study selects the ratio of total annual logistics cost to GDP as the indicator. The ratio of annual logistics cost to GDP is declining, from 23.5% in 1991 to 14.2% in 2019 (Fig. 4). It can be concluded that with the increase of annual logistics amount and revenue, the logistics operation will become more and more efficient as time goes.

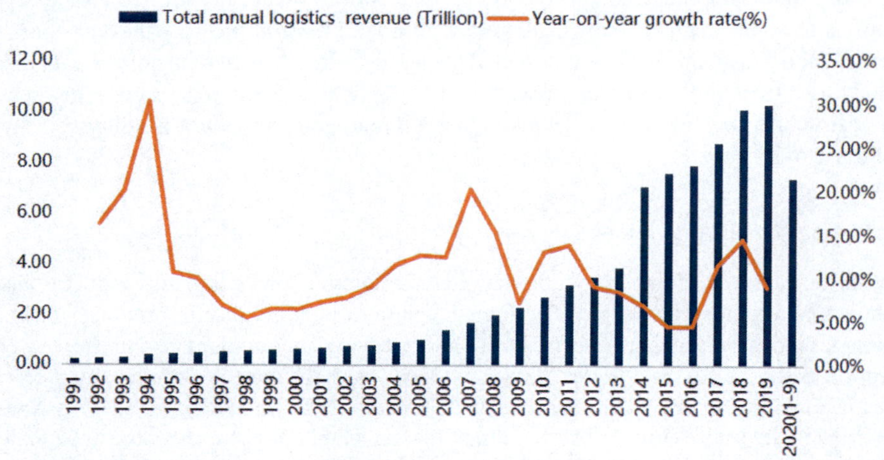

Fig. 2. Total annual logistics revenue in China from 1991to 2020 (Jan.-Sep.), in trillion CNY, comparable prices. Source: China Logistics Yearbook.

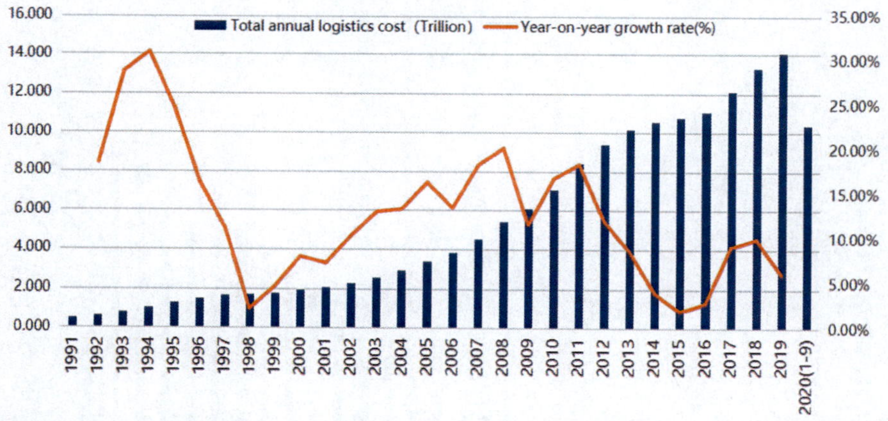

Fig. 3. Total annual logistics cost in China from 1991to 2020 (Jan.-Sep.), in trillion CNY, comparable prices. Source: China Logistics Yearbook.

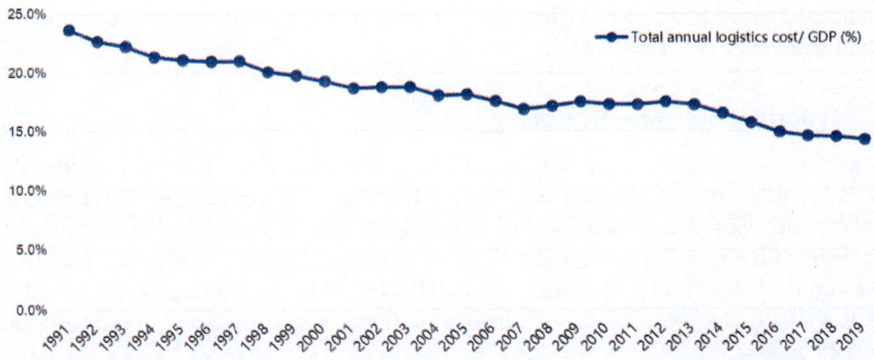

Fig. 4. Ratio of total annual logistics cost to GDP from 1991 to 2019. Source: China Logistics Yearbook.

2.2 Infrastructure Construction

In recent decades China has been famous for infrastructure construction in the world. In the past decades, China was the biggest beneficiary of moderately advanced infrastructure construction. Infrastructure construction is the foundation and precondition of logistics development. Since the reform and opening up, transportation infrastructure has become the frontier of China's economic development. With the support of government policies and funds, transportation infrastructure construction has been gradually strengthened and achieved great results. By 2019, China's total railway mileage is 139900 km; highway mileage is 5010000 km; expressway mileage is nearly 150000 km;

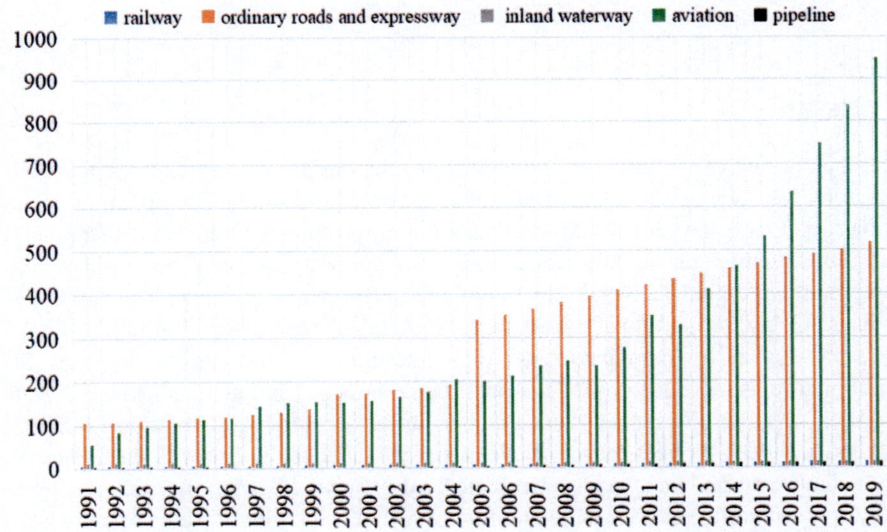

Fig. 5. Overview of transportation infrastructure construction in China from 1991 to 2019. Source: China Logistics Yearbook.

inland waterway mileage is 127000 km; civil aviation mileage is 9480000 km; and pipeline mileage is 126600 km (Fig. 5).

2.3 Industrial Structure Upgrading

To measure the level of industrial structure upgrading in China, this study divides the GDP into three parts of primary, secondary and tertiary industry and selects the proportion of the added value of the secondary and tertiary industries in GDP as the indicator. It is presented in Fig. 6. It can be seen that the ratio of annual added value of the secondary and tertiary industries to GDP was just 75.97% in 1991, and continually increased to 92.45 in 2019. Therefore, the China's industrial structure is continuously optimized from 1991 to now.

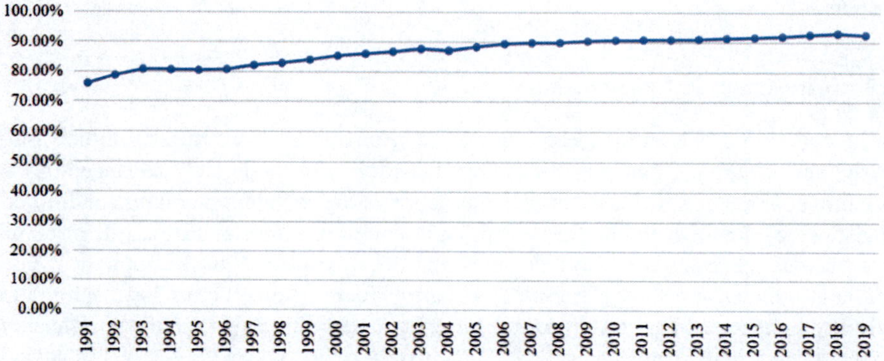

Fig. 6. The indicator of industrial structure upgrading in China from 1991 to 2019. Source: China Statistical Yearbook

3 Results

Based on the above analysis, Fig. 7 presents the relationship between China's industrial structure, logistics development, and infrastructure construction. The direction of the arrows represents the China's industrial structure upgrading, the logistics development, and infrastructure construction respectively over the past decades. The outside circle represents the "digital platform" which combined all those new information technologies such as 5G, big data, artificial intelligence, blockchain, cloud computing.

Smart logistics is combination of "digital platform" with traditional logistics. So traditional logistics industry plus the outer circle in Fig. 7 mutually constitute "smart logistics". As a service industry using transportation infrastructure, traditional logistics industry provides services for the whole industrial value chain. But with outside circle of "digital platform" and "new technological infrastructure", smart logistics can serve the whole industrial value chain better. According to the theories of New Structure Economics, the relation between smart logistics and industrial structure can be summarized as follows: on the one hand, smart logistics is endogenous in the industrial structure; on the other hand, smart logistics also can promote industrial structure upgrading.

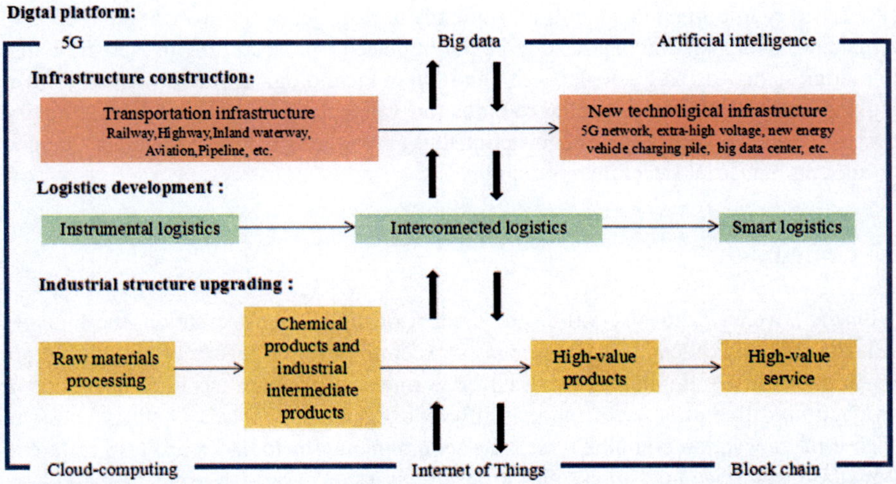

Fig. 7. The relationship between China's industrial structure, logistics development, and infrastructure construction.

4 Discussion

The different industrial structure also means the different scope of production, different scale of market, different organization forms of firms, different degree of transaction complexity. Firms in labor-intensive industrial structure are usually relatively small, and scope of market and production is mainly regional, has limited economies of scale. Transaction among firms is very simple, limited to local markets with familiar people. During this period, logistics just played an instrumental role, referring to the physical flow of goods from the place of supply to the place of receiving, including the basic activities of transportation, storage, loading and unloading, order picking, packaging and so on. Firms in capital-intensive industrial structure became bigger, the scope of production and market expanded and transaction among firms turned more complicated. Logistics that are needed must comply with the necessities of national and global markets where business transactions are long distance and large in quantity and value. Interconnected logistics combines those basic activities integrally to form a complete supply chain, and connects highway, railway, aviation, inland waterway and maritime transportation to develop multimodal transportation and provides users with multi-functional and integrated comprehensive services. In technology-intensive industrial structure, the organization forms of firms are diversified. Many manufacturing enterprises mainly focus on two sectors in order to reduce costs: one is the sector of research and development, the other is marketing, which do not include logistics. This phenomenon leads to a very high requirement for logistics. Smart logistics is emerged as a further development on the basis of interconnected logistics.

Smart logistics represents the development trend of logistics and also can promote the upgrading of industrial structure. It can channel more capital to the whole industrial chain, especially for SMEs. SMEs often face financial constraints, such as the lack of personal funds, the complexity of access to bank loans, insufficient state financial support [13].

IPO is a very important channel for a company to raise funds for growth, but there still remains some legislation that need to be put in place [14]. According to New Structure Economics, due to “digital platform” and “new technological infrastructure”, smart logistics can channel more capital to industrial value chain and reduce the transaction cost, serve the whole industrial chain better, thus accelerating factor endowment changes, promoting industrial structure upgrading.

5 Conclusion

With the advent of industry 4.0, in the context of digital transformation, the logistics industry will take a leap to smart logistics. According to the theory of New Structure Economics, this study indicates that different industry structure has different scope of production, different scale of market; different organization forms of firms, different degree of transaction complexity, and its requirements for logistics are also different. With the development of China’s industrial structure from labor-intensive, capital- intensive to technology-intensive, logistics has taken the same stages of development which starts from instrumental logistics, interconnected logistics to smart logistics. Therefore, smart logistics is endogenous in the industrial structure.

This study also indicates that with “digital platform” and “new technological infrastructure”, smart logistics can provide better services to the entire industrial chain, thus accelerating factor endowment changes, promoting industrial structure upgrading.

This study points out that infrastructure construction has an important role in the relationship between smart logistics and industrial structure. Different logistics development also requires different infrastructure construction. Infrastructure construction also has the imprint of time. While the infrastructure supporting the development of instrumental logistics and interconnected logistics referred to transportation infrastructure before, such as high-speed railways, expressways and airports, the infrastructure supporting the development of smart logistics in the future will be infrastructure in technological innovation fields such as 5G network, extra-high voltage, new energy vehicle charging pile, big data center.

The development of smart logistics is consistent with latent comparative advantage and has a great externality to the national economic development. In order to address externality problems, government should play a proactive, facilitating role. Therefore, the government should provide smart logistics with appropriate industry policy. Furthermore, logistics policy should be deployed or adjusted timely in consistency with logistics development. In the future, logistics policies should focus more on strengthening logistics informatization, promoting the construction of logistics platform ecosystem.

References

1. Arkhipov, A.Yu., Fomicheva, E.V.: Human development and structure: the main directions of the economic breakthrough of Russia. In: Human and Technological Progress Towards the Socio-Economic Paradigm of the Future. Part 1, pp. 129–142. Walter de Gruyter Gmbh, Berlin, Munich, Boston (2020). https://doi.org/10.1515/9783110636147-015

2. Issaoui, Y., Khiat, A., Bahnasse, A., Ouajji, H.: Toward smart logistics: engineering insights and emerging trends. Archiv. Comput. Methods Eng. **28**(4), 3183–3210 (2020). https://doi.org/10.1007/s11831-020-09494-2
3. Guimares, V.D.A., Skroder, G.C., Ribeiro, G.M.: Rev. Facult. Ingen. **98**, 104–116
4. Jiang, Z.Z., Wan, M.Z., Pei, Z.: Comp. Oper. Res. **125**, 105091 (2020). https://doi.org/10.1016/j.cor.2020.105091
5. Liu, W.H., Wei, W.Y., Yan, X.Y.: J. Clean. Prod. **276**, 124189 (2020). https://doi.org/10.1016/j.jclepro.2020.124189
6. Pan, X., et al.: Transp. Res. Part E Log. Transp. Rev. **137**(1), 101939 (2020). https://doi.org/10.1016/j.tre.2020.101939
7. Shu, B., Lu, S., Hu, H.: Int. J. Serv. Sci. Tech. **9**(11), 381–398 (2016). https://doi.org/10.14257/ijunesst.2016.9.11.34
8. Miguel, F.L.: Industry 4.0 and the digital society: concepts, dimensions and envisioned benefits. Proc. Int. Conf. Bus. Excell. **12**(1), 386–397 (2018). https://doi.org/10.2478/picbe-2018-0034
9. Zheng, P., et al.: Smart manufacturing systems for Industry 4.0: conceptual framework, scenarios, and future perspectives. Front. Mech. Eng. **13**(2), 137–150 (2018)
10. Douaioui, K., Fri, M., Mabroukki, C., Semma, E.A.: The Interaction between Industry 4.0 and smart logistics: concepts and perspectives. In: Proceedings of 2018 International Colloquium on Logistics and Supply Chain Management, pp. 128–132. Institute of Electrical and Electronics Engineers (2018)
11. Bag, S., et al.: Res. Pol. **68**(5–8), 101776 (2020). https://doi.org/10.1016/j.resourpol.2020.101776
12. Lin, J.Y.: New structural economics: the third generation of development economics. Asian Educ. Dev. Stud. **9**(3), 279–286 (2020)
13. Shevchenko, D., Arku, H.K.: State institutions of financial support for small business in volatile economies: possible application of Russian experience in Ghana. In: Carvalho, J., Sabino, E. (eds.), Strategy and Superior Performance of Micro and Small Businesses in Volatile Economies, pp. 1–17. IGI Global (2019). https://doi.org/10.4018/978-1-5225-7888-8.ch001
14. Shevchenko, D., Njoroge, P.K.: Effects of people's IPO on the Russian Financial Market and Corporate Governance Praxis. In: Agyemang, O.S., Ansong, A., Agyei-Mensah, B.K. (eds.), Corporate Governance Models and Applications in Developing Economies, pp. 83–101. IGI Global (2020). https://doi.org/10.4018/978-1-5225-9607-3.ch005

Modern Social Process of Goods and Services Promotion and Formation

Valery V. Nikishkin, Maria D. Tverdokhlebova, Elena V. Aleksandrova(✉), and Vladimir N. Kovnir

Plekhanov Russian University of Economics, Moscow, Russia

Abstract. This article focuses more specifically on studying and meeting market needs, especially relationship management. Over the past three years, people's buying behavior has changed dramatically in almost all market segments, and this applies to all segments of the population, regardless of age or income level. The relevance of the research is to study and understand the needs and desires of people, to implement and support the implementation of socially significant projects. The task is to make production financially attractive, consumption constant, income stable. Support active participation in the formation of a civilized market for marketing services. The work aims to study and understand market trends, correctly assess situations, promote economic growth, and increase the welfare of the population by improving the quality of goods and services. As well as the development of non-standard management mechanisms corresponding to the unfolding market transformations. A special place in this process is occupied by complex and purposeful work in marketing management. The main research methods are comparative analysis, archival search, deduction, theoretical and methodological approaches, and the study of modern domestic and foreign literature in the field of marketing.

Keywords: Marketing · Sales potential · Company · Advertising · Market

1 Introduction

Marketing is a market activity. Marketing is the company's activity to meet market needs [1]. Marketing is a constant movement of perception. This is an understanding of the needs and desires of the people of the world. Marketing is the most important thing in developing a given industry, product, or consumer response [1]. The art of marketing in its most essential and elementary form comes down to an understanding of what the consumer wants to buy at a price that allows the manufacturer or the plant to sell in large quantities at a minimum profit so that it gives the plant a gross profit that allows it to buy raw materials, capitalize production machinery, lend money to finance equipment and products, pay employees real and fair wages, and realize sufficient income to reward employees for outstanding efforts and contributions [2, 3]. Moreover, ultimately, to make the entire production from start to finish financially attractive enough for the plant to have an interest in continuing this production.

P. V. Trifonov and M. V. Charaeva (Eds.): DITEM 2021, LNNS 380, pp. 406–410, 2022.
https://doi.org/10.1007/978-3-030-94245-8_55

2 Materials and Methods

The main research methods are scientific abstraction, system analysis, deduction, information structuring, analysis of archival materials. The historical-genetic and historical-typological methods were used. Review of research and recent publications. In the process of working on the article, we used the research of the famous scientists-economists Baker [1], Lichtenstein [3], Kumar [4], Grether [5], as well as publications of such modern marketing researchers as Nikishkin [6, 7], Zhuravleva [2], Skorobogatykh [8], etc.

3 Results

So, a factory or a plant or a nation as a whole can conceptualize a product or a need or a desire and assume that by producing that product and selling it to the public, they, in turn, will respond favorably and continually buy it. At the same time, the population becomes a consumer of this particular product or type of service. Many factories can produce finished products. Many businesses can offer a variety of services. Failure usually affects those who do not carefully study their product or type of service to honestly determine whether there is a genuine need or desire for it. One of the aspects of marketing is honest reflection when determining what a consumer will buy. Moreover, most importantly, when buying this product or service, how much he wants to pay.

It is pointless to produce finished products and then find out that the price of these products does not allow them to be sold in an amount sufficient to pay for the costs of the plant for its production. Market research would allow the manufacturer to understand and estimate what would be marketed at a price that would allow the supplier, after selling the product at that price, to realize a profit sufficient to cover all the costs of producing and delivering the product to the consumer. When assessing expenses, it is necessary to keep careful records from the beginning because for the plant to work successfully for a long time; all expenses must be taken into account from the beginning to the end to correctly determine the selling price of the product. While the manufacturer of a product or service offers a product that the consumer willingly purchases, the factory must be reasonable enough to make sure that the selling price is fair and the profit is acceptable. Otherwise, watching your success, a potential competitor will very quickly realize that he can produce a similar product cheaper and, by doing so, will push you out of the market. Competition between manufacturers for life and death occurs worldwide, and those who know the needs of the market and have a fine sense of market prices succeed [4].

When evaluating a particular product, it is necessary to analyze how the product or type of service comes to market, whether a consumer is a private person, a housewife, the family's breadwinner, or the whole country. Marketing analysis will allow determining exactly how many consumers will buy this product, and according to this data, the head of the enterprise will be able to determine whether the profit from the sale of products will cover its production costs [6, 9].

Another part of marketing is the study and understanding of market trends. This is a key element in the successful operation of the enterprise. When the product will be delivered to the consumer, how long it will remain fashionable. How long this item will sell quickly to maintain production capacity [7, 10]. For example, if you follow the Parisian fashion designers who annually dictate what well-dressed women of the world will wear next spring, you can very quickly understand what is fashionable this spring and will go out of fashion next. The Parisian trendsetters change models of clothing from year to year to maintain production at a constant level. Changing styles of clothing and its length are their tactics for ensuring the sale of products. It is an interesting marketing phenomenon, and our task is to go into detail to show how it works and what happens when it does not work. To make big production profitable and keep employment levels high, Parisian fashion designers, as well as shoe and cosmetics manufacturers, have for many years, through enormous spending on advertising, inventive design, and, in addition, a clever sales technique, convinced women that clothes from last year, still quite usable, are no longer fashionable and should be thrown away. Cosmetic companies have now reached the point where women spend huge amounts of money on creams, hair dye, lipstick, deodorants, products that have no utilitarian value and essentially do nothing to enhance a woman's natural, God-given beauty. If they are young, they want to look older; if older – they desperately want to revive the lost look of youth. This is an example of marketing in the most elegant form when factories have adapted the consumer's will to their needs through marketing. This is a risky business, and every step can lead to an unpredictable result [5].

For a long time, French designers have been dictating fashion to the women of the world. One year they decided that women's skirts should be long and lush to give women a Victorian style. Another year, the skirts had to be two centimeters below the knee to liberate the female appearance. Year after year, the length of the skirt rose and fell. Women all over the world obediently followed the instructions of Parisian designers. Two years ago, Parisian designers decided to raise the length by 7–15 cm above the knee instead of lengthening the skirt from the knee to the ankles. Without conducting an appropriate marketing analysis, they started a multimillion-dollar advertising campaign for this fashion, persuading women worldwide to expose their flesh above the knee. Designers invited very slender girls to show their work. Women who were persuaded not to question designers' wisdom tried to prepare for these very uncomfortable, tight, and little decorating products. Suddenly it turned out that a very high percentage of women have thick, loose hips, and they look bad in these Parisian fashions. However, the women went on a diet, started exercising, and desperately tried to lose weight. At the height of the season, when the sale of new products was supposed to reach a maximum, everything suddenly stopped. Women who could not follow a diet, who did not lose weight, eventually rebelled and decided that Parisian fashions were created for teenagers and children and not for them at all. Clothing manufacturers and factories immediately reduced production, laid off thousands of employees. The moral of this story is that the excellent designers who set the tone in fashion forgot to conduct a marketing analysis. They did not ask themselves what percentage of the female population is comfortable showing their hips, how men will react to this, especially if the hips are unattractive. This is an example of people who adjusted marketing to their desires. A thorough analysis of the market would save everyone from this unfortunate experience [8].

Another example of a flourishing dynamic industry is the production of women's shoes. After a thorough study of the market, manufacturers have determined that women do not want to wear well-made, comfortable, practical flat shoes. Manufacturers making such products for years have realized that with the right approach, there is a huge market for women's shoes that are not practical, not durable, not functional, and do not meet any of the requirements of regular shoes. Having discovered this, shoe manufacturers, especially in Italy, began to produce very narrow shoes to give women a sense of elegance. The webbed shoes came into fashion because laces were unelegant and, finally, the most important innovation was the four-inch heels, which made women look taller and slimmer, giving a beautiful shape to their calves and ankles, thus making an even greater reason for women to wear them. Shoes that had no practical value were presented to the female population with all the above attributes. The Italian shoe industry, which has been experiencing difficulties for years, unable to compete with the American one in terms of economic indicators, has now highly appreciated a piece of leather and has begun to make millions of profits over the past decades. Despite the outstanding success, which testifies to talented marketing, shoe manufacturers almost ruined themselves, deciding that the heels of women's shoes should be even higher and should reach almost the limit. Again, women obediently bought shoes, not considering that it would be difficult to walk in shoes. The narrow, sharp heels swayed and did not provide the required support for the foot. Feet began to swell, heels clung to the carpets, and a string of accidents followed. Despite this discomfort, the new shoes of fashion designers were quickly sold out until it was discovered that sharp heels spoil the floors in buildings. It is clear that after that, the sale of shoes fell sharply, factories, warehouses, and stores suffered heavy losses, and practical shoes came back into fashion.

This is an example of marketing taken to the extreme, again, an example of how people have not conscientiously studied marketing.

4 Discussion

Simple production management and product creation are not enough to implement the plan. It is necessary to improve the management technology in general, constantly improve the quality of goods and services, and the use of modern innovative and information technologies, the continuous development of supporting resources (primarily intellectual). This is the task of marketers.

5 Conclusion

In this article, the authors have tried to explain what marketing is and show its impact on the consumer and what contribution marketing has made to a given plant, factory, or service company's success or failure. Marketing strategies, customer relationship management are discussed.

The marketing analysis as a denominator for establishing the accurate and fair price of a given product or service was carried out. Marketing research and its use are applied to assess buyers' whims, market trends, local, regional, and national habits.

The article is recommended for students of educational institutions of this profile.

References

1. Baker, M.: Marketing Encyclopedia. Peter Publishing House, Saint-Petersburg (2002)
2. Zhuravleva, G.P., Aleksandrova, E.V.: Institutional foundations for searching the ways out of the modern global permanent crisis. Indian J. Sci. Technol. **9**(42), 98637 (2016)
3. Lichtenstein, S., Slovic, P.: Reversal of preferences between bids and choices in gambling decisions. J. Exper. Psychol. **89**, 46–55 (1971)
4. Kumar, N.: Marketing as a Strategy. Pretext, Moscow (2008)
5. Grether, D., Plott, C.R.: Economic theory of choice and the preference reversal phenomenon. Am. Econ. Rev. **69**(4), 623–638 (1979)
6. Nikishkin, V.V.: On the question of the manifestation of passionarity of modern marketing. Pract. Mark. **1**(95), 17–20 (2005)
7. Nikishkin, V.V.: Development of the concept of marketing management of a large shopping center. In: Digital Economy: Trends and Prospects of Development. Collection of Abstracts of Reports of the National Scientific and Practical Conference, vols. 151–153 (2020). https://www.elibrary.ru/item.asp?id=44290899&pff=1
8. Skorobogatykh, I.I.: Marketing: from transactions to values. The evolution of marketing and the evolution of marketing definitions in the world and Russia. Marketing in Russia 2017. In: Handbook of the Guild of Marketers. Publishing House: “School Press” LLC, Moscow, pp. 178–199 (2017). https://www.elibrary.ru/item.asp?id=29833976
9. Marketing in Russia. In: Handbook of the Guild of Marketers. Berezin, I.S. (eds). Moscow (2017). https://www.elibrary.ru/item.asp?id=29833951&selid=29833976
10. Tverdokhlebova, M.D., Nikishkin, V.V.: The role of marketplaces in the retail trade services market. Pract. Mark. **6**(268), 3–8 (2019)

Innovation Management in RUSNANO Corporation

Natalia B. Safronova[1](✉), Roman S. Volodin[2], Lyudmila L. Shtofer[3], and Shaozhui Ma[4]

[1] The Moscow State University of Civil Engineering, Moscow, Russia
safronovanb@mgsu.ru
[2] Southern Federal University, Rostov-on-Don, Russia
[3] Rostov State University of Economics, Rostov-on-Don, Russia
[4] Don State Technical University, Rostov-on-Don, Russia

Abstract. Innovative management is a certain coordination and economic method and directions of management of all stages and types of innovation processes at an enterprise with maximum efficiency. The latter is part of the investment management, which studies the tools for investing resources in the development of the innovation sphere, the expediency of their investing and profitability. State regulation of the innovation sphere is carried out primarily through normative and legislative forms. The paper considers the problems associated with the regulation of the legal aspect of financing innovative activities in RUSNANO JSC, depending on the legal support of this environment at the government level. The issues of increasing the effectiveness of the process of managing the innovative activities of enterprises are becoming more and more urgent, the need for an effective response to the changing national and international challenges of external conditions is growing. Therefore, a detailed structural study of the area is required, covering all components associated with innovation processes. The purpose of the study is to analyze the successful experience of RUSNANO JSC in innovation management. The experience of the largest innovation corporation in the Russian Federation, RUSNANO JSC, has been studied. The estimates of the innovative potential of enterprises by the group of cost indicators for 2019 have been provided, the data on the effectiveness of innovation management have been obtained. Some proposals have been formulated to improve the efficiency of innovative activities based on the results of the research.

Keywords: Innovation activity · RUSNANO JSC · Efficiency assessment · Dynamic indicators · Legislative base

1 Introduction

As is known, the main goal of an organization's innovation activity is to provide the most effective ways to apply the innovation corporate strategy at certain stages of development. In the process of implementing this management tool, they are guided by the solution of the following major tasks: ensuring high growth rates of enterprises and

P. V. Trifonov and M. V. Charaeva (Eds.): DITEM 2021, LNNS 380, pp. 411–416, 2022.
https://doi.org/10.1007/978-3-030-94245-8_56

their competitiveness; maximizing profits through effective innovation; minimization of risks in financing innovative and nano-technology projects; maintaining the stability and solvency of the business in the implementation of new processes; search for ways to accelerate the creation of innovative projects.

At the current stage of economic development, the importance of innovation is recognized at both the micro and macro levels [1].

According to some scholars [2, 3], the main restrictions on the innovative development of business are:

- low level of state funding for innovative development and insufficient internal funds;
- low investment activity of foreign and domestic investors;
- shortcomings in tax regulations;
- undeveloped legal legislation on innovation;
- lack of motivation of domestic enterprises to initiate scientific research and implement their results;
- underdeveloped relationships between science, scientific and innovative developments and production.

The role of RUSNANO JSC as a transformational leader can be assessed based on the position of the authors, who note: "the results show that transformational leaders can not only promote innovation within the organization, but also ensure the market success of innovations" [4].

Therefore, strengthening the innovative development of enterprises should be associated, firstly, with the renewal of fixed production assets, which should become the main source of increasing production output and creating conditions for the manufacturing of domestic products at the level of the best world models [5].

This study selected the financial indicators and indicators of business activity to assess the effectiveness of management of innovation activities of the corporation. It should be noted that a Delphi survey of 264 experts with diverse professional and academic backgrounds examined the perceived importance of each factor at different stages of the innovation implementation [6].

The success of a firm in the modern economy is impossible without the creation, development and diffusion of innovations. The introduction of innovations requires investing in capital assets subject to aggregate actions of investors. At the initial stage of the project, each investor wants to know whether its participation in a particular project will be successful, and whether the products expected to be manufactured will be in demand. In this regard, the economic assessment of investment projects is an important aspect of the activities of the firm's staff.

2 Methods

The company monitors how effectively the mechanisms by which innovative activities are organized and provided financially are functioning [7]. When developing the Program, the company analyzed the possibility of using various tools for investing in innovative activities, including: financing the development of new equipment and its production through the conclusion of future contracts, creation of special-purpose companies; creation of the Fund for financial support of R&D; venture financing; implementation of projects related to innovation, in accordance with public-private partnerships.

The study of the tools according to which the provision of financially innovative activities through the organization of the above-mentioned Fund enabled to reveal the presence of a number of inherent activities carried out by the company, which exclude the advisability of using this tool:

- a significant number of the company's facilities are strategically important for the government, and information about them is a state and commercial secret;
- the company acts as the sole consumer of the engineering solutions that it uses. In this regard, opportunities for joint investment with investors who might be interested in financing the relevant projects and their implementation are limited;
- functions related to the preparation of R&D plans, and funding R&D in the company are carried out on a centralized basis.

To increase the efficiency of financial support for innovative activities, the company is analyzing the possibility of attracting financial resources for the implementation of innovative projects within the framework of public-private partnerships. This tool could be realized via government subsidies for the implementation of projects related to the creation of high-tech industries, in which state scientific institutions and universities participate [8].

The study of proposals related to participation in venture projects was carried out in accordance with the Principles of Co-investment in these projects regulating investments and innovation policy of the company, developed by the Ministry of Economic Development.

Several groups of indicators are used as indicators of RUSNANO's innovative activity:

- dynamic indicators;
- technological indicators;
- cost indicators.

On their basis, an integral indicator of the level of innovative potential is calculated and compared with the normative data.

By implementing the Program, the company will continue to monitor how appropriate it is to use possible mechanisms to ensure financial innovation.

3 Results

Analyzing the indicators of the effectiveness of innovation activities of RUSNANO JSC and comparing them with the standard indicators [9] (Table 1).

Table 1. Assessment of the level of innovation potential of JSC RUSNANO for 2019.

Indicator	Standard value	Actual value
Group of dynamic indicators	0.4	0.36
Group of technological indicators	1.61	1.42
Group of value indicators	1.6	1.43
Integral indicator of investment potential level	1	0.92

Analyzing the indicators, one can say that RUSNANO JSC conducts stable operations, which means that the level of management efficiency at this enterprise is higher. It should be noted that all indicators are at a fairly high level, but still slightly inferior to the standard values. Based on the calculations, each group of indicators will be analyzed separately (Table 2).

Table 2. Results of the assessment of the innovative potential of RUSNANO JSC by the group of dynamic indicators for 2018–2019.

Dynamic indicators	Standard value	RUSNANO JSC
Revenue growth rate	0.170	0.169
Profit growth rate	0.1	0.11
Change in profitability ratio	0.08	0.05
Change in asset turnover ratio	0.05	0.028
Total	0.4	0.357

Based on the results of the table, one can conclude that the growth of revenue and growth of profit are as close as possible to the standard values, while the change in the profitability ratio and the asset turnover ratio is below the standard thresholds (Table 3).

Table 3. Results of assessing the innovative potential of RUSNANO JSC by the group of technological indicators for 2019.

Technological indicators	Standard value	RUSNANO JSC
Number of patents and copyright certificates	0.01	0.06
Technology novelty	0.9	0.55
Management efficiency level	0.6	0.68
Own pilot production	0.1	0.36
Total	1.61	1.65

Having considered the group of technological indicators, one can conclude that the number of patents and owned innovative production facilities exceed the regulatory threshold, which indicates a good level of the company's innovative activity, however, such an indicator as the degree of innovation novelty is below the standard, which evidences that RUSNANO JSC is borrowing foreign technologies and developments (Table 4).

Table 4. Results of assessing the innovative potential of RUSNANO JSC by the group of value indicators for 2019

Value indicators	Standard value	RUSNANO JSC
Return on investment in innovative developments	0.5	0.62
Investment in R&D	0.3	0.41
Growth in the value of net assets	0.05	0.02
Total	0.85	1.05

Based on the results obtained, it can be concluded that RUSNANO is a company that is primarily engaged in investing in innovative activities and innovative projects, therefore related indicators exceed the standard threshold values.

4 Discussion

It should be noted that the entry of an enterprise into the innovation market should be understood as an integral system, as a set of planning, organization and control activities, in which the dynamic indicators analyzed in this study play an important role [10].

Thus, the participants in the innovation process need an effective financial and legal mechanism that formalizes the relationship between the actors and other players in the economic market. The entire set of regulations on innovative activities should form a favorable legal environment for innovative entrepreneurship, including for RUSNANO JSC.

An analysis of legal acts in the field of innovation allows stating an incompleteness of regulations in this area. The absence of the main consolidated regulatory legal act at the federal level is the key negative side of the current legislation of the innovation industry. Also, there is some "blurring" in connection with the directions of government policy in the field of innovation. RUSNANO JSC is a leading innovative corporation with a high level of management and strategic planning. Nevertheless, an analysis of the dynamic indicators of innovation activity made it possible to reveal a number of hidden problems, such as an orientation towards foreign developments. In general, the conducted research enables to confirm that RUSNANO corporation is a leading innovative company whose innovative potential exceeds the standard economic indicators.

5 Conclusions

Based on the current situation, RUSNANO JSC can be offered the following measures:

- development of new types of services that do not depend on the costs of marketing;
- development of new types of products, including vital ones;
- optimizing the budgeting structure for new innovative projects;
- efficiently distributing financial resources by sources of funding: own funds and government loans;
- dynamic management of innovation activities based on the indicators obtained in this study.

References

1. Diego, A.-L., et al.: Eur. J. Manag. Bus. Econ. **18**(3), 266–284 (2018). https://doi.org/10.1108/ejmbe-05-2018-0056
2. Vasyaicheva, V.A.: Voronezh State Univ. Bul. **4**, 54–67 (2020). https://doi.org/10.17308/econ.2020.4/3195
3. Bryskina, E.O.: Int. Res. J. **2**, 91–109 (2020). https://doi.org/10.23670/IRJ.2020.94.4.023
4. Mokhber, M., Khairuzzaman, W., Vakilbashi, A.: J. Manag. Org. **24**(1), 108–128 (2018). https://doi.org/10.1017/jmo.2017.26
5. Christensen, J.F.: Ind. Corp. Change **11**(2), 263–288 (2002). https://doi.org/10.1093/icc/11.2.263
6. Pichlak, M.: J. Manag. Org. **22**(4), 476–494 (2016). https://doi.org/10.1017/jmo.2015.52
7. Czarnitzki, D., Kraft, K.: J. Econ. Behav. Org. **71**(2), 372–383 (2009). https://doi.org/10.1016/j.jebo.2009.03
8. Kosareva, E.N., Safronova, N.B.: Serbian J. Manag. **11**(1), 129–140 (2016). https://doi.org/10.5937/sjm11-9090
9. Information disclosure – RUSNANO group. https://www.rusnano.com/about/highlights
10. Zhu, L., Cheung, S.O.: J. Manag. Eng. **33**(5), 04017020 (2017). https://doi.org/10.1061/(ASCE)ME.1943-5479.0000534

Value Orientation for Marketing Customer Experience Management in Companies in a Digital Transformation

Yulia V. Syaglova[1](✉), Ekaterina S. Bozhenko[2], Natalia G. Larkina[2], Elena Yu. Polyakova[2], and Irina V. Stefanova[2]

[1] The Russian Presidential Academy of National Economy and Public Administration, Moscow, Russia
Syaglova-yv@ranepa.ru

[2] Southern Federal University, Rostov-on-Don, Russia

Abstract. This publication substantiates aspects of marketing management of the client experience in modern companies that are analogue, multi- and omnichannel companies, as well as digital platforms characteristic of the conditions of digital transformations in economic systems. The prerequisites for the study are based on the evolution of transformations in marketing tools for the formation and transformation of customer experience, as well as on the evolution of transformations of the economic systems themselves, associated with the transition of companies in various industries and spheres of economic activity to digital forms of functioning. The study aims to develop recommendations on the need to transform marketing tools for managing customer experience in modern digital companies and platforms of the economic system. The research methods were classification, systematisation, generalisation, critical and factor analysis. These methods allowed the authors to substantiate the requirements and recommendations for managing customer experience in a digital environment. The result of the study was a list of parameters and recommendations aimed at improving the performance of modern companies using an effective mechanism for managing customer experience. The novelty of the study is represented by the scientific validity of a set of measures for managing client experience in modern digital companies. The main practical significance of the study is determined by the possibility of extrapolating the presented set of measures for managing the client's experience into the activities of modern digital companies in various industries and spheres of economic activity. The theoretical significance of the study is to systematise, classify and generalise the factors and parameters of modern digital economic activity in order to form an effective toolkit of measures for the formation and management of client experience in companies.

Keywords: Digital transformation · Digital platforms · Customer Journey Map · Digital marketing

1 Introduction

In the context of digital transformations, most companies in various industries and business areas are faced with the task of forming, adapting, "setting up" or adjusting the

P. V. Trifonov and M. V. Charaeva (Eds.): DITEM 2021, LNNS 380, pp. 417–422, 2022.
https://doi.org/10.1007/978-3-030-94245-8_57

strategy of interaction with their customers [1–3]. This task is due to global trends that influence, on the one hand, the behaviour of the client in the digital environment, and, on the other hand, the formation of an optimal, effective business model of the organisation. The novelty of the study is represented by the scientific validity of a set of measures for managing client experience in modern digital companies. The main practical significance of the study is determined by the possibility of extrapolating the presented set of measures for managing the client's experience into the activities of modern digital companies in various industries and spheres of economic activity. The theoretical significance of the study is to systematise, classify and generalise the factors and parameters of modern digital economic activity in order to form an effective toolkit of measures for the formation and management of client experience in companies. The hypothesis of this study was the possibility of the influence of parameters, factors and trends of modern economic systems on the change in the marketing mix aimed at shaping and managing the client's experience. The study aims to develop recommendations on the need to transform marketing tools for managing customer experience in modern digital companies and platforms of the economic system. The main objectives of the study are as follows: 1. Systematisation of scientific and analytical materials in order to identify the main trends in the functioning of modern economic systems. 2. Based on the identified trends, the formulation of requirements for the transformation and transformation of economic systems themselves into digital platforms. 3. Justification and classification of types of digital platforms. 4. Justification of the need and feasibility of transformations in the methodology of marketing management of client experience in modern digital companies.

2 Methods

The research methods presented in the study, such as classification, systematisation, generalisation, critical and factor analysis, allowed the authors to review scientific and analytical sources of information which present the conditions, parameters, factors and trends of the digital economy on the basis of which it was possible to substantiate the requirements and recommendations for managing customer experience in a digital environment. According to the authors, analytical types of sources make it possible to identify factors characteristic of modern companies in the digital economy, due to the determination of the influence of which it is possible to justify the tools of the marketing mix in order to effectively manage the client's experience. Scientific sources allow us to trace the evolution of change and transformation of elements of the marketing concept associated with the parameters of the formation of customer behaviour in companies, as well as the route, or Customer Journey Map, with which you can manage the customer experience using marketing tools.

3 Results

After analysing scientific sources, as well as analytical material from global research and consulting companies such as Deloitte, Gartner, PWC [4, 5] and others, the authors systematised the information and formulated two types of trends characteristic of the digital business environment:

Technological Trends

This category includes such important parameters of digital business as the use of immersive technologies and artificial intelligence; technologies of virtual and augmented reality and blockchain; Big Data technologies and cloud storage systems; technologies that ensure the security of data functioning and the "digital footprint" left by stakeholders.

The presented list of technologies, of course, is not finalised, it is constantly expanding, since the conditions for the formation and functioning of digital platforms are also subject to rather rapid changes and require adaptation and adjustment to socio-cultural trends, which will be discussed below.

Sociocultural Trends

This group is associated with changes in the requirements and needs of the client, who acquires a new experience of interacting with a digital platform in the modern digital space and forms new points of contact in its route – the Customer Journey Map (CJM). This group should include the requirements of customer focus and customer centricity when building a company's business model; transparency and traceability of all interactions with the digital platform; "seamlessness" and customer experience when moving between points of contact in CJM; speed and accuracy when interacting with the platform; omnichannel experience in transactions.

The formulated technological and sociocultural trends allow the authors to come to the conclusion that in the conditions of the increasing penetration of Internet technologies, both in Russia and in international economic spaces (the penetration rate is about 60%), the increasing willingness of customers to make an increasing number of transactions using e-commerce (from 50% to 80% depending on the geographic location of the client, the type of product purchased, the generation of clients, the availability of the technical ability to access the Internet, and so on), new digital platforms are being formed in the digital economic system, which enable the client to realise everything the above requirements for the formation of a new customer experience. We are talking about such digital platforms as marketplaces at a global, national or niche level of coverage, as well as ecosystems that include many stakeholders and areas of economic activity which are united by a single space and create an integrated effective business model [6, 7]. Successful digital business models include Alibaba, Amazon, Sberbank, Wildberries, OZON, Yandex Market, Sbermegamarket and many others.

In such digital spaces, clients of companies that, according to the parameters of customer focus and customer centricity, can be divided into external – ultimate customers of products or goods and internal – team, personnel, talents within the company, get the opportunity to form the so-called omnichannel customer experience, organising a special CJM, including offline and online touchpoints. Consequently, the global goal of such digital platforms is to form an optimal and effective strategy for managing all points of contact for all types of customers in order to achieve the best customer experience, on the one hand, and the highest profitability of the company [8].

Such an omnichannel strategy for managing customer experience in a digital environment should be based on the requirements of "seamlessness", regardless of the type of platform interaction point; the speed and accuracy of performing the required operations – when placing an order or application, purchasing, organising logistics, marketing operations, and so on; convenience and comfort when interacting with the platform – high-quality usability and design; the presence of technically advanced digital spaces, namely, a website and/or a mobile application with fast content loading; availability of various options for delivery and payment for goods; availability of a relevant stock confirmed by the warehouse suitable for placing an order; availability of customer support and service which is able to quickly solve emerging problems for the client; monitoring and control of risks arising in the digital environment; digital business performance metrics and KPI requirements.

The authors of the study are confident that the strategy for managing customer experience in new digital platforms can be successful, while ensuring effectiveness for all stakeholders along the entire chain of origin of touchpoints, starting from manufacturers and suppliers of products and goods; the teams and talents of the digital company, which are the internal links of the system; logistics and marketing service providers; ultimate consumers; and the digital platform itself providing the opportunity to establish relationships between these stakeholders as an intermediary.

4 Discussion

Based on data from scientific and analytical sources, the authors analysed the main parameters of modern economic systems, as a result of which they identified technological trends that affect the formation and functioning of companies in the digital environment. The analysis made it possible to identify the following technological trends: the use of immersive technologies and artificial intelligence; technologies of virtual and augmented reality and blockchain; Big Data technologies and cloud storage systems; technologies that ensure the security of data functioning and the "digital footprint" left by stakeholders. In addition to technological trends, on the basis of the analysis [9], the authors of the study substantiated the impact of socio-cultural trends on the aspects of marketing management of the client's experience, such as: customer-focusedness and customer-centricity in building a company's business model; transparency and traceability of all interactions with the digital platform; "seamlessness" and customer experience when moving between points of contact in the customer route – CJM; speed and accuracy when interacting with the platform; omnichannel experience when performing transactions [10–12]. Technological trends, of course, affect the penetration of customer traffic and the number of transactions carried out using Internet communications giving them a greater volume, coverage and speed of operations [13, 14]. As a result, falling under these trends, new digital companies are being formed, namely, platforms, ecosystems, marketplaces which include an increasing number of participants requiring adjustment and adaptation of their customer experience [15]. Sociocultural trends form new parameters, features of customer behaviour in digital companies, i.e. platforms, which requires the analysis, categorisation and structuring of maps or routes for directed customer promotion along the chain of relationships with companies at various points of contact.

5 Conclusion

As a result of the analysis, the authors came to the following conclusions. 1. Currently, there is a transformation of the economic system and an active immersion of companies in various industries and spheres of economic activity in the digital environment. 2. Global technological and socio-cultural trends affect both the forms and sizes of modern companies in the digital environment, and the parameters of the marketing management of the marketing mix in these companies. 3. Among the main types of companies in the digital economic environment, one should single out digital platforms, marketplaces and ecosystems which can be represented by global, national, or niche coverage. 4. In these digital platforms, there is a transformation of the experience of customers which are represented both by the external segment, i.e. consumers of products, goods and services of companies, and internal segments, represented by the team – by the company's personnel, as well as by other stakeholders included in the infrastructure of the digital ecosystem or marketplace. Research prospects are further detailing scientific and analytical data in the field of digital economy formation with the aim of clustering companies in the digital environment, transforming their marketing mix, categorising customers and participants of digital platforms, transforming or adapting marketing tools in terms of managing the experience of clients of various categories and segments in order to increase the effectiveness of the functioning of the company, i.e. the platform. The limitations of this study are presented by the limited amount of scientific sources justifying the formation of digital economic platforms, as a result of which it is rather difficult to structure and categorise the factors of influence on aspects of marketing management of the client experience.

References

1. Dunakhe, K., Panse, C.: Int. J. Innov. Sci. (2021). https://doi.org/10.1108/IJIS-11-2020-0263
2. Farhat, K., Aslam, W., Mokhtar, S.S.M.: J. Internet Commer. (2021). https://doi.org/10.1080/15332861.2021.1905474
3. Krishen, A.S., Dwivedi, Y.K., Bindu, N., Kumar, K.S.: J. Bus. Res. **131**, 183–195 (2021)
4. Gartner Top Strategic Technology Trends for 2021. Analytical Report. Burke, B. (ed.). Gartner Inc. (2020)
5. Tech Trends 2021. Analytical Report. Buchholz, S., Bechtel, M., Briggs, B. (eds.). Deloitte Development LLC (2021)
6. El-Darwiche, B., Herzog, M., Singh, M., Maalouf, R.: Understanding Digital Content and Services Ecosystems. Analytical Report. PwC (2019)
7. Li, C., Fengxia, Z.: J. Mark. Manag. **37**, 1191–1218 (2020). https://doi.org/10.1080/0267257X.2021.1900895
8. Ayawickramarathna, W., Mulye, R., Rahman, K., Fry, T.: J. Cons. Aff. (2021). https://doi.org/10.1111/joca.12364
9. Woods, R., Doherty, O., Stephens, S.: Ind. High. Educ. 09504222211009180 (2021). https://doi.org/10.1177/09504222211009180
10. Hu, L.: J. Internet Commun. (2021). https://doi.org/10.1080/15332861.2021.1907170
11. Malyy, M., Tekic, Z., Podladchikova, T.: Tech. Forec. Soc. Change **169**, 120794 (2021)
12. Wei, J., Lu, J., Chen, W., Xu, Z.: J. Theor. App. Elec. Comm. Res. **16**(3), 445–465 (2021). https://doi.org/10.3390/jtaer16030028

13. Dolega, L., Rowe, F., Branagan, E.: J. Retail. Cons. Serv. **60**, 102501 (2021). https://doi.org/10.1016/j.jretconser.2021.102501
14. Hanninen, M., Smedlund, A.: J. Manag. Stud. (2021). https://doi.org/10.1111/joms.12701
15. Mabarsi, A.R., Njotoprajitno, R.S., Hadianto, B., Wiraatmaja, J.: J. Asian Fin. Econ. Bus. **8**(4), 475–482 (2021). https://doi.org/10.13106/jafeb.2021.vol8.no4.0475

Modern HR Analytics: Digital Opportunities in Assessing the Effectiveness of Personnel Management

Ekaterina V. Batovrina(✉), Natalia N. Oparina, and Galina V. Cherniaeva

Lomonosov Moscow State University, Moscow, Russia
batovrinaev@spa.msu.ru

Abstract. The issues of assessing business performance, including assessing the effectiveness of personnel management, were actualized in the 1980s with the introduction and active development of strategic management. In the same period, theoretical and methodological grounds for conducting complex multi-method assessments were formed, which served as a starting point in the formation of modern HR analytics. The use of HR analytics is seen as a new stage in assessing the effectiveness of personnel management, the need for which arose with the emergence of talent management and individualization of approaches to personnel management. The extreme demand for HR analytics in the practice of management activities has led to the need to understand the possibilities of its application from the standpoint of an interdisciplinary approach. The objective is identification of the leading areas of modern HR analytics, opportunities, limitations and priorities for the use of modern digital tools for assessing the effectiveness of personnel management. Research methods are general scientific theoretical research methods; comparative analysis; SEO analysis. The leading directions of modern HR analytics have been identified and their high importance in assessing the effectiveness of personnel management has been shown; the opportunities, limitations and priorities of using modern digital capabilities for assessing the effectiveness of personnel management have been identified; on the basis of comparative analysis and SEO analysis, the priority practices of modern HR analytics have been identified and the current directions of its development have been shown.

Keywords: HR analytics · Leading areas of HR analytics · Personnel management effectiveness · Digital tools for assessing personnel management effectiveness

1 Introduction

HR analytics is one of the most popular and demanded areas in the field of personnel management. According to Leonardi and Contractor, "more than 70% of companies now say they consider people analytics to be a high priority" [1]. Similar data are presented in the report on the results of an international study conducted by Deloitte [2]. A study

P. V. Trifonov and M. V. Charaeva (Eds.): DITEM 2021, LNNS 380, pp. 423–428, 2022.
https://doi.org/10.1007/978-3-030-94245-8_58

by the consulting company KPMG revealed that 70% of HR leaders are aware of the need for analytics in personnel management, while 60% of the corporate human capital investment budget is expected to be spent on developing digital HR analytics capabilities [3].

The range of HR analytics tools is quite wide: it includes both economic tools that allow measuring the economic effectiveness of personnel management, and sociological and psychological tools used to assess the social effectiveness of personnel management and intra-organizational diagnostics of employees. Until recently, many HR analytics tools have been difficult to apply. Conducting a full-fledged HR analysis required too much time and money and special training of HR staff. Digitalization of HR analysis enabled to obtain the maximum amount of relevant information about the personnel and the state of personnel management in the organization in a short time, significantly simplified the procedures for collecting, analyzing, interpreting economic, social and personal data necessary for making personnel decisions.

The objective of the research is to identify the leading areas of modern HR analytics, opportunities, limitations and priorities for the use of modern digital tools to assess the effectiveness of personnel management.

2 Methods

In preparing this paper, general scientific theoretical research methods, comparative analysis, SEO analysis were used.

3 Results

Evaluation of the effectiveness of personnel management is "a systematic, clearly formalized process aimed at measuring the costs and benefits associated with programs of personnel management to correlate their results with those of the base period, the indicators of competitors and the goals of the enterprise" [4]. The use of HR analytics has become a breakthrough in assessing the effectiveness of personnel management and has introduced such economic tools for assessing effectiveness as HR metrics, key performance indicators, a balanced scorecard, and benchmarking techniques [5, 6].

The use of economic tools for assessing the effectiveness of personnel management is one of the most common ideas about HR analytics, which significantly narrows the scope of its application, but, nevertheless, is reflected in the practice of personnel management and training programs for HR specialists. This, in particular, is evidenced by the results of a study randomly selected 10 h analytical courses from 150 currently operating posted on one of the Internet portals [7]: HR analytics, HR Analytics Master Course with Excel, Python and R, The Most Comprehensive Course on Data Analytics and HR Analytics developed in 2016–2020.

Based on the use of the content of the headings "What you will learn", it proved possible to obtain information about the HR analytical competencies formed in students. Thus, an array of data was formed and a frequency SEO analysis was carried out, which enabled to identify priority areas of training (see Fig. 1).

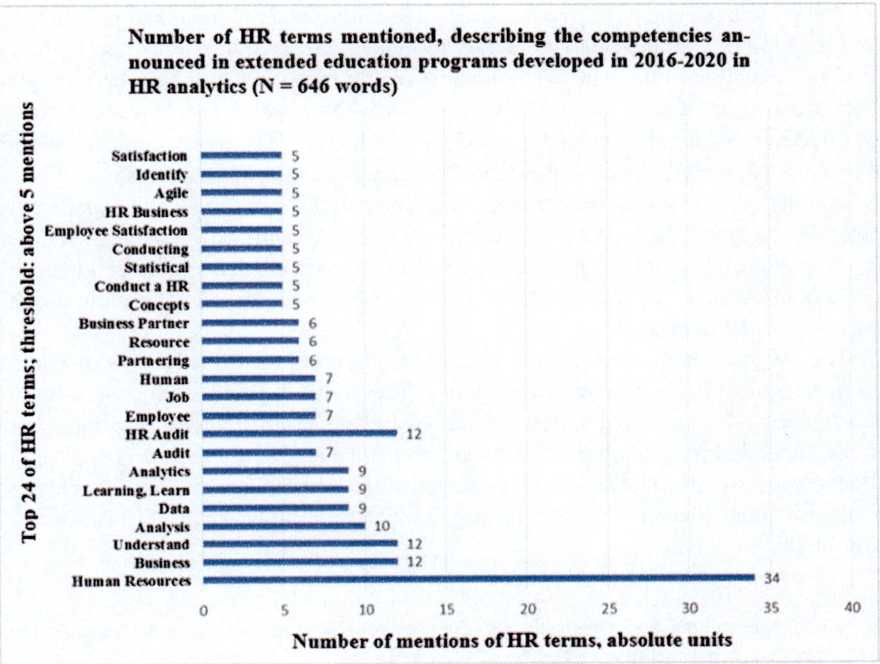

Fig. 1. Results of SEO-analysis of HR analytical competencies formed in students, reflected in 10 existing HR analytical courses, developed in 2016–2020. *Source:* Developed by the authors.

An honorable place in the data array, which is quite logical, was taken by the term "business", which indicates that the economic direction of HR analytics is especially in demand. This is partly caused by the fact that it provides an assessment of the key (and, importantly, understandable for commercial organizations) economic effects of personnel management, the activities of HR departments and particular HR specialists.

Such topical areas of personnel management as: selection and adaptation of personnel; assessment and certification; development and loyalty; talent and innovation management were not reflected in the studied data array. Sociological and psychological directions of HR analytics were also underrepresented.

Nevertheless, the tendency to consider the use of sociological and psychological tools in assessing the effectiveness of personnel management as full-fledged areas of HR analytics is becoming more and more popular. This is not surprising: sociological and psychological tools have been used for these purposes for more than a dozen years [8–13].

Sociological tools (first of all, questionnaires and interviews) are widely used in assessing the effectiveness of certain areas of personnel management – HR branding, recruitment, motivation, assessment and training of personnel, corporate culture management. A range of indicators has been formed, measured mainly using sociological instruments. These include the attractiveness of an employer in the labor market, loyalty, involvement, commitment of staff, staff satisfaction with jobs and remuneration system in the company, impact of staff training on the results of their performance, etc.

The growing demand for psychological research in modern organizations has led to the formation of an independent direction of modern HR analytics – the psychological vector of personnel research. In the context of the transition to a new technological order, such a task of psychological HR analytics as the search for resources to increase the effectiveness of personnel activities comes to the fore. The psychological direction of modern HR analytics uses well-established, recognized, somewhat classic methods of psychology: psychological observation; biographical method; survey and experimental methods; diagnostic techniques; documents analysis. Meantime, the range of requests by the current HR practices and HR specialists has expanded. Variable and available methods of quantitative and qualitative content analysis can become an indispensable assistant for HR managers.

The range of application of digital tools in assessing the effectiveness of personnel management is already quite wide today. The digital capabilities for assessing the effectiveness of personnel management can include: data-driven approach; application of analytical digital technologies for processing and analysis of big data [14]; artificial intelligence [15]; machine and deep learning; use of sophisticated methods for analyzing the data obtained (data mining, mind maps, decision tree); cloud services and some other innovative digital tools.

Table 1. Opportunities and prospects for the use of HR analytics in order to assess the effectiveness of personnel management.

	Digital capacities	Current and prospective areas of assessment	Types of assessing the effectiveness of personnel management
Current HR analytics	• Digital technologies • Digital platforms and add-ons • Digital analytical methods of data processing • Digital tools for assessing effectiveness • Digital formats for presenting HR analytics results	Actual current directions for assessing the effectiveness of personnel management:	Assessment of economic effectiveness
			Assessment of social effectiveness
			Assessment of psychological effectiveness
			Assessment of intra-organizational effectiveness
		Actual promising directions for assessing the effectiveness of personnel management:	Assessment of information technology effectiveness
			Assessment of management content effectiveness
			Assessment of digital HR brand effectiveness

Source: Compiled by the authors.

The digital technologies for HR analytics are applied via special digital analytical platforms (Business Intelligence Platforms), among which the most in-demand ones are Power BI, Oracle BI, Statistica, Tableu, which allow presenting analytics results in a compact, dense, highly informative dashboard format, which clearly demonstrates the investigated HR metrics and key performance indicators of personnel management and visualizes trends and prospects for the dynamics of personnel management effectiveness.

Currently, digital capabilities for assessing the effectiveness of personnel management are widely used in the economic direction of HR analytics. In the mainstream of the sociological and psychological directions of HR analytics, digital platforms and digital analytical methods of data processing are primarily used. Meantime, as can be seen from Table 1, the potential for using digital capabilities in order to assess the effectiveness of personnel management is by no means exhausted. Promising areas are the development of a matrix and system tools for HR analytics, the identification of scientifically based principles and criteria for HR analytic research.

4 Discussion

The use of HR analytics tools is aimed at assessing the economic, social and psychological effectiveness of personnel management. The success of their application in practice is largely determined by the use of digital capabilities, as acknowledged by the results of the discussion of the potential and prospects for the development of modern HR analytics with personnel managers of Russian companies – students of the additional education program "Practical tools for building an effective HR analytics system", implemented at the School of Public Administration of the Lomonosov Moscow State University since November 2020. From the point of view of HR specialists, the use of digital tools to assess two areas of personnel management – personnel motivation and corporate culture management seems to be a priority. At the same time, it is extremely attractive to independently develop research tools and conduct an assessment based on specialized Internet platforms with a built-in data processing function and graphical display of the results obtained in the form of graphs and diagrams, enabling one to initiate and assess the effectiveness of personnel management promptly, in a short time, using a unique toolkit that is focused on specific goals and relevant context of the organization.

5 Conclusion

Modern HR analytics is a set of tools that provides a multi-method systematic assessment of the effectiveness of personnel management. The leading areas of HR analytics are economic, sociological and psychological. The digital capabilities of HR analytics lie in the use of digital technologies, platforms, add-ons, digital analytical methods of data processing and other digital tools to assess economic, social, psychological and intra-organizational effectiveness. A promising direction for the development of HR analytics is a significant expansion of the range of digital tools and areas for assessing the effectiveness of personnel management.

References

1. Leonardi, P., Contractor, N.: Better people analytics. Harvard Bus. Rev. 70–81 (2018). https://hbr.org/2018/11/better-people-analytics
2. Deloitte Global Human Capital Trends. https://www2.deloitte.com/us/en/pages/human-capital/articles/introduction-human-cap-ital-trends.html
3. HR Future 2019. KMPG Report. https://assets.kpmg/content/dam/kpmg/ru/pdf/2019/04/ru-ru-future-of-hr-2019.pdf
4. Ivantsevich, J.M., Lobanov, A.A.: Human Resources for Management. Delo, Moscow (1993)
5. Fernandez, V., Gallardo-Gallardo, E.: Tackling the HR digitalization challenge: key factors and barriers to HR analytics adoption. Compet. Rev. An Int. Bu. J. (2020). https://doi.org/10.1108/CR-12-2019-0163
6. Zeidan, S.: HR analytics and organisational effectiveness. Int. J. Emerg. Technol. **11**(2), 683–688 (2020)
7. Udemy.com. https://www.udemy.com/course/hr-analytics-course/
8. Shcherbina, V.V.: Factory sociology and management consulting in the Soviet and post-Soviet periods. Sociol. Res. **6**, 115–124 (2008)
9. Popova, N.V.: Sociology at an enterprise in the urals: notes by a factory sociologist. Sociol. Stud. **7**, 60–66 (2017). https://doi.org/10.7868/S0132162517070078
10. Kachainova, N.B., Popova, N.V.: Factory sociology: origins and prospects. Bull. Tyumen State Univ. Socio-Econ. Legal Res. **2**(3), 29–36 (2016). https://doi.org/10.21684/2411-7897-2016-2-3-29-38
11. Subochev, N.S., Lenskaya, I.Yu.: Sociological methods in the study of personnel management processes. Sci. Bull. Volgograd Branch RANEPA. Ser. Polit. Sci. Sociol. **1**, 40–46 (2014)
12. Tikhonov, A.I.: Application of HR analytics tools in Russian companies. Moscow Econ. J. **1**, 540–546 (2020)
13. Kasparova, A.: Organization of effective interaction with personnel: the experience of JSC "Russian Railways." Hum. Cap. Prof. Educ. **1**, 55–58 (2016)
14. Agarwal, R., Dhar, V.: Big data, data science, and analytics: the opportunity and challenge for IS research. Inf. Syst. Res. **25**(3), 443–448 (2014). https://doi.org/10.1287/isre.2014.0546
15. Davenport, T.: Is HR the most analytics-driven function? Harvard Bus. Rev. (2019). https://hbr.org/2019/04/is-hr-the-most-analytics-driven-function

Development of Methodology for Identifying the Target Clients of Medical Tourism Using Internet Technologies

Irina Yu. Stolyarova, Dmitry Yu. Stolyarov, Marina V. Rossinskaya(✉), Evgenia V. Gordeeva, and Larisa A. Kamenskaya

Sochi State University, Sochi, Russia

Abstract. Medical tourism is developing dynamically as a new direction formed at the intersection of medicine and tourism. The country has the necessary prerequisites for development, especially in the segment of medical resort. Sanatorium-and-health-resort services are in demand. The number of tourists who choose medical treatment and rehabilitation as the main purpose of their trip is growing. The rehabilitation program for those who have had COVID-19 became especially relevant. The majority of sanatoriums have developed a special treatment program for such clients, which is in high demand and is successfully implemented through tour operators and travel agencies as a special type of tourism. Little attention is paid to the marketing of medical tourism and consumer segmentation. The analysis of this area and the identification of the target tourism segment are complicated by the lack of statistics and methods for assessing the target clients of medical tourism. The study aims to develop a methodology for identifying the target clients of medical tourism by evaluating Internet statistics data on the medicine and tourism segments and identifying the corresponding average characteristics of medical tourist. To identify the characteristics inherent in the target clients of medical tourism, it is proposed to analyze characteristics of the medical and tourist website visitors. Then to obtain average characteristics of both groups for users, which may be interested in medical tourism. As a result of the study, a new methodology for assessing the target clients of medical tourism was obtained based on the analysis of website traffic.

Keywords: Medicine · Tourism · Medical tourism · Medical tourist · Internet statistics · Internet technologies · Target clients · Evaluation method

1 Introduction

Medical tourism is a modern direction of the Russian tourism market [1, 2]. The country has the necessary prerequisites for its development, especially in the segment of medical resort (sanatorium-and-health-resort) tourism [3].

The sphere of medical resort tourism is one of the dynamic segments of the tourism market in the South of Russia [4–6]. Sanatorium-and-health-resort services are in

P. V. Trifonov and M. V. Charaeva (Eds.): DITEM 2021, LNNS 380, pp. 429–441, 2022.
https://doi.org/10.1007/978-3-030-94245-8_59

demand. The number of tourists who choose medical treatment and rehabilitation as the main purpose of their trip is growing.

There is a large number of publications on the development of medical tourism, analysis of development prospects in the Russian Federation and in the South of Russia, in particular; management issues and organizational aspects of the medical tourism development [7–9]. Little attention is paid to the marketing aspects of medical tourism development and consumer segmentation. The research works talk about the segmentation of competitors [10], the segmentation of the international medical service market [11]. Using the traditional method, the works demonstrate segmentation of medical service consumers carried out on the example of a number of health care institutions and the strategic grouping of competitors in the market of sanatorium-and-health-resort services [3, 7–9, 12], the criteria of segmentation [13] and its features in tourism [14], the segmented approach to the selection of target markets [15]. The possibilities of the Internet are considered from the perspective of promoting a tourism product [16–19], Internet advertising [20–23]. Conclusions are drawn about the ability of Internet marketing to reach a large percentage of potential customers by expanding the target audience [24]. The segmentation of medical tourism consumers and the use of Internet technologies in this area remain unaffected. The development of methodology for identifying the target clients of medical tourism using Internet technologies is actualized.

The study aims to develop a methodology for identifying the target clients of medical tourism by evaluating Internet statistics data on the medicine and tourism segments and identifying the corresponding average characteristics of a medical tourist.
The main study objectives:

- Study of theoretical basis.
- Analysis of website traffic regarding the Medicine and Tourism fields followed by the allocation of key segments according to the selected criteria.
- Identification of the average medical tourist characteristics based on comparative analysis.

2 Methods

Medical tourism is at the intersection of medicine and tourism. To identify the characteristics inherent in the target clients of medical tourism, the authors propose to analyze individual characteristics of the medical and tourist website visitors. The next step is to collect the average characteristics of both groups that are inherent in the Internet users that may be interested in medical tourism services.

The study uses methods of statistical analysis, graphical interpretation of data, and content analysis. The results of the website traffic analysis by the following characteristics are given:

1. Attendance.
2. Regions.
3. Countries.
4. Gender and age characteristics.
5. Devices.
6. Sources of traffic.

3 Results

The analysis of the medical website visits statistics collected from the Liveinternet system is presented. The segment of Medicine according to the selected characteristics.

The average attendance for 3 months – 68,604,347 people per month; the average number of views – 5.7; the average duration – 1.9 min.

The traffic sources were analyzed by the number of clicks from search engines and social networks.

Figure 1a contains the number of referrals from search engines. Figure 1b shows similar characteristics of referrals from social networks.

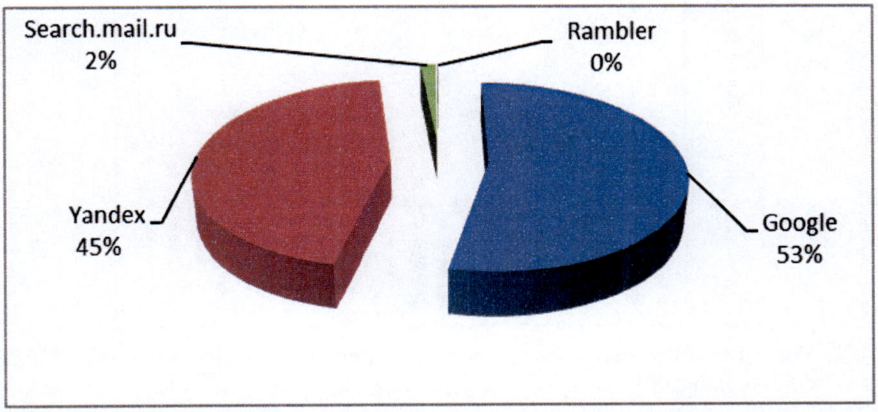

a

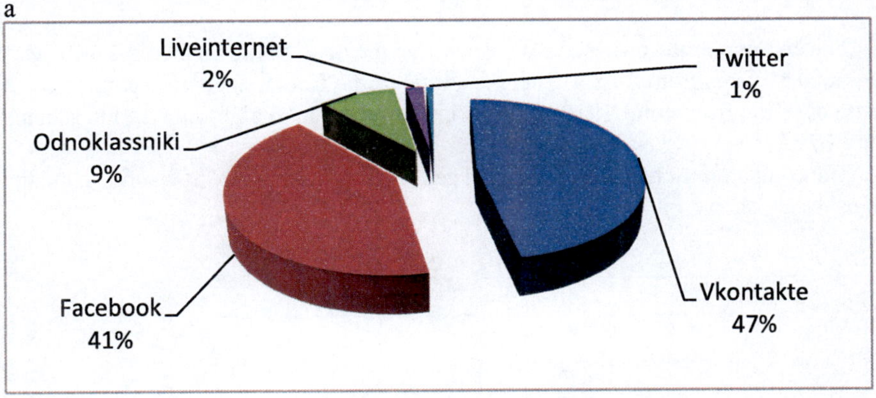

b

Fig. 1. Referrals from search engines regarding the Medicine group (a); Referrals from social networks regarding the Medicine group (b). *Source:* compiled by the authors

The main traffic on the medical websites is generated by search engines. There are about a hundred times fewer referrals from social networks.

Among social networks, the greatest activity is noted among the Vkontakte and Facebook users. Comparative characteristics of visitors to medical websites in the regional context are shown in Fig. 2.

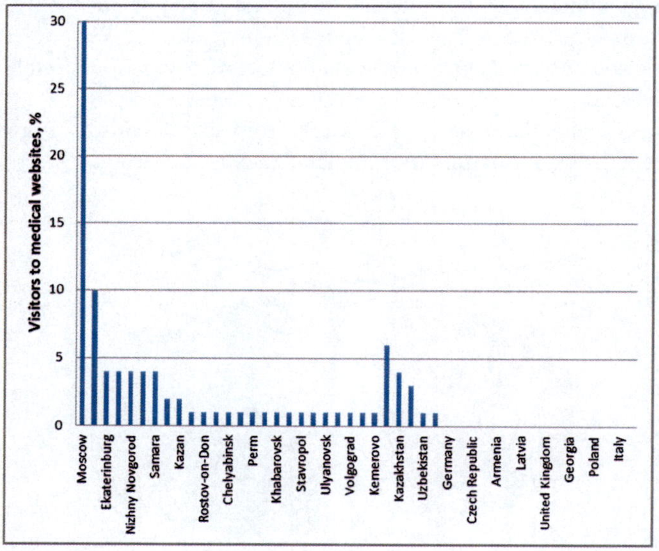

Fig. 2. The structure of visitors to medical websites in Runet by the region of residence. *Source:* compiled by the authors

The analysis of the obtained data shows that the main traffic on medical websites is provided by users from Russia. There is a little traffic from neighboring countries. The share of traffic from non-CIS countries is insignificant. In total, foreign traffic accounts for 20.7%.

The comparative characteristics of the gender and age structure of visitors to medical websites are shown in Fig. 3.

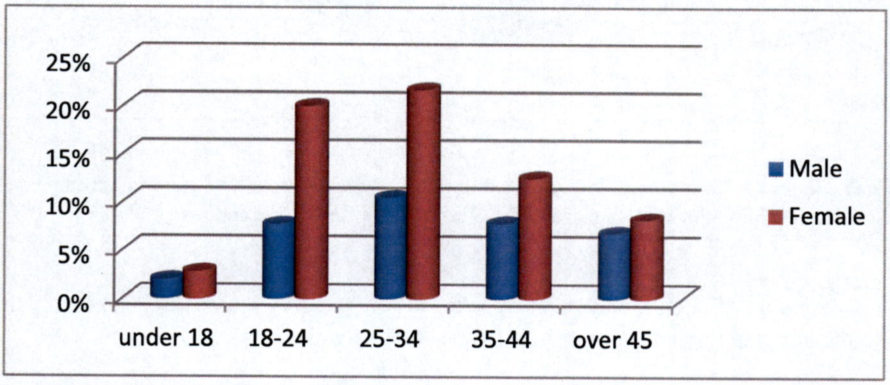

Fig. 3. Age and gender structure of visitors to medical websites. *Source:* compiled by the authors

The analysis of website traffic statistics showed that in the total share on traffic to medical websites, there are 1.87 times more women than men.

In order to identify the devices, from which users browse medical websites, statistical data has been collected regarding the operating system used by visitors and the type of device it belongs to. The structure of attendance by devices used is shown in Fig. 4.

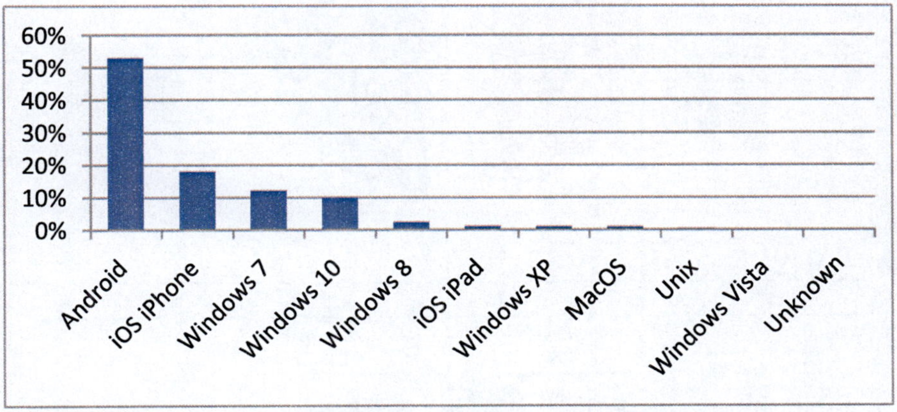

Fig. 4. Operating systems of medical website users. *Source:* compiled by the authors

Thus, 72.3% of the medical website visitors use mobile devices, and only 27.4% use computers and laptops.

The second stage of the analysis is the analysis of traffic statistics for tourism websites according to the same characteristics.

The average attendance for 3 months is 50,867,409 people per month. The average number of views is 6.8. The average duration is 2.77 min.

Figure 5a shows the characteristics of referrals from search engines. Figure 5b shows similar characteristics of referrals from social networks.

The main traffic on the travel websites is generated by search engines. There are approximately 30 times fewer referrals from social networks. A significant portion of the traffic comes from the Google and Yandex search engines.

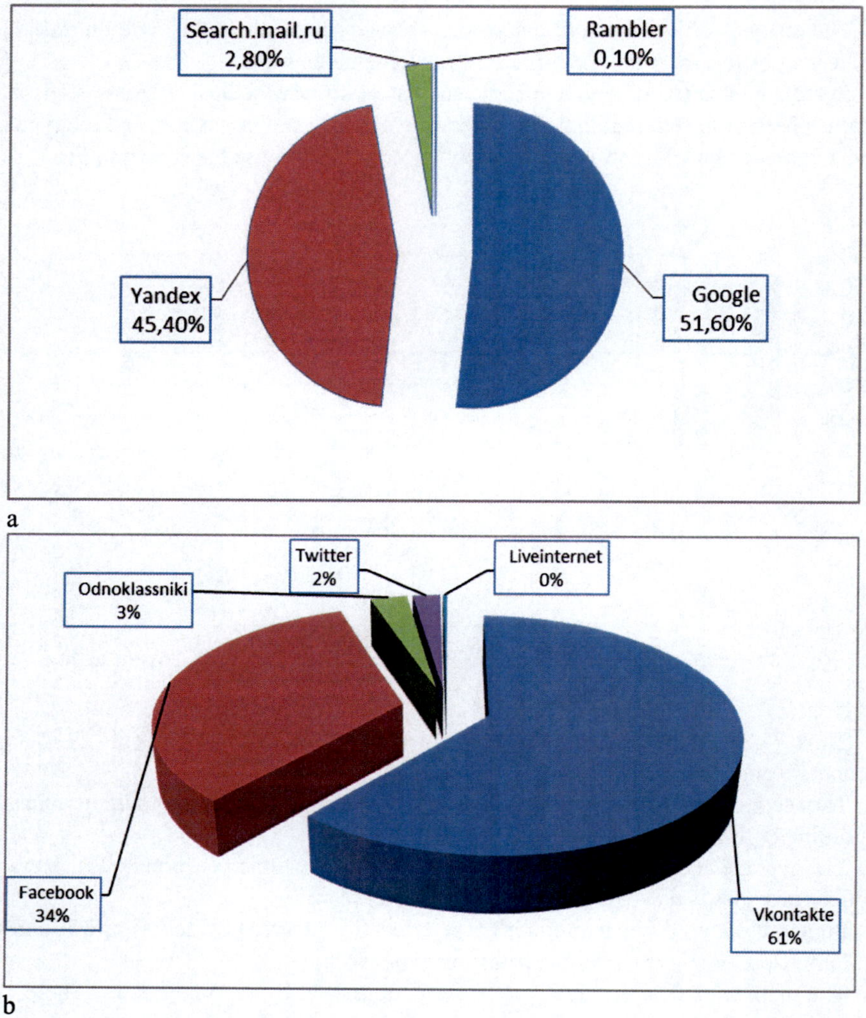

Fig. 5. Referrals from search engines regarding the Travelling group (a); Referrals from social networks regarding the Travelling group (b). *Source:* compiled by the authors

Among social networks, the greatest activity is noted among the Vkontakte and Facebook users. Comparative characteristics of visitors to tourism websites in the regional context are shown in Fig. 6.

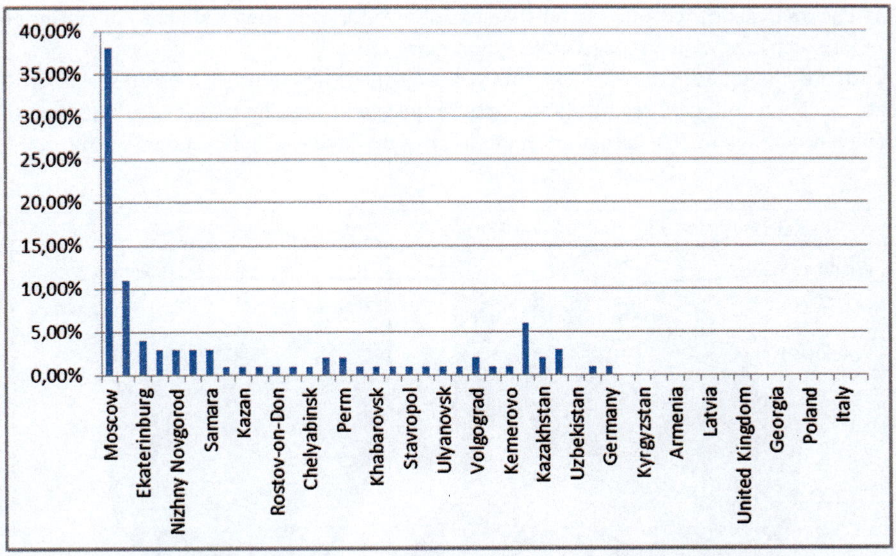

Fig. 6. The structure of visitors to tourism websites in Runet according to the residence region. *Source:* compiled by the authors

The main traffic on the tourism websites is provided by users from Russia. There is a little traffic from neighboring countries. The share of traffic from non-CIS countries is insignificant. In total, foreign traffic accounts for 19.6%.

The comparative characteristics of the gender and age structure of visitors to tourism websites are shown in Fig. 7.

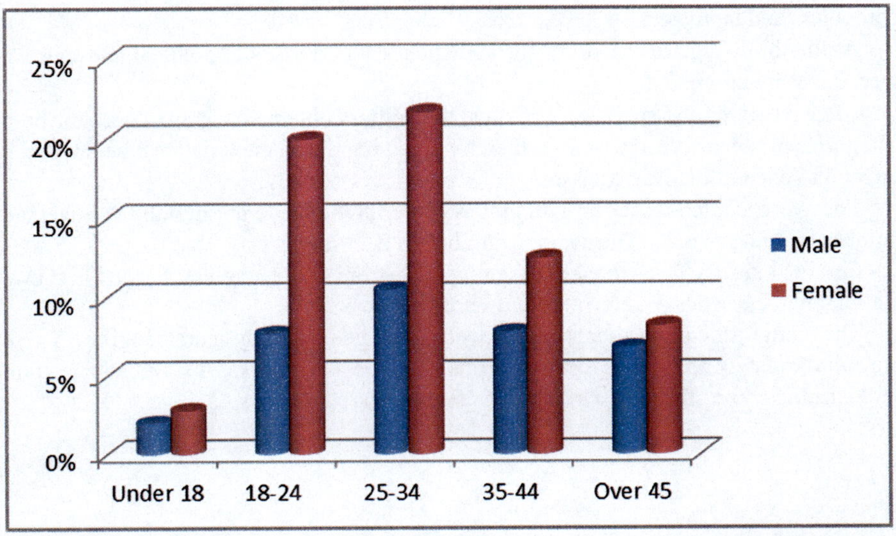

Fig. 7. Age and gender structure of visitors to tourism websites. *Source:* compiled by the authors

The analysis of website traffic statistics showed that in the total share of traffic on tourism websites women make 48.9%, and men – 51.1%.

In order to identify the devices, from which users browse tourism websites, statistical data has been collected regarding the operating system used by visitors and the type of device it belongs to. The structure of attendance by devices used is shown in Fig. 8.

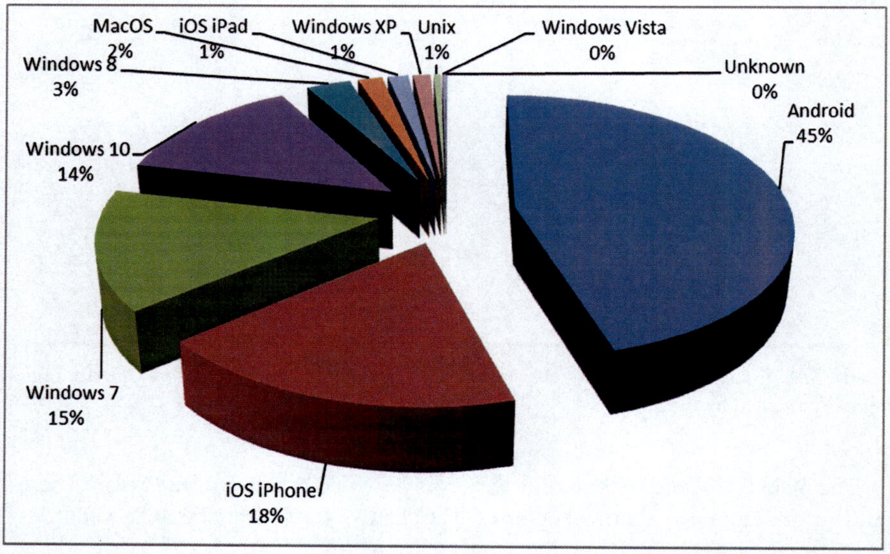

Fig. 8. Operating systems of tourism website users. *Source:* compiled by the authors

65.2% of travel websites are visited from mobile devices and only 34.7% – from computers and laptops.

At the third stage of the study, the statistical data on visits to medical and tourism websites were compared.

After receiving the specified statistics, the authors obtained averaged data for both categories of the analyzed websites, thereby obtaining the characteristics inherent in the target audience of medical tourism.

The average attendance for 3 months is 59,735,878 people per month. The average number of views is 6.25. The average duration is 2.34 min.

Figure 9a shows the characteristics of referrals from search engines. Figure 9b shows similar characteristics of referrals from social networks.

The main traffic on medical tourism websites is generated by search engines. There are approximately 55.9 times fewer referrals from social networks. A significant portion of the traffic comes from the Google and Yandex search engines.

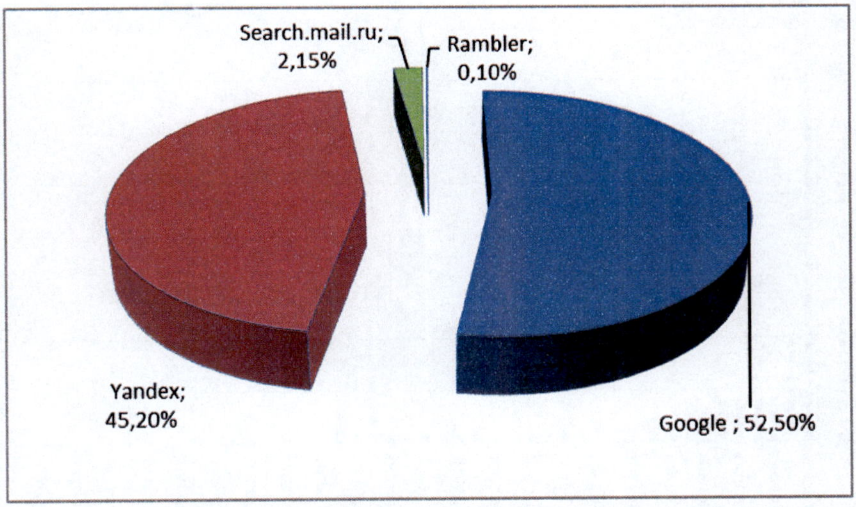

a

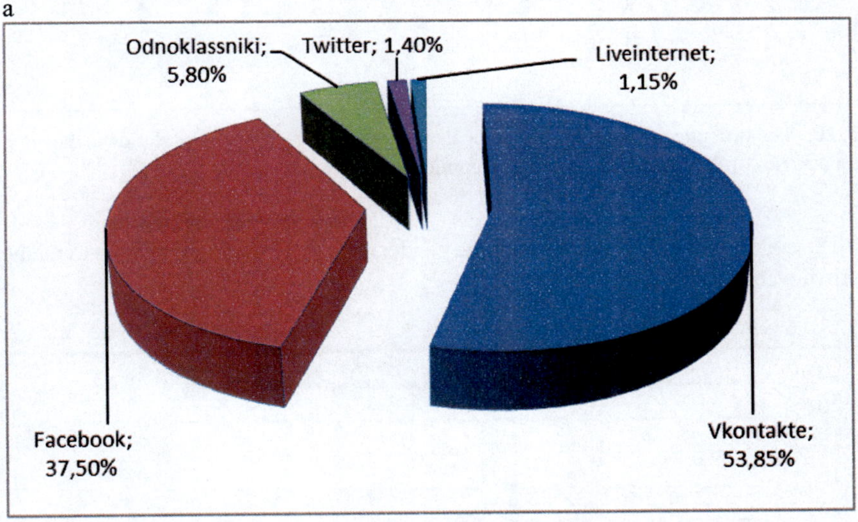

b

Fig. 9. The average number of referrals from search engines in the Medical Tourism group (a); Referrals from social networks in the Medical Tourism group (b). *Source:* compiled by the authors

Among social networks, the greatest activity is noted among the Vkontakte and Facebook users. Comparative characteristics of the website visitors interested in medical tourism services in relation to the regions and countries of their residence are shown in Fig. 10.

The main traffic on the medical tourism websites is provided by users from Russia. There is a little traffic from neighboring countries. The share of traffic from non-CIS countries is insignificant. In total, foreign traffic accounts for 20.15%.

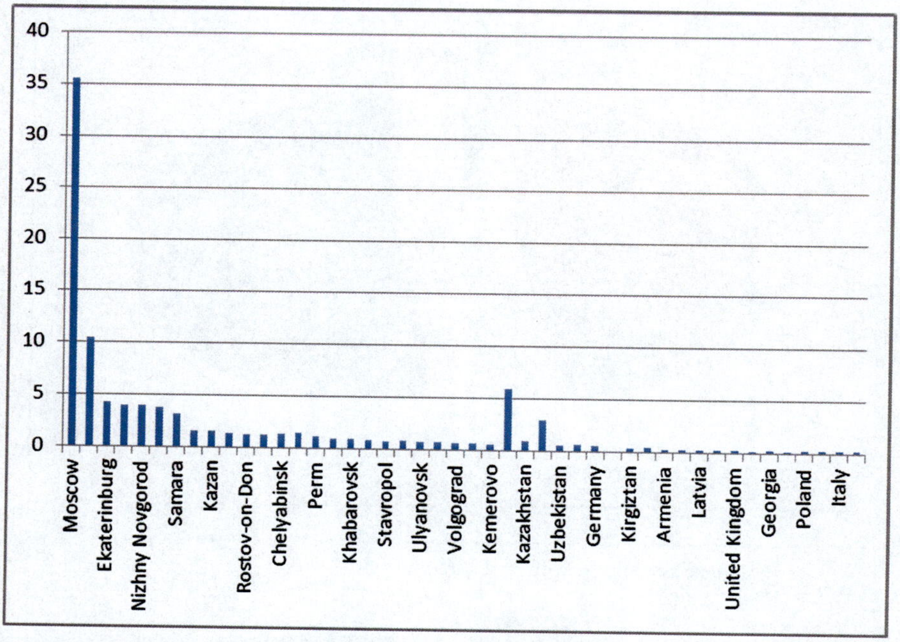

Fig. 10. The structure of visitors to the medical tourism websites in Runet in relation to the regions and countries of their residence. *Source:* compiled by the authors

The comparative characteristics of the gender and age structure of visitors to medical tourism websites are shown in Fig. 11.

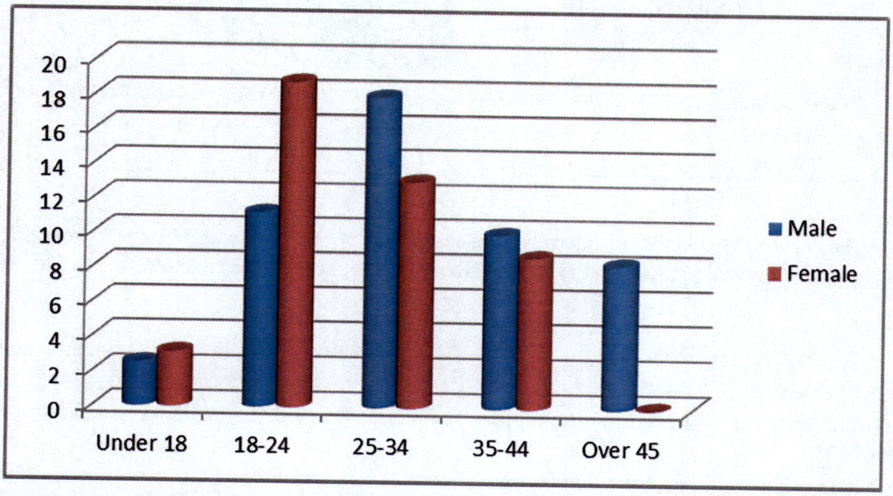

Fig. 11. Age and gender structure of visitors to medical tourism websites. *Source:* compiled by the authors

The analysis of website traffic statistics showed that in the total share of traffic on medical tourism websites women make 50.25%, and men – 49.75%.

In order to identify the devices, from which users browse medical tourism websites, statistical data has been collected regarding the operating system used by visitors and the type of device it belongs to. The structure of attendance by devices used is shown in Fig. 12.

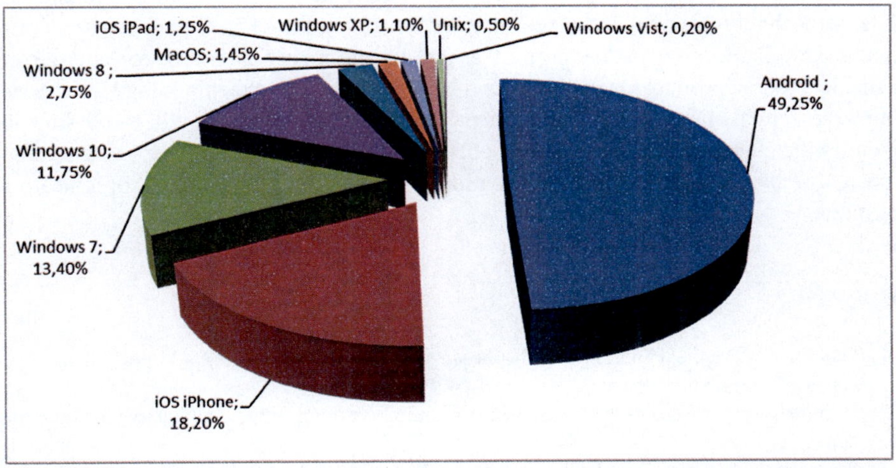

Fig. 12. Operating systems of medical tourism website users. *Source:* compiled by the authors

68.75% of travel websites are visited from mobile devices and only 31.25% – from computers and laptops.

4 Discussion

Over the past five years, there has been a discussion on the terminology and directions for the development of medical tourism in general and in the context of individual countries and regions, management, and organizational aspects of the development of medical tourism. The segmentation of competitors, the international market of medical services, and the strategic grouping of competitors in the market of sanatorium-and-health-resort services are touched upon. Little attention is paid to the marketing aspects of medical tourism and the analysis of the Internet possibilities. The segmentation of medical tourism service consumers and the development of methodology for identifying the target clients of medical tourism through Internet technologies are not touched upon.

This study serves to further develop research on medical tourism and is aimed at developing a methodology for identifying the target clients of medical tourism by assessing Internet statistics on the segments of medicine and tourism and identifying the corresponding average characteristics of a medical tourist. In the future, the research results can be used to promote medical tourism services for different target requests.

5 Conclusions

Based on the analysis the following main statements can be made.

The current state statistics does not allow obtaining real data on the characteristics of the target clients of medical tourism (social and demographic characteristics, regions and countries of residence of potential medical tourists, search engines used by potential medical tourists, etc.) and the further use of this data for the promotion and implementation of medical tourism services. A way out in this situation can be the use of various Internet technologies aimed at identifying and assessing the desired characteristics of a medical tourist, forming his/her portrait, as well as optimizing the promotion and sale of a medical tourist product via the Internet. This study proposes a methodology for assessing the potential clients of medical tourism, with the help of which it will be possible to identify the characteristics of the target clients based on traffic statistics for tourism and medical websites, and, if necessary, to build predictive values for a number of indicators of interest.

References

1. Galkin, V.V.: Medical Tourism as a Business Area (2011). http://vadim-galkin.ru/articles/medical-business/medical-tourism/
2. Patients Beyond Borders: Medical Tourism and Consulting. https://patientsbeyondborders.com/
3. Saveleva, N.A., Toporov, S.I.: Sochi. J. Econ. **13**(2), 175–190 (2019)
4. Vasilieva, S.O., Seleskerova, E.F.: Bul. Med. Internet Conf **7**(1), 187–188 (2017)
5. Medical Tourism Opportunities in Chinese Healthcare Market. http://www.medicaltourismmag.com/medical-tourism-opportunities
6. The Concept for the Development of Sanatorium, Health Resort and Tourist Complex of the Krasnodar Territory until 2030. https://min.kurortkuban.ru
7. Savelyeva, N.A., Kovalenkova, O.G., ZhYa, K.: Sci. Edu: Manag. Econ. Entrep; Law Manag. **11**(114), 21–26 (2019)
8. Savelyeva, N.A., Kovalenkova, O.G., Shafieva, Y.: Sci. Edu. Manag. Econ. Entrep. Law Gov. **10**(113), 32–38 (2019)
9. Savelyeva, N.A., Shmeleva, T.V., Toporov, S.I.: Sci. Edu. Manag. Econ. Entrep. Law Gov. **7**(110), 15–19 (2019)
10. Savelyeva, N.A., Shmeleva, T.V., Losevich, D.V.: Sci. Edu. Manag. Econ. Entrep. Law Gov. **10**(113), 39–44 (2019)
11. Shmeleva, T.V., Toporov, S.I., Savelyeva, N.A.: Sci. Edu. Manag. Econ. Entrep. Law Gov. **8**(111), 39–45 (2019)
12. Kareva, O.S.: Market segmentation of medical service consumers. In: III (I National) All-Russian Scientific and Practical Conference on Management in the Healthcare: Challenges and Risks in the 21st century. Volgograd State Medical University Publishing House, Volgograd (2018)
13. Khatskelevich, A.N., Yasyreva, A.A., Antineskul, E.A., Lekomtseva, A.A.: Marketing. Perm State National Research Institute, Perm (2019)
14. Penkina, N.V.: Tourism and Recreational Complex: Theoretical and Methodological Approaches to the Analysis. Nizhnevartovsk State University Publishing House, Nizhnevartovsk (2017)

15. Goncharova, I.V., Rozanova, T.P., Morozov, M.A., Morozova, N.S.: Tourism Marketing. Federal Agency for Tourism, Moscow (2014)
16. Dunets, A.N.: Design and Promotion of a Regional Tourism Product. AltSTU Publishing House, Barnaul (2014)
17. Konovalenko, N.V., Pigina, T.N.: Promotion of a hotel enterprise through internet resources (on the Sayanogorsk hotel example). In: Second International Scientific Student Conference on Strategy for the Development of the Hospitality and Tourism Industry. State University – UNPK, Oryol (2015)
18. Moisenko, I.V., Usova, M.V.: Azimuth. Sci. Res. Econ. Manag. **8**(3(28)), 261–265 (2019)
19. Stolyarov, D.Yu.: Formation of a System for the Promotion and Sale of Tourism Products through the Internet. Abstract of PhD Thesis in Economic Sciences. Optima Printing House, Sochi (2011)
20. Bayramukova, V.Z.: Digital marketing as a tool to improve the competitiveness of a modern enterprise. In: International Scientific and Practical Conference on Scientific Potential of Higher Education – the Future of Russia. DSTU, Rostov-on-Don (2020)
21. Arakelova, I.V., Bramnik, I.V.: New aspects of medical service marketing. In: IV All-Russian Scientific and Practical Conference on Management in Health Care: Challenges and Risks of the XXI Century. Volgograd State Medical University Publishing House, Volgograd (2019)
22. Gubanova, A.Yu.: Internet for Children: Social Functions, Audience Specifics, Content Requirements. Ph.D. Thesis in Sociological Sciences. Moscow (2016)
23. Mukharsky, T.A., Anikina, Zh.P.: Innovative tools for marketing analysis of the market. In: International Scientific and Practical Conference on Actual Problems of Marketing of the 21st Century. Institute of Business, BSU, Minsk (2019)
24. Korzina, M.I., Trapeznikov, A.D.: Internet marketing technologies in promoting small businesses in the confectionery sector. In: International Scientific and Practical Conference of Young Scientists, Ph.D. Candidates and Students on Actual Problems of Modern Science: The View of Young Scientists; Panel Discussion on Axiological Crisis of Human Existence: Humanism – Transhumanism – Post-transhumanism. Chechen State University Publishing House, Grozny (2020)

Employee Motivation and Engagement Projects as an Effective HR Management Tool

Alena K. Rybakova[1,2](✉) and Anna V. Lukyanova[2]

[1] METRO Cash and Carry, Moscow, Russia
[2] Financial University under the Government of the Russian Federation, Moscow, Russia

Abstract. The main prerequisites for the research are the need to search for optimal project implementation concepts in order to assess involvement in large retail companies. The research is aimed at finding effective tools for the project implementation to assess motivation and engagement by the example of retail companies. The research methods were comparative analysis, assessment, in-depth interviews and a survey of the retailer's HR processes. The research results consist in identifying tools for assessing engagement in large retailers; in addition to conducting surveys, it is proposed to conduct in-depth interviews and work with line managers – HR managers – to create a microclimate in teams of personnel of different levels. The research confirms that the higher the frequency of surveys and those activities, the higher the representativeness of the sample and the results obtained reflect more accurately the company value for shareholders in terms of non-financial indicators, such as the percentage of engagement and the loyalty index. Some technical issues concerning the project implementation were also considered and process digitalisation for assessing involvement was proposed on the basis of a single platform solution and line manager training.

Keywords: Project management · Personnel engagement · Engagement research · Personnel management

1 Introduction

The topic of personnel motivation and engagement management is a cornerstone in the effective human capital management [1]. Scientists, both theorists [2–5] and practitioners [6, 7], have been studying the issue of employee motivation for more than a decade, which has been especially relevant for the last 20 years, since the labour market model has changed; and employers enter into fierce competition among themselves for not only the best minds but also mass personnel recruiting [8].

A large number of tools have been proposed to study projects for managing employee motivation and engagement [9, 10]. One of the most highly competitive markets in the labour market is the trade and retail market. It is the companies of the retail group that meet the challenge in how to motivate hundreds of managers and thousands of cashiers, salesroom employees and produce stockers as part of a single business [11]. The number of staff, the heterogeneity of the assessment and the scale of engagement

P. V. Trifonov and M. V. Charaeva (Eds.): DITEM 2021, LNNS 380, pp. 442–447, 2022.
https://doi.org/10.1007/978-3-030-94245-8_60

projects lead to the search for optimal approaches to assessing engagement in large companies, which was a prerequisite for researching engagement projects in the retail sector. The international METRO Cash & Carry company was selected for the analysis.

The main hypotheses of the research are as follows: non-financial indicators of growth in the value of a large retailer company (percentage of involvement and employee loyalty index) depend on the assessment, frequency and continuity of prompt response to changes in these indicators by developing activities at the microclimate level by line managers and HR managers, as well as on the level of digitalisation of assessment, which can be implemented in large companies in the form of a single large-scale HR project or a portfolio of projects.

The research aims to search for methodological tools for the implementation of projects to assess the motivation and involvement of employees in large retail companies. The key objectives of the research were analysis of the methodological approaches to the project implementation in order to assess personnel involvement in large retail companies; analysis of the possibilities of digitalisation of the personnel engagement assessment system and assessment of the impact of its level on the results; analysis of the influence of line managers and HR managers on the microclimate and the results of the retail company employee involvement including the company's non-financial indicators, percentage of involvement and loyalty index.

2 Methods

Projects for managing employee motivation and engagement are an important part of modern HR management, since there is a direct correlation between the level of employee engagement and the business performance of companies [11].

Motivation management projects are developed taking into account the types of motivation the key ones being tangible and intangible incentives.

The overseas project management experience based on staff motivation and involvement offers the market a number of project models [12–14].

The main difference between the Russian experience in project management in terms of motivation and employee engagement from the foreign one is the focus on tangible incentives, as noted in a number of studies [15, 16].

In this research, employee motivation and engagement projects were considered by the example of retail companies. In Russia, these projects can be characterised by complexities of the implementation, which is associated with a large difference in the motivation of the management of these companies and the staff, as well as a high level of competition in the labour market for the latter HR category [11].

In our today's retail company practices, the main project for motivating and involving employees is payroll system that combines wage and bonus parts, as well as a percentage of profits [17].

However, there are other statistics, for example, a survey of the motivation and engagement of retail industry employees was carried out by HAYS in 2018: the survey lasted two months, and 486 employer companies and 3,114 normal staff of the retail industry took part in it. The survey showed that 94% of respondents chose a decent salary as a motivation factor, while 71% are looking for professional development opportunities in the company [18].

The personnel motivation and involvement projects of METRO Cash & Carry were selected as the target of this research. In Russia, the company is represented by 93 shopping centres in 51 regions, 3 office entities including the Moscow headquarters, as well as the logistics platform MGL (METRO Group Logistics) in Noginsk. The number of employees is about 12,000.

The company identifies the following employee target groups for assessment:

- employees of the headquarters;
- employees of shopping centres – normal staff;
- managers of shopping centres;
- youth with a high potential;
- Sales Line Force.

Based on the heterogeneity of the audience, the company implemented several projects at once to manage motivation and engagement by units at different times. This approach had a significant drawback, the results of comparing the employee satisfaction levels of different personnel groups for different projects were not representative, nor could be compared.

Since 2012, the company has been implementing an employee engagement research project, in which employees of all levels take part simultaneously four times a year (the quarterly format has been introduced since 2018), the results of which make it possible:

- to identify common trends in the mood of employees;
- to measure the level of staff involvement at all levels;
- to measure the employee to employer loyalty index (eNPS);
- to measure the extent to which employees share the behavioural indicators of the company's core values;
- to receive answers to relevant open questions;
- to receive responses in terms of potential improvements in employee-oriented processes;
- to handle results and implement improvements in existing processes.

In addition, for nine years of surveying in all 26 countries of the company's presence, it was found that there had been a statistically proven correlation of 81% between the percentage of engagement and the employee loyalty index (hereinafter eNPS). In other words, the methodology assumes that with a high level of involvement, the level of employee loyalty to the company is also high. For several years in a row, both indicators in METRO Cash & Carry Russia were high, the deviation as a result in the downward percentage of engagement was no more than 5%. However, at the end of 2020, the survey results showed that the engagement rate was 78%, while the eNPS was only 30 points out of 100.

During the research and additional surveys in 2020, the company found that line managers had not worked with teams in the context of discussing results and developing plans for implementing improvements that could provide the basis for projects to motivate teams and individual employees.

Three hypotheses have been put forward as to why managers do not work with teams on research and survey findings:

- No understanding why the company is conducting employee engagement research and survey.
- No understanding why so often.
- No understanding of how to work with research and survey findings.

Based on HR research and in-depth interviews in the company, all the three hypotheses were confirmed. In addition, additional manager "resistance" zones were identified when processing the employee engagement survey findings, which, in turn, were caused by the lack of skills in working in the information system, inability to interpret the findings, lack of skills in providing feedback to direct subordinates and inability to work on the involvement assessment findings along with the team.

The work of line managers with HR managers resulted in a separate sub-project, the creation of an e-learning course that closed all "zones of resistance". The course was focused on two main areas, such as the development of skills to work with the platform and numerical results as well as the skills of managers to work with teams on research results. It should be noted that this approach of METRO Cash & Carry showed high efficiency: the NPS of the project was 9.8 out of 10 (the level of loyalty to learning), while the company's eNPS increased from 30 points to 68 out of 100.

However, it is important to take into account that such tools would not be highly effective without the company's having the technical facilities for implementation. From a technical point of view, it should be possible to pass surveys on any devices, including mobile ones, taking into account the possibility of identification and authentication; at the same time, the results should be impersonalised and of a non-attribution nature. In addition, it should be possible to obtain analytical summary findings for the operation of line managers with their employees and the work of HR managers with line managers.

After an in-depth analysis of the technical and methodological project implementation to increase staff motivation and engagement, some areas for improvement were identified. In 2018, METRO abandoned the annual survey format and the services of the third-party technical provider AON Hewitt in favour of a quarterly survey format and created its own platform for conducting research and handling results consisting of a survey page and three analytical modules: for setting up a survey, for setting module with results and the module itself for working with results.

3 Results

Based on the example of METRO Cash & Carry, the research results confirmed the hypothesis about the influence of assessing the involvement of employees of large companies on intangible indicators (percentage of employee involvement and employee loyalty index) including a direct relationship between in-depth interviews and the work of HR managers with line managers to create a microclimate in personnel teams of a different level.

The research also confirmed the hypothesis about the impact of the level of digitalisation of the processes of working with company employees in assessing engagement on the company's non-financial performance. Many information systems automate the processes and results of assessing the level of involvement. Research has shown that engagement assessment results in large retail companies require high frequency to correlate data with current performance. It is no longer relevant to assess engagement in a large retail company 1–2 times a year. A more frequent assessment is required, which in turn is impossible without the digitalisation of those processes.

4 Discussion

Engagement research is an ongoing process that is proposed to be implemented within the framework of individual projects or a portfolio of projects. Such initiatives are often viewed from a process perspective. At the same time, the effectiveness of project management in involvement in small retail companies remains a controversial issue.

Also controversial is the issue of assessing the intrinsic value of the proposed solutions for a company in relation to non-financial indicators.

5 Conclusion

The main work of HR managers with line managers should be aimed at developing skills in interpreting the results of assessing the level of employee involvement, feedback from them and the ability to use the opportunities for assessing involvement in current activities to improve the business performance of employees. In turn, these processes (multi-level assessment of personnel involvement, multi-level interviews, and work with line managers and assessment findings) can be implemented in large companies within the framework of one project or a portfolio of projects, which increases the effectiveness of those activities.

METRO Cash & Carry carried out a sequential integration of information systems for conducting surveys and preparing an employee database, as well as analytical information systems into a single platform for handling research findings, which greatly simplifies such project implementation, communication with project participants and creation of opportunities for in-depth and continuous analysis of the results.

References

1. Furnham, A.: Motivating your staff. In: The Engaging Manager, pp. 114–116. Palgrave Macmillan (2012). https://doi.org/10.1057/9781137273871_32
2. Rahman, M.S., Mannan, M., Hossain, Md.A.: Int. J. Edu. Manag. **32**(5), 761–782 (2018)
3. Welch, J., Byrne, J.A.: My years at GE. Mann, Ivanov and Ferber (2011)
4. Egorova, A.S.: Personnel Involvement: 7 Steps to Understanding. Infra, Moscow (2014)

5. Pulyaeva, V.N., Ivanova, I.A., Valishin, E.N.: Theory and Practice of Human Resource Management. KnoRus, Moscow (2018)
6. Human Factors. https://human-factors.ru/trendy/assessment-employee-engagement
7. State of the global work place: Employee engagements insights for business leaders worldwide. www.gallup.com/file/strategicconsulting/164735/State%20of%20the%20Global%20Workplace%20Report%202013.Pdf
8. Employee Engagement Overview Brochure. Gallup Consulting (2016)
9. Delaney, M.L., Royal, M.A.: Indust. Org. Psy. **10**(1), 127–140 (2017). https://doi.org/10.1017/iop.2017.2
10. Employee Engagement Management: Actual Practices and Tools. www.planetahr.ru
11. Labour market research: trends, articles (2021). https://www.kellyservices.ru/blog/tag/issliedovaniia-kelly
12. Krugliko, M.E., Guryeva, I.Y., Shatokhina, A.O.: Fund. App. Res. Mod. World **11**(2), 134–136 (2015)
13. Mamalimova, R.A., Rabtsevich, A.A.: Vectors Mod. Sci. Devel. **1**, 147–149 (2015)
14. Moiseeva, V.O.: Econ. Manag. Innov. Tech. **3**(30) (2014)
15. Pyankova, L.A.: Newsl. Sib. State Indust. Univ. **1**(31), 71 (2020)
16. Helving, J.: Employee Engagement in Russia: How to Build a Corporate Culture Based on Employee Engagement, Customer Focus and Innovation. Russia Advisory Group Oy, Helsinki (2012)
17. Lozovaya, I.V.: NovaInfo. Ru. **3**(53), 95–99 (2016)
18. Hays motivation & job satisfaction report. http://vneftegaze.ru/content/wp-content/uploads/2018/05/Hays-Motivation-Guide-2018.pdf

Digital Infrastructure of the Commercialization Process of Innovations in the Russian Economy

Andrey P. Savchenko(✉), Mikhail R. Zakaryan, Darya V. Lanskaya, Darya A. Detkina, and Dmitry V. Urmanov

Kuban State University, Krasnodar, Russia

Abstract. In the innovation policy of many countries, much attention is paid to the development of the digital infrastructure of innovation activity. Researchers see the reason for the low level of innovation in the Russian economy in the inefficiency of the innovation infrastructure. The purpose of the study is to assess the adequacy of the model of the national innovation infrastructure to the goals of innovation development and the specifics of innovation activity. The research used the systemic method, the method of empirical observation, logical analysis, and content analysis. The conducted analysis of the state of elements of the Russian national digital infrastructure enabled to draw the following conclusions: (1) the existing platforms solve only part of the tasks at the stages of R&D, at the stage of commercialization of the R&D results, the infrastructure is ineffective; (2) work on the design and implementation of the digital infrastructure for innovation development is haphazard, fragmented; (3) the main reason for the problems is the discrepancy between the principles of centralized management embedded in state information systems, the decentralized, non-linear nature of innovation activity. The development of the architecture of the national innovation infrastructure of innovation activity based on the network model of interaction will eliminate the disadvantages of the existing systems and implement the advantages that network structures acquire.

Keywords: Scientific innovation · Network models · Innovation management · Digital transformation

1 Introduction

Innovation activity is a priority in ensuring sustainable development in most countries of the world. The results of the innovation development of the Russian economy cannot be considered satisfactory: according to Rosstat data, the level of innovation activity of organizations has been steadily decreasing since 2011 and in 2019 was only 9.1%. Among the reasons for this state of affairs, many researchers cite the low efficiency of the innovation infrastructure. The existing infrastructure elements are weakly connected to each other, the principle of the "triple helix" has not been implemented, which implies the effective interaction of the participants in the innovation process (hereinafter referred to as IP) [1]. Pilipenko highlights the discrepancy between the existing elements

P. V. Trifonov and M. V. Charaeva (Eds.): DITEM 2021, LNNS 380, pp. 448–455, 2022.
https://doi.org/10.1007/978-3-030-94245-8_61

of the Russian innovation infrastructure (technoparks, business incubators, etc.) and the peculiarities of higher technological orders [2]. Khasanshin, exploring the dialectical problems of the development of information provision for the national innovation system, points to the lack of formation of organizational and economic mechanisms for implementing the innovation strategy, the lagging behind of the regulatory framework of the digital economy, a decrease in the quality of information resources against the background of their rapid growth [3].

The theoretical basis of the study was publications on the management of innovations and innovative structures [4, 5], classification of innovation structures [6–8], models of the innovation process [9]. Many scientists agree that the key factor for the success of innovation activity is the effective interaction of participants in the innovation process based on a network model [10–15].

Thus, in Russia, there is an acute problem of building an integrated model of an innovative development management system, which will ensure productive cooperation of all IP subjects.

Research hypothesis: the existing elements of the digital infrastructure of innovation activity (hereinafter – DIIA) are not effective due to the inadequacy of the conceptual model underlying the architecture of the national innovation management system, the structure and specifics of innovation activity.

The purpose of the study is to assess the adequacy of the model of the national innovation infrastructure to the goals of innovation development and the specifics of innovation activity.

Research objectives: analysis of the structure, functionality and effectiveness of elements of the Russian DIIA, assessment of strengths and weaknesses, and research into the reasons for the low efficiency of infrastructure.

2 Materials and Methods

The first, theoretical stage of the research is devoted to the analysis of scientific publications and expert interviews for 2011–2020, dedicated to the development of DIIA in the context of forming the digital economy. The authors used systemic and general scientific methods, as well as the method of logical analysis.

At the second stage, the authors used methods of empirical research and content analysis of large Russian digital platforms created within the framework of federal projects and Internet portals focused on information provision and support for innovation activity. The analysis of their structure, functionality, compliance with the requirements of participants in the innovation process at various stages of its implementation is conducted. When summarizing the results, a systemic method and comparative analysis were used.

3 Results

At the first stage of the study, it was shown that the innovation process is characterized by a network type of interaction between subjects [10–16]. Modern IP models of the 5th and 6th generation are also of a network nature [17], reflecting the following features:

1. strategic network interaction of participants in the process;
2. decentralized nature of innovation activities;
3. asynchronous parallel execution of processes within the framework of innovation activities;
4. the existence of many non-linear scenarios for the IP development.

Networking is especially important at the stage of commercialization of innovations, when it is required to ensure effective multilateral communication of the participants in the innovation network (universities, business, government). In this case, two tasks are solved: (a) the search for new technologies that meet the requirements of the business; (b) search for consumers for newly developed technologies. The effectiveness of these processes largely depends on the quality of information provision for technology transfer processes.

State information systems created to solve these problems, as a rule, implement a *centralized* approach to design and management, which has proven itself in the management of *linear processes* [3], but does not correspond to modern views on the nature of innovation activity.

Features such as strict regulation and coordination of the activities of participants, as well as their lack of economic independence contradict the key features of innovation activity. Attempts to implement innovation systems with command-and-control management lead to a lack of healthy competition, a narrowing of the area of responsibility of each subject, formal fulfillment of their obligations and a lack of interest in the final result of the IP [18].

At the second stage, Russian digital platforms and Internet portals, which support innovation activity, were examined. The effectiveness of the elements of the innovation infrastructure and the quality of interaction between IP subjects were assessed. The main functional capabilities of these resources were analyzed in the context of the stages of the innovation process, their advantages and disadvantages were formulated (see Table 1). In addition to the resources listed in Table 1, a number of large projects launched in 2005–2018 were identified, which had been phased out by the time of the study.

As it can be seen from Table 1, the elements of the DIIA are quite clearly divided according to the stages of the innovation process, within the framework of which they are involved.

Thus, the *AS of the Foundation for the Promotion of Innovations* and *IS of the Russian Foundation for Basic Research* are intended only for information support of procedures for R&D grant support and are of a closed nature. At the same time, they have been actively used by researchers for more than 10 years and can be considered effective.

The only fully functional platform in the authors' sample is the *STP "My Business"*. The platform was created by the order of the Ministry of Economic Development of the Russian Federation in 2019, at the time of the study it was in experimental operation. The declared functions of the platform cover the information needs of all stages of the IP, and also provides support in the form of electronic services for registering and launching a business. The services are classified into 6 main categories. In the section "Services and support", among other things, innovative, information and consulting support for business is declared. Unfortunately, content analysis of the platform showed a low quality of implementing these services.

Table 1. Characteristics of the elements of the Russian digital infrastructure for innovation development.

Infrastructure elements	Functions for supporting the stages of the innovation process, including			General business support	Advantages (+)/disadvantages (–)
	R&D	Introduction of new technologies into production	Marketing of innovations		
AS "Fond-M" (Foundation for the Promotion of Innovations)	Grant support and maintenance of the register of R&D reports (about 12 thousand)	–	–	–	+ effectively solves its tasks; – is of a closed nature, full texts of reports are not available via the Internet; – in the process of preparing the article, the website was under technical maintenance
Information system of the Russian Foundation for Basic Research	Grant support and maintenance of the register of R&D reports (over 20 thousand)	–	–	–	+ effectively solves its tasks; – does not solve the problems of commercialization, access is provided only for Russian scientific organizations upon application
Russian Technology Transfer Network (RTTN)	–	Identifying the potential of the subject of innovation activity for technology transfer Maintaining a register of innovative technological proposals Maintaining a register of innovative technological requests Negotiating and concluding a technology transfer agreement		–	+ detailed profiles of developers and consumers of innovation technologies + convenient search system + focus on critical stages of the innovation process – small number of participants

(*continued*)

Table 1. (*continued*)

Infrastructure elements	Functions for supporting the stages of the innovation process, including			General business support	Advantages (+)/disadvantages (–)
	R&D	Introduction of new technologies into production	Marketing of innovations		
State digital platform "My business"	Maintaining a register of R&D services	Maintaining a register of production services, legal and financial services	Maintaining a register of services in the field of marketing, consulting	Business training Online office Electronic services for registration and business start-up Checking counterparties	+ coverage of most of the stages of the innovation process + integration with the offices "My Business" – formalist approach to implementation – weakly filled with services for supporting the innovation process

Thus, in the sections related to innovation activities (R&D execution, advanced training, high technologies, etc.), a very small number of services (from 5 to 40 items) from a small number of suppliers are presented.

But the main problem of the platform is an absolutely formalistic approach to implementing support services, which is expressed in a number of features:

- in the card of the service – there are more than 60 fields, many of which are redundant, making the service catalog inconvenient for work;
- for many services, the tab with the description is not filled or is filled with template SEO texts, which makes the card virtually useless and negatively affects the perception of the platform;
- the description of the services is of a general, non-specific nature and is rather a declaration of the supplier's capabilities than a specific commercial proposal.

In the authors' opinion, the only Russian resource that effectively solves the most important task of the commercialization of innovations and technology transfer is the *Russian Technology Transfer Network (RTTN)*. The network is based on a system of technological profiles of suppliers and consumers of innovations with a search and feedback system. The disadvantages of RTTN include the local nature and the small number of participants in the system (there are only 60 items in the catalog of technological proposals). However, even with such a modest scale, several large successful international innovation projects have been implemented within the network.

4 Discussion

Based on the results of the first, theoretical stage of the research, the following conclusions can be formulated.

1. The process approach based on operations research methods does not correspond to the content of innovation activity, which is focused on achieving unique goals and in this sense is project-oriented. For project management, network models and methods of network planning are more effective [19].
2. The dynamism of the modern economic environment makes linear IP models ineffective. The model of an innovative network structure that generates and transforms knowledge in the process of creating innovations is more adequate to the IP structure. At the same time, innovativeness becomes a characteristic of the business processes of the organization as a whole.

The second stage of the study showed that the existing elements of the DIIA provide effective support only at the R&D stages; information support for the commercialization of R&D results is practically absent, and the few existing systems are ineffective.

As a result of the research, a characteristic feature of the "innovative" period of development of the Russian economy (2011–2020 – the period of the Strategy for the Innovative Development of the Russian Federation) was manifested in the form of a large number of state and private-state projects for the construction of DIIA, many of which did not receive development and were withdrawn after 3–5 years of work. This leads to the conclusion that there is no systematic approach to the development of DIIA. Activities are fragmented and spontaneous, projects do not achieve their goals, and the created platforms are closed as ineffective.

The fundamental reason for this state of affairs lies in the inadequacy of the models underlying state information systems, the structure and principles of IP. Other researchers come to similar conclusions [3, 17]. The attempts to build a centralized system for supporting innovation development have not given tangible results for more than 10 years, despite the significant resources allocated by the state.

Thus, the hypothesis of the study on the ineffectiveness of using centralized linear control models when creating DIIA is confirmed. From the point of view of the modern theory of innovation management, the network model of interaction between IP participants is more adequate.

It is interesting that the digital platforms have long been successfully developing on the market built on the basis of this model and satisfying the specified requirements of network interaction – these are social online networks. The characteristic features of networks include:

- possibility of stable multilateral communication due to redundancy of connections in the network;

- independence of network entities, which ensures competition, free search for partners and increases the efficiency of interaction;
- filling the network with user-generated content, the content and form of which is determined only by the information requirements of the network participants;
- stable long-term nature of interaction, which allows to minimize transaction costs;
- the possibility of forming associations, collaborations based on the principles of community of interests, which makes interaction more effective.

Moreover, in the West, there are specialized networks aimed at communication in the framework of research-scientific and innovation activities. The largest scientific social networks include Academia, ResearchGate, Mendeley, etc. There are no such specialized networks on the Russian Internet. The two largest scientific electronic libraries eLibrary and Cyberleninka, for all their relevance, also have a centralized architecture, which systematically leads to failures in their work and negative side effects associated with the high complexity of centralized moderation of such large-scale systems.

5 Conclusions

The development of an innovation economy requires creating an effective information support system for innovation activity, an important element of which is digital infrastructure, including digital communication platforms, electronic libraries and databases, as well as specialized information and analytical systems.

The conducted analysis of the state of the DIIA elements in Russia allows to draw the following conclusions:

- the existing platforms solve only part of the tasks at the R&D stage, information support at the stage of commercialization of R&D results is ineffective;
- at the federal and regional levels, the work on the development of the DIIA is unsystematic and fragmentary;
- the main reason for the above-mentioned problems is the discrepancy between the principles of centralized management embedded in state digital platforms, and the decentralized, non-linear nature of innovation activity.

The development of the architecture of the national DIIA based on the network model of interaction will allow to eliminate the disadvantages of the existing systems and implement the advantages of network structures, which is especially important in the context of forming the digital economy in Russia.

References

1. Degtyareva, I.V., Marina, A.V., Shalina, O.I.: Mod. Econ. Quest. **1**, 49–58 (2013)
2. Pilipenko, E.V., Batalov, Yu.V.: Bull. Baikal State Univ. **1**, 8–12 (2013)
3. Khasanshin, I.A.: Bull. Acad. Knowl. **2**(31), 224–228 (2019)
4. Ermolenko, V.V., Lanskaya, A.E., Yakovlenko, D.V.: Bull. Acad. Knowl. **3**(32), 125–133 (2019)

5. Amitrano, C.C., et al.: Sustainability **10**(10), 3744 (2018). https://doi.org/10.3390/su10103744
6. Meyers, P.W., Tucker, F.G.: J. Acad. Mark. Sci. **17**(1), 73–82 (1989). https://doi.org/10.1007/BF02726356
7. Chandy, R.K., Tellis, G.J.: J. Mark. **64**(3), 1–17. https://doi.org/10.1509/jmkg.64.3.1.18033
8. Miller, L., Miller, R.: Int. J. Innov. Tech. Manag. **9**(1), 1250004 (2012). https://doi.org/10.1142/S0219877012500046
9. Rothwell, R.: Int. Mark. Rev. **11**(1), 7–31 (1994). https://doi.org/10.1108/02651339410057491
10. Badaraev, T.D.: Bull. Tomsk State Univ. Econ. **2**(34), 203–211 (2016)
11. Karlik, A.E., Krechko, S.A., Platonov, V.V.: Lab. Econ. **4**(4), 295–308 (2017)
12. Makoveeva, V.V.: Bull. Tomsk State Univ. **354**, 163–166 (2012)
13. Savchenko, A.P.: State Munich Manag. Sci Notes North Caucasus Acad. Public Adm. **1**, 129–135 (2017)
14. Shaposhnikova, T.L., Kotlyarova, E.A., Ternovaya, L.N.: Sci. Notes. Lesgaft Univ. **12**(154), 305–311 (2017)
15. Piazza, M., Mazzola, E., Abbate, L., Perrone, G.: Eur. Plann. Stud. **27**(9), 1857–1878 (2019). https://doi.org/10.1080/09654313.2019.1642856
16. Bek, N.N., Gadzhaeva, L.R.: Bull. Moscow Univ. Ser. 6: Econ. **1**, 140–159 (2018)
17. Pakhomov, A.V., Pakhomova, E.A., Rozhkova, O.V.: Nation Interests: Prior Secur. **14**(3), 520–540 (2018). https://doi.org/10.24891/ni.14.3.520
18. Savchenko, A.P.: Econ.: Theory Pract. **4**(48), 49–55 (2017)
19. Zhikharev, K.L.: Sci. Notes. E Sci. J. Kursk State Univ. **3–1**(15), 86–94 (2010)

Flows in the Digital Economy: New Approaches to Modeling, Analysis and Management

Ruslan Kh. Ilyasov(✉)

Chechen State University, Grozny, Russia

Abstract. The digital economy development is transforming modern management models for socio-economic systems. The main condition for improving the economic system management in the digital economy is the analysis of information about exact, real states at the moment, as well as about real changes in the system. In conditions when digitalization has provided almost continuous access to services and goods in streaming mode, the research purpose is to develop methods of operational real-time management of economic processes. This requires continuous monitoring, accounting and management of processes in dynamic economic systems. The elements of dynamic economic systems that describe states and changes in states are stocks and flows. The article proposes to assess the efficiency of flow management in the economic system, in their analytical description by methods putting aside the distortion of real data. This makes it possible to take into account even small changes in states in the system. As shown in the article, the best quality of analytical flow description is achieved by interpolating by splines. The effectiveness of the proposed method is assessed by modeling the dynamics of housing loans by cubic splines, with the transformation of the dynamics of states into a flow through differentiation. The analysis showed that the spline model analytics turns out to be relevant to the tasks of taking into account the real states of the process under study, with the possibility of transforming successive changes in the model – flow, rate and acceleration of the flow.

Keywords: Digital economy · Economic systems · Stocks · Flows · Modeling · Management · Interpolation · Splines

1 Introduction

The main condition for effective economic system management in the digital economy is the analysis of information about exact, real states at the moment. It should be borne in mind that the analysis of generalized or averaged data will a priori contain unacceptable errors of discrepancy with reality. Certainly, the results of the operational economic system management based on the analysis of generalized data will also be unsatisfactory. Under the new conditions, a management model should take into account the dynamics of real continuous-successive changes in economic systems. Digitalization has provided almost continuous access to services and goods – today it happens in streaming mode. Consequently, there are tasks of flow accounting, modeling, analysis and management

P. V. Trifonov and M. V. Charaeva (Eds.): DITEM 2021, LNNS 380, pp. 456–463, 2022.
https://doi.org/10.1007/978-3-030-94245-8_62

in real time. For flow modeling in economic systems, this creates a requirement for an exact match between model and actual data at all nodal points.

Flows as dynamic indicators of changes in economic systems appear to be a source of more valuable information for decision-making than stocks. Stocks characterize a static part of the economic system, determining states at specific points in time, while flows determine changes in stocks per unit of time. Some examples will be given to illustrate flows in the economy: these are changes in the money supply in the economy for the year, an increase in the number of unemployed for the quarter, a change in loan debt for the month, etc. In modern conditions, changes in stocks have to be taken into account not only for relatively equal time periods – month, quarter, year, etc., but also for intervals of random length. For example, one may need to provide the ability to repay a loan debt at any time – for time intervals of random length and not for a whole month. Flow modeling, in this case, must satisfy an important requirement – an analytical description by functions that are non-vital for the variability of an interpolation step.

Another important requirement for analytical flow modeling in the digital economy is accuracy – the coincidence of model and real processes at nodal points. Obviously, in the digital economy, making effective decisions in real time requires analysis and local changes in the economic system. Moreover, it becomes relevant to a search for patterns in successive changes in system states through flows – finite differences or derivatives. This allows formulating the research purpose – the development of new approaches to analytical flow modeling and analysis that increase the efficiency of real-time process management [1].

2 Materials and Methods

The need to manage flows in real time firstly makes it necessary to drop smoothing procedures that distort real dynamics. The second argument is a deterministic mechanism of mutual transformation of stocks and flows, which also requires the preservation of accuracy in the model representation of input data. For example, Sharda, Abdulelah Al-Sudani and other scholars point out the need to maintain accuracy when studying flows [2, 3]. Requirements for the accuracy of physical flow management models are also relevant in the study of flows in economic systems. Among the known functions, the selected requirements are simultaneously satisfied by third-degree spline functions. Cubic splines are known for their flexibility, the best interpolation behavior, the continuity of both the function and the first two derivatives [4–6].

In the proposed new concept for modeling and analyzing processes in economic systems, residuals, first differences or derivatives become a source of accurate information about successive changes – flows. Then the spline interpolation model of the state dynamics can be "decomposed" through differentiation into models of successive changes in states – flow models, flow rate and acceleration model.

To identify shortcomings of the classical approach to process modeling, the authors will consider the results of approximation by the least square method (LSM) using the example of the dynamics of debt on housing loans – an indicator of the type of stocks. The conceptual feature of LSM-models is their smoothness and the best approximation in terms of the minimum sum of squared deviations between the model and actual data at nodal points.

3 Results

A polynomial curve that describes well enough the dynamics of states (stocks) – housing loan debt will be constructed through LSM-modeling:

$$\text{LS\_DL} = \frac{19778489}{19448}t^3 + \frac{1198430353}{136136}t^2 + \frac{789309267}{6188}t + \frac{1665}{77}. \quad (1)$$

As can be seen in Fig. 1, a third-degree polynomial has a smooth model trajectory of dynamics of debt on housing loans, which can be useful for extrapolating the revealed trend. However, the transition from the LSM-model of the dynamics of states (housing loan debt) to the model of changes in states – the flow – becomes meaningless.

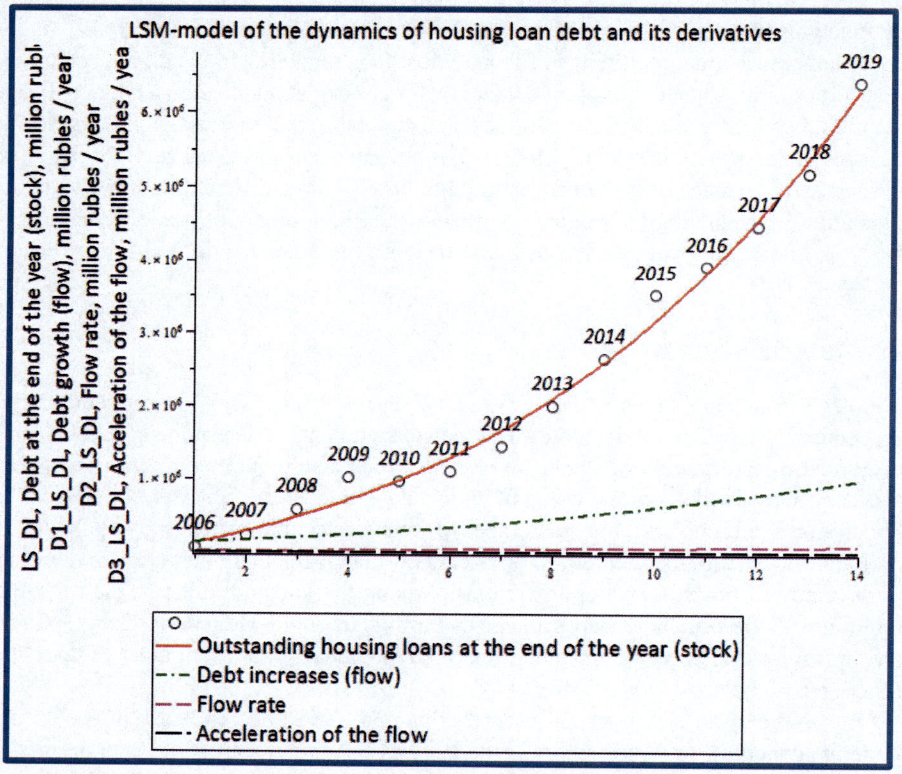

Fig. 1. LSM-model of the dynamics of housing loan debt and its derivatives. *Source:* Compiled by the author.

By differentiating the polynomial model (1), the authors will perform the transition from the state (stock) model to the rate of change (flow) model:

$$\text{D1\_LS\_DL} = \frac{59335467}{19448}t^2 + \frac{1198430353}{68068}t + \frac{789309267}{6188}. \quad (2)$$

The model of the instantaneous rate of changes in housing loan debt (flow model), obtained by differentiating the stock function, demonstrates a parabolic trajectory of uniform changes. The second-order derivative retains even less real information about local changes, demonstrating the linear nature of the flow rate:

$$\mathrm{D2\_LS\_DL} = \frac{59335467}{9724}t + \frac{1198430353}{68068}. \tag{3}$$

In turn, the acceleration of the flow of housing loan debt in Russia, described by the third derivative, turns out to be constant:

$$\mathrm{D3\_LS\_DL} = \frac{59335467}{9724} \tag{4}$$

As the analysis has shown, LSM-models, abstracting from "random" fluctuations in the dynamics under study, lose information about real states of the economic system at specific points in time.

Spline interpolation of zero error dynamics at nodal points of the process allows acquiring a new understanding of DataMining as a way to extract knowledge about the development patterns from accurate numerical, analytical and graphic images. In successive changes in states of the economic system, the authors find an opportunity to study the development rate and acceleration, the flow correlation, to conduct a phase analysis of flow cyclicality and seasonality [7, 8]. Using the example of spline modeling of the dynamics of housing loan debt, the authors will demonstrate some of the analytical advantages of the proposed methodology.

$$\mathrm{SPL3\_DL} = \begin{cases} 30478,1\,t^3 - 91434,3\,t^2 + 222921,2\,t - 84569,0 & t < 2 \\ 25673,5\,t^3 - 62606,7\,t^2 + 165266,1\,t - 46132,3 & t < 3 \\ -212743,1\,t^3 + 2083142,9\,t^2 - 6271982,9\,t + 6391116,8 & t < 4 \\ 237158,0\,t^3 - 3315670,5\,t^2 + 15323270,6\,t - 22402554,7 & t < 5 \\ -59576,8\,t^3 + 1135352,1\,t^2 - 6931842,0\,t + 14689299,7 & t < 6 \\ 25145,4\,t^3 - 389648,0\,t^2 + 2218158,3\,t - 3610700,9 & t < 7 \\ -53996,7\,t^3 + 1272335,4\,t^2 - 9415725,3\,t + 23535027,6 & t < 8 \\ 104340,3\,t^3 - 2527753,5\,t^2 + 20984985,7\,t - 57533535,3 & t < 9 \\ -259891,7\,t^3 + 7306512,7\,t^2 - 67523410,1\,t + 207991652,2 & t < 10 \\ 246664,6\,t^3 - 7890176,0\,t^2 + 84443476,8\,t - 298564637,6 & t < 11 \\ -96921,5\,t^3 + 3448164,6\,t^2 - 40278269,3\,t + 158748431,8 & t < 12 \\ 144967,5\,t^3 - 5259840,2\,t^2 + 64217787,6\,t - 259235796,0 & t < 13 \\ -131297,5\,t^3 + 5514495,0\,t^2 - 75848570,0\,t + 347718420,3 & t < 14 \end{cases}$$

As Fig. 2 shows, cubic spline interpolation stores in the model the information about the states of an indicator of a type of stocks – housing loan debt with absolute accuracy at all nodal points. Consequently, converting the dynamics of stocks into an analytical flow model is also feasible with zero error.

The transformation of the stock spline model into a flow occurs automatically by differentiation. The first derivative of the stock function, by analogy with the physical interpretation, is interpreted as the instantaneous rate - in this case, as the instantaneous rate of change in stocks or the flow. The broken second derivative can be used to observe

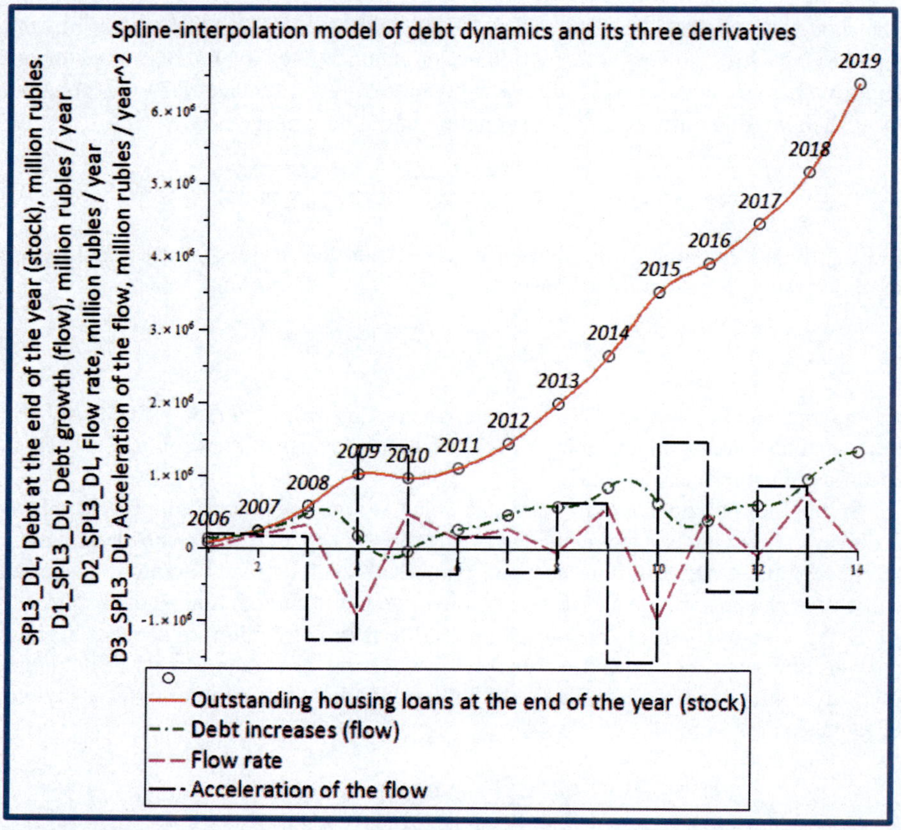

Fig. 2. Spline-interpolation model of housing loan debt dynamics and its three derivatives. *Source*: Compiled by the author.

changes in the instantaneous flow rate. The flow acceleration is described by the third derivative – in Fig. 2 one can observe it along a broken line of piecewise constant changes.

Below is the comparison of the quality of displaying annual changes in housing loan debt by spline interpolation D1_SPL3_DL and polynomial D1_LS_DL flow models. If D1_SPL3_DL is the instantaneous rate of change in stocks, the change in stocks (flow) over a certain time interval from t-1 to t can be determined by numerical integration:

$$\int_{t-1}^{t} D1\_SPL\_DL(t)dt \quad - \text{ for the spline model;}$$

$$\int_{t-1}^{t} D1\_LSM\_DL(t)dt \quad - \text{ for the LSM – model.}$$

The authors compared the values of the housing loan debt flow calculated through integration with the annual chain of debt increments. According to the calculation results, housing loan debt flows, determined by the integration of the spline interpolation model, coincided with absolute accuracy with the annual chain debt increments. The sum of all consecutive changes in debt stocks from 2006 to 2019 made up as follows:

$$F_{2006/2019} = \int \sum F_i(t)dt = 6321729\, \textit{million rubles.}$$

The accumulated debt at the end of 2019, taking into account the debt at the end of 2006, amounted to:

$$S_H = S_L \pm \int \sum F_i(t)dt = 77396 + 6321729 = 6399125\, \textit{million rubles},$$

where SL = 77,396 million rubles – debt at the end of 2006:

SH = 6,321,729 million rubles – debt at the end of 2019;
$\sum$ – for all i-th flows in the interval from 2006 to 2019;
Fi(t) – one of the i-th flows.

4 Discussion

LSM-modeling of debt dynamics resulted in noticeable flow distortions. Thus, the absolute deviations of the modeled flow values from the real ones are observed in the range from 881.665 to 319,946.1 million rubles per year. In relative terms, the deviations range from 0.5% to −625.8% per year. The total deviation between the calculated and actual values of housing loan debt at the end of 2019 was relatively small – 157,282 million rubles or 2.5%.

The analysis shows that the analytics of the spline model turns out to be relevant to the tasks of taking into account the real states of the process under study, with the possibility of transforming successive changes in the model – flow, rate and acceleration of the flow. Continuity in the analytical flow representation allows it to be estimated for time intervals of any length, which is important for the efficiency of real-time process management [9, 10]. Mathematically, the measurement of flows (changes in stocks) over irregular time intervals can be performed by numerical integration by parts. For example, the amount of housing loan debt flows was calculated for the following irregular intervals:

from 31.12.2006 to 30.06.2010 – lasting 3.5 years;
from 01.07.2010 to 30.06.2015 – lasting 5 years;
from 01.07.2015 to 31.12.2019 – lasting 4.5 years.

$$F_{1/13} = \int_1^{14} F_i(t)dt = \int_1^{3.5} F_i(t)dt + \int_{3.5}^{9.5} F_i(t)dt + \int_{9.5}^{14} F_i(t)dt = 758919.9678$$
$$+2271041.920 + 3291767.112 = 6321729.0000\, \textit{million rubles.}$$

5 Conclusion

Calculations confirm the ability of splines to model the dynamics of states (stocks) in economic systems with high accuracy, to perform analytical transformations of state models into accurate flow models, including in the flow model at time intervals of random length [11, 12]. The ability to analytically and quantitatively represent states at randomly chosen times, as well as represent changes in states over time intervals of random length becomes the basic condition for the real-time management of the flows in dynamic systems.

Derivatives are becoming effective tools for detecting, visualizing and analyzing small changes in states. As mathematically accurate indicators of slowdown, growth acceleration, as well as the achievement of extreme levels in the process development, derivatives become effective tools for detecting and analyzing crises in the economy.

Applying accurate mathematical flow models allows managing economic systems with due account for their continuous development, adapting functioning modes to local changes [13, 14]. Knowledge about the regularities of the system functioning and development should now be obtained not only by smoothing and averaging the dynamics of states, but also by studying real successive changes, growth acceleration or deceleration. This is required by the tasks of studying transformation processes in the world economy, flow integration and complication [15]. The study showed that spline functions simulate the complex dynamics of states with accurate preservation of all values of a real process, which allows transforming stock models into flows without errors. The analytical potential of the method is fulfilled through spline derivatives – the ability to quantitatively, analytically and graphically study the flows.

References

1. Markova, V.D., Markov, A.A.: Innov **7**(249), 83–87 (2019)
2. Sharda, V.N., et al.: Hydrol. Sci. J. **53**(6), 1165–1175 (2008). https://doi.org/10.1623/hysj.53.6.1165
3. Abdulelah Al-Sudani, Z., Salih, S.Q., Sharafati, A., Yaseen, Z.M.: J. Hydrol. **573**, 1–12 (2019). https://doi.org/10.1016/j.jhydrol.2019.03.004
4. Spline functions in economic and statistical research. In: Rosin, B.B. (ed.) Collection of Articles. Siberian Branch, Institute of Economics and Organization of Industrial Production of the Siberian Branch of the Russian Academy of Sciences. Nauka, Novosibirsk (1987)
5. Ilyasov, R.Kh.: Econ. Anal.: Theory Pract. **19**(496), 173–187 (2020). https://doi.org/10.24891/ea.19.1.173
6. Reich, B.J., Storlie, C.B., Bondell, H.D.: Technometrics **51**(2), 110–120 (2009). https://doi.org/10.1198/TECH.2009.0013
7. Tsai, T.-L., Yang, J.-C.: Adv. Water Res. **28**(7), 661–670 (2005). https://doi.org/10.1016/j.advwatres.2005.01.005
8. Wijnands, J.S., Qian, G., Kuleshov, Y.: Appl. Math. Model. **40**(19–20), 8685–8707 (2016). https://doi.org/10.1016/j.apm.2016.05.013
9. Odoardi, I., Muratore, F.: Adv. Intell. Syst. Comput. **373**, 385–392 (2015)
10. Hung, L.D.: J. Econ. Integr. **35**(3), 503–518 (2020). https://doi.org/10.11130/jei.2020.35.3.503

11. Schmitt-Grohé, S., Uribe, M.: J. Econ. Theory **192**, 105195 (2021). https://doi.org/10.1016/j.jet.2021.105195
12. Chang, L.-Y.: J. Air Transp. Manag. **34**, 123–130 (2014). https://doi.org/10.1016/j.jairtraman.2013.09.003
13. Chen, Y., Zhu, L., Gonder, J., Young, S., Walkowicz, K.: Transp. Res. C: Emerg. Tech. **83**, 134–145 (2017). https://doi.org/10.1016/j.trc.2017.08.003
14. Grüne, L., Semmler, W.: J. Econ. Dyn. Control **28**(12), 2427–2456 (2004). https://doi.org/10.1016/j.jedc.2003.11.002
15. Franco-Villoria, M., Scott, M., Hoey, T.: Environmetrics **30**(2), e2522 (2019). https://doi.org/10.1002/env.2522

Issues of Digital Business Transformation in the Current Context

Oleg N. Misko[1(✉)], Artem V. Smirnov[2], and Yuriy V. Kotelkin[1]

[1] North-West Institute of Management, Branch of the Russian Presidential Academy of National Economy and Public Administration, St. Petersburg, Russia
[2] St. Petersburg State University, St. Petersburg, Russia
a.smirnov@spbu.ru

Abstract. The issues of digital transformation of business in the current context are more pressing than ever. Most entrepreneurs and scientists have two views on the digital economy. In their eyes, it can be an impetus for further business development, but for many of us it is an incomprehensible value, and, therefore, a potential danger. Digital transformation carries many risks that need to be identified and dealt with. In response to contemporary challenges, it raises the question of creating new models of business management because "old" traditional methods have reached their limits in the digital world. A huge positive and explosive growth in cyberspace does not mean a more prosperous future for the global economy. They go hand in hand with the greatest risks and cyber threats that are growing as the digital economy expands. In addition, while studying the latest issues of digitalization in business, it is also necessary to consider the opportunities and advantages that arise when using new technologies in order to combat the COVID-19 pandemic. The development of advanced technologies, neural networks and convergent innovations will ultimately allow the global economy to overcome the economic crisis more painlessly, which has arisen due to the pandemic. This paper proposes the analysis of the essence of the digital economy, examines some of its aspects, and identifies possible risks that may carry the transformational processes of transition to the digital economy.

Keywords: Digital transformation of business · Digitalization · Risks of digital economy · Fourth Industrial Revolution · "Digital" investments · Automated management system · Cybersecurity · "Digital" management

1 Introduction

In January 2016, at the World Economic Forum (WEF) in Davos, the beginning of the Fourth Industrial Revolution was announced, and digitalization becomes a global trend. The new revolution is in its infancy – the transition to a new era has just begun where the digital, biological and physical worlds merge. In this revolution, opportunities and growth are built on a supportive regulatory and business environment, the readiness of both Information and Communication Technologies (ICT) to embrace new technologies and society's willingness to accept and use them. For the country to thrive in this century,

P. V. Trifonov and M. V. Charaeva (Eds.): DITEM 2021, LNNS 380, pp. 464–470, 2022.
https://doi.org/10.1007/978-3-030-94245-8_63

digital intervention is inevitable. The WEF has proposed a Networked Readiness Index (NRI) to assess and calculate the readiness of countries to receive and benefit from new technologies in the digital economy. According to NRI ranking from publication the top ten countries are Singapore, Finland, Sweden, Norway, the United States, the Netherlands, Switzerland, the United Kingdom, Luxembourg and Japan. It seems natural that half of them are small countries of Western Europe.

Researchers estimate that the share of the digital economy is already about 5% of world GDP and covers 3% of the labor market [1]. According to a report by Oxford Economics, "the global digital economy will reach $ 23 trillion by 2025, and every dollar invested in digital technologies will increase global GDP by $ 20. The average return on digital investments is 6.7 times higher than that of investments in traditional sectors of the economy" [2]. The digital economy becomes the main driver of economic development around the world. It "generates" high value in both developed and developing countries but faces a number of significant regulatory challenges. The pace of technological change is accelerating, and technological change often outpaces regulation [3].

The expected development is associated primarily with the implementation of the possibilities of "breakthrough technologies", which, like any real innovation, destroy traditional business, change the market and, in the short term, introduce an imbalance in the economy. It is useful to mention that the term "breakthrough technologies" in Russia has the same meaning as "disruptive technologies".

K. Schwab, Forum's President, notes that "the development and implementation of the latest technologies are associated with uncertainty and mean that society has no idea how the transformations caused by the Fourth Industrial Revolution will develop in the future" [4].

To date, most business leaders are acutely aware of the fact that the digital economy can be either an opportunity or a threat. The question is not whether they should involve their business in it. Rather, it is about how to address the opportunities of digital technologies while supporting a healthy business, using existing IT investments and innovation without disrupting them. However, it should also be noted that, despite the obvious advantages, many business leaders have refrained from this task. According to a recent study by MIT Sloan and Capgemini, only 15% of CEOs are pursuing a digital strategy, although 90% agree that the digital economy will affect their industry. At the same time, those who adopted digital transformation at the early stages achieve a 9% increase in revenue, a 26% impact on profitability and a 12% increase in the market [5].

Nevertheless, despite the caution of experts in long-term macroeconomic forecasts, the business sector focuses primarily on the "advantages" of digitalization: reducing the uncertainty of the external environment in the short term, increasing the speed of decision-making and operations, and reducing costs. According to the study by Oxford Economics and SAP, 84% of the interviewed representatives of global holdings believe that "digital transformation, or the introduction of modern technologies into business processes, will become crucial for the survival of companies in the next five years" [6].

2 Materials and Methods

Having analyzed the most problematic issues (according to a number of researchers) that digitalization addresses to business community, and based on the statistical data during

Russian and international practices over the past five years, the authors of the paper set out their views on changes in business processes that inevitably arise during digital transformation of the economy, and also offer possible directions and ways of adapting business entities to these changes. In this paper, the authors tend to accept the following definitions of the digital economy: first, it is "an economy which functions primarily by means of digital technology, especially electronic transactions made using the Internet" [7]. Second, it is "the economic activity that results from billions of everyday online connections among people, businesses, devices, data, and processes. The backbone of the digital economy is hyper-connectivity which means growing interconnections of people, organizations, and machines that results from the Internet, mobile technology and the internet of things (IoT) [8].

3 Results

Scaling up a business with the help of "numbers" and the use of breakthrough technologies involves radical changes in business processes and the creation of new conditions in the relationship of the company with the external business environment. This concerns not only the creation of new industries and products, but the emergence of new services, methods and forms of communication between all participants in business processes as well. UNCTAD Digital Economy Report (2019) notes that "an entirely new "data value chain" has evolved, comprising firms that support data collection, the production of insights from data, data storage, analysis and modelling. Value creation arises once the data are transformed into digital intelligence and monetized through commercial use [9].

The ability of "old firms" to adapt to new conditions, "fit" into this "value chain" largely depends on their ability to digital transformation. This will be difficult to achieve based on traditional models that use the paradigm of "one-size-fits-all means" of management, regardless of the specifics of the business. Flexible management should be accompanied by the ability to develop, refine and adapt to the specifics of a particular company. Flexible planning should be accompanied by an understanding and analysis of all business processes of the company. When using a flexible management methodology, an infrastructure is created that allows you to quickly respond to changes in the market and adapt to innovations. Manufacturing processes used to end in the factory workshop; now a challenge has been thrown down to them by the business model of incomplete products with digital applications or 3D components in the consumer market; supply chains end up at a supermarket or other store [10].

One of the features of current business processes is that they are based on large and mega-large databases for all channels of the company. In this regard, in the field of business management, there are some critical issues as the prompt collection of relevant information, the correct identification of data by unique ID-codes, accounting of transactions and the possibility of their recording in a single access window. This means that solving the problems of digital business transformation will require significant additional costs for IT support. Along with the purchase of powerful hardware and its support in mode 7/24, it will be necessary to implement training or retraining programs for the company's employees. These requirements for changing key IT competencies should apply to all employees – from PC users to ICT developers and e-business experts.

Common challenges in the areas of management for many Russian companies are the fragmentation of databases, the absence of centralized system for collecting information, and the high labor intensity of the existing methods of collecting and processing data. However, the creation of a fully automated system for managing internal processes serviced by qualified personnel will not automatically remove the risks associated with the introduction and application of new business models. Risk management that is combined with the possibilities of the digital economy, when setting and solving specific business problems, becomes the main challenge for a company. Therefore, the digital economy raises questions of security, privacy and trust. Countries and governments should use new technologies for benefiting and improving cybersecurity to reduce cyber threats. Cybersecurity is a key factor contributing to the development of the digital economy and society.

Opportunities and boom in cyberspace combine with risks and cyber threats that grow as the digital economy expands. In 2014, global cybercrime cost more than $ 400 billion a year; in 2016 it grew to $ 450 billion [11, 12].

According to another report, the cost of global cybercrime reaches $ 575 billion annually. For the four largest economies (the United States, China, Japan and Germany), the total loss from cybercrime reached $ 200 billion [11].

Losses from cybercrime are not limited to actual losses due to attacks – they include recovery costs and missed opportunities. The study has showed that in Italy, losses from cybercrime amounted to $ 875 million. The cost of reconstruction and missed opportunities is $ 8.5 billion [11].

Cybercrime also affects the company's reputation, goodwill and share prices. 2013 is the year of the beginning of cybercrime actions (data leaks) against mega-level databases (mega data), including 8 super leaks (mega-hacks, in which more than 10 million personal data were stolen each time) [13].

In 2016, the total number of identity theft amounted to 1.1 billion, including 15 mega-hacks [14].

In the digital world, where both the defenders and the attackers participate in an online battle, skills and knowledge determine battlefield success. Demand for manpower to ensure cybersecurity is constantly growing. It is difficult to find qualified personnel in this area of activities, but already hired specialists require time and training to fully meet the needs of a particular business. As we could see, the number of cybercrimes and attacks is growing in the hostile environment of the Internet; the demand for cybersecurity specialists is growing proportionally, although according to the international professional IT association ISACA, the growth rate of personnel in this area has slowed down [15].

Another problem in the development of human resources in the field of cybersecurity is the length of maturity time required for specialists. Achieving peak performance requires years of IT knowledge, including at least a year of security experience. In addition, technologies are constantly and dynamically developing, so that cybersecurity often faces new cyber threats. The online struggle against them can only be carried out by qualified cybersecurity specialists with the assistance of advanced users who are active in cyberspace [16].

The problems associated with the latest cyber threats are taken into account in the economic development plans of a number of states, as well as companies that intend to reduce the risks of digitalization through cybersecurity strategies [17].

According to the researchers "the typical goals" of such strategies are:

- detection of cyber attacks and timely response to them;
- prevention of threats, support of products and services that are reliable from the point of view of cybersecurity for the state and business entities;
- support of state institutions and infrastructure operators;
- promoting the development of education in the field of digital
- technologies" [18].

Considering the current issues of digital transformation in the business sphere, it should also be noted the opportunities provided by new technologies in the fight against the COVID-19 pandemic. An interesting approach to the development of the digital economy from this point of view is offered by S. Lee and S. Trimi, who believe that sustainable innovation is necessary for the survival and success of an organization in the market environment of the digital age, especially in the current COVID-19 – pandemic crisis [19].

The authors explore a new category – *convergent innovations* (CI), defining them as innovations based on the effect of merging various objects, technologies, ideas and strategies, and consider them as a new sustainable core competence of organizations. CI is an autonomous ecosystem provided by advanced technologies, unique life cycle characteristics, relationships with other innovative approaches, as well as the goal of creating value for consumers. CI is much more dynamic than digitalization or automation, because it uses the power of merging various objects, ideas, people, functions, technologies, organizations, industries and societies. More detailed information about the idea of CI can be found in the monograph called "Living Innovation: From Value Creation to the Greater Good" [20].

The key feature of CI is a system that is designed to make the necessary decisions or actions autonomously, by scanning the environment with support for intelligent sensors, Internet technologies, the Internet of Things, big data analytics, machine learning and related computing. The relevant information is then sent to the next level to evaluate innovative ideas obtained from both internal and external sources for implementation.

4 Discussion

Showing interest in the CI concept put forward by Lee and Trimi [20], the authors of the paper believe that CI can be of great help both in the fight against the COVID-19 pandemic and in the post-pandemic period. In the first case, these are such activities as scanning the environment in real time; creating a transparent database, analytics and information for decision-making; forming international business cooperation to overcome the pandemic, etc. In the second case, these are: the creation of new digital operating systems to overcome huge disruptions in global supply chains for organizations; the creation of remote or "contactless" business services; digital transactions not only by young people, but also by older generations; digital transactions with not only everyday goods but with cars, and real estate as well.

5 Conclusion

Having summarized all the presented views of the authors of this paper it should be stated that during the digital transformation of business, owners and firms' CEOs will no longer be able to achieve the required competitive position using traditional management methods. In the face of the Fourth Industrial Revolution, in order to identify the risks of digitalization and manage them, flexible management should be based primarily on the specifics of the internal and external environment of a particular company. This means, among other things, the ability to organize the collection, processing and use of mega-large databases for business information support, as well as the ability to create and maintain the necessary cybersecurity system for the company. The approach to business organization from the standpoint of convergent innovations, in our opinion, makes it possible to manage the company more efficiently in the context of the implementation of global systemic risks.

References

1. Bukht, R., Heeks, R.: Informatics Working Paper no. 68 (2017). https://doi.org/10.2139/ssrn.3431732
2. Smirnov, A.: The future of the digital economy was discussed in Moscow. https://rg.ru/2020/09/28/budushchee-cifrovoj-ekonomiki-obsudiat-na-konferencii-v-moskve.html
3. Dahlman, C., Mealy, S., Wermelinger, M.: OECD Development Centre's Working Papers no. 334 (2016). https://doi.org/10.1787/4adffb24-en
4. Schwab, K.: The Fourth Industrial Revolution. Moscow (2016)
5. Bazzoun, M.: Int. J. Soc. Sci. Econ. Invent. **05**(09) 117 (2019)
6. Digital transformation in business: what it is and why changes are inevitable. https://blog.ingate.ru/detail/digital-transformatsiya-v-biznese-chto-eto-takoe-ipochemu-izmeneniya-neizbezhny
7. Oxford Dictionary: Digital Economy. Oxford University Press, Oxford
8. Bukht, R., Heeks, R.: Int. Org. Res. J. **13**(2), 143 (2018)
9. UNCTAD: Digital economy report. https://unctad.org/system/files/official-document/der2019_en.pdf
10. Ng, I.: J. Rev. Pricing Manag. **13**, 149–155 (2014). https://doi.org/10.1057/rpm.2013.27
11. Net Losses: Estimating the global cost of cybercrime. https://csis-website-prod.s3.amazonaws.com/s3fs-public/legacy_files/files/attachments/140609_McAfee_PDF.pdf
12. Graham, L.: Cybercrime costs the global economy $450 billion: CEO. https://www.cnbc.com/2017/02/07/cybercrime-costs-the-global-economy-450-billion-ceo.html
13. Internet Security Threat Report: https://docs.broadcom.com/doc/istr-14-april-volume-19-en
14. Internet Security Threat Report: https://docs.broadcom.com/doc/istr-21-2016-en
15. State of Cyber Security 2017: https://www.cybersecobservatory.com/wp-content/uploads/2017/06/state-of-cybersecurity-2017-part-2_res_eng_0517-1.pdf
16. Assante, M., Tobey, D.: IT Prof. **13**(1), 12–15 (2011)
17. Gruber, H.: J. Ind. Bus. Econ. 44(3), 327–342 (2017). https://doi.org/10.1007/s40812-017-0073-x

18. Van Caenegem, B., Skordas, T.: Telecommun. Syst. **35**(3–4), 89–97 (2007). https://doi.org/10.1007/s11235-007-9043-3
19. Lee, S., Trimi, S.: J. Bus. Res. 123, 14–22 (2021). https://doi.org/10.1016/j.jbusres.2020.09.041
20. Lee, S., Lim, S.: Living Innovation: From Value Creation to the Greater Good. Emerald Publishing Ltd., Bingley (2018)

Modern Trends of Development of Energy Saving Management in Organization

Aleksey S. Budakov[1](✉), Alla A. Lysochenko[2], Irina A. Pryadko[2], and Chzhan Minyuy[3]

[1] International Institute for Energy Policy and Diplomacy, MGIMO University, Moscow, Russia
[2] Southern Federal University, Rostov-on-Don, Russia
[3] Don State Technical University, Rostov-on-Don, Russia

Abstract. The modern agenda of energy saving management in the management of organizations dictates the need to involve automated systems for utility monitoring and metering. As the information sources, the data on the use of closed methods for assessing the effectiveness of a transition between price categories in the implementation of complex solutions in connection with the creation of automated information and measurement systems for utility monitoring and metering at industrial enterprises have been analyzed. Within the framework of this paper, the main opportunities of introducing automated information and measuring measurement systems for utility monitoring and metering have been presented, as well as their key advantages over traditional metering devices. This study explores the technology of smart metering and real-world measurements derived from the activities of a commercial company. And the article also presents the application of the proposed methodology to determine the basic results related to energy saving and improving the economic efficiency of the company.

Keywords: Energy saving · Energy efficiency · Organization management · Electricity metering

1 Introduction

Domestic and foreign literature pays significant attention to the issues of organizing the management of energy saving and increasing energy efficiency.

This paper is related to the study of local problems of energy management in connection with the fact that the study of energy conservation management issues is associated for the most part with the description of technological capabilities, their impact on potential cost-effectiveness.

The result of the issue under consideration is the reflection of the economic effect from the introduction of local energy saving mechanisms – the installation of automated systems for utility monitoring and metering without developing an expensive energy saving system and assessment of the current energy balance of the enterprise.

P. V. Trifonov and M. V. Charaeva (Eds.): DITEM 2021, LNNS 380, pp. 471–478, 2022.
https://doi.org/10.1007/978-3-030-94245-8_64

As noted in industry research papers [1], accurate and cost-effective large-scale energy forecasting is a vital step towards next-generation energy efficiency initiatives.

2 Methods

Smart metering is a critical method for organizing more accurate asset management and for developing a new awareness of energy use. The smart meter is solely designed to check the energy consumption and control each electronic equipment [2].

Zykov [3] puts a need to create effective energy metering systems both for utilities and for industrial enterprises. An attempt to find a comprehensive solution to the problem of energy metering implies an increase in the metering efficiency using automation methods, modernization and implementation of up-to-date engineering equipment and energy metering systems.

Innovations in the energy sector lead to an increase in the efficiency and quality of generation, conversion, transport, storage and distribution of electric and thermal energy, to a decrease in the negative environmental impact and, in general, provide opportunities for economic growth [4].

The first step in reducing energy costs is to recognize the importance of energy as a costly resource rather than as an overhead for an enterprise. It is necessary to exclude from practice the attitude to energy saving as to a specialized engineering activity, since energy is an object of production management, not just a technical element [5].

Improving energy efficiency is an indispensable condition for ensuring competitiveness and the main motivation for achieving energy efficiency indicators through the modernization of equipment, the use of new technological solutions, as well as the introduction of organizational measures [6].

Having analyzed the profile studies [7–10] it is proposed to identify the following some economic barriers to the implementation of energy-saving technologies:

- non-market barriers (heterogeneity of economic agents, economic market failures);
- technological market failures;
- asymmetry of information;
- floating technology costs;
- inconsistency of motives and conditions for the introduction of energy-saving technologies;
- hard access to credit capital;
- high level of investment risks.

Utility monitoring and metering, in turn, can help to manage an enterprise more efficiently, in a timely manner and provide full accountability for activities financed from borrowed capital.

Energy consumption can be monitored and analyzed for various reasons. The root cause may be related to operating costs, building's performance, customer expectations, rental commitments, production efficiency, continuity of service, sustainability goals, or management reporting.

Energy metering software provides analysis of energy use and distribution of costs for individual generating units, areas and the entire system.

Reports are segmented by energy cost analysis, profit making, production billing, energy consumption and can be made out for generators and intersystem exchanges.

To identify the general need for metering devices, their types and areas with the highest energy consumption, it is necessary to conduct a complete inventory of the capacities and equipment involved in manufacturing.

After the formulation of motives and a preliminary assessment of the goals of introducing electricity metering and monitoring, the next step is to consider possible costs and draw a preliminary budget.

Actual capital costs and subscription fees for the electricity metering and monitoring system varies depending on the nature of the facility, the needs of the facility and the selected equipment supplier.

An overview of energy management and monitoring platforms ranges from the simplest to the most complex.

Specialized energy survey and analysis service providers offer platforms that collect data from multiple sources including gauges, temperature sources, other sensors and billing data, provide real-time monitoring, visualization and reporting data, interpret data in analytical systems and provide active monitoring and optimization of the system.

Modern realities of the regulatory framework in terms of the implementation of intelligent energy metering systems affect to a greater extent three main market players:

- guaranteeing suppliers of electrical energy;
- power grid companies;
- developers.

The need to install metering devices for guaranteeing suppliers and grids occurs under one of the following conditions:

- a device is out of order;
- a device was absent;
- within a technological connection.

Metering devices must comply with the rules for providing access to a minimum set of functions:

- transmission of meter readings;
- providing information on the parameters of energy efficiency;
- limitation/resumption of the consumption mode;
- setting/changing zones of the day for summing up the volumes of energy efficiency;
- display (transfer) of settings;
- transfer of reference information;

- transfer of data archives;
- notification of emergency situations.

Implementation of measures to control energy saving can be aimed at:

1. Improving energy management practices, including:
- setting and achieving energy efficiency targets – compared with other facilities;
- establishing an energy background and using it to measure savings from energy conservation measures;
- detecting unexpected increases in energy consumption;
- knowledge of the impact of each piece of equipment or area;
- monitoring the impact of efficiency measures.
2. Cost management, including:
- control of electricity costs;
- charging tenants for the use of electricity;
- control of energy consumption by individual units;
- energy consumption forecast for various operating conditions;
- reduced demand (for example, by managing peak loads and saving costs for paying for network power;
- increasing the efficiency of implementing business cases for the projects seeking to improve efficiency.
3. Quality of corporate reporting in the field of sustainable development, in particular through participation in carbon reduction programs and voluntary sustainability reporting schemes to comply with corporate social responsibility.
4. Supporting funded solutions and grants that assess and verify guaranteed savings under contracts for energy efficiency, including as part of the reconstruction of buildings, etc.
5. Measurement and validation using an internationally recognized method (e.g., The International Performance Measurement and Verification Protocol to check organization's performance by performance ratings) to quantify energy efficiency cost savings.
6. Maintaining compliance with the rules for charging tenants or subleasers for electricity or the use of metering devices for other operations.
7. Troubleshooting and diagnostics, including identification of equipment misalignment and component failure to optimize maintenance and replacement times.
8. Ensuring interaction with any existing control and management systems, for example, BMS, PLC, SCADA – identifying inefficiencies in the control settings and equipment operation and extending its life.

3 Results

Well-designed energy saving programs provide customers of all types with the opportunity to take energy saving measures and cut electricity bills.

To reflect the local effect of smart metering devices without developing an expensive energy saving system and assessing the current energy balance of the enterprise, the experience of their implementation was analyzed at an industrial enterprise in Goryachy Klyuch city.

As part of an analysis of the effectiveness of the implementation of automated information and measuring systems for monitoring and accounting of energy resources, 8 objects were analyzed (Table 1).

Table 1. Distribution of objects by price category.

Object	Voltage level	MMEU group	Optimal price category (Pc)
Water intake 1–2	CH-2	Less than 670 kW	4/6
			4/6
Water intake 1–2	CH-2	Less than 670 kW	4/6
			4/6
Water intake 2	CH-2	Less than 670 kW	3/5
			4/6
Gas pumping station	CH-2	Less than 670 kW	4/6
			4/6
Pump room 3	HH	Less than 670 kW	3/5
			3/5
Pump room 4	CH-2	Less than 670 kW	4/6
			4/6
Sewage treatment	CH-2	Less than 670 kW	4/6
			4/6
Local pumping station Lead-in-1	CH-2	Less than 670 kW	4/6
			4/6
Local pumping station pump room-Dubzavod	CH-2	Less than 670 kW	4/6
			4/6

The costs of installing information and measuring systems for utility monitoring and metering at 8 facilities were distributed by 6 different price categories (Table 2).

Table 2. Distribution of costs by price category.

Object	PC1 Costs, Rub ths	PC2 Costs (3 Day Zones), Rub ths	PC2 Costs (2 Day Zones), Rub ths	PC3 Costs, Rub ths	PC4 Costs, Rub ths	PC5 Costs, Rub ths	PC6 Costs, Rub ths
Water intake 1–2	380.0	408.1	397.6	341.1	260.0	340.2	259.1
Water intake 1–3	366.1	393.2	383.1	328.7	250.5	327.8	249.7
Water intake 2	393.4	426.1	428.6	344.8	346.7	344.0	345.8
Gas pumping station	173.6	186.5	181.7	154.6	116.2	154.2	115.8
Pump room 3	44.4	42.9	44.0	39.8	60.1	39.7	60.0
Pump room 4	101.4	115.9	112.7	95.0	77.1	94.8	76.9
Sewage treatment	418.7	449.7	438.2	375.9	286.5	374.9	285.6
Local pumping station Lead-in-1	133.4	143.3	139.6	119.8	91.3	119.5	91.0
Local pumping station pump room-Dubzavod	48.3	52.5	51.7	43.6	35.9	43.5	35.8

The cost for price category 1 amounted to 277,463.68 rubles excluding VAT. The cost for price category 3 amounted to 275,395.22 rubles excluding VAT, including savings for the reporting period in the amount of 2,068.46 rubles excluding VAT, which is 0.75% (Table 3).

Table 3. Distribution of savings by price category.

Object	Saving (PC2 to PC1), Rub ths	%	Saving (PC2 to PC1), Rub ths	%	Saving (PC3 to PC1), Rub ths	%	Saving (PC4 to PC1), Rub ths	%	Saving (PC5 to PC1), Rub ths	%	Saving (PC6 to PC1), Rub ths	%
Water intake 1–2	−28.1	−7.4	−17.6	−4.6	38.9	10.2	120.0	31.6	39.8	10.5	120.9	31.8
Water intake 1–3	−27.1	−7.4	−17.0	−4.6	37.5	10.2	115.6	31.6	38.3	10.5	116.4	31.8
Water intake 2	−32.6	−8.3	−35.1	−8.9	48.6	12.4	46.8	11.9	49.5	12.6	47.7	12.1
Gas pumping station	−12.8	−7.4	−8.0	−4.6	19.0	11.0	57.4	33.1	19.4	11.2	57.8	33.3
Pump room 3	1.5	3.3	393.8	0.9	4.6	10.4	−15.7	−35.3	4.7	10.6	−15.6	−35.1
Pump room 4	−14.5	−14.3	−11.3	−11.1	6.4	6.3	24.3	23.9	6.6	6.5	24.5	24.2

(continued)

Table 3. (*continued*)

Object	Saving (PC2 to PC1), Rub ths	%	Saving (PC2 to PC1), Rub ths	%	Saving (PC3 to PC1), Rub ths	%	Saving (PC4 to PC1), Rub ths	%	Saving (PC5 to PC1), Rub ths	%	Saving (PC6 to PC1), Rub ths	%
Sewage treatment	−31.0	−7.4	−19.4	−4.6	42.9	10.2	132.2	31.6	43.8	10.5	133.2	31.8
Local pumping station-Lead-in-1	−9.9	−7.4	−6.2	−4.6	13.7	10.2	42.1	31.6	14.0	10.5	42.4	31.8
Local pumping station Pump room-Dubzavod	−4.2	−8.8	−3.4	−7.0	4.7	9.7	12.3	25.5	4.8	10.0	12.4	25.8

The payment to the energy utility amounted to 55% of the resulting savings: 1,137.65 rubles excluding VAT.

4 Conclusion

Automated systems for commercial metering of electricity are designed to automate utility metering and ensure control of metering indicators in accordance with the established requirements, as well as for automated data transmission to the information collection centers of the energy supplier.

Such systems perform the following functions:

- automated commercial and engineering (controlling, technological) metering of the amount of electrical energy, thermal energy (heat carrier), cold and hot water;
- automation of the process of commercial and engineering utility metering;
- increasing the reliability, efficiency and accuracy of metering due to up-to-date metering devices and increasing the degree of protection of equipment and commercial information from unauthorized interference.

References

1. Jain, R.K., Smith, K.M., Culligan, P.J., Taylor, J.E.: Appl. Energy **123**, 168–178 (2016). https://doi.org/10.1016/j.apenergy.2014.02.057
2. Kurde, A., Kulkarni, V.S.: Int. J. Sci. Res. Publ. **6**(9), 411–415 (2016)
3. Zykov, D.D., Urazaev, D.R., Khabibulina, N.Yu.: Rep. TUSUR **20**(4), 131–134 (2017). https://doi.org/10.21293/1818-0442-2017-20-4-131-133
4. Mikheev, A.V.: Inf. Math. Tech. Sci. Manag. **4**(20), 158–167 (2020). https://doi.org/10.38028/ESI.2020.20.4.014
5. Kharchenko, V.V., Panchenko, V.A.: E-Scio **5**(44) (2020)
6. Ilyina, M.N., Ilyina, G.F.: TPU Bull. 91–98 (2018)

7. Agybayev, M.Sh., Rakhimbekova, Zh.S.: Eur. J. Econ. Manag. Sci. **2**, 96–100 (2020). https://doi.org/10.29013/EJEMS-20-2-96-100
8. Choi, K.S., Hong, S.: New design approaches and implementation of smart metering system. In: The 18th IEEE International Symposium on Consumer Electronics, 2014, pp. 1–2 (2014). https://doi.org/10.1109/ISCE.2014.6884309
9. Živic, N.S., Ur-Rehman, O., Ruland, C.: Evolution of smart metering systems. In: 2015 23rd Telecommunications Forum Telfor (TELFOR) 2015, pp. 635–638 (2015). https://doi.org/10.1109/TELFOR.2015.7377547
10. Liu, X., et al.: J. Clean. Prod. **251**, 119693 (2020). https://doi.org/10.1016/j.jclepro.2019.119693

Financial Economy: Priorities, Trends, New Solutions and Areas

The Rationale for the Implementation of Investment Projects of the Fuel and Energy Complex Enterprise

Miroslava G. Salko(✉), Elena M. Deberdieva, Olga V. Lenkova, Larisa S. Kovalzhina, and Ekaterina A. Funtikova

Industrial University of Tyumen, Tyumen 625000, Russia
salkomg@tyuiu.ru

Abstract. Currently, many companies pay special attention to the organization and implementation of activities within the framework of their developed investment policy, as they see their responsibility in maintaining the stability of the development of the state's fuel and energy complex. The leaders in the volume of investment in the national economy are industry enterprises with activities associated with risks to human health and the environment. They support directions related to everyday areas of society, such as health care, sports, education, culture, and relevant to their professional activities. Socially-oriented projects are based on the attributes of industrial companies: a significant spread of subsidiaries across the Russian Federation, the age structure of personnel, the presence of subdivisions in the areas of nutrition, medical care, and health and recreation and cultural institutions, etc. The support of many social investment projects by the oil and gas company allowed the authors to consider the problem of effective implementation of social programs. The authors systematized the procedure for assessment of investment projects by calculating the integrated index, considered various methods of transformation and aggregation of indices, and formed the base of indices, which can be used to fully determine the social, economic, and financial efficiency of social programs. The formed complex integrated assessment of investment projects allowed justifying the choice of the most promising investment projects for further implementation of the largest oil and gas producing company in Russia.

Keywords: Investment project · Oil and gas production company · Project evaluation methods · Integrated index

1 Introduction

The implementation of positive changes in public life is achieved through investment design, which aims to design efficient production facilities and create social value.

Important state tasks in the social policy include the protection of public health, assistance to social structures of the population, the formation of a social service system, etc. [1].

P. V. Trifonov and M. V. Charaeva (Eds.): DITEM 2021, LNNS 380, pp. 481–487, 2022.
https://doi.org/10.1007/978-3-030-94245-8_65

State authorities emphasize implementing housing and educational programs; they are engaged in developing health care, agriculture, and support for large families and young professionals in rural areas [2, 3].

In the field of investment design, the implementation of social projects is actively engaged in sectoral enterprises. The oil and gas production enterprise, which is the object of the study, annually allocates money for social investment design as shown in Fig. 1 [4].

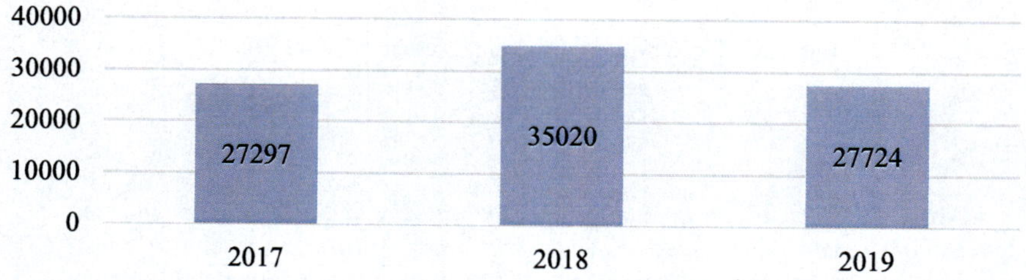

Fig. 1. Dynamics of oil and gas production company investments in social projects, RUB mln.

Oil and gas companies allocate funds for the most important social projects: construction of sports and recreational facilities, support for social events, funding housing programs for low-income citizens, and so on. Special attention should be paid to projects aimed at professional training and professional development of existing and potential enterprise personnel, support for socially-oriented programs, ensuring occupational safety and health [5].

The effective implementation of investment projects largely depends on a high-quality socio-economic assessment. The existing socio-economic phenomena are caused by a multitude of parameters, which creates difficulties associated with identifying the structure of relationships between these parameters. To solve this problem, the authors recommend the use of integral assessment, methods of multivariate statistical analysis, the implementation of which will justify the selection of the most promising projects in terms of the economic and social importance of their implementation [6–11].

2 Methods

The use of multidimensional statistical methods aims to increase the validity of the choice of project implementation options that will allow assessing the authenticity of the results obtained in conditions of limited information.

Integral assessments aim to determine the overall indicator, which provides methodological integrity of all private indicators and uniqueness of the assessment of the object under study. It is recommended to start with selecting and calculating indicators to calculate the integrated index for various investment projects, followed by the transformation of the results to ensure their comparability. The use of methods of transformation of indicators forms the content side of the integrated index, for example, the rating method, the "maximum-minimum" method, etc. [12, 13].

Next, the aggregation of indicators is carried out using various methods of summarizing private indicators into common ones. It is possible to carry out the objects' ranking, provided their number is small based on integral evaluations. Each object is assigned a rating that corresponds to a place in the ranked row [14].

3 Results

The authors have chosen three investment projects of oil and gas production company social orientation, for which integrated indices are calculated to justify the choice of the most effective and expedient for implementing the project. Project 1 is the opening of a sports and recreation complex in the village of Udmurtia. Project 2 is an update of the material and technical base in three schools in the Amur region. Project 3 involves implementing major repairs of two departments of the city hospital in the Komi Republic [15–17].

It is proposed to use three groups of indicators to determine the integral indicator for social projects. The first group includes evaluation criteria that measure social effectiveness, which determines the project's impact on the lives of the population. The second group of indicators includes those covering the social orientation of the project. The third group is a set of financial and economic indicators that assess the macroeconomic content of investment projects [18–21].

At the stage of transformation of indicators, the authors used the "maximum-minimum" method since it avoids the impact of a particular indicator on the integrated one. The method of summing up the weighted average arithmetic group indicators was chosen to carry out the aggregation of indicators. The authors used the hierarchy analysis method, and, thus, the values of the significance ratios were determined to identify the value of the integrated index.

Table 1 shows the calculations of the integrated index for each social project. In the presence of statistical data in the sources of determining the values for the indicator, the base was carried out using direct calculation. In the absence of the possibility of objective measurements of the parameters of social projects, the expert method was used, through which specific evaluation results were obtained [22–24].

The highest value of the integrated index belongs to project 2; therefore, it is the most attractive for implementation, as it has low implementation costs. The first and second projects have lower values of the integrated index, as their implementation requires high initial investments, and project 1 has costly annual maintenance and low financial income. Despite this, all three considered investment projects are socially effective and allow solving important social problems.

Table 1. Evaluation of socially oriented investment projects (fragment).

Indicator	Feature/value			Changed indicator			Aggregated value		
Social performance indicators							0.20		
Project	1	2	3	1	2	3	1	2	3
Indicator reflecting the standard of living	5,304 RUB	3,075 RUB	3,410 RUB	0.829	0.604	0.549	0.56	0.20	0.56
Indicator reflecting the level of population employment	0.54%	0.89%	0.46%	0.484	0.188	0.433	0.26	0.56	0.26
Social orientation index							0.42		
Target audience	All village residents	Students teaching staff	Middle-aged and elderly people	0.864	0.657	0.608	0.20	0.20	0.20
Project priority indicator in the strategy for socio-economic development of the region	Developing the potential of children through sports; maintaining a healthy lifestyle	Unlocking the children potential the learning process is more exciting and productive	Favorable conditions for the work of medical staff and patient care	0.750	0.735	0.784	0.30	0.30	0.30
Growth rate of social tension in society	Residents will have the opportunity to engage in sports and lead a healthy lifestyle	Students will be more involved in the educational process and sports activities	Patients will obtain medical care in a comfort-able and supportive environment	0.333	0.420	0.493	0.10	0.10	0.10
Financial and economic indicators							0.38		
Social profitability ratio	0.079	0.88	0.296	0.078	0.880	0.295	0.41	0.41	0.41
Overall project profitability	0.089	1.32	0.309	0.088	1.320	0.308	0.31	0.31	0.31
Net present social value	−83.8 mln	4.87mln	−1.3 mln	– 0.840	0.482	−0.143	0.17	0.17	0.17
Total net present value	−78.2 mln	10,0 mln	10.4 mln	– 0.784	1.004	1.042	0.11	0.11	0.11
Integrated index							0.353	0.751	0.510

4 Discussion

The methodical approach proposed by the authors to evaluating investment projects allowed justifying the implementation of the most promising project. Adjustment of the system indicators is possible to justify the implementation of projects with the allocation of environmental importance, taking into account the energy efficiency of design solutions and other criteria reflecting the specifics of investment projects. It is also possible to perform for the industry as a whole to identify the share of the potential of investment decisions of individual economic units.

The methodological approach recommended by the authors is attractive in its use of several techniques to justify the system of indicators, assess their significance, and integrate the evaluation criteria into a single indicator. The scientific potential of the study is expressed in the possibility of a reasonable choice of methodology, taking into account the peculiarities of the integration of various indicators.

Oil and gas companies investing in the development of education, health care, sports, and recreational projects ensure the qualitative growth of human capital. The measurement of labor potential during the implementation of projects allows assessing the long-term prospects and potential benefits of companies participating in social investment.

It is possible to expand the evaluation criteria system for investment projects with the allocation of sectoral and macroeconomic indicators as a development of the author's recommendations. In this case, it is possible assessing the impact of investment projects on the development of the industry and the state, which will increase the validity of the development of investment policies for oil and gas companies.

5 Conclusion

As a result of scientific research, methodological approaches to calculating the integrated index for evaluating projects were systematized. In general, this allowed justifying the methodological tools for implementing the most effective investment projects of social orientation.

Consideration of the methods and their organization into integral estimates allowed the authors to recommend the transformation of indicators. The use of various methodological tools allowed for improving the quality and reliability of solutions to implement socially successful investment projects.

The scientific approach proposed by the authors to the assessment of social investment projects will allow to:

- improve the methodological tools for evaluating socially-oriented projects;
- justify the feasibility of evaluating social investment projects in various industries and regions;
- determine the need for the implementation of investment projects;

The evaluation of socially-oriented projects is specific and does not always determine the priority of implementing more commercially effective options. However, for the most part, the social efficiency of the project implementation is fundamental for the oil and gas company, which aims to contribute to improving the quality of life in the Russian Federation.

References

1. The Constitution of the Russian Federation: No. 11-FKZ from July 01 2020. In: Collection of Legislation of the Russian Federation. Moscow, Russia (2020)
2. Edronova, V.N., Solovyova, N.V.: Reg. Econ.: Theory Pract. **40**, 2–7 (2010)
3. Kuzmina, N.L., Tonysheva, L.L.: Reg. Probl. Econ. Transform. **9**, 118–126 (2018)
4. Our social initiatives. https://sustainability.gazpromreport.ru/2019/5-social/
5. Employee development and support. https://www.gazprom-neft.ru/social/employee-development
6. Azizov, Sh.S., Nurov, A.G., Azizov, R.Sh.: Tajikistan Mod. World **2**(57), 85–94 (2017)
7. Andreeva, E.I., Gorshkova, I.D., Kovalevskaya, A.S.: Recommendations for assessing the socio-economic effectiveness of social programs. Definitions, approaches, practical experience. Prospekt, Moscow (2014)
8. A sports and recreation complex was opened in Udmurtia under the "Gazprom for Children" program. http://www.gazprom.ru/about/subsidiaries/news/2018/december/article46952
9. Manokhin, P.E., Shishkina, M.L.: Determining the social effect of the reconstruction of public buildings. In: Manokhin, P.E., Shishkina, M.L. (eds.) Materials of the 3d International Scientific Conference Ser. Young Scientist: Technical Sciences: Theory and Practice. Limited Liability Company "Molodoi Ucheny" Publishing House Chita (2016)
10. Maryganova, E.A.: Social efficiency: concept and evaluation. In: Maryganova, E.A. (ed.) Materials of the International Scientific and Practical Conference: Values and Interests of Modern Society. Moscow State University of Economics, Statistics, and Informatics, Moscow (2015)
11. Denscombe, M.: The Good Research Guide for Small-Scale Social Research Projects. Open University Press, Berkshire (2010)
12. Klyushnikova, E.V., Shitova, E.M.: Infocenter **1**(10), 4–18 (2016)
13. Shekova, E.L.: Economics and management of a non-profit organization. Lan, Saint Petersburg (2004)
14. Smarchkova, L.V.: A systematic approach to determining the socio-economic efficiency of enterprises. In: Smarchkova, L.V. (ed.) Materials of the International Scientific and Practical Conference: Innovative Entrepreneurship: Socio-economic and Marketing Aspects. Nauchnaya kniga, Voronezh (2017)
15. Buzmakova, M.V.: Innov. Invest. **8**, 32–37 (2019)
16. The opening of two renovated departments in the city hospital took place in Ukhta. https://www.gazprom.ru/about/subsidiaries/news/2019/december/article496939
17. Gazprom processing Blagoveshchensk helped three schools of the Svobodnensky district to update the facilities and equipment. https://www.gazprom.ru/about/subsidiaries/news/2020/january/article498891
18. Avtonomov, A.S., Khananashvili, N.L.: Social Projects Evaluation. Yurist, Moscow (2014)
19. Alaev, A.A., Kozlova, S.V., Malyutin, K.M., Perova, I.T.: Res. Financ. Inst. Financ. J. **4**(26), 41–52 (2015)
20. Nagornykh, V.N., Butenko, T.V.: Socio-cultural and economic aspects of the development of physical culture and sports in the subjects of the Russian Federation. In: Nagornykh, V.N., Butenko, T.V. (eds.) Materials of the 11th All-Russian Scientific and Practical Conference: The Development of Physical Culture and Sports in the Context of Human Self-realization in Current Socio-economic Conditions. Lipetsk State Pedagogical University, Lipetsk (2018)
21. Fedorova, M.V.: Transp Syst. Tech. **3**(4), 179–203 (2017)
22. Galtsova, A.A., Dudkin, A.I.: Ural Sci. Bull. **12**(1), 23–24 (2017)

23. Ludwig, E.S.: Socio-economic content of mass sports events. In: Ludwig, E.S. (ed.) 30th International Plekhanov Readings, pp. 330–334. Plekhanov Russian University of Economics, Moscow (2017)
24. Funtikova, E.A., Glukhova, M.G.: On methods of evaluating social projects of gas-producing enterprises. In: Plenkina, V.V. (ed.) Materials of the All-Russian Scientific and Practical Conference With International Participation: Innovations in Managing Regional and Sectoral Development. Tyumen Industrial University, Tyumen (2019)

Manufacturing Industry Financial Condition of the City of Rostov-On-Don: Examine by the Altman Models

Oscar S. Gasanov[1(✉)] and Marina A. Chirskaya[2]

[1] Don State Technical University, Accounting, Analysis and Audit, Rostov-on-Don, Russia
[2] Rostov State University of Economics, Financial Management, Rostov-on-Don, Russia

Abstract. In the context of continuing population growth in regional centers, one of which is the city of Rostov-on-Don, industrial enterprises of the city are becoming important as centers of job creation. The reduction in the number and volume of output of industrial enterprises in the city has significant dimensions. This study is scoped at assessing the financial stability of industrial enterprises that continue to operate in the city using the Altman Z-score models, which have proven their effectiveness in the probability of bankruptcy at the global level. The study was conducted on a sample of ten industrial enterprises in the manufacturing industry. Two Altman Z-score models were used as tools: for private enterprises in the industrial sector, and for enterprises from emerging markets. The results of the analysis suggest that the Z-model for private industrial companies is better suited for assessing the financial stability of the manufacturing industry companies. The use of the classic Z-score model and the emerging market Z-score model has some limitations. Assessment of financial stability based on Altman models indicates a good and satisfactory financial condition of the majority of industrial enterprises in the city. As a result of the Altman Z-score models assessment, it was found that the main variables that negatively affect the final results of the Z-score are the long-term functioning of economic entities with negative financial results and negative net working capital.

Keywords: Financial stability · Bankruptcy · Manufacturing industry · Altman's Z-score

1 Introduction

The preservation and development of industrial enterprises are considered today as the urgent problems of economic well-being for Russian regions [1]. For the development of conditions for the realization of intellectual, labor, and resource potential, the preservation of industrial potential for large regional centers is especially important [2]. Regional and interregional interaction, which is considered by economists as an element of intensifying economic growth and development of innovations, cannot be imagined without regional industrial centers [3]. Rostov-on-Don is a regional center of the South of Russia; historically the city was considered its industrial center. In the past two decades, the

P. V. Trifonov and M. V. Charaeva (Eds.): DITEM 2021, LNNS 380, pp. 488–499, 2022.
https://doi.org/10.1007/978-3-030-94245-8_66

decrease in industrial production in the city is noticeable without statistical calculations. Suffice it to say that during this period more than 10 large industrial enterprises of the manufacturing industry were closed down in the city.

Industrial enterprise's bankruptcy is a worldwide problem and is believed to cause economic damage around the world [4]. The main reasons for the industrial manufacturing crisis in the city, of course, were the same factors as throughout Russia: lack of experience in managerial work in the transition to a market economy, inability to adapt to the conditions of a market economy, devaluation, loss of working capital, loss of established markets and technological ties, and, most importantly, low competitiveness and lack of demand for products.

Nevertheless, some industrial enterprises of the city were able to more or less adapt to the new economic conditions and continue to function at present. Adaptation does not mean that there will be no problems in the future since the loss of financial solvency and bankruptcy are the main problems in the context of current events, which can lead to price imbalances, disruptions in the supply of components and finished products. In these conditions, the assessment of financial stability and the ability to withstand shocks look especially relevant. Knowing the likelihood of bankruptcy with reasonable accuracy in advance, companies can better protect themselves and take action to reduce the risk of financial loss or avoid bankruptcy. Besides, the further development of industrial enterprises in the city cannot be imagined without the introduction of innovations and modern technologies that require large investments and investors should have access to qualified information about the financial condition of the companies in which they are interested in investing. For regional economic authorities, such an assessment can help in making decisions on the structural policy of the region, which is recognized as an important component of a balanced economy [5, 6]. Therefore, it will be useful to assess the financial health of industrial enterprises in the city using proven and recognized methods.

Therefore, this study is aimed at assessing the financial sustainability of manufacturing enterprises in the city of Rostov-on-Don using Altman's Z-models. The classical model and its variations are designed to assess the probability of default (bankruptcy) of a company based on the analysis of several indicators of its financial and economic activities and the calculation of the consolidated indicator of the financial condition [7].

Our study contains several differences from previous studies in this area of assessing the financial problems of business entities. First, we apply Altman's methodology in the study of industrial companies in the mechanical engineering segment, located in a separate zone, which is an example of a Russian industrial city.

Secondly, in the Russian-language segment of the scientific literature, we have found no previous works directly related to our research. In the scientific electronic library "eLibrary" for the keywords "Rostov-on-Don" and "Altman" no single article can be hunted. The only work that addresses the problem of socio-economic development of the city is [8]. At the same time, there is a lot of applied research on assessing the financial stability of companies using the Altman Z-model. In existing studies based on different variants of the Altman Z-model the financial stability of individual companies [9, 10], branches of the economy [7, 11, 12] or countries [13–17] are assessed.

Thirdly, taking into account the existing organizational and legal forms of ownership and the absence of circulating securities of companies in the engineering industry of Rostov-on-Don on the stock market, the study was carried out based on an assessment of data from non-public companies. At the same time, with certain adjustments for the non-public nature of companies from the sample, we assessed all three Altman models.

Fourthly, we hope that this study will be useful for investors, as well as for CEOs of companies in the sample, the Ministry of Economic Development of the region, and the City Administration. This hope is justified by the fact that the assessment using Altman's models provides signals of financial distress before it occurs with high efficiency [7]. Investors can get information to analyze industrial companies operating in the city and make decisions about their presence in the region and the city.

2 Materials and Methods

2.1 Methodology

The Z-score model was first proposed by E. Altman in 1968 using a linear combination of five weighted coefficient indicators [12]. The coefficients for the indicators were estimated by discriminant analysis and comparison of two data sets of public companies: survivors and those who filed for bankruptcy with a comparison by industry and approximate size (assets). After more than fifty years, researchers still widely view the Altman Z-model as an indicator of a company's financial classification. Several authors estimate the probability of forecast accuracy for this model at the level of up to 90% [17].

We use two Altman models, a model for private companies in the manufacturing sector whose shares are not listed on the stock exchange (Z_{NP}-score) and a model for companies from emerging markets (Z_{EM}-score). The calculation of integral indicators is carried out according to the formulas (1) and (2) [13, 14]:

$$Z_{NP} = 0.717 * X_1 + 0.847 * X_2 + 3.107 * X_3 + 0.420 * X_4 + 0.998 * X_5 \tag{1}$$

$$ZEM = 3.25 + 6.56X1 + 3.26X2 + 6.72X3 + 1.05X4 \tag{2}$$

where X_1 – Working capital/Total assets;
X_2 – Retained Earnings/Total assets;
X_3 – Earnings before interest and taxes/Total assets;
X_4 – Equity /Total Liabilities (long-term and current);
X_5 – Sales/Total assets.

For Z_{NP} interpretation of the results is carried out according to the scale:

$Z_{NP} > 2.99$ – "Non-bankrupt sector" (the probability of bankruptcy is low);
$1.23 < Z_{NP} < 2.99$ – "Gray area" (area of uncertainty);
$Z_{NP} < 1.23$ – "Bankrupt sector" (high probability of bankruptcy, risk area).
For Z_{EM} interpretation of the results is carried out according to the scale:
$Z_{EM} > 2.6$ – "Non-bankrupt sector" (the probability of bankruptcy is low);
$1.1 < Z_{EM} < 2.6$ – "Gray area" (area of uncertainty);
$Z_{EM} < 1.1$ – "Bankrupt sector" (high probability of bankruptcy, risk area).

2.2 Data

In the study, the financial stability of the sample of 10 manufacturing enterprises in Rostov-on-Don was assessed using all three described models. The selection of manufacturing enterprises located in the city of Rostov-on-Don was made from the list presented on the portal "Encyclopedia of Industry of Russia" (WikiProm) [18]. The data for the analysis were obtained on the information disclosure portals Rusprofile [19] and Interfax [20]. The number of enterprises equal to ten is not a choice of a round number, but a random coincidence of a choice from enterprises located in the city according to two criteria: (1) the main activity is related to the production of industrial products (except for building materials) and (2) the volume of annual revenue is at least 100 million rubles based on the results of two reporting periods out of five studied ones (from 2014 to 2018 inclusive). The sample list is in Table 1. There are also the main parameters of the financial performance of companies for 2018.

Table 1. The sample of industrial enterprises in Rostov-on-Don (data as of 01.01.2019).

Full name	Code	Main activity	Assets	Revenue	Net profit
PJSC "Rostvertol"	RVP	Helicopters production	95 583 228	93 988 806	15 911 956
LLC "Rostselmash"	RSM	Agro-industrial equipment production	40 207 278	39 704 836	4 751 434
JSC "Prodmash"	PDM	Food-industrial equipment production	1 011 120	272 125	1 023
JSC "Rostovgasoapparat"	RGA	Gas heating equipment production	382 020	364 225	525
LLC "Rostov foundry"	RFP	Foundry production	354 751	1 019 780	273
LLC "Rostov Compressors Plant"	RCP	Compressors (pumps) production	273 732	446 374	8 063
JSC "Sevkavelektroremot"	NCE	Electrical transformer production	249 721	489 978	35 636
LLC "Yujtechmontaj"	YTM	Metal structures production	1 955 96	364 225	525
LLC "Rosvelektroremot"	RER	Electrical machines equipment production	70 748	100 291	441
JSC "Elektrotechnika"	ELT	Technological equipment production	70 656	86 255	4 393

3 Results and Discussion

3.1 Descriptive Statistics of the Models

For both models, calculations were carried out based on the results of 2014–2018 financial years. The results of descriptive statistics of variables, the annual results of which are presented according to model (1), can be found in Table 2. In general, we can state a good homogeneity of the financial performance of the sample, since the mean values of the variables and the standard deviation (SD) are rather densely grouped. Significant deviations from the mean values are observed only for variable X4, especially for

Table 2. Descriptive statistics of variables in model (1).

Variables	Year	Min	Max	Mean	SD
X1	2014	– 0.6377	0.7663	0.2905	0.4329
	2015	−0.7523	0.7848	0.2918	0.4517
	2016	−0.7510	0.7660	0.2991	0.4409
	2017	−0.5676	0.7042	0.2947	0.4038
	2018	−0.6600	0.7142	0.1002	0.4927
X2	2014	−0.0563	0.7663	0.2578	0.3337
	2015	−0.0463	0.7800	0.2669	0.3280
	2016	−0.0291	0.7903	0.3171	0.3236
	2017	−0.0237	0.8127	0.4026	0.3109
	2018	−0.0563	0.8155	0.2745	0.3591
X3	2014	−0.0969	0.3101	0.1198	0.1410
	2015	−0.0630	0.1812	0.0660	0.0812
	2016	−0.0204	0.3901	0.1295	0.1377
	2017	−0.0182	0.4599	0.1630	0.1557
	2018	−0.0247	0.2078	0.0628	0.0758
X4	2014	−0.0532	4.4973	0.8035	1.4672
	2015	−0.0405	4.6107	0.8273	1.4596
	2016	0.0000	4.2791	0.9186	1.3615
	2017	0.0000	11.6635	2.2199	3.5775
	2018	−0.0532	11.8934	2.2073	3.6995
X5	2014	0.2961	3.2518	1.5352	1.0190
	2015	0.2056	2.7389	1.3594	0.8263
	2016	0.3932	3.1204	1.5993	0.8204
	2017	0.5946	2.9225	1.4474	0.7110
	2018	0.2691	2.8746	1.3782	0.6948

2017 and 2018. For model (2) we do not present data: results of calculations completely coincide with model (1).

Table 3 provides descriptive statistics for the X1–X5 and Z-score variables for each model. Calculations show that the minimum value for almost each of their variables demonstrates negative values; the exception is variables X5 in models (1).

Table 3. Descriptive statistics of variables and Z-score for the sample for the entire period.

Score	Min	Max	Mean	SD
X1	−0.7523	0.7848	0.2874	0.4125
X2	−0.0563	0.8155	0.3243	0.3173
X3	−0.0969	0.4599	0.1113	0.1232
X4	−0.0532	11.8934	2.2986	3.1600
X5	0.2056	3.2518	1.4702	0.7910
Z_{NP}	**0.11**	**8.80**	**3.26**	**2.12**
X1	−0.7523	0.7848	0.2874	0.4125
X2	−0.0563	0.8155	0.3243	0.3173
X3	−0.0969	0.4599	0.1113	0.1232
X4	−0.0532	11.8934	2.2986	3.1600
Z_{EM}	**−1.71**	**23.24**	**9.35**	**6.38**

The mean values of most of the variables are at comfortable levels. Calculations based on the indicators of return on assets and equity in the sample (Table 8) demonstrate results that are higher than the average for the Russian industry in the aggregate. According to Rosstat, the profitability of industrial enterprises assets (excluding SMEs) in the period 2014–2018 amounted to respectively: 2.5; 3.7; 5.9; 3.8 and 4.7% [21]. Fluctuations in variables for the period under study are within fairly moderate limits. The widest scatter is observed here for the X4 variable: the standard deviation is 3.16.

According to the results of the Z-account obtained for the entire period of estimate, one can also judge the sufficient stability of the financial condition of the sample. The results of statistics of model (1) are not alarming, the results for model (2) are slightly different: when evaluating by the model for emerging markets, both negative values of the composite indicator are observed, as well as the largest spread (24.95 units) with a standard deviation of 6.38 units. The average Z-score for both models is in the safe area. Therefore, based on the results of the analysis of descriptive statistics of variables and resulting indicators of the Z-score, it can be argued that the financial situation of industrial enterprises of the city of Rostov-on-Don for the entire period locates in a safe area.

Next, we will analyze the statistical results of evaluating models for each year of the study period separately. The obtained data of descriptive statistics are in Table 4.

Table 4. Descriptive statistics of Altman's Z-score by years.

Score	Year	Min	Max	Mean	SD
Z_{NP}	2014	0.75	5.53	3.03	1.89
	2015	0.58	7.62	3.09	2.37
	2016	1.19	8.80	3.48	2.33
	2017	0.59	8.59	3.50	2.39
	2018	0.11	7.52	2.79	2.09
Z_{EM}	2014	−0.32	15.68	8.56	5.31
	2015	−1.71	20.73	9.20	6.94
	2016	−1.24	23.24	9.61	6.86
	2017	−0.40	22.98	10.08	6.96
	2018	−0.81	22.26	7.51	7.58

At the annual assessment, there are both similarities and differences with the summary data for the entire period (Table 3). The negative minimum value of the Z-score was found only in the results by model (2). Here, the largest spread of the resulting indicator is observed. Z-score volatility for emerging markets ranges from 5.31 to 7.58, which is approximately three times higher than that for model (1). The presence of a 3.25 constant in the formula for calculating Z_{EM}, which should "elevate" the results obtained at the expense of the remaining elements of the formula, is superfluous for our sample. Should we conclude that this model is not suitable for assessing the financial soundness of our sample? The answer can be obtained only after detailing the results for each of the companies included in the sample and evaluating other financial indicators.

3.2 The Results of the Z-score for Each Enterprise

The results of calculating the variables and Z-score are detailed for each model separately and are presented in Tables 5, 6, respectively. According to the results of the assessment using model (1), six of the ten (60%) industrial enterprises in Rostov-on-Don are in a safe financial area. It is important to note that the financial condition of all enterprises in the sample has improved over the period.

According to the Emerging market model, eight out of ten elements (80%) of our sample received the maximum score. One enterprise (YTM) gets the lowest score: it is in the "risk area". The main reason for which the company receives negative Z-Scores is the negative amount of net current assets, insignificant volumes of operating, and retained earnings. Separately, we should dwell on the results of another outsider of model (1), "PDM", which in model (2) receives Z-scores (S) and (G). Therefore, we can conclude that this object gets an overestimated Z-score due to the (3.25) constant in formula (2).

Table 5. Estimate results for every enterprise – model (1): for non-public companies.

Enterprises	2014		2015		2016		2017		2018	
	Z_{NP}	Class.	Z_{NP}	Class.	Z_{NP}	Class.	Z_{NP}	Class.	Z_{NP}	Class.
RVP	0.85	D	1.08	D	1.90	G	2.10	G	2.64	G
RSM	1.72	G	2.72	G	3.76	S	5.12	S	4.21	S
PDM	0.75	D	0.75	D	1.19	D	0.59	D	0.11	D
RGA	3.47	S	3.38	S	3.42	S	3.08	S	3.56	S
RFP	4.38	S	3.28	S	4.97	S	5.10	S	3.70	S
RCP	n/d	–	0.99	D	1.25	G	1.88	G	1.72	G
NCE	4.66	S	4.85	S	4.34	S	3.97	S	3.65	S
YTM	1.24	G	0.58	D	1.33	G	0.85	D	1.50	G
RER	4.71	S	5.63	S	3.85	S	3.69	S	3.13	S
ELT	5.53	S	7.62	S	8.80	S	8.59	S	7.52	S

$Z_{NP} > 2.99$ – "Non-bankrupt sector" (the probability of bankruptcy is low) – (S);
$1.23 < Z_{NP} < 2.99$ – "Gray Area" (Area of Uncertainty) – (G);
$Z_{NP} < 1.23$ – "Bankrupt sector" (high probability of bankruptcy, risk area) – (D).

Table 6. Estimate results for every enterprise – model (2): for emerging market companies.

Enterprises	2014		2015		2016		2017		2018	
	Z_{EM}	Class.	Z_{EM}	Class.	Z_{EM}	Class.	Z_{EM}	Class.	Z_{EM}	Class.
RVP	7.21	S	7.56	S	9.13	S	9.26	S	9.56	S
RSM	7.13	S	9.12	S	11.54	S	16.68	S	15.18	S
PDM	3.65	S	3.06	S	3.61	S	2.49	G	2.41	G
RGA	15.68	S	15.81	S	15.33	S	13.61	S	13.95	S
RFP	5.19	S	5.52	S	9.15	S	10.70	S	6.45	S
RCP	n/d	–	3.89	S	3.07	S	3.49	S	3.21	S
NCE	11.82	S	11.88	S	10.84	S	10.81	S	10.15	S
YTM	−0.32	D	−1.71	D	−1.24	D	−0.40	D	−0.81	D
RER	12.72	S	16.19	S	11.47	S	11.17	S	9.99	S
ELT	13.92	S	20.73	S	23.24	S	22.98	S	22.26	S

$Z_{EM} > 2.6$ – "Non-bankrupt sector" (the probability of bankruptcy is low) – (S);
$1.1 < Z_{EM} < 2.6$ – "Gray Area" (Area of Uncertainty) – (G);
$Z_{EM} < 1.1$ – "Bankrupt sector" (high probability of bankruptcy, risk area) – (D).

3.3 The Estimate of Profitability and Financial Stability Indicators

We do not evaluate the level of defaults (due to the lack of reliable information), which would allow us to objectively assess the suitability of the employed models for applied use, therefore, the effectiveness of the financial results of the sample was assessed using traditional indicators of return on assets (ROA) and on equity (ROE). In addition to them, we evaluate the Margin of Financial Stability (MFS) estimated as the ratio of the number of own funds and long-term loans to total assets.

The choice of this indicator is due to the fact that a rather high requirement of – 75% and higher is imposed on its standard value. All three indicators were evaluated for all elements of the sample for the entire period of the analysis. Descriptive statistics of the ratios are provided in Table 7. Large variation is observed only in the indicators of ROE. This is due to the fact that one element of the sample (PDM) has a loss in the first two periods, and another (YTM) has a negative value of equity. These two objects showed the worst marks for all three Altman models. For the rest of the sample, there are quite good levels of profitability. Thus, when comparing estimates by Altman's models and indicators of profitability, we can state that all three models identified those two enterprises which have the worst indicators of profitability as unfavorable.

Table 7. Descriptive statistics of profitability and financial stability.

Variables	Year	Min	Max	Mean	SD
Return of assets (ROA)	2014	−0.0461	0.2635	0.1083	0.1043
	2015	−0.0123	0.1434	0.0634	0.0622
	2016	0.0079	0.3122	0.1130	0.1045
	2017	0.0012	0.3648	0.1342	0.1202
	2018	0.0008	0.1665	0.0520	0.0596
Return of equity (ROE)	2014	−2.3896	0.7138	−0.2725	1.0914
	2015	−5.0922	0.9507	−0.3060	1.7099
	2016	0.0147	6.8457	1.0441	2.0594
	2017	0.0034	0.8258	0.3440	0.2301
	2018	−1.9211	0.4722	−0.0146	0.6934
Margin of financial stability (MFS)	2014	−0.0562	0.8181	0.5440	0.2876
	2015	−0.0422	0.9069	0.5253	0.3377
	2016	0.0075	0.9163	0.5364	0.3320
	2017	0.0116	0.9210	0.5280	0.3286
	2018	-0.0562	0.9224	0.3984	0.3841

Finally, we compare the results of the models with the Margin of Financial Stability (MFS), which is not included in the variables of Altman's models. Full details of the ROA, ROE, and MFS are given in Table 8.

The sample is divided into five groups according to the MFS. The first group consists of three elements (RGA, RER, and ELT), in which the coefficient values above the recommended level (75%) are observed throughout the period. Two elements of the sample (RVP and RSM) show values at the level of 65–70%. For the next two objects

Table 8. Ratios of return on assets, equity and financial stability

Firm	MFS					ROA					ROE					
	14	15	16	17	18	$\overline{X}$	14	15	16	17	18	$\overline{X}$	14	15	16	17	18	$\overline{X}$
RVP	0.66	0.66	0.72	0.70	0.62	0.67	0.03	0.06	0.14	0.14	0.17	0.11	0.21	0.37	0.54	0.43	0.35	0.38
RSM	0.57	0.65	0.68	0.81	0.82	0.71	0.04	0.13	0.20	0.37	0.12	0.17	0.11	0.26	0.33	0.47	0.15	0.26
PDM	0.32	0.18	0.24	0.16	0.16	0.21	−0.05	−0.01	0.02	0.00	0.00	−0.01	−2.39	−5.09	0.87	0.07	0.06	−1.30
RGA	0.82	0.82	0.81	0.77	0.78	0.80	0.02	0.00	0.01	0.00	0.00	0.01	0.03	0.00	0.02	0.00	0.00	0.01
RFP	n/d	0.38	0.42	0.56	0.67	0.51	n/d	0.26	0.11	0.22	0.18	0.20	n/d	0.71	0.27	0.40	0.28	0.42
RCP	0.61	0.12	0.01	0.06	0.06	0.17	0.00	0.00	0.01	0.05	0.03	0.02	0.00	0.95	0.86	0.83	0.47	0.62
NCE	0.61	0.65	0.63	0.57	0.53	0.60	0.22	0.14	0.09	0.17	0.14	0.15	0.37	0.23	0.15	0.31	0.28	0.27
YTM	−0.06	−0.04	0.01	0.01	0.01	−0.01	0.11	0.01	0.05	0.01	0.00	0.04	− 1.92	−0.25	6.85	0.47	0.20	1.07
RER	0.80	0.89	0.79	0.61	0.57	0.73	0.19	0.05	0.07	0.20	0.11	0.12	0.23	0.06	0.09	0.33	0.18	0.18
ELT	0.79	0.91	0.92	0.92	0.92	0.89	0.15	0.14	0.31	0.23	0.06	0.18	0.20	0.15	0.34	0.25	0.07	0.20

(RFP and NCE), the coefficient is in the range of 40 to 60%. For PDM, the MFS fluctuates within 20%, and for YTM, about 0%. Thus, the assessment both according to Altman's models and in terms of profitability and financial stability ratio shows two outsiders. At the same time, the results that are closest to alternative calculations are shown by the second Z-score model (Z_{NP}), i.e. a model designed to examine the probability of bankruptcy of industrial enterprises whose shares are not traded on the market. Actually, our entire sample consists of such enterprises, which allows us to declare a high degree of efficiency of using model (2) when analyzing enterprises of the manufacturing industry in Russian conditions.

Our findings are consistent with other studies based on Russian data [11, 16], which showed that model (2) overcomes the limitations of the first model and can be used to assess financial sustainability both at the level of individual economic entities and at the level of over industry data.

4 Conclusion

Our analysis of the financial stability of the sample of manufacturing enterprises located in the city of Rostov-on-Don using three versions of the Altman Z-model indicates that the model for private (non-public) industrial enterprises is able to most objectively assess the financial sustainability of industrial enterprises. The classical model requires the public nature of the subject of estimate; therefore, instead of the market price of enterprises, one has to use the net asset value of the balance sheet. The model for emerging markets gives too high values of the resulting indicator due to the presence of the (3.25) constant in the formula. This allows us to state the need for further research of this model to examine this constant and, possibly, to select more correct coefficients for the model variables.

The results of the analysis allow us to assert that most of the manufacturing enterprises of the city of Rostov-on-Don have good or satisfactory financial stability. Data for those enterprises that show a high level of bankruptcy risk indicate the importance of variables X2 and X3 which are formed with the participation of indicators of net working capital and retained earnings, respectively. The effect of the presence of a positive result of retained earnings on the data of our sample is especially highlighted: the long-term operation of a business with a negative financial result is a reliable indicator of its inefficiency and impending bankruptcy. Long-term operation with negative financial results also becomes the reason for negative equity. And equity is an indicator that is involved in calculating variable X4 in the model for non-public companies.

Summing up, we note that, in accordance with the calculations, it looks ideal when all the variables in the models are positive, in order to have a higher resulting Z-score.

References

1. Khanin, G.I., Fomin, D.N.: J. Inst. Stud. **11**(1), 155–175 (2019). https://doi.org/10.17835/2076-6297.2019.11.1.155-175
2. Khanin, G.I.: World Econ. Int. Relat. **60**(9), 105–108 (2016)

3. Averina, L.M., Sirotin, D.V.: Econ. Reg. **16**(1), 268–282 (2020). https://doi.org/10.17059/2020-1-20
4. Kaufman, G.G.: Cato J. **16**(1), 17–45 (1996)
5. Nikulina, E.V., Ovchinnikova, O.P.: Belgorod State Univ. Sci. Bull. Ser.: Econ. Inf. Tech. **13**(210), 27–35 (2015)
6. Gizatullin, K.N., Garipov, F.N., Garipova, Z.F.: Econ. Reg. **14**(1), 43–52 (2018). https://doi.org/10.17059/2018-1-4
7. Khanin, G.I.: Terra Econ. **17**(2), 124–145 (2019). https://doi.org/10.23683/2073-6606-2019-17-2-124-145
8. Noviczkij, A.G., Noviczkaya, A.I.: State Munich Manag. Sch. Notes NCACS **2**, 115–122 (2016)
9. Bamadio, B., Lebedev, K.A., Shevchenko, I.V.: Multi. Internet El Sci. J. Kuban State Agr. Univ. **108**, 334–356 (2015)
10. Khanin, G.I.: Terra Econ. **16**(2), 17–26 (2018). https://doi.org/10.23683/2073-6606-2018-16-2-17-26
11. Savel'eva, M.Yu., Majorko, E.A., Vagajceva, V.P.: Sib Financ. Sch. **4**(123), 82–87 (2017)
12. Khasanov, R.X., Kashtanov, N.N., Margaryan, L.G.: Bull. Univ. Financ. **5**(77), 44–53 (2013)
13. Altman, E.I.: J. Financ. **23**(4), 589–609 (1968). https://doi.org/10.1111/j.1540-6261.1968.tb00843.x
14. Altman, E.I., Hartzell, J., Peck, M.: Emerging market corporate bonds – a scoring system. In: Levich, R.M. (ed.) Emerging Market Capital Flows, pp. 391–400. Springer, Boston (1998). https://doi.org/10.1007/978-1-4615-6197-2_25
15. Altman, E.I.: Emerg. Markets Rev. **6**, 311–323 (2005). https://doi.org/10.1016/j.ememar.2005.09.007
16. Lvova, N.A.: Fin. Anal.: Prob. Sol. **7**(241), 37–45 (2015)
17. Sasivimol, M., et al.: Applying emerging market Z-Score model to predict Bankruptcy: a case study of listed companies in the stock exchange of Thailand (Set). In: Postorozi Slovakia International Conference, 25–27 June 2014, pp. 1227–1237 (2014)
18. Factories and industrial enterprises in Rostov-on-Don. https://www.wiki-prom.ru/city/116city.html
19. Quick and convenient verification of counterparties. https://www.rusprofile.ru
20. Corporate information disclosure center. http://www.e-disclosure.ru
21. Federal State Statistics Service. Finances. https://www.gks.ru/folder/11192

Today's Contradictions and Priorities of Financial Technology Development

Petr V. Akinin(✉), Valentina P. Akinina, Yulia O. Boldareva, Kumsiyat R. Aliyeva, and Igor I. Kravchenko

North Caucasus Federal University, Stavropol, Russia

Abstract. Having burst into the financial accounting sphere, financial technology ("FinTech") is contributing to its revolutionary transformations. This process is characterized by the dominance of infocommunications over the financial essence. In this regard, many FinTech development scenarios, forecasts and programs are fragmented, unstructured and lack fundamentality. The research purpose is to conduct a systematic analysis of financial technology development, to identify its priorities and contradictions. The main research methods are a systemic analysis and logical modeling. Based on the created structural-logical FinTech model, its current image was formed, the consequences were determined, challenges and contradictions were identified. The most pressing issues include the construction of institutions of the future; creation of a theoretical FinTech platform; ensuring cybersecurity; formation of adequate models and competencies of specialists. The scientific novelty of this study implies the identification of these problems and the presentation of recommendations for their implementation and resolution. It should be borne in mind that FinTech development takes place under conditions of a fundamental civilization matrix reformatting characterized by the collapse of a liberal model in general and a dollar-centric model in particular, the change of world leaders, the growing role and place of Russia in the world space.

Keywords: Theoretical platform · Institution design · Cybersecurity · Global actors

1 Introduction

One of the phenomena of modern civilization is a digital economy, which is based on infocommunications. This industry has reached its greatest development in the financial accounting sphere [1].

In a generalized form, computer technology in this area has been defined as financial technology, or FinTech.

Having burst into the financial accounting sphere, FinTech is contributing to its revolutionary transformations. This process is characterized by the dominance of infocommunications over the economic-financial essence. It appears that this postulate contains the main FinTech contradictions. There are many other problems as well, but the race of infocommunication technologies leaves no time and space for their solution. Today

P. V. Trifonov and M. V. Charaeva (Eds.): DITEM 2021, LNNS 380, pp. 500–505, 2022.
https://doi.org/10.1007/978-3-030-94245-8_67

the Russian authorities are making a decision to introduce a digital ruble in 2023. This is done virtually without deep theoretical study. The main argument for making this decision is to keep up with the developed technological payment systems.

Many FinTech development scenarios, forecasts and programs are fragmented and unstructured. In this regard, an attempt is made to conduct a systematic analysis of development of this industry: its characteristics, directions, consequences, contradictions and threats, as well as to substantiate the main priorities.

2 Methods

The main research methods include a systematic approach, logical modeling, comparative analysis and expert review.

The information database of this study included Internet resources, reports of the Central Bank of the Russian Federation and other financial institutions, reviews of communities of foreign and domestic scientists, their articles in peer-reviewed journals and scientific collections, authors' personal publications (over the past 5 years).

3 Results

The research results are presented in Table 1.

Table 1. Analysis of current FinTech development.

Indicators	Quantitative and qualitative characteristics
1	2
1. Current FinTech image	In 2017, 250 IT organizations were involved in this industry in Russia. The leading directions are as follows: – lending, including microfinance institutions – 10.3% of the market, P2P lending – 15.7%. – cryptocurrency sector – 15.1%, including digital wallets, exchange offices, exchanges, etc. related to the production and movement of digital coins; – selection of financial products and/or their purchase – 14.7%; – enterprise financial management – 13.2%; – payment sector – 11.2%; There are 2 digital banks registered in the Russian Federation operating under their own banking license: Tinkoff Bank and Modulbank [2]. FinTech is growing by 20% annually. In terms of financial technology penetration, Russia is among the leaders – 82% (in third place). China is in first place – 87%, followed by India – 87%, South Africa – 82% and Colombia – 76%. Therefore, the USA occupies one of the last places in this indicator – 46%, followed by France – 32% and Japan – 34% [3]. According to expert estimates, by 2035, FinTech penetration into the payment and funds transfer segment in Russia will exceed 96%. In the financing segment – 37%

(*continued*)

Table 1. (*continued*)

Indicators	Quantitative and qualitative characteristics
	In the sector of private wealth management and financial consulting – 46%. Operators of insurance FinTech services will pay about 10% of all insurance premiums [3]
2. Directions	1.Remote service; 2.Artificial intelligence; 3.Electronic money; 4.Mathematization; 5.New technology (such as blockchain technology). 6.Institutions of the future [1, 4–6]
3. Consequences	Institutional transformations: from 2008 to 2018 the number of commercial banks decreased twice, their branches – by 3.9 times, the volume of the number of accounts increased by 6.9 times, including the ones with Internet access – by 38.2 times, the share of Internet accounts increased from 16 to 91% [7]. Transparency, increased interface to the real sector, integration with other institutions of the economy [8]
4. Challenges and contradictions	Lack of a theoretical platform, which is a source of tension in the labor market [9]. Damage from cyberattacks (in the world it amounts to 8–10 trillion US dollars annually, in Russia – more than 1 trillion rubles) [10]. Dictate of global actors; the threat of specialists' degradation, their transformation into an appendage of a computer

4 Discussion

The following aspects from Table 1 should be studied in more detail:

design of institutions of the future;
creation of a theoretical FinTech platform;
ensuring cybersecurity;
formation of adequate models and competencies of specialists;
elimination of global actors' dictatorship.

As for the design of institutions of the future, it should be noted that the intensification of integration processes led to significant transformations in financial engineering, which, starting with the creation of structured tools – products, spread to the design of new integrated institutions of the future. Figure 1 shows a model of a bank of the future as an example [7].

In the authors' opinion, the key problem in FinTech development is the lack of a profound theoretical basis [11–16]. This primarily concerns the theory of money. Fundamentally, modern civilization is in a collapse in this sphere. Having recognized the US dollar as the global currency (Bretton Woods 1944), mankind has retained its material content – the gold standard. In 1976 in Jamaica, it was canceled and replaced with a floating exchange rate. To some extent, it could be justified as a material asset by the US economy; today this concept is greatly blurred.

Firstly, the US economy is already greatly inferior to the Chinese economy. China's gross domestic product (GDP) based on purchasing power parity (PPP) is 18.5% globally versus 15% in the case of the USA.

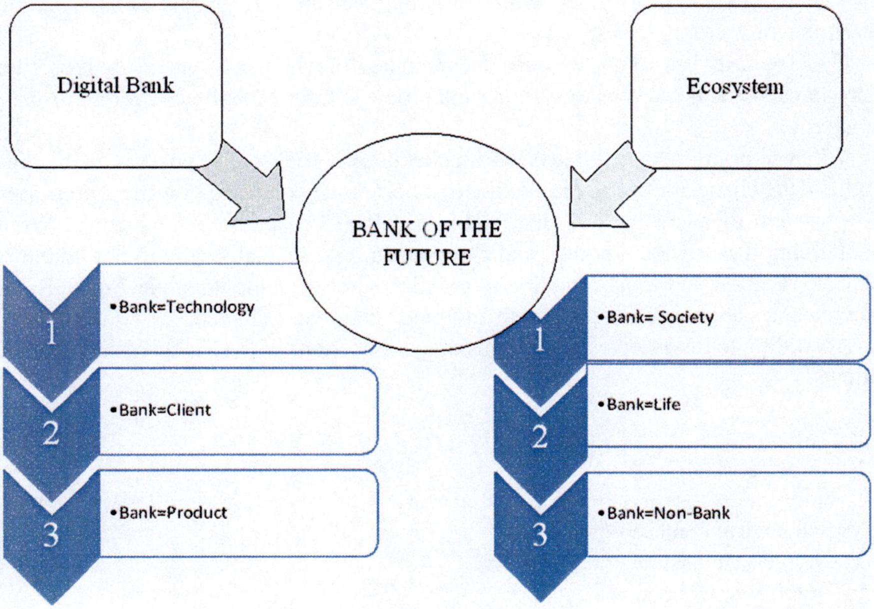

Fig. 1. Model of a bank of the future.

Secondly, the unjustified printing of American banknotes continues. The US national debt has already approached 28 trillion US dollars and continues to grow.

Thirdly, the situation with derivatives should also be mentioned. In the 1960s, the Nobel Prize was awarded for their invention, while in 2008, they basically "blew up" the financial world, becoming a trigger for the global crisis [17].

Thus, today the dollar-centric system is a financial pyramid, the collapse of which threatens world civilization.

Without understanding the above-mentioned problem, they offer a cryptocurrency unsecured by any tangible asset (except for the cost of electricity for mining) and the economy of any state. The main advantage of cryptocurrencies is high infocommunication technology [18].

Thus, Russian authorities rush to introduce the digital ruble into use in 2023, without an adequate organizational-legal basis, putting forward the desire to keep up with world technology as an argument.

It should be noted that attempts to create an information theory of money were made in Russia before, in particular by V. Yurovitsky, but they have not been well developed [19].

Summing up the above, it should be stated that today humanity is experiencing an urgent need for a solid theoretical basis of money circulation in the context of FinTech development, its legal aspects.

Damage from cyberattacks has already been mentioned. In this area, a clear definition of cybersecurity should be given, which provides answers to many questions, including development vectors [4, 5].

Cybersecurity is a set of measures, the main task of which is to ensure the protection of systems, networks, software systems and products from cyberattacks and third-party penetration.

When committing cyberattacks, an attacker usually tries to take possession of confidential data, change or destroy it in order to extort money or disrupt business processes. The problem of adequate and clear staffing in FinTech sphere is very acute. While maintaining the previous models and competencies of specialists, there is a threat of their degradation and transformation into appendages of computers. New models and competencies are needed to prevent this threat from being executed.

According to research results, promising qualifications in the FinTech market are as follows:

- IoT Architect
- Bioinformatist
- Data journalist
- Virtual environment designer
- Voice interface designer
- IoT interface designer
- Data security engineer
- Data scientist
- IT lawyer
- Computational linguist
- Neurointerface developer
- Digital logistics specialist
- Digital marketer

These specialists should have the following competencies:

- enhanced understanding of the professional area, as well as knowledge and experience in related areas (T-shaped specialist);
- understanding of the opportunities and risks associated with the use of new technology;
- knowledge of project management methods;
- possession of tools for working with large data and visualization tools;
- understanding of cybersecurity basics;
- skills in working with databases;
- emotional intelligence;
- lifelong learning ability;
- ability to solve problems to the fullest extent;
- adapt to work in conditions of uncertainty [6].

Finally, the last contradiction is the dictatorship of global actors (Google, Twitter, Instagram). Today, it can be seen that they block websites and, conversely, retain sites with negative and distorted information. As is known, the latter affects ratings, stock indices, etc. In response, Russia started to impose penalties, but as practice shows, they are ineffective. There is only one way out – to create domestic social networks.

5 Conclusion

The main findings of the study are as follows. The most acute problem is the creation of a theoretical basis to abandon the use of the US dollar in international settlements, withdrawal from payment systems that use this tool; the use of electronic money.

The continuous process of scientific-technical development is the provision of cyber-security, the creation of adequate models and competencies of specialists, as well as the design of institutions of the future.

References

1. Keshelava, A.V. (ed.): Introduction to the Digital Economy. VNIIGeosystem, Moscow (2017)
2. Fintech: Annual survey of the financial technology market in Russia (2019). https://ict.moscow/research/fintekh-2019-godovoe-issledovanie-rynka-finansovykh-tekhnologii-v-rossii/
3. Serykh, I.: How Russian fintech lives. https://iq.hse.ru/news/334262759.html
4. Udalov, D.V.: Econ. Secur. Qual. **1**(30), 12–18 (2018)
5. Malakhova, I.V.: Bul REA by Plekhanov GV, **6**(108), 183–190 (2018)
6. Basic competencies of the digital economy. https://digital.ac.gov.ru/poleznaya-informaciya/4166/
7. Akinin, P.V., Akinina, V.P., Korolev, V.A.: Modern transformation of the Russian banking sector in the context of digitalization. In: Penkova, I.V. (ed.) Russia, Asia, Europe: Digitalization of the Global Space, pp 28–31. SEKVOYA, Stavropol (2020)
8. Implications of fintech developments for banks and bank supervisors. https://www.bis.org/bcbs/publ/d431.pdf
9. Digitalization sets trends in the labor market. https://www.iot.ru/promyshlennost/
10. Lavrov, S.V.: Losses of the global economy from cyberattacks could amount to $8 trillion in 2022. https://www.finversia.ru/publication
11. Eskindarov, M.A., et al.: The directions of FinTech development in Russia: expert opinion of the financial university. World New Econ. **12**(2), 6–23 (2018). https://doi.org/10.26794/2220-6469-2018-12-2-6-23
12. Inshakova, A.O., Goncharov, A., Kazachenok, S.Y.: Adv. Intell. Syst. Comput. 726, 68–377 (2018)
13. Lunyakov, O.V.: Moscow Univ. Bull. **6**(2), 28–45 (2020)
14. Lavoie, M., Reissl, S.: Further insights on endogenous money and the liquidity preference theory of interest. SSRN Electron. J. (2018). https://doi.org/10.2139/ssrn.3193027
15. Sawyer, M.: Endogenous money and the tyranny of demand and supply. In: Rochon, L.-P., Rossi, S. (eds.) Advances in Endogenous Money Analysis, pp. 227–244. Edward Elgar Publishing (2017). https://doi.org/10.4337/9781783472246.00020
16. Konovalova, M.E., Kuzmina, O.Y., Salomatina, S.Y.: Adv. Intell. Syst. Comput. 908, 315–328 (2020)
17. Belobrodko A Derivatives. A story that teaches nothing. http://aszk.ru/finansy/item/2831
18. Ancri C Fintech innovation: An overview. https://pubdocs.worldbank.org/
19. Yurovitsky, V.M.: Evolution of Money: Money Circulation in an era of Change. GROSSmedia, Moscow (2004)

On the Methodology for Assessing the Fiscal Effectiveness of Tax Incentives for Private Investors in the Securities Market

Olga S. Belomyttseva(✉)

National Research Tomsk State University, Tomsk, Russia

Abstract. When identifying the effectiveness of tax incentives for private investors in the securities market, the use of standard effectiveness methods is not possible. At the same time, tax incentives for individual and institutional investors are quite widespread both in the Russian and international practice. The purpose of this study is to develop elements of the methodology and present the author's technique for assessing the effectiveness of tax incentives for private investors in the securities market. To achieve this goal, the methods of logical, comparative and statistical analysis were used. Methodologically, the work focuses on justifying the circulation of capital due to the provision of tax incentives to private investors. The concept of tax expenditures is also mentioned, the interpretation of the terms "economic effect", "fiscal effect" and "fiscal effectiveness" in relation to tax incentives is given. The author also deciphered other effects of private investors' investments in the securities market. The main methodological finding of the author is the use of stock indices in calculating the fiscal effect and fiscal effectiveness. Stock indices are introduced into the methodology for identifying the sectoral structure of investments and distribution of investments in stocks and corporate bonds of particular enterprises. The author's methodology for assessing the effectiveness of tax incentives for private investors in the securities market consists of ten stages. It is proposed to test the author's methodology using the example of individual investment accounts and non-government pension funds. The author's technique has some controversial points, which are also noted in this study.

Keywords: Tax incentive · Effectiveness of tax incentives · Tax expenditures · Fiscal effectiveness · Stock index

1 Introduction

The provision of tax incentives to private investors for the purchase of securities is widespread both in domestic and international practice. So, for example, in the Russian Federation, there are individual investment accounts, in the USA there are widespread IRA accounts, in the UK – ISA, in Australia – Superannuation. At the same time, investors significantly increase the profitability of their essentially tax-free investments, while the state incurs quite large expenses in the form of lost tax payments. Therefore, the

P. V. Trifonov and M. V. Charaeva (Eds.): DITEM 2021, LNNS 380, pp. 506–511, 2022.
https://doi.org/10.1007/978-3-030-94245-8_68

provision of tax incentives to private investors in the financial market requires analysis and evaluation.

Standard approaches to assessing the effect of tax incentives normally mean analyzing the relationship between the increase in tax payments of a taxpayer (enterprise) and the amount of tax incentives granted. In the case of a tax credit for the purchase of securities, investments are distributed among a wide range of industries, businesses and securities, which precludes the use of standard methodologies.

The question of the effectiveness of tax incentives for private investors has been limitedly researched in foreign literature, mainly American and mainly studying individual investors. Thus, studies on the effectiveness of tax incentives in different years were carried out by Poterba, Venti and Wise [1], Attanasio and DeLeire [2], Engen, Gale and Scholz [3], Hubbard and Skinner [4]. At the same time, the work of Feldstein [5] is noteworthy, as he noted the influence of tax incentives for the population on the growth of equity capital and tax payments of corporations. Feldstein's research was continued by Rudjeri and Fougere [6], claiming that such effects take place only in a closed economy. In the future, similar studies, also in terms of income taxation, were continued by Rudjeri and Vincent [7].

In the domestic literature, the issue of the effectiveness of private investments has been studied superficially, probably due to the insufficient period of existence of these tax benefits. Exceptions are the studies by Belomyttseva [8, 9], Maimulov [10]. The analysis of tax incentives for individual investors, in general, is carried out in Tenetnik [11], Karpova and Panova [12].

The purpose of this study is to develop individual points of the methodology for assessing the fiscal effectiveness of assessing private investments in the securities market and to present the author's assessment methodology.

The theoretical hypotheses of the research are formulated as follows:

H1: Tax incentives for private investors in the securities market contribute to the growth of debt and equity capital of Russian corporations;

H2: Tax incentives for private investors in the securities market boost corporate tax payments;

H3: The fiscal effect and fiscal effectiveness of tax incentives for private investors in the securities market can be calculated using stock indices.

2 Methods

2.1 Methodological Foundations for Assessing the Effectiveness of Private Investments in the Securities Market

The presentation of the author's methodology for assessing the effectiveness of private investments is preceded by a discussion of some methodological aspects.

1. The author interprets the notion "tax expenditures" as lost tax revenues of the budgetary system in connection with the application of various deviations from the regulatory structure of taxes, which provide benefits to certain groups of taxpayers. Lost tax revenues can be estimated as the amount of tax revenues missed by the budget as a result of the introduction of tax benefits [13, 14].

2. The author's methodology is based on the disclosure of the stimulating function of taxes. Accordingly, the author is interested in private investments in the corporate sector, namely, in stocks and corporate bonds.
3. The author highlights the following effects resulting from the provision of tax incentives to private investors:

- an increase in the debt and equity capital of corporations due to the acquisition by investors of their securities;
- an increase in corporate tax payments due to the attraction of additional financing;
- a decrease in public debt as a result of an increase in budget revenues due to an increase in corporate tax payments.

Per se, an increase in equity and corporate debts are not indicative and do not evidence investment stimulation. At the same time, an increase in corporate profits, tax payments and a decrease in public debt are already indicative and effective. Thus, the provision of tax incentives to private investors forms a kind of capital circuit. The increase in corporate tax payments that exceeds the governmental tax expenditures on tax benefits will be a measure of the fiscal effect and effectiveness.

4. In the interpretation of the terms "economic effect" and "fiscal effect" in relation to tax benefits, the author adheres to the point of view of Maiburov et al. [13, 15]. The economic effect of a tax incentive, in theory, is to increase the taxpayer's financial resources. Within the framework of the author's methodology, the economic effect of a tax incentive is also present in two versions:

- as an effect for the taxpayer, who receives, due to the tax incentive, a high return on investment and thus maximizes capital;
- as an effect of capital gains for issuers whose securities were acquired by investors.

By the fiscal effect, the author means an increase in subsequent tax revenues over current tax expenditures. In fact, in the author's methodology, fiscal and economic effects and, accordingly, fiscal and economic effectiveness, are intertwined and combined.

Fiscal effectiveness, in turn, by analogy with the profitability indicator, is interpreted as an indicator of the ratio of subsequent annual tax revenues to current annual tax expenditures in percent per annum.

5. An important question is which securities of which issuers are in the portfolios of private investors. There is no such official information. It is needed to identify the sectoral structure of investments broken down by shares / corporate bonds and particular issuers. For this purpose, the author introduced into her methodology the stock indices reflecting the average price of the so-called "basket of securities". Moreover, the index includes the most popular or traded securities. What matters is not the values of the indices, but their structure or the composition of the basket. By operating the structure of the stock/bond index at a particular date, one can roughly identify the structure of private investments.

2.2 Author's Methodology for Assessing the Effectiveness of Private Investments in the Securities Market

The author's methodology for assessing the effectiveness of private investments includes the following stages:

6. Analysis of the dynamics of private investments separating investments in stocks and corporate bonds. At the same time, it is desirable to have at least five-year quarterly data. At this stage, it is possible to conclude the contribution of investors to the equity/debt capital of corporations.
7. Calculation of the amount of government tax expenditures for the provision of tax incentives, broken down by types of securities (shares/corporate bonds).
8. Analysis of the structure of the stock index: shares, corporate bonds or pension savings indices. At this stage, the amount of investment in the sectors/subsectors is identified.
9. Based on the data on the sectoral return on assets, the amount of profit for the subsectors is identified at the expense of investors.
10. Calculation of the increment in income tax of subsectors at the expense of investors.
11. Collecting data on assets and revenues of particular enterprises and consolidated groups of enterprises.
12. Based on the share of investors' funds in assets, identification of the increase in revenue of individual corporations and consolidated groups of enterprises at the expense of investors' funds.
13. Calculation of the fiscal effect and the fiscal effectiveness of tax incentives:
14. Calculation of reducing the impact of public debt due to the fiscal effect.
15. Reforming tax incentives for private investors in order to achieve greater impact.

Testing of the author's methodology for tax incentives for individual investors is possible within the framework of individual investment accounts. This tax incentive has extensive statistical material since 2015. There are other tax incentives for individual investors, but they are not widely used. With regard to institutional investors, the analysis can be carried out for non-state pension funds that have tax incentives for income tax when investing pension reserves and pension savings. The data of the Bank of Russia, the Federal Tax Service, and the Moscow Exchange can be used as statistical data.

3 Results and Discussion

The author's methodology for assessing the fiscal effectiveness of tax incentives for private investors in the securities market has 2 limitations:

- tax incentives are provided to investors only for the purchase of securities of domestic issuers. In other words, the economy is "closed", tax incentives are provided to stimulate only domestic corporations;
- investors are reasonable and rational; calculate the effect of using tax incentives and declare tax deductions on time, if possible.

In practice, in the Russian Federation, these restrictions are observed, but not fully. So, both individual and institutional investors are allowed to purchase foreign securities, however, there are some percentage restrictions for funds. As for the rationality of investors, this fact requires additional study. In some cases, investors fail to declare tax deductions for individual investment accounts.

As a disadvantage of the author's technique, one can note its high labor intensity. Thus, the analysts will have to evaluate the financial statements of dozens of securities issuers. For example, the Moscow Exchange index as of 05/07/2021 includes 44 shares, the broad market index of the Moscow Exchange includes 100 shares, and the corporate bond index contains 123 bonds.

The following controversial aspects of the author's methodology for assessing the effectiveness of private investments can be highlighted:

16. The use of stock indices in the methodology covers only a certain average portfolio of the investor. More accurate data can be obtained, however, only for institutional investors, for example, non-state pension funds. It is not possible for individual investors to analyze the real structure of all portfolios.
17. The question of which corporate reporting standards – Russian or international – to analyze when calculating the fiscal effect and fiscal effectiveness is controversial. It is more reasonable to carry out the analysis in both options, which also significantly increases the complexity of testing the author's methodology.
18. The most controversial point of the methodology is the assessment of the influence of the additional capital of the corporation on the growth of its profits. Obviously, other factors can also influence the growth of profits. This issue can be solved by factor analysis or by introducing the EPS (earnings per share) indicator into the methodology.
19. Also controversial is the question of the analysis of the capital gains of a particular corporation as a result of the purchase of its securities by investors or a consolidated group of corporations.

4 Conclusion

Thus, one can state that:

1. The author's methodology for assessing the fiscal effectiveness of tax incentives for private investors in the securities market requires testing;
2. testing is possible in the following options:

   - for individual investors (via individual investment accounts) and for institutional investors (via non-state pension funds);
   - based on RAS and based on IFRS;
   - for individual corporations and for consolidated groups of corporations;
   - by an increase in corporate income tax and by an increase in tax payments in general.

3. the true indicators of the effectiveness of tax incentives for private investors can be an increase in profits and tax payments of corporations, as well as a decrease in public debt due to an increase in tax payments of corporations;
4. Application of the author's methodology is possible not only in connection with tax incentives for private investors, but also for any other forms of subsidies, grants, as well as for assessing other widely distributed public investments among recipients, for example, national projects of a federal scale.

References

1. Poterba, J.M., Venti, S.F., Wise, D.A.: How retirement saving programs increase saving. J. Econ. Perspect. **10**(4), 91–112 (1996). https://doi.org/10.1257/jep.10.4.91
2. Attanasio, O.P., DeLeire, T.: The effect of individual retirement accounts on household consumption and national saving. Econ. J. **112**(481), 504–538 (2002). https://doi.org/10.1111/1468-0297.00728
3. Engen, E.M., Gale, W.G., Scholz, J.K.: Brookings Papers Econ. Act. **25**(1), 85–151 (1994)
4. Glenn Hubbard, R., Skinner, J.S.: Assessing the effectiveness of saving incentives. J. Econ. Perspect. **10**(4), 73–90 (1996). https://doi.org/10.1257/jep.10.4.73
5. Feldstein, M.: The effects of tax-based saving incentives on government revenue and national saving. Q. J. Econ. **110**(2), 475–494 (1995). https://doi.org/10.2307/2118447
6. Ruggeri, G., Fougere, M.: The effect of tax-based savings incentives on government revenue. Fiscal Stud. **18**(2), 143–159 (1997). https://doi.org/10.1111/j.1475-5890.1997.tb00258.x
7. Ruggeri, G.C., Vincent, C.: An Economic Analysis of Income Tax Reforms. Routledge (2018). https://doi.org/10.4324/9780429453489
8. Belomyttseva, O.S.: Tax Taxation **5**, 47–60 (2020). https://doi.org/10.7256/2454-065X.2020.5.33585
9. Belomyttseva, O.: Determination of the effects of tax-privileged private investors investments in the securities market. In: Global Social and Economic Challenges and Regional Development. In: Proceedings of the 16th International Conference of the Association of Economic Universities of South and Eastern Europe and the Black Sea Region (ASECU), pp. 228–231. Novosibirsk (2021)
10. Maimulov, M.S.: Fin Bus **6**(208), 59–65 (2020)
11. Tenetnik, O.S.: Citizen's savings as a tool for the systemic development of the economy. Finance: Theory Pract. **22**(3), 22–35 (2018). https://doi.org/10.26794/2587-5671-2018-22-3-22-35
12. Karpova, D., Panova, S.: The study of investment preferences of Russians on individual investment accounts. J. Rev. Global Econ. **7**, 626–644 (2018). https://doi.org/10.6000/1929-7092.2018.07.58
13. Mayburov, I.: Soc. Econ. **4**, 71–91 (2013)
14. Mayburov, I.A., Leontyeva, Y.V.: Indian Econ. **1**(65), 5–16 (2014)
15. Ivanov, Y., Mayburov, I., Nazarenko, A.: Soc. Econ. **10**, 80–96 (2016)

The New Silk Road: Current Financial Trends and Vulnerabilities in the Transport Sector

Zhaxybay I. Ismailov(✉)

Institute of Control Sciences of Russian Academy of Sciences, Moscow, Russia

Abstract. Studies of the main elements of the development of the New Silk Road project continue. The current situation in the financial management strategy, including its transport component, is considered. A model of international logistics system development in a graph model (cognitive map) is proposed. Based on the scenario analysis methods of complex systems, a number of scenarios are considered, and the most vulnerable components of the system are identified. The current state of transport integration and financial aspects of the development of the New Silk Road in terms of the global logistics system is considered. A comparative analysis of financing processes is carried out, taking into account the countries' national interests participating in the New Silk Road. The prospects for strengthening economic and transport ties between China and the countries of the New Silk Road in the development of transport infrastructure, taking into account the specific model of rail transport, the development of national consumption markets, and the political influence of the subsequent inflow of foreign investment. This paper also assesses the financial and economic risks of China's partnership system initiative. The tasks of China and the participating countries in the light of the many problems and the security of attracting foreign investment in the construction of numerous infrastructures in the event of a failure to return the foreign investment. The impact of COVID restrictions on rail transit along the New Silk Road route, along with increased security threats, is assessed.

Keywords: New Silk Road · Railway · Development scenarios · Vulnerability

1 Introduction

This paper presents the latest results of the studies initiated in [1–3]. The main purpose of the research is to develop proposals for the design of nodal points of the New Silk Road (hereinafter – NSR) in various strata of its socio-economic development.

In 2013, in the city of Nur-Sultan, the Republic of Kazakhstan, the President of the People's Republic of China Xi Jinping introduced the world to the Chinese initiative "One Belt – One Road", as it later became known as the "New Silk Road", aimed at strengthening China's economic and transport ties with the continents of Eurasia and Africa.

According to the World Bank documents "Belt and Road Economics: Opportunities and Risks of Transport Corridors in 2019" [4], the implementation of the NSR initiative

P. V. Trifonov and M. V. Charaeva (Eds.): DITEM 2021, LNNS 380, pp. 512–517, 2022.
https://doi.org/10.1007/978-3-030-94245-8_69

will require about $575 bln. of investments, including for transport infrastructure for $144 bln. About half of the investment from the total investment of the NSR will be in Indonesia, Malaysia, Pakistan, and the Russian Federation.

In the overland route of the NSR, passing through Central Asia, transit is structurally divided into the northern route – through Kazakhstan and Russia to Europe, and the central route – through Afghanistan, Iran, and Pakistan to the Persian Gulf. At the initial stage, the Chinese authorities allocated about a trillion dollars to develop their infrastructure to promote the NSR and strengthen their political influence. However, over time and circumstances, China's relations with many countries have changed, and the views of participating countries on the NSR have changed. Although they offered investment projects worth billions of dollars, there was also concern that participating countries would not justify the foreign investment made, which threatens the loss of independence and sovereignty.

Therefore, many experts see in the NSR the prospects for business and trade development and serious risks for servicing infrastructure investments due to the excess of foreign investment over the potential of trade volume, as well as social, demographic, and environmental consequences. Of course, the transport corridors of the NSR are aimed at increasing the speed of delivery of goods, increasing trade turnover, and attracting investment. According to the author, the initial implementation of the NSR project will allow the delivery speed to increase by 12% for the participating countries and the rest of the countries by an average of 3%.

2 Methods

The amendments to the Federal Law "On railway transport in the Russian Federation" and certain legislative acts adopted by the State Duma of the Russian Federation aimed at implementing projects for the construction and reconstruction of railway transport infrastructure facilities" are aimed at integrating Russia into the NSR initiative and at a deeper study of the consequences of the initiative on the environment, during the modernization and expansion of the main infrastructure.

According to the adopted amendments, it can be concluded that the Russian Federation in the future relies on international transit transportation, along with natural resources. At the same time, the emphasis is on the construction of high-speed railways from the eastern borders to the western ones for the transportation of goods from China to Europe and back, since the Trans-Siberian Railway and BAM cannot develop high speed. Therefore, it is planned to lay new high-speed tracks.

3 Results

The Law sets targets to increase the carrying capacity of the Baikal-Amur and Trans-Siberian railway lines to 180 million tons by 2024. The same Law provides measures to increase the capacity of transit transportation of containers by four times to ensure the growth of volumes and reduce the time of transportation of containers by rail to 7 days from the Far East to the western border of the Russian Federation.

The cost of the first stage, which is currently being implemented by Russian Railways, is estimated at more than $7 bln. Of this amount, $2 bln. are the funds of the National Welfare Fund. With such a volume of project financing, in May 2020, the General Director of Russian Railways announced a deficit of the investment program by more than $6.5 bln. in equivalent.

According to experts, China's "non-market" practice in relation to subsidizing the transportation of NSR by rail, aimed at conquering the markets of Europe and Russia, has achieved its goal. However, the costs of the budget of China and its provinces for these benefits have become enormous, so that the Chinese authorities had to urgently cancel the subsidization of rail transport from March 1 of this year. Before that, Beijing subsidized railway transportation by covering from 30% to half of the total railway tariff for transportation on their territory, which allowed Chinese commodity producers to ensure competitive conditions in the market.

The Rules adopted by the Government of the Russian Federation for granting subsidies from the federal budget to Russian Railways to compensate for lost revenues arising from the establishment of preferential tariffs for the transit transportation of goods in containers on the territory of the Russian Federation by public railway transport is aimed at subsidizing tariffs for the transit transportation of goods in the West-East and East-West directions and is, to some extent, compensation for the canceled subsidization of transportation by rail by China and the existing risk of redistribution of the cargo base to the transport capacities of alternative NSR routes.

According to the Decree, subsidies are provided to Russian railways regardless of the transportation scheme (container train, single or group shipments of containers) and the specialization of the container. Russian Railways, in turn, will charge a tariff for transit transportation, reduced by the amount of the subsidy.

The amount of subsidy in the East-West direction per 40-foot container is about $1,000 per container, in the opposite direction about $650.

According to the calculations of the "Analytical Credit Rating Agency" JSC, the revenue of Russian Railways from the use of subsidies for container transit through the territory of the Russian Federation may amount to up to $600 mln.

And what can the NSR Russia initiative offer to the growing consumer market?

According to the structure of loading of the main types of cargo of Russian Railways in 2020 (Fig. 1), it is possible to see the export potential of the Russian industry. As can be seen from the figure, the main cargo volume is coal, oil and petroleum products, iron ore, and forest cargo.

In 2021, "Russian Railways" JSC plans to increase loading to 1.26 billion tons, by 1.6% more than of 2020, the volume of coal loading will increase to 30%, but the result will depend on many factors including the restrictions related to COVID-19.

In addition to COVID restrictions, rail transport was affected by a week-long closure of the Suez Canal; because of it, the cost of delivery of 40-foot containers from China to Europe has increased four times compared with the same period last year, with a price up to $8,000 and the cost of accumulated cargo at the entrance to the canal for that week was estimated at $10 bln. Moreover, according to expert estimates, the approximate cost of goods that pass through the Suez Canal per hour is $400 mln.

Shortly, considering the growth of sea rates and the shortage of empty containers in China, prices are unlikely to return to the previous level.

At the same time, the situation may be negatively affected by a series of forced "convention prohibitions" in connection with checks due to the coronavirus through the border crossing at railway junctions with China, mainly for the export of raw materials to the PRC and for the import of containers from the PRC.

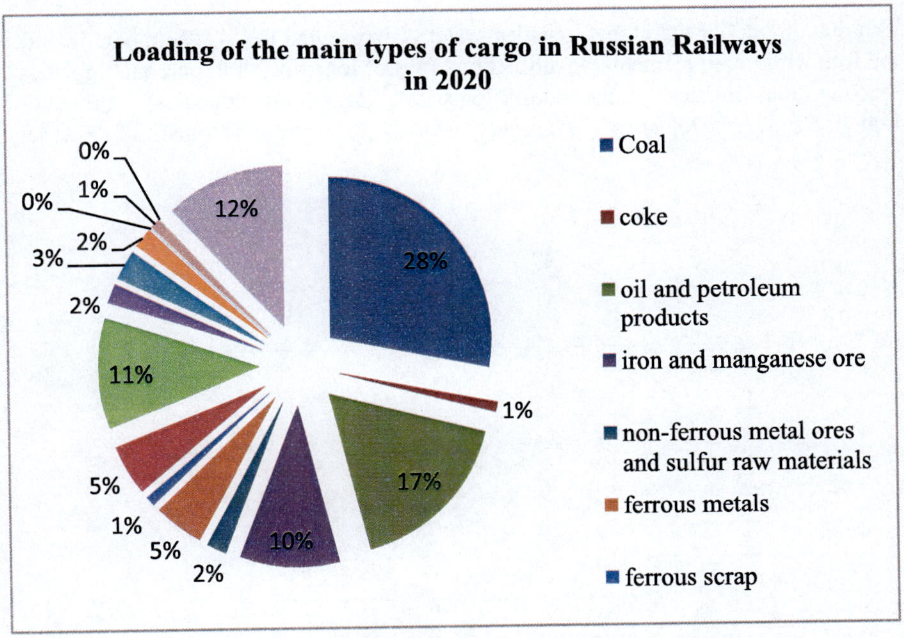

Fig. 1. The share of individual goods in the total loading.

In 2020, despite the pandemic, the volume of cargo transportation in containers via the NSR transit through the territory of Russia and Kazakhstan amounted to 517,500 TEU, which is 65% more than in 2019. The volume of cargo transportation between Kazakhstan and Russia in 2020 exceeded 85.6 million tons. Furthermore, in the first quarter, container traffic on the northern route of the NSR increased by 57%, compared to the same period in 2020.

Therefore, from a financial point of view, the NSR is beneficial to all participants of the NSR project integrated into the global supply chains to consume raw materials and materials with the production of goods with high added value. At the same time, the development of NSR can activate measures for social security, increase the competitiveness of human resources and increase the mobility of the population, contributing to an increase in passenger traffic.

4 Discussion

Based on identifying the main types of international logistics, a method for studying the development of international logistics relations is proposed, based on the methodology of scenario analysis of complex systems.

4.1 International Logistics System Model

The main model of the global logistics system is presented in [5]. The model includes the following main elements, identified by types of logistics: "International logistics", "Information logistics", "Distribution logistics", "Warehouse logistics", "Production logistics", "Service logistics", "Transport logistics", "Procurement logistics", "Financial logistics".

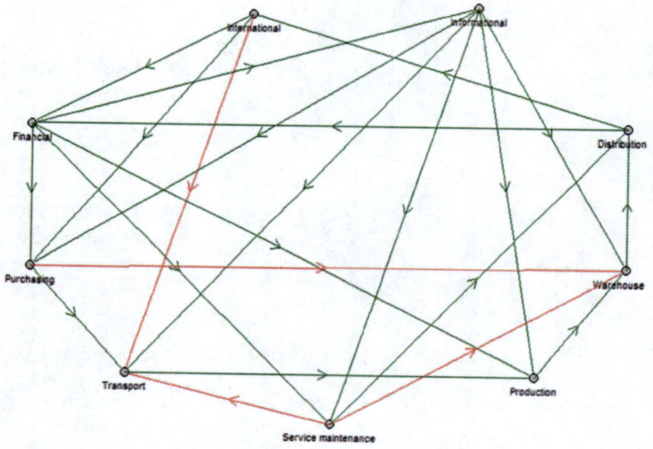

Fig. 2. Global logistics linkage model.

In the figure, the dots mark the vertices of the model graph, and the green arrows represent the arcs of the graph, which reflect an increase (decrease) in the parameters at one vertex, leads to an increase (decrease) in the parameter values at the vertex directly associated with it, the red lines reflect a decrease, that is, the opposite effect (Fig. 2).

4.2 Scenarios for the Development of the Logistics System

The constructed model (cognitive map) allows studying scenarios for the development of the global logistics situation. The report considers the following scenarios:

Scenario 1. Normal mode of operation.
Scenario 2. Reduction of financial logistics.
Scenario 3. Reduced warehouse productivity due to the epidemic.
Scenario 4. Reduced productivity of transport logistics.

The characteristics of the vulnerability of the considered types of logistics in comparison with the normal mode of operation are determined.

5 Conclusion

In this paper, the research of the main elements of the development of the New Silk Road project is continued. The current situation in the financial management stratum, including in railway transport, is considered. Based on the proposed model of developing the international logistics system in the form of a graph model (cognitive map), a number of scenarios for the development of the situation are considered, and the most vulnerable components of the system are identified.

It seems that it is advisable to involve interested persons, from interstate bodies to private investors, to develop the proposed digital model.

References

1. Ismailov, Z., Kononov, D.: Integrated management system for rail transport: planning of cargo turnover in conditions of uncertainty. In: Proceedings of the 11th International Conference "Management of Large-Scale System Development" (MLSD). IEEE Catalog Number CFP18GAE-ART, Denver (2018)
2. Ismailov, Z., Kononov, D.: New silk way: effective management of container transportations in the conditions of uncertainties. In: Proceedings of the 12th International Conference "Management of Large-Scale System Development" (MLSD). IEEE, Denver, pp. 1–5 (2019)
3. Ismailov, Z., Kononov, D.: IFAC-PapersOnLine **52**(25), 311–315 (2019)
4. Belt and Road Economics: Opportunities and Risks of Transport Corridors. https://openknowledge.worldbank.org/bitstream/handle/10986/31878/211392RU.pdf?sequence=12&isAllowed=y
5. Ismailov, Z., Kononov, D., Ponomarev, N.: Scenario analysis of development of complex logistics systems. In: Proceedings of the 13th International Conference "Management of Large-Scale System Development" (MLSD). IEEE, Moscow, pp. 1–5 (2020)

Financial Instruments for Recovery and Growth of the Russian Economy

Valeria E. Krolivetskaya(✉)

Saint Petersburg State University of Economics, St. Petersburg, Russia
kroliveckaya.v@mail.unecon.ru

Abstract. The global economic crisis of 2020–2021 caused by the pandemic coronavirus, having forcibly limited the economic activity, required special anti-recessionary measures for restoration of the Russian economy and also set the task to find due funding sources for sustainable economic growth. The purpose of the study is to identify the sources and mechanisms of financing the economic growth in the country in the post-crisis period. The paper uses the methods of historical and logical evaluation, analysis, synthesis, comparison and correlation. Following the research, the author concluded that the recent crisis has led to a dramatic decline in the global economy and its less-than-prompt recovery due to its unprecedented nature. In virtue of this, some previously inactivated fiscal and monetary policy instruments have been widely used to provide financial support to the economic entities affected by the pandemic. Considering the possible measures to ensure further post-crisis economic growth in Russia, the author systematised a number of financial sources and realisation tools, proposed independently and by other academic economists within the last five years, in the context of the economic development strategy adopted in Russia involving restriction on external sources of funding.

Keywords: Economic crises · Methods to support economic entities · Economic growth

1 Introduction

The problem of finding financial sources securing sustainable economic growth in Russia which belongs to the category of countries with emerging markets has been given special attention throughout the history of market relations development in the country. This issue was studied at different times by a number of scholars-economists: Abalkin, Aganbegyan, Andryushin, Burlachkov, Glazyev, Ershov, Larionova, Manevich, Krolivetskaya, Neshitoi, Senchagov and others. Some authors in their research, [1–4] pointed to the situation of monetary restriction within the budget channel of money supply to the economy, as well as the underdevelopment of the credit channel in terms of the refinancing schemes of commercial banks. A number of proposals were made to change the main objective of the monetary policy, to increase the role of the national debt monetisation instruments in economic regulation [4–6], to turn the refinancing system of the Bank of

P. V. Trifonov and M. V. Charaeva (Eds.): DITEM 2021, LNNS 380, pp. 518–522, 2022.
https://doi.org/10.1007/978-3-030-94245-8_70

Russia towards active investment, to develop the investment corporate bonds market and other instruments aimed at funding the innovative growth [1, 4, 5, 7]. The historical analysis has shown that until recently, in order to overcome the negative effects of financial crises in Russia, a number of monetary policy instruments were mainly used, including unconventional ones (mainly credit easing). Unlike in the previous crises, the priority in supporting the economic actors during the current coronary crisis was given to fiscal policy instruments, without using them to capitalise the banking sector. In the conditions of the shortfall in oil and gas revenues in the federal and regional budgets in 2020, as well as reduction of their taxable base, the Russian Federation Ministry of Finance actively used the funds of the RF National Wealth Fund to provide financial support to the economy and citizens; in addition, in the conditions of growth of the federal budget deficit due to the increased spending, it extended the scope of borrowings by additional placement of government securities in the financial market for periods exceeding 5 years. The Bank of Russia, in order to stimulate the demand and the economic growth and to overcome the consequences of the pandemic coronavirus, in 2020 proceeded to realising a variant of unconventional monetary policy – credit easing.

In the conditions of economic recession in Russia due to the coronavirus pandemic, a number of Russian academic economists (Aleksashenko, Sonin) in April 2020 suggested that the Bank of Russia should launch a quantitative easing programme as an anti-recessionary measure, along with credit easing, by realising monetary emission through the purchase of government bonds by the RF Central bank. However, the Bank of Russia, unlike a number of other central banks, refused to purchase them on both the primary and the secondary market in order not to be exposed to pro-inflationary risks. To what extent was this decision substantiated? And what are the prospects for using the fiscal channel of money supply to the economy for its recovery and growth, as well as using other innovative development funding sources? At present, it is expedient to give answers to these questions in order to get out of the current crisis and to ensure the further growth of the Russian economy.

2 Materials and Methods

The author substantiated, in order for the country to enter the trajectory of sustainable economic growth within the framework of a new innovative model of economic development by 2020, the use of optimal combination of financial instruments of lending and budgetary channels of "long" money supply to the Russian economy; devoted her PhD degree thesis [4] to this purpose, using the banking and budget statistics for 12 years, the relevant data grouping, analysis and comparison in the historical aspect, as well as the materials of articles and monographs of Russian scholars published in academic journals on this subject. However, the delays in realisation of the measures planned by the Russian Government for the transition to the sixth wave of innovation, as well as the sequent economic crisis in the country in 2014–2016, followed by the coronavirus crisis, made the author reconsider some previously proposed financial sources for Russian economy growth, relying mainly on internal ones, which was reflected in the author's articles [8, 9] and those of her associates [5, 10–13].

3 Results

In the author's opinion, for the country's economy to be reliable and sustainable, it is necessary to actively develop and improve the industrial production and knowledge-intensive sectors of the national economy on a new technological basis, providing Russian economy actors' access to both debt and equity financing, using actively a variety of financial market instruments for this purpose, especially the institution of "long" money. An important source of "long" money and a key instrument of liquidity formation in the world economy at present are "long" governmental securities that are issued by national ministries of finance and are purchased by central banks at the time of their primary placement, with entry on respective balance sheets before their maturity and simultaneous issue of national currency [12]. Issuance of national currency on the basis of direct purchase of government bonds by the central bank is excluded in Russia according to article 22 of the Federal law "On the Central Bank of the Russian Federation (Bank of Russia)", although the law has a postscript: "except for cases when it is provided by the federal law on federal budget". However, the Bank of Russia still disengages itself from the technology of forming "long" resources with its participation, while the purchase of government bonds by commercial banks means just withdrawal of money from the economy [4, 12]. To secure fully-featured operation of the budget channel of "long" money supply to economic entities for investment and innovation purposes, only political will of the RF Government is needed.

In the future, it is the fiscal policy in accordance with the modern monetary theory that is considered by Western economists as the main factor maintaining economic growth, being the source of its financing, because monetary policy measures alone are not sufficient to stimulate the economic growth. In these terms, in Russia it is possible to increase the volume of federal budget funds placed on deposits of Vnesheconombank, as well as the funds of the RF National Wealth Fund, which will make it possible to develop more actively the Project Financing Factory (PFF) programme developed jointly with the Government, which provides that the project funding is based on the combination of syndicated lending and stock financing by issuing bonds of a specialised project finance company [9, 11].

Given the geopolitical and economic situation in the country, it is necessary to revive rather than curtail (as fixed, starting from 2018) the lending channel of "long" money supply to the Russian economy. As early as in 2014, in connection with the imposition of economic sanctions by the USA and EU countries and the closure of external sources of long-term funding, the Bank of Russia proceeded to refinancing of large investment projects on concessional terms for up to 3 years secured by government guarantees and financed by the leading banks; in addition, it turned to a special programme of selective refinancing, for up to 3 years, of commercial bank loans provided to small and medium businesses, enterprises engaged in import phaseout as well as leasing companies for investment purposes. But the volume of these loans is small, since they are artificially restrained by the Bank of Russia for the fear of non-compliance with the inflation targeting policy which is rightly recognised by some economists [5, 10, 12] as exhausted at the present stage. According to the author, economic growth targeting should be treated as the main goal of the monetary policy [4, 12].

The enhanced lending activity aimed at ensuring due economic growth should also be facilitated by more active development of syndicated lending of objects that do not fall under the Project Financing Factory programme, in particular, through the institution of other domestic loan syndicates, especially since the opportunities to use this type of lending have significantly extended with the adoption of the federal law "On Syndicated Credits (Loans)" in December 2017, owing to the extended membership in loan syndicates due to admission of not only credit institutions, but also a number of institutional investors, Vnesheconombank and some international financial organisations.

The step-up of investment activity in the country and, accordingly, the economic growth can be facilitated by securitisation of investment loans and leasing assets, which will ensure due supply of "long" money to the economy through the issue of securities circulating in the financial market against the collateral of these assets. The securitisation of commercial organisations' assets is currently effected in accordance with the Russian legislation enforced on 01.07.2014; the author's analysis thereof showed the advantages of such securitisation in comparison with the practice based on the English law [8].

In order to replace the bank loans with the borrowed money raised in the securities market for the purpose of funding of economic entities, the Bank of Russia as a mega-regulator envisaged a number of measures for active development of the securities market since 2015. In particular, in order to finance the investment projects implemented in Russia on the basis of project financing, the bonds of Vnesheconombank's specialised project finance company under the Project Financing Factory programme appeared in 2018, as well as the structured bonds designed to diversify the investment portfolio of aggressive qualified investors; in 2019, the green bonds market, designed to finance exclusively environmental projects, started developing in connection with the transition from the traditional to "green" economy meeting the environmental demands; this was supplemented by the development of the bond market focusing on sustainable economic growth (combination of "green" and social bonds) [14]. In 2020, the market of federal loan bonds issued by the Russian Ministry of Finance for an extended period of 10 years acquired a longer-term character. In the author's opinion, the active use of such sources promoting the Russian economy development, along with the instruments for their realisation, should cumulatively yield the desired effect.

4 Discussion

Since 2014, the author has not changed her point of view concerning the factual change of the principal goal of the Bank of Russia's monetary policy – inflation targeting – for economic growth targeting. At the same time, a number of economists [10] suggest that the Bank of Russia switches to the policy of targeting the GDP nominal prices level. The author is also remaining committed to the expansionary budgetary policy aimed at recovery and growth of the Russian economy, which has been observed in recent years in Russia, though without the participation of the Bank of Russia. The criticism of such position of the regulator was expounded by the author in a number of published academic articles. The author's point of view is shared by many scholars [5, 9–12, 15], as concerns the active use of a number of financial instruments in the conditions of hampered access to external financial sources, namely syndicated investment lending, securitisation of assets of financial institutions, medium-term refinancing of commercial banks' investment loans by the Bank of Russia, the implementation of project funding mechanism through different options.

5 Conclusion

The author concluded, following the undertaken research, that, first, the coronavirus crisis has surpassed all previous economic crises that Russia has gone through – in terms of scope and structure of financial sources (with priority of budget ones) used to overcome the negative phenomena, as well as the tools for their realisation. This enabled the author and her associates to raise the issue of changing the principal goal of the monetary policy in Russia – inflation targeting – for economic growth targeting, which is relevant within the framework of the present line of economic policy in the country. Second, some possible sources of funding the further economic growth in Russia were identified and systematised, proceeding from the geopolitical situation and the chosen strategy of economic modernisation based on new technological grounds involving digital innovation, with specification of their prospective financial instruments.

References

1. Agangebyan, A.G.: Investment credit as a main link in overcoming the decline in Russia's socio-economic development. Money Cred **5**, 11–18 (2014)
2. Burlachkov, V.K.: Monetary Theory and Dynamic Economy: Conclusions for Russia. Editorial URSS, Moscow (2003)
3. Glazyev, S.: Strategy for Russia's Advanced Development in the Conditions of the Global Crisis. Economics, Moscow (2010)
4. Krolivetskaya, V.E.: Money as a Factor of Economic Growth. Aletheia, St. Petersburg (2014)
5. Andryushin, S.A.: Financial policy of Russia in the conditions of macroeconomic uncertainty. Banking **11**, 12–20 (2019)
6. Afanasyeva, O.N.: The Impact of Monetary Policy on Economic Development. Ru-Science, Moscow (2019)
7. Dvoretskaya, A.E.: Financial mediation in modern Russia: structural aspect. Banking **5**, 44–54 (2016)
8. Krolivetskaya, V.E.: Specific features of the new stage of securitisation in Russia. SPb Econ. J. **1**, 52–56 (2019)
9. Oro, S.: Financial development and economic growth: evidence from Russia, Europe. Asia Stud. **64**(2), 247–256 (2012)
10. Andryushin, S.A., et al.: Expert opinion regarding the draft document "main directions of the unified state monetary policy for the year 2020 and for the period 2021–2022." Banking **10**, 8–17 (2019)
11. Dvoretskaya, A.E.: Project funding: new trends and mechanisms. Banking **5**, 4–13 (2019)
12. Ershov, M.V.: The world in 2020: new challenges reflect systemic changes in the economy. Econ. Issues **12**, 5–25 (2020)
13. Parusimova, N.I., Krolivetskaya, L.P., Krolivetskaya, V.E.: Credit relations of Russian banks with the borrowers of the real economy sector. Intell. Innov. Invest. **6**, 75–87 (2019)
14. Usoskin, V.I.: Unconventional monetary policy: methods and results. Money Cred **11**, 12–18 (2016)
15. Lavrushin, O.I., Valenceva, N.I., Dadasheva, O.: The Role of Credit and Modernisation of Lending Activities. Knorus, Moscow (2012)

Improvement of Methods for Measuring the Fair Value of Securities in Credit Institutions

Larisa I. Yuzvovich[1(✉)], Natalia N. Mokeeva[2], Oksana A. Boytush[1], Gulmira A. Nasyrova[2], and Elena A. Smorodina[1]

[1] Ural State University of Economics, Ekaterinburg, Russia
[2] L.N. Gumilyov, Eurasian National University, Astana, Kazakhstan

Abstract. The article is devoted to methods for measuring the fair value of securities in credit institutions. The main attention is paid to certain parameters of a business model of a credit institution. The research subject is the economic relations arising in the process of improving methods for measuring the fair value of securities in credit institutions within the framework of sustainable development. The research purpose is to develop proposals for improving methods for measuring the fair value of securities traded both on the organized market in the Russian Federation and on the unorganized (over-the-counter) securities market, in order to increase the compliance of fair value measurement methods with IFRS 13 requirements. The fair value of assets/liabilities and securities in particular is currently one of the most controversial valuation types. It is still actively discussed by both Russian and foreign scientists. Due to significant changes in banking legislation associated with the transition of credit institutions to IFRS 9 in terms of mandatory classification of assets into three categories based on a business model and characteristics of cash flows on financial assets, in the authors' opinion, the issues of adequacy of measuring the fair value of securities are of particular importance. Based on the conducted analysis, the authors have developed a number of proposals, the main purpose of which is to detail (clarify) the approaches of credit institutions to measuring the fair value of securities that meet IFRS 13 requirements, but simultaneously preserve the variability of fair value measurement methods.

Keywords: Securities · Fair value

1 Introduction

The relevance of the research topic is caused, on the one hand, by the increased attention of the Bank of Russia to the issues of reliability of asset disclosure by credit institutions [1, 2] at fair value, on the other hand, by the multiplicity of valuation methods, quotes and assumptions used by different banks for the same financial instruments, i.e. the current lack of uniform regulatory requirements containing quantitative criteria for the input data taken to measure the fair value of securities. The problems of practical fair value application are considered by many scientists [3–5].

P. V. Trifonov and M. V. Charaeva (Eds.): DITEM 2021, LNNS 380, pp. 523–529, 2022.
https://doi.org/10.1007/978-3-030-94245-8_71

In addition, the importance of fair value measurement methodology [4], including the degree to which the input data used by a credit institution for the purpose of fair value measurement are consistent with the nature of assets, current market conditions, input data and assumptions, was influenced by the introduction of new financial asset classification and assessment rules provided for by IFRS 9 into banking practice from the beginning of 2019 [6].

Improvement of the quality and objectivity of methods used for measuring the fair value of financial instruments directly affects the amount of the bank's own funds (capital), the values of capital-containing ratios, financial performance, including the amount of profit/loss for the period, the reliability of the bank's financial statements.

2 Materials and Methods

Based on the research results, it was determined that securities revalued at fair value and the nature of transactions with financial instruments should satisfy a business model, which goal is achieved in one of two ways:

- through the sale of securities (regular purchase and sale transactions);
- in a combined way (both receiving the security payments stipulated by the terms of issue and through its sale).

Taking into account IFRS 9 approaches [7–9] to the asset classification, which stipulate that asset valuation methods should correspond to the bank's business model and contribute to the achievement of its goal, the authors proposed to conduct regular tests of the business model used by a credit institution to manage securities in order to generate cash flows.

During testing, in the context of securities issue, it is proposed to assess the compliance of transactions with them and cash payments formed on them. In case of identification of securities that do not meet the parameters of the business model, the authors consider it expedient to promptly inform the management of the Risk Management Service, the Internal Control Service, the Internal Audit Service and the bank's Board of Directors about this in order to raise the issue of business model revision and possible securities reclassification.

To date, the Bank of Russia regulations give credit institutions "freedom of choice" in terms of development and application of methods for measuring securities at fair value, the only requirement is that the fair value of securities shall be measured in accordance with IFRS 13 [10].

These circumstances entail the banking industry accepting increased risks in terms of recognizing and measuring the value of securities, including initial and subsequent measurement, due to the fact that IFRS 13 establishes fairly broad principles and methods for measuring assets/liabilities at fair value, which may be interpreted by each bank in its own way and does not contain specific quantitative criteria for fair value measurement.

The importance of the quality of securities measurement (revaluation) at fair value and the compliance of the methodology used by banks for fair value measurement with IFRS 13 requirements is due to the following:

- firstly, the direct impact of revaluation results on the amount of the bank's own funds (capital) and, as a consequence, on all capital-containing ratios.
- secondly, the impact on the amount of the bank's market risk, i.e. the risk of financial losses for a credit institution due to changes in the fair value of financial instruments and goods and in foreign exchange rates and/or discount prices for precious metals.

3 Results

Based on the results of consideration of internal methods for measuring the fair value of securities of two credit institutions [11] and their deficiencies in terms of incomplete compliance with IFRS 13 criteria, the authors formulated the following proposals to improve the measurement of securities at fair value:

1. Fix in the internal methods the rules on the need to use as the main source of information only those prices/quotes that were calculated by the Moscow Exchange PJSC based on the results of trading in financial instruments in the main (zero-address) modes (order books T0 and T+) – for securities traded on the organized market in the Russian Federation [12].
2. When measuring the fair value, take into account the ratio between the volume of an open position on the securities issue (without revaluation and coupon income) and the volume of exchange trading, on the basis of which a trade organizer calculated the market/weighted average price, since in the case of a market sale of a significant block of securities characterized by low activity and liquidity, there will definitely be a change in market prices. If a significant excess of the volume of the bank's position over the volume of exchange trades is revealed, it is considered expedient to apply adjustment factors lowering the initial price.
3. However, it should be noted that in this case, IFRS 13 permits the use of a price quotation classified as Level 1 in the hierarchy even if the regular daily trading volume in that market is not sufficient to absorb the available stock of such securities, and placing orders to sell the corresponding position within one transaction could have an impact on the price quotation [10, 13].
4. Provide in the methods the need to assess the presence of market-specific factors and their influence on the formation of exchange quotations, taking into account as follows: the identification of forced, amicable or contractual transactions concluded with affiliated and other related parties, as well as between market participants of a limited group. If such factors are identified, it is proposed to make adjustments to the chosen fair value measurement method, i.e. to the initial quotations of securities.
5. In order to implement a conservative approach to fair value measurement, provide in the methods the priority of using bid (buyer) quotes as of the measurement date disclosed by the Bloomberg news agency in BGN, CBBT or BVAL data sources – for securities traded on the unorganized (over-the-counter) market. Considering that most of the quotes on the OTC market are indicative, using the bid price within the spread between the seller's price and the buyer's price will allow to more reliably determine the price that can possibly be obtained when selling eurobonds.

6. Use the so-called "stop factors", in case of activation of which it is not allowed to revalue securities at fair value. It means that securities are valued by creating reserves for possible losses.

According to the authors, stop factors include:

- significant deterioration in the issuer's financial situation;
- commencement of the issuer's insolvency (bankruptcy) procedure;
- default or presentation of proposals for the restructuring of the issuer's debt obligations;
- other significant events in the issuer's activities.

As part of the study of completeness and quality of fair value measurement methods in two banks, the authors developed an adjustment system for measuring the fair value of securities using a market method and a method based on market quotations (Table 1).

Table 1. Adjustment factors for price quotations of securities.

Indicator name	Risk assessment ratio
1. Adjustment for low activity in organized and OTC markets. The total volume of concluded transactions:	
– 10% of the issue volume and more;	0
– from 8% to 10% of the issue volume;	– 0.04
– from 6% to 8% of the issue volume;	– 0.05
– from 4% to 6% of the issue volume;	– 0.06
– from 2% to 4% of the issue volume;	– 0.07
– from 1% to 2% of the issue volume;	– 0.08
– from 0.5% to 1% of the issue volume;	– 0.10
– from 0.1% to 0.5% of the issue volume;	– 0.12
– less than 0.1% of the issue volume	– 0.15
2. Adjustment for low activity. Number of transactions:	
– 10 and more;	0
– from 7 to 9;	– 0.03
– form 5 to 7;	– 0.04
– from 3 to 5;	– 0.05
– less than 3	– 0.06
3. Adjustment for low activity. Number of trading days from the previous thirty (30) trading days:	
– 8 days and more;	0
– from 6 to 7 days inclusive;	– 0.03
– from 4 to 5 days inclusive;	– 0.04
– from 2 to 4 days inclusive;	– 0.05
– less than 2 days	– 0.06

(*continued*)

Table 1. (*continued*)

Indicator name	Risk assessment ratio
4. If the BVAL Score is low – for securities that are not traded on the Russian market and are priced using Bloomberg quotes:	
– Score ≥ 8; – Score 7; – Score 6; – Score 5; – Score 4; – Score 3; – Score 2; – Score 1	0 – 0.02 – 0.03 – 0.04 – 0.05 – 0.06 – 0.07 – 0.08
5. Place of securities accounting (storage):	
– property rights to securities are certified by a depositary; – property rights to securities that do not meet the criteria are certified by a depositary	0 – 0.15

Using the example of securities portfolios of the credit institutions under consideration, testing of securities value measurement was carried out using different price indicators and adjustments proposed both in Table 1 and provided by internal methods of the banks.

Hypothetical securities portfolios were formed by the method of random non-repeated selection from debt and equity securities traded on Moscow Exchange PJSC [12] and included in the Composite Bond Index (RUABITR), the Index Moscow Exchange (IMOEX), respectively. The carrying value in terms of issues is distributed in proportion to the randomly selected values and was equal to the total carrying value of portfolios of the banks under consideration, respectively.

On the basis of the weighted average prices (WAPPRICE) disclosed by the Moscow Exchange PJSC [11], the number of acquired securities of each issue was furthermore calculated. The preliminary fair value for each release was defined by multiplying price quotations (the weighted average price, the minimum and maximum prices for the previous thirty (30) days) by the number of securities.

4 Discussion

Thus, after applying adjustment factors to the initial data on the method used by the first bank in relation to bonds DOM.RF B-5 and KAMAZ BO14 – factor 0.01, on the method used by the second bank in relation to bonds FGC UES-4-bob – two factors with a total value of 0.06, as well as for all four of the above bonds on the author's method, the revaluation of securities, taking into account adjustments, will take the following value (Table 2).

Table 2. Revaluation of securities using adjustment factors, thousand rubles.

Issuer	Short name	Revaluation at weighted average price	Revaluation on the banks' methods	Revaluation on the author's method
Lenta LLC	Lenta BO-3	−11,786	−11,786	−33,475
DOM.RF JSC	DOM.RF B-5	−7,638	−11,025	−31,341
KAMAZ PJSC	KAMAZ BO14	9,388	6,178	−16,293
FGC UES PJSC	FGS UES-4-bob	−4,997	−10,154	−23,045
Total:		−15,033	−26,786	−104,154

The use of stricter adjustment factors proposed by the authors will lead to a greater decrease in the amount of equity (capital), if adjustments are made to the initial data of securities measured at fair value on the banks' methods and proposed by the author, respectively.

Thus, in the studied example, making adjustments for low activity in organized trades will not lead to a violation of the H1.0 standard, since the values recalculated based on different methods will exceed the minimum permissible numerical value of this standard, set at 8.0%, but will allow to assess more conservatively the actual level of market risks assumed by banks.

Professional judgments should be based on the following core principles:

- the principle of reliability and objectivity;
- the principle of precaution;
- the principle of materiality;
- the principle of priority of the economic content of operations/transactions over their legal form.

5 Conclusions

Thus, in order to control the risks assumed by banks, when revealing signs or facts of incorrect recording of securities at fair value on the basis of professional judgment containing the analysis and conclusions about the inconsistency of methods with specific IFRS 13 clauses, the Bank of Russia [14] will be able to take measures aimed at objectively recording the results of securities revaluation in the bank's accounting and financial statements, namely:

1. Send a letter to the bank with the detected signs/facts of unreliability of securities valuation and recommendations for their elimination;
2. Conduct a meeting with the bank to discuss the reasons for unreliability of securities valuation at fair value and develop measures to eliminate them, including amending the current valuation method.

In addition, in order to summarize and save the results of securities revaluation at fair value, to provide information on the valuation of assets at fair value to the management of a credit institution and to the Bank of Russia, the authors consider it appropriate to record the results of fair value measurement in the form of professional judgments (consolidated ones or on a specific issue) with the disclosure of the valuation method, the level of hierarchy of the source data, the quotes used, indicators on the need to adjust the source data and the adjustment factors applied.

References

1. Direction of the Bank of Russia dated 08.10.2018 No. 4927-U "On the List, Forms and Procedure for Compiling and Submitting Reporting Forms of Credit Institutions to the Central Bank of the Russian Federation" (Registered with the Ministry of Justice of Russia on 13.12.2018 Under No. 52992, as Amended). https://base.garant.ru/72074044/
2. Official Website of the Central Bank of the Russian Federation. http://www.cbr.ru/
3. Druzhilovskaya, T.Y.: The use of fair value to measure accounting items: practical issues. Int. Account. **21**(9), 1086–1099 (2018). https://doi.org/10.24891/ia.21.9.1086
4. Kozlova, A.S., Taraskin, D.S.: Bull. Saratov. State Soc. Econ. Univ. **1**, 104–109 (2018)
5. Frolova, E.E., Karmadonova, E.V.: MID **4**(20), 20–28 (2014)
6. Borisov, A.: IFRS and ISA Credit. Inst. **1**, 53–62 (2019)
7. International Financial Reporting Standard (IFRS) 9 "Financial Instruments" (Put into Effect by Order of the Ministry of Finance of the Russian Federation No. 98n Dated 27.06.2016, as Amended). http://www.consultant.ru/document/cons_doc_LAW_202060/
8. Dolotova, M.A.: The convergence of accounting approaches: IFRS 9 financial instruments for purposes of the Russian accounting system. The market economy compliance test. Int. Account. **22**(11), 1271–1284 (2019). https://doi.org/10.24891/ia.22.11.1271
9. Kuvaldina, T.B., Lobachev, E.V.: IFRS 9: Estimation of expected credit losses and their recognition in financial reporting. Int. Account. **22**(4), 364–378 (2019). https://doi.org/10.24891/ia.22.4.364
10. International Financial Reporting Standard (IFRS) 13 "Fair Value Measurement" (Put into Effect by Order of the Ministry of Finance of the Russian Federation No. 217n dated 28.12.2015, as Amended). http://www.consultant.ru/document/cons_doc_LAW_193740/
11. Volkov, V.A.: Methodology for measuring the fair value of securities in Lanta-Bank JSCB (JSC). Approved by the First Deputy Chairman of the Board of Lanta-Bank JSCB (JSC). Minutes of the Board Dated 06.06.2019 No. 772. https://www.lanta.ru
12. Official Website of the Public Joint Stock Company "Moscow Exchange MICEX-RTS". https://www.moex.com/s249
13. Druzhilovskaya, T., Dobrolyubov, N.A.: Problems of conceptual framework formation in accounting of financial instruments of organizations. Int. Account. **22**(5), 484–499 (2019)
14. Regulation of the Bank of Russia Dated 03.12.2015 No. 511-P "On the Procedure for Calculation of Market Risk by Credit Institutions" (Registered with the Ministry of Justice of Russia on 28.12.2015 Under No. 40328, as Amended). https://base.garant.ru/71283076/

Promises of Private Pension Funds in Slovakia and Reality

Vladimir M. Bachishin[1](✉), Maria B. Vladimirova[2], and Ivana Lenerová[3]

[1] Comenius Analytica, Ltd., Bratislava, Slovakia
[2] Lomonosov Moscow State University, Moscow, Russia
[3] University of Economics in Bratislava, Bratislava, Slovakia

Abstract. The pension system for the population of the 27 countries of the European Union is based on the fact that today's population pays pensions to those who have retired. It can hold up well only in several ways: either by an increase in the retirement age or by the creation of a private savings system. The article analyzes the state of various pension reforms in the world, including in Slovakia.

Keywords: Pensions · Savings · Funds · Capital markets

1 Introduction

The most important demographic change in Europe will probably be the apparent shift towards a much older population structure, a development of which can already be seen in several of the 27 EU member states.

As a result, the share of working-age people in the 27 EU countries is falling, while the relative number of retirees is growing. The proportion of the elderly in the total population will increase significantly in the coming decades. In turn, this will increase the burden on working-age people who will have to finance the social spending required for a range of services to support an aging population.

Currently, 20.3% of the population of Western Europe is over 65 years old. However, in 2050, this category will account for up to thirty percent of the total population.

All of these have serious implications, especially in terms of the cost of pension systems, especially in Europe, where individual pension schemes are rather generous. But will they be socially and financially sustainable in the medium or long term?

It turns out that the most viable way, in addition to constant funding, is also the accumulation of funds paid by employees and employers and their investment in a promising active business. This trend seems to be more transient compared to what is happening today – simple "reuse" of the funds raised and their payment to today's retirees will not be enough to maintain the current level for several years.

The demographic arithmetic is relentless: in the European Union, there are currently more than four working-age people per person aged 65 and over. However, this ratio will change rapidly. By 2030, two working-age people will "work" for one pensioner, while in 2050; this ratio will be equal to at least 1.5:1.

P. V. Trifonov and M. V. Charaeva (Eds.): DITEM 2021, LNNS 380, pp. 530–537, 2022.
https://doi.org/10.1007/978-3-030-94245-8_72

Thus, it is clear that maintaining the current level of retirement benefits in relation to average earnings will become extremely costly or financially unacceptable.

Social security benefits and their high standard in continental European countries can significantly weaken the competitiveness of their economies in the global market. They can also jeopardize public expenditure projects if they lead to budget deficits.

The current form of unfunded financing of PAYG (pay-as-you-go) pension schemes no longer has the desired effect. Where they looked for funded pension schemes, for example, in the Netherlands, Switzerland and the UK, the assets acquired grew up to 100% of the gross domestic product. Suppose that there are no such schemes in France and Italy. However, the fact is that PAYG and other unfunded platforms in the European Union contribute up to 84% to funding for retirement benefits [1]. This situation is also alarming, since the EU has a common labor market and free movement of workers, which is very disappointing and irritating for employers in multinational companies, for which maintaining pension standards is an increasingly serious problem.

Pensions should perform three main functions – insurance against certain risks (longevity, disability), savings (income transfer from the youth to the elderly) and redistribution (from the rich to the poor). It is reasonable to divide the provision of these different services from each other into different "pillars". Redistribution is best ensured by the pillar of the state, while savings can accumulate well in the pillar of private capital. Insurance is best performed by all the components together. The real situation is putting serious pressure on reforming pension models in various combinations.

This is not only about Europe, since this trend is global in nature. In the recent past, 125 countries have undertaken partial or fundamental pension reform. Basically, these were partial steps – most often a change in the amount of contributions (51 cases), then a change in the pension calculation coefficients (24 cases) and a change in the retirement age (15 cases).

A significant number of less developed countries, from Mexico to Poland, have introduced or are constantly preparing a mandatory fund system that more or less follows the Chilean model of the 1980s. These countries have a particular interest and motivation to obtain long-term resources and prevent volatility in foreign investment flows. If the capital market in their countries stabilizes, it can be assumed that the long-term economic growth will accelerate, which will have a positive impact on the pension sector.

However, it is currently controversial whether only a pension saving scheme capitalizing on "old-age contributions" will contribute to the financing of pensions in the developed part of Europe. This is not to say that the unfunded system in France puts the country's economy and growth at a disadvantage compared to the UK, where financial schemes work to a large extent.

However, this scheme aims to "settle down" in the long term, which is relevant and, above all, is of paramount importance in the demographic context. In addition, it has several comparative advantages: it shifts the pension burden from the population to the public sector, it is a tool for establishing international pension schemes, it has an incentive to harmonize tax systems and, above all, it protects pensioners in countries with a shrinking population. It is also characterized by long-term growth prospects.

In some other European countries, such as Italy and Spain, pension saving schemes are developing flexibly, but very slowly.

The need to reform the pension system is also evident in Germany, although there are the most sophisticated and complex pension schemes.

2 Methods

2.1 Pension Market in the USA

The problem of pension funds in the USA has diametrically opposite dimensions. Their expansion will develop into the creation of global structures, which will further increase the capacity of the domestic market. An example is the recent acquisition by the Nationwide Mutual Insurance Company (Columbus, Ohio), which paid over £1 billion for Gartmore, one of the UK's leading retirement businesses.

This is the first international platform of the mentioned expansion. The second one is the predation of US pension funds in the domestic capital market. The share of these funds in the portfolio of ordinary domestic shares is constantly growing. Due to the fast-emerging market with a steep rise in share prices, the strongest US pension funds have achieved average returns of 6–7%, and this level has stabilized at least in recent years.

However, US pension fund managers are also cautious. In recent years, the investment in common stock has been very profitable, but some of the proceeds are usually invested in less risky government bonds.

It should be added that US pension fund managers manage a large amount of assets – according to estimates, 18.8 billion dollars [2].

There are over 3,000 private and public pension funds operating in the USA – this "club" of large funds includes the funds with assets of more than $100 million.

The American pension model can simply be characterized by the fact that an employee contributes to a fund, and these resources are subsequently invested in the market. This means that pension contributions are well defined, but the final yield depends on the market.

The old model is a defined income scheme, in which retirement benefits are paid by the employer and the pension amount is defined as a portion of an employee's real wages until the end of his working career.

However, these pension schemes are also becoming too expensive due to the demographic aspect, and the aforementioned "undefined retirement income", the amount of which depends on the market mechanisms, comes to the fore.

The existing "defined pension scheme" stems from the freedom of decision and choice of the employee, who decides for himself what benefits he will pay for his retirement age, choosing from the pension proposals where to invest the money.

2.2 Japan

The aging of the Japanese population is the most significant of all industrialized countries, which has a very strong impact on the financing of pension funds.

In Japan, the demographic arithmetic is even more ruthless than in Europe. In the coming decades, one employee will fund two or three retirees! In the past decade, this ratio has been 1: 1. A similar demographic trend may have prevailed in the USA, but it has been eliminated there by the immigration policy. The United Nations Organization has pointed to the negative demographic development of several Japanese governments, but without a more tangible effect. Japan is proud of its ethnic homogeneity, which is an important factor in social cohesion. Probably, in the future, Japan will not allow a liberal immigration model, which, however, will negatively affect the structure of the pension plan. The PAYG (pay-as-you-go) model, which assumes that the current generation of working-age people provides financial resources to pay pensions to retirees, is unacceptable in times of labor market shortages.

The government and employers are looking for a way out, but they constantly face the same problem – Japanese pension funds are severely undercapitalized [3].

Despite the efforts of some banks (Sakura and Sanwa Bank) to place temporarily spare financial resources in pension funds, there is still no systemic solution. The government wants to raise the retirement age and hastily introduce a new law to the parliament to define the 401K pension scheme for the first time. This is a model similar to the American one: an employee pays remuneration to the fund, which is then evaluated as part of the investment in the capital market. Fixed contributions and uncertain income are in stark contrast to the level of retirement benefits, at which the Japanese know exactly what their pension will be. The new law strongly opposes Japanese trade unions, which argue that it is highly inappropriate for employees who have worked hard for many years to be exposed to possible risk capital market shocks.

Thus, the 401K pension scheme can be expected to be significantly modified, but even then the current situation in Japan is ripe for a new fund boom. Proof of this is an agency "PASONA" that offers different combinations of insurance products and investment options within a graduated monthly fee.

2.3 Chile

Forced saving by Chilean citizens through pension funds has become one of the cornerstones of long-term economic prosperity and extraordinary domestic saving dynamics. Today's pension funds are more than double the foreign exchange reserves of the Central Bank of Chile. The economic situation of the Chile pension fund management company (AFP – Administradoras de Fondos de Pensiones) is extremely important for the above reasons, and every decision they make has a short-term and long-term macroeconomic effect. For example, 60% of the mortgage loans for social housing come from the sources accumulated in pension funds. A kind of social pact and trust in pension funds has been created, as they focus on both the individual future of contributors and the financial needs of the national economy [4].

Chilean pension funds are waiting for some adjustments in the short term. They will aim at enhancing individuality and easing the tight government oversight that was needed to lay the foundations. The younger generation of policyholders with a longer retirement period will be given the opportunity to make more risky investments with a higher return, and thus the overall return will increase the achieved pensions. The autonomy of contributors in choosing investment options, up to the possibility of exit and independent maintenance of the pension account, is a growing trend. Before the cancellation, a so-called profitability zone is also held, which must contain all AFPs under threat of recall.

Pension funds need to address this new issue with the utmost caution, since they are overseen not only by the government, but also, for example, by the Social Investment Forum, which assesses carefully how pension funds perform their obligations in the context of socially responsible investment.

3 Results and Discussion

3.1 Private Compulsory Pension Funds in Slovakia

What is the main characteristic of the pension system in the Slovak Republic?

It has three components [5]. The first component is pay-as-you-go. The second component is compulsory private savings in funds.

What are old-age pension savings? Old-age pension savings (Component II) are savings which purpose – along with pension insurance – the so-called first component, is to provide income to contributors in old age or to survivors in the event of a contributor's death. Old-age pension savings represent a capitalization component, which is a fixed contribution, i.e. the amount of the retirement benefit depends on the contributions paid to this component and their increment.

Entry into Component II is voluntary, but after the transition to this level, old-age pension savings become mandatory, i.e. the exit from it is impossible and along with the first component, it forms the basis of pension benefits.

A contributor is an individual who has a concluded funded old-age pension agreement (hereinafter referred to as the "Agreement with the Management Company"), entered into the register of agreements with companies managing private savings, or has entered into an agreement on the payment of an old-age or early retirement pension. The old-age pension depends on the chosen program.

A contributor is also an individual whose pension fund management company has been appointed by the Social Insurance Agency. The first agreement can be concluded by a person who has entered into the first pension insurance agreement and has not reached the age of 35 on the date of signing the first agreement.

Contributions are paid to the contributor's personal pension account. The funded contributions for old-age pensions are compulsory and voluntary.

The collection of compulsory contributions is handled by the Social Insurance Agency, which is obliged to transfer them to the contributor's pension management company, which then credits them to contributors' respective personal pension accounts held in pension funds.

The amount of compulsory contributions is determined as a percentage of the tax base achieved in the relevant period. From 01 September 2012 to 31 December 2016, this rate was 4% of the tax base. After this period, the rate increases by 0.25% annually. In 2024 and in subsequent years, the rate will remain at 6% of the tax base.

Contributors can also send voluntary contributions to their personal pension accounts to increase their pension savings, the amount of which is not limited.

Compulsory contributions and voluntary contributions of a contributor shall be paid to his personal pension account, which is managed by the chosen pension management company. A contributor can choose a pension fund management company, as well as a pension fund, in which the paid contributions will be distributed in accordance with the investment strategy of this pension fund.

To better protect allocated savings from unexpected negative events in financial markets, the gradual and automatic transfer of assets to conservative bond funds begins at the age of 52. This transfer only applies to the assets consisting of compulsory contributions.

The assets qualifying for voluntary contributions can be invested by a contributor in any pension fund, regardless of his age and in accordance with his investment risk profile.

A pension recipient shall be paid an old-age pension from assessed contributions, an early retirement pension or a survivor pension can be paid to survivors. At the decision of a pension recipient, the pension shall be paid to him either by the chosen insurance company (life and temporary pension) or by a pension fund management company (choice of a pension scheme).

In the event of the contributor's death, the authorized person specified by the contributor in the agreement with the savings management company shall become entitled to payment of the corresponding amount from the personal pension account of the deceased contributor. If the contributor has not indicated an authorized person in the pension agreement in the event of his death or does not have such a person, his property shall become an inherited item after his death.

3.2 Pension Savings

At the end of 2019, there were 9.37 billion euros in accounts of pension fund management companies, which is 10% of the volume of the Slovak economy. There were 1,564,152 contributors in the system, which is 57% of the economically active population of Slovakia. Bond-backed pension funds had the largest share of pension fund assets (the most conservative investment option) at 72%.

The pension system in Slovakia has been operating for 15 years. It is possible to solve the above problem only partly, because the system is significantly inferior to its capabilities and does not really appreciate the Slovakian savings.

The system as such does not achieve its potential. There are several reasons. The first is political interference. Frequent political intervention does not benefit the pension system, which aims to ensure the long-term sustainability of the system and provide pensions that are enough for people to keep alive. The rules change and the settings change frequently.

The second negative phenomenon is the ineffective investment strategies of pension fund management companies. Compared to other countries in Europe and beyond, the performance of Slovakian pension fund management companies is significantly lower than in other countries.

How much do private pension funds earn? Not much for conservative investors. The average appreciation for the studied period reached 1.58%, which, unfortunately, is below the average inflation rate, and therefore savings are not valued. A fee charged by management companies, which is also deducted from the amount saved, should also be mentioned.

Those who invested in index funds got an increase of 5.3%. For equity funds, it is 2%, for mixed funds – 1.5%, for bond funds (the most conservative option) – 1.1% [6].

4 Conclusion

Can this be improved so that the Slovaks have more money in savings accounts? If so, in what way? Yes, it is possible and can be achieved through the principles underlying the reform of compulsory and voluntary retirement savings.

The first principle is to change the basic "system setup". This means automatically logging in, choosing an investment strategy and being able to log out at a specified time.

The second one is a change in the distribution of current contributors to support the transition of the Slovaks to significant risks, long-term investments to compensate for short-term losses.

The third one is the introduction of investment performance benchmarks – they are needed to compare whether a chosen strategy is profitable and how effectively a management company invests savings.

The fourth one is a reduction in fees – the current payment system is difficult to justify compared to compatible systems elsewhere in the world. Currently, the fees for actively and passively managed funds are the same, this needs to be changed.

The fifth one is to take into account the life cycle – it is necessary to automatically adjust the system to a gradual decrease in equity capital – a more risky and simultaneously more profitable component of the portfolio. The older a person is, the less risky his/her portfolio is.

A political risk cannot be completely eliminated from pension schemes, but it can be reduced and transformed. It is necessary to strengthen the pension system in the form of a constitutional law to minimize the negative impact. This change will allow eliminating the risk of poor performance and contributors will receive a significant pension.

Ethics should also be taken into account when investing. Fund managers are also required to fulfill their obligations to employees and employers who are exposed to investment risks in the fund's operations. This means that if a certain profit cannot be obtained through the fund's fault, the number of recipients shall not be reduced any way.

Fund managers will need to ensure that their managers and advisors clearly manage all the matters related to socially responsible investments to be able to clearly assess social, environmental and ethical issues, as well as the changing business and entrepreneurial climate in the long-term increase in the cost of financial resources of pension funds.

The relationship between the pension fund and the savings rate has attracted the attention of economists for decades. The main question was whether a pay-as-you-go pension system reduces aggregate national savings or the greater importance of funded pension schemes leads to an increase in national savings. If a corresponding strong correlation were found between these indicators, this would mean that changes in the pension policy could be used to increase overall national savings and, as a consequence, higher economic indicators per capita.

Analytical reports usually conclude that there is insufficient evidence to hypothesize that introducing a rolling system would lower national saving rates. There may be some indications, but they are either insignificant based on empirical evidence or their impact is hidden by other factors.

It has been similarly shown, mainly based on data from the USA, that the accumulation of assets in pension accounts will lead to an increase in total household savings, albeit to a lesser extent than an increase in balances in these accounts. The positive effect of developed pension funds seems to be in a shadow or replaced by the "behavior" of other components of national savings. At the very least, in OECD countries, there was virtually no correlation between the growth rate of pension assets and the growth rate of total savings in the economy.

References

1. Huges, G., Stewart, J.: Reforming Pensions in Europe. Edward Elgar, Chelteham (2004)
2. Pension Fund Assets Rose to USD 32 Trillion in 2019 But COVID-19 Impacts Are Set to Reverse Some of These Gains. https://www.oecd.org/pensions/Pension-Funds-in-Figures-2020.pdf
3. Noriyasu, W.: Private pension plans in Japan. In: Bodie, Z., Mitchell, O.S., Turner, J.A. (eds.) Securing Employer-Based Pensions: An International Perspective, pp. 121–147. University of Pennsylvania Press (1996)
4. Chile Pension Fund Sector: Technical Note (English). http://documents.worldbank.org/curated/en/951971468014980189/Chile-Pension-fund-sector-technical-note
5. Horbulák, Z.: Current situation of pension reform in Slovakia. Tourism Econ. Entrepreneur. **2–3**, 16–34 (2009)
6. Financial Stability Report (2019). https://www.nbs.sk/_img/Documents/ZAKLNBS/PUBLIK/SFS/protected/SFS_112020.pdf

Demand Factor and Bubbles in the Market of Bank Loans to the Population

Elena N. Alifanova(✉), Ludmila I. Nivorozhkina, Yulia S. Evlakhova, Elena N. Makarenko, and Aleksandra A. Tregubova

Rostov State University of Economics, Rostov-on-Don, Russia
alifanovaen@mail.ru

Abstract. The dynamics of market bank loans to individuals is subject to market laws, including the impact of supply and demand factors and the formation of market bubbles. The interaction of these phenomena significantly affects the prospects of this market. The objective is on the basis of an analysis of the dynamics of the Russian market of bank loans to individuals and identification of market bubbles, as well as of the risk appetite and deviant financial behavior of future generations of borrowers, to identify how market bubbles and the demand factor can affect the development of this market. To study bubbles the method of exponential smoothing was used. The study of the demand factor, based on the data of a survey of students of the largest universities in the south of Russia according to the original authors' questionnaire, was investigated on multidimensional probit models. In the Russian market of bank loans to individuals for 2005–2019, bubbles were revealed, since the credit bubble was replaced by a bubble of overdues on retail loans and vice versa. The thesis about the leading role of the demand factor in this market has been confirmed. The scale of the propensity of future borrowers to various types of deviant financial behavior has been identified. The novelty of the results obtained lies in the identification of the development options for the Russian market of bank loans to individuals, considering the interaction of the demand factor and the formation of market bubbles.

Keywords: Credit cycles · Lending boom · Risky financial behavior · Future generations of borrowers

1 Introduction

In many countries of the world, including Russia, the population is a donor of resources for the banking sector. Even in the sphere of retail bank lending, the balance of cash flow is in favor of banks [1, 2]. The size of the transfer of funds from the people to banks reaches its peak values during crisis periods, when banks receive more in the form of repayment of old debts than they issue new loans.

The dynamics of the market for bank loans to individuals is subject to market laws: it is not free from the formation of market bubbles [3], and is also subject to the influence of supply and demand factors [4]. Reviews [5, 6] are devoted to the interaction of these phenomena and their influence on the prospects of the credit market.

P. V. Trifonov and M. V. Charaeva (Eds.): DITEM 2021, LNNS 380, pp. 538–544, 2022.
https://doi.org/10.1007/978-3-030-94245-8_73

In contrast to existing research, this paper is devoted to the study of the Russian market of bank loans to individuals. On the one hand, in the dynamics of this market for 2005–2019, the authors tried to identify some episodes of market bubbles (or in-between states leading to them).

On the other hand, the authors focused on the study of the demand factor in the market for bank loans to individuals, by which they mean the needs of the people for extra funding, satisfied via loans.

The demand (rather than supply) factor turned out to be a constant driver of growth in the Russian market of bank loans to individuals [7], supporting it even in the face of a drop in economic activity due to the coronavirus pandemic.

When studying the demand factor in the credit market, it is traditional to refer to the analysis of the financial condition of the borrower, to assess the risks of his/her solvency. In this study, the authors remain within the framework of the traditional approach, clarifying it in connection with: a) the target audience (not all population groups, but young people, university graduates, who are the "future" generations of borrowers; b) a broader approach to a set of risks (not so much the risk of insolvency, which may be difficult for graduates to assess as they are unemployed yet and have no own income, how many are the risks of lavishness and financial misbehavior, risk appetite in making decisions).

The purpose of this study is to analyze the dynamics of the Russian market of bank loans to individuals, to identify market bubbles, and to study the risk appetite and deviant financial behavior of "future" generations of borrowers to find out how market bubbles and the demand factor can affect the development of the Russian market of bank loans to individuals.

2 Methods

To identify the episodes of credit bubbles (or in-between states causing them, the authors used an approach different from the existing one [8, 9], successfully tested by the authors in assessing the level of threats to the financial security of a country. The key idea of this approach is a quantitative assessment of the gap between the actual and forecast values of particular macroeconomic indicators. The forecast values of indicators are found using the exponential smoothing model [10]. The general view of the model can be represented as follows:

$$\hat{y}_{T+1} = \alpha \cdot y_T + (1 - \alpha) \cdot \hat{y}_T \ or \ \hat{y}_{T+1} = \alpha \cdot y_{T+1} + (1 - \alpha) \cdot \hat{y}_T,$$

where α is the smoothing constant parameter (usually from 0 to 1), $y_T(y_{T+1})$ and $\hat{y}_T(\hat{y}_{T+1})$ is the actual and predicted values of the series level at the moment/for the period T and (T + 1).

Note that empirical studies [11, 12] have shown that simple exponential smoothing in most cases gives a fairly accurate forecast.

The information base for the study of bubbles in the market of bank loans to individuals was formed by the data of the Central Bank of the Russian Federation on the development of the banking sector for the period from 01.01.2005 to 01.01.2019.

The methodology for assessing the propensity of the population to risky financial behavior is based on the authors' questionnaire [13, 14], the concept of which implies the presence of a system of basic personal, behavioral, socio-economic, demographic characteristics and attitudes, which together form a typology by the propensity to manifest deviations in financial behavior.

In the study, the authors proceed from the need to assess the risk of manifestation among young people of deviations in financial behavior, as well as lavishness and financial misbehavior, risk appetite in making decisions.

Risk assessment, depending on its objectives, can be carried out with varying degrees of depth and complexity. The easiest option is to identify to which group an individual will belong (spenders/risk inclined/financially misbehaving/financially deviant), on the basis of his/her questionnaire, and thus find out whether he/she belongs to a risk group. More accurate can be considered an approach that makes it possible to assess the risk that a given individual will be inclined to engage in illegal financial activities, based on a scoring assessment [15], which enables to obtain a fairly reliable answer in the conditions of a small amount of customer data and a limited time.

As a rule, subject that credit histories are available as information database, modern scoring cards are made using the apparatus of multiple logistic regression. However, assessing the risks of students manifesting certain types of financial behavior by evaluating logistic or polynomial regression models may give an incorrect result, since the estimates of the model parameters may turn out to be inconsistent. Therefore, as an alternative to one-dimensional logit models or multinomial logit models, the authors used multidimensional probit models [16, 17]. Evaluation of multidimensional probit models made it possible to identify the determinants of all personal characteristics, which act as indicators of the propensity to financial deviations, in a single system, taking into account their potential relationship with each other.

The information base for the study of the demand factor in the market of bank loans to individuals was made up of the data from a survey of students of the largest universities in the south of Russia. The sample set was formed on the basis of a probabilistic multistage sample with a proper random non-repetitive selection method. The survey involved 912 students.

3 Results

The following estimates of the forecast and actual values of the indicators of the market of bank loans to individuals were obtained (Fig. 1).

According to the data in Fig. 1, going beyond the upper boundaries of the "stability corridor" of the loans/assets indicator, which indicates the appearance of a credit bubble in the banking sector, was observed on 01.01.2013 – 01.01.2014. The boom in lending (primarily consumer lending) during this period was primarily due to the post-crisis economic recovery.

Going beyond the upper bounds of the "stability corridor" of the arrears/loans indicators, indicating the emergence of a bubble of arrears in the banking sector, was observed on 01.01.2010 and 01.01.2016. Taking into account the lagging (in relation to economic dynamics as a whole) nature of the credit market indicators, the growth in the volume of overdue debt beyond the boundaries of the "stability corridor" is a consequence of the crises of 2008 and 2014, accompanied by a drop in household income.

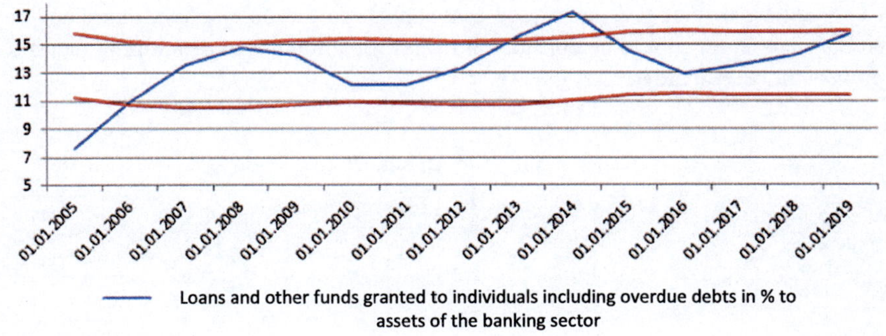

a

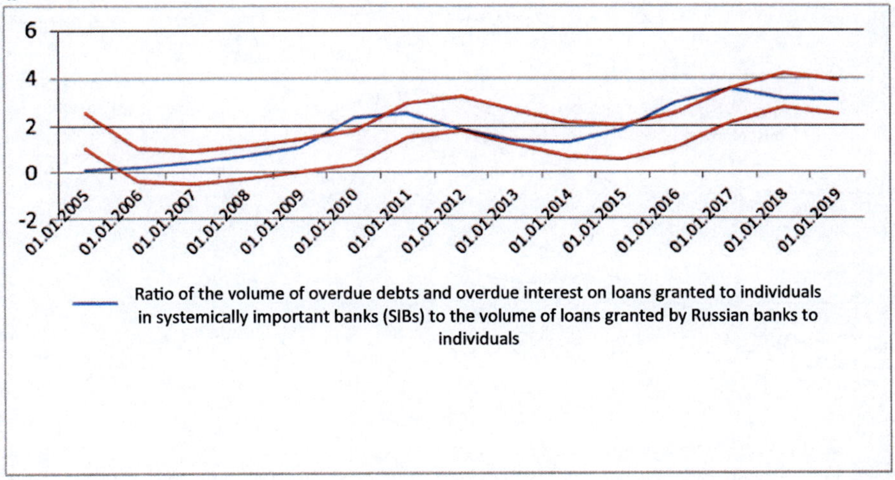

b

Fig. 1. Actual values **a** and forecast interval of values **b** of indicators of the market of bank loans to individuals. *Source:* developed by the authors.

According to the results of a survey of university graduates, the number of groups of respondents was estimated, in which, both individually and in combination, the following features were manifested:

1. Risk inclination,
2. Attitude towards money – spenders,
3. Financial misbehaving,
4. Tendency to deviate in financial behavior.

Figure 2 shows the results of such an assessment. Estimated shares are obtained as the ratio of the number of respondents who have manifested combinations of the corresponding types of deviations to the total number of respondents.

The authors do not give normative assessments to the obtained values, since any deviation risk factor should be considered from the standpoint of developing measures to prevent and eliminate it.

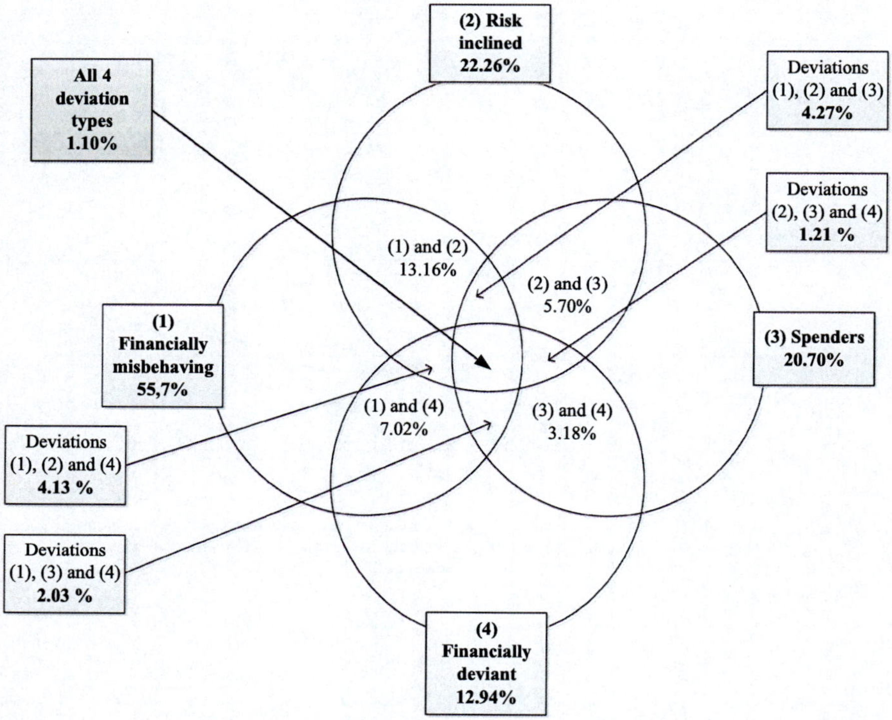

Fig. 2. Estimates of the scale of deviant behavior of the people.

Based on the results of evaluating multidimensional probit models, the main socio-demographic characteristics of individuals who are potentially more inclined to implement risky financial strategies were found [18]. These are young men who live with families whose family income is more than 50 thousand rubles, studying in any areas, except for economics and management, whose average spending on personal needs per month will be in the range from 10 to 50 thousand rubles, whose mothers have specialized secondary education. Note that in this case the authors are talking about projective strategies, a tendency to possible deviations in financial behavior, and not about the actual implementation of such strategies.

4 Discussion

Thus, in the market of bank loans to the individuals of the Russian Federation in 2005–2019, as the exponential smoothing model shows, market bubbles were observed, while there was a certain sequence and recurrence of the phenomena, since the credit bubble was replaced by a bubble of overdue debts in connection with retail loans and vice versa. This suggests that Russian banks as a whole are pursuing an aggressive policy in the market of bank loans to individuals (which, of course, may not be like that for a particular bank), and in crisis conditions they tend to reduce risk appetite and carry out credit rationing on a scale that does not significantly affect positive dynamics of the market of bank loans to individuals.

The identified episodes of bubbles in the market of bank loans to individuals also confirm the thesis about the leading role of the demand factor in this market. In this regard, it is important to understand that the influence of the demand factor will continue in the future, since with the development of national financial ecosystems, the scale of the client database becomes a competitive advantage, and the ways to increase it are the accumulation of a large amount of data, the formation of a customer portrait, a combination of the supply of financial and non-financial services. along with their customization according to the needs and the seamlessness of their receipt.

New generations of borrowers are entering the market, the prospects for the development of the market for bank loans to individuals depend to a certain extent on what their characteristics are.

A study of the scale of the propensity of future borrowers to various types of deviant financial behavior showed that although all 4 types of deviations are observed in 1.1% of respondents, the share of financially misbehaving (55.7%), risk inclined (22.26%) and lavish (20.7%) turn out to be quite significant in the analyzed sample. In the authors' opinion, such categories of the population are potentially candidates for borrowers with overdue debts, and their number (taking into account the representativeness of the sample under study) suggests that in the future the factor of demand for bank loans by individuals will be burdened by problem debts. In this regard, the socio-demographic characteristics of individuals who are potentially more inclined to implement risky financial strategies identified in the course of the study are of great importance. The formed portraits allow obtaining a preliminary assessment of the client's risk, being guided by which financial institutions can determine whether in this case it is necessary to organize a specialized underwriting procedure or not.

5 Conclusion

On the Russian market of bank loans to individuals, the credit bubble is being replaced by a bubble of overdue retail loans. This cyclicality, the authors believe, will continue in the future due to the support of the demand factor for loans on the part of individuals.

The future demand for loans is being shaped by new generations of borrowers. Studies of the scale of the propensity of university graduates as future borrowers to various types of deviant financial behavior show three main differences (characteristics) that are significant for the credit market – financial misbehaving, risk appetite and lavish attitude to money. This suggests that future generations of borrowers will not be able to mitigate the overdues problem, which will require banks to remain focused on scoring and underwriting procedures.

Acknowledgments. The paper was prepared with the financial support of the Russian Foundation for Basic Research (agreement No. 18-010-00657).

References

1. Huifen, Y.: Int. Insolvency Rev. **27**(1), 58–76 (2018). https://doi.org/10.1002/iir.1296
2. Zinman, J.: J. Bank. Finance **34**(3), 546–556 (2010). https://doi.org/10.2139/ssrn.1335438
3. Virtanen, T., Tölö, E., Viréna, M., Taipalus, K.: J. Fin. Stab. **36**, 66–81 (2018). https://doi.org/10.1016/j.jfs.2018.02.008
4. Mamonov, M.E.: Econ. J. High. Sch. Econ. **21**(2), 251–282 (2017)
5. Livshits, I.: J. Econ. Surv. **29**(4), 594–613 (2015). https://doi.org/10.1111/joes.12119
6. Garmaise, M.J., Natividad, G.: J. Fin. **72**(5), 2331–2368 (2017). https://doi.org/10.1111/jofi.12522
7. Kuzina, O.E., Krupensky, N.A.: Econ. Issues **11**, 85–104 (2018). https://doi.org/10.32609/0042-8736-2018-11-85-104
8. Alessi, L., Detken, C.: J. Fin. Stab. **35**, 215–225 (2018). https://doi.org/10.1016/j.jfs.2017.06.005
9. Harvey, D.I., Leybourne, S.J., Whitehouse, E.J.: J. Emp. Fin. **58**, 226–246 (2020). https://doi.org/10.1016/j.jempfin.2020.06.004
10. Brown, R.G.: Exponential Smoothing for Predicting Demand. Arthur D. Little Inc., Cambridge, MA (1956)
11. Makridakis, S., et al.: J. Forecast. **1**(2), 111–153 (1982)
12. Makridakis, S., Winkler, R.L.: Manage. Sci. **29**(9), 987–996 (1983)
13. Alifanova, E.N., Nivorozhkina, L.I., Evlakhova, Y.S.: Int. J. Econ. Bus. Admin. **VII**(2), 3–15 (2019). https://doi.org/10.35808/ijeba/366
14. Sinyavskaya, T.G.: Account. Stat. **3**, 47–53 (2014)
15. Sadretdinov, I.E.: J. New Econ. Assoc. **2**, 30–36 (2009)
16. Ashford, J., Sowden, R.: Biometrics **26**(3), 535–546 (1970)
17. Cappellari, L., Jenkins, S.P.: Stata J. **3**, 278–294 (2003)
18. Nivorozhkina, L.I., Alifanova, E.N., Evlakhova, Y.S., Tregubova, A.A.: Int. J. Econ. Bus. Admin. **VIII**(4), 551–562 (2020). https://doi.org/10.35808/ijeba/606

Green Bonds: Peculiarities and Directions of Development

Yulia A. Vlasova(✉), Tatyana M. Rogova, Natalya V. Ivolgina, Zhanneta I. Gerzelieva, and Artem M. Lyukshin

Plekhanov Russian University of Economics, Moscow, Russia
{Vlasova.YA,lyukshin.am}@rea.ru

Abstract. The article considers the problems associated with the peculiarities of current development, and the prospects for the global and national markets of green bonds, which are a tool for financing projects to improve the environment. The main problematic aspects of issuing green bonds are low investment appeal, high risks and low liquidity of these securities, as well as the high politicization of such markets, if the projects relate to strategically important sectors of the economy. This demonstrates the urgent need to develop forms of state support for green bonds, especially at the regional and municipal levels. Moreover, the green investment market in Russia also needs government help to create favorable conditions both for issuers of green bonds and incentives to invest in these instruments. To increase the demand for such bonds the authors suggest introducing special benefits and supporting measures for the regions, highlight that the access to the purchase of green bonds by individuals would have a positive impact on the development of this segment at the sub-federal level.

Keywords: Green bonds · Borrowing · Debt · Investment · Financing · Environment

1 Introduction

The relevance of the topic of the research is due to the lack of funding for projects aimed at improving the environment of the country compared to other countries.

The forms and instruments of state participation in such projects deserve special attention in solving the problem of financing green bonds. Most of the scientific research is related to the problems of inadequate financing of green investment projects [1–7]. One of the obstacles is that the green securities segment does not have a high investment appeal. This is due to the high risks of investing in industrial intellectual property at the stage of design and development of objects of exclusive rights (technologies, inventions, industrial designs, allowing to improve the state of the environment). Certain works on this subject are of interest. Kaminker and Stewart [6] in their work touched upon the problems with credit risks of borrowers and issuers. In study [7], the need to use various hybrid instruments to reduce financial risks was noted. Despite the fact that the chosen topic is sufficiently scientifically advanced, there is a need for constant updating

P. V. Trifonov and M. V. Charaeva (Eds.): DITEM 2021, LNNS 380, pp. 545–551, 2022.
https://doi.org/10.1007/978-3-030-94245-8_74

of knowledge and theoretical developments, since international and national green bond markets are constantly in flux and subject to change.

The purpose of this research is to assess the current features of the financing of the green economy in Russia and to offer recommendations on the directions of its development. The objectives of the research assume the study of global and national experience in the field of green bond market financing, analysis of forms and tools for the implementation of green projects and assessment of the prospects for the development of international financial and national green bond markets.

2 Materials and Methods

The methods of our research were comparative, expert and statistical research methods based on officially published information of issuers, governments, methods.

Global green bond issuance rebounded in the second half of 2020 to a record $290.1 billion by the end of December, up from the previous record of $266.8 billion set in 2019. Developed markets accounted for 80% of total green bond issuance in 2020, up from 73% in 2019. Emerging markets accounted for 16% versus 22% in the previous year, as Chinese issuers shifted their focus to issuing social bonds. The contribution of supranational organizations was 4% versus 5% in 2019.

The share of European countries in the total volume of emissions was 156 billion U.S. dollars, or 48% [8].

The main issuers were state organizations and non-financial corporations. Government support provides more opportunities for private sector investment in Europe A wider range of issuers enters the market, in addition to utilities, real estate companies and banks [9].

North American issuance trends remained broadly unchanged from the previous year, with $61.5 billion in green bond issues compared with $60 billion in 2019. Latin America and the Caribbean region showed growth of nearly 65% over the previous year, reaching $7.9 billion in 2020. More than half of the total came from Chile, including four sovereign bond issues totaling $3.8 billion. African countries have issued green bonds to the amount of $1.2 billion. The issuers were South African non-financial corporations with FirstRand Bank loan ($225 million) and Standard Bank of South Africa with 10-year bonds ($200 million); Ghana's $41 million (€35 million) bilateral loan and Egypt's debut $750 million sovereign green bond issue [8].

The drop in green bond issuance by Asia-Pacific and supranational issuers can largely be attributed to the COVID-19 pandemic and the subsequent increase in social and sustainable development bonds. Such bonds were issued to support health care, drug coverage, and other urgent needs related to the pandemic.

3 Results

Based on the worldwide trend of growing attention to green finance around the world, interest in the green bond market has been growing in Russia as well. At the beginning of 2021, the number of green bond issues was 16 from more than six issuers. Since 2018, the annual volume of green bond issues has shown high growth rates (Fig. 1).

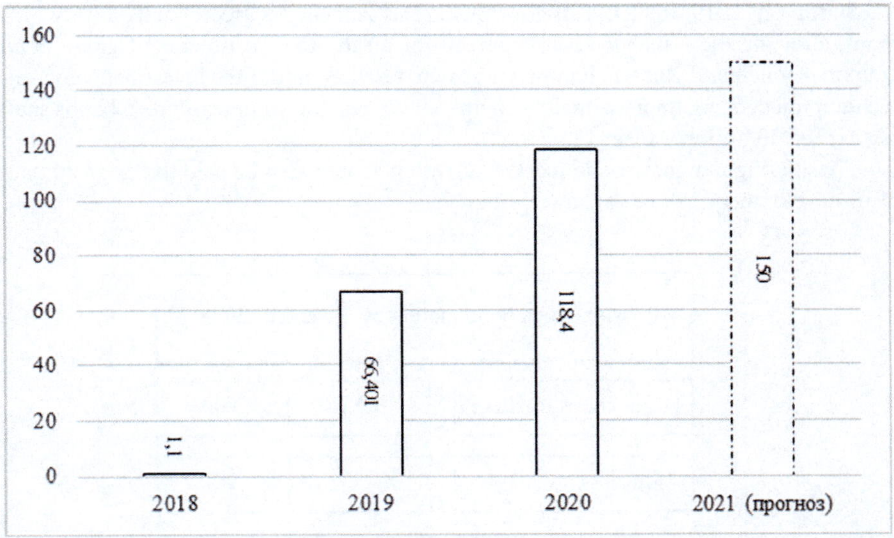

Fig. 1. Volume of green bonds issued by Russian issuers, billion rubles. *Source:* Authors' calculations based on data [10].

Despite the difficult economic conditions that arose in 2020 due to the aggravation of a number of economic problems in Russia and the manifestation of the global coronavirus pandemic, the Russian green bond market is showing positive growth rates. For example, only in 2020 the volume of green bonds issued increased by 78%. According to calculations and taking into account the current economic situation, the volume of green bonds issuance in 2021 will exceed 150 billion rubles. The growth potential of domestic green bonds could be realized by the end of 2030 and the market could reach 1.5 trillion rubles [11].

The key growth drivers of the green bond market during this period could be:

- global growth in the green finance market;
- growing investor interest in environmental investment projects;
- "ecologization" of the consciousness of society as a whole;
- government support for the green bond market;
- Reducing the level of uncertainty in the Russian economy.

Speaking of government support for the domestic green bond market, it is important to note that in May 2021, the Ministry of Finance of the Russian Federation together with the Bank of Russia developed a roadmap for the development of the green investment market. The Ministry of Economic Development of Russia has submitted a package of necessary legal documents to the Russian government for the launch and development of the green investment market in Russia [11]. Among the stimulating mechanisms for the development of the green bond market, the Ministry of Economic Development of the Russian Federation suggested reducing taxes on green bonds [12].

Starting in 2019, the Russian government has launched a green bond subsidy program, allowing the issuer to be reimbursed for up to 70% of the cost of green bond coupon payments. If Russian equipment is purchased as part of the investment environmental project being implemented, then up to 90%, and the volume of green bonds shall not exceed 30 billion rubles [13].

Considering the green bond market, attention should also be paid to the geography of its development, which is shown in Fig. 2.

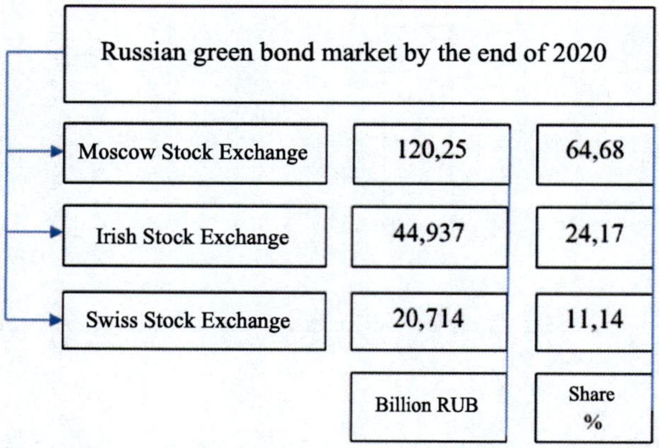

Fig. 2. Geography of green investment market in Russia at the beginning of 2021. *Source:* Authors' calculations based on data [10].

As indicated in Fig. 1, the main share of green bonds issued is traded on the Moscow Stock Exchange is 120.25 billion rubles, or 64.68%. Russian Railways placed its green bonds on the Irish and Swiss stock exchanges for 500 million euros (44.937 billion rubles at the exchange rate of the Bank of Russia at the end of 2020) and 250 million francs (20.714 billion rubles at the exchange rate of the Bank of Russia at the end of 2020), respectively. The main types of represented in the Russian green investment market are shown in Fig. 3. The largest in terms of volume (100 billion rubles) are perpetual bonds, the issuer of which is JSC "Russian Railways" [14]. The purpose of issuing these bonds is to promote the development of green technologies in rail transport.

The second largest issue is the Eurobonds placed by a company related to JSC "Russian Railways". Eurobonds were placed on the Irish and Swiss stock exchanges. (Fig. 2).

Project green bonds are issued by LLC "Transport Concession Company", which is engaged in the development of public transport. The main purpose of the issue of these bonds is to create and develop a streetcar network in St. Petersburg.

Thus, we can conclude that thanks to green bonds in Russia, environmental projects for the development of streetcar and rail transport, the development of environmental energy (the issue of green bonds LLC SFO RuSol 1), housing and public utilities and environmental waste management, real estate construction.

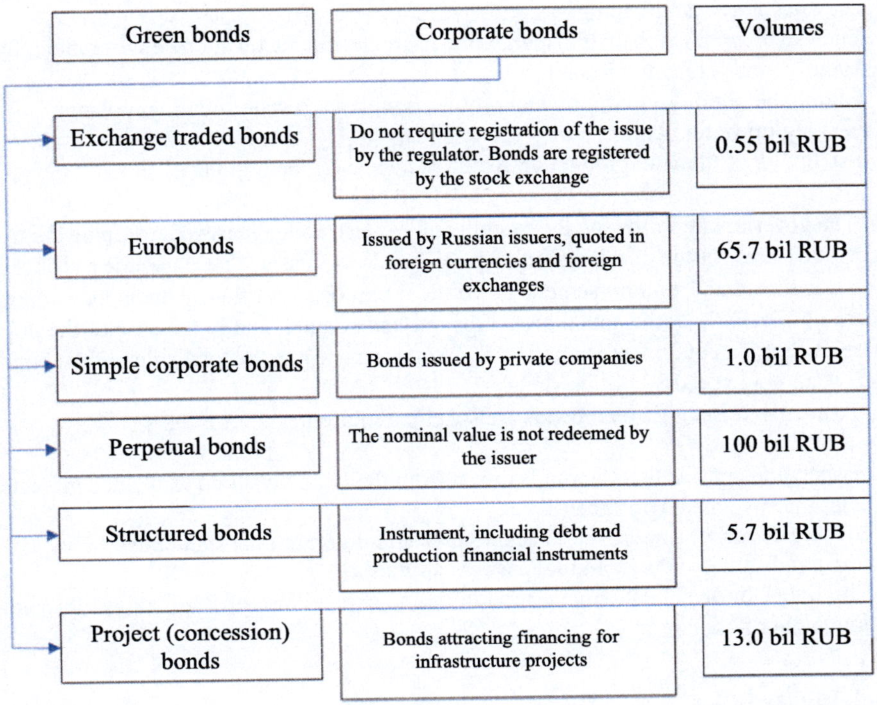

Fig. 3. Structure of the Russian green bonds market by type at the beginning of 2021. *Source:* Developed by the author on the basis of data [10].

The Russian authorities are also showing interest in the green bond market. For example, Moscow will be the first subject of the Federation to enter the market with green bonds. The total volume of green bonds will amount to 70 billion rubles is 70 million bonds at face value of one thousand rubles each. Maturity of the bonds is 7 years. According to the concept of green bonds in Moscow, the issue will comply with the principles of green bonds in 2018, the International Capital Markets Association, as well as methodological guidelines for the development of investment activities in the field of green finance in the Russian Federation, developed by VEB.RF. The issuance of green bonds would require the region to publish annual green bond reports on how the funds were spent.

4 Discussion

Given the limited revenue potential of the constituent territories of the Russian Federation and the reduction by many regions of spending on ecology and housing and communal services in the last 5 years, the issuance of green bonds would allow the constituent territories to:

- carry out active landscaping
- renewal of public urban transport with its replacement by more environmentally friendly modes (electric buses)
- solving the region's environmental problems with wastewater treatment plants
- cleaning up water bodies
- installation of filters at production facilities owned by the region.

The government, following the example of Western colleagues, is developing stimulating measures for the development of green bonds in Russia. These include a reduced tax rate; subsidized reimbursement of verification costs and the payment of coupon income (up to 70% of the actual cost of paying the coupon, and in the case of the purchase of Russian equipment is up to 90%). At the same time, the total value of the bond issue shall not exceed 30 billion rubles [15, 16].

Problems in issuing green bonds are specific requirements for issuers:

1. issuers must prove that the funds raised from the bonds. Will go to finance projects that improve the environment;
2. the bonds must be approved in accordance with international standards;
3. the bonds must undergo an independent appraisal;
4. the issuer must publish annual reports on the expenditure of funds raised through the issue of bonds.

5 Conclusion

To increase the demand for such bonds, we believe it is necessary to develop special benefits and support measures for the regions, including monetary ones in terms of reimbursement of part of the cost of organizing the emission. However, given the limited range of investors and, at the same time, the growing investment activity of individuals and the growing interest in environmental issues, access to the purchase of green bonds by individuals would have a positive impact on the development of this segment at the sub-federal level.

At its current stage of development, the green investment market in Russia needs government support to create favorable conditions both for issuers of green bonds and incentives to invest in these instruments. Moscow's issuance of green bonds could serve as an example for other regions of Russia and contribute to the further development of the green investment market and the implementation of major environmental projects [17, 18]. In order to develop this market, the Moscow Stock Exchange needs to conduct an active information campaign on green investments. Issuers should work on attractive bond yields, possibly with government support. The state should continue to work to support the green bond market in terms of lowering taxes, creating additional incentives for issuing these bonds.

References

1. Noh, H.J.: Strategies of Developing Green Finance. Korea Capital Market Institute, Seoul (2010)

2. Noh, H.J.: Financial Strategies to Activate Green Company Eco System. Korea Capital Market Institute KCMI, Seoul (2014)
3. Noh, H.J.: Climate Finance. Park Young Sa, Seoul (2014)
4. Noh, H.J.: Financial strategies to accelerate green growth. In: Sachs, J., Woo, W., Yoshino, N., Taghizadeh-Hesary, F. (eds.) Handbook of Green Finance. Sustainable Development. Springer, Singapore (2019)
5. Heinkel, R., Kraus, A., Zechner, J.: The effect of green investment on corporate behavior. J. Financial Quant. Anal. **35**, 431–449 (2001)
6. Kaminker, C., Stewart, F.: The role of institutional investors in financing clean energy. In: OECD working papers on finance, insurance and private pensions, no. 23. OECD, Paris (2012)
7. Gouldson, A., Kerr, N., Millward-Hopkins, J., Freeman, M.C., Topi, C., Sullivan, R.: Innovative financing models for low carbon transitions: exploring the case for revolving funds for domestic energy efficiency programmes. Energy Policy **86**, 739–748 (2015)
8. Sustainable Debt Global State of the Market 2020. Climate Bonds Initiative. https://www.climatebonds.net/system/tdf/reports/cbi_sd_sotm_2020_04d.pdf?file=1&type=node&id=56395&force=0
9. Rodenkova, T.N., Gerzelieva, Z.I., Borisova, I.V.: Public-private partnership in science, technology and social sphere in the digital economy. In: Proceedings of the International Scientific and Practical Internet Conference "Public and Private Finance in the Digital Economy", pp. 217–224. Moscow, February 12–08 April 2018 (2018)
10. Register of green and social bonds of Russian issuers as of December 28 (2020). https://infragreen.ru/frontend/images/PDF/INFRAGREEN-green-social-bonds-281220.pdf
11. By 2030 the volume of green bonds market in Russia could reach 1.5 trillion rubles, predict the Ministry of Economic Development. http://www.finmarket.ru/news/5474199
12. Ministry of Economic Development discusses tax cuts on green bonds. https://www.vedomosti.ru/economics/articles/2021/05/21/870680-minekonomrazvitiya-nalogov
13. Green light. https://rg.ru/2019/05/12/pravitelstvo-utverdilo-pravila-subsidirovaniia-zelenyh-obligacij.html
14. "Green" financing. JSC "Russian Railways". https://company.rzd.ru/ru/9972
15. Moscow government approved the decision to issue green bonds. https://www.mos.ru/mayor/themes/7299/7308050/
16. Moscow will be the first Russian region to issue green bonds. https://www.vedomosti.ru/economics/articles/2021/02/28/859580-zelenie-obligatsii
17. Vlasova, Y.A.: Espacios **38**(52), 63–69 (2017)
18. Vlasova, Y.A.: J. App. Econ. Sci. **11**(7), 1434–1441 (2016)

Methodological Approaches to Measuring Financial Development

Konstantin V. Krinichansky[1(✉)], Oleg V. Lunyakov[1], and Mikhail Y. Makovetsky[2]

[1] Financial University under the Government of the Russian Federation, Moscow, Russia
kkrin@fa.ru
[2] Moscow Witte University, Moscow, Russia

Abstract. Despite significant progress in the designing of financial development indicators, empirical studies reveal the problems of using this kind of indicators as descriptors of financial development and predictors of economic growth. This, for example, manifests itself when such key indicators as private credit to GDP turn out to be neutral or even negatively associated with economic growth, as well as in cases where the nonlinearity of the relationship between finance and growth is revealed, the nature of which remains unclear. The purpose of this study is to develop an approach to modifying indicators of financial development in order to eliminate distortions arising from the fact that increase in the indicator value in some systematic part does not anticipate long-run economic growth, and even, under certain conditions, precedes a financial crisis or a recession and ultimately negatively affects long-term growth. The authors use correction factors to calculate modified indicators of financial depth, ensuring that the growth of the corresponding indicators reflects financial development and meets the criterion of linearity and progressiveness of this development. The result of the study is the modification of five indicators of financial depth related to the categories of financial institutions and markets. The use of these modified indicators in empirical studies is intended to provide more objective assessments of financial development at the national level, to conduct a comparative cross-country analysis, to find more substantiated estimates of the relationship between financial development and long-term economic growth, and to formulate policy recommendations more accurately.

Keywords: Financial system · Financial market · Economic growth · Benchmarking

1 Introduction

Financial development today is a topic of discussion and a core theoretical notion in several fields of economics. The most extensive of these should be recognized as the area that discusses the relationship between financial development and economic growth. Over the 30-year history of this discussion, many indicators have been proposed to

P. V. Trifonov and M. V. Charaeva (Eds.): DITEM 2021, LNNS 380, pp. 552–557, 2022.
https://doi.org/10.1007/978-3-030-94245-8_75

operationalize the concept of financial development. For brevity, all of them for the purposes of this study will be denoted as FDIs (Financial Development Indicators).

Using FDIs as a kind of benchmarks for financial development helps to describe the financial sectors of countries [1], regions [2], administrative parts of individual countries [3] in terms of statics and dynamics and analyze the relationship between financial development and economic growth. It should be noted, however, that based on those theoretical assumptions that imply a consistent change in FDIs and economic growth variables [4], these indicators do not always help to confirm empirically the importance of financial development for long-term growth. In this paper, the authors discuss how this might be related to the disadvantages of FDIs. Ultimately, the authors aim to develop methodological approaches to measuring financial development by, firstly, improving the methods for calculating certain types of FDIs, and secondly, expanding the range of indicators classified as FDIs. The achievement of this goal constitutes the expected scientific novelty of this research.

2 Methods

The main method used in this study to achieve the goal of improving the methodology for calculating specific FDIs is to modify their models in order to consider the dual nature of the effect of increasing these indicators, in particular, when this increase after overcoming a certain threshold is associated with the activation of a systemic risk and threatens with negative consequences for economic growth.

Let us describe in more detail the problem covered here. The most widely used FDI is Private Credit by Deposit Money Banks to GDP (hereinafter referred to as PCB). There are several reasons for this, which emphasize the relevance of using this indicator as an FDI and its advantages over other metrics. First, this indicator affects the activity of the banking sector, which is the backbone sector of the financial system in all countries. Second, bank credit is an important source of external resources for firms, both facilitating the current operating cycle of enterprises and providing resources to cover their capital expenditures, which form the basis for long-term growth. Third, historical data for this indicator cover the largest number of countries over a significant period [5]. Fourth, a significant number of influential studies find that this indicator and long-term economic growth are linked directly to a positive relationship. See, for example, a review in [6].

At the same time, this indicator has a number of shortcomings. Let us dwell on two of them. First, it aggregates all issued loans, regardless of their quality. In cases where the deterioration in the quality of loans with an increase in their volume lead to the accumulation of systemic risks, it would not be correct to associate the increase in the amount of loans with the expectation of economic growth on this basis. Rather, on the contrary, a credit boom accompanied by a decline in credit quality creates the threat of a recession, while financial crises and recessions lead to a subsequent decline in long-term GDP growth. Second, this indicator, being a stock variable, does not necessarily grow linearly with financial development. Thus, the expansion of the use of securitization vehicles (reflecting the fact of financial development) leads to loans substitution and a decrease or slowdown in the volume of loans on bank balance sheets. At the same time, the issuing of loans by banks does not weaken at all, while a transformation of the financing method in the process of delivering funds from the lender to the borrower occurs. The same phenomenon may be the cause of undesirable heterogeneity in the input data given that different countries have achieved different success in promoting securitization.

Now let us concentrate on the methods by which the authors propose overcoming the above disadvantages. The first of these refers to loans quality. It is proposed to overcome this disadvantage by means of an additional adjustment procedure. In its current form, the PC^B indicator is calculated according to the formula [7]:

$$PC^B = \frac{0.5 \times \left({}^{PC_t}/_{CPI_t} + {}^{PC_{t-1}}/_{CPI_{t-1}} \right)}{{}^{GDP_t}/_{CPI_t^{av}}}, \quad (1)$$

where PC is the indicator value presented in the IMF International Financial Statistics database; CPI_t is the value of the consumer price index at the end of the year t; $CPI^{av}{}_t$ is the average annual value of the consumer price index in year t; GDP_t is the value of GDP in year t.

Here, introducing an additional adjustment to the formula for calculating this indicator, which is addressed to PC_t:

$$PC_t^{adj} = PC_t \times (1 - NL_t), \quad (2)$$

where NL_t is the share of non-performing loans in the total loan portfolio in year t.

Thus, in formula (2), the authors propose the use of an adjustment via a well-known credit characteristic included in the already existing range of statistical observations by country [7].

An alternative approach could be to adjust the amount of loans for the amount of non-performing loans shown on the balance sheet and overdue debts.

3 Results

The proposed methodology for adjusting FDIs can be successfully extended to several indicators belonging to the category of financial depth. The table below shows our suggestions for improving FDIs (Table 1).

The proposed adjustments will enable to more accurately approximate the data for some FDIs and smooth out the fluctuations explained by the business cycle, so that the resulting data series better reflect the dynamics of financial development and can be used as regressors (or predictors) to explain long-term economic growth.

Table 1. Financial depth indicators adjustment methods.

Indicator not adjusted	Indicator adjustment procedure
Financial institutions	
Private credit to GDP	Application of two adjustments: 1) adjustment for non-performing loans; 2) reverse adjustment of securitized loans
Financial institutions' assets to GDP	Application of three adjustments*: 1) volatility of stock prices; 2) critical yield spread of bonds**; 3) shares of investment-grade bonds, for which a default is declared
Financial markets	
Domestic private debt securities to GDP	Application of two adjustments: 1) critical yield spread of bonds**; 2) shares of bonds, for which a default is declared
Public debt securities to GDP	Application of the public debt critical threshold adjustment***
Stock market capitalization to GDP	Application of the volatility of stock prices adjustment

Notes:
* the adjustment requires the publication of time series characterizing the structure of assets of financial institutions of countries – the share of investments in equities, non-financial corporative bonds and government debt securities in the total amount of assets;
** these spreads indicate about "spread crises"; they are calculated within the framework of the concept of "debt crises without default". For this concept and the results of assessing the critical levels of interest rate spreads for bonds, see [8];
*** means such a critical threshold or debt limit, upon reaching which the government's solvency becomes questionable, and the risk of default increases sharply. See on this topic, for example, in [9, 10].

4 Discussion

The evolution of the literature on the relationship between finance and growth has gone from irrepressible enthusiasm and optimism in the 1990s, when most researchers strongly argued for a positive relationship between financial variables and growth indicators [11, 12], to the contradicting findings of the 2010s, skepticism and uncertainty, when researchers either found that finance negatively affects growth [13] or found the role of finance in growth to be "exhausted" [14, 15], or found that such a relationship has the property of nonlinearity [1, 16].

Against this background, the authors continue to express the opinion that the very concept of financial development needs to be clarified, which predetermines the urgency of solving the problem of its operationalization.

The modern academic studies have made significant headway through the efforts of the World Bank and the IMF, which have provided extensive data sets on financial development. However, these data still only fragmentarily cover some of the most important functions of financial systems, have short observation periods for a number of countries, need to be improved due to discrepancies in calculation methods, as well as need additional editing and construction of adapted and combined indicators that could more sensitively capture changes in financial development [17]. The approach proposed in this study to improve the financial development data set through the construction of additional modified FDIs is one of the attempts to overcome the existing drawbacks of financial development benchmarking.

At first glance, the proposed approach is fully integrated into the already existing recognized taxonomy of financial development indicators, formed under the auspices of the World Bank and the IMF [4, 18], since it allows integrating the newly obtained indicators into it (as mentioned above, the authors referred the resulting modified indicators to the category of depth). However, in reality, the new indicators are combined, because, including in the calculation algorithm the indicators of credit risk, volatility, the share of non-performing loans, one can combine indicators of two categories at once – depth and stability. The authors consider this approach promising, given that a large number of researchers have recently found some vulnerabilities in the use of known depth indicators.

5 Conclusion

This study proposes an approach to the modification of financial depth indicators, which consists in the use of adjustments that smooth out the time series of the corresponding indicators (including the most frequently used in private credit to GDP) using the measures showing the accumulation of risks with the growth of the considered depth indicators. Using the proposed approach (combining indicators of depth and stability) allows obtaining new data on the financial development of countries, better reflecting not only the quantitative, but also the qualitative aspect of financial development.

As a result, the authors expect that the use of new data in empirical research will lead to a better understanding of the relationship between financial development and economic growth. Presumably, the previously discovered nonlinear properties of this relationship will come under attack, being difficult to explain within the framework of economic concepts and generated, possibly, by the shortcomings of the operationalization of the financial development concept and the imperfection of the applied assessment methodology. For errors in estimates arising from the shortcomings of widely used methods for analyzing panel data [19].

Regarding the limitations of the implementation of these assumptions, the authors emphasize that they are mainly determined by the limited data that are supposed to be used to construct the adjustment factors. To a lesser extent, this applies to the data describing the banking sector. At the same time, the data related to the volatility of stock markets, bond spreads, etc., require unification of calculation methods and the establishment of an order for collecting and publishing relevant information.

References

1. Botev, J., Égert, B., Jawadi, F.: Int. Econ. **160**, 3–13 (2019). https://doi.org/10.1016/j.inteco.2019.06.004
2. De la Torre, A., Ize, A., Schmukler, S.: Financial development in Latin America and the Caribbean: the road ahead. The World Bank Latin American and Caribbean Studies (2011). https://doi.org/10.1596/978-0-8213-8847-1
3. Krinichansky, K.: Econ. Quest **10**, 94–108 (2015). https://doi.org/10.32609/0042-8736-2015-10-94-108
4. Beck, T., Feyen, E., Ize, A., Moizeszowicz, F.: Benchmarking Financial Development. Policy Research Working Paper; No. 4638. World Bank, Washington, DC (2008)
5. World Bank: "Global Financial Development" statistical dataset. (2021) https://databank.worldbank.org/source/global-financial-development
6. Levine, R.: Finance and growth: theory and evidence. In: Aghion, P., Durlauf, S.N. (eds.): Handbook of Economic Growth, pp. 865–934. Elsevier, Amsterdam (2005). https://doi.org/10.1016/S1574-0684(05)01012-9
7. World Bank Data Catalog. https://datacatalog.worldbank.org/private-credit-deposit-money-banks-gdp
8. Mitchener, K.J., Trebesch, C.: Sovereign Debt in the 21st Century: Looking Backward, Looking Forward, CESifo Working Paper No. 8959. CESifo, Munich (2021)
9. Ghosh, A.R., et al.: The Econ. J. **123**(566), 4–30 (2013). https://doi.org/10.1111/ecoj.12010
10. Cheron, A., et al.: J Money. Cred Bank **51**(2–3), 655–682 (2018). https://doi.org/10.1111/jmcb.12543
11. Levine, R., Zervos, S.: Amer. Econ. Rev. **88**, 537–558 (1998)
12. Beck, T., Levine, R., Loayza, N.: J. Fin. Econ. **58**(1, 2), 261–300 (2000) https://doi.org/10.1016/S0304-405X(00)00072-6
13. Cecchetti, S., Kharroubi, E.: Why does financial sector growth crowd out real economic growth?.(2015). https://www.bis.org/publ/work490.pdf
14. Rousseau, P., Wachtel, P.: J. Econ. Inq. **49**, 276–288 (2011). https://doi.org/10.1111/j.1465-7295.2009.00197.x
15. Arcand, J.L., Berkes, E., Panizza, U.: Too much finance? J. Econ. Growth **20**(2), 105–148 (2015). https://doi.org/10.1007/s10887-015-9115-2
16. Samargandi, N., Fidrmuc, J., Ghosh, S.: World Devel. **68**, 66–81 (2015). https://doi.org/10.1016/j.worlddev.2014.11.010
17. Svirydzenka, K.: Introducing a New Broad-based Index of Financial Development. IMF Working Paper WP/16/5, International Monetary Fund (2016). https://doi.org/10.5089/9781513583709.001
18. Čihák, M., Demirgüç-Kunt, A., Feyen, E., Levine, R.: J. Fin. Persp. **1**(2), 17–36 (2013)
19. Fajeau, M.: Int. Econ. **165**, 14–36 (2021). https://doi.org/10.1016/j.inteco.2020.10.003

Assessment of the Economic Security of an Entity

Elena N. Makarenko[1], Yulia G. Chernysheva[1(✉)], Gary I. Shepelenko[2], Svetlana S. Galazova[2], and Tatyana V. Makarenko[1]

[1] Rostov State University of Economics, Rostov-on-Don, Russia
[2] North Ossetian State University after K.L. Khetagurov, Vladikavkaz, Russia

Abstract. The concept of security, including the economic security, was first widely studied in the 20th century without losing its relevance at present. This concept is considered from various points of view, at both macro and micro levels. The most acute problem of assessing economic security is characteristic of an individual entity. What is the essence of economic security and how to correctly assess its level in an entity? To understand the essence of this category and create methods for its assessment, it is necessary to study these and other issues in more detail. The article investigates, identifies and summarises the key characteristics of the essence of economic security in Russia and abroad. The authors use tools and methods of business analysis, consider the qualitative characteristics and features of the category of economic security in relation to the country as a whole and an individual entity in particular. The authors' position is based on the fact that nowadays it becomes increasingly difficult for each entity to withstand the competition. This problem has been exacerbated by the pandemic crisis. Each entity has its own unique features taking into account which increases the objectivity and accuracy in assessing its economic condition. The authors found that the category of economic security of an entity should be considered taking into account the individual factors affecting it that is unique for each entity. A scenario approach was proposed for assessing the state of economic security and the possibility of changing it.

Keywords: Economic security · Uniqueness of entity · Scenario approach

1 Introduction

Today, the world as a whole and countries individually are undergoing constant changes as a result of technological revolutions and crisis situations, especially if we take into account and highlight the 2020 pandemic [1]. The modern economy can be characterized by a wide range of challenges, threats and risks [2]. The realization of those threats and risks leads to a disruption in the stability of the functioning of both the national economy as a whole and that of an individual entity.

Economic security is a widely recognised and relatively new subject area, for example, in contrast to military security, which has been a long-established area of security

P. V. Trifonov and M. V. Charaeva (Eds.): DITEM 2021, LNNS 380, pp. 558–563, 2022.
https://doi.org/10.1007/978-3-030-94245-8_76

research. It became increasingly popular after the oil shocks of the 1970s, when most security research experts recognised the importance of economics for the well-being of the nation state and citizens. Economic security for each country provides the basis for stable and sustainable socio-economic development, as well as the well-being of the nation [3]. By analysing the world's experience in terms of assessing economic security, we can say that there is still no single approach in both determining the essence of economic security in countries and in the assessment methods at both the macro and micro levels [4].

2 Methods

In the scientific literature, the fundamental approaches to assessing the economic security of an entity and methodological tools for the practical application of analytical tools in that area are insufficiently presented, since the interaction of such categories as "economic security", "the risk of its loss", individual characteristics inherent in a particular entity and influencing its development.

The article discusses the scenario method as a tool that has the greatest value in assessing threats and risks for a project (problem), and in this case, in assessing the economic security of an entity. The scenario method is widely considered in the global practices [5, 6]. It involves the developing scenarios (alternative options for the development of events in the future), predicting the probabilities of their occurrence, assessing the investment acceptability of the action plan and a developed action plan in the event of a certain development scenario, which is consequently, saving time) forms an effective action plan consistent with its capabilities, allowing to minimise losses and restore economic security [7].

The development of scenarios begins with the identification of the key stakeholders and analysis of the external environment of an entity (with an emphasis on individual factors), and ends with the development of an entity's action strategy depending on the described scenario.

3 Results

The Economic Security Strategy of the Russian Federation up to 2030 approved in May 2017 by the Decree of the President of the Russian Federation involves tackling issues of the Russian economy, one of which is to ensure the economic security. Economic security is "the state of protection of the national economy from external and internal threats, which ensures the country's economic sovereignty, the unity of its economic space, conditions for the implementation of the strategic national priorities of the Russian Federation" [8]. In the world, economic security is considered as a combination of the current state, conditions, and factors reflecting the stability, sustainability and progressive development of the economy of a territory, a certain independence and integration with the national economy [9].

Unlike overseas countries, understanding the economic security in Russia is special; it is not only considered on the scale of the country as a whole but also within the framework of regions and individual entities, since ensuring economic security at the level of each individual entity will contribute to ensuring the economic security of a country. Therefore, for any business, both large and especially small, the role of implemented strategies is important in order to survive in the competition and maintain economic viability.

The term economic security began to be widely used in the 1970s of the 20th century in Western Europe. Until the 20th century, the protection of economic security was mainly relevant for peripheral countries, which saw the competition of other, more developed countries among the most dangerous economic threats. At the beginning of the 20th century, during the Great Depression (world economic recession), developed countries found themselves on the verge of an economic catastrophe and the economic security issues became especially urgent. Since that time, under the protection of economic security, countries have come to understand the counteraction to destructive influences for the economy, from not only outside but also from within.

This new approach to understanding the national economic security in the 1930s was formulated by the English economist John Maynard Keynes [10].

For modern Russia, both the neoclassical and Keynesian versions of the national economic security are relevant in many ways. At the same time, our country has special national economic characteristics and problems preventing us from fully fitting into one concept of the national economic security or another.

As noted, the concept of economic security is distinguished in the context of a country and an individual entity. Let us consider this concept on a national scale (Table 1).

By analyzing the positions of Russian researchers, the typical features of economic security are independence should not be absolute, since when the economy is divided, they still remain dependent on each other; economic security is aimed at protecting

Table 1. Definitions of the "economic security" concept at the macro level.

Author	Definition
L.I. Abalkin	"Economic security is a community regime (from the family to the state as a whole) to ensure that interests in preventing threats and their consequences can be met" [11]
V.K. Senchagov, E.A. Ivanov	"Economic security is a state of public life in which the protection of interests, the focus of politics, a defence regime can be ensured even in the context of internal and external processes" [12]
O.R. Rakhimov	"Economic security is an economic benefit in interpreting this concept as follows: economic security appears as a benefit provided on a paid or free basis to various economic entities ranging from personal security to the security of the state as a whole" [13]
V.V. Krivorotov A.A. Kalina N.D. Edriashvili	"The economic security of a territory is such a state of its economy, in which the territory can always maintain acceptable values of pre-determined criterion indicators – indicators of economic security" [14]

(*continued*)

Table 1. (*continued*)

Author	Definition
I.Ya. Bogdanov	"Economic security is the state of the national economy, which, firstly, in terms of volumetric and structural parameters, is sufficient to ensure the existing status of the state, its political and socio-economic development independent of external pressure, and, secondly, is able to maintain the level of legal income providing the absolute majority of the population with well-being that meets the standards of a civilised country" [15]

the economy from threats and risks, self-development and progress; creating favorable conditions for development and achievements;

Despite a significant number of positions in defining the economic security of an entity, there is no single point of view on its essence. When summarising them, we can distinguish the state of security, sustainable development and efficient use of resources. The most capacious from our point of view is the following definition "ensuring the most efficient use of an entity's resources to prevent threats and create conditions that would ensure the stable functioning of the key elements of a company" [12]. It considers an entity as a structural unit of the national economy, which, in our opinion, is more logical to ensure the safety of parts first, and then the entire system as a whole.

In foreign scientists' definitions of the economic security, common features can also be distinguished, e.g. the state pursues the economic development and economic protection, maintains the achieved standard of living and contributes to an increase in the economy.

By summarising the definitions of the economic security of an entity by both Russian and overseas scientists, common aspects can be called, such as the survival of an entity, threats to its activities, operational efficiency, external and internal risks, damage from the loss of economic security, stability and freedom and sustainability of economic development, "business continuity" or "economic business security".

4 Discussion

To develop approaches to a reasonable assessment of the economic security of an entity, it should be considered as a need for the entity and to identify the stakeholders to whom it is important.

At present, the emphasis on uniqueness for each entity is relevant in its assessment, since each of them has its own developmental characteristics, which are not identical to others, even of one type of activity. Uniqueness allows you to develop more effective solutions to any problems of the entity, since they are designed for only a given case and a specific entity. This makes it necessary, when assessing the economic security of an entity, to highlight the individual (unique) factors of its economic security, those that can affect it to the greatest extent and reduce the level of economic security. Each entity should determine this group of factors independently in order to develop in advance measures to prevent or mitigate their influence.

The most effective method for this will be the scenario approach. Key stakeholders are owners, managers and personnel. The interest of the owners is in ensuring a sufficient level of economic security of an entity in a crisis, the interest of managers, in addition to the same interest as the owners, is the ability to formulate an action plan in advance in the event of a risk and the implementation of the worst-case scenario. Personnel motivation is while maintaining economic security and sustainable development, they will not lose their jobs.

As a necessary scenario for assessing the economic security of an entity, we propose to assess the possibilities of promoting certain products or services to meet changing consumer demand; identify new ways of working that can benefit both employees and customers. It is also essential to understand which features are critical to achieving a competitive advantage, as their protection should be a priority. Other functions can be postponed, adapted, or re-formed.

It is proposed to set up a list of the main aspects of an entity's activities for developing a scenario and risk assessment:

- customers (contractual obligations, the need to retain customers, preserve the brand);
- sales and distribution channels;
- relationships with suppliers, customers, partners etc.;
- key resources (equipment, premises etc.);
- personnel (non-compliance with the work schedule (absenteeism), stress during a crisis, illness, family responsibilities, financial obligations, moral climate);
- income streams;
- expenses (loan obligations, utilities, wages, etc.);
- personal situation related to the business owner (health, financial obligations, personal availability).

The scenario method also assumes an assessment of the likelihood of the risk of these aspects. It can be ranked on a scale: 0–5%, 6–20%, 21–50%, 51–80%, 81–100%. You also need to assess the "speed of the situation" (at the moment, after 1 week, after 2–3 weeks, after 1 month, after 2–3 months) and assess the positive or negative consequences of each development option.

In the subsequent development of contingency plans, it becomes possible to identify new opportunities for the organization or to generate ideas for the creation or distribution of a product (service) produced in new ways.

5 Conclusion

The proposed scenario approach to assessing the economic security of an entity is based on two fundamentally new positions, an emphasis on unique factors and the development of scenarios in order to restore/maintain/improve economic security with the development of a specific action plan based on the assessment of its current state.

References

1. Government of Canada announces investment that will help improve women's economic security in northern British Columbia. https://www.canada.ca/en/status-women/news/2019/08/government-of-canada-announces-investment-that-will-help-improve-womens-economic-security-in-northern-brit-ish-columbia.html
2. World Bank: The World Bank Development Report: Conflict, Security, and Development, pp. 5–6 (2011)
3. Ocepek, A.L.: Economic Security and the European Dream. https://ru.scribd.com/document/266628141/EU-economic-security
4. Economic Security Strategy 2020–2023. https://www.icrc.org/en/download/file/166099/ecosec_strategy_2020-2023.pdf
5. Ogilvy, J.: Scenario Planning and Strategic Forecasting. https://www.forbes.com/sites/stratfor/2015/01/08/scenario-planning-and-strategic-forecasting/?sh=2cb4ad51411a
6. Mietzner, D., Reger, G.: Int. J. Tech. Intellig. Plan. **1**(2), 220–239 (2005). https://doi.org/10.1504/IJTIP.2005.006516
7. Chernysheva, Y.: Sci. Edu. J. Econ. Econom. Entrep. Law Gov. **4**(83), 12–15 (2017)
8. Decree of the Russian Federation President dated May 13. No. 208. Economic Security Strategy of the Russian Federation for up to 2030. https://www.garant.ru/products/ipo/prime/doc/71572608/ (2017)
9. What is Economic Security? International committee of the red cross. https://www.icrc.org/en/document/introduction-economic-security
10. Keynes, J.: The General Theory of Employment. Interest and Money, Uright, Moscow (2018)
11. Abalkin, L.I.: Econ. Quest **12**, 4–16 (1994)
12. Senchagov, V.K., Ivanov, E.A.: The Structure of the Mechanism of Modern Monitoring of the Economic Security of Russia. Report at the meeting of the sectional Academic Council of the scientific area Economic policy of the RAS Institute of Economics, Moscow (2015)
13. Rakhimov, O.R.: Sci. Newsl. MSITI **5**, 76–81 (2001)
14. Krivorotov, V.V., Kalina, A.V., Erypalov, S.E.: Newsl URFU. Econ. Manag. Ser. **5**, 121–132 (2014)
15. Bogdanov, I.: Economic Security of Russia: Theory and Practice. Institute for Socio-Political Studies of the Russian Academy of Sciences, Moscow (2001)

The Role of the Analytical Function in the Financial Control

Aleksandra I. Novitskaya[1](✉), Anna A. Bakulina[2], Ekaterina A. Nesterenko[3], and Elena A. Ivanova[4]

[1] Southern Federal University, Rostov-on-Don, Russia
achalova@sfedu.ru
[2] MGIMO University, Moscow, Russia
[3] Yuri Gagarin State Technical University of Saratov, Saratov, Russia
nesterenko.67@bk.ru
[4] Don State Technical University, Rostov-on-Don, Russia

Abstract. The article focuses on the role of analytical function in the financial control. It justifies the need for analytical activities during financial control at both the state and the corporate level. To achieve the goal, the following tasks were identified and solved. Firstly, an analysis of the legislation in the field of financial control was conducted. Secondly, terms were defined in relation to control activities. Thirdly, the concept of financial control from the functional point of view was revealed. Fourthly, the relationship between the concepts of analysis and financial control is substantiated within the described conditions. The analysis of methodological foundations of the analytical function in financial control, generalization of theoretical and practical fundamentals of financial control, as well as methods of classification in determining the interaction between the analytical function and the control function were used as methods of research. Information was selected, processed, and systematized using methods of comparison, classification, expert evaluation, and graphical presentation of data based on analysis of published books, articles, and periodicals. The work performed has shown that there is a direct link between the analytical function and financial control, because analysis is a method of financial control, and control is one of the tasks of analysis. In this case, on the one hand, financial control ensures the reliability of information sources in the analysis and the adoption of conclusions and decisions based on it. On the other hand, financial control is not possible without using analytical methods. The research makes it possible to clearly articulate in the scientific world the role of the analytical function in the financial control. In addition, on a practical level it is necessary to improve the effectiveness of the financial control environment at the level of the state and individual organizations, pointing to the need for a higher degree of accuracy and control.

Keywords: Financial control · Analysis · Analytical function · Interaction

1 Introduction

The importance of the analytical function in carrying out financial control in Russia is enshrined at the legislative level in the Federal Law of 05.04.2013 No. 41-FZ "On

P. V. Trifonov and M. V. Charaeva (Eds.): DITEM 2021, LNNS 380, pp. 564–569, 2022.
https://doi.org/10.1007/978-3-030-94245-8_77

the Accounts Chamber of the Russian Federation" [1]. The Accounts Chamber performs financial control (audit) to verify the reliability of financial transactions, budget accounting or reporting, the intended use of funds. To this end, the Accounting Chamber performs analytical type of activity and expert-analytical activities, which emphasizes the role of the analytical function in the implementation of financial audit.

Analytical function in financial control is to identify the causes and patterns that lead to violations of legislation in the management of financial resources. Thus, the analysis of a phenomenon or subject in order to study the patterns is the basis of financial control.

Analytical functions can be used to improve the effectiveness of the financial control environment, not only external [2], but also internal at the corporate level [3]. Traditional accounting, reporting and control methods have become labor-intensive because they involve analyzing thousands of transactions to form conclusions. However, through analytics, auditors can optimize the analysis of documents and data, providing a higher degree of accuracy and control, reducing the risk of errors.

2 Materials and Methods

The article is focused on the study of the role of analysis and analytical function in financial control. Many researchers are exploring how analytics helps solving real-time financial control and accounting problems [4, 5].

For the purpose of the study, it is necessary to clarify the terms used in relation to control activities. The processes for obtaining and analyzing information regarding control activities include the collection, processing and analysis of information used for control purposes. Some studies distinguish between financial and non-financial information [6, 7]. Financial information includes basic concepts used in research applicable to control, such as planning, forecasting, budgeting, goal setting, analysis, and accounting reports. Obviously, these processes overlap and interact. Thus, the analysis of factual data provides a basis for action and planning. Financial information can be obtained not only from analysis of internal financial data, but also from external sources. For example, in the case of organizations, using the accounts of other organizations in the analysis of competitors.

Non-financial information can be presented in the form of feedback/evaluation of stakeholders, innovation criteria, reputational factors, etc. As practice shows, non-financial information correlates with financial information.

The basis of analysis were the legal acts of Russia, as well as the legal acts of developed and developing countries, and official sites of national organizations of public audit system. To determine the analytical role in the implementation of the control function, various aggregate studies were studied: "The financial function: the basis for analysis" [8], "Business Model Report: New Practices and Future Trends" [9], "Digital Control" [10].

3 Results

Based on the study it may be concluded that financial control from a functional point of view is a set of tasks [11]:

- Collection, processing and analysis of information on actual performance;
- Comparison of indicators with targets, identification of deviations, analysis of reasons for identified deviations;
- Development of measures to achieve the goals set.

As can be seen, the tasks (analysis, comparison) are directly related to the performance of the analytical function. Also, the analytical function is separately manifested in the process of gathering information. For example, it is possible to obtain information with the help of analytical forms. Such forms ensure an objective assessment and analysis of the causes of deviations and allow facilities to independently evaluate and analyze their work. This creates a kind of "bridge" between one' own control and external financial control.

Thus, the use of analytical forms and the implementation of analytical measures expands the horizon of financial control.

In general, the relationship of the concepts of analysis and financial control is obvious.

On the one hand, without financial control one cannot unfailingly rely on the sources of information that will be used to analyze and make decisions based on it [12]. The presence of financial control allows you to detect undesirable deviations in time and take any action.

On the other hand, financial control uses in its work analytical methods. Analytical and control functions are similar in terms of tasks to ensure the reliability of information for decision-making [13]. Thus, analysis is a method of financial control, and control is one of the tasks of analysis.

At the same time, as practice shows, it is necessary to improve the mechanism of information analysis. For an objective assessment of any object of control it is necessary to study data not for one year, but for 5–7 or even 10 years, in order to assess the dynamics of indicators and see the full picture of socio-economic results. It is assumed that information databases with elements of passportization of objects under control and reaching a new level of interaction between all subjects of budgetary legal relations will serve as a basis.

The system of passportization covers not only individual ministries and departments (organizational information, the normative-legal field of their action, analysis of financial-economic activity over a long period, information about the activity of subordinate organizations), but also industries. Such a system can also be applied to the corporate level. The advantages of the system are the simplification of the financial control process and the ability to carry out qualitative analysis of financial and economic activities of facilities in a short time.

At the corporate level, there is also an interaction between the analytical function and the control function (see Fig. 1). The variety of operational processes creates difficulties in control. Combining analytical functions with control processes makes it easier to detect inconsistencies. The interaction is carried out by means of informing. The flow of information comes from one activity for use by another, or by direct influence. There is a direct use of information to influence.

For example, analysis and accounting are interconnected with the compliance-control function. On the one hand, analysis-based accounting provides the necessary information about compliance control in the form of data, financial reports. On the other hand, compliance control affects accounting and reporting by guiding accounting activities, determining the analysis necessary for compliance.

In turn, compliance control interacts with management when financial results can be measured on the basis of legal requirements, such as the return on risky assets in bank regulation. Compliance control also affects the activities and risks of the company through management and control, imposing restrictions on its actions.

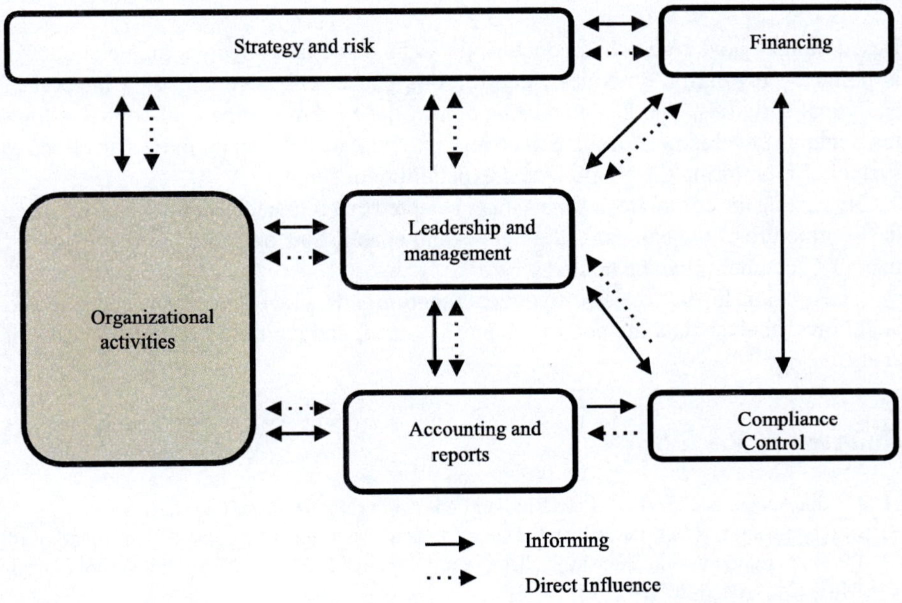

Fig. 1. Interaction of financial control with the analytical function at the organizational level. *Source:* compiled by the author.

4 Discussion

Cases of dishonest reporting around large companies in the absence of proper financial controls (Enron, Tyco International, WorldCom) stressed its necessity [14]. There is an increased understanding of the importance and reliability of the financial information obtained in the analysis.

But at the same time, the financial information needed for control is based on the analysis and aggregation of the financial implications of organizational activities. However, the content of information, methods of analysis and classification, timeliness will be different, which can lead to a number of problems.

Researchers have noted [15] that "management accounting lost its value when it began to focus on using information only for financial control that was irrelevant to management and performance control". That is, systems that perform an analytical function, such as accounting and operating systems, must meet both control and performance management and control needs.

It should also be noted that the analytic function has institutional pressures that appear in the ever-changing ecosystem of data analytics [16], which emphasizes the need for an improved information analysis mechanism.

5 Conclusion

Based on the above research, a conclusion can be made that effective financial control is based on regularity, timeliness, due to compliance with a certain time interval of analytical activities. The analytical function in the system of financial control allows the dynamic development of financial control at the state level and provides an effective system of monitoring the planning and expenditure of funds.

Regarding the corporate level, it should be noted that if financial controls have a place in the structure of the company, the figures and conclusions obtained in the analysis of financial statements can be trusted.

The relationship between the analytical function and financial control can be reflected as follows: analysis is a method of financial control, and control is one of the tasks of analysis.

References

1. On the Accounts Chamber of the Russian Federation (No. 41-FZ of 05.04.2013)
2. Mukhametshin, R.F.: State (financial) control as an instrument of management function of the state. Economics and Business: Theory and Practice **2**, 57–61 (2020). https://doi.org/10.24411/2411-0450-2020-11025
3. Gaining value from advanced analytics in finance and controllership. https://www2.deloitte.com/us/en/pages/advisory/articles/advanced-analytics-in-controllership-blog.html
4. Krahel, J.P., Titera, W.R.: Consequences of Big Data and formalization on accounting and auditing standards. Account. Horiz. **29**(2), 409–422 (2015). https://doi.org/10.2308/acch-51065
5. Sun, Z., Sun, L., Strang, K.: Big Data analytics services for enhancing business intelligence. J. Comput. Inf. Syst. **58**(2), 162–169 (2018). https://doi.org/10.1080/08874417.2016.1220239
6. Henttu-Aho, T.: The role of rolling forecasting in budgetary control systems: reactive and proactive types of planning. J. Manag. Control. **29**(3–4), 327–360 (2018). https://doi.org/10.1007/s00187-018-00273-6
7. Eskindarov, M.A., Fedotova, M.A. (eds.): Estimation of Business Value, 2nd edn. KNORUS, Moscow (2018)
8. The Finance Function: A Framework for Analysis. Retrieved from https://www.icaew.com/-/media/corporate/files/technical/business-and-financial-management/finance-direction/finance-function-a-framework-for-analysis.ashx?la=en (2011)
9. Reporting your business model. Emerging practices and future trends. Retrieved from https://integratedreporting.org/wp-content/uploads/2017/01/corporate-reporting-business-models.pdf (2016)
10. Digital Controllership: Analytics & Insights. https://www2.deloitte.com/content/dam/Deloitte/us/Documents/risk/us-rfa-analytics-insights.pdf

11. Grishin, V., Nadareishvili, G.: Financial control as an instrument of managerial decision-making. Upravlenie **1**(26), 37–46 (2013)
12. Sheremet, A.D., Negashev, E.V.: Methodology of Financial Analysis of Commercial Organizations. INFRA-M, Moscow (2021)
13. Demina, T.A.: Financial stability and paying capacity as a subject of economic analysis. In: Larionova, A.A., Chernikova, L.I., Kaurova, O.V. (eds.) Problems and Experience of Management, Finance, Accounting and Taxation of Enterprises, Industries, Complexes: Collection of Articles, pp. 142–145. Stepanenko Publisher, Moscow (2012)
14. WorldCom, Tyco, Enron–R.I.P. Retrieved from https://www.forbes.com/2002/07/01/0701topnews.html?sh=4ee51b405397 (2002)
15. Kaplan, R.S., Johnson, H.: Relevance Lost: The Rise and Fall of Management Accounting. Harvard Business School Press, Boston (1987)
16. Kaya. I., Akbulut D.H.: Big Data analytics in financial reporting and accounting. PressAcademia Procedia. **7**(45), 256–259 (2018). https://doi.org/10.17261/Pressacademia.2018.892

Responsible Financing Practices in the Transition to Digital Economy

Sung Laibin[1], Sergei Yu. Abramovich[2](✉), Aleksander V. Paleev[3], and Olga V. Fateeva[4]

[1] Southern Federal University, Rostov-on-Don, Russia
[2] RANEPA, Rostov-on-Don, Russia
[3] Military Training Center, Rostov-on-Don, Russia
[4] Don State Technical University, Rostov-on-Don, Russia

Abstract. The article is devoted to the problem of developing responsible financial practices in the transition to the digital economy and their role in achieving sustainable development goals. Having analyzed the dynamics of the growth of responsible investment in world practice, the authors reveal the reasons why the Russian financial and credit market rejects important international initiatives in the field of socially responsible investment. Responsible financing can be considered the next important step in the development of corporate finance, which allows not only to bring high returns in the long run, but also to count on being included in various indices and funds, and, consequently, on investor cash inflows. It can be said with a high degree of certainty that companies which generally ignore or pay too little attention to responsible financing in their long-term strategy may simply find themselves outside the broad investment field and will lose to their more responsible competitors. The authors link the effectiveness of responsible management with the level of development of automated IT decision support systems. They show that the introduction of new methods of corporate finance management and tools based on digital information technologies will allow integrating the principles of sustainability into development strategies, making responsibility management the main element of corporate culture, and responsible investment one of the key performance indicators. The implementation of such systems could become a significant area both in the "digital economy" project and in the processes of improving systems for responsible financing and corporate finance management.

Keywords: Responsible financing · Digital economy · Sustainable development · Corporate finance

1 Introduction

On September 25, 2015, in New York, 193 member states of the United Nations (including Russia) unanimously adopted a new Agenda for Sustainable Development. 17 Sustainable Development Goals (SDGs) until 2030 were formulated [1, 2]. In the world today, sustainable development is associated with these goals. Modern conditions in the field of sustainable development require new approaches to solving the problems

P. V. Trifonov and M. V. Charaeva (Eds.): DITEM 2021, LNNS 380, pp. 570–576, 2022.
https://doi.org/10.1007/978-3-030-94245-8_78

of monitoring, analyzing, assessing and preventing the emergence of catastrophic risks and presuppose the introduction of new methods of corporate finance management and tools based on digital information technologies that provide systemic and innovative interdisciplinary approaches.

Currently, work is being carried out on global and Russian sites to introduce projects, conventionally called "digital economy" into business practices. According to many experts, the period of extensive growth in developed economies ended in the 1st decade of the 21st century and an active transition to the 4th technological order began. A high-tech economy is a harmonious combination of technologies of the physical, digital and biological realms, opening up new opportunities on the one hand, and on the other, creating new threats to economic, social and, above all, ecological systems [3]. It is the financial support of environmental safety, the preservation and improvement of the human life environment that is the priority task of today [4], which is directly related to the ability of financial institutions to effectively use the digital economy tools, to adapt their financial strategies in the new reality. An effective solution to these problems has not yet been found, and the urgency of this issue is confirmed by the corresponding statistics reflecting the trends of the past 25 years (Fig. 1).

The statistics backed up by more than 15 thousand scientists from 184 countries of the world clearly show that significant growth in global carbon emissions and average temperatures continues. Since 1992, the average global temperatures have increased by 0.5 degrees Celsius, and the average annual carbon dioxide emissions, by 62 percent. Over the past quarter century, the amount of fresh water per capita has decreased by 26%. During the same period of time, there has been a significant increasing pollution of water with industrial waste. The number of "dead" zones in the ocean has increased by 75%. Since 1992, 1.2 million square kilometers of forest land have been lost, most of which have been converted for agricultural use. Together, these factors have led to a 29% reduction in the number of mammals, reptiles, amphibians, birds and fish worldwide. Extinction of species is happening thousands of times faster than natural rates. Plastics have filled the oceans to so much extent that fish became addicted to eating them.

In these conditions, the most important role in achieving sustainable development is assigned to financial systems and digital technologies which must be responsible, socially oriented [6] and implemented on a global scale. To do this, business practices need to widely implement digital platforms to support the adoption of responsible financial decisions based on modern information technologies.

2 Methods

In the course of the study, methods of economic and logical analysis, expert assessments, and a systematic approach were used.

3 Results

If we analyze the dynamics of socially responsible investment in different regions of the world, we can see that even under the conditions of instability, responsible financing is a steady trend. Such conclusions are confirmed by the materials presented at the Paris

Conference on Climate Change which became a "historic turning point" on the path to reducing the rate of global warming [7], and the World Economic Forum in Davos, Switzerland.

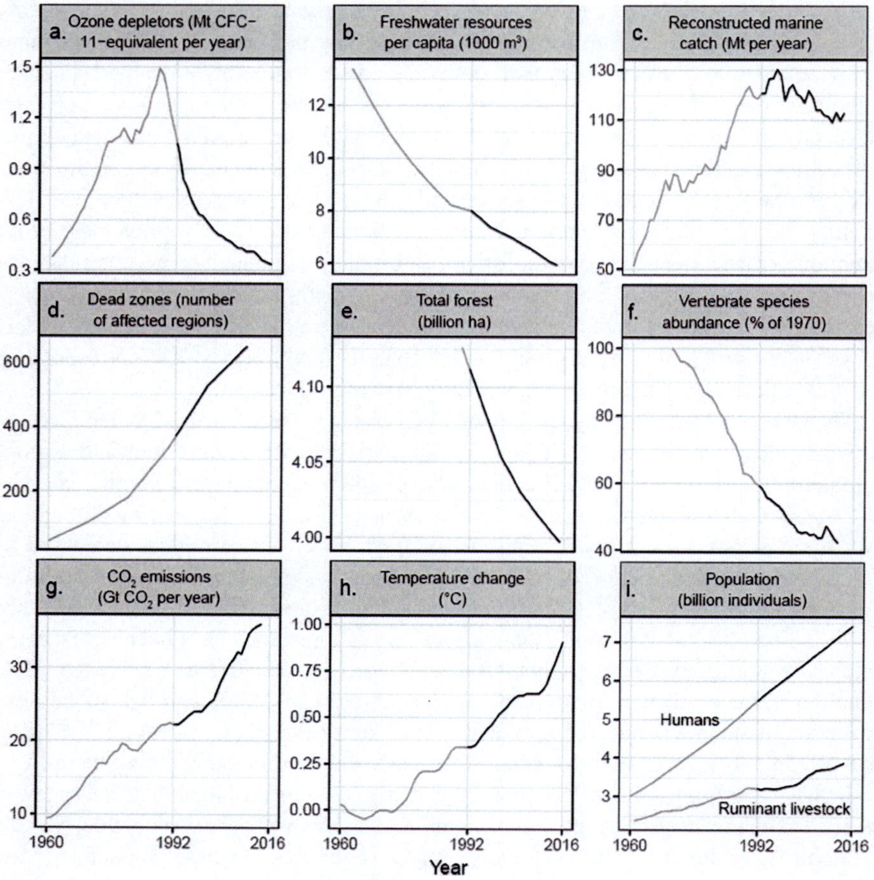

Fig. 1. Global environmental change [5].

Responsible financing can be considered the next important step in the development of corporate finance, which allows not only to bring high returns in the long run, but also to count on being included in various indices and funds, and, consequently, on investor cash inflows. It can be said with a high degree of certainty that companies that generally ignore or pay too little attention to responsible financing in their long-term strategy may simply find themselves outside the wider investment field and lose to their more responsible competitors.

The Equator Principles were signed by the 21st largest banking group in the world (39 countries, 90 banks, 80% of the world financial market [8]). The number of financial institutions that signed the UN Principles for Responsible Investment increased from

100 in 2006 to 1488 in 2016, and assets under management are estimated at more than 60 trillion US dollars, which is about half of the total assets of institutional investors in the world [9]. Unfortunately, Russia does not fall into these statistics, as such a system of socially responsible investment is still absent, leading international initiatives in this area are not widespread, each financial organization acts independently choosing its own path [10].

There are no unambiguous approaches to solving these issues yet, mainly due to the lack of information bases that can accumulate a sufficient amount of information for an objective assessment of the processes taking place in the systems of responsible financing.

The formation of the necessary information resources for a wide range of parameters, most of which are currently not taken into account, will require a long period of time and complex unified IT platforms, which will help to achieve the implementation of operational monitoring, analysis, control, objective assessment, as well as responsibility for management decisions, including environmental ones.

Environmental and social components are present in almost all investment projects (Table 1).

Table 1. Dynamics of socially responsible investment in different regions of the world in 2014–2018 (billion dollars) [11].

Countries	2014	2016	2018
Europe	10.775	12.040	12.075
USA	6.572	8.723	11.995
Canada	729	1.086	1,699
Australia and New Zealand	148	516	734
Japan	7	474	2.180
Total	18.3	22.9	24.9

In 2014–2018, there was an increase in socially responsible investment: in Europe since the beginning of 2014 by 12%, in the USA and Canada, by almost 50%. In general, the volume of SRI increased by 36% compared to 2014 and reached $24.9 billion. Statistics of the last 10 years also speak in favor of responsible investment. Thus, the MSCI World and USA ESG Leaders indices outstripped MSCI World and S&P 500 by 40% and 23%, respectively, in terms of total return including dividends. Almost all large, medium and even small capitalization companies included in these indices have taken care of this topic, but responsible funds now differ little in their structure from their more popular counterparts, such as the S&P 500 index [12].

Of course, one cannot ignore the fact that socially oriented activities are not specifically widespread in Russian banking practice, which is primarily due to the lack of a clear understanding of CSR. According to the Artisan Group Public Relations study [13], 32% of Russian banks consider CSR as part of the company's initiatives in the role of an employer. At the same time, more than a quarter of banks interpret CSR as the implementation of charitable projects. It is remarkable that only 11% of bankers consider CSR as a separate area.

In general, the above factors indicate that, unlike in developed countries, the Russian reality does not stimulate the achievement of global sustainable development goals, including the most important tool of this system, responsible financing. The priority in the implementation of financial strategies is still focusing on short-term goals of making a profit in short-time periods.

4 Discussion

What are the reasons behind the fact that the Russian financial and credit market rejects important international initiatives in the field of responsible financing? Let us designate the main ones.

First of all, it should be noted that banks are afraid of losing borrowers if they set conditions to comply with environmental and social requirements. It is no secret that most business structures do not bother with the obligation to comply with the necessary requirements in the field of ecology and labor relations.

It should also be noted that there is a certain lack of incentives to induce the financial services sector to actively implement the practice of taking into account environmental and social factors in investment activities. This is a manifestation of the lack of due attention to the problems of socially responsible investment both on the part of the state and on the part of unions and associations operating in the financial sector.

The structural crisis also exerts a negative influence on the economy of our country, as well as its financial sector which has lost the ability to obtain the required amount of foreign loans due to sanctions imposed for political reasons and used as an economic instrument of external pressure (7 largest Russian banks (more than 50% of assets [14]) are still under sanctions).

Is the Russian economy ready to make a transition to innovative methods and instruments of responsible financing?

In the Russian version of the "digital economy" program developed by the government of the Russian Federation in July 2017, a plan for the development of the digital economy until 2024 is proposed, which defines five basic directions: legal regulation, personnel and education, the formation of research competencies and technical groundwork, information infrastructure and information security. The country's leadership has set a rather ambitious strategic goal: to transform management systems from managing money and hydrocarbon export flows to managing technologies that shape the future.

It is worth noting that the Russian version of the project for changing economy (transition to digital economy) is less ambitious than a similar western project, which is considered to be represented by the recommendations of experts from the Davos Economic Forum, an international discussion club where the development of the world economy and society is annually examined [15].

The national document is mainly aimed at changing the technological base, which makes it possible to automate routine operations and significantly accelerate implementation of many processes, as well as to increase the efficiency of the activities of structures that ensure public administration.

As before, priority is given to development which ensures the growth of "gross domestic product" (GDP), "macroeconomic stability," "lower inflation," and so on. To a lesser extent, it stipulates the development of directions that determine the quality of human life, the person's social development and security, including: medicine, science and education, a high-quality living environment, etc. It is envisaged to annually allocate more than 100 billion rubles for these purposes, if the need significantly exceeds this amount.

What measures must be taken to change the landscape on the credit and financial market in Russia, making it financially responsible and turning it to face the problems of sustainable development?

In particular, an effective impact could have the creation and dissemination of the Code of Socially Responsible Investment which defines the conditions, principles and rules for regulating socially responsible investment business practices, including the environment, in the context of the digital economy and the development of automated IT platforms to support socially responsible management decisions, including those ensuring the implementation of environmental goals of sustainable development, which will allow in a practical plane to solve vital environmental problems [16].

Another major issue of our time is the creation of appropriate IT platforms fully supporting the adoption of socially responsible financial management decisions.

The most "expensive" mistakes are strategic ones. As a rule, they cannot be corrected at the higher levels of management. Computer systems have an undeniable ability to avoid making erroneous decisions, which ultimately implies additional economic, environmental and social benefits.

5 Conclusion

In this regard, the development of automated systems (IT platforms) to support the adoption of responsible socially oriented financial decisions should be the main part of corporate governance procedures. They primarily contribute to the minimization (optimization) of risks and prevention of irretrievable risks, due to the fact that they take into account environmental and social issues, along with the financial and economic component, and also allow an objective assessment of the effectiveness and efficiency of the participants' activities and the degree of their responsibility.

The key point is that, despite the large amount of software, today's leaders, when making a decision, in most cases rely on their experience and the experience of their team, and also widely practice manual management, which increases the risk of making ineffective or erroneous decisions. Cognitive technologies of digital economy make it possible to foresee the most probable consequences of the decisions made and "tell" the manager which parameters are most consistent with the set goals and objectives. The development and implementation of such systems could become a significant area both in the "digital economy" project and in the processes of improving systems for responsible financing and corporate finance management.

References

1. Sustainable Development Goals. 17 goals to transform our world. http://www.un.org/sustainabledevelopment/ru/sustainable-development-goals/
2. Global Sustainable Investment Review 2016. http://www.gsi-alliance.org/wp-content/uploads/2017/03/GSIR_Review2016.F.pdf
3. Nankervis, A., Connell, J., Montague, A., Burgess, J. (eds.): The Fourth Industrial Revolution. Springer, Singapore (2021). https://doi.org/10.1007/978-981-16-1614-3
4. Fabius, L.: With landmark climate accord, world marks turn from fossil fuels. http://www.reuters.com/article/us-climatechange-summit-idUSKBN0TV04 L20151212#gVKudBATCD0EGdxL.97
5. Ripple, W.J., et al.: Bioscience **67**(12), 1026–1028 (2017). https://doi.org/10.1093/biosci/bix125
6. Rockström, J., Sukhdev, P.: State of the union style adress. EAT Stockholm Food Forum. https://www.youtube.com/watch?v=tah8QlhQLeQ (2016)
7. Paris Climate Agreement. http://unfccc.int/resource/docs/2015/cop21/eng/l09r01.pdf
8. Equator Principles. http://www.equator-principles.com
9. What is responsible investment? Principles for Responsible Investment. https://www.unpri.org/about/what-is-responsible-investment
10. Abramovich, S.Yu: Econ Law Iss. **03**(93), 116 (2016)
11. Investing to Curb Climate Change: A Guide for the Individual Investor. http://www.ussif.org/files/Publications/SRI_Climate_Guide.pdf
12. Is responsible investing becoming the norm? https://www.finam.ru/analysis/newsitem/otvetstvennoe-investirovanie-stanovitsya-normoiy-20200215-11000/
13. Romanova, E.: Kommersant. "Bank". App 1. http://www.kommersant.ru/doc/2427127 (2014)
14. The Impact of Sustainable and Responsible Investment. http://www.ussif.org/files/Publications/USSIF_ImpactofSRI_FINAL.pdf
15. Barmuta, K.A., et al.: Entrep. Sust. Iss. **8**(1), 945–959 (2020)
16. Charaeva, M.V.: Formation of structural and logical model of company's financial standing analysis. In: 2nd International Conference on Social, economic, and academic leadership (ICSEAL 2018), Prague, Czech Republic, pp 127–133, 16–18 June 2018

ESG Investing as a Corporate Sustainability Factor

Nataliia N. Muraveva[1] (✉), Elena A. Chumachenko[2], Maria P. Glyzina[3], and Evgeny A. Zhabin[1]

[1] Southern Federal University, Rostov-on-Don, Russia
[2] Rostov State University of Economics, Rostov-on-Don, Russia
[3] Don State Technical University, Rostov-on-Don, Russia

Abstract. This study is a companion study to research projects conducted by Amundi Financial Company on socially responsible investing (SRI) in the stock market. The main idea of the new research is to study the impact of ESG investments on asset prices in the financial market. The methods used by domestic and foreign scientists to test ESG in active and passive management are applied to analyze them. The ideal ESG model is an extended five-factor model that includes both the momentum and stability factors. This extended multi-factor model is based on value maximization principles to determine the cost of capital, where social and environmental risks, in particular, determine the consumption cost; thus, they affect the cost of capital and the cost of equity for firms. Our findings confirm the literature data that sustainability indicators reduce risk and reduce the equity cost for the firm. In addition, the results suggest that the appropriate cost of capital used in the evaluation of investment projects using net present value or internal rate of return methods should be adjusted downward when making strategic decisions on the implementation of sustainable projects.

Keywords: Socially Responsible Investment · Ecology · Social sphere · Sustainable development

1 Introduction

Stakeholder demands for more sustainable business practices have led to a change in the context of strategic asset allocation decisions. Longitudinal data show that company sustainability performance is becoming increasingly important to major investors (retail investors, asset managers, institutional investors) and that investors are now using sustainability-related, environmental, social, and governance criteria in their investment decisions. In a survey of more than 3,000 global executives, 75% believe that sustainability metrics improve revenue performance and reduce capital and risk costs. Investors' belief that sustainability is positively correlated with return on investment (ROI) is supported by studies that collectively show a positive return on sustainability in revenue, stock performance, brand reputation, and long-term organizational performance. Given that sustainability is now rapidly becoming a competitive necessity for organizations

P. V. Trifonov and M. V. Charaeva (Eds.): DITEM 2021, LNNS 380, pp. 577–583, 2022.
https://doi.org/10.1007/978-3-030-94245-8_79

and firms that have changed their core business models to include sustainability, it is clear that executives need better asset valuation models to more accurately assess assets when making asset allocation decisions.

A couple of decades ago, only a few socially conscious investors expressed concern about ethics in business practices. This trend has given rise to socially responsible investments (SRI), which have evolved from marginal to mainstream. Investors, stakeholders, numerous regulatory changes, and ample media coverage fuel the craze for responsible investing. This investment form is different as it allows investors to align their portfolios with their beliefs. The global pandemic has re-emphasized the importance of integrating environmental, social, and governance considerations into financial investments. Responsible corporations have indeed proved more resilient to this recent bear market.

The financial system constantly evolves, and investors pay more attention to sustainability criteria affecting asset prices. This phenomenon is likely to continue as demand for responsible financial instruments grows faster than supply. For example, practice shows that the relationship between ESG and the financial portfolio ranking and stock market performance is direct and characterized by a positive trend of "best stocks" compared to "worst stocks" due to the growing demand for highly ranked securities. Recently, ESG portfolios in developed countries have surpassed traditional capitalization-weighted benchmarks.

Around the world, sustainable investment has proven to be an ideal practice, and many companies want to see if they can understand its components and adjust accordingly [1]. As an investment procedure, sustainable investments potentially influence sustainable development by coordinating long-term social, environmental, management criteria, and monetary considerations into investment decisions. Thus, sustainable investments are green and ethical investments that use both Socially Responsible Investments (SRI) and ESG investment procedures [2]. It is not surprising that the current growth of research in sustainable investment has prompted both scientists and policymakers to assess the existing relationships in the above-mentioned factors related to sustainable investment.

2 Results and Discussion

This research work seeks to build on existing research by conducting a systematic review of sustainable investment research [3]. Over the past few decades, traditional profit-seeking companies have shown an increased interest in understanding and caring about the broader impact of their business. However, non-governmental organizations (NGOs), governments, and corporations have failed in their attempts to solve some global problems, such as climate change, inequality, and poverty. Thus, sustainable investment has proven to be a key solution to environmental and social problems, making financial markets increasingly responsible for the ESG effects.

ESG investing is a moral choice between risk and efficiency in the use of modern financial and digital technologies to support the long-term development of society through the creative use of quantitative assessments and continuous innovation.

Since ESG investing is a new topic for discussion by the scientific community in the context of corporate bonds, this management style is less mature than the others in the stock market. It is still too early to call ESG investing a new factor in the corporate bond market.

The use of criteria based on environmental, social, and governance (ESG) considerations is becoming an increasingly important aspect of investment decision-making, especially for large institutional investors. On February 8, 2019, Justina Lee of Bloomberg reported that "about $12 trillion is allocated for sustainable investing in Europe alone".

In recent years, "green bonds" have received increasing attention. Corporate green bonds direct their income to finance environmental and climate projects. It immediately raises the question of why the company needed to issue such bonds. By forcing itself to use the proceeds in a certain way, the company loses the flexibility to use the funds more generally if circumstances change.

In addition, if the "green" goal is the best use of funds raised from general obligation bonds, which can always be allocated to "green" projects.

As noted, three rationales have been put forward to explain why companies issue green bonds. First, green bonds can serve as a reliable signal of a company's commitment to environmental protection. Second, issuing green bonds can be a form of "greenwashing" of public relations. Third, if investors are willing to pay more for green bonds, the company should issue them at a lower yield. This is a bond option for a lower cost of equity. However, unlike stock returns, bond yields are directly observable, making it easier to assess whether green is important enough for investors to accept lower yields.

An empirical study by foreign researchers has shown that the yield on green bonds is almost identical to the yield on other comparable non-green bonds. This information suggests that a large part of the focus of green bonds may indeed be related to public relations.

In terms of corporate governance, the ESG movement has gone beyond analyzing investor concerns to question the goal of value maximization that underlies much of financial theory. Of course, some criticism is wrong in that it confuses value maximization with short-term profit maximization. If a company fails to treat its customers or employees properly or ignores the impact of its activities on society and the environment, future cash flows – and, consequently, the cost of equity – are likely to be negatively affected.

As a result, there may be no need for a broader corporate goal to solve social problems. However, experienced managers, who undoubtedly understand the above difference, strongly advocate expanding corporate tasks.

We will review the theoretical context of ESG model development by briefly reviewing the literature on sustainability and equity value indicators, environmental economics, sustainable strategic management, and sustainable finance. Research by foreign scientists has shown that firms with socially and environmentally sustainable business practices have significantly lower equity costs. They found that only environmental sustainability and governance indicators reduced the equity cost, probably due to the more direct correlation between these indicators and risk reduction and financial performance. Social sustainability was not significantly related to the equity cost. However, using the generalized ESG sustainability indicator, the authors found a significant negative relationship between ESG and the equity cost. They concluded that the strong ESG sustainability indicators strengthen the link between financial performance and equity cost.

While research shows that sustainable business performance reduces the equity value, most traditional asset pricing models do not consider the value of social and natural capital and the resulting systematic risks associated with climate change, resource depletion, or civil unrest. The ESG model includes risks associated with social and natural capital, which are additional assessed assets that have not yet been covered in any CAPM extensions. Given that the ESG model predicts a sustainability factor based on the microeconomic principles of maximizing a firm's social and natural capital value, we can better estimate the benefit of the sustainability factor and its factor loadings. Accounting for these assets is becoming critical as research shows that the global value of both natural and social capital is rapidly declining, affecting the quality of life of both current and future generations.

Ecosystem services are now estimated to have declined by about $20 trillion a year [4], and global social capital has trended downward since 1978, as measured by the GPI, a true progress indicator [5].

Although the risks associated with the loss of social and natural capital cannot be diversified, managers can make strategic decisions to use their resources more sustainably, increase social capital, and reduce social and environmental externalities, thereby reducing the firm's risks and lowering the cost of equity through improved sustainability performance.

The development of the ESG model requires a change in the theoretical paradigm underlying traditional multi-factor CAPM models, which are based on the neoclassical economic assumption of a closed system economy. Economic theory informs all business disciplines, which is reflected in the misplaced specificity of modern financial and strategic management models, which misconceive abstract models as accurate representations of the current business environment. The closed economic system perspective of neoclassical economics also provides a theoretical context for a managerial philosophy that narrowly defines key stakeholders for shareholders. However, open systems models of ecological economics [6] provide managers with a framework for broadly defining stakeholders, including the creation of social and environmental value and economic benefits. Ecological economics is a theoretical economic construct that embeds sustainable strategic management and the ESG model.

Sustainable strategic management is the next coevolutionary level in developing the field of strategic management, reflecting the transition from the assumptions of the neoclassical economic paradigm to the open system assumptions underlying ecological economics [7]. While traditional models of strategic management, like traditional financial models, implicitly assume that the economy is a closed system concerning society and the ecosystem, sustainable strategic management views the organization and the economy as open, coevolutionary subsystems of the larger society [8]. This paradigm shift allows managers to manage their economic and social, and environmental performance by developing sustainability strategies focused on greater performance. Therefore, the economic value and the cost of natural and social capital are taken into account in sustainable strategic management. We believe this management philosophy provides managers with a framework for strategic management to develop tools, such as the ESG model, that consider the value of natural and social capital when making asset allocation decisions [9].

As investors increasingly demand ESG criteria when making investment decisions, it is not surprising that interest in integrating sustainability into financial markets and financial products is attracting much attention. At the beginning of 2018, the U.S. sustainable investment market was $8.72 trillion, a fifth of all investments, with a growth rate of 33% over the two years. According to Merrill Lynch, millennials could invest between $15 trillion and $20 trillion in sustainable investments in major U.S. financial companies such as Blackrock, Goldman Sachs and Morgan Stanley, which have implemented ESG-conscious investment productsover the next two decades [10].

The exact meaning of sustainability for the financial sector is the subject of much debate. While the 2015 U.S. Department of Labor ruling allowed pension fund managers and 401(k) plans to incorporate ESG criteria into investment decisions, the ruling did not clarify or consolidate portfolio managers' various definitions for sustainable investing [11]. One of the biggest problems for sustainable investment is the lack of a single agreed definition. In general, sustainable finance/investment refers to creating social and environmental value along with economic value in financial models, products, and markets that align investors with their sustainability investment goals. It improves economic well-being both today and, in the future, while protecting and restoring ecological systems and increasing social capital. Currently, the terms "sustainable finance" and "sustainable investment"' are usually used as an umbrella concept for all assets based on sustainable development. In other words, the field of finance seems to be moving toward profit performance measurement, thereby creating a demand for better pricing and asset valuation models.

3 Conclusion

Thus, the ESG model is embedded in the theoretical constructor of an open system ecological economy, where sustainable strategic management is the dominant management philosophy. A paradigm shift is required to create an environment conducive to the development of the ESG model. Traditional financial models, such as CAPM, are implicitly based on the neoclassical economic assumption of a closed economic system. However, financial theorists are beginning to rethink their traditional financial models based on the company's assumption (and, obviously, its financial models) as a coevolutionary subsystem of the economy, society, and ecosystem. New concepts of sustainable finance, sustainable investment, ESG criteria, socially responsible investing, etc., all indicate that financial theorists have begun to question the dominant neoclassical economic paradigm underlying their models and recognize the value of environmental and social capital. Thus, the purpose of this paper is to develop an extended multi-factor asset pricing model that values environmental and social capital and to test whether sustainability performance is measured in the firm's equity cost. In the next section, we will discuss the methodology used in the study.

The ideal ESG model is an extended five-factor model that includes both the momentum and stability factors. This extended multi-factor model is based on value maximization principles to determine the cost of capital, where social and environmental risks, in particular, determine the consumption cost; thus, they affect the cost of capital and the cost of equity for firms. We assume that the equity cost will be reduced when firms practice sustainable strategic management that mitigates social and environmental risks. Thus, the price of social and environmental risk will be positive. We also assume that smaller firms will benefit less than larger firms from spending on environmental and social risks due to costs associated with their mitigation. Thus, small firms are exposed to higher social and environmental risks than large ones.

The ESG model uses the Fama-MacBeth two-pass regression analysis method to determine whether sustainability is estimated in the firm's equity cost. The returns on 50 portfolios sorted by size, the book value of equity, and operating income are used as test assets. Empirical measurement shows that the price of social and environmental risks is positive.

Overall, due to negative factor burdens, reflecting policies to mitigate the effects of sustainable strategic management, firms worldwide have reduced the equity cost by 1.6–2.9%. Based on the results of the factor loading, most of this benefit went to larger firms. As the system's overall level of social and environmental risks increases, firms practicing less sustainable management that do not consider social and environmental costs when making asset allocation decisions have a higher equity cost than firms engaged in sustainable strategic management. The ESG model shows that this is due to the costs of replacing factors as overall environmental and social risks increase [12].

In terms of asset allocation decisions, the sustainability factor is a haphazard risk factor, meaning that managers have strategic choices about whether to internalize the costs of social and environmental risks when evaluating their assets and making asset allocation decisions. Given our findings that sustainable strategic management reduces the cost of capital, the obstacle level for sustainable projects should be valued at a lower equity cost than other riskier, unsustainable projects. Our findings confirm the literature data that sustainability indicators reduce risk and reduce the equity cost for the firm. In addition, the results suggest that the appropriate cost of capital used in the evaluation of investment projects using net present value or internal rate of return methods should be adjusted downward when making strategic decisions on the implementation of sustainable projects.

References

1. Bai, J., Zhou, G.: Fama-MacBeth two-pass regressions: Improving risk premia estimates. Financ. Res. Lett. **15**, 31–40 (2015). https://doi.org/10.1016/j.frl.2015.08.001
2. Balvers, R.J., Du, D., Zhao, X.: Temperature shocks and the cost of equity capital: capital: Implications for climate change perception. J. Bank. Finance. **77**, 18–34 (2017). https://EconPapers.repec.org/RePEc:eee:jbfina:v:77:y:2017:i:c:p:18-34
3. Azevedo, V., Santos, A., Campos, L.: Corporate sustainability and asset pricing models: Empirical evidence for the Brazilian stock market. Production **26**(3), 516–526 (2016). https://doi.org/10.1590/0103-6513.201115
4. Kassel, M.: How much do you know about ethical investing?. Wall Street Journal. (2018). https://www.wsj.com/articles/how-much-do-you-know-about-ethical-investing-1499349815
5. Voronkova, O.Y., Petrova, S.N., Ivanova, E.A., Klochko, E.N., Logvina, E.V., Penkova, A.N.: Assessment of the regional potential of recreation and health-promoting natural resources. J. Exp. Biol. Agric. Sci. **8**(Special Issue 2), S314–S321 (2020). https://doi.org/10.18006/2020.8(Spl-2-AABAS).S314.S321

6. Kolbel, J., Busch, T., Jancso, L.: How media coverage of corporate social irresponsibility increases financial risk. Strateg. Manag. J. **38**, 2266–2284 (2017). https://doi.org/10.1002/smj.2647
7. Kiron, D., et al.: Corporate sustainability at a crossroads. MIT Sloan Management Review. (2017) https://sloanreview.mit.edu/projects/corporate-sustainability-at-a-crossroads/
8. Charaeva, M.V., Naumova, O.A., Kosyakova, I.V., Denisov, A.D.: Financing – the basis of organization and realization of the investment policy of Russian enterprises. The International Journal of Environmental and Science Education (IJESE). **14** (2016). http://www.ijese.net/makale/982
9. Brigham, E., Houston, J.: Fundamentals of Financial Management, 9th edn. Cengage Learning, Boston (2015)
10. Burton, J.: Revisiting the capital asset pricing model [Interview with William Sharpe] (1998). https://web.stanford.edu/~wfsharpe/art/djam/djam.htm
11. Crifo, P., Forget, V.: The economics of corporate social responsibility: A firm-level perspective survey. J. Econ. Surv. **29**, 112–130 (2015). http://hdl.handle.net/10.1111/joes.12055
12. Kubiszewski, I., Constanza, R., Anderson, S., Sutton, P.: The future value of ecosystem services: Global scenarios and national implications. Ecosyst. Serv. **26**, 289–301 (2017). https://doi.org/10.1016/j.ecoser.2017.05.004

Best Practices in Public Finance Management

Elena E. Tolikova[1], Elena V. Chernyavskaya[1], Vladislav V. Koltyshev[1](✉), Matvey D. Gvozdev[1], and Tamara B. Solomatina[2]

[1] Russian Customs Academy, Lyubertsy, Russia
vv.koltyshev@customs-academy.ru
[2] Russian New University, Moscow, Russia

Abstract. The purpose of this article is to develop five best practices in public finance management to shape new approaches in customs administration. The article developed the following best practices in the field of public finance management: assessment of the efficiency of customs authorities; identification of barriers to trade and economic cooperation, as well as on the part of customs regulation; assessment of the impact of information technology on the development of public services, tools and infrastructure; calculation of the "administrative pressure" index and reduction of business costs, including logistics risks. The elaborated best practices in the field of public finance management are based on the currently existing methodologies for improving public finance management in the fields of science and practice related to customs. The result of each best practice is the development of optimal solutions to improve customs administration and public finance management. A set of measures proposed by the authors in terms of applying the best practices will improve the financial activities of the federal executive authorities on the example of the Russian customs service, reduce the financial burden on participants in foreign economic activity, form a new approach to the formation of a mechanism for assessing the effectiveness of customs authorities and assess the degree of development of digitalizing customs authorities.

Keywords: Customs administration · Efficiency of customs authorities · Trade barriers · Administrative pressure · Customs infrastructure · Global supply chains · Business environment

1 Introduction

In the activities of the federal executive bodies of the Russian Federation, as in the public sector of the economy as a whole, the work continues to update the approaches and form the best practices in public finance management.

The order of the Government of the Russian Federation dated January 17, 2019 No. 20-r "On the approval of the plan "Transformation of the business climate" and the recognition of the acts of the Government of the Russian Federation" as invalid (hereinafter referred to as the Order) provides for the development of new approaches in customs administration:

P. V. Trifonov and M. V. Charaeva (Eds.): DITEM 2021, LNNS 380, pp. 584–593, 2022.
https://doi.org/10.1007/978-3-030-94245-8_80

1. Creating conditions for accelerating the transition to electronic document flow between participants in foreign economic activity and state control bodies.
2. Reducing the time required to complete all procedures related to the import of goods and vehicles into the Russian Federation and their export from the Russian Federation and others [1].

The work aims to develop 5 best practices in the field of public finance management in the activities of customs authorities for developing the institution of customs administration.

The theoretical basis of the study was the scientific works of Russian and foreign economists, and specialists in the field of finance and customs administration, as well as logistics. The research methodology is based on a systemic approach. The study used general scientific methods (analysis, synthesis, comparison, observation, and extrapolation), expert methods, and methods of mathematical modeling.

2 Methods

2.1 The First Best Practice in Public Finance Management "Evaluating the Effectiveness of Customs Authorities"

In 2021, the World Customs Organization (hereinafter – WCO) is faced with the urgent task of unifying the systems for assessing the activities of customs authorities in all WCO member countries, which will allow for a transparent analysis and assessment of work efficiency, and, on the basis of these studies, to form ratings of the level of development of customs administration by WCO member states.

Due to this purpose, the WCO Strategic Development Plan for 2019–2022 was developed [2–5], which is a new and improved approach to the management and strategic planning of the activity of customs authorities [6].

Currently, under the chairmanship of the FCS of Russia [7], the work is underway to form a national mechanism for assessing the effectiveness of the activity of customs authorities.

Based on the data of the final report of the FCS of Russia for 2020, Table 1 was compiled on the implementation of the Strategic Development Plan of the WCO for 2019–2022 by the FCS of Russia.

Table 1. Key performance indicators of the planned values of the FCS of Russia [8].

No.	Strategic direction	Indicators	Results %	Planned results %
SO1	Facilitating international trade	The share of automatically registered electronic declarations for exported goods	99.1	99

(continued)

Table 1. (*continued*)

No.	Strategic direction	Indicators	Results %	Planned results %
		The share of automatically registered electronic declarations for imported goods	99.3	99
		The share of automatically issued low-risk electronic declarations for exported goods	93.7	80
		The share of automatically issued low-risk electronic declarations for imported goods	86	80
SO2	Collectability of customs payments	The share of customs and other payments made using personal accounts of legal entities opened in a unified information resource of the FCS of Russia	100	100
SO3	Protecting the society	The share of automatic risk profiles in the total number of active risk profiles	90.03	90
		The share of crimes detected by the operational units of the customs authorities in the total number of criminal cases initiated by the customs authorities, referred to the competence of the customs authorities	96.5	90

Source: compiled by the authors

As Table 1 shows, the FCS of Russia successfully conducts activities that are customer-focused and are aimed at increasing the level of conscientiousness of participants in foreign economic activity, improving the trading conditions, increasing the efficiency of customs control and reducing fiscal risks for the country's budget, which is a confirmation of implementing the first best practice.

2.2 The Second-Best Practice in Public Finance Management "Identifying Barriers to Trade and Economic Cooperation"

To develop integration processes, the Eurasian Economic Commission, together with the member states of the Union, has developed the terminology of obstacles (Table 2) for the free movement of goods, services, capital, and labor in the internal market of the Union and determined methodological approaches to dividing obstacles into barriers, derogations, and restrictions taking into account the specifics of the Eurasian economic integration.

Table 2. Definition of the terms "barriers", "derogations", "restrictions" [9, 10].

Obstacles		Definition
Type	Term	
Forbidden	Barriers	Obstacles to the free movement of goods, services, capital, and labor within the framework of the functioning of the Union's internal market, which arose as a result of the discrepancy between the legislation of the member states or the existing law enforcement practice in the member states with the Union law
Allowed	Derogations	Exceptions provided for by the Union law in terms of non-application by a member state of the general rules for the functioning of the Union's internal market
	Restrictions	Obstacles to the free movement of goods, services, capital, and labor within the framework of the functioning of the Union's internal market, arising from the lack of legal regulation of economic relations

Source: compiled by the authors

The dynamics of the number of obstacles in the internal market of the Union is presented in Table 3.

Table 3. The number of obstacles in the internal market of the Union [11].

Obstacles	2017	2018	2019	2020
Barriers	9	11	16	19
Restrictions	34	37	38	37
Derogations	17	17	17	14
Total	60	65	71	70

Source: compiled by the authors

When implementing the second-best practice, it should be noted that the member states adhere to integration sentiments, this is regularly supported by political statements and constructive initiatives at the supranational and national levels. At the same time, economic contradictions between countries and the imperfection of the legal field remain the key reasons for the preservation and emergence of barriers and obstacles in the Union.

2.3 The Third Best Practice in Public Finance Management "Assessing the Impact of Information Technology on the Development of Public Services, Tools and Infrastructure"

The main indicator characterizing the development of information technologies in every state of the world and determining its place among all countries is the index of the development of electronic government.

With regard to managing public finances in the activities of customs authorities, the authors will adapt the electronic government development index in relation to state customs services.

Below, the formula is provided for calculating the "comprehensive index" of electronic customs (E – Customs Service Index – ECSI):

$$CSI = \frac{1}{3} \times (OSI + TII + EDI), \tag{1}$$

where OSI (Online Service Index) is the indicator of the development of online customs services;

TII (Telecommunication Infrastructure Index) is the indicator of the development of the telecommunications infrastructure of the customs authorities;

EDI (Electronic Declaration Index) is the indicator of electronic declaration of goods.

In total, from 2018 to 2020, the FCS of Russia created 8 electronic customs and 16 electronic declaration centers. Electronic declaration centers conduct their activities taking into account the types of transport, categories of goods, as well as the place of tax registration of the declarant.

Table 4. Dynamics of the transition to electronic declaration.

Indicators / year	2018	2019	2020
The share of customs declaration of goods in the EDC, %	32	67	97
Number of places for customs declaration of goods, pcs	500	120	16

Source: compiled by the authors

Based on the data in Table 4, it can be concluded that the EDI indicator in Russia is 0.97, which makes the FCS of Russia one of the world leaders in this category.

The other two indicators are identified on the basis of the expert review method, which makes it impossible to instantly calculate the ECSI index. Based on this index, it is possible to compare not only the dynamics of the development of information technologies in the customs service, but also to determine the place of one or another federal executive body in the all-Russian rating of federal executive authorities.

2.4 The Fourth Best Practice in Public Finance Management "Calculation of the Index of "administrative Pressure" in the Field of FEA"

To use the fourth practice of improving financial management in the activities of customs authorities, the authors adapt the administrative pressure index used in the investment sphere to the activities of the customs service.

The administrative pressure index is calculated according to four parameters (P1-P3, P5):

1. The share of warnings from the total number of punishments (P1).
2. The share of FEA participants subject to control and supervision (P2) of the total number of controlled ones.
3. The share of fines imposed without scheduled and unscheduled inspections, of the total number of fines (P3).
4. The number of imposed fines (P5).

Within the framework of the regional customs offices, for each profile of a state authority interacting with a FEA participant, a general index of administrative pressure is determined. And then the place of each regional customs office is determined in the general index of administrative pressure.

After calculating this index by profiles, it is required to analyze the results and draw a conclusion regarding the methodology for reducing administrative pressure in order to improve the business environment (Fig. 1).

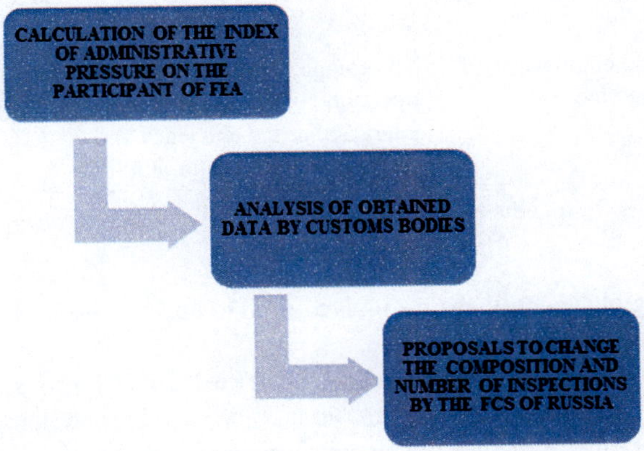

Fig. 1. Algorithm of actions to reduce administrative pressure using adaptive methodology.

Thus, the profiles will be able to cover both external and internal subjects of control of the FCS of Russia, while the results of calculating the index can become the basis for making proposals on adjusting the activities of control and supervisory bodies of general and special competence.

2.5 The Fifth Best Practice of Public Finance Management "Reducing the Costs of FEA Participants"

In order to identify regional risks of global supply chains, the collection of input data is the first and most important step [12]. There is the Logistics Performance Index, calculated by the World Bank for 6 indicators based on the method of expert assessments (hereinafter – LPI) [13]. Each indicator is rated up to 5 points. This index shows the state of supply chains in a particular country.

Table 5. LPI indicators for the Russian Federation in 2018.

No.	Indicator	Brief description	Indicators in Russia
1	Organization of international transportation	Possibility of organizing international transportations at competitive prices	2. 64
2	Quality of trade and transport infrastructure	Quality of ports, railways, highways, development level of logistics infrastructure	2. 78
3	Trackability of the passage of goods	The ability to identify and track the passage of goods in real time	2. 65
4	Efficiency of customs and border operations	Border crossing process and customs clearance process (speed, simplicity)	2. 42
5	Timeliness of cargo deliveries	Delivery of cargo exactly on schedule	3. 31
6	Quality and competence of logistics services	The competencies of logistics operators, which serve as a guarantee of the efficiency and safety of cargo transportation	2. 75

Source: compiled by the authors

To optimize global supply chains, it is required to improve each of the indicators of this index, presented in Table 5.

To reduce the financial costs of the FCS of Russia and business, it is proposed to modernize the customs infrastructure: temporary storage warehouses and customs warehouses – in terms of the use of special warehouse technologies – the introduction of special warehouse technologies – Warehouse Management System (hereinafter – WMS). With the help of the WMS-system, it is possible to optimize the functions of the operational management of a warehouse with targeted storage and thereby increase the efficiency of business processes and reduce the financial costs of all participants in foreign economic activity.

Thus, in order to implement the fifth best practice of reducing business financial costs, it is required to modernize supply chains and introduce special warehouse technologies.

3 Results

The presented best practices in the field of public finance management in the activities of customs authorities will help to improve the activities of the Russian customs service and obtain significant financial effects in the activities of FEA participants.

The first best practice focuses on optimizing customs performance using national and international assessment techniques.

The practice "Identifying barriers to trade and economic cooperation" will help to determine the existing barriers in foreign economic activity and the prospects for Russian exports in conjunction with the deepening of Eurasian economic integration.

The calculation of the electronic customs index from the practice "Assessing the impact of information technologies on the development of public services, tools and infrastructure" helps to increase the efficiency of the provision of customs services by the FCS of Russia to businesses, which entails an increase in the traceability and transparency of transactions with monetary funds and the timeliness of transferring customs payments to the federal budget.

The practice of "Calculating the index of administrative pressure" will help to improve the management of public finances of federal executive authorities, as this will reduce the financial and time costs of participants in FEA and the FCS of Russia, which will contribute to the growth of economic activity and improve the conditions for conducting business.

Using the "Reducing business costs" practice, the LPI index can be improved as one of the tools to enhance the quality of global supply chains through the following measures: updating software and hardware; developing the infrastructure of road checkpoints; optimizing the work of national special logistics providers providing a full range of services using information technology.

4 Discussion

In order to improve the business environment, government bodies conduct an analysis and assess the possibility of the practical use of the best practices.

The best practices in the public sphere will help to reduce budgetary risks, which is very important in the post-pandemic period.

Some scientists are inclined to believe that the indicator of the electronic customs index is influenced not only by the work of electronic government itself, but also by the modern methodology for calculating this index, comparable to the international methodology used by the UN.

When calculating the adaptive index of administrative pressure in the activities of the customs service, it is necessary to take into account that sometimes inspections are replaced by other forms of control and supervision, while state statistics does not keep a systematic record of such forms and does not assess their impact.

The development of customs infrastructure through the introduction of special customs warehouse technologies and global supply chains are closely linked. Improving customs infrastructure has a positive impact on the entire transport and logistics system, and, as a result, on the supply chain. In addition, participants in foreign economic activity must be confident that the work of the customs authorities will not cause serious disruptions in the cargo flow, and, perhaps, and must even be ready to invest in improving places of customs control. This explains why public-private partnerships have become a major topic of discussion in the business community and government agencies dealing with national security issues [14, 15].

5 Conclusion

To form the best practices in the field of public finance management, it is required to apply a set of measures aimed at the financial activities of the customs service, namely:

- Continue the work towards improving the national mechanism for assessing the efficiency of customs authorities based on the Strategic Development Plan of the WCO (until 2022).
- Continue the work to optimize the use of trade and economic barriers in Russia's international trade, as well as trade with the EAEU member states.
- Consider the issue of using a "comprehensive index" of electronic customs.
- In order to improve the business environment, use the "adaptive index" of administrative pressure in the field of FEA.
- Apply improved approaches to managing customs infrastructure facilities in terms of implementing an automated warehouse management system.

These best public finance management practices will help to create a favorable environment for effective interaction between the FCS of Russia and FEA participants.

References

1. Order of the Government of the Russian Federation dated January 17, 2019 No. 20-r "On approval of the plan "Transformation of the business climate" and invalidation of acts of the Government of the Russian Federation". http://www.consultant.ru/document/cons_doc_LAW_316129/
2. Strategic Plan of the WCO for 2019–2022. http://www.wcoomd.org/-/media/wco/public/global/pdf/about-us/administrative-documents/wco-strategic-plan-2019-2022.pdf?db=web
3. WCO Implementation Plan for 2019–2020. http://webcache.googleusercontent.com/search?q=cache:CVvK3RoWWd4J:www.wcoomd.org/-/media/wco/public/global/pdf/about-us/administrative-documents/annex-to-the-implementation-plan-2019_2020_concrete-deliverables_publication_version.pdf%3Fdb%3Dweb+&cd=1&hl=ru&ct=clnk&gl=ru
4. Appendix to the implementation plan for 2019–2020. Exact results. http://webcache.googleusercontent.com/search?q=cache:CVvK3RoWWd4J:www.wcoomd.org/-/media/wco/public/global/pdf/about-us/administrative-documents/annex-to-the-implementation-plan-2019_2020_concrete-deliverables_publication_version.pdf%3Fdb%3Dweb+&cd=1&hl=ru&ct=clnk&gl=ru
5. WCO Annual Report for 2019–2020. http://www.wcoomd.org/-/media/wco/public/global/pdf/about-us/annual-reports/annual-report-2019_2020.pdf
6. Official website of the WCO. http://www.wcoomd.org
7. Official website of the FCS. https://customs.gov.ru

8. Tolikova, E.E.H., Zubavlenko, E.A.: Econ. Bus. Theory Prac. **3–2**(73), 175–179 (2021). https://doi.org/10.24412/2411-0450-2021-3-2-175-179
9. Methodology for dividing obstacles in the internal market of the Eurasian Economic Union into barriers, derogations and restrictions and recognition of barriers, approved by the EEU Board dated February 2, 2021 No. 10. https://eec.eaeunion.org/upload/medialibrary/820/Strategy_2025.pdf
10. Register of barriers of the EAEU (as of April 17, 2021). https://barriers.eaeunion.org/en-us/Pages/obstacles.aspx
11. Tolikova, E.E.H., Chernyavskaya: Econ. Bus. Theory Prac. **4–2**(74), 198–203 (2021). https://doi.org/10.24412/2411-0450-2021-4-2-198-202
12. Chu, C.-Y., Park, K., Kremer, G.E.: Adv. Eng. Inf. **45**, 101053 (2020) https://doi.org/10.1016/j.aei.2020.101053
13. Official website of the World Bank LPI International. https://lpi.worldbank.org/international/global
14. Pourakbar, M., Zuidwijk, R.A.: Euro. J. Operat. Res. **271**(1), 331–340 (2018). https://doi.org/10.1016/j.ejor.2018.05.012
15. Mourtzis, D., Samothrakis, V., Zogopoulos, V., Vlachou, E.: Procedia. CIRP **79**, 574–579 (2019). https://doi.org/10.1016/j.procir.2019.02.097

Analytical Aspect of the Profit-Forming Indicators Control in the Context of Strengthening Financial Business Security

Elena A. Sharovatova(✉), Ekaterina V. Parada, Elena N. Karpova, and Radmila N. Magomedova

Rostov State University of Economics, Rostov-on-Don, Russia

Abstract. This article discusses the need to calculate the indicators of an enterprise in the context of strengthening the financial security of the business. The financial security of an organization is characterized by a number of indicators of the financial company condition (qualitative and quantitative), reflecting the protection degree from various internal and external risks. The level of financial security is provided by – ensuring stable economic development of the company, minimizing the impact of external negative influences, control of the debt load level, and creation and implementation of a system for tracking the financial company condition. In order to identify crisis phenomena and prerequisites for future bankruptcy the monitoring is used. In the study, the authors consider the mail profit-forming factors, the impact of each factor on profitability is calculated separately. The overall results and efficiency of the production and economic activity of the enterprise are determined by analyzing the formation of profit indicators, which can be carried out by means of using horizontal analysis of financial results. The authors consider the algorithm for conducting this type of analysis on the example of a specific enterprise. It is determined that by conducting a factor analysis of the profit from sales for a specified period of time it is possible to identify the reasons for the profit growth; and the more often factor analysis is carried out, the easier it is for a company to stay afloat in the conditions of market changes without a sharp loss of income.

Keywords: Financial enterprise security · Qualitative and quantitative indicators · Factor analysis of profit

1 Introduction

Financial enterprise security is the most important part of its economic security. Among the economic security elements, the financial part is of leading importance for ensuring enterprise activity. The financial security of an organization may be characterized by a number of qualitative and quantitative indicators of the financial company condition, reflecting the protection degree from various internal and external risks. The main tasks of maintaining the necessary level of financial security include:

P. V. Trifonov and M. V. Charaeva (Eds.): DITEM 2021, LNNS 380, pp. 594–602, 2022.
https://doi.org/10.1007/978-3-030-94245-8_81

- ensuring stable economic development of the company;
- minimizing the impact of external negative influences. The impacts include both global, for example, economic crises, and the actions of competing organizations directed against a particular enterprise;
- control of the debt load level;
- creation and implementation of a system for tracking the financial company condition. The monitoring purpose is to identify crisis phenomena and prerequisites for future bankruptcy [1].

Working in these areas will help the company to form a safety margin necessary and sufficient to overcome the consequences of crisis situations. It is noteworthy that many of these tasks are directly interrelated with the financial results of the company. Planning is impossible without factor analysis of various indicators that affect the company's activities. These criteria make it possible to justify plans, to assess the quality of accounting and control systems. Factor analysis of profit from sales is especially important, which makes it possible to identify the influence of a wide variety of factors on the amount of profit generated. Therefore, it means that on the basis of such an analysis it is possible to make effective management decisions related to pricing policy, assortment adjustment, cost structure optimization, etc.

Net sales revenue is the difference between revenue and the amount including cost, management and commerce costs [2]. The main factors on which the increase or decrease in income depends are considered to be changes of:

- sales volume;
- assortment range;
- cost price;
- price characteristics.

2 Methods

Factor analysis of the profit from the sale involves the study of the structure of the receipt of financial resources by the enterprise in its pure form, as well as the conditions that influenced the growth of profitability and losses. Therefore, research is required in order to make the work of the organization more efficient and to choose the right management strategies.

The main purpose of the analysis of profit-forming factors is to identify strengths and weaknesses in the sales chain while making a profit. In order to realize this, the following is needed:

- determination of fluctuations by influential criteria;
- study of changes in indicators;
- assessment of the profitability of the organization for the selected period [3].

For the analysis of profit-forming factors, the method of chain substitutions is popular, which allows to identify not only general, but also conditional changes. To do this, the indicators in the reporting period are multiplied by the difference between the planned profit and costs, and fixed planned expenses are deducted from the amount received.

It is advisable to use this method of analysis for those companies where profit fluctuations are determined by seasonal demand, since in this case the analysis of changes in realized volumes is carried out for each of the seasons, and the results obtained are compared in order to identify an imbalance. Profit values are studied in relation to other parameters – the flow of buyers, price changes, which allows to determine how much seasonal fluctuations have affected the decrease or increase in profit, and whether it is possible to increase profitability in the "low" season by influencing certain factors.

3 Results

The structure and composition of profit, fluctuations in comparison with similar reporting periods, and the effect of the accounting strategy for each revenue line, tax and dividend payments are investigated.

The easiest way to analyze the factors on which profit depends is by entering the indicators in a summary table. How factor analysis of profit from sales is carried out by this method can be seen in the following example as well (Table 1).

Table 1. Profit-forming factors. Example 1.

Criteria	Investigated time periods, Rubles		Difference	
	Prior	Reporting	Rubles	Percentage
Revenue	11500000	12000000	500000	4.35
Cost price	8000000	7700000	−300000	−3.75
Commercial costs	1300000	1500000	200000	15.38
Management costs	600000	750000	150000	25
Profit from sales	1600000	2050000	450000	28.13
Indexation of price changes	1	1.25	250	25

The impact of each factor on profitability is calculated separately. For example, in order to calculate the impact of the volume of products sales at base prices, revenue in the reporting period is divided by the index indicator, then by previous data, and multiplied by one hundred percent. If you remove the extra zeros, the result is as follows: $12000/1.25/11500 \times 100\% = 83.5\%$. Therefore, it may be concluded that the commodity volume decreased by 16.5%, which negatively affected the profitability. By converting interest into real money, it is possible to see that the company's profit decreased by 264 thousand rubles (1600 thousand $\times$ (−0.165)).

In order to identify the impact of cost, recalculate the data of the base period in relation to the changed quantity of goods sold. That is, 8000 thousand is multiplied by 0.835. It turns out to be 6680 thousand rubles. Comparing with the current indicator, they conclude: the cost increased by 1.020 thousand rubles, which led to a decrease in net income. Comparing current and previous cost data, it can be revealed that costs have increased. This also led to a decrease in profits. But the price increase had a positive impact, leading to an increase in net income by 2.400 thousand rubles. This figure is revealed by calculating the volume of sales in basic prices: 12.000/1.25 = 9.600 thousand rubles and then subtracting it from the amount of reported revenue.

It is important to determine the marginal profit of the enterprise, which shows the income from the sale of a commodity unit. By calculating this indicator, it is possible to identify as accurately as possible the degree of impact on the amount of profit received by such factors as sales volume and the value of variable costs – expenses for raw materials, materials, fuel, utilities, etc. Margin profit is determined by multiplying the number indicating the number of products sold by the difference between the commodity value and variable costs, after which the resulting value is reduced by the amount of fixed costs associated, for example, with the payment of rent of premises, depreciation of equipment, etc. Conducting such a study is necessary to calculate the minimum quantity of goods required for sale in order to cover the costs of the enterprise [4].

4 Discussion

Here is an example of how such a calculation is made. The company produces one type of product from one type of material; the planned production volume is 400 pcs and the actual one is 500 pcs. As for one unit of product, the price per gram is 200 rubles. With a standard of 50 g per product, in fact the price is increased to 220 rubles, but the waste per product was reduced, and the material consumption was 40 g per product. So it is possible to assume that the general production and general economic costs of the plan and the fact coincided. At the same time, there was an improvement in sales figures – instead of the planned 400 units, 480 units were sold, and the selling price increased from the planned 12.000 rubles by 200 rubles. Also, due to optimization measures, it was possible to reduce labor costs (280 rubles per hour with the planned 300 rubles), but this had a negative consequence – the labor intensity increased to six hours instead of five for the manufacture of one unit of the product.

Table 2 shows calculations of deviations by material, Table 3 – by remuneration, and Table 4 demonstrates the calculation of deviations by revenue.

According to the calculated indicators, margin profit can be determined by reducing revenue by the amount of direct production costs: $480 \times 12.200 - 480 \times 220 \times 40 - 480 \times 280 \times 6 = 825.600$ rubles, while the planned figure is much less than actually achieved: $400 \times 12.000 - 400 \times 200 \times 50 - 400 \times 300 \times 5 = 200.000$ rubles. Thus, it can be concluded that the deviation of 625.000 rubles of planned indicators from the actual ones is a positive factor and indicates the successful development of the enterprise.

It is important to note that the calculation of deviations should be a mandatory element of the control of profit-generating indicators within the framework of the enterprise planning and budgeting system. If it is necessary to implement such an analysis in a small enterprise, then no special software is required – the tables calculated in the Excel program provide a full-fledged opportunity to perform an analysis of the impact of any factors on profit in advance and adjust the strategy of purchases and sales in accordance with the result obtained [4].

Table 2. Calculating deviations for materials based on three factors. Example 2.

Type of deviation	Calculation formula	Calculation	Deviations, rub.
Deviations in output volume – Dv	Dv = (Q – Q) × Pн × Nн	Dv = (500 – 400) × 200 × 50	1 000 000
Deviations in the price of the material – Dp	Dp = Qф × (Pф – Pн) × Nн	Dp = 500 × (220 – 200) × 50	500 000
Deviations in the material consumption rate – Dn	Dn = Qф × Pф × (Nф – Nн)	Dn = 500 × 220 × (40 – 50)	−1 100 000
Total deviation in material – Dm	Dm = Dv + Dp + Dn		400 000

Table 3. Calculating deviations in piecework wages by three factors. Example 2.

Type of deviation	Calculation formula	Calculation	Deviations, rub.
Deviations in output volume Do	Do = (Qф – Qн) × Зн × Тн	Do = (500 – 400) × 300 × 5	150 000
Salary rate deviations Ds	Ds = Qф × (Зф – Зн) × Тн	Ds = 500 × (280 – 300) × 5	–50 000
Deviations in labor intensity Di	Di = Qф × Зф × (Тф – Тн)	Di = 500 × 280 × (6 – 5)	140 000
Total deviation on piecework payment Dp	Dp = Oз + Oc + Oт		240 000

Table 4. Calculating revenue deviations based on two factors. Example 2.

Type of deviation	Calculation formula	Calculation	Deviations, rub
Revenue deviations Dr	Dr = (Пф – Пп) × Цп	= (480 – 400) × 12000	960 000
Deviations by sale price Dsp	Dsp = Пф × (Цф – Цп)	= 480(12200 – 12000)	96 000
Total deviation by revenue			1 056 000

The overall results and efficiency of the production and economic activity of the enterprise are determined by analyzing the formation of profit indicators, which can be carried out by means of using horizontal analysis of financial results [5]. The following data allow considering the algorithm for conducting this type of analysis on the example of a specific enterprise (Table 5).

Table 5. Analysis of the factors of profit formation of the enterprise in 2018–2020.

The name of the indicator	2018	2019	2020	Deviations from 2019 to 2018 (+;–)	Deviations from 2020 to 2019 (+;–)	Deviations of 2019 to 2018 in %	Deviations of 2020 to 2019 in %
Revenue	209365	242906	264517	+33541	+21611	116.02	108.90
Cost price	191658	215723	223997	+24065	+8274	112.56	103.84
Gross profit	17707	27183	40520	+9476	+13337	153.52	149.06
Commercial expenses	–	–	–	–	–	–	–
Management expenses	–	–	–	–	–	–	–
Profit from sales	17707	27183	40520	+9476	+13337	153.52	149.06
Interest payable	790	1638	1584	+848	–54	207.34	96.70
Income from participation in other organizations	–	–	–	–	–	–	–
Other income	8577	12148	9899	+3571	–2249	141.63	81.49
Other expenses	12528	10403	11708	–2125	+1305	83.04	112.54
Profit before tax	12966	27290	37127	+14324	+9837	210.47	136.05
Net profit (loss) of the reporting period	12966	27290	37127	+14324	+9837	210.47	136.05

As can be seen from Table 5, there is a trend of growth in all profit indicators. It is possible to identify the reasons for the profit growth by conducting a factor analysis of the profit from sales for a specified period of time.

First of all, it is important to determine the impact on the financial result of such a factor as the "revenue level", for which the amount of revenue will be adjusted by bringing it to a comparable form to the prices of the previous year. In order to do this, an inflation index for the manufactured product or a price index is needed, as it shows how much the price of products sold increased in 2019 and 2020 compared to the base year. So, in 2019, the price increased by 2.77%, so the price index will be 1.03, and in 2020 – by 6%, i.e. the price index, in this case, is 1.06. The revenue from the sale of products at comparable prices will be calculated using the following formula:

$$B = \frac{B_1}{B_p},$$

where B_1 is revenue from sales;
B_p – price index;

$$B_{2009} = \frac{242906}{1.028} = 236289,88\,\text{(thousand roubles)};$$

$$B_{2010} = \frac{264517}{1.06} = 249544,34\,\text{(thousand roubles)}.$$

This means that if there were no price growth of 2.77% in 2019 and 6% in 2020, the revenue would be 236289.88 thousand rubles and 249544.34 thousand rubles, respectively.

Consequently, in 2019, due to the price, revenue was increased by:

$$\Delta B_p = B_1 - \frac{B_i}{B_p} = 242906 - 236289.88 = 6616.12\,\text{(thousand roubles)}.$$

In 2020, due to the price, revenue was increased by:

$$\Delta B_p = B_1 - \frac{B_i}{B_p} = 264517 - 2349544.34 = 14972.66\,\text{(thousand roubles)}.$$

Then the number of goods sold changed in 2019 compared to 2018:

$$\Delta_{sales} = 236289.88 - 209365 = 26924.88\,\text{(thousand roubles)}.$$

The number of goods sold changed in 2019 compared to 2018:

$$\Delta_{sales} = 249544.34 - 242906 = 6638.34\,\text{(thousand roubles)}$$

Based on this, the change in profit in the reporting year compared to the previous year due to a decrease in revenue (excluding external prices) was:

$$\Delta \Pi_{\text{n}}^{B} = \frac{[(B_1 - B_0) - \Delta B_p]R_0^{\Pi}}{100},$$

where $\Delta \Pi_{\Pi}^{B}$ – profit from sales due to revenue,
B_0 – revenue from the products of the previous year,
R_0^{Π} – profitability of sales of the previous year.
In 2019 compared to 2018:

$$\Delta \Pi_{\Pi}^{B} = \frac{[(242906 - 209365) - 6616.12]0.084}{100} = 22.62\,\text{(thosand roubles)}.$$

In 2020 compared to 2018:

$$\Delta \Pi_{\Pi}^{B} = \frac{[(264517 - 242906) - 14972.66]0.112}{100} = 7.43\,\text{(thosand roubles)}.$$

Due to the increase in sales revenue, the profit from sales increased in 2013 by 22.62 thousand rubles and in 2014 by 7.43 thousand rubles.

Then it is possible to calculate the influence of the "Price of sold products" factor:

$$\Delta\Pi_{\text{п}}^{p} = \frac{\left(B_1 - \frac{B_1}{B_p}\right)R_0^{\Pi}}{100} = \frac{\Delta B_p R_0^{\Pi}}{100}$$

where ΔB_p – changes in revenue due to price.

In 2019:

$$\Delta\Pi_{\Pi}^{p} = \frac{6616.12 \cdot 0.084}{100} = 5.55\,(\text{thousand roubles}).$$

In 2020:

$$\Delta\Pi_{\Pi}^{p} = \frac{14972.66 \cdot 0.112}{100} = 16.77\,(\text{thousand roubles}).$$

Due to the increase in prices for products sold the profit from sales increased by 5.55 thousand rubles in 2019 and by 16.77 thousand rubles in 2020.

The calculation of the influence of the factor "Cost of sales" is carried out according to the formula:

$$\Delta\Pi_{\Pi}^{C} = \frac{B_1(\Delta C_1 - \Delta C_0)}{100},$$

where $\Delta\Pi_{\Pi}^{C}$ – change in profit from sales due to cost;

ΔC_1, ΔC_0 – the share of cost in revenue, respectively, of the base and reporting period.

In 2019:

$$\Delta\Pi_{\Pi}^{C} = \frac{249606 \cdot 2.69}{100} = 6534.17\,(\text{thousand roubles}).$$

In 2020:

$$\Delta\Pi_{\Pi}^{C} = \frac{264517 \cdot 4.13}{100} = 10924.55\,(\text{thousand roubles}).$$

The level of cost in 2019 in relation to sales revenue was increased by 2.69%. Therefore, the profit from sales was increased by 147217.01 thousand rubles. In 2020, the level of cost in relation to sales revenue was increased by 4.13%. Therefore, the profit from sales was increased by 10924.55 thousand rubles.

The overall impact of the factors is as follows:

$$\Delta\Pi_{2009} = 22.62 + 5.55 + 6534.17 = 6562.34\,(\text{thousand roubles});$$

$$\Delta\Pi_{2010} = 7.43 + 16.77 + 10924.55 = 10945.75\,(\text{thousand roubles}).$$

5 Conclusion

How often you need to analyze sales depends on the specifics of the market, seasonality, competitiveness. Usually, the analysis is carried out every two to three months. But there are industries where monthly research is required. This usually refers to enterprises with a high level of competition. The more often factor analysis is carried out, the easier it is for a company to stay afloat in the conditions of market changes without a sharp loss of income.

References

1. Financial security of the enterprise + useful formulas. https://www.gd.ru/articles/10974-finansovaya-bezopasnost-predpriyatiya
2. Charaeva, M.V.: The main directions of increasing the profitability of industrial enterprises in Russia. PSU Bul. Ser. Econ. **4**, 35–44 (2012)
3. Factor analysis of profit: features of conducting by example. https://www.kom-dir.ru/article/2772-faktornyy-analiz-pribyli
4. Malikova, S.: Deviation management: types of deviations and methods of analysis. Econ. Life **32**, 14 (2019)
5. Shumilina, E.V., Khaustova, G.I.: Analysis of financial results of a commercial organization. Econ. Bus. Theory Pract. **1–2**, 231–236 (2021)

Crowdfunding (Crowdinvesting) as an Alternative to Securitization in the Real Estate Market

Sergey G. Sternik[1,2,3](✉), Yuliya R. Mezentseva[4], Yuliya S. Zharkova[5], and Byan Tszinpin[6]

[1] Institute of Economic Forecasting of Russian Academy of Sciences, Moscow, Russia
[2] Financial University under Government of Russian Federation, Moscow, Russia
[3] Moscow State University of Civil Engineering, Moscow, Russia
[4] Don State Technical University, Rostov-on-Don, Russia
[5] Rostov State University of Economics, Rostov-on-Don, Russia
[6] Southern Federal University, Rostov-on-Don, Russia

Abstract. A comparative analysis of various mechanisms for attracting investment in the real estate market is carried out. The theoretical and practical aspects of attracting to the market investments from non-qualified investors are considered. The practice of crowdfunding in the real estate of the American market is studied. The best practice cases of successful projects for attracting non-institutional investors in different countries are given. A review of Russian crowdfunding platforms offering warehouse and retail real estate as investment objects is made. The analysis of the legislative conditions constraining the development of crowdfunding as an alternative investment tool is carried out.

Keywords: Industrial policy · Legislative system · State program · Federal target program · Megaproject

1 Introduction

Crowdinvesting is one of the youngest areas of alternative investments. For example, in Russia, the first platforms (StartTrack, VCStart) appeared in 2014. Currently, in Russia, in different business niches, the platforms operate at Vdolg.ru (vdolg.ru), Loanberry (loanberry.ru), S miru po nitke (smipon.ru), Rufander, Boomstarter.ru, Planeta.ru, Kroogi, ThankYou.ru, Together, Naparapet, StartTrack, CrowdPress, etc.

Crowdfunding in real estate is a relatively new concept that is actively developing in different countries of the world. The "people's" investments attracted in this way are used for the purchase, development, and renovation of residential and commercial real estate for sale or for obtaining rental income.

According to the world practice, crowdfunding in real estate can bring more income than a number of other financial instruments (securities, deposits). The most common crowdfunding type used in the world is debt crowdfunding, which brings a fixed return on capital (for example, from renting premises) in the amount of 8–12%.

P. V. Trifonov and M. V. Charaeva (Eds.): DITEM 2021, LNNS 380, pp. 603–607, 2022.
https://doi.org/10.1007/978-3-030-94245-8_82

When financing the construction or renovation of real estate objects with subsequent sale, one can get up to 20% per annum, but such projects carry more risks.

Abroad there are quite a lot of examples of successful crowdfunding platforms that collect money from collective investments and direct them, among other things, to the development of various segments of real estate, startups, and so on. According to the World Bank's forecast, the volume of the crowdfunding industry may grow to $96 billion by 2025 [1].

There are hundreds of such investment platforms in the world. Fundrise, Realty Mogul, RealtyShares, and RealCrowd are popular in the American market. Hundreds of millions of dollars have been invested through such platforms. In Europe, these are Exporo, Property Partner, and The House Crow with investments of up to 200 million euros through each site. "As a rule, foreign sites attract money for a short period – up to two or three years, receive profitability either from speculative transactions, or working with development/redevelopment, bank leverage is often used, and the entire project is transferred to the bank as collateral", explains Olga Sharygina, Becar Asset Management consulting company vice-president [2].

Among the largest projects that have been implemented based on crowdfunding platforms, Becar Asset Management calls the Weissenhaus Grand Village project in Germany. The author of the project bought a village on the Baltic Sea coast and in 2014–2015 raised money to turn it into a luxury spa resort through the reconstruction of 40 buildings. Through the Companisto platform, the author of the project managed to attract $8.1 million from 1,800 investors.

Another successful project is the Golden Hill Fort in the UK. The old fort was divided into 18 separate blocks, which can be rented through Booking.com. The funds were raised through the Property Partner platform, and now the asset is estimated at more than $2 million.

2 Methods

Crowdfunding is the practice of collective funding for different projects by founders/lenders. By means of the Internet platforms, it allows raising funds in the form of so-called "fiat" (or fiduciary) money [3] from a large number of people for the purpose of implementing a project, developing a product or service, or financing a creative process.

A distinctive feature of crowdfunding as compared to other forms of raising funds from a large number of people is the payment of financial remuneration that the investor receives in exchange for support. Such remuneration can be paid in three main forms: as part of the project's revenue or profit, as a percentage of the invested funds, or in the form of a transfer of part of the company's assets or shares.

3 Results and Discussion

Crowdfunding in Russia is mainly used to raise funds for the development of a business or a startup. At the same time, crowdfunding in real estate is practically non-existent. There are several sites in the market that attract funds for commercial real estate, among them PNK rental [4], Aktivo [5], and Credo Capital. These sites can hardly be called crowdfunding in its pure form. They follow the principle of closed-end mutual investment funds (CEIFs), which are formed for a specific project. These are investment platforms in terms of functionality: selection of facilities, setup of investment mechanisms, varying by goals set and the facility specifics, and the facility management. Thus, the platforms are aimed at obtaining the result announced to potential investors. Their similarity to crowdfunding platforms is that they attract investors' funds through their platform on the Internet.

Recently, the Russian developer of industrial facilities PNK Group offered its new product aimed at generating income from rental payments – investments in warehouse real estate Rent PNK. The developer itself, acting as the main fund co-investor, will give the fund a preferential right to purchase already pre-rented facilities.

The yield offered to investors starts at 11.5% per annum. The group is ready at the initial stage to invest in industrial buildings valued at more than 10 billion rubles. The platform will be focused on private investment and will offer new buildings already leased out for a long term. The fund's management company is A class Capital. The minimum investment is 5000 rubles. Income is paid quarterly; the recommended investment period is 5 years with the possibility of an early exit. Attracting "people's" investment will allow the company to speed up the turnover of its facilities and get a quicker return on investments in construction.

The Aktivo platform was created in 2015 by the founder of KupiVIP, Oskar Hartmann, with partners [6]. Now the company has more than 700 investors, and the value of the assets (commercial real estate) is estimated at 3 billion rubles. The starting share price is 300,000 rubles and higher. In 2019, more than 450 million rubles income was distributed to investors. According to Aktivo data, the commercial real estate payback in Russia is 7–8 years due to the indexation of lease agreements, and the income (after taxes) in the first year of ownership is 9–10%. Now Aktivo is focused on low-risk investments. All of them offer investors indirect ownership of commercial real estate in Russia with professional management. Due to the fact that no financial leverage such as bank credit is used for facility purchase, the ownership risks are significantly reduced, explains Egor Klimenko, CEO of the Aktivo crowdfunding platform.

Another well-known platform for investing in commercial real estate, Credo Capital, was created in 2017 by Alexander Ruchjev, the founder of the Osnova development group [7]. Here, a separate CEIF was established for each commercial real estate facility offered to the investor. At the start, the minimum entrance threshold for the project was 900,000 rubles. At the same time, the platform took care of all maintenance issues, tenants' search and operation, and all facilities offered for investment were pre-leased and had professional management, which allowed investors to start earning at once. The platform is still operating today, but it is now re-focused onto big investors. Thus, today it can no longer be called a crowdfunding platform in the usual sense.

"So far, there is not a single crowdfunding platform for investment in residential real estate in Russia, and perhaps they will not appear soon. The current legislation regulating crowdfunding prevents the development of "people's" investments in residential real estate as an alternative to shared-equity construction", – says the head of the Petersburg

Real Estate consulting center (a member of the Setl Group development company), Olga Trosheva [1].

The “Law on crowdfunding” [8] (Article 8, paragraph 2) states that “digital utility rights may not constitute a right to demand property, subject to state registration, and (or) transactions with which are subject to state registration or notary certification”. This point, according to Trosheva, excludes the possibility to invest and get housing in return. In addition, ordinary citizens (non-qualified investors) cannot invest more than 600,000 rubles a year through online platforms (Article 7, paragraph 1).

At the same time, the use of crowdfunding to generate additional income, and not the housing itself, is quite attractive against the background of risk-free rates decrease, Trosheva believes. However, at the moment, the legislation regulating crowdfunding is not sufficiently developed and does not fully protect the investor, she adds.

“Previously, many of the residential real estate developers used creative financing redistributing cash flows between projects, financing new construction projects with the funds of existing projects. Therefore, the legislator stopped such schemes by introducing escrow accounts. This caused an increased need for capital from developers for the implementation of new projects. This situation opens up new opportunities for collective investment. We are considering the development of Aktivo in this direction” [5].

In Russia, according to the developer A. Ruchjev, several basic conditions indispensable for the development of crowdfunding in real estate are absent: a regulatory legal framework, legal mechanisms for protecting the rights of investors, and an investment culture.

In Western countries, there exists a longstanding investment culture, all this is transparent and regulated by law. “While in European countries all this is clearly regulated by legislation and there are ways of additional protection in the form of a blockchain, today we have neither one nor the other. Hence the distrust of potential investors. Therefore, I assess the prospects for the development of crowdfunding in real estate as extremely low” [6].

The segment of collective investments in commercial real estate has more prospects. With the key rate and the deposits yield decrease, the interest in alternative investment opportunities will obviously grow, Klimenko believes. According to him, mutual funds grew in 2019 and hit the maximum: the share of mutual funds’ assets in GDP was 4.7%, the number of shareholders increased by 240,000.

In the future, in Russia, not only ready-made commercial real estate facilities organized based on the principle of closed-end funds can be offered for investment through crowdfunding platforms, but also facilities with a higher risk that will be focused on obtaining a higher income, Klimenko predicts [5].

Since the crowdfunding mechanism, as a rule, attracts a wide range of investors (mostly non-qualified) to participate, and is also technically implemented via the Internet, one can talk about a high degree of similarity between the mechanisms of token placement and crowdfunding. The key difference between these mechanisms at the moment is the deep connection of token placement models with blockchain technology, as well as the fact that real (fiat), rather than digital currency, is used for crowdfunding financing.

4 Conclusion

Thus, in the authors' opinion, taking into account the current dynamics of blockchain technology development and the further spread of crowdfunding in the future as a competitive tool in relation to classical securities issued on the basis of real estate market assets (objects and projects), a gradual transition of crowdfunding to blockchain technology with the subsequent significant displacement of classical securitization tools from the market seems inevitable.

References

1. Crowdfunding's Potential for the Developing World. https://www.infodev.org/crowdfunding
2. Velesevitch, S.: The whole world. Why there is no crowdfunding in real estate in Russia. https://sber.pro/publication/pochemu-v-rossii-net-kraudfandinga-v-nedvizhimosti
3. Amosov, A.O.: Advantages and disadvantages of the bitcoin cryptocurrency compared to fiduciary money. Econ **11**(32), 82–85 (2017)
4. Investments in industrial real estate with rental income. https://pnkrental.ru/
5. Collective investment in commercial real estate. https://www.aktivo.ru/
6. KupiVIP founder launches a crowdfunding project for real estate investments. https://aktivo.ru/blog/sf-crowdfunding-project
7. The founder of "Morton" Alexander Ruchev invests in commercial real estate. https://www.vedomosti.ru/realty/articles/2017/03/21/681975-osnovatel-mortona-investiruet-nedvizhimost
8. Federal Law "On Attracting Investments Using Investment Platforms and on Amendments to Certain Legislative Acts of the Russian Federation" dated 02.08.2019 No. 259-FZ. https://www.consultant.ru/document/cons_doc_LAW_330652/

Optimization of Tax Payments and Their Accounting in the Accounting and Analytical System of the Corporation

Galina E. Krokhicheva[1](✉), Evgeny A. Zhabin[2], Irina V. Koltyreva[3,4], and Sun Shutsyuan[1]

[1] Don State Technical University, Rostov-on-Don, Russia
[2] Southern Federal University, Rostov-on-Don, Russia
[3] Interdistrict Inspectorate of the Federal Tax Service No. 23 for the Rostov-on-Don Region, Rostov-on-Don, Russia
[4] Rostov State University of Economics, Rostov-on-Don, Russia

Abstract. Optimization of the tax burden includes the reduction of tax liabilities as part of the taxpayer's applying instruments enshrined in the existing law. One of such tools for reducing the tax burden is accounting policy. The relevance of this topic lies in the improvement of tax regulation tools and the use of new information technologies in the area of accounting and economic policy in view of the tax optimization. As the methodology, the following research methods were used: theoretical (analysis, synthesis and induction), empirical (analysis of the volume of non-cash transactions according to the Federal Tax Service reporting data and interpretation of expert opinions on the issue), methods for the graphical use of data. The research is aimed at considering the main areas of tax optimization. An accounting policy, as a tool for optimizing taxation, can be effective in view of the current legislative and regulatory concepts, which allow for several accounting methods, the use of various methods and approaches, permission for an economic entity to independently develop accounting methods, the variable choice of accounting elements and no strict regulation of possible accounting methods, which provides ample opportunities to reduce the tax burden for an entity. The scientific research outcomes obtained as a result of the research are aimed at supplementing the theoretical basis of economic disciplines related to financial management at an enterprise and accounting policy arrangements.

Keywords: Accounting policy · Taxation · Optimization · Tax liabilities · Economic entity · Tax policy · Accounting · Economic activity · Economic effect · Modeling

1 Introduction

In economic science, the term tax optimization has been embodied as a reduction in the scope of tax liabilities in connection with the lawful behavior of a taxpayer, which primarily concerns the fulfillment of the prescriptions established by tax legislation. Let us take a closer look at the tax optimization components of contemporary entities:

P. V. Trifonov and M. V. Charaeva (Eds.): DITEM 2021, LNNS 380, pp. 608–612, 2022.
https://doi.org/10.1007/978-3-030-94245-8_83

- fulfillment of tax obligations;
- control over the assessment of taxes and fees;
- timely payment of taxes to the state treasury;
- some others.

At the same time, it should be noted that in the taxation optimization there are tools that allow you to effectively implement it, so one of the most important tools is an accounting policy, the observance of which allows an entity not to violate tax legislation [1].

Today, against the backdrop of the information technology development, there is an improvement in accounting policy, which determines the relevance of the research into the stated accounting policy implementation problem. Indeed, due to competent approaches to accounting, system analysis and risk management, as well as the rationality of management decisions, they can reduce costs and optimize taxation. All this together enables entities to improve their accounting policies.

Today, the organization and improvement of accounting policy are considered by not only economic science but are also vital for research in such areas as jurisprudence, public and municipal administration, and at the same time become important for practitioners including employees of the Russian tax service. In support of this, we will give an illustrative example that the subjects of civil law, especially entrepreneurs and legal entities (that means businesses), at the time of drawing up a civil law transaction should focus on not only the provisions of civil law but also specific tax consequences that may occur in the event of such a transaction. In turn, legal practitioners should also conduct an examination of the transaction for the implementation of tax law by taking into account the existing provisions of the Tax Code of the Russian Federation [2].

In considering the accounting policy issue, we consider it necessary to explore theoretical approaches to the problem posed. In examining the accounting policy, O.V. Nagovitsyna notes that the business entities need to analyze the specifics of their activities in order to shape successful management decisions, which will determine the accounting policy of the entities. To complement this approach, A.A. Zhidikin notes that it is necessary to find the best option for an enterprise; this will subsequently create a competent accounting policy.

We also note the theses of A.O. Volokhov, N.N. Makarova and O.V. Shakina, who also consider the effective tools for accounting policy, in particular, by calling the optimization of taxation for individual accounting items of the organizational and legal forms of firms.

In addition, it should be noted that tax optimization and its accounting policy tool often become a topic for research in scientific articles, monographs and other educational literature.

2 Results

As already noted, contemporary regulatory instruments of the Russian Federation provide some tax benefits for the entities that makes it possible to reduce the tax burden. As such regulatory instruments, the Tax Code of the Russian Federation, Federal Law No. 402 "Accounting" and a number of bylaws, including instruments of the Russian

Ministry of Finance, can be singled out. The said regulatory documents provide economic entities with preferential conditions for paying taxes, or tax exemptions and other concessions. That is why it is very difficult to consider all the tax optimization tools in a unified manner; therefore, it was decided to study tax optimization based on drafting the accounting policy of enterprises [3].

First of all, in economic science, it is enshrined that accounting policy is a significant component of accounting for all business transactions; in addition, accounting policy allows in practice to use the methods that an enterprise needs to account for its own financial activities. It is due to the developed accounting policy rules that accounting can be implemented.

Analysis of clause 2 of the Russian Accounting Regulations PBU No. 1/2008 consolidates the concept of accounting policy, which means the following: the totality and form of accounting, including primary observation, value measurement, implementation of stock-taking and workflow, application of accounts, and assessment of an entity's activities.

In the continuation of this provision, it is important to note that since January 1, 2008, the Russian Federation has ratified an international agreement in accordance with which it pledged to apply the provisions of the International Financial Reporting Standards. In particular, it is necessary to note the eighth part of the specified standard, which regulates the concept of accounting policy, changes in accounting estimates and errors. This provision of the standard has become necessary for entities that have already begun to develop their accounting policies based on the international standard. For example, the eighth section of the International Standard defines accounting policies as follows: specific principles, frameworks, agreements, rules and practices adopted by an enterprise for the preparation and presentation of financial statements [4].

Modern Russian legislation allows economic entities several optimal ways to represent their accounting, which is important for the enterprises themselves, as it allows them to choose, and it is also effective for optimizing taxation. Indeed, due to the choice given to entities, businesses can optimize taxation, which is also important for determining the income of an entity.

Let us consider an example: some real estate can be leased out by any entity, regardless of its type of activity, which means that lease relations can be in any tax treatment and will be qualified as other income of the entity. Due to various instruments, a company can independently determine in the accounting policy its properties with a useful life within a year (12 months) by taking it into account as part of fixed assets. It should be noted that it is also possible to calculate depreciation for various groups of means of production using different methods, but it should be remembered that after an item is put into operation, it will be impossible to change the method for calculating depreciation. Thus, an entity can choose a single method for calculating depreciation for all items, but also distribute the methods its own method of calculation per each individual item, this proves the effective optimization of taxation.

Based on the provisions of the Russian Accounting Regulations (PBU), it is important to note that each enterprise independently organizes and applies accounting methods; in addition, an enterprise should take into account the current domestic legal regulation. For example, the analysis of PBU No. 6/015 allows us to conclude that an entity can independently establish a cost criterion for inventory (hereinafter referred to as inventory). At the same time, in 2000, the Order of the Russian Ministry of Finance stated that the entry of debit record 10 "Materials" was not provided, and code 08 "Investments in non-current assets". Despite this, it is sometimes necessary for entities to have such accounting entries. A striking example can be two sets of furniture, where their total cost was 100,000 rubles (50,000 each, respectively) excluding VAT, and the assembly of those is also provided at a cost of 5% of the cost of furniture sets (that is, 5,000 for each set) [5].

It is also important to note that in the policy of an enterprise, it is possible to mark individual sub-accounts on account 10 "Materials", at the same time fix the debit record of account 10 – credit record of account 08. So, when implementing such an accounting policy, it can be noted that the acceptance of fixed assets that the initial cost of up to 50,000 rubles, is recorded by posting on debit account 10 and credit 08. It is for this reason that the definition in the accounting policy of attributing tax items to fixed assets, or to inventories is an important aspect.

Of course, when drafting an accounting policy, each company should independently mark items of taxation, which should correspond to not only its economic interests but also the regulatory framework of the Russian Federation. When giving an example for a manufacturer it should be noted that it needs to focus on the location of construction-in-progress objects in a certain line of expenses, and also there is a variability in the definition and assignment of completed construction objects. For a commercial undertaking, it is necessary to make careful accounting for the purchase and write-off of goods.

Optimization of taxation is an important element for the formation of the accounting policy of an enterprise, based on specific ideas what plans this enterprise has and what strategies are being implemented at the moment; and it is also necessary to take into account what goals this entity has set for itself. That is why it is important, when drafting an accounting policy, an entity should take into account not only immediate wants, but this will not lead to the achievement of its goals for itself. That is why an entity as a taxpayer should take into account long-term goals in order to understand financial resources, strengthening competitiveness.

3 Conclusion

When drafting an accounting policy in order to optimize taxation, it is necessary to apply the following approaches:

- contemporary Russian legislation enables an enterprise to implement several accounting methods, that implement various methods and approaches, which allows you to save tax deductions by modeling the accounting policy of a business (entity);
- there are issues in respect of which no regulatory methods of accounting are established;
- accounting policy should include some important elements that can be varied, however, this is directly provided for by regulatory documents.

Thus, the implementation of these approaches allows economic entities, despite the difference in activities, organizational and legal forms and scales, to shape their own accounting policies, which will optimize taxation in accordance with the current legislation.

References

1. Alekseeva, I.V., Evstafieva, E.M., Charaeva, M.V., Mosentseva, V.A.: Trajectories of the internal control system in the management of agricultural organizations. Int. J. Econ. Bus. Adm. **VII**(Special Issue 1), 521–534 (2019) https://www.ijeba.com/journal/298
2. Aleshko, R., Petrova, L., Ivanova, E., Plotnikova, A., Melnikov, M., Antonov, V.: Human capital in the digital economy format. Int. J. Eng. Adv. Technol. **9**(1), 7517–7523 (2019). https://doi.org/10.35940/ijeat.A2201.109119
3. Zotova, A.S., Kandrashina, E.A., Ivliev, A.D., Charaeva, M.V.: Evaluation methods basis of strategy development effectiveness of the enterprise. Int. J. Environ. Sci. Educ. (IJESE). **14** (2016) http://www.ijese.net/makale/954
4. Evstafyeva, E.M., Charaeva, M.V., Krokhicheva, G.E., Paleev, A.V.: Business processes' reengineering in the context of increasing the firm's value. Eur. Res. Stud. J. **XXI**(Special Issue 1), 125–135 (2018) https://www.ersj.eu/index.php?option=com_content&task=view&id=1212
5. Voronkova, O.Y., Petrova, S.N., Ivanova, E.A., Klochko, E.N., Logvina, E.V., Penkova, A.N.: Assessment of the regional potential of recreation and health-promoting natural resources. J. Exp. Biol. Agric. Sci. **8**(Special Issue 2), S314–S321 (2020). https://doi.org/10.18006/2020.8(Spl-2-AABAS).S314.S321

Development of the Audit Services Market in Russia

Elena N. Makarenko, Nikolay T. Labyntsev, Natalia A. Andreeva(✉), and Tatiana V. Makarenko

Rostov State University of Economics, Rostov-on-Don, Russia
rector@rsue.ru

Abstract. Improving the conceptual framework for the provision of audit services in crisis conditions caused by the impact of the COVID-19 pandemic should be aimed at expanding both the subject of audit activities and diversification of the services provided by audit organizations and individual auditors. The implementation of measures in this area will be able to provide reliable and reliable information about the activities of economic entities for making effective management decisions. The purpose of the study was to identify directions for the development of the audit services market in the Russian Federation, to substantiate the feasibility of increasing the digital component in auditing and consulting, as well as to further improve the legislative framework for auditing. To achieve this goal, the following methods of scientific knowledge were used: observation, abstract-logical, historical, comparison, systematization of theoretical and practical material, methods of generalizing statistical information, including using information resources of the global Internet and modeling. Based on the results of the study, the following results were obtained: the main trends in the development of the modern Russian market of audit services were determined, the need to strengthen the digital component in the activities of individual auditors and audit companies in post-pandemic conditions was substantiated, as well as the author's model of audit in the digital economy was proposed, the objective need to improve the legislative basis of the sphere of audit activities and diversification of audit services by expanding the practice of in-demand tasks was established.

Keywords: Audit · Audit activity · Audit services market · Audit examination

1 Introduction

The market economy is characterized by the functioning of numerous economic entities, united by numerous financial and economic ties, therefore, one of the necessary conditions for their effective functioning is reliable information about each other's activities, which is necessary for making correct and informed management decisions. The source of such information for various groups of interested users is the organization's reporting, the formation of which is the final stage in accounting. To ensure user confidence in the information contained in the accounting (financial) statements, an audit is used, the purpose of which is to express an opinion on its reliability. Moreover, the need to

P. V. Trifonov and M. V. Charaeva (Eds.): DITEM 2021, LNNS 380, pp. 613–624, 2022.
https://doi.org/10.1007/978-3-030-94245-8_84

obtain reliable information about the activities of an economic entity arises not only from external users of financial statements, primarily from partners of the company, investors and creditors, to assess the prospects for cooperation, but also from its internal users in the person of shareholders and employees of the organization to assess the value and the stability of their income. An audit of the reliability and reliability of reporting information is the only correct solution for all groups of its users. That is why this institution, having emerged several centuries ago, does not lose its relevance to the present day.

The Government of the Russian Federation is pursuing a course towards improving the system of ensuring information transparency of the economic activities of organizations. This is evidenced by the application of International Standards on Auditing (ISA) in Russia from January 1, 2017, which is due to the need to increase confidence in the audit results on the part of users of financial statements and society as a whole, as well as the concentration of efforts on the riskiest areas where there is a possibility of significant distortion of information in the financial statements.

Scientists from all over the world have been and continue to be engaged in the development of auditing. A significant contribution to the solution of these aspects in the field of auditing has been made by such foreign and Russian scientists as Toropova [1], Kutera and Soliman [2], Kizilov and Bogataya [3], Miroshnichenko [4], Bezdolnaya and Malakhova [5], Dvoretskaya [6], Shvyreva, Kruglyak and Petukh [7, 8], Alekseeva and Evstafieva [9], Werner and Gehrke [10, 11], Manita et al. [12], etc.

However, despite the significant level of development of the audit field, in practice, the types of audit services and the requirements for the results and conditions of the audit are entirely dependent on the realities of the market economy, therefore, the actualization of unexplored aspects in the form of new challenges due to the strengthening of the digital component of economic processes, as well as local crisis situations generated by various factors, one of which was coronavirus infection. The new economic realities not only did not weaken the position of the audit, but even more indicated their relevance, and not so much in the audit itself of the reliability of the reporting information, as in the services associated with the audit and consulting.

The working hypothesis of this study is based on the assumption that responding to the new challenges caused primarily by the COVID-19 pandemic and the introduction of the self-isolation regime, the only possible option for the development of the audit services market is to strengthen the legislative and digital determinants, while the current legislation in the field of providing audit services will increase the transparency of auditors' activities and trust in them from stakeholders, and a high degree of automation and digitalization will increase the value of audit services by reducing the time factor of their provision, as well as increase the efficiency and quality of the audit due to the ability to cover all operations in the audit evidence collection process, not just within the audit sample.

The purpose of this study is to study the state and identify directions for the development of the audit services market in the Russian Federation, substantiate the feasibility of increasing the digital component in the process of conducting an audit and consulting, as well as further improving the legislative framework of auditing.

2 Methods

To obtain new generalized data on the state and development prospects of the audit services market in Russia, we used such research methods as observation, abstract-logical, historical, comparison, systematization of theoretical and practical material, methods of generalizing statistical information, including the use of information resources of the global Internet and modeling.

In the process of applying the above-mentioned research methods, critically scientific and special literature was analyzed and practice-oriented material was summarized regarding the historical aspects of the emergence and formation of audit activity in the Russian Federation, the current state of the audit services market, highlighting the main problems and trends of its further development, including simulation of an audit using IT technologies.

The study is based on statistical information, databases of the Ministry of Finance of the Russian Federation and analytical studies of the RAEX agency (RAEX-Analytica), as well as on previously published results of studies by domestic and foreign authors in this subject area.

3 Results

The Russian market of audit services, which arose in the process of the formation of market relations, in contrast to the international market, has only about 30 years in its development. However, in spite of its young age, the Russian market of audit services is a well-functioning structure, which is based on a fairly effective system of legislative regulation and, most importantly, in accordance with international requirements, which is confirmed by the application of international audit standards in the Russian Federation and the transition from a rigid state regulation of audit activities to self-regulation [1]. In 2019, a significant event took place for the auditors of the Russian Federation, consisting in the creation and operation from 2020 of a single self-regulatory organization based on the SRO (self-regulatory organization) of the Auditing Association "Sodruzhestvo", which, in the opinion of practitioners in this area, will positively affect the activities of audit companies, so how to develop a unified strategy for the development of the audit services market [13].

In accordance with the concepts of audit and audit activity, defined by the Federal Law of 30.12.2008 No. 307-FZ "On Auditing", it becomes obvious that they are not identical [14], and the goals and objectives of audit activities are very multifaceted, and therefore the spectrum audit services provided by audit organizations on the market is quite extensive (Fig. 1).

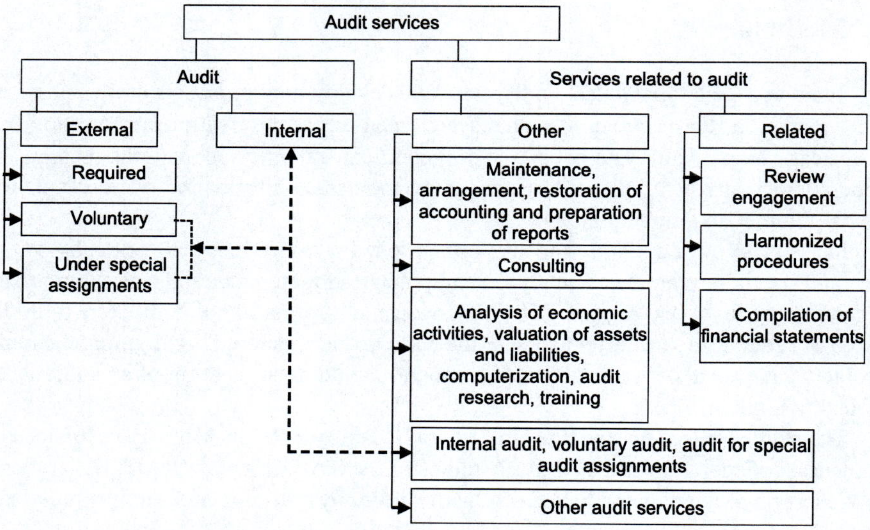

Fig. 1. General classification of audit services in the Russian Federation.

The main participants in the market of audit services in the Russian Federation are audit organizations and individual auditors who directly provide such services, and audited entities, legal entities and individual entrepreneurs, and regulatory authorities and regulatory structures represented by a single self-regulatory organization based on the SRO of Auditors of the Sodruzhestvo Association.

Currently, the profession of an auditor is quite in demand in the labor market, it is generally recognized that the services of individual auditors and audit companies will be in demand, but with the condition of expanding the range of competencies necessary for these specialists. According to official data, the dynamics of the quantitative composition of audit organizations and individual auditors is shown in Fig. 2.

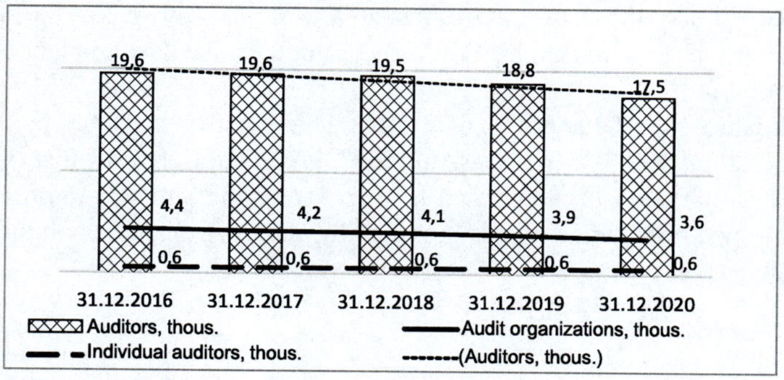

Fig. 2. Number of audit organizations and auditors in the Russian Federation for the period 2016–2020 [15].

It is necessary to note a noticeable reduction in the scale of the audit market in Russia over the past two years. The reason is undoubtedly the crisis situation caused by the COVID-19 pandemic, which negatively affected the activities of users of audit services in the direction of reducing their number and solvency, which led to withdrawal from the market and takeovers among audit companies.

In addition, the number of auditors has also steadily declined during the period under review and especially over the period 2019–2020. This happened, firstly, in connection with a decrease in the number of audit companies, and, secondly, in connection with an increase in the qualification requirements for the auditor profession (availability of an auditor's qualification certificate, membership in a self-regulatory organization).

For a more in-depth analysis of the state of the audit services market, we will diagnose the income of audit organizations in the Russian Federation for the period 2017–2020 in the form of a histogram in Fig. 3.

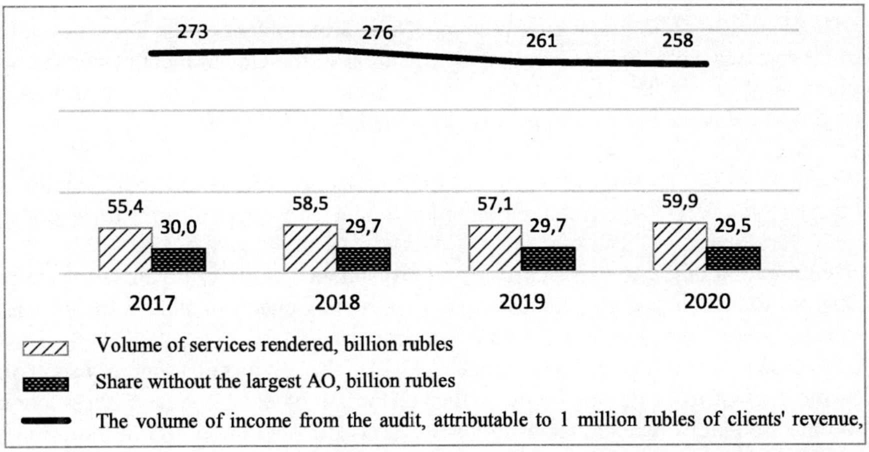

Fig. 3. Dynamics of income of audit organizations in the Russian Federation for the period 2017–2020 [15].

Assessing the dynamics of revenues, we note, on the one hand, the receipt of stable volumes of services provided by audit organizations for the period 2017–2020, and on the other hand, the reduction of audit revenues per 1 million rubles of client revenue at audit companies, especially in the last two years of the period under review. The current trend is explained as an objective price dumping of audit services due to a decrease in the solvency of clients operating in crisis conditions, which have developed under the influence of the threat of the spread of a new coronavirus infection.

The resulting final analytical conclusions are confirmed by the dynamics of the distribution of income of audit organizations, shown in Fig. 4, which indicates a decrease in the share of auditors' income from conducting an initiative audit due to a reduction in the costs of clients faced with harsh conditions of survival due to the risk of uncertainty in the economic environment.

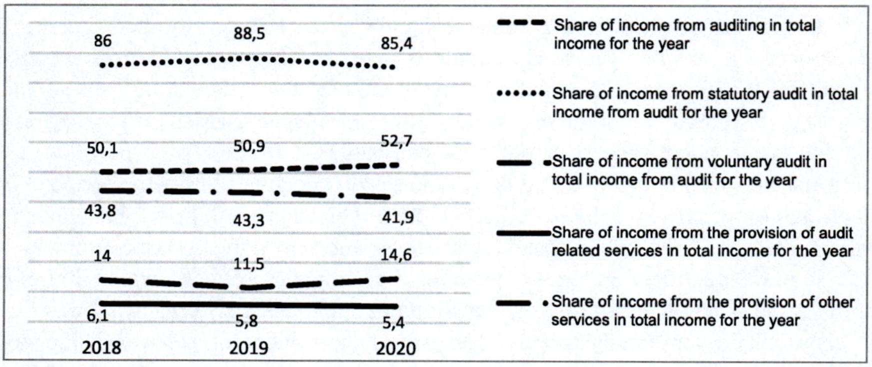

Fig. 4. Distribution of income of audit organizations in the Russian Federation for the period 2018–2020 (in %) [15].

Thus, let us outline the main problems of the development of the Russian market of audit services, noted in the approved by the order of the Government of the Russian Federation of December 31, 2020 No. 3709-r "Concept for the development of audit activities in the Russian Federation until 2024" consisting of [16]:

1 Decrease in the clients' solvency due to the crisis conditions of their functioning caused by COVID-19 and their reorientation to audit companies offering their services at lower prices;
2 Deterioration of the business climate and investment activity of organizations in the market as a whole, leading to the departure of existing clients of audit companies and the outflow of new ones in the field of proactive audit;
3 Decrease in the volume of audit services related to the conduct of a statutory audit due to the increase in the criteria for its conduct on the part of legislators, so the restrictions on the amount of revenue received were increased from 400 to 800 million rubles, and in terms of asset value – from 60 to 400 million rubles;
4 Underestimation of the services of auditors by business entities and low interest in them from the state.

To overcome the identified problems and increase the importance of the audit institution in accordance with the approved by the order of the Government of the Russian Federation of December 31, 2020 No. 3709-r "Concept for the development of audit activities in the Russian Federation until 2024", the following main trends in the development of the audit services market can be formulated, presented in Fig. 5 [16].

In continuation of the measures outlined in the Concept for the development of the audit services market in relation to expanding the format of the audit report, the Ministry of Finance of the Russian Federation in the Recommendations to the subjects of audit activity regarding the audit of the annual accounting (financial) and consolidated financial statements of organizations for 2020 in the context of the spread of the new coronavirus infection, approved by The Audit Council on December 18, 2020, Minutes No. 56, draws the attention of the auditors to the need to assess, among other things, the legality of the management of the audited entity's use of the going concern assumption

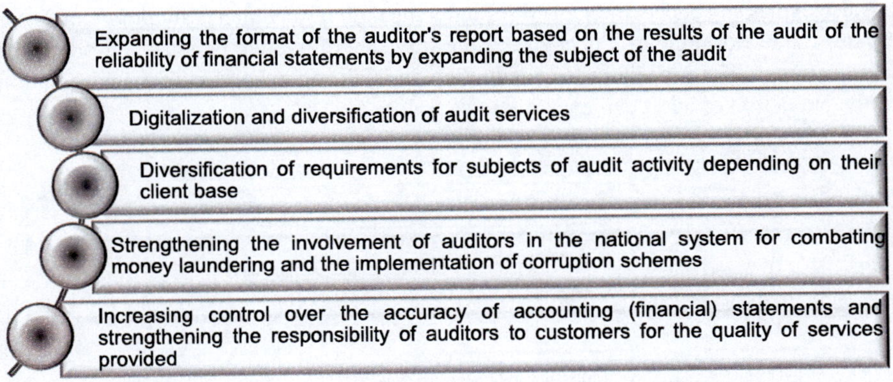

Fig. 5. Main trends in the development of the audit services market in Russia.

under the influence of COVID-19 conditions in accordance with the requirements of ISA 570 Going Concern (Revised) [17].

We will focus on the statement that despite the outlined decrease in demand for auditor services in the field of proactive audit in its classical form, strengthening of fiscal control and its digitalization will act as a catalyst in the process of increasing demand for proactive tax audit. The second factor contributing to the growth of the needs of economic entities in tax audit is the adoption and implementation of new Federal Accounting Standards in accounting practice, in connection with which the likelihood of errors that can have negative tax consequences for organizations increases and the value of audit services that contribute to their prevention becomes especially valuable for clients of auditors and audit organizations.

In order to increase confidence in the audit market, its transparency and suppression of illegal activities of audit organizations, the adoption of measures to strengthen control, in accordance with the adopted Concept in Russia, will follow the path of increasing the responsibility of audit organizations to customers and other interested parties, as well as consolidation at the legislative level the need for audit organizations to disclose information about their activities on the relevant official websites, especially when it comes to their relationship with socially significant economic entities.

Attention is paid to these issues in foreign practice, as Kutera believes that it is necessary to provide a legislative basis for the need to change audit companies by audited persons to reduce the corruption component, the basis of which lies in the establishment of closer ties between auditors and their clients in the provision of services related to audit and consulting, as a result of which the independence of auditors is lost. According to this researcher, such a rotation of auditors will help to increase the reliability of the results of the audit [2].

In the context of the digitalization of the economy, special attention in the development of the audit services market will be paid to the processes of transformation of auditor services in the digital environment. Only the mastery of digital tools by auditors can ensure their relevance in modern realities. The introduction of IT technologies into the practice of providing audit services will allow not only preserving their entire range, but also expanding and, at the same time, reducing their cost, which is an important factor against the background of a decrease in the solvency of audited persons and prices for audit services. The procedure for conducting an audit in the context of the automation of the collection of audit evidence is shown in Fig. 6.

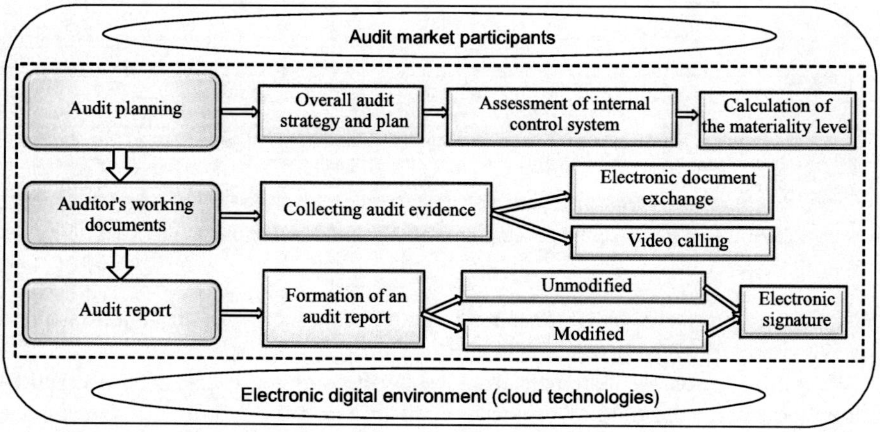

Fig. 6. Model of an audit in the context of digitalization.

Automation of procedures for collecting audit evidence in the process of conducting an audit can be implemented on the basis of programs used in the practice of audit companies, such as AIS "Audit", IT Audit, ExpressAudit, AuditXP, 1C:Mobile, EZ-R Stats for Windows, as well as cloud technologies including blockchain technology.

Thus, the proposed author's model of audit in the context of digitalization with the use of electronic data analysis technologies that increase the intensification of the process of providing audit services and consulting is relevant not only in the context of the self-isolation regime associated with the COVID-19 pandemic, which was a major challenge that business entities encountered en masse in the first half of 2020 and a certain part of them is still facing today, but also in terms of optimizing the costs of auditing activities and, consequently, the possibility of making auditing services cheaper and more accessible to a wide range of users even in the general reduction of solvency of the auditing companies' clients.

In addition, increasing the importance of IT technologies in the provision of audit services in the digital environment of the modern economy is a solution that will eliminate the contradiction, consisting, on the one hand, in the need to optimize and minimize the costs of organizations in crisis conditions of their functioning, and, on the other hand, in increasing the request for obtaining reliable information about their own activities and the activities of partners with confirmation of the possibility of the continuity of their activities in the near future, due to the presence of a high degree of uncertainty in the market environment, which is potentially possible through the development of the audit services market.

4 Discussion

A lot of scientific and practical research and analytical developments by leading Russian and foreign scientists and practitioners are currently devoted to the development of the audit services market. All of them had one goal – to investigate the importance of the profession of an auditor, to identify problems in the provision of high-quality audit services and to propose directions for the development of audit activities in order to increase confidence in its results on the part of customers of audit services and stakeholders.

Turning to the issue of substantiating the importance of audit, many scientists draw their attention to the historical stages of its emergence and development, thus emphasizing the invariably high degree of relevance of auditing throughout the history of mankind, from ancient times to modern days. So, according to Kizilov and Bogataya, the prehistory of the emergence and development of audit dates back to Ancient India. In the 3rd century B.C. the political and economic treatise Arthashastra mentioned the keeping of accounts, which contained provisions regulating the keeping of accounts in the accounting office, which in ancient India was headed by a supervisor, under whose authority were the accountants [3]. The system of control over the expenditure of state funds was also developed in Ancient Rome and Ancient Greece. The above statements confirm and testify to humanity's awareness of the need to control the movement of financial resources in the ancient world.

The term "audit" itself, which in Latin means hearing, appeared in Great Britain in 1825, and here in the 18–19 centuries the modern international market of audit services began to form, the impetus for the emergence of which was the industrial revolution, which led to the emergence of numerous companies, the reliability of the reports of which had to be checked and confirmed.

In her research, Toropova established that in Russia the first mention of auditing dates back to 1711, when Peter I issued a decree establishing the post of auditor, which at that time belonged to the military ranks, along with quartermaster, adjutant and reviewer. Auditors were called upon to resolve property disputes, often combining the positions of secretary, clerk, investigator and prosecutor [1].

We state that the audit in Russia has been mentioned since the 18th century, but the audit services market in its modern sense is very young in comparison with the foreign world, and in its development, according to Miroshnichenko, has gone through three main stages. The first – from 1987 to 1993, characterized by the spontaneous emergence of auditing. The second – from December 1993 to 2001, when the activities of auditors were based on the temporary rules of auditing. And the third – from August 2001 to the present, is marked by the adoption of the Federal Law dated 07.08.2001 No. 119-FZ, which was replaced by the law dated 30.12.2008 No. 307-FZ, and 34th FPSAD (FEDERAL RULES (STANDARDS) OF AUDITING ACTIVITIES), which since 2017 replaced by ISA, in connection with which it can be argued that the Russian market of audit services in its current state meets international requirements [4].

Over the past five years, especially during 2019–2020, the Russian as well as the international market for audit services has been characterized by similar problems associated with reforming legislation in order to improve the quality of audit services and trust in auditors from interested users, on the one hand, and mastering auditors with professional digital tools, on the other. A particularly problematic field entered a phase of exacerbation during the COVID-19 pandemic, which is confirmed by studies by Russian and foreign scientists such as Kutera and Soliman [2], Bezolnaya and Malakhova [5], Dvoretskaya [6], Shvyreva, Kruglyak and Petukh [7, 8], Alekseeva and Evstafieva [9] and leading practitioners – Borodina [18], Shapiguzov [19]. Due to the increased likelihood of unreliability of the reporting information of economic entities operating in conditions of an extreme degree of uncertainty. To solve the problems of ensuring the use of IT technologies, it is necessary to improve audit standards in order to expand the format of the auditor's report, including, in addition to the results of standard audit procedures, also the result of assessing the risks of going concern of organizations, as well as conducting audits and providing other audit services offline, or becomes impossible, or is reduced to a minimum, which necessitates the active implementation of IT technologies in the work of auditors and audit companies and their transition to online mode.

Note that the area of scientific research presented by Werner and Gehrke [10, 11], Manita et al. [12] puts emphasis on the need for auditors to use cloud technologies, blockchain technologies, including process mining of audit evidence, which will make it possible to cover the entire set of operations in the process of assessing the internal control system of the audited entity, and not only their part in the redistribution of the audit sample, will increase efficiency audit and will ensure a high degree of reliability of the auditors' conclusions and will add credibility to the quality of the audit services provided by customers and stakeholders.

5 Conclusion

So, in the process of analyzing the state of the audit services market, it was established that the key factors of its development will be:

- improving the legal framework and auditing standards in order to increase the transparency of the relationship between auditors and audit organizations with audited persons, which will increase confidence in the results of the audit. An important positive point is the fact that legislators pay close attention to audit issues in the future, as the only possible tool for confirming the reliability of the reporting information of economic entities, expressed in the approval of the Concept for the development of audit activities until 2024 and an action plan for its implementation;
- expansion of the information component of the audit report, which would allow not only to draw a conclusion about the degree of confidence in the reporting information of the audited entity, but also contain an assessment of the likelihood of continuity of its activities in the next 12 months after the reporting date, which becomes a vital condition for making effective management decisions by stakeholders of this reporting, especially in the context of a high degree of instability of the modern economic environment. Legislators pay attention to this in the recommendations for conducting an audit in the context of the COVID-19 pandemic, but there is currently no standardization of the reflection of this information in reporting;
- the widespread introduction of IT technologies, as one of the directions of the development of the audit services market in the current crisis conditions of the functioning of economic entities, and not only in the process of consulting, but also

during audits, since practice has shown that in the conditions of self-isolation and its consequences – this is the only way to avoid the risk of receiving inaccurate information.

Thus, based on the results of the research conducted by the authors of the article on the improvement of the conceptual framework for the provision of audit services in crisis conditions caused by the impact of the COVID-19 pandemic, it can be concluded that the approaches to the format in the field of audit activities of audit organizations and individual auditors have objectively changed. In this regard, all the designated areas of development of the audit services market in Russia are in demand from the standpoint of improving the professional activities of audit organizations and individual auditors.

References

1. Toropova, I.S.: Audit in the Russian Federation: history of development and current state. Sci. Bul. Fin, Banks, Invest. **4**, 153–157 (2017)
2. Kutera, M., Soliman, K.S. (Eds.) Changes in the European audit services market. In: Sustainable Economic Development and Advancing Education Excellence in the Era of Global Pandemic. In: Proceedings of the 36th International Business Information Management Association Conference (IBIMA), pp. 1285–12944, 5 November 2020, Granada, Spain. IBIMA, Norristown (2020)
3. Kizilov, A.N., Bogataya, I.N.: Evolution of audit as a science and practical activity. Econ. Analysis: Theory Prac. **22**, 43–57 (2012)
4. Miroshnichenko, T.A.: Audit: Tutorial. Part 1. Fundamentals of Audit. Donskoy GAU, Novocherkassk (2018)
5. Bezdolnaya, T.Yu, Malakhova, T.Yu: Assessment of the effectiveness of audits in the context of digitalization of auditing. Bul. Inst. Friendship Peop. Caucasus. Econ. Sci. **1**(49), 146–151 (2019)
6. Dvoretskaya, V.V.: Digitalization in tax audit. Econ. Manag: Prob, Solut **8**(3), 62–66 (2019)
7. Shvyreva, O.I., Kruglyak, Z.I., Petukh, A.V.: Features of the auditor's report on the accuracy of the accounting (financial) statements for 2019: COVID-19 as qualitatively significant information (part 1). Int. Account. **23**(6(468)), 701–722 (2020). https://doi.org/10.24891/ia.23.6.701
8. Shvyreva, O.I., Kruglyak, Z.I., Petukh, A.V.: Features of the auditor's report on the accuracy of the accounting (financial) statements for 2019: COVID-19 as qualitatively significant information (part 1). Int. Account **23**(7(469)), 777–799 (2020). https://doi.org/10.24891/ia.23.7.777
9. Alekseeva, I.V., Evstafieva, E.M.: Theoretical and practical aspects of the formation of predictive financial statements and their audit in the context of digitalization. Bul. IPB **6**, 25–33 (2018)
10. Werner, M., Gehrke, N.: Identifying the absence of effective internal controls: an alternative approach for internal control audits. J. Inf. Sys. **33**(2), 205–222 (2019). https://doi.org/10.2308/isys-52112
11. Werner, M.: Financial process mining – Accounting data structure dependent control flow inference. Int. J. Account Inf. Sys. **25**, 57–80 (2017). https://doi.org/10.1016/j.accinf.2017.03.004
12. Manita, R., et al.: The digital transformation of external audit and its impact on corporate governance. Techn. Forecast Soc. Change **150**, 119751 (2020). https://doi.org/10.1016/j.techfore.2019.119751

13. Efremova, L.: Interview with project partners: "Russian audit". https://raex-a.ru/editions/Efremova/2020
14. Federal Law No. 307-FZ dated 30.12.2008 "On Auditing Activity" (as amended on 09.03.2021). http://www.consultant.ru/document/cons_doc_LAW_83311/
15. Key indicators of the audit services market in the Russian Federation in 2019–2020. https://minfin.gov.ru/ru/perfomance/audit/audit_stat/MainIndex/
16. Concept for the development of audit activity in the Russian Federation until 2024. https://www.garant.ru/products/ipo/prime/doc/400070454/
17. Clarifications and recommendations. http://www.consultant.ru/document/cons_doc_LAW_371705/
18. Borodina, N.: Interview with project partners: "Russian audit". https://raex-a.ru/editions/borodina/2020
19. Shapiguzov, S.: Interview with project partners: "Russian audit". https://raex-a.ru/editions/Shapiguzov/2020

Digitalization of Corporate Financial Management

Elena M. Evstafyeva[1](✉), Marina V. Charaeva[2], Sergey A. Konovalenko[3], and Yan Shentsyuan[4]

[1] Rostov State University of Economics, Rostov-on-Don, Russia
[2] Southern Federal University, Rostov-on-Don, Russia
[3] Ryazan Branch of V.Y. Kikot Moscow University of the Ministry of Internal Affairs of Russia, Ryazan, Russia
[4] Don State Technical University, Rostov-on-Don, Russia

Abstract. The article reveals the methodological aspects and key issues of financial management development in the digital economy. Digital financial management in corporate management system brings economic activity to a new effective level. Digitalization system allows corporate employees to perform their functions remotely. In this case the work process does not affect the quality of work and the results achieved. Digitalization also allows for financial oversight of the corporation's financial resources, for example, as part of tools to improve the effectiveness of internal controls or the "e-Budget" financial management system. Numerous works of Russian and foreign scientists confirm the increasing effect of modern technologies and digitalization processes on the financial management of companies. This effect is referred to as synergistic. The purpose of this article was to study the positive and negative effects on financial management of the digital economy. Theoretical and methodological basis of research were the works of both Russian and foreign specialists in the field of management and administrative activities, as well as periodicals of leading academic economists. The main methods of research were the method of theoretical system analysis, monographic (descriptive) method.

Keywords: Digitalization · Financial management · Informatization · Digital infrastructure · Digital transformation

1 Introduction

The strategic direction of the modern concept of financial management has become the creation of smart enterprises and a new type of management mechanism. Informatization makes it possible to increase labor efficiency on the basis of the intensive factor of the available resource base, to increase the managerial efficiency of decisions, to diversify the existing horizontal production and management connections. The digitization of production expands the possibilities of effective management. This includes fast results and growth of investments in production, expansion of sales channels, new forms of payments, the possibility of forming production clusters remotely. As part of digitalization,

P. V. Trifonov and M. V. Charaeva (Eds.): DITEM 2021, LNNS 380, pp. 625–631, 2022.
https://doi.org/10.1007/978-3-030-94245-8_85

the problem of using relevant information and new knowledge has become particularly important. A very important attribute of the digital program for the development of new technologies is the control of the life cycle of wealth creation, the organization of new forms of payments and payments in virtual space. Under this model of financial management, the development of digital infrastructure and the digital environment become the key vectors of its development [1].

Modern scientific literature highlights both positive and negative phenomena associated with the digitalization of financial management. New digital technologies lead to changes in:

1) Economic phenomena, processes and their dynamics;
2) Emergence of new forms of payments and means of payment (payment technology with cryptocurrencies, electronic money, etc.);
3) Formation of electronic budgets, development of electronic services and trading platforms;
4) Institutional and structural changes in the digital economy.
5) Introduction of innovative technologies in management levels out the human factor and increases automation of the production process and, as a consequence, requires a qualitatively new workforce. It requires higher engineering education, education in IT-technology, availability of modern economic knowledge, programming skills [2].
6) On the one hand, management technologies act as a rule (having a compelling force for agents). On the other hand, they lead to the transformation of already existing standards, rules, norms.
7) Growth of cybercrime, namely information and computer crimes associated primarily with the theft of financial resources from electronic accounts and wallets, loss of funds in the form of cryptocurrency;
8) Increase in the costs required to protect information and information resources;
9) Growth of additional costs aimed at preventing crimes on the Internet, namely legal changes, cryptocurrencies and standards;
10) Economic agents experience additional stress and effort due to constant changes and developments in information technology, which puts a serious burden on them;
11) Information, new knowledge, and information technology become a product. One thing this "triad" has in common is their rapid obsolescence over time. A review of the economic literature on the reasons for the digital transformation of financial management reveals:

   - Improvement and development of information infrastructure. This position requires the use of digital technology to increase productivity and speed of dissemination of knowledge and information;
   - Increasing the technological level of specific subsystems of the digital infrastructure;
   - Digitalization leads to serious social consequences, which in turn causes changes in the behavior of economic agents [3].

2 Materials and Methods

As a significant methodological problem of financial management in the digitalization of the economy is to identify further ways of digital transformation of financial management. These include:

1. Availability of 80% of own funds necessary to implement modern digital technology [4].
2. The introduction of digital technology [5].
3. New digital technology increases productivity by about 15% or more. It is new compared to its previous counterpart, while it is not an improved innovation [6].
4. The new digital technology increases the amount of information by about 5% of the previous amount of information that was with the previous technology [7].
5. The new digital technology should cover the value of current needs, not exceed it [8].
6. The new digital technology should use existing information platforms that economic agents are accustomed to and do not require additional funds to purchase [9].
7. The new digital technology must be clear and easy to use, and adaptation to its use must be as quick and low-cost as possible [10].
8. The new digital technology must be implemented with the modern development of information systems [3].

Based on this comes the understanding that in today's environment management must implement new tools and technologies, so T. Copeland and A. Dolgoff introduced the concept of financial management based on expectations. The following features of financial management based on expectations can be distinguished:

- There is a correlation between the growth of the company's value and market expectations;
- The correct methodology for assessing the effectiveness of the company increases the chances of improving its performance and sustainability. At the same time, it must allow for control over the creation of value for the owners;
- Two key elements for top managers are forecasting and assessing expectations;
- Top managers are primarily investor-oriented;
- Companies and management on the basis of expectations are more aimed at establishing the most effective exchange of information, especially with potential investors;
- Performance standards are developed taking into account the opinion of counterparties [11];
- Continuous improvement of the company's planning and budgeting system [12];
- Financial incentives for employees are closely related to the performance of the company using expectation-based management. This tool allows you to improve the compensation model [13];
- The company knows the mechanism of formation of the market rate of its shares and understands market expectations [14].

Four areas will be affected by the digital transformation of financial management in companies:

- Social sector;
- Information changes;
- Management system;
- Innovation sector [15].

Thus, we can formulate and describe the author's approach to identifying the positive and negative effects of digital transformation in relation to financial management (Table 1).

Table 1. Positive and negative effects of digital transformation.

Consequences	Characteristics of effect
Increase in labor efficiency	Positive effect
Organizational and managerial effectiveness	Positive effect
Reducing the labor intensity of human labor in the volume of output	Positive effect
The emergence of new means of payment and forms of settlement	Positive effect
Expansion of trading platforms, sales markets, and logistics capabilities	Positive effect
Effective management on the basis of replication and, as a consequence, obtaining fast results	Positive effect
Improving the comfort of life	Positive effect
Additional time from employees for personal matters	Positive effect
Transactional productivity	Positive effect
The speed of dissemination of knowledge and information becomes faster	Positive effect
Increased unemployment, increased competition for a place in the company	Negative effect
Increased requirements for employees, the need for higher engineering education, education in IT-technology, the availability of modern economic knowledge, programming skills, etc.	Negative effect
Additional costs for employee training, training becomes continuous throughout the working life	Negative effect
Constant change in technology reduces the sustainability of business units and economic agents	Negative effect
Cybercrime and the additional cost of preventing such crimes	Negative effect

(*continued*)

Table 1. (*continued*)

Consequences	Characteristics of effect
Social consequences, which include: opportunistic behavioral patterns of economic agents, emotional and physical stress, high dependence on social networks and information technology, increasing financial and information inequality	Negative effect
The workforce has the following characteristics: poor physical and moral health, detachment from real life, loneliness, increased apathy and ambition	Negative effect
Additional costs for information security	Negative effect
Growth of "Big Data" information, difficulties in processing and additional processing costs	Negative effect
Information noise, duplication of information, duplication, presence of "junk" information	Negative effect
Decrease in the intellectual level of economic agents	Negative effect
Increase in the legal framework and, consequently, the number of court cases and proceedings, increased transaction costs and reduced innovation potential	Negative effect
Increase in the legal framework and, consequently, the number of court cases and proceedings, increased transaction costs and reduced innovation potential	Negative effect
Building up the debt economy for companies	Negative effect
Speculation in the IT sector, the emergence of mortgage and technology bubbles	Negative effect
Rapid digital obsolescence	Negative effect
A New Model of Behavior of Economic Agents called "Irrational Optimism"	Negative effect
Increasing financial and digital inequality	Negative effect

Source: Compiled by the authors.

3 Results

The modern concept of financial management is based on the digital transformation of economic activity of a modern company. The transition to a new technological paradigm provides both positive changes in the company's activities and forms a number of key problems. From the author's point of view, the key methodological problems of digitalization of financial management in the foreseeable future may be the following:

Firstly, the difficulty of extracting relevant information from the flow of information.

Secondly, there is a gap between equity and debt capital of companies, a constant shortage of own funds for the economic activity of organizations.

Third, the threat of cybersecurity and complete zeroing of company financial resources in the digital space.

Fourthly, the difficulty of recruiting staff, high requirements in terms of engineering education, education in IT-technology, the availability of modern economic knowledge, programming skills, etc.

Fifth, a significant reduction in the life cycle of companies. Digitalization and poor management decisions in financial markets and trading venues can lead to instant bankruptcy and insurmountable financial instability. The problems of digitalization of financial management are largely offset by competitive advantages and the possibility of excess profits companies in the market. In particular, competitive advantages may include:

- a jump in labor productivity;
- expansion of trading platforms, markets, and logistics capabilities of companies;
- extensive investment promotion;
- emergence of new means of payment and forms of settlement, increased productivity of transactions;
- reducing the labor intensity of human labor in the volume of output, reducing the dependence of the production process on the state of the resource base.

Today we can distinguish four areas that will be affected by the digital transformation of financial management in companies: social sphere, information changes, management system, innovation sphere [11].

4 Discussion

The modern concept of financial management defines four main directions for the digitalization of the company management activities (social sphere; information systems and databases; production process management system; innovation sphere). The question remains open as to the extent and quality of the impact of digitalization on management systems. A number of researchers consider this process only positive, and the effect of digitalization synergistic. In turn, some representatives of academic schools consider these trends mainly negative and speak of the need to form protective mechanisms in the field of digital legislation, developing payment systems, creating a legal framework for digital trade, and ensuring effective mechanisms for cybersecurity.

5 Conclusion

The digitalization of corporate financial management is now an objective and irreversible process of modern companies. Leading researchers point to both the positive effects of digitalization and informatization and the negative impact on business operations of corporations. On the one hand, digitalization opens up additional opportunities in the field of management, expressed in a reduction in the use of material resources and the expansion of corporate markets. On the other hand, the challenges of digitalization are primarily related to the cybersecurity of assets and funds and the shortening of the lifecycle of companies. The cost of error for making incorrect management decisions in the management of financial assets and funds of payment increases.

References

1. Sukharev, O.S.: Industrialization 4.0 and models of technological development to overcome the effect of “2D”. Bulletin of the South Russian State Technical University (NPI). Series: Socio-Economic Sciences 1:4–23. https://vestnik.npi-tu.ru/index.php/vestnikSRSTU/article/view/497 (2020)
2. Samaniego, R.M., Sun, J.Y.: Productivity growth and structural transformation. Rev. Econ. Dyn. **21**, 266–285 (2016). https://doi.org/10.1016/j.red.2015.06.003
3. Sidorova, E.Y., Timokhova, G.V.: Digital management and process approach in business process management system. International Accounting **3**, 316–337 (2021). https://doi.org/10.24891/ia.24.3.316
4. Yangulbaeva, L.S., Merzho, M.S.: Digitalization of the process of production in agriculture as part of financial management. Moscow Economic Journal **13**, 42 (2019). https://doi.org/10.24411/2413-046X-2019-10352
5. Arasteh, A.: Combination of real options and game-theoretic approach in investment analysis. J. Ind. Eng. Int. **12**(3), 361–375 (2016). https://doi.org/10.1007/s40092-016-0144-z
6. Akhobadze, M., Zumburidze, O., Kurtskalia, E.: Analysis of spatial economic processes for defining investment policy. J. Math. Sci. **218**(6), 703–708 (2016). https://doi.org/10.1007/s10958-016-3052-2
7. Bai, C., Sarkis, J.: Supplier development investment strategies: a game theoretic evaluation. Ann. Oper. Res. **240**(2), 583–615 (2014). https://doi.org/10.1007/s10479-014-1737-9
8. Qian, X., Liu, X., Pei, J., Pardalos, P., Liu, L.: A game-theoretic analysis of information security investment for multiple firms in a network. J. Oper. Res. Soc. **68**(10), 1290–1305 (2017). https://doi.org/10.1057/s41274-016-0134-y
9. Begiazi, K., Asteriou, D., Pilbeam, K.: A multivariate analysis of United States and global real estate investment trusts. IEEP **13**(3), 467–482 (2016). https://doi.org/10.1007/s10368-016-0349-z
10. Rossi, A., Vismara, S., Meoli, M.: Voting rights delivery in investment-based crowdfunding: a cross-platform analysis. J. Ind. Bus. Econ. **46**(2), 251–281 (2018). https://doi.org/10.1007/s40812-018-0109-x
11. Gadzhiev, N., Konovalenko, S., Trofimov, M., Leontieva, J., Akhmedova, K.: Management mechanism of human capital formation under a postindustrial economy. In: E3S Web of Conferences. Topical Problems of Green Architecture, Civil and Environmental Engineering, TPACEE 2019. p. 10005 (2020). https://doi.org/10.1051/e3sconf/202016410005
12. Ezhei, M., Tork Ladani, B.: Interdependency analysis in security investment against strategic attacks. Inf. Syst. Front. **22**(1), 187–201 (2018). https://doi.org/10.1007/s10796-018-9845-8
13. Farzan, F., Mahani, K., Gharieh, K., Jafari, M.A.: Microgrid investment under uncertainty: a real option approach using closed form contingent analysis. Ann. Oper. Res. **235**(1), 259–276 (2015). https://doi.org/10.1007/s10479-015-1929-y
14. Riad, D., Hattaf, K., Yousfi, N.: Mathematical analysis of a delayed IS–LM model with general investment function. J. Anal. **27**(4), 1047–1064 (2019). https://doi.org/10.1007/s41478-018-0161-y
15. Gao, X., Zhong, W., Mei, S.: A game-theoretic analysis of information sharing and security investment for complementary firms. J. Oper. Res. Soc. **65**(11), 1682–1691 (2014). https://doi.org/10.1057/jors.2013.133

Control in the Accounting System: The Experience of Russia and China

Tan Chzhen[1], Anna V. Nazarenko[2], Irina A. Kislaya[3], and Oksana V. Chukhrova[3](✉)

[1] Don State Technical University, Rostov-on-Don, Russia
[2] Penza State Technological University, Penza, Russia
[3] Rostov State University of Economics, Rostov-on-Don, Russia

Abstract. The article is devoted to a comparative analysis of controls in the accounting system in Russia and China as harmoniously developing countries, which are strategic partners in economic cooperation. The purpose of the research is a comparative analysis of the theoretical and practical foundations of the functioning of the accounting system controls in Russia and China. The methods of analysis, synthesis, comparison and ranking were used to achieve this goal, allowing us to identify the peculiarities in each country. Academic novelty of the research lies in the analysis of international experience in the control implementation in the accounting system in Russia and China, identification of similarities and differences to recognize current trends and areas of improvement. The study, based on the study of the regulatory framework and scientific literature, has established the importance of control. Control consists in eliminating the possibility of potentially dangerous damage, inappropriate waste of funds, elimination of financial risks associated with incompetence or deliberate criminal actions of accountants. Its goals, objectives, supervisory bodies, principles, and peculiarities have much in common. At the same time, the Chinese system is more orderly and transparent. It provides for more restrictions on abuse and abuse of power, built on trust. The Chinese system is, however, limited by the clear framework of the law. It can become a certain example in the construction of these principles in the Russian state. Control in the Russian state is an evolving mechanism that incorporates the principles of imperative and structured.

Keywords: Accounting · Control · Economic entity · System

1 Introduction

The accounting system is a complex mechanism that requires control and organized external and internal control in order to improve the relationship of subjects of economic activity. External control is represented by compulsory and initiative inspections. External initiative supervision in the accounting system is carried out at the initiative of the head with the involvement of third-party independent experts from an auditing company. It is used to prevent or correct errors, miscalculations in the accounting system.

P. V. Trifonov and M. V. Charaeva (Eds.): DITEM 2021, LNNS 380, pp. 632–637, 2022.
https://doi.org/10.1007/978-3-030-94245-8_86

If the external control is regulated and formally defined, the internal control has not yet received a precise and complete definition, which leads to different interpretations. Hence, there is appropriateness of considering internal control. Under internal control, most often, we understand the process of data management of the subjects of companies to achieve their goals with minimal costs and the prevention of distortions and risks during the activities of the organization [1].

The relevance of the research of the stated issues due to the purpose of internal control in the accounting system to obtain reasonable assurance that the economic entity ensures the effectiveness and efficiency of its activities [1].

The accounting systems of different countries are quite varied. At the same time accounting is becoming a means of international communication in the active interaction of countries in connection with the intensification of the processes of globalization and economic integration. In this regard, the experience of two dynamically developing countries with similar types of economic system: Russia and China is of interest [2, 3]. Control issues in the accounting system are actively considered in the modern scientific domestic, foreign literature. In this case, of particular interest is the comparison of accounting systems in countries with a similar economic platform, which are Russia and China [4].

The legal basis for the regulation of relations in the Russian accounting system are the Federal Law "On Accounting" No. 402 from 06.12.2011, Law practice 11/2013 "Organization and implementation of internal control over the facts of economic life, accounting and preparation of accounting (financial) statements by an economic entity", Tax Code of the Russian Federation, etc.

In China among the main legislative sources are the Law "On Accounting" adopted on January 21, 1985, the Law of the People's Republic of China of September 4, 1992 (as amended on February 28, 1995), "On the administration of the collection of taxes", which contains a special section II "Control over the keeping of accounting records and other business records", Law "On Certification of Professional Accountants", "Enterprise Reporting Standards", approved by the State Council of the PRC on November 16, 1992, Notification of the Ministry of Finance of the PRC and the State Economic System Reform Committee of May 25, 1995 "On the Reporting System for Experimental Mutual Enterprises" etc.

The problem of control in the accounting system in Russia and China is reflected in the works of Alaverdova, Arzybaeva, Babaeva, Kovaleva, Saparbaeva, Suprunova, Chang [5–9]. However, a comparative characterization of the control of these countries is not widespread. Parameters of comparison of the organization of accounting in China and Russia are described in the works of Petrov. He highlights the management of the rules for the preparation and presentation of accounting statements, the objects of accounting, monetary units and accounting language, the composition and timing of accounting statements. The experience of controlling the accounting system in China is widely represented in the works of Babaeva, Haibo, Hongtao, Jkan, Fugui, who describe the principles and fundamentals of accounting. The aim of the study is a comparative analysis of the theoretical and practical foundations of the functioning of the accounting system controls in Russia and China. A number of tasks serve to achieve the goal: analytical review of the experience of control mechanisms in the accounting system in Russia and China; study of the regulatory framework and scientific literature of the stated sphere; identification of similarities and differences of control in the accounting system.

Academic novelty of the research lies in the analysis of international experience in the implementation of control in the accounting system in Russia and China, identifying similarities and differences to identify current trends and areas for improvement.

2 Methods

The research was conducted with the help of general theoretical methods, including comparison, systematization, ranking, analysis and synthesis. The method of comparative analysis of qualitative indicators of accounting system control allowed to identify common and different features of the system of China and Russia. This comparison made it possible to establish regularities in the functioning of the systems in question.

3 Results

The results obtained during the analysis of regulatory and scientific literature will be presented in the form of a table (Table 1).

Table 1. Comparative characteristics of internal control in the accounting system of Russia and China.

Comparison criterion	Russia	China
The concept of internal control	There is no unambiguous interpretation of "internal control"	
	"A set of organizational structure, methods and procedures approved by an organization for the orderly and effective conduct of financial and economic activities" [10]	Mechanism of legal, obligatory observation of the financial statements of the organization, which consists in checking and analyzing the objects of the accounting system for compliance with the standards established by the state
Regulatory framework	1) Federal Law "On Accounting" No. 402 – FZ from 06.12.2011, which entered into force on January 1, 2013. [11] 2) Letter No. PZ-11/2013 of the Ministry of Finance of the Russian Federation "Organization and implementation by an economic entity of internal control over the facts of economic life, accounting and preparation of accounting (financial) statements" [1] 3) Tax Code of the Russian Federation [10] 4) Federal Law dated 30.12.2008 No. 307-FZ (revised on 09.03.2021) "On Auditing Activities" [11]	Quaijifa (accounting laws) is the general name of the legal norms regulating economic relations 1) Law "On Accounting" adopted on January 21, 1985. (as amended on December 29, 1993) [13] 2) Law of PRC of 04.09.1992 (amended on 28.02.1995) "On the management of tax collection" (section "Control over the accounting records and other economic documentation") 3) The law on auditing, which came into force on January 1, 1995 [14]

(*continued*)

Table 1. (*continued*)

Comparison criterion	Russia	China
Subjects of internal control	Since 2013, all economic entities must have their own internal control service [11] The subjects of internal control include: the management bodies of the economic entity; the audit commission (auditor) of the economic entity; the chief accountant or other officer of the economic entity charged with keeping accounting and other [5]	The subjects of internal control include: -accounting authorities and accounting personnel; - financial department; -auditors
Elements of internal control	- control environment; - risk assessment; - internal control procedures; - information and communication; - assessment of internal control [5]	
Functions of internal controllers and internal auditors	1) Timely identification and assessment of the significance of business process risks using the forms of preliminary, current and follow-up control 2) Compliance with current legislation in improving the efficiency of management systems of economic entities 3) Organization and implementation of internal control over the facts of economic activity [15]	1) Checking accounting records 2) Obtaining testimonial materials by checking accounting records 3) Maintaining proper accounting records in accordance with the Company and Internal Revenue Ordinance
Supervisory authority	Ministry of Finance of the Russian Federation	Ministry of Finance of the PRC
Requirements for internal control	No opportunities to exceed official authority, rapid mobility in the face of possible changes in the law, the timely elimination of errors in the accounting system. An important requirement for financial reporting in China, as well as in Russia, is the reflection of information on cash and receipts from securities; receipt, disposal and use of valuables; debt obligations and settlements; capital increase and expenses incurred; calculation of income, expenses and cost of sales	

The comparative analysis has shown that the internal control in the accounting system of Russia and China have similarities and differences. It may be noted that in China the control system is more centralized and transparent. The accounting calculations here are based on actual economic actions, truthfully reflecting the financial situation and economic results. Accounting information meets the requirements of state macroeconomic regulation and provides an opportunity for its comprehensive understanding. The method of accounting during the accounting period must be uniform; accounting calculations must be made in a timely manner; accounting records and reports must be clear and usable. There is a dependence on international accounting and financial reporting standards, for which the China Accounting Standard Committee (CASC) is responsible. A distinctive feature of Chinese accounting is the presence of organizational stamps and seals. These stamps with the imprint of the organization replace the signature of a legally competent person. Often documents are not signed, but marked with a seal. Thus, all organizations and companies in China should develop sound internal control systems.

In general, control in China is characterized by the preservation of historicity, ancestral memory, symbolism, which can be borrowed in the Russian system of control. In turn, Russia has also introduced streamlined and mandatory accounting controls.

4 Discussion

The analysis of the regulatory framework and scientific literature showed an insufficient degree of understanding of the meaning of internal control, the lack of clear requirements for the internal control service of the economic entity in the accounting system of Russia and China. Accordingly, the subject of control has to collect disparate material, interpret the provisions of laws and standards, which can lead to inconsistencies and costs.

The analysis of internal control in the accounting system of Russia and China has revealed certain regularities: Chinese organizations and enterprises have more autonomy and freedom in choosing their own methods of accounting, a high degree of historicity of its development and formation, and a high degree of trust in employees on the part of the economic entity. In general, it should be emphasized that there is a need for consistency between the accounting system and the control system [12].

5 Conclusion

The processes of economic globalization have set all countries a common task to find a common business information language, namely, the language of financial reporting. This has also become a major challenge for China, which is one of the world's major powers. Reforming the regulatory framework for accounting in China has a long historical development. Russia at the present stage is at the stage of development of the control system.

Thus, internal control contributes to the achievement of the objectives of the economic entity. It must ensure the prevention or detection of deviations from established rules and procedures, as well as distortions of accounting, accounting (financial) and other reporting data.

References

1. Information N PZ-11/2013 of the Ministry of Finance of the Russian Federation "Organization and implementation by an economic entity of internal control over the facts of economic life, accounting and preparation of accounting (financial) statements". http://www.consultant.ru/document/cons_doc_LAW_156407/
2. Audrey, S., Paterson, F., Changwony, P., Miller, B.: The British Acc. Rev. **51**(5), 100844 (2019). https://doi.org/10.1016/j.bar.2019.100844
3. Kiswanto, R., Fajar, C.M., Adi, S.: J. Ekonomi. BisEntrep **13**(1), 61–73 (2019). https://doi.org/10.5281/zenodo.3522807
4. Yu, X.: Research on internal accounting control in colleges and universities. In: Proceedings of the 2018 5th International Conference on Education, Management, Arts, Economics and Social Science (ICEMAESS 2018). https://doi.org/10.2991/icemaess-18.2018.196
5. Alaverdova, T.P.: Naukovedenie. **6**(25). (2014). https://doi.org/10.15862/190EVN614

6. Arzybayev, A.A., Saparbayeva, S.S.: Bul. Eurasian Nat. Univ. **1**, 25–29 (2012)
7. Babaeva, Z.S.: Int. Acc. **29**(179), 61–64 (2011)
8. Kovalev, A.E.: Ideas Ideals **1**(11), 175–196 (2019). https://doi.org/10.17212/2075-0862-2019-11.1.1-175-196
9. Suprunova, E.A.: Int. Acc. **1**(439), 54–64 (2018). https://doi.org/10.24891/ia.21.1.54
10. Tax Code of the Russian Federation. Part Two. http://www.consultant.ru/document/cons_doc_LAW_28165/
11. Federal Law dated 30.12.2008 N 307-FZ (revised on 09.03.2021) "On Auditing Activities". http://www.consultant.ru/document/cons_doc_LAW_83311/
12. Federal Law of 06.12.2011 N 402-FZ "On Accounting". http://www.consultant.ru/document/cons_doc_LAW_122855/
13. The Law on Accounting in the People's Republic of China. https://asia-business.ru/law/law2/accounting/accounting/
14. Makkaeva, R.S.A., Pataeva, S.A., Mazhetova, T.S.: Prob. Sust. Devel. Soc. **4**, 88–94 (2021)
15. Mukhina, E.R.: Apriori. Ser. Hum. **5**, 43–49 (2015)

Regional Development Budgets as an Investment Policy Tool

Karina N. Samoylova(✉), Svetlana N. Rukina, Irina P. Denisova, and Aida S. Takmazyan

Rostov State Economic University, Rostov-on-Don, Russia

Abstract. The research purpose is to substantiate the need to develop a development budget as an independent monetary fund, contributing to the achievement of efficiency of the investment policy of a region. The next methods were used: the use of a systematic approach, logical-semantic and comparative analysis of theoretical and practical materials made it possible to study measures of regional support for investment activities. The study is based on the works of domestic and foreign scientists devoted to the issues of state support for investment activities, as well as open data from public authorities and governing bodies. The necessity of forming a development budget in Russian regions as an independent monetary fund as part of the consolidated budget of a constituent entity of the Russian Federation has been substantiated. An analysis of the sources of fixed investment financing showed that investment is made mainly at the expense of enterprises' own funds, budgetary funds are extremely unevenly distributed. The constituent entities of the Russian Federation strive to create comfortable conditions for attracting investment using a variety of tools, including fiscal ones. In order for a development budget to function as a regional investment policy tool, it is necessary to legislatively consolidate its status, expand the tax independence of regions, maintain the stability of tax legislation. The research novelty lies in the systematic presentation of a regional development budget through analyzing the content of its components.

Keywords: Investment · Forms of state support · Tax exemptions · Tax expenditures · Economic growth · Employment recovery

1 Introduction

As is known, economic growth directly determines the level of the socio-economic development of a state and the well-being of its citizens. Fulfillment of the country's economic development potential is impossible without taking into account the spatial organization. To achieve annual growth rates at the world average level (at least 3.0–3.5% per year), it is necessary to eliminate disparities in the economic and social development of regions and increase the contribution of industrial federal districts to economic growth. Obviously, it is virtually impossible to create equal conditions for economic and social development in all 85 constituent entities of the Russian Federation, but it is clearly possible to bring

P. V. Trifonov and M. V. Charaeva (Eds.): DITEM 2021, LNNS 380, pp. 638–644, 2022.
https://doi.org/10.1007/978-3-030-94245-8_87

the levels of development of territories closer, to ensure equal rights of citizens to buy quality goods and receive services, thereby increasing the standard of living in Russia. Based on the Pareto theory, it can be assumed that the regional development depends on positive macroeconomic dynamics by 80% and on the existing structure of regions' economy and features of regional development by 20%. The factors aimed at leveling the parameters of socio-economic development in the regions can be investment and income of the population living in their territory. Investment is undoubtedly the main driver of economic growth. In the post-COVID period, they will ensure the restoration of the economy, employment of the population and their real income, thereby implementing the national Action Plan of the Government of the Russian Federation, and in the long term – structural changes in the economy, its competitiveness and resilience to external shocks.

Numerous studies of state and regional investment policy reveal its various aspects: factors and directions of investment policy in the regions (Emelyanova, Kharchikova [1], Sochulakova, Igazova [2]), forms of state support (Soldak, Shamileva [3], Troshin [4], Korotkevich, Shparun, Xu Zi-Ming [5]). The work by V. Raghupathi and W. Raghupathi can be noted among foreign publications on the development of national investment and innovation ecosystems, integrating disparate strategies to stimulate economic growth [6]. However, there is no detailed examination of the content of a development budget, its role in achieving the goals of regional investment policy in the scientific literature, and this confirms the relevance of this study.

The research purpose is to substantiate the need to develop a development budget as an independent monetary fund, contributing to the achievement of efficiency of regional investment policy, which reflects the link between tax revenues, tax benefits, tax expenditures. To achieve this goal, the dynamics of investment financing sources in federal districts is analyzed, the definition of a regional development budget is formulated, two options for the formation of its income are proposed, and the content of forms of state support for investment activities is disclosed on the example of the Rostov region.

2 Methods

The study uses statistical data from the Federal State Statistics Service, official information from the Government of the Rostov region, domestic and foreign scientific publications, and Internet resources. The study of the legislation of the Rostov region made it possible to group the forms of state support for investment activities, highlighting tax support (built-in investment stimulation tools) as significant for business and research, as well as to analyze their relationship with tax expenditures, components of the regional development budget. The research methodology is based on a systematic approach, statistical and comparative analysis.

3 Results

Below is the analysis of the structure of fixed investment financing sources by federal districts.

The data shown in Table 1 show that from 2014 to 2019 the largest share in the structure of investment financing sources was occupied by organizations' own funds (profit, depreciation), and their share was growing steadily in all federal districts, especially in the Volga, Ural and Siberian federal districts. Budget investment primarily addresses the issues of infrastructure development in those areas where business efforts are not enough, or the motives for the commercial implementation of investment projects are not clear.

Budget investment is provided at the expense of the federal budget and budgets of constituent entities of the Russian Federation. According to the analysis of Table 1, funds from the federal budget prevail in the structure of the budget fixed investment. They are distributed extremely unevenly among the federal districts, despite the implementation since 2019 of the national projects included in the state programs of the Russian Federation and determining the key areas of spatial development.

A significant share of federal budget investment falls on the North Caucasian and Southern Federal Districts: in 2014 – 31.9%, in 2016 – 49.7%, in 2019 – 54.6%, while in 2019, in the Ural Federal District this figure was only 1.5%, in the Volga Federal District – 6.9% and in the Central and Siberian Federal Districts – 6.1% each. As for the share of investment financed from budgets of constituent entities of the Russian Federation, it is insignificant and exceeded 10% in 2019 only in the North Caucasus and Central Federal Districts.

Thus, the limited influence of intraregional factors on positive economic dynamics should be noted, therefore, in today's situation, it is necessary to search for new tools and solutions aimed at enhancing investment processes in Russian regions. in the authors' opinion, this tool can be a regional development budget, which will ensure the formation of a favorable environment for private investors and activate investment processes in the post-pandemic period, will allow to achieve the national development goals set by the President of the Russian Federation in the Decree "On National Development Goals Through 2030".

In today's budgetary legislation there is no concept of "development budget". Based on the provisions of Article 6 of the Budget Code of the Russian Federation, the following definition can be formulated: "A development budget is a form of education and spending of funds to finance production and infrastructural capital expenditures (construction and modernization of the real sector of the economy, construction and repair of educational institutions, highways, healthcare institutions, sports and cultural facilities, etc.). In the current practice, it is not only about spending funds from the regional development budget. Thus, in the Rostov region, the development budget is planned for 2021 in the amount of 47.9 billion rubles, which is 19% of the total revenues of the regional consolidated budget. Compared to 2020, its volume increased by 2.2 billion rubles. About 320 regional and municipal property items, including highways, will be designed, built and repaired with the use of budget resources (development budget) [8]. In the authors' opinion, a development budget is important for the regional socio-economic and investment development, including the implementation of national projects.

It appears necessary to determine the sources of its formation, since the expenditure has developed in practice; it can be adjusted taking into account tax expenses. There are two possible options for the formation of development budget revenues. The first option

Table 1. The structure of fixed investment (in %) by federal districts of the Russian Federation in the period of 2014–2019 [7].

Federal districts	Own funds			Bank loans			Federal budget funds			Funds from budgets of the constituent entities of the Russian Federation			Other		
	2014	2016	2019	2014	2016	2019	2014	2016	2019	2014	2016	2019	2014	2016	2019
Central federal district	44.6	49.4	50.5	10.7	10.0	14.4	12.4	9.8	6.1	11.3	12.4	12.6	21.0	18.4	19.4
Northwestern federal district	43.6	51.9	53.3	7.2	3.6	5.0	13.5	12.8	10.2	6.3	5.0	5.3	29.4	26.7	26.2
Southern federal district	32.5	38.5	41.4	17.1	7.6	9.3	6.8	22.0	24.8	3.2	3.9	5.8	40.4	28.0	18.7
North Caucasus federal district	30.8	35.8	34.8	9.8	7.3	11.0	25.1	27.7	29.8	9.6	8.8	10.7	24.7	20.4	13.7
Volga federal district	52.9	57.7	62.8	13.4	13.5	9.1	6.3	7.5	6.9	5.4	4.8	6.7	22.0	16.5	14.5
Ural federal district	52.1	54.1	65.5	8.3	17.8	12.8	2.9	2.0	1.5	4.1	2.0	3.6	32.6	24.1	16.6
Siberian federal district	46.5	54.8	62.6	8.4	5.9	5.3	7.1	7.9	6.1	4.3	3.5	4.6	33.7	27.9	21.4
Far eastern federal district	40.6	44.2	48.8	9.6	6.1	9.3	15.3	10.8	7.1	7.5	6.0	3.5	27.0	32.9	31.3

implies tax and non-tax revenues credited to the budget revenues of a constituent entity of the Russian Federation in accordance with the provisions of Chapter 8 of the Budget Code of the Russian Federation: excise taxes on oil products and transport tax, which are the road fund sources of a targeted use; 10–15% of return of corporate income tax, personal income tax, special tax regimes, income from the use of state-owned property, etc. Consequently, the availability of one's own tax and non-tax revenues implies the independent and responsible use of funds from the regional development budget. In case of the second option of development budget formation, budget revenues include not only tax and non-tax revenues, but also interbudgetary transfers from the federal budget (subsidies), which is of great relevance in the implementation of national projects. Regions receive additional investment resources through intergovernmental transfers. In times of coronavirus crisis, due to the provision of interbudgetary transfers to the Russian regions in the amount of 3.1 trillion rubles, the growth of capital investment in the consolidated budgets of constituent entities of the Russian Federation was ensured by more than 10%. In the authors' opinion, this practice reduces the independence of sub-federal authorities, since they implement federal programs and projects, as well as infrastructural developments for their functioning, which account for the bulk of subsidies from the federal budget. Therefore, it is necessary to restructure interbudgetary relations towards the creation of mechanisms for the economic recovery growth based on stimulating investment processes.

Certainly, development budget expenditures should reflect the volumes of tax expenditures that correspond to the amount of stimulating tax benefits provided by the sub-federal government to investors. It should be pointed out that "tax expenditures of a constituent entity of the Russian Federation are shortfall in budget revenues of a constituent entity of the Russian Federation, due to tax exemptions for taxes and special tax regimes provided as state support measures in accordance with objectives of state programs of the constituent entity of the Russian Federation and objectives of socio-economic policy of the constituent entity of the Russian Federation" [9]. Since 2020, financial authorities of constituent entities of the Russian Federation have evaluated the effectiveness of tax expenditures and their contribution to the change in the value of an indicator of achieving state program objectives, the budgetary efficiency of tax expenditures, overall budget efficiency, which may lead to the cancellation of ineffective tax exemptions [10]. This technique was developed by the Ministry of Finance of the Russian Federation.

4 Discussion

As noted in the works, investment attractiveness of constituent entities of the Russian Federation directly depends on comfortable business conditions, which determine the vector and dynamics of private investment in various economic and infrastructural facilities [11–13]. Thus, each region seeks to offer competitive investment conditions to investors who are ready to invest their own financial resources in promising projects [14, 15].

The set of measures of state support for investment activities implemented by the authorities of the Rostov region is shown in Fig. 1.

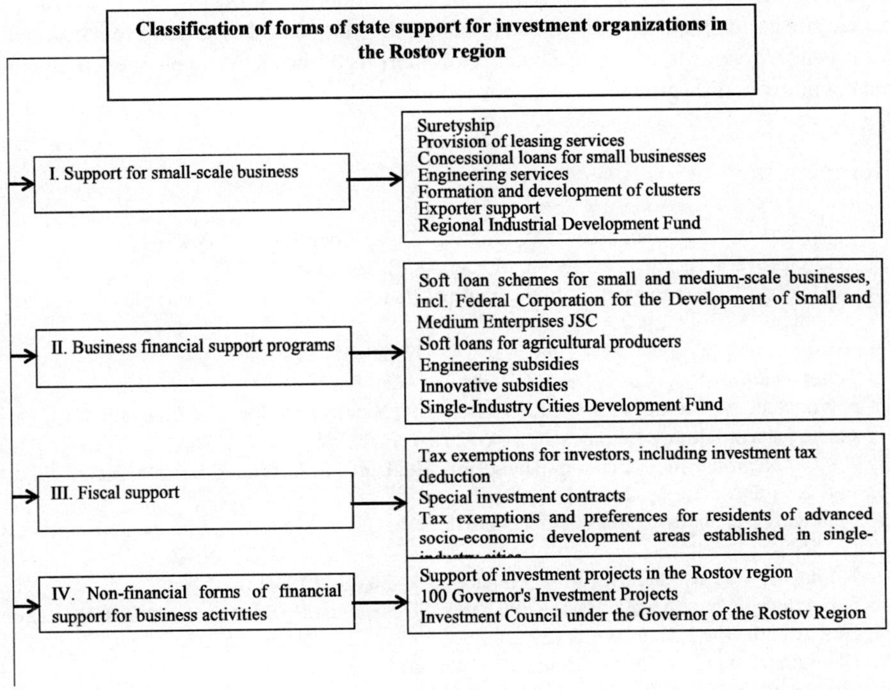

Fig. 1. Classification of forms of state support for investment organization in the Rostov region.

As is obvious, the regional profile of incentives for corporate investors in the Rostov region is represented by various demanded forms.

5 Conclusion

Sustainable economic growth, increased employment of the population, its real income and savings, stable receipts to regional budgets can be ensured by boosting investment processes in Russian regions. In modern conditions, an increase in investment in the regions is associated with the implementation of state programs of the Russian Federation and national projects. There is no doubt that budget expenditures are an important element in stimulating private investment, since they demonstrate the vector of state socio-economic policy and form the infrastructure conditions for the implementation of commercial projects. Taking into account the peculiarities of socio-economic development, each constituent entity of the Russian Federation offers investors various forms of investment support that characterize its investment profile.

It appears that under current conditions the independence of sub-federal authorities should not be limited, since the formation of positive economic dynamics, the restoration of employment, the strengthening of financial and investment capacity is of paramount importance to them. The resolution of this issue lies in ensuring the stability of tax legislation for the period up to 2030, simplifying the investment tax deduction mechanism, expanding tax powers of sub-federal authorities, which will allow the regions to form a development budget, the main objective of which is to create an attractive environment for private investment, ensuring a balance of interests of the state and business to restore the economy and its growth, investment inflows.

References

1. Emelyanova, E.V., Kharchikov, N.V.: Manag. Rus. Abroad **2**, 40–47 (2020)
2. Sochulakova, J., Igazova, M.: Econ. Manag. **18**(3), 501–508 (2013)
3. Soldak, M.O., Shamileva, L.L.: Econ. Industry **1**(81), 21–42 (2018). https://doi.org/10.15407/econindustry2018.01.021
4. Troshin, A.S.: Sci. Edu. New Time **5**(16), 40–47 (2016)
5. Korotkevich, A.I., Shparun, D.V., Zi-Ming, X.: New Econ. **2**(72), 53–58 (2018)
6. Raghupathi, V., Raghupathi, W.: Exploring science-and-technology-led innovation: a cross-country study. J. Innov. Entrep. **8**(1), 1–45 (2019). https://doi.org/10.1186/s13731-018-0097-0
7. Russian Regions: Socio-Economic Indicators: Statistical Collection. Rosstat, Moscow (2020)
8. Fedotova, L.V.: Budget **2**, 50–53 (2021)
9. Samoylova, K.N., Rukina, S.N., Gerasimova, K.A., Takmazyan, A.S.: Fin. Res. **2**(67), 19–28 (2020)
10. Klimanov, V.V., Mikhailova, A.A.: Finance **3**, 26–30 (2020)
11. Samoylova, K.N., Muzaev, M.Z., Rukina, S.N., Denisova, I.P., Takmazyan, A.S.: Int. J. Econ. Bus. Admin. VII **I**(4), 691–699 (2020)
12. Dzwigol, H., et al.: J. Entrep. Edu. **22**(1S) (2019)
13. Khasanova, S., et al.: Espacios. **39**(31), 20 (2018)
14. Explanatory information on "The development of investment potential of the Rostov region through 2025 and the implementation of the regional investment program in 2020". http://www.donland.ru/result-report
15. Federal Law "On the Protection and Encouragement of Investment in the Russian Federation" dated 01.04.2020 (as amended). https://www.consultant.ru/document/cons_doc_LAW_38145

Analysis of Management Paradigms in the Contemporary Finance Theory

Lyudmila N. Usenko(✉), Anastasia M. Usenko, Victoria A. Guzey, and Alina S. Bidzhieva

Rostov State University of Economics, Rostov-on-Don, Russia

Abstract. The structure of the market economy is undergoing major changes in connection with the fundamental development of the global market in various areas. Globalisation makes modern enterprises to quickly respond to changes in the external environment and adapt to them. As a result, the existing financial relations in the areas of production, promotion, settlements and interaction with partners and consumers are being transformed. There is an increasing scientific interest in the study of management paradigms and their influence on the peculiarities of the financial system creation and business entity operation. The study reviewed the interpretation of the concept of "paradigm" and proved the ambiguity of this concept and the need to develop your own approach to this term from the perspective of financial management, aimed at achieving a balanced cash flow of an enterprise based on the development of relationships within and outside the enterprise. The study focuses to identify and highlight conceptual changes in management processes that affect the establishment of financial relations in organisations. It has been established that the creation of well-coordinated relations within one enterprise can allow building effective relationships with other actors in the market economy. The research results proved that globalisation is leading to a change in management paradigms in the contemporary finance theory. The value of this research is to show the feasibility of the creation of sectoral paradigms in the management of enterprises in various sectoral fields for structuring and identifying general trends in their functioning in markets of various sort.

Keywords: Financial management · Paradigm analysis · Financial relations · Organisation finance

1 Introduction

The modern finance theory presupposes building steady relationships among enterprises, populations, or countries.

The finance of an enterprise, like any financial relationship, is built on concepts, theories and paradigms.

Initially, the term "paradigm" began to appear from the time of the creation of science. For the first time, the concept was applied by Plato, who defined the paradigm

P. V. Trifonov and M. V. Charaeva (Eds.): DITEM 2021, LNNS 380, pp. 645–650, 2022.
https://doi.org/10.1007/978-3-030-94245-8_88

"as the structure and form of material things as something eternal and an unchanging transcendental pattern" [1].

In the 20th century, Bergman introduced this concept into the philosophy of science as "the definition of slowly changing methodological standards of research" [2].

The general changed concept of "paradigm" in 1962 was suggested by Kuhn "this is a conceptual scheme that serves as a regulatory framework for the practical activities of the scientific community in a historically limited period of time" [3].

2 Materials and Methods

The above definitions show the ambiguity of this concept and can characterise the development of their approach to this term. This circumstance is explained by the fact that the formation of the definition of "paradigm" took place at different times, which left its imprint through the events that took place at that time.

The paradigm is a "universal method of making evolutionary decisions, an epistemological model of evolutionary activities" [4].

There is another definition of this concept. "A paradigm is an initial conceptual framework, a model for posing problems and solving them, research methods. Moreover, the constructed model refers to a specific period of time" [5].

In other words, a paradigm is understood as certain truths that imply everything that is proven, known and accepted by the society. This definition assumes a certain model that has been accepted by the society and has been in effect for a fairly long period of time, or is outdated and has been replaced by a new model.

It is important to understand the basic assumptions of each paradigm because without understanding the basics it would be difficult to relate each with financial theories [6].

Research shows that when a pre-determined time period of particular value for a particular state ends, a major paradigm shift takes place. At this point, the current paradigm becomes obsolete and ceases to operate, and a completely different new paradigm begins to gain momentum in its application [7].

As market relations developed in the 20th century, there was a complication of the external and internal environments of the company operation in the market. This circumstance provided the basis for finding ways to build optimal and sustainable relationships among organisations through the creation of a different kind of management. The change in management has led to the need to create a new kind of paradigm that defines management at various levels within and outside organisations.

Based on such changes, there is a need for a different definition of the term "management paradigm". It is understood as a system of methods and forms of management at various levels.

This issue has been studied by various scientists and some of them believe that the "management paradigm is a system of views" [8].

Some authors understand this definition as a "concept" [9].

According to Kuhn, "a paradigm in a scientific concept implies a system of means, methods and judgments" [10].

3 Results

The management paradigm in the practice of its application at enterprises of various fields of activity can be considered from several positions:

- asset and liability management of a company implies maintaining the optimal ratio of the considered positions. This makes it possible to achieve the stability of the company in the market with an optimal level of liquidity and solvency. This fact can be explained by the fact that the greater the degree of liquidity and solvency of a company, the faster it can cope with the occurrence of possible crisis situations with the need to quickly cover debts;
- cash flow management in organisations makes it possible to reasonably plan the incomes and expenditures of the organisation.

When creating an optimal system of planning service and creating plans for the coming period, it becomes possible to optimise activities from a financial point of view. This concept allows you to plan the activities of a company in the market and, in comparison with other counterparties, get a greater financial advantage.

Moreover, the above two provisions of the management paradigm are interrelated and complementary to each other. This is due to the fact that a non-optimal structure of the asset/liability ratio of a company will never lead to its financial stability and the availability of the required amount of free cash in circulation. Consequently, the company will not be able to achieve the required level of liquidity and solvency and, as a result, the level of competitiveness of the company will be significantly lower compared to an organisation in which the management of cash flow and asset/liability ratio is optimised.

The key area in the management paradigm implementation is to correctly organise a company's cash flow, which is based on building the relations inside and outside the company [11].

4 Discussion

An organization can build relationships of a various nature in order to achieve a specific goal pursued by the company.

Building relationships among several organisations in the domestic or international markets is based on the establishment of effective relationships within the organisation [12]. This circumstance provides the basis for the purposeful promotion of the organisation's activities in the market. Relationships within the organisation should be built in order to arrange for the activities of various departments. Given that the organisation functions as a single system, maintaining the effectiveness of the relationship is the key to the effectiveness of its activities [13].

If we consider separately manufacturing products, performing work or rendering services, then well-coordinated relations in the organisation's team will allow streamlining this process, regardless of any critical situation [14].

For example, with continuous production at an enterprise, there may be a shortage of workers for various reasons, such as illness, vacation or other.

If an organisation has well-established relationships between the teams of various departments and within one specific department, then in a similar situation, when conducting additional training, self-replacement of personnel at workplaces can be organised. Most often, this can be seen within the same department, where the personnel activities may not differ much from each other but only have some peculiarities. In this case, the correct organisation of the self-replacement of personnel will improve the operation of both one department and the organisation as a whole. However, this can only be the case if a similar style of building relationships is maintained not only within one department but also in other departments, that is, in the organisation as a whole. This method of building relationships within an organisation will provide the basis for the effective use of the organisation's resources.

In addition, the promotion of well-coordinated relationships within one company can help build effective relationships with other companies in the market. This can be achieved through the use of various methods of building relationships to achieve specific goals. The key methods of building relationships with other companies can be as follow:

- analysis of the demand for the products of that organisation;
- identification of the interest of other organisations in the activities of one organisation, and so on.

The above positions can be achieved by conducting additional market research and conducting additional social surveys of the closest counterparties of the company and other stakeholders. The implementation of these social surveys will allow the company to establish the positive and negative aspects of its activities. This will make it possible to identify more significant areas of activity and, if possible, strengthen the activities of the company in those areas.

Based on this, the management paradigm implementation redirects the organisation's activities in the market, among both similar entities and in the external markets, with absolutely different enterprises.

With this approach to building relationships, the main goal of an entity can change, e.g. not maximising market value but maximising customer satisfaction with products, works and services of its own manufacture. This circumstance allows you to make every effort to increase the productivity of the entity's personnel.

In recent years, there has been a tendency to create an international market for the relationships between different countries through the relationship of various enterprises and organisations. Some of the elements of such relationships are:

- the branches of the operation of enterprises, where in one way or another the enterprises interact as inter-related;
- areas of knowledge. Recently, various knowledge has been developed, e.g. innovation, which involves the promotion of innovative projects and, on their basis, the innovative product or service will be developed, and they will be sold in the market. You can also see the significant development of various types of infrastructure, which is aimed at maintaining and further developing the activities of various industries.

5 Conclusion

The relevance of the application of the practical paradigm is increasing more and more due to the promotion of innovative operation of organisations, and the creation of various information and communication technologies, such as robotics and artificial intelligence. The creation of these types of information and communication technologies increases the relevance of additional research into the effectiveness of human knowledge embedded in machinery. With the development of science and technology, physical work has been transformed into brainwork. In addition, experts predict that in the future, all tasks will be completed by robots [15].

Until now, additional research into the use of machine knowledge is being carried out; and new developments are being sought to improve it. This is done in order to identify evidence that robotics and artificial intelligence, at some stage of their activities, can completely replace human labour in many operations of organisations. This aspect will maximise the automation of creating an innovative project and an innovative product or service based on it with the aim of further selling it in the market and enhancing the competitiveness of enterprises.

At the moment, the promotion and development of sectoral paradigms are under way. This is necessary in order to create general guidelines for the development of the operation of a particular industry and its member enterprises for the uniformity of their activities.

The promotion of various types of paradigms in the management of the activities of enterprises in various areas of activity will allow a more structured process to carry out the development of their functioning in various types of markets.

References

1. Plato, T.: Plato. Selected Dialogues. In: Skoda, V.V. (ed.) Plato. AST, Moscow (2004)
2. New philosophical encyclopedia. In: Stepin, V.S., Guseinov, A.A., Semigin, G.Yu, Ogurts, A.P. (eds.) In 4 volumes, V 3, IP RAS, National social and scientific fund. Mysl, Moscow (2001)
3. Kuhn, T.: The structure of scientific revolutions. In: Kuznetsov, V.Yu. (ed.) AST, Moscow (2003)
4. Paradigm/Academician. Dictionaries and encyclopedias on Academician. https://clck.ru/XSjpD
5. The paradigm of financial management. Studopedia. https://studopedia.ru/10_56573_paradigma-finansovogo-menedzhmenta.html
6. Sultana, F.: J. Fin. Account Res. **2**(1), 94–113 (2020). https://doi.org/10.32350/JFAR/0201/04
7. Soufian, M., Forbes, W., Hudson, R.: Crit. Persp. Account **25**(8), 724–742 (2014). https://doi.org/10.1016/j.cpa.2013.08.009
8. Polukratov, V.L.: Fundamentals of Management. Knorus, Moscow (2007)
9. Kovalev, V.V.: Introduction to Financial Management. Finance and Statistics, Moscow (2001)
10. Paradigm: these are simple words. Development stages. https://zajmy-onlayn.ru/paradigma-eto-prostymi-slovami
11. Usenko, A.M.: Econ. Human. Sci. **4**, 69–74 (2015)
12. Martinez-Sanchez, A., Vicente-Oliva, S., Pérez-Pérez, M.: Europ. J. Innov. Manag. (2020). https://doi.org/10.1108/EJIM-10-2019-0314

13. Usenko, A.M.: Sci. Works Free Econ. Soc. **6**(226), 280 (2020). https://doi.org/10.38197/2072-2060-2020-226-6-280-293
14. The Forrester – International Independent Analytical Company. https://www.forrester.com
15. Shkarlet, S., Dubyna, M., Shtyrkhun, K., Verbivska, L.: WSEAS Trans. Environ. Devel. **16**, 413–422 (2020). https://doi.org/10.37394/232015.2020.16.41

Research of Features of Payment Technologies in the Payment Industry

Ilya A. Nozhenkov[1] (✉), Albina N. Sokolova[2], Tatiana I. Demidenko[3], and Tatiana N. Michnenko[3]

[1] Southern Federal University, Rostov-on-Don, Russia
[2] Don State Technical University, Rostov-on-Don, Russia
[3] Rostov State University of Economics (RINH), Rostov-on-Don, Russia

Abstract. Retail payments research is on the rise in the digital economy. However, due to logical, conceptual similarity, the notion of "payment technology" is blurred in the studies of payment systems. Hence, there is a need to form the conceptual apparatus of payment technologies and to identify the features in the functioning and implementation of payment technologies. Due to the analysis of structural elements of the payment industry and identifying areas of payment technology used in the mechanism of retail settlements, it seems possible to form the features of payment technology. Their comparison with a number of principles and criteria of functioning in the future seems possible expediency of development of new technologies, which characterizes the practical significance of the research. The purpose of the research is to form a conceptual framework in the field of payment technology, namely, the definition of the concept of "payment technology" and the identification of features of their functioning, as well as the classification by analyzing the opinions of scientific researchers in the field of payment systems, banking and sciences related to financial technology. The originality of the research is that we formed a theoretical and methodological base of research in the field of payment systems, payment technologies and financial management of companies, industries that directly depend on the level of development of payment technologies, including retailers and payment service providers.

Keywords: Digital payments · Non-Cash money turnover · Payment systems · Digital banking · Mobile wallets · Fintech · Paytech

1 Introduction

In the last decade, the role of non-cash payments in the daily life of the population has reached the greatest social and economic significance. According to the National Retail Federation's Consumer Demand and Behavior Survey, 58% of consumers mention mobile payments among the various technologies that they use in daily life [1]. For example, in Sweden, researchers note that by 2030 society is expected to fully transition to cashless payments [2]. Given the epidemiological situation in the world in 2020, the dynamics of non-cash payments transformed in quantitative and qualitative form [3]. The criterion of qualitative transformation implies the transformation of technologies

P. V. Trifonov and M. V. Charaeva (Eds.): DITEM 2021, LNNS 380, pp. 651–656, 2022.
https://doi.org/10.1007/978-3-030-94245-8_89

associated with the implementation of non-cash payments and qualitative form. However, the fundamentals of payment technologies have not been sufficiently studied in the practice during payment systems research. On the one hand, researchers distinguish payment technologies in the role of payment instruments. On the other hand, payment technologies as a service are a source of cash inflows for companies providing payment services, i.e. a source of income. And then again they define the role of the financial technology industry. For reasons of logical, technical and conceptual similarity of the term "payment technologies" and "payment instruments" in the array of economic, legal and technical sciences is blurred. Hence there is a need for functional or role division of payment technologies in the payment system.

In the works by Belousova and Zemlyakova, the scientists explore the issue of terminology and features of payment innovations, originally defined by the Committee on Payment and Settlement Systems of the Bank for International Settlements. Researchers mention two types of innovations: Process-oriented innovation and product-oriented innovation [4]. The application of cloud computing technologies and improvement of payment technologies (from the aspect of improving payment security). As well as the integration of payment technologies in the concept of "Internet of things" was considered by a group of researchers from China [5]. The list of features of the above-mentioned technologies forms the main pool of payment technology development directions. The paper "Impact of Blockchain Technology on Business Models in the Payment Industry" by Murman highlights the peculiarities of blockchain technology and the role of its impact on the payment services market. However, the paper does not note the conceptual features of "payment technology" [6]. The particular degree of influence of payment technologies and the type of interaction between them and consumers are highlighted as the socio-economic importance of the impact of payment technologies, based on the theory of consumer demand [7]. The most significant developments in the field of the payment market are reflected in the work "Actualization of the conceptual model of the payment market in the digital economy" by Russian scientist Korobeinikova [8]. This model characterizes the payment market on the scale of the payment system of the country. Partial elements of this model will reveal key features of payment technologies.

Considering the payment industry in detail, it is possible to note a number of common features in the studies of Russian and foreign researchers. The conceptual features of payment services of Russian and foreign studies are formed mainly from the term payment service defined in the glossary of the Committee on Payment and Settlement Systems of the Bank for International Settlements. The linguistic peculiarities of research in the field of payment technologies in Russian practice are mostly in the use of the term "non-cash payments", in European practice the term "payments" is used more often. In the projection of this study, in our opinion, the two concepts can be identified. At the same time, identifying these concepts, in our opinion, it is necessary to clearly differentiate the concepts of payment technologies in the industry of payment services, compare them with the principles of organization of cashless turnover and define the structure of the industry with the classification of these technologies.

2 Materials and Methods

Theoretical (analysis, synthesis, induction), empirical (analysis of the volume of non-cash transactions, interpretation of expert opinions on the issue), methods of graphical representation of data were used as the main methods of research. By applying the method of induction in the consideration of the mechanism of non-cash payments on the example of the life cycle of mobile payment can form a number of functional features and key areas of application of payment technology.

3 Results

First of all, it is worth referring to the conceptual and linguistic background of the term "Technology" in the implementation of the task of separating conceptual features. Turning to the ancient Greek terms "Techno" and "Logos", which can be translated literally as ability to think, we can understand the fundamental basis of the term. With the development and digitalization of the payment market, the term "payment technology", in our opinion, is closest to describing the term "information technology". This characterizes a set of methods, software and hardware and technological means that ensure the collection, accumulation, processing, storage, presentation and dissemination of information [9]. In our opinion, compliance with the principles of the organization of cashless circulation, as well as compliance with normative and regulatory documents (laws, directives, standards) is a key criterion for the creation or introduction of payment technologies. Comparing features of payment technologies to principles of the organization of a cashless turnover, it is possible to note the principles most applicable to payment technologies. The legal regime of settlements, maintenance of liquidity and solvency of participants in the settlement relations, consent (acceptance) of the payer to the payment, control of participants of settlements to meet the deadlines and the accuracy of the settlements, the urgency of payment come to the fore [10]. Cash equivalents and quasi-money can also be referred to the principle of non-cash settlements on bank accounts.

In our opinion, the concept of "payment technologies" can be defined as a set of methods, software and hardware and technological means aimed at the development and improvement of the mechanism of non-cash settlements in the retail payment system in the payment infrastructure and functioning on the basis of the principles of organization of non-cash turnover and regulatory acts. On the example of the mechanism of POS (Point of sale) acquiring functioning it is possible to note zones of payment technologies application [11]. Considering this process, we can divide it into technological elements in the breakdown of categories similar to those of payment innovation. Dotted areas in Fig. 1 represent payment technologies in use, with the potential for transformation or development of technologies at each stage of settlement highlighted here.

Considering the practice of foreign studies of payment services market, we can note the significant role of payment services providers in the mechanism of settlements with customers. Due to state regulation of trade concessions in European countries, the cost of non-cash payments in retail outlets, which is carried by the trading company on average 10 times less compared to the payment market in Russia. Hence, there is a large supply of payment services by clearing and non-cash settlement companies. Therefore, in

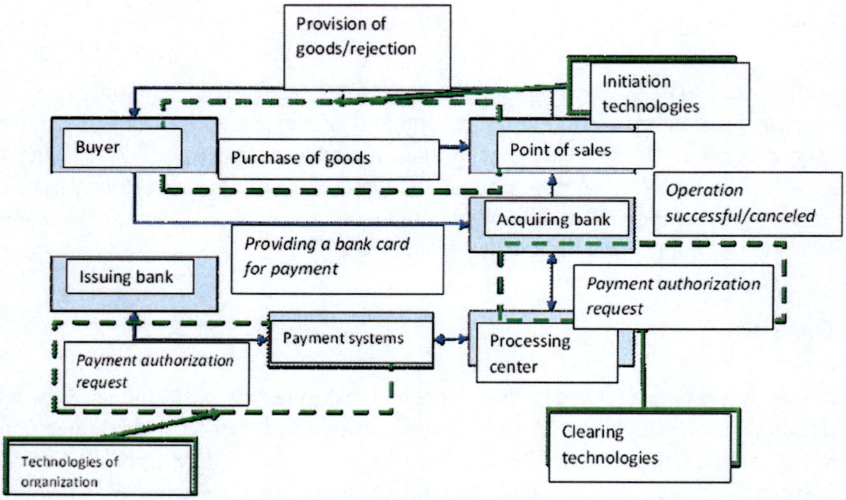

Fig. 1. Acquiring settlements (in terms of payment technologies).

order to achieve their competitive advantage, payment service provider services (PSPS) develop and improve payment technologies in their processes. For example, achieving a competitive advantage is possible by reducing the speed of the transaction or by providing additional customer loyalty services. Analyzing market trends, interpreting reports of information agencies in the field of research of cashless payments, it is reasonable to introduce the classification of different trends in payments and types of technologies used, as well as taking into account the target reference point of application of these technologies.

Figure 2 presents payment technologies in terms of focus of application, current platforms and technologies (including early-stage technologies and regulations depending on the country of implementation). Additionally it is worth mentioning that in accordance with the Strategy of national payment system of Russia one of the target objectives of the Bank of Russia is the implementation of international standard ISO 20022, an international standard to ensure seamless payments and interoperability of participants in the payment market [12].

Separately, it is worth noting the potential environment for the introduction of payment technology. As noted earlier on the example of acquiring the process can be divided into technological elements. The development of these elements on the part of payment service providers in the context of extensive development of demand for payment services, is determined by such concepts as PAAS, BAAS, Payment factories, IHB. According to foreign studies in the field of retail payments in the fintech industry distinguish the concept of PAAS (Payment-as-a-service) [13]. Adhering to this concept, payment technology is accepted as a software architecture that provides secure two-way communication with the payment system of the country. This concept reflects the study of payment technologies from the side of a set of tools that provide cashless turnover. On the other hand, as a result of the transformation of operational models and new technologies in banking in foreign practice and with the last 5 years of Russian practice introduced

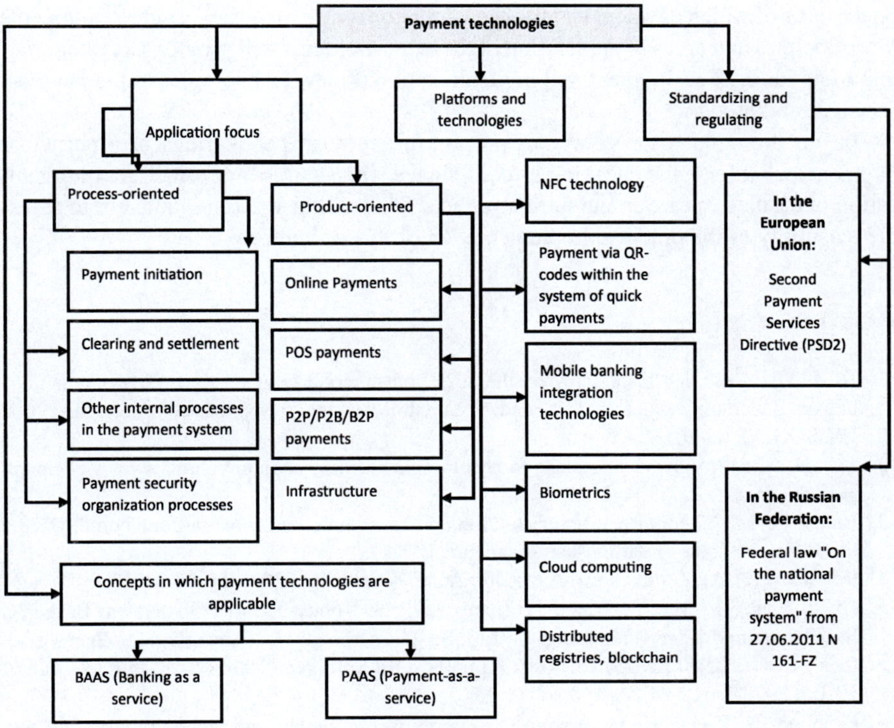

Fig. 2. Structure of payment technology industry.

the approach BAAS (Banking as a service), A key feature of which is cloud technology and architecture based on API (Application Programming Interface) [14] and flexible business processes (Agile, Scrum) [15]. The application of PAAS and BAAS concepts reflects the technological component of the process and, in our opinion, opens a prospect for the transformation of the current mechanisms of operation of non-cash payments and payment technologies.

International scientific studies in the field of payments and corporate finance also separately distinguish the practice of "In-house banking" (IHB) [16]. In-house banking automates the clearing of settlements of companies within the group, including consolidation of cash balances and programming of automatic payments. The concept of "payment factories" functions similarly, with a slight difference in the direction of automated payments. With the high rate of development of information technology, the concepts described above form a favorable environment for the development or implementation of payment technologies.

4 Conclusion

Under conditions of the growing importance of payment technologies in the payment industry for the development of theoretical foundations a clear interpretation of the concept of payment technologies is necessary, and systematization of features and principles on the basis of which payment technologies function is also necessary. Understanding the key principles and areas of application of these technologies will provide a methodological basis for the development and implementation of new technologies in the payment services market.

In our opinion, in the course of purposeful study of peculiarities of a particular payment technology and checking its compliance with a number of principles of organization of cashless turnover and normative acts. This will be a certain indicator to assess the feasibility of the practical implementation of this technology.

References

1. IBM, NRF Research Insights: Meet the 2020 consumers driving change. Why brands must deliver on omnipresence, agility, and sustainability. https://www.ibm.com/downloads/cas/EXK4XKX8 (2020)
2. Savage, M.: Why Sweden is close to becoming a cashless economy. https://www.bbc.com/news/business-41095004 (2017)
3. Beresford, M.: The future of luxury retailing postpandemic. https://edgardunn.com/2021/02/the-future-of-luxury-retailing-post-pandemic (2021)
4. Belousova, A.A., Zemlyakova, A.V.: Bul. Acad. Knowledge **4**(23), 50–53 (2017)
5. Hu, B., Xie, H., Ma, Y., Wang, J., Zhang, L.-J.: A Robust Retail POS System Based on Blockchain and Edge Computing. In: Liu, S., Tekinerdogan, B., Aoyama, M., Zhang, L.-J. (eds.) EDGE 2018. LNCS, vol. 10973, pp. 99–110. Springer, Cham (2018). https://doi.org/10.1007/978-3-319-94340-4_8
6. Holotiuk, F., Pisani, F., Moormann, J.: The impact of blockchain technology on business models in the payments industry. In: International Conference on Wirtschaftsinformatik, St. Gallen, vol. 13 (2017)
7. Xiao, X., Hedman, J., Runnemark, E.: Use of payment technology: a perspective based on theory of consumption value. In: ECIS 2015 Research-in-Progress Papers, vol. 40 (2015)
8. Korobeinikova, O.M.: Theory Prac. Soc. Devel. **11**, 77–80 (2017)
9. Information Technologies. Great Russian Encyclopedia. https://bigenc.ru/technology_and_technique/text/2016059
10. Burlakov, D.V.: Finance, Money and Credit. Yurayt Publishing House, Moscow (2019)
11. Vibha, R.K., Pandey, U.S.: Research objectives and overview of wireless technologies in payment systems. In: Proceedings of International Conference on Reliability, InfoCom, Technology and Optimization (IRCITO) (2010)
12. Central Bank of the Russian Federation: Development strategy of the national payment system for 2021–2023 economy. http://www.cbr.ru/Content/Document/File/120210/strategy_nps_2021-2023.pdf (2020)
13. Ponamorenko, V.E.: Legal Sci. **5**, 113–120 (2017)
14. Skinner, C.: Digital Bank. Strategies for launching or becoming a Digital Bank. Mann, Ivanov and Ferber (2015)
15. Bruno, P., Townsend, Z., Zell, J.: The emerging era of PaaS: "payments as a service". McKinsey Global Payments Report 2019: Amid sustained growth, accelerating challenges demand bold actions. https://ru.scribd.com/document/433880893/McKinsey-Global-Payments-Report-2019 (2019)
16. Wölfing, D., Moormann, J.: J. Pay. Strategy Sys. **11**(4), 343–354 (2018)

Integrating Pension Funds into Financial Ecosystems

Yulia Yu. Finogenova(✉)

Plekhanov Russian University of Economics, Moscow, Russia
finogenova.yy@rea.ru

Abstract. The article developed a new topic – the possibility of integrating pension products offered by pension funds with ecosystems. They gain greater reach of potential customers, faster information exchange, and a better understanding of current customer needs while integrating with ecosystems, which for non-state pension funds are primarily financial marketplaces. National platforms-marketplaces should connect participants at the level of providing information about services and value creation processes, hence the efforts of funds to transfer relevant data to a suitable platform are justified in implementing long-term strategies. The most significant result of the study was the conclusion that the standardization of pension products offered to the market, which will be based on the guarantee of the return of the originally invested capital, is the optimal solution for pension funds, with which they could enter the marketplaces. Speaking about the entire world, it is necessary to mention that the basis for the development of such products is the European standardized Pan-European private pension product, and in Russia – the guaranteed pension plan. The incentive for the development of these products can be tax incentives, through the provision of tax deductions, as well as the development of integrated pension proposals, where in addition to the pension product, insurance protection against related risks of individuals will be offered.

Keywords: Non-state pension funds · Standardized pension products · Pension funds' investment strategy · Financial marketplaces

1 Introduction

An ecosystem is an interconnected set of services that allows users to meet different needs on a single marketplace.

According to [1], "the mission of ecosystems has shifted from access to innovative funds to the joint creation of industry 4.0 solutions, and then to the joint creation of intelligent business solutions".

If traditional industry limitations disappear, ecosystems and digital platforms will provoke a significant impact on the future of financial institutions. Different studies have shown that ecosystems will generate $60 trillion in revenue by 2025, since 30% of global sales [2].

P. V. Trifonov and M. V. Charaeva (Eds.): DITEM 2021, LNNS 380, pp. 657–663, 2022.
https://doi.org/10.1007/978-3-030-94245-8_90

The popularity of marketplaces is promoted by the opinion that consumer risk on marketplaces is reduced, due to the triggering of the signal role of information about the product and the seller [3, 4]. According to Meents, Verhagen, 2020, "although information about the product and the seller is indeed an important signal of risk reduction and, as such, can play an important role in stimulating purchases, the potential for the risk-reducing of these information forms differs depending on the types of risks studied" [5].

This study is novel in that the authors speak about the prospects for entering the marketplaces of pension funds, with a focus on the Russian pension market, which is waiting for the next round of reforms of the funded pension. The study hypothesis allows us to expect that marketplaces should be considered as a way to monetize the classic components of the value chain of financial institutions: product development, pricing, service provision and claims management.

The purpose of the study is to study the role of pension funds within the marketplaces development, as well as tools that need to be developed in order to maximize the coverage of the customer base.

The actual tasks are as follows: to study the trends within the development of ecosystems and financial marketplaces in the Russian Federation and the world; to present the structure of the marketplace that fits into the Russian regulatory framework; to determine the main components of pension products that would be more in demand on financial marketplaces.

2 Methods

In order to solve the tasks, the authors of this article chose the following methods: analysis, synthesis, induction, observation. The methodological basis of this study is the general scientific research methods: logical and causal analysis, generalization, classification, comparison, induction and deduction, grouping.

2.1 Identification of the Pension Funds Role in the Ecosystems Context

Within the framework of ecosystems, funds may perform two functions: first, they can be an ecosystem participant and, secondly, its organizer. All in all, participation means offering an additional pension provision as a service by integrating with existing platforms that offer new distribution channels and new access to customers.

2.2 Integration of Pension Funds with Digital Ecosystems

During the integration process, it is necessary to follow a certain algorithm:

- to develop a development strategy taking into account the available resource base in comparison with other industry players in order to determine their role in the digital space (which ecosystems they would like to join, and what their services may consist of except selling pension products);

- to identify the most significant options for using marketplaces that benefit both the client and the fund at the same time;
- to clearly identify the value of the products and tools they offer for the ecosystem, to create a platform that facilitates integration and identify incentives for joining other partners, to work out the process of monetization of cooperation.

As for most of the funds, connecting to the marketplace is really a serious step, so it is urgent to join forces and create industry platforms and, possibly, to exchange services with each other and other participants.

2.3 Standardization of Pension Products

Standardization will allow the exchange of services between funds and between industry players, and now there is a process of "initial rules formation". In the conditions of increased volatility of the capital market and uncertainty about the amount of future income received among the population with low and middle income, savings products that guarantee the return of capital have become increasingly popular. The McKinsey study shows that groups of respondents (at least 1,000 citizens in each country) with low and middle income levels, compared with more affluent groups of citizens, prefer guarantees of return on invested funds to the detriment of profitability [2].

3 Results

3.1 Features of Russian Marketplaces

The Bank of Russia plans to increase the availability of financial products and services for the majority of citizens. If the competition between financial service providers increases, then the commissions and the spread between the rates of loans and deposits in different banks will decrease.

Not only deposits, but also insurance products and mutual fund units will be traded on the created marketplaces, because as the financial instruments offered grow, the attractiveness of the marketplace for potential users will grow as well.

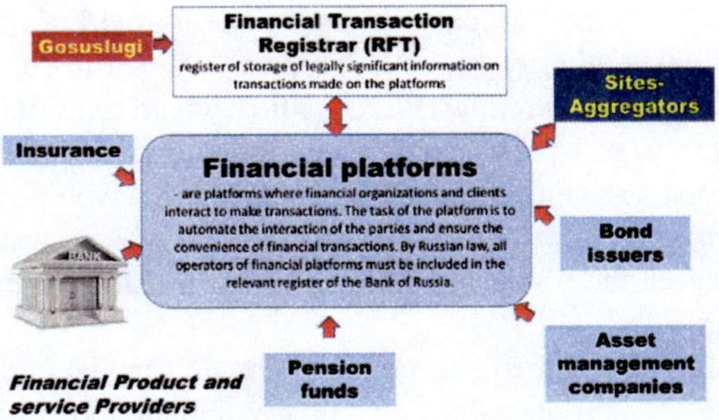

Fig. 1. Bank of Russia marketplace – structure [6].

As it can be seen from Fig. 1, the marketplace is a system that combines platforms for financial transactions, storefronts for collecting and presenting information about financial products/services and bots, specialized algorithmic consultants for selecting products/services for end-users – individual persons [6].

The system's infrastructure will be combined with a platform for remote identification, which will allow transferring financial services to a digital environment and increasing their accessibility for consumers. While building the marketplace, such innovative developments as open interfaces (Open API) and the fast payments platform will also be used [7].

The main advantages of the concept implemented by the Bank of Russia are: (1) improving the situation with accessibility; (2) remote access of users to financial services 24 h a day; (3) a variety of financial products (bank deposits, government and corporate bonds, insurance products). In the nearest future, this list can be expanded at the expense of credit, investment and pension financial products.

Along with the advantages of developing marketplaces, there are also a number of disadvantages. For example, large financial organizations are developing their own ecosystems in which customers can order the same products, but from one supplier [8]. Thus, there is a monopolization not only of the financial market, but also of other economic sectors by financial organizations. For example, in Russia, Sber Bank is actively developing its own ecosystem.

Possible risks of developing marketplaces for the National Pension Funds (NPFs):

1. NPFs will pay a commission to the marketplace for the provision of services, which should cover the operating costs of the platform. Therefore, this will lead to an increase in the cost of the service.
2. The platform will identify the user by biometric data that he needs to submit in advance in order to use the service. Biometrics implies extremely sensitive information that cannot be replaced in the event of media leak: documents and passwords are replaceable, but in the case of biometrics, it is problematic to replace the data.
3. And the last question – will all the financial platforms be able to guarantee the reliability and security of customers' personal data?

3.2 Standardization of Pension Products in the Russian Federation and in the EU

1. EU practice. In 2013, the European Commission and the European Insurance and Occupational Pensions Authority (EIOPA) proposed the formation of a standardized pan-European private pension product (PEPP), which would be available in the 28 member States of the European Union. The potential purchasers of PEPP can be not only employed citizens, but also self-employed and even unemployed persons.

The PEPP suppliers or distributors are obliged to provide citizens with appropriate recommendations about the product, to inform about investment options and the level of capital protection.

The advantage of PEPP is the focus on preserving the capital of depositors at retirement, taking into account contributions at the accumulation stage after deducting all taxes and fees.

One of the key features of PEPP is to limit the costs for providers and include them in the marginal cost of the service – in the amount of no more than 1% of the accumulated capital per year. In addition, each client will receive a key information document (KID). This document is formed by an asset management company and contains information about how the risks of sustainable development are integrated into the investment decisions that underlie the proposed product [9].

Thus, the advantages for depositors in comparison with traditional products are as follows:

– PEPP will contain such investment options, using which the depositor at least receives a guarantee of return on the invested capital, along with a certain number of alternative investment options (up to 6 options).
– Service providers will have the opportunity to offer PEPP throughout the EU, which will allow depositors to move across borders without necessities to change the product. PEPP will also allow switching between service providers periodically, every five years.

2. The Russian Federation practice. Currently, one of the reasons for the dynamics lack in the development of non-state pension provision is the low level of Russian citizens' salaries and the lack of necessary financial resources for their participation in voluntary pension programs.

In 2019, the creation of a new accumulative system was announced in the Russian Federation. It was a guaranteed pension product (GPP), although its introduction was not formalized by law. The concept of a guaranteed pension plan provides a number of benefits. In particular, for individuals, a tax deduction is proposed for the contribution amount to the GPP, but not more than 6% of income. One of the important points of the concept is the possibility for citizens to convert their pension savings into GPP within the framework of mandatory pension insurance. The GPP scheme is demonstrated in Fig. 2.

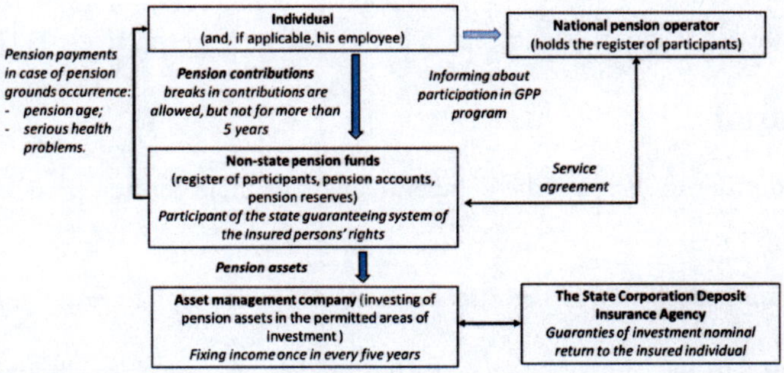

Fig. 2. An illustration of the proposed structure of the Guaranteed Pension Plan (GPP) in Russia. *Source:* Compiled by the author on the basis of [10].

If a Guaranteed Pension Plan is adopted, the role of non-state pension funds (NPFs) is not completely clear, since the investment of pension savings will be carried out by management companies, accounting accounts and pension schemes, as well as by the pension operator.

Nevertheless, the GPP can become an attractive product for the non-state pension insurance industry, allowing citizens to carry out voluntary pension savings and similar to the PEPP pension plan, which is planned to be implemented in the EU countries in the nearest future.

4 Discussion

In order to increase the attractiveness of pension products while promoting through marketplaces in Russia, pension funds need, on the one hand, to consider the possibility of integration with insurance products, and on the other hand, to review the principles of investing pension assets. It is necessary in order to increase the product attractiveness for potential customers.

1. Cooperation between NPFs and life insurance companies. As foreign experience shows, the activities of these financial institutions are quite organically combined, therefore, options for their joint cooperation are also potentially possible in Russia.

In addition, in many countries of the world there is a practice of transferring the formed pension savings to insurance companies at the time of pensions payment through the purchase of a lifetime or term annuity.

Such cooperation will be facilitated by the adoption of the law on the guaranteed pension plan (GPP), as well as the decision of the Bank of Russia to launch individual investment accounts (IIA) of the third type (designed for long-term investments for a period of more than 10 years and assuming accompanying tax deductions and benefits for investors).

2. Investing pension assets. In Russia, the profitability of investing pension savings and reserves in 2020 remained above the level of inflation, but decreased relative to 2019. One of the reasons for the decline was the negative results of the first quarter of 2020, caused by the negative stock market conditions due to the spread of COVID-19 in the world and in Russia. The gross return on investment of pension savings and reserves at the end of 2020 was 7.2 and 5.8%, respectively (−3.5 p.p. and −2.2 p.p. for the year). Net profitability indicators also decreased to 5.2 and 4.6%, respectively [11].

5 Conclusion

Therefore, in order to increase the competitiveness of their pension products, pension funds need:

1. To develop a portfolio of investment assets to create alpha returns and generate attractive returns for clients.
2. To reconsider and simplify the operating model, as well as to digitize the main business processes, which will allow to reduce administrative costs at least twice.
3. To actively use the opportunities of growing ecosystems that provide contact with potential customers, to form standardized products that will be sold to individuals through financial marketplaces.

References

1. Benitez, G.B., Ayala, N.F., Frank, A.G.: Industry 4.0 innovation ecosystems: An evolutionary perspective on value cocreation. Int. J. Prod. Econ. **228**, 107735 (2020). https://doi.org/10.1016/j.ijpe.2020.107735
2. McKinsey: Global Insurance Industry Insights. An in-Depth Perspective. McKinsey Global Insurance Pools (2017)
3. McKinsey: Competing in a world of sectors without borders. McKinsey Quarterly (July 12, 2017) (2017). www.mckinsey.com/industries/financial-services/our-insights/ecosystems-and-platforms-how-insurers-can-turn-vision-into-reality
4. Rawlins, J.M., De Lange, W.J., Gavin, C.G., Fraser, N.: Ecosystem service value chain analysis Framework: a conceptual paper. Ecol. Econ. **147**, 84–95 (2018). https://doi.org/10.1016/j.ecolecon.2017.12.023
5. Meents, S., Verhagen, T.: Reducing consumer risk in electronic marketplaces: the signaling role of product and seller information. Comput. Hum. Behav. **86**, 205–217 (2018). https://doi.org/10.1016/j.chb.2018.04.047
6. The Bank of Russia. The "Marketplace" project. https://www.cbr.ru/fintech/market_place
7. Haddad, C., Hornuf, L.: The emergence of the global fintech market: economic and technological determinants. Small Bus. Econ. **53**(1), 81–105 (2018). https://doi.org/10.1007/s11187-018-9991-x
8. The Bank of Russia: A report for public consultations. New approaches of the Bank of Russia to the development of competition in the financial market. Moscow, Bank of Russia (2019)
9. European Federation of Insurance Intermediaries (BIPAR). PEPP Background. https://www.bipar.eu/en/page/pepp
10. Government of the Russian Federation: Draft of federal law "On amendments to certain Legislative Acts of the Russian Federation on non-State pension provision" (September 17, 2019) (2019). https://release.minfin.gov.ru/ru/document/npa_projects/?id_4=5993-o_vnesenii_izmenenii_v_otdelnye_zakonodatelnye_akty_rossiiskoi_federatsii_po_voprosam_negosudarstvennogo_pensionnogo_obespecheniya
11. The Bank of Russia: Review of Key Indicators of Non-State Pension Funds. Bank of Russian, Moscow (2021)

Business Education and New Economic Realities: Challenges or Opportunities

Modelling of Economic Clusters in the Russian Regions in the Context of Formation of the Industry 4.0

Natalya G. Vovchenko[1](✉), Olga V. Andreeva[1], Valeria D. Dmitrieva[1], Svetlana S. Galazova[2], and Terziev V. Krstev[3,4,5]

[1] Rostov State University of Economics, Rostov-on-Don, Russia
[2] North Ossetian State University named after K.L. Khetagurov, Vladikavkaz, Russia
[3] Georgi Ravoski Military Academy, Sofia, Bulgaria
[4] Kanchev University Hospital, Ruse, Bulgaria
[5] Ruse University, Ruse, Bulgaria

Abstract. The Russian cluster system includes more than 140 clusters. No one of them is calculated and no one checked. Some of them function in reality but others are paper items. Taking into account the peculiarities and problems of the Russian clusters the authors notice that the digital transformation and formation of the Industry 4.0 becomes a serious challenge to all the enterprises and a threat to those who cannot "catch" the trends. The focus of the Industry 4.0 is neo-industrialization based on digital, smart-, nano- and biotechnologies, quantum computing and robotics, resource saving. The Russian clusters are formed in the traditional branches, they use mainly the common system of contracts as the search of a new partner leads to the growth of transaction costs. The question arises if such clusters could exist in the context of the Industry 4.0. Methodologically while there are no actual statistics the research is based upon the institutional approach at which the revealing of interconnections between the individual organizations inside and outside the structure has a priority: the workers share their knowledge and experience with each other in the evident and non-evident manner, and the enterprises and organizations in the cluster share their contracts, standards and instructions. The authors conclude of the necessity to revise and restructure current cluster system. On the contrary, it is very important for the innovative activity to have a flexible structure, a new specialization and to invest into the new knowledge.

Keywords: Cluster policy · Neoindustrialization · Knowledge economy · Digitalization

1 Introduction

During the last two decades, the cluster approach is actively applied within the frameworks of some concepts of socio-economic, branch and spatial development, it presupposes the segmenting of the territories of the region or a group of regions according to the branch feature taking into account the interfirm connections and localization of the sale markets.

P. V. Trifonov and M. V. Charaeva (Eds.): DITEM 2021, LNNS 380, pp. 667–673, 2022.
https://doi.org/10.1007/978-3-030-94245-8_91

The innovation clusters, and there were created more than a hundred of them on the territory of Russia, scientific, industrial, IT- and technoparks, technological valleys (the latter exist in the form of projects only) represent close structures that are so close in type, that sometimes there are no discrepancies between them in understanding. It is difficult to separate the notions when there is no clear methodological guideline and regulatory documentation is quite complicated. Thus, Silicon Valley in the USA is considered to be a cluster and a technological valley; in Russia the technological valley "Vorobyovy gory" on the base of the Moscow State University has been projected for several years already and it is also called a cluster by the authors of the project. However, the substitution of notions would not be critical if these structures no matter how they are called could bring the result they were created for.

Despite the long term of implementation of the cluster approach to the projecting of the innovative systems of the regions of the Russian Federation the efficiency of such structures in most cases is quite low (with rare exception) and the activity itself is actually invisible. In reality such structures exist often on paper only or in the form of associations or consortiums that does not correspond to the formal requirements specified to the cluster initiatives.

In the context of the global transformation of the economic systems under the impact of such significant external forces that change the arrangement on the political map and transform the whole branches, only those complicated structures can survive and exist that respond to the challenges of the time and are capable to generate the real innovations.

The success of the organization in the modern realities depends upon how fast it adapts to the changes, generates the new knowledge and includes into the global chains of creation of innovations. Neither state nor region or enterprise can "Be smart and alone" [1]. Theoretically the network innovative structures were created for joint activity through the achievement of synergistic effect. Moreover, it was presupposed that this synergetic effect would create not only the additional value for the organizations being the participants of such structures but also the favorable socio-economic conditions for the development of the territories where they are located.

A new focus of the modern transformation of the economic systems is neoindustrialization the base of which is the made up of digital, smart-, nano- and biotechnologies, quantum computing and robotics, resource saving. Together with the development of technologies the image of the modern production shall change as well.

The main objective of this research is to determine the possibilities of re-modelling of the cluster system in Russia in the context of transformation 4.0. As a result, the fair question arises: can the Russian clusters correspond to the requirements of the Industry 4.0?

2 Methods

The Russian cluster system is rather complicated for understanding. Its complicity is not in the peculiarities of the structure but in its "unclear" character: clusters registered in the register of the Russian cluster observatory are not always recognized by the Government and the regional authorities. Furthermore, the formal indicators of the activity of clusters are not available for public or are absent. Thus the data collected for analysis are like 'patchwork" quilt.

Methodologically in the absence of the actual statistics the research is based upon the institutional approach where the revealing of the interconnections between the individual organizations inside and outside the structure is a priority. In such a sense any cluster is not only the organizations and enterprises but the people who work there [2]. The interpersonal relations, exchange of experience and timely information, staff turnover gives a dynamic to the cluster. The more numerous the connections are the larger and stronger the cluster is.

3 Results and Discussion

3.1 Why Do not Economic Clusters Work Properly in Russia?

Approximately 140 economic clusters function on the territory of regions of the Russian Federation. In most cases, their functioning cannot be taken literally as they do not perform their functions and exist on paper only. This happens due to many reasons, and those guilty are government, business and society.

The first reason of inefficiency of the domestic cluster (and generally, network) structures is in the failure to understand the essence of innovation as it is. The innovative character of functioning of a cluster organization is its key parameter. In this respect in the legislation and scientific literature, there is no clear definition of innovations and innovative activity. The cluster itself is an innovative structure and all its activity shall be innovative by definition.

Here is the second reason: the definition of a cluster and the purpose of its creation are often unclear in the national economic practice. It was mentioned above that the notions are substituted and the cluster is understood as industrial or techno-park (or vice versa) and sometimes the cluster is identified by mistake as one or another branch in the regional context.

The uncertainty of the definition of clusters creates the problem of their identification and recognition. As a result Ministry of Economic Development of the Russian Federation, Ministry of Industry and Trade of the Russian Federation, Russian cluster observatory, National informational and analytical centre for monitoring of innovative infrastructure of the scientific and technological activity and regional innovative system create the individual lists of cluster initiatives. Specifically these lists rarely coincide or intersect with each other.

And one more problem comes from it: difficulties in identification of economic clusters lead to the absence of control from the government bodies and authorities as it is not clear who should supervise it. And what is more important – what criteria for estimation are. The absence of estimation criteria of the functioning efficiency of clusters and external control show the non-transparency of activity of innovative structures. In 2020 the Government of Astrakhan Region stopped the activity of the local business-incubators in Astrakhan, Akhtubinsk, Kamyzyak and Olya village after many inspections during which the spending of federal funds and use of the property for unauthorized purposes were revealed. As the local mass media say in one of the buildings of business-incubators the Olya village the hotel was organized and in Kamyzyak the facilities of the business-incubator were leased to the ordinary self-employed entrepreneurs (tourist agencies, beauty salons, etc.) [3]. This case is a bright example of all that was said above:

failure to understand the essence of innovations and economic clusters as they are, aims of their existence and absence of constant control from the side to the authorities led to the connivance and non-coverable financial losses.

The third reason is the absence of the public control the reason of which is a simple unawareness: an average Russian citizen is hardly interested in the problems of formation of cluster systems, he does not even know about their existence in his region.

In the opinion of the authors of the research, the main objective of functioning of the economic clusters is the increase of welfare level of the territory at the expense of the positive social (synergetic) effect from carrying out the coordinated policy. Other tasks written in many programs of the cluster development (increase of competitiveness of the Russian enterprises, attraction of investments into the region, development of the regional infrastructure, exit mobility of labor, development of small and medium-sized entrepreneurship, saturation of domestic market, import phase-out) just assist the achievement of the main objective [4].

3.2 How to Create an Economic Cluster V4.0?

The concept of cluster development of the economy and the Industry 4.0 is quite rarely considered in conformity to each other. However, it does not mean they are incompatible.

The main features of the forming Industry 4.0 are:

- Digitalization and virtualization;
- Application of Big Data and technologies IoT;
- sharing economy;
- decentralization and distribution (dissipation) of processes;
- vertical integration and network integration;
- creation of value chains;
- predominance of tertiary sector;
- investments into development of skills and life-long learning, etc.

Actually, all these features shall be peculiar to the cluster structures with the exception of the circumstance that organization of the Industry 4.0 related with the help of ICT does not require geographical proximity, at the same time the clusters, on the contrary, are meant to create the agglomerative effect. This ability to "relate" the territory of the region is one of the key abilities for the economic cluster. Its other important ability is to enforce the regional specialization and competitiveness.

However, actually the cluster can relate not only one region but several. And not only in one country. The cluster can be distributed around the world. Thus the boundaries of the agglomerative clusters are much wider than the limits of the real interfirm interaction.

The economic clusters are an advanced format of network interaction. Traditionally the following peculiarities are typical for clusters as networks structures:

- restricted by a certain branch [5];
- includes vertical and horizontal relations in the branch [6];
- participants of the cluster manufacture the homogeneous or related products and consequently function in the same market conditions [7–11];

- branch market remains competitive, development of cluster is at the expense of the high productivity based upon the specialization [12];
- all participants of cluster have a public access to the labor market and infrastructure [6];
- community of infrastructure promotes the accelerated sharing of knowledge and technologies [5];
- territorial proximity of enterprises grows strong and trust relations between the participants of the cluster [7–10, 13, 14];
- support of non-commercial organizations, universities and research and development centers [15];
- within the frameworks of the cluster transport expenses and delivery chains can be decreased [15];
- high concentration of companies increases the sale market for each of them [8–10].

Clusters have a significant impact on competitiveness and cooperation of firms in the region and the branch, while excluding any tendency to monopolization of the branch and the firms participating in the cluster preserve its legal independence.

Technological innovations in the rapidly growing branches speed up the business-cycle, extend the production opportunities, create the sustainable competitive media and include the certain regions and countries into the global chains of creation of value. The specific character of the Russian cluster system is than the clusters are created mostly in the traditional branches and not in the new ones. In this case the creation of the new mechanisms and models of interaction between the organizations and enterprises leads, as a minimum, to the growth of transactions costs.

The Industry 4.0, presupposing the large-scale digital transformation of the branches constitutes a threat for those organizations that for some reasons cannot or have no time to make the digital modernization [16]. In the conditions of selection of the traditional branches for clusterization and inertia of the Russian economy in general it can become a critical factor [17]. On the other hand, the ability to adapt to the new digital realities shall be considered as one of the criterion of the cluster efficiency.

It can be definitely declared that the "cluster 4.0" shall have two new characteristics:

- to possess a flexible adaptive and, what is extremely important, structure distributed in the space;
- to invest into the development of own knowledge and competencies, especially competencies in demand by the Industry 4.0.

The effectiveness of clusters depends mainly upon the activity of the organizations managing them that shall promote the transfer of technologies, provide the inflow of the necessary knowledge and availability of the support measures [16]. This happens in reality but a lot of questions arise:

- shall a cluster self-organize its activity?
- in the conditions when business, science and authority do not understand the demands of each other, can a cluster "understand" what knowledge is required?

- what mechanisms shall be used for distribution of knowledge between the participants of the cluster?

The form of the economic "cluster 4.0" shall be determined by the mechanisms of distribution of knowledge and competencies [18].

4 Conclusion

This research is a consideration inspired by the article of M. Götz, "The Industry 4.0 Induced Agility and New Skills in Clusters". The author explains the idea of "compatibility" of the models of the Industry 4.0 and economic clusters.

In the work, it is emphasized that in the scientific literature the connection between the clusters and the Industry 4.0 [16] is almost not described and that initially suggests an idea that they cannot coexist: industrial character, local restrain and niche specialization of clusters cannot be related to the territorial and activity distribution and diversity of the Industry 4.0. However it is just at first sight.

Continuing the idea of M. Götz about "compatibility" of clusters and the Industry 4.0 the authors of the present research study the problem from the point of view of the institutional approach. The relations and contracts inside the clusters as the network structures are at all levels and differ by the diversity of forms: workers share their knowledge and experience with each other in the evident and non-evident manner, and the enterprises and organizations in the cluster share their contracts, standards and instructions. The distribution of knowledge outside the cluster is also possible. The establishment of such multilevel system of relations within the frameworks of the cluster corresponds to a certain extent to the characteristics of the Industry 4.0.

On the other hand, the analysis of intra- and intercluster connections brings the authors to the conclusion of the necessity to revise and restructure the existing cluster system: traditional branches, traditional contracts and subcontracts cannot create innovations. On the contrary, it is very important for the innovative activity to have a flexible adaptive structure, a new specialization and to invest into the new knowledge. The competencies to generate new knowledge qualitatively and rapidly are the skills determining the value of organization in the context of the Industry 4.0.

References

1. Kutsenko, E., Islankina, E., Kindras, A.: Foresight STI Gov. **12**(1), 25–45 (2018). https://doi.org/10.17323/2500-2597.2018.1.25.45
2. Lindqvist, G., Sölvell, Ö.: Organising Clusters for Innovation: Lessons from City Regions in Europe. Clusnet final report (2011)
3. Brovkina, M.: Why are Astrakhan business incubators empty. https://rg.ru/2018/10/30/reg-ufo/pochemu-pustuiut-astrahanskie-biznes-inkubatory.html
4. Dmitrieva, V.D.: Does Russia need economic clusters?. In: Proceeding of the International scientific conference "Science and Society – 2019". Moscow Witte University, Moscow (2019)
5. Krugman, P.: Geography and Trade. MIT Press/Leuven UP, London (1991)

6. Bergman, E., Feser, E.: Industrial and Regional Clusters: Concepts and Comparative Applications. (1999). http://www.rri.wvu.edu/WebBook/Bergman-Feser/contents.htm
7. Feser, E.: Old and New Theories of Industry Clusters. In: Steiner, M. (ed.) Clusters and Regional Specialisation: On Geography, Technology and Networks. Pion, London (1998)
8. Porter, M.: Harvard Bus Rev Nov-Dec 1998 (1998)
9. Porter, M.: Econ Devel Quart. 14(1) (2000). doi:https://doi.org/10.1177/089124240001400105
10. Porter, M.: Clusters and Economic Policy: Aligning Public Policy with the Economics of Competition. ISC White Paper (2007)
11. Simmie, J., Hart, D.: Euro. Plan Stud. **7**(4), 445–462 (1999). https://doi.org/10.1080/09654319908720529
12. Rosenfeld, S.: Industrial Strength Strategies: Regional Business Clusters and Public Policy. The Aspen Institute, Aspen (1995)
13. Morosini, P.: World Devel. **32**(2), 305–326 (2004). https://doi.org/10.1016/j.worlddev.2002.12.001
14. Swann, G.M.P., Prevezer, M.A.: Comparison of the Dynamics of Industrial Clustering in Computing and Biotechnology. Oxford University Press, Oxford (1998)
15. Boja, C.: Int. J. Econ. Prac. Theor. **1**(1), 34–43 (2011)
16. Götz, M.: Foresight STI Gov. **13**(2), 72–83 (2019). https://doi.org/10.17323/2500-2597.2019.2.72.83
17. Natalia, V., Olga, A., Valeria, D., Valeriya, K.: The National Technological Initiative as a Digital "Lego Bricks" for Industries: Some Prospects of the Modernization New Wave in the Russia's Economy. In: Aliev, R.A., Kacprzyk, J., Pedrycz, W., Jamshidi, M., Babanli, M., Sadikoglu, F.M. (eds.) ICAFS 2020. AISC, vol. 1306, pp. 346–353. Springer, Cham (2021). https://doi.org/10.1007/978-3-030-64058-3_43
18. Fallah, M.H., Ibrahim, S.: Knowledge Spillover and Innovation in Technological Clusters. In: Proceedings, IAMOT-2004 Conference, Washington, D.C (2004)

Business Education as a Component of the General Algorithm of Postgraduate Education

Tatiana S. Markova(✉)

Russian State Social University, Moscow, Russia
markowats@mal.ru

Abstract. The article discusses possible ways of postgraduate education for graduates of various specialties, mainly with the prospect of entering the global labor market. Particular attention is paid to business education as the most demanded by modern standards component of higher education in general. The role of business education is not only to prepare graduates for practice in the fourth industrial revolution, particularly in a crisis but also to make them able to take advantage of the potential opportunities offered by the business education they receive, such as mastering new prestigious professions such as HR business partner. The relevance of the article lies in the fact that it evaluates various criteria of possible options of additional education, in particular: the content of education, quality of education, recognition in the global labor market, compliance with the goals of the student, time and cost of training, additional stress, including psychological. The main methods used were analyzing articles, dissertation research, monographs, and Internet resources containing material on this topic; the method of content analysis, i.e., the study of texts and documents to select the optimal options; the classification method, i.e., grouping the collected data array by individual criteria; the method of comparing possible education options. The background of the study is the identification of specific factors influencing the choice of an appropriate algorithm for postgraduate training, taking into account the additional workload factor. The practical implications of the work are that it can really help a graduate navigate the options offered by Russian and foreign educational institutions of postgraduate education and choose the best option, taking into account various factors.

Keywords: Market demand · Factor · Business education · Qualification

1 Introduction

Our time is characterized as the era of the fourth industrial revolution. Twenty years have passed since the beginning of the millennium, so experts are already noting the new realities of the economy. These include the emergence of financial and technological platforms such as the Bitcoin cryptocurrency, Blockchain technology, etc., which can reduce transaction costs between the lender and the borrower and reformat the world of

P. V. Trifonov and M. V. Charaeva (Eds.): DITEM 2021, LNNS 380, pp. 674–683, 2022.
https://doi.org/10.1007/978-3-030-94245-8_92

finance. New business models of financial and technological start-ups determine their higher estimated value than the capitalization of traditional investment banking groups [1]. This is confirmed by German experts, who expect that a new asset class will be observed all over the world due to the infrastructure and regulatory framework created for digital assets, which are still the only marketable products on the market, and the Bitcoin market – in terms of market capitalization – is leading by a sufficient margin [2]. At the same time, there are many negative events in the crypto-asset sector-stolen bitcoins, thousands of deceived customers, market manipulation, money laundering, etc. In this regard, in Germany, for example, several steps have already been taken by the German legislative bodies to comprehensively regulate the crypto-asset sector [3].

Naturally, it is not enough for specialists to have knowledge obtained at the university to navigate in this situation. First of all, it concerns obtaining additional knowledge in the field of business, which raises the question of which way of postgraduate education is the most appropriate in terms of financial and time costs. In other words, it is necessary to build an algorithm of postgraduate education, one component of which should be business education. Without it, it is impossible to correctly navigate the current situation on the market in general and the labor market in particular.

Modern literature has repeatedly emphasized that one of the areas of postgraduate education is becoming business education, aimed primarily at obtaining practical skills. The reasons for this are not only the desire to create independent start-ups, but also dissatisfaction on the part of employers with the training of university graduates, namely the content of education and lack of practical readiness to make independent decisions, particularly in crises, while the modern market requires practitioners who can think independently and competently from an economic perspective, and most importantly can make effective decisions and take responsibility for decisions that are observed in the collective monograph "Increasing the innovative activity of industrial enterprises" [4]. Business development entails the emergence of new prestigious professions, one of which is HR-business partner or strategist, who understands the company's business, helps resolve conflicts in the team, and competently uses employees' potential. The HR Business Partner Training Seminar, held in Vienna May 18–20, 2017, discussed the differences between an HR Business Partner and an HR Manager and Human Capital Management Specialist. Dr. Christoph Wolf, in his report at the seminar, stressed that an HR business partner must have, among other things, basic knowledge in the field of labor legislation, contract law, working time standards, administrative control systems, the structure of work contracts, etc. [5].

Dissertation study of Anisimova, which notes the role of business education as one of the most important components of innovative education, characterizes the main differences between traditional economic education aiming to learn and understand general economic laws and processes, which is basic education and the target group of which is mostly young people, from business education, which, being an interdisciplinary field, is aimed at in-depth study of material, financial, and informational tools with which the administrative and managerial staff of institutions and entrepreneurs can influence the resources used to run a particular business [6]. The authors of the work "Improving the level of professional competencies using a virtual educational environment" offer information and practical web quest "In the world of IT professions", including tasks with elements of role-playing games, contributing to the formation of an idea about the professions of this sphere, about the requirements for specialists of this profile, as well as preparing for the choice of the path of their professional development [7]. The use of a virtual educational environment is certainly useful, but this web quest is intended mainly for pre-university education and aims to prepare applicants for the choice of the profession they are studying.

In the work of Sharova, the influence of the experience of previous crises on the existing market of Russian business education is noted. The author notes a decrease in demand for long-term educational training programs, an increase in demand for short training and the development of specific applied skills, and an increase in demand for innovative educational formats. In addition, changes in legislation require the creation of programs that reflect changes in legislative acts and regulatory documents [8].

Sergey Myasoedov, Director of the Institute of Business and Business Administration (IBBA) of RANEPA, President of the Russian Association of Business Education, speaking about current issues of business education quality assessment, underlines that the leading international accreditation holders and participants of internationally recognized business education rankings have an advantage when choosing a business school, which requires Russian business schools to take a course on international accreditation and participation in rankings, especially since the highest world accreditation AACSB International (International Association for the Development of University Business Schools) has only one business school in Russia so far, namely the Institute of Business and Business Administration (IBBA) of RANEPA. This means that it is necessary to revise the indicators that characterize the quality of domestic business education. Such indicators include the connection of the learning process with innovative technologies and the digital economy, which can ensure the competitiveness of companies in the world market; the orientation of business schools to the sustainable development of the ecosystem and the social responsibility of business [9].

2 Materials and Methods

The material for our analysis was articles devoted to this topic in periodical Russian and foreign publications, dissertation research, monographs, and books, as well as Internet resources, mainly over the past five years. The methods used are descriptive, analysis of scientific and methodical and periodical literature, and comparative-analytical methods in choosing the best ways to solve the problem.

3 Results and Discussion

The main ways to improve education after graduation are getting a second higher education; advanced training with various options; master's degree, residency training, MBA, self-education.

When choosing a specific option, i.e., when building an appropriate algorithm for postgraduate education, several factors should be considered. We include: 1) the presence of a specific first higher education, which determines the main goal – professional development in the received specialty or retraining; 2) motivation; 3) time factor; 4) financial factor; 5) psychological factor, additional load factor; 6) quality factor; 7) factor of official recognition of education received in the labor market, where we distinguish two options: a) the national labor market, b) the international labor market.

Based on these factors, the algorithm can have different options. In the case of first specialized education not directly related to economics and/or business, the question arises as to whether the goal is to a) improve skills in the chosen specialty; b) complete retrain; c) get a business education in addition to the chosen specialty. In the case of advanced training in an existing specialty (medicine, pedagogy), if this does not pursue the goal of obtaining additional business education, it is advisable to choose the master's degree, residency training, internship, and advanced training in the workplace, as an addition – self-education. The next step may be doctoral studies. If we are talking about complete retraining, i.e., a change of profession, it is advisable to prefer a second higher education since it gives fundamental knowledge of the newly chosen profession. When choosing a business education, professional development at the workplace, an MBA program may be recommended as a necessary additional component. In any case, self-education is necessary as an additional component.

An important factor when choosing one of the ways of postgraduation education is motivation. Recently, there has been a tendency in Russia for adults to receive higher education [10]. This is mainly due to obtaining a second higher education. At the same time, the main motives are increasing competitiveness in the labor market; the need for a second education for career growth, for example, to occupy a higher position or a position requiring knowledge in a related field; the desire to change the unsuccessfully chosen first profession, if it was chosen under pressure from parents, lost its prestige, etc.; the need to master new competencies under the influence of modern trends; the desire to acquire an additional circle of communication and connections; the desire to meet market demands; the desire to deepen knowledge in their profession. Analyzing the structure of motivation to obtain higher, including second, education, Krushelnitskaya O. I. et al. distinguish two clusters of motives, grouped around the motives of material well-being and self-actualization and the motive of professionalization [11].

Similarly, although not in all respects, the motivation to receive a second (or first) higher education, regardless of specialty, is assessed in Germany. The works of German psychologists and sociologists, including several works by Hans Dietrich, discuss the reasons that encourage young people to get a first or second education after the age of 25, starting almost from scratch. The author notes that this fact was overlooked by a number of sociologists studying the processes of education in Germany [12, 13]. According to the author himself, one of the reasons is the desire to improve one's social status and stabilize one's career [14]. Another reason is the social origin, which is relevant for the environment of emigrants in Germany [15]. Besides, as the choice of the first place of work influences the subsequent career growth, the motive for getting the second education can be a mistake in choosing the first education, for example, on the advice of parents and friends [16].

The motive of striving to improve the social or material status inclines to choose an MBA program, a master's degree, a doctoral degree, advanced training at the place of work. The initial mistake in choosing a profession involves getting a second higher education. In both cases, especially when trying to master new competencies that meet modern trends and market demands, it is logical to choose business education as one of the components of obtaining additional education.

An important factor determining the choice of the option of postgraduate education improvement is the financial factor. At first glance, one of the most attractive ways is self-education, since it has several advantages: it allows planning your time easily, you only have to pay for books and the use of Internet resources. Self-education helps to develop self-discipline skills. In addition, there are no restrictions on the learning pace. At the same time, self-education has its drawbacks. One of them is the motivation factor, which requires an iron will not to deviate from the planned schedule. The disadvantages of self-education include, depending on the chosen field of study, the lack of practice, which is especially necessary for engineering and medicine, but in principle, is required in any field of activity. Often you need the assistance of a more competent person who can help correct mistakes, give the right advice, help highlight the essentials and the minor, etc.

In self-education, it is important to choose the right sources, the main of which are books and articles in scientific journals, which can help almost free of charge to get the knowledge that, from the user's point of view, is most needed at a given time. You can find articles offering, for example, an analysis of the innovative digital business models needed to accelerate the growth of start-ups, describing the basic configurations of successful business models, etc. The content of the article can be found in its abstract [17]. However, to read most of these articles, it is desirable to know the original language, since even if there is a translation or an opportunity to get an Internet translation, it should be taken into account that the translation is not always perfect and not even because of insufficient knowledge of the language or specialty of the translator, but because of a certain "knowledge conflict" and confrontation of concepts in the contexts of the source and the target languages [18]. Thus, in the fields of law and accounting, articles by foreign authors are often useless since the knowledge here is mainly purely national. However, it is also necessary to get acquainted with foreign basic legal acts if you have the prospect of entering the global labor market.

However, the biggest disadvantage of self-education in Russia is the attitude of employers to this method of professional development. The company's management, no matter how advanced and loyal it may be, requires confirmation of the qualifications of employees with appropriate educational documents. Therefore, following the results of the self-education process, it is advisable to take courses or pass exams externally, for example, at a university, to get a diploma, certificate, or another document. Thus, self-education is good as an additional component of the algorithm of postgraduation education.

Another component in the algorithm for obtaining additional knowledge and, most importantly, skills in the specialty can be participation in internships, which, incidentally, is possible not only at the end of high school but also during studentship. As the author of the article notes, "It is not fashionable to be an amateur!" Natalia Stoyan, summing up the results of the All-Russian student Olympiad "I am a professional," participation in internships allows students to gain work experience in their specialty and thereby reduce the discrepancy between university graduates and the requirements of the labor market, which is expensive for the state economy [19]. An internship abroad in a foreign company is especially valuable, which means not just another line in the CV of an applicant for a position, but the necessary experience gained by an intern, familiarity with technologies, and methods of working in a foreign company. After completing an internship program related to business education, graduates often open their businesses or change their profession. With a favorable impression made during the internship on the management of a foreign company, it is possible to sign a contract at the place of internship. Since an internship abroad involves daily communication in a foreign language, replenishing the stock of specialized vocabulary, and improving conversational skills in a language environment, often also obtaining a document on successful completion of language courses and working in a foreign company, this provides graduates with additional advantages in the labor market.

In some cases, though not always, it is possible to get a free graduate degree. The advantages of obtaining a master's degree are the multidisciplinary nature of master's programs, as well as the availability of highly specialized programs [20], and a draft determent. However, master's programs are chosen mainly by bachelors who plan to engage in research subsequently, teaching and/or management activities [21]. The majority of bachelor graduates have no motivation to study for a master's degree since, at present, a master's degree does not give any obvious advantages when applying for a job. Therefore, it is not considered a necessary stage of professional education [22], especially since it is associated with a certain time spent on training (2 years). In addition, the most prestigious master's programs are not free. The price of paid training varies from 40 thousand to 200–300 RUB thousand per year. For example, training under the program Business Management in the hospitality industry, which promises to teach how to bring new companies and projects to the market, manage start-ups, develop business concepts and business models in the hospitality sector, costs 300,000 RUB/year; the cost of training under the program Development and real estate management, which introduces the formation of concepts and business plans, is 300,000 RUB/year for full-time, 240,000 RUB/year for part-time education. However, at the end of the training, graduates receive a state diploma with a master's degree. However, it is necessary to pass entrance tests for admission, particularly written testing consisting of two blocks of questions (management theory, marketing theory) associated with additional psychological stress, which will be discussed below.

When considering the financial factor, one cannot ignore the result of receiving additional education, and the most important in this respect is its recognition in Russia and the global labor market. In this regard, the most effective, although the most expensive, is participation in MBA programs. Before choosing the appropriate program, you should get acquainted with the expert opinion. In this regard, the assessment of existing MBA programs given by the Dean of the Said Business School at the University of Oxford, Peter Tufano, is of interest. He writes about the obsolescence of the traditional model of business education and that innovations in business schools lag behind innovations in the business itself, resulting in graduates of traditional MBA programs being poorly prepared for the complex problems of leading companies. The author divides modern challenges into three main categories: first, a multitude of stakeholders, which encourages the creation of training programs that allow students to understand the needs and problems of all stakeholders but are limited to the interests of shareholders, economic benefits for shareholders. In this environment, companies have to compromise, balance different stakeholders, and decide which stakeholders to prioritize and make judgments about what is fair and what is not.

Moreover, since the traditional tools of the MBA, which are based on economics, data analysis theory, and psychology, are not suitable for preparing program graduates to solve questions about justice, since justice, unlike discounted value, is a much more subtle matter, this means that business schools should revise and expand their curricula [23]. Secondly, the author calls on business schools to include science in the expanded curricula to take into account environmental problems and study real climate change. According to the author, the third factor is the need for diversity, in particular ethnocultural diversity.

Svetlana Tretyak, a specialist of the Netology Online Educational Center, is also trying to understand the subtleties of obtaining a modern business education with the help of an MBA, according to whom the updating of MBA programs consists mainly in teaching program participants the competent use of digital technologies and the possession of digital tools since the age of digitalization dictates new business rules [24]. Since we are primarily interested in choosing an MBA program as one of the components of postgraduate education, we need to find out the pros and cons of its choice. This paper indicates that training in the MBA business schools programs for middle and senior managers in the commercial, state, and municipal management sectors usually lasts two years. The basis of training is the methodology of business cases. The main requirements for candidates are a degree in higher education, at least 2–3 years of practical work experience, the results of the MBA aptitude test, letters of recommendation, the TOEFL or IELTS language test. The advantage of studying at international business schools is that they are taught by company managers, consultants, i.e., practitioners with extensive professional experience, which is important for implementation on the world market. However, Russian MBA schools often include practical modules in Europe or the United States in the course of training. There are special programs for different levels and areas of training, both general management (General MBA) and highly specialized. Upon graduation, the graduate receives an MBA qualification degree, a diploma of professional retraining, and an MBA diploma from a business school. Successful completion of the program will require high dedication and serious work, which means much time and a high concentration of attention, so choosing such a program is advisable only if there is a specific learning goal and motivation to change yourself and your business.

The ratings of the best foreign business schools are published by the Financial Times, US News, and Business Week. The most prestigious business schools in the world are Stanford University Graduate School of Business, Harvard Business School (the first MBA program in history appeared here), the Wharton School of Business at the University of Pennsylvania, the London Business School, the INSEAD French Business School. The standard of MBA programs' quality is their international accreditation. Accreditation indicates compliance with internationally recognized quality standards of business education. The rating of the school and its accreditation affect the cost of education. In the most prestigious business schools in America, a two-year MBA program costs about $160,000. In the European INSEAD, the cost of a 10-month program is

€87,000. This price does not include accommodation, meals, and other expenses [25]. In some cases, the employer may pay for the training in full or in part, setting certain conditions. You can try to get a grant or scholarship.

The DAAD (Deutscher Akademischer Austauschdienst – German Academic Exchange Service) scholarship for postgraduate studies is very attractive; it is available to graduates of Russian universities of all fields of study who received their Specialist, Master or Bachelor diploma no earlier than 2014, as well as to final-year students of Russian universities who will receive their diploma before they start studying in Germany (01.10.2021). The duration of the scholarship is from 10 to 24 months, depending on the duration of the chosen postgraduate study program. The scholarship is 861 euros. The scholarship holder is entitled to medical insurance, accident insurance, individual civil liability insurance, partial compensation for travel expenses, and a one-time student allowance. The main selection criteria are professional qualification (based on academic standing), motivation and justification for the master's program selection, knowledge of German and/or English to the extent necessary for the education in the chosen postgraduate program, confirmed by the TestDaF certificate, Goethe-Certificate C2, TOEFL or IELTS (Scholarships for 2019). A certificate (diploma) of higher education obtained at a German state university or a university with state accreditation means it is recognized on the world market. It should be noted that the knowledge obtained as a result of this training is mainly theoretical (academic) in nature, i.e., it prepares the graduate in subsequent research work, while most people seeking additional business education are aimed more at practice-oriented learning, which allows implementing the knowledge and skills obtained in business [26].

It is interesting to know what are the student preferences when choosing a particular way of postgraduate education. According to the survey, the content of the program (51.5%) and the composition of teachers (31.15%) prevail in the "Business Overheard" (BO) group, and depending on the goals of training; we are talking about the presence of academic degrees or practical work experience. Next come the prestige and rating of the business school, the university, the center, the program itself (40.35%), the cost of education (30.09%), the educational document (13.81%), the format (online or offline) (12.74%), and the duration of education (10.27%) [27].

Unfortunately, it should be noted that in the Russian literature, little attention is paid to such a factor associated with obtaining additional education as an additional workload factor. This factor is very important since it means the need for additional time costs; mental stress associated with passing exams, tests; responsibility to the family, since postgraduate education is already received by adults who often have their own families, who are paid less attention, especially if additional education is associated with the need for travel. As Robert Schleicher notes, the speed of technological development leads to a rapid aging of professional qualifications: "lifelong learning" is becoming a requirement for many. However, systematic professional development at the enterprise is available only to about 20 percent of employees [28]. At the same time, even in the case of advanced training at the enterprise, there may be a certain psychological tension within the team caused by internal competition [29]. In the case of on-the-job training, additional mental stress arises, caused by fear for the preservation of the workplace, and in the case of its loss – the stress associated with finding a new job. This also means additional financial expenses.

4 Conclusion

From the above in this article, we conclude that business education is necessary for almost every modern professional, especially those seeking to enter the global market, to assess the situation competently and realistically. A number of factors influence the choice of the algorithm for postgraduation education, in which the main component is business education. Because of this, several options may be offered, depending on the goal and motivation. The goal involves either retraining in general or choosing a business education as a supplement to an existing first degree if it includes business and economics training to some extent. In the first case, it is advisable to get a second higher education since it provides basic knowledge in the business field that was previously absent. Then you can proceed to the second option, which assumes at least a theoretical knowledge of business models, business processes, etc.

The choice of the second option of the algorithm depends on the motivation of the graduate. If the motive is further self-fulfillment in research activities, it makes sense to get postgraduate education, choosing the appropriate DAAD scholarship, because it meets several requirements: saving money, the acquisition of a document with international status, improving language skills, under certain conditions, travel with family, which is important because it is adult education, for whom this factor is essential, the prospect of later work abroad. There is also such an option as a master's degree, and subsequently, a doctoral degree.

If the motive is further career growth or creating your own business, when choosing the path of postgraduate education, it plays a role whether the goal is to enter the national or global market. In both options, the factor of training time, financial costs, and the training program's content should be taken into account. In the first case, advanced training and/or internal internships are preferable since they are more likely to guarantee job retention. In the second case, an internship abroad or MBA program is preferable, especially in a foreign business school since they allow to get practical knowledge necessary for activity in the world market, to get a diploma or certificate of international status, help to improve knowledge of a foreign language, necessary for a specialist of the international level. In any case, self-education is a necessary component.

References

1. Kobyakov, A.: Challenges of the 21st century: how the fourth industrial revolution is changing the world. https://www.rbc.ru/opinions/economics/12/02/2016/56bd9a4a9a79474ca8d33733
2. Sandner, Ph., Groß, J., Blockchain, K.M.: DLT und Crypto Assets: Trends für 2020. https://www.capital.de/wirtschaft-politik/blockchain-dlt-und-crypto-assets-trends-fuer-2020
3. Sandner, Ph.: Deutschland reguliert Krypto – Assets ab dem 1. Januar 2020: Was sind die besten Strategien für Blockchain-Startups, FinTechs, Banken, Krypto-Börsen und auch Industrieunternehmen?. https://medium.com/@philippsandner/deutschland-reguliert-krypto-assets-ab-dem-1-2211dff8ed72

4. Veselovsky, M.Y., Kirova, I.V., Yunusov, L.A.: Increasing the Innovative Activity of Industrial Enterprises. Nauchny Consultant, Moscow (2018)
5. Wolf, Ch.: Arbeitsrechtliches Know-how für den HR-Business Partner. https://cms.law/de/aut/events/ausbildung-zum-zert.-hr-business-partner
6. Anisimova, S.V.: Entrepreneurial activity in the Russian business education market. PhD thesis in Economic Sci. Moscow (2017)
7. Nemtinov, V.A., Borisenko, A.B., Morozov, V.V., Nemtinova, Yu.V.: Higher Edu. Rus. **30**(3), 104–113 (2021). https://doi.org/10.31992/0869-3617-2021-30-3-104-113
8. Sharova, S.V.: Creative Econ. **12**(5), 641–650 (2018). https://doi.org/10.18334/ce.12.5.39146
9. Myasoedov, S.: Assessment of the quality of business education in Russia. https://academia.interfax.ru/ru/analytics/business/3435/
10. Kosyakova, Y.: Europ. Soc. **20**, 65–88 (2018)
11. Krushelnitskaya, O.I., Polevaya, M.V., Tretyakova, A.N.: Psych. Pedag. Res. **11**(2), 43–57 (2019). https://doi.org/10.17759/psyedu.2019110205
12. Maaz, K., Ordemann, J.: Bildungsprozesse im Lebensverlauf: Der kontinuierliche Erwerb von allgemeinbildenden Bildungsabschlüssen. In: Quenzel, G., Hurrelmann, K. (eds.) Handbuch Bildungsarmut, pp. 435–465. Springer Fachmedien Wiesbaden, Wiesbaden (2019). https://doi.org/10.1007/978-3-658-19573-1_17
13. Virdia, S., Schindler, S.: Res. Soc. Stratif. Mobil. **60**, 29–38 (2019)
14. Dietrich, H., Patzina, A., Kretschmer, S.: Kölner Zeitschrift für Soz Sozialpsy **71**, 357–383 (2019)
15. Kalter, F., Granato, N.: Migration und ethnische Ungleichheit auf dem Arbeitsmarkt. In: Abraham, M., Hinz, T. (eds.) Arbeitsmarktsoziologie, pp. 355–387. Springer Fachmedien Wiesbaden, Wiesbaden (2018). https://doi.org/10.1007/978-3-658-02256-3_10
16. Dietrich, H., Martin, A.: Übergänge in ausbildung und arbeitsmarkt. In: Abraham, M., Hinz, T. (eds.) Arbeitsmarktsoziologie. Probleme, Theorien, Empirische Befunde, pp. 77–116. Springer VS, Wiesbaden (2018)
17. Koch, S., et al.: HMD Praxis der Wirtschaftsinformatik **56**(1), 468–484 (2019)
18. Engel, N.: Zeitschrift für Pädagogik **5**, 730–747 (2019)
19. Stoyan, N.: It is not fashionable to be an amateur!. https://nark.ru/news/12-10-2020-diletantom-byt-ne-modno-.php
20. Barinova, N.V.: Innov. El. Sci. J. **3**(32), 1–11 (2017)
21. Mikidenko, N.L., Storozheva, S.P.: Voc. Edu. Modern World **8**(2), 1886–1895 (2018). https://doi.org/10.15372/PEMW20180217
22. Kogan, E.A.: Higher Edu. Today **3**, 64–67 (2017)
23. Tufano, P.: How Should Business Education Change in the 21th Century. https://hbr-russia.ru/biznes-i-obshchestvo/nauka/826207
24. Dolinsky, A.: The Sovereign Market: How Does Business Education in Russia Differ from the World One? https://education.forbes.ru/authors/mba-v-rossii
25. Tretyak, S.: What is an MBA and why Digital MBA is Gaining Popularity: Key Features of Training. https://netology.ru/blog/03-2020-mba-i-digital-mba
26. Scholarships for Postgraduate Study for Graduates of Universities of all Specialties. https://www.daad.ru/ru/stipendien/ctipendii-dlya-poslediplomnogo-obuchen/
27. Kochkina, O.: Where they Teach Business: Formats and Programs of Business Education with Comments from Graduates. https://texterra.ru/blog/gde-uchat-biznesu-formaty-i-programmy-biznes-obrazovaniya-s-kommentariyami-vypusknikov.html
28. Schleicher, R.: Psychische Belastungen in der Arbeitswelt – Stand, Entwicklungen, Handlungsmöglichkeiten. Hans-Böckler-Stiftung-Industriegewerkschaft Metall, Vorstand Funktionsbereich Sozialpolitik. Frankfurt/Main (2002
29. Blum, U., Gabathuler, J., Bajus, S. (eds.): Weiterbildungsmanagement in der Praxis: Psychologie des Lernens. Springer, Heidelberg (2021). https://doi.org/10.1007/978-3-662-62631-3

Prospects for the Development of Higher Education in the Conditions of Practice-Oriented Student Training

Lyudmila V. Bogoslavtseva(✉), Oksana Yu. Bogdanova, Oksana I. Karepina, Svetlana N. Meliksetyan, and Vera V. Terentyeva

Rostov State University of Economics, Rostov-on-Don, Russia
b_ludmila@bk.ru

Abstract. The most important feature of modern higher education is its innovative orientation based on the intellectual level of the teaching staff, as well as information and digital educational services which predetermine the need for constant updating and increment of new knowledge. The problem of providing a practice-oriented approach in obtaining higher education has been actively studied in the works of domestic and foreign authors for more than 10 years. Purpose of the research: identifying the features of the development of higher education and determining measures to improve the quality of practice-oriented training of students in conditions of limited financial resources. The methodological basis of the study was the determinism of the process of providing educational services and their financial support, as well as the methods of systemic and structural-functional approaches. In the case of strengthening the practice-oriented training of graduates of higher educational institutions, an increase in the quality of mastering professional competence, obtaining practical skills and experience necessary for starting practical activity is expected. Achievement of these results will ensure the growth of competitiveness not only of graduates, but also of universities.

Keywords: Higher education · Vocational training · Innovations in education · Financing of education

1 Introduction

At the present stage, human potential is defined as a factor in the socio-economic development of society which predetermines the need to increase the efficiency and practical orientation of higher education in accordance with the conditions of digitalization.

However, in the context of the pandemic, the socio-economic infrastructure has changed significantly, which has led to a significant reduction in the bases of practices against the background of growing employers' requirements for the professional competencies of graduates.

In the new conditions, each country introduces its own innovations in the created models of higher education. For example, Malaysia has established a Programme (KTP, Knowledge Transfer Partnership) to facilitate knowledge transfer, collaboration and

P. V. Trifonov and M. V. Charaeva (Eds.): DITEM 2021, LNNS 380, pp. 684–691, 2022.
https://doi.org/10.1007/978-3-030-94245-8_93

interaction between academics in public higher education institutions and other parties interested. [1] In the studies of domestic scientists and practitioners, it is also concluded that it is necessary to strengthen the role of national personnel through innovations in the organization of the educational process in the changed conditions [2, 3].

In Russian practice, the interaction of universities with potential partners and employers is being strengthened, project methods of teaching and mastering practical skills and abilities are being introduced, and special attention is being paid to improving the qualifications of the teaching staff.

Thus, the hypothesis of the study is to substantiate the provision on the objective need for financial support for practice-oriented training of graduates of higher educational institutions, which is designed to improve the quality and effectiveness of educational services and the demand for graduates in the labour market.

The scientific novelty of the study can be attributed to the identification of specific professional functions and their multiplier effect on the development of a practice-oriented approach to training students in higher educational institutions in the context of the digitalization of the economy and the epidemiological situation in the world.

The purpose of the study is to identify the prospects for the development of higher education in the context of practice-oriented training of students and limited financial resources.

To achieve the goal, the following tasks are being solved: identifying the features of the implementation of practice-oriented educational services in higher education institutions in modern conditions; identification of measures to expand practice-oriented training of graduates in conditions of limited financial resources.

The theoretical and practical significance of the research results lies in substantiating the need for practice-oriented higher education, the development of which ensures recruitment of applicants for traditional and innovative professions, the demand for graduates in the labour market, and the competitiveness of higher educational institutions.

2 Methods

The levels of education and the organization of the educational process are not unchanged; their transformation depends on the changing factors of the provision of educational services, as well as on the methods of the systemic and structural-functional approach to the provision of educational services.

Note that since 1976 UNESCO, in the International Standard Classification of Education (ISCED), has been setting the levels of education that are transformed taking into account the changing socio-economic conditions [4]. So, for example, in Russia until 2015, the level of higher education was called higher professional education, but at present, according to international standards, it is called higher education.

In ISCED UNESCO identifies not only levels of education, but also organizational approaches to learning: formal, non-formal, informal and non-systemic. In our opinion, these approaches are reflected in educational standards and requirements for the educational process and the procedure for its financing. Depending on the founder, higher education institutions use budgetary and non-budgetary sources of funding which ensure the implementation of constitutional guarantees of the general availability of higher education on a competitive basis [5]. The attraction of extra-budgetary sources reflects the demand for students to master certain areas of higher education preparation in accordance with the approved educational programmes, that is, formal rules and standards. At the same time, there is still an opinion about the relevance and informal order of education.

In this regard, it is difficult to disagree with V.V. Krasnoshchekov that "… there is no reason to recognize non-formal education as the only possible form of education in the future… Also, the opinion of T. O'Driscoll, an American expert, seems to be fundamental, that without the basis of formal education, the system of non-formal education will collapse like a brick wall without cement mortar." [6].

Practical classes and all types of practice, in our opinion, have advantages in formal education, since they are carried out according to approved work programmes, but are always proactive, creative, independent (informal) in the disclosure of issues.

In order to further develop the practical component of training in the new edition of the Federal Law No. 273-FZ dated December 29, 2012 (as amended on April 20, 2021) "On Education in the Russian Federation" in Article 2, Clause 24 "practice" was replaced by "practical training" as a form of organizing educational activities in the development of an educational programme.

Thus, practical training includes practice itself, as well as practical classes, workshops, laboratory classes and even lecture-type classes, if their goal is to form, develop, and consolidate practical skills and competencies related to future professional activities.

3 Results

The use of ISCED is based on a methodology or an algorithm for intercomparability of curricula of various countries in an internationally comparable register of categories for establishing levels (stages) of education, which will allow for the exchange of students, and expand the conditions for mastering professional competencies. Though, in this case, the question arises about the procedure for financing additional educational services.

In domestic practice, financing of higher educational institutions is carried out by subsidizing the state assignment and attracting extra-budgetary sources and is aimed, first of all, at the implementation of vocational education of students. Based on the analysis of the typology of higher education functions presented in the economic literature [7], we will define a specific professional function the implementation of which, in our opinion, is of particular relevance and predetermines the prospects for innovative development of society within the framework of the issue under study. See Table 1.

The data in Table 1 reflect that the implementation of the professional functions of higher education is designed to ensure the development of practical knowledge, skills and abilities and is achieved through the interaction of all participants in the higher education system with other social structures and sectors of the economy, including financial provision of educational services.

Table 1. Specificity of professional functions implemented by institutions of higher education.

Functions	Contents
Mastering competencies	Shaping the necessary professional knowledge, skills and abilities in accordance with educational standards in the field of professional activity, that is, shaping of relevant competencies and preparation for practical activity
Communicative	Rendering, using and processing the information obtained in the course of mastering scientific knowledge, including information based on digital resources, services and platforms
Discipline-specific	Training of highly qualified personnel for all sectors of the economy, taking into account new promising areas
Innovative	It is closely related to the professional function, since the innovative development of society depends on the quality of training in traditional and new specialties
Formation of a worldview	It develops objectively, but for a specific person, allows the complex use of knowledge, skills and abilities in practical activities
Development of personality and thinking	It allows the learner to think independently, analyse practical situations and make decisions
Implementation of lifelong learning	It orientates a person to professional development, retraining, and active participation in professional and social activities
Redistributive	It provides funding for educational services of higher education through state subsidies for educational assignments, targeted subsidies, payment of individuals and (or) legal entities for tuition, grants, income from other statutory activities

In conditions of limited own funds, higher educational institutions use extra-budgetary sources in the form of subsidies for the fulfillment of state educational assignments, including paid educational services [8] and maternity capital [9]. Unification of tuition fees for similar educational institutions or educational services should also lead to effective competition in educational institutions.

At the same time, budgetary or contractual training bases should not be a priority factor when choosing a practice base. Practice-oriented training assumes full coverage of students with real bases of practice and their mastery of professional competencies which are designed to ensure the competitiveness of graduates and universities in all socio-economic areas.

4 Discussion

The problem of the quality of training of specialists and their further successful employment raises the issue of the advisability of developing a mechanism for interaction between universities and potential employers. The implementation of this problem is seen by some authors through the introduction of practice-oriented learning [10].

Other authors propose the creation of university-wide institute practice-oriented platforms (incubators) allowing to implement practice-oriented learning in the process of performing real tasks by students in the mastered training profile with the participation of professionals commissioned by enterprises and organisations [11].

The use of practice-oriented technologies, according to a number of authors, should be considered as a set of means and methods of teaching and development of students and can be classified into the following groups: interactive learning; contextual competence training; modular training; self-regulating teaching [12].

Taking into account the problems of modern education, over the past two decades, foreign authors (Jacinto Jardim, Universidade Aberta • Department of Social Sciences and Management (DCSG) PhD) have developed, applied and tested EPEC, an Educational Project for Entrepreneurship and Citizenship, the model of comprehensive education for students from preschool to higher education which, in our opinion, lays down a practice-oriented model of education [13].

It is difficult to disagree with the authors that universities should prepare students to overcome the many difficulties that they will have to face, both in the academic world and in the future period of integration into the working career. In the context of the development of practice-oriented education, the conclusions of foreign authors that "within the framework of policy, government agencies must provide students with access to an entrepreneurial atmosphere are of interest [14].

Within the framework of the implementation of the Priority Project on Universities as Centres of Space for Creating Innovations, in the domestic practice of higher education, university centres for innovative, technological and social development of regions are being created, as well as "the educational sphere is the centre of attraction for public-private partnerships projects in the regions" [3]. One of the important values of the activities carried out is the creation of conditions for the development of professional competencies by students.

The implementation of Modern Digital Educational Environment in the Russian Federation, another priority project, approved for the period 2016–2021 within the framework of the Development of Education state programme, made it possible by the beginning of 2020 to organise an electronic information educational environment (EIE) in higher education institutions and get the first practical experience of its use even before the start of the pandemic [15]. At the heart of the created EIOS, domestic and foreign digital services and tools of online education portals are widely used, which actually created conditions for practice-oriented student learning.

Based on the above, we will present an interconnected chain of participants in the educational process: university-student – employer which ensures the achievement of the goal of practice-oriented learning only if they interact. See Fig. 1.

The data in Fig. 1 proposed by the authors reflect the logical scheme for providing practice-oriented learning at the university: from goal-setting to the risks of achieving the set goal [16], which confirms, on the one hand, the complexity of providing practice-oriented training of students; on the other hand, the reality of this process is subject to taking into account the problems and risks that hinder its development.

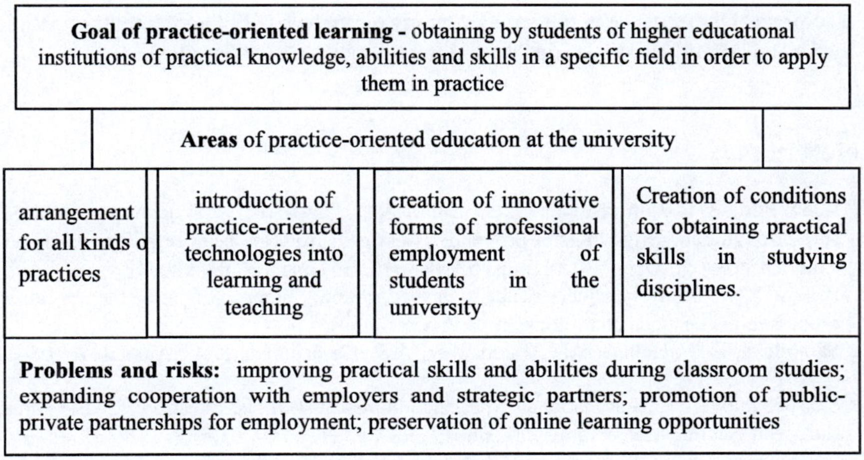

Fig. 1. Logical scheme for providing practice-oriented learning at a university.

Thus, the directions for achieving the goal of preparing graduates for professional activity are designed to create a comprehensive learning process at the university with the implementation of term paper and final qualification paper based on the materials of the practice base, which helps to motivate students of higher educational institutions to master professional competencies, improve the quality of education and increase the competitiveness of graduates on the labour market [17].

In turn, the financial transparency of budgetary and extrabudgetary flows of the university is an important condition for clustering the achievement of the goals of increasing the number of applicants in specialties in demand in society and training competence specialists [18].

5 Conclusion

The practical orientation of the study made it possible to substantiate the need to develop a practice-oriented approach to teaching in order to attract applicants to higher educational institutions and increase the competitiveness of both graduates and universities by:

- transformation of the organisation of knowledge transfer: from the rendering of lecture material to educational technologies for mastering new material, for example, a lecture-discussion, a lecture on problematic issues;
- practice-oriented training in the process of traineeship in accordance with the curriculum for the mastered profile of training;

- the development of club activities at the graduating departments with the development of creative projects used in the implementation of term papers and final qualifying paper;
- attracting students to the implementation of grants and contract-based issues;
- systematic advanced training of teachers in accordance with the subject of the taught discipline at the expense of university funding or on the basis of agreements concluded by the university on cooperation with potential partners and employers.

References

1. Radin Firdaus, R.B., Mohamad, O., Mohammad, T., Gunaratne, M.S.: Community partnership through knowledge transfer program: assessment from the perspectives of academics' experience. SAGE Open (2020). https://doi.org/10.1177/2158244020980742
2. Falkov, V.N.: On the readiness of the higher education system to organize the educational process in the changed conditions. https://www.mgri.ru
3. Samoilova, K.N., Rukina, S.N., Takmazyan, A.S., Gerasimova, K.A.: Audit J. **2**, 105–111 (2019)
4. Skorobogatova, V.I.: International experience in educational classification: ISCED and the European Qualifications Framework. https://clck.ru/XHQsY
5. Bogoslavtseva, L.V., Karepina, O.I., Meliksetyan, S.N.: Development of Financial Support for Education in the Context of Budgetary Reforms. AzovPechat, Rostov-on-Don (2016)
6. Krasnoshchekov, V.V.: Sci. Rev. Pedag. Sci. **3**, 24–28 (2018)
7. Dzhamaludinov, G.M., Dzhamaludinova, Z.G.: Bul. Dagestan State Pedag. Univ. Psy. Pedag. Sci. **3**(32), 23–26 (2015)
8. Romanova, T.F., Andreyeva, O.V., Takmazyan, A.S., Otrishko, M.O.: Financial support of the education system: theory and practice. AzovPechat, Azov (2016)
9. Bogdanova, O.: Maternity (family) capital as a form of social support for mothers and children in modern conditions. In: Actual Problems of Taxation and Development of Key Areas of the Economy: Collection of Scientific Articles of the International Correspondence Scientific and Practical Conference V, Penza, November 21–28, pp. 175–179. Penza State university, Penza (2014)
10. Vyatkina, I.V.: Practice-oriented training as a means of professionalizing the training of future specialists at the university. In: Pudov, E.Yu. (ed.) A New Look at the Education System. Digest of Materials of the International Scientific and Practical Conference II, pp. 007.1–007.5 (2019)
11. Polisadov, S.S.: Practice-oriented teaching at the university. In: Level training of specialists: E-Learning and Open Educational Resources. Proc. I All-Russian Scientific-Methodical Conf., pp. 349–352. TPU Publishing House, Tomsk (2014)
12. Olesova, M.M.: Phil Sci. Ques Theory Prac **7**(73), 201–204 (2017)
13. Jacinto, J.: Project: EPEC – Educational Project for Entrepreneurship and Citizenship: Socio-Educational Impact Assessment (2020). https://doi.org/10.13140/RG.2.2.16080.87042
14. Anjum, T., Heidler, P., Amoozegar, A., Anees, R.T.: Adm. Sci. **11**(2), 45 (2021). https://doi.org/10.3390/admsci11020045
15. Artemenko, D.A., Goncharenko, L.I.: Digital Economy and Education: Pros and Cons. In: Penkova, I.V., Koroleva, V.A. (eds.) Digital Economy and Online Education: European Experience, pp. 334–346. North Caucasus Federal University, Stavropol (2020)
16. Karepina, O.I., Bogdanova, O.Yu.: Risk-based approach to financial management in higher educational institutions. In: Statistics, the language of digital civilization. Reports Digest of the Open Russian Statistical Congress II: in 2 volumes, pp 159–163. AzovPrint LLC, Rostov-on-Don

17. Bogoslavtseva, L.V.: On the question of motivating students of higher educational institutions to master professional competencies. In: For the Rights of Workers! Improving the Quality of Working Life as a Strategic Task of the State and Trade Unions. Materials of the Fifth International Scientific and Practical Conference, pp 48–51 (2019)
18. Ivanova, O.B., Vovchenko, N.G., Kostoglodova, E.D., Bogoslavtseva, L.V., Rukina, S.N., Karepina, O.I.: Financial transparency in budget sector of economy as a necessary condition of clustering. International Journal of Trade and Global Markets. T. 10. № 2–3, C, 207–216 (2017). https://www.elibrary.ru/item.asp?id=31061813

Digital Educational Ecosystem: Transformation and Development Trends of Online Education

Alzhanat E. Suleimankadieva[1,2](✉), Maxim A. Petrov[1], Olga A. Popazova[1], Maria A. Gridneva[1], and Eleonora B. Molodkova[1]

[1] St. Petersburg State University of Economics, Saint Petersburg, Russia
[2] St. Petersburg Electrotechnical University "LETI", Saint Petersburg, Russia

Abstract. The authors' research focuses on the transformation of the online education market in the current context and the creation of the concept of a digital educational ecosystem. The relevance of the topic can be substantiated by reasons that are associated with the overall digitalisation of the economy and the COVID-19 pandemic. Determination of the reasons for the transformation of the conventional hierarchical education system, the rapid development of the online education market and, as a consequence, the development of the concept of a digital educational ecosystem as a present-day approach to arranging for the educational environment in the context of the society's transition to a new technological set-up, networkisation of the world educational space and active implementation of digital technologies in the educational process is defined by the authors as the key goal of the research. Based on the empirical data analysis, a conclusion is made about the accelerated rates of development of the online education market, the predicted value of the growth rate of online education in the future in the short term. The author's vision of the digital educational ecosystem as a network infrastructure, which shapes a single technological educational platform with digital technologies, creates conditions for effective interaction between the stakeholders through the provision of tailored educational services based on taking into account the interests of consumers, is formulated.

Keywords: Transformation of education · Digital educational technologies · Educational ecosystem · Network educational infrastructure · Trends in the development of online education

1 Introduction

The reasons for large-scale digitalisation are associated with the development of Industry 4.0 technologies [1, 2], which were supplemented by the objective reality of COVID-19 which accelerated the digitalisation processes. Together, they substantiated the relevance of replacing the current model of human capital development with a new one, based on the ideas of inter- and trans-disciplinary approaches to learning [3, 4].

With regard to the digital education system, the ecosystem approach is currently actively used [5, 6]. The studies offer many definitions of this term [7, 8], but they all

P. V. Trifonov and M. V. Charaeva (Eds.): DITEM 2021, LNNS 380, pp. 692–697, 2022.
https://doi.org/10.1007/978-3-030-94245-8_94

pursue their own local goals and objectives. It is advisable for us to choose the actual, justify it and expand the concept taking into account the purpose of the research.

The concept of a digital educational ecosystem appeared relatively recently and is associated with the following phenomena: 1) the intensive development of the digital economy is actively changing the requirements for the training of graduates; 2) COVID-19 turned out to be the main reason for the transition to distance learning which accelerated the processes of intensive development of digital educational technologies and the formation of the online education market; 3) the rate of socio-economic changes, has proved the feasibility of developing a network infrastructure of the educational environment based on an intelligent platform and digital tools for the transfer of knowledge; 4) the active development of online education has led to an increase in competition between IT companies, educational organisations and concerned parties, has led to the personalisation of the educational process and the creation of multivariate educational tracks. Is the formation of the concept of a digital educational ecosystem a consequence of the transformation processes taking place in education in the context of the active introduction of digital technologies and the networkisation of the world educational space?

2 Materials and Methods

The aim of the study is to determine the reasons for the transformation of the traditional education system, the rapid development of the online education market and trends in the development of the concept of a digital educational ecosystem as an approach to organizing a modern educational space. The authors solved the following tasks: 1. Studying the reasons for the transformation of the traditional hierarchical education system into a digital educational ecosystem; 2. Determination of the main trends and forecasting the development trends of the global and Russian online education market; 3. Discourse analysis of the digital educational ecosystem as a new approach to the organisation of the modern educational environment.

The methodological basis of the study was the combined use of general scientific and empirical methods: a comparative analysis of the development trends of the world and Russian educational spaces; grouping and systematisation of empirical data on the Russian online education market, identifying the features of the development of the digital educational environment; extrapolation of the research results which made it possible to determine the prospects for the development of online education in the context of intensive digitalisation of educational products and the formation of the concept of an ecosystem educational environment; the discourse analysis of the digital educational ecosystem made it possible to form the author's vision of this phenomenon.

3 Results

An innovative way to transfer and control knowledge is online education, the large-scale development of which began in 2001 [9]. The projected volume of the global online education market (Research of the Russian Online Education Market) in 2021 will amount to $241 billion, and by 2023 this figure will grow by another 17% (Fig. 1).

There is an obvious linear trend towards an increase in the volume of the global online education market, the average annual growth rate of which is 8.2%. In 2017, the share of online education in the volume of the world market of educational services amounted to 3.0% [9], and by 2025, according to our calculations, this indicator may grow to 4% and more.

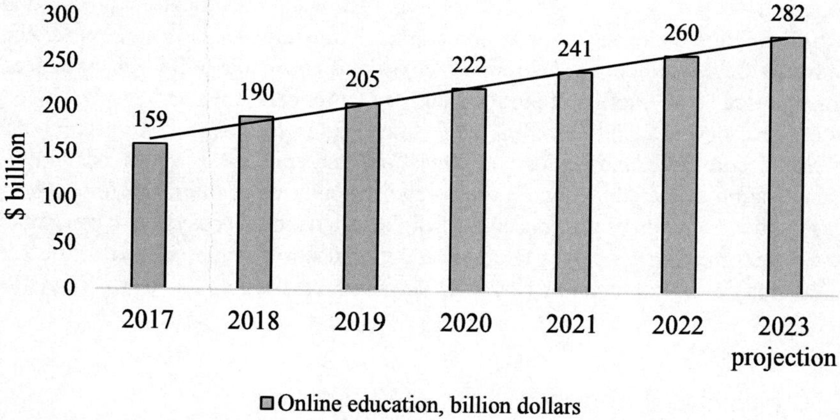

Fig. 1. The volume of the global online education market, billions of dollars. *Source:* EdTechX-Global, Global Market Insights, Education International.

A comparative analysis of the Russian education system (in the format of traditional education and online education) by segments over five years of development allows us to note that in 2021, compared to 2016, there is an increase in the volume of education in general, and the share of online education has accelerated growth trend (Table 1).

In 2021, the growth in the volume of the online education market will be 2.6 times, which indicates the insaturation of this segment of Russian education. Higher education is the most developed where the value of digital education in 2021 will amount to 15 billion roubles against 6.8 billion roubles in 2016. The market of Russian online education in 2021 will continue to grow, the growth rate will be 20–25% compared to 2020 [11].

The main trends in the development of Russian online education for the period of 2019–2020: gamification; interactive video; microlearning; use of artificial intelligence; adaptive corporate training; project training; entry of Russian edTech companies into the world market of online educational services; shortage of qualified specialists; the emergence of new services; development of the Hobby Education segment by more than 19% [12].

The positions in the market of online educational services in the USA, the countries of the Asia-Pacific region (APAC) and the countries of Latin America are strong. The APAC countries were the most successful in 2020, accounting for 54% of the market. If in 2016 the US share of the global online education market was 50–55%, then by 2018 this figure dropped to 40%. According to experts, this is due to the fact that the American online education market is currently at a saturation stage. The growth rates

Table 1. Trends in the development of Russian education by segments and forms of education in 2021 compared to 2016, billion roubles.

	2016		2021	
Segments of education	Educational services market value	Online education market value	Educational services market value	Online education market value
Pre-school education	462	0.6	548	1.7
General secondary education	572	0	699	10
Extra school education	130	3.6	149	10
Higher education	386	6.8	336	15
Secondary vocational education	146	0.6	175	1.8
Extra vocational education	105	7	103	11
Language learning	26.8	1.6	24.6	3.9
TOTAL	1827.8	20.2	2034.6	53.4

Source: Research of the online education market in Russia[10]

of the online education market of the world leaders are different. Thus, the average annual CARG indicator for the United States for the period from 2017–2022 is 5%, for European countries it is 14%, for APAC countries it is 19.4%, and the global online education market has a growth of 10.26% [13].

Currently, the ecosystem approach is actively used in the organisation of the online education system. The digital educational ecosystem is defined as "a socio-cultural environment where the process of personalised training of students is based on the process of learning the needs in the interaction of all participants in educational relations using variable and adaptive digital content" [5]. Given the current conditions, the digital educational ecosystem can be characterized as a network infrastructure that forms a single digital platform that creates conditions for the most effective interaction of concerned parties, capable of providing personalised educational services to students. The digital educational ecosystem acts as an open set of information systems designed to ensure the solution of a complex of tasks of the educational process, which makes it possible for concerned parties to access their content when certain conditions are met.

The main systemic elements of the digital educational ecosystem are: digital platforms which are various interactive modules, applications, services that help teachers place educational materials in an electronic environment allowing developers to create educational products and interact with other concerned parties; concerned parties (participants in the open digital educational space); unlimited consumer segments.

All of these elements are subsystems that interact with each other. Consequently, the digital educational ecosystem can be objectively characterised as an infrastructure that creates certain conditions for ensuring the interaction of participants in the digital educational space.

4 Discussion

The content analysis [8] allows you to determine the prerequisites for the development of the ecosystem approach in relation to the educational environment:

1 The acceleration of the process of obsolescence of knowledge has led to the fact that the average period of relevance of new knowledge is shrinking every year.
2 The digital economy is constantly making new demands on the competencies of graduates. The traditional education system does not keep pace with the development of the economy, which leads it to a crisis. A knowledge transfer system that can quickly capture changes in the environment, adapt to these conditions, analyse large amounts of data, and identify general trends is needed.
3 The transformation process of the traditional education system and the development of a digital ecosystem format are due to several reasons, among which COVID-19 has played a significant role which accelerated the digitalisation process. More than half of respondents (56%) note that it is COVID-19 that has made a great contribution to the development of online education, and they see new opportunities for business development in online education; 43% of respondents note that they had to adapt the product to new conditions. 22% of respondents changed their business model due to the pandemic, and 11% of respondents were forced to master an online project due to a drop in sales [14]. If in 2016 the share of online education in the entire Russian education system was 2.2% (or 0.6 billion roubles), then by 2021 this figure increased to 5.5% (or 1.7 billion roubles) [15]. According to our calculations, the volume of Russian online education increased 2.8 times over this period.
4 The digital ecosystem of education functions as a network infrastructure that is supported by digital technologies and creates conditions for effective interaction on a single technological platform where each of the participants has access to common ecosystem resources
5 The technological revolution is actively converting all communications of the educational system into digital ones. The forecast of the annual growth rate of the EdTech segment in Russia in the next 5 years will be about 20–25% [15].
6 The outstripping development of the online education market has caused intense competition between players, which leads to an increase in investments in the online sector and an increase in the quality of the digital products offered.

5 Conclusion

The digital education ecosystem must be understood as a network infrastructure that includes a set of intensively interacting concerned parties that:

1. includes an educational platform, developers, students, teaching materials, business representatives, creators of new knowledge and creates conditions for their effective interaction and its regulation;
2. organises and carries out its activities either with the participation of only digital technologies, or on the basis of their preferential use, which allows the successful development of online education where the educational process is digitised and is the quintessence of the development of IT technologies in education;
3. speeds up the process of generating intellectual capital and significantly increases the market capitalisation of both the education system itself and the digital network infrastructure.

The main task of the digital educational ecosystem is to improve the quality of life, create conditions for increasing the availability of knowledge from concerned parties.

References

1. Schwab, K., Davis, N.: Technologies of the Fourth Industrial Revolution. Eksmo, Moscow (2018)
2. Zakharov, N., Britvina, I., Kuznetsov, A.: The dilemma of job creation and loss before of industry 4.0. In: Proceedings of Scientific Papers from the International Scientific Conference The Impact of Industry 4.0 on Job Creation 2019, pp 457–462. Alexander Dubček University in Trenčín, Faculty of Social and Economic Relations (2020)
3. Suleimankadieva, A.E., Petrov, M., Popazova, O.: IOP Conf. Series: Mat. Sci. Eng. **940**, 012092 (2020). https://doi.org/10.1088/1757-899X/940/1/012092
4. Maksimtsev, I.A., et al.: Theory and Practice of Bioeconomy Development: Innovation, Digitalisation, Transformation. Publishing House of SPbGEU, St. Petersburg (2019)
5. Kondakov, M.A.: Ecosystem of Digital Education. https://firo.ranepa.ru/files/docs/cifr_didactika/plenar/kondakov_plenar.pdf
6. Koroleva, D.O., Naushirvanov, T.O.: Facts Edu. **6**(31) (2020)
7. Ivanov, V.V.: Mod. Econ. Probl. **4**, 334–337 (2010)
8. Katrich, A.: Digital Financial Ecosystems. https://asros.ru/upload/iblock/4c5/19421_katrich.pdf
9. Kuznetsov, N.V.: E-Manag. **2**(1), 19–25 (2019)
10. Research of the Online Education Market in Russia. http://fgosvo.ru/uploadfiles/presentations/Educ_online_2020.pdf
11. Kolesnikova, K.: Learning-site. https://rg.ru/2021/01/12/v-2021-godu-rynok-onlajn-obrazovaniia-v-rossii-prodolzhit-vzryvnoj-rost.html
12. Top Trends in Online Education in 2019. https://vc.ru/marketing/58813-5-glavnyh-trendov-v-onlayn-obrazovanii-v-2019-godu
13. Research of the Russian Online Education Market. https://innoagency.ru/files/Issledovanie_rynka_rossiyskogo_online_obrazovania_2020.pdf
14. Pandemic of Digital Education. https://vc.ru/education/153144-pandemiya-cifrovogo-obrazovaniya
15. Research of the Russian Market of Online Education and Educational Technologies. https://estars.hse.ru/mirror/pubs/share/211448255

Integration of Higher and Vocational Education: Trends and Challenges

Sergey P. Anofrikov(✉), Yulia V. Dyubanova, Maria V. Oblaukhova, Larisa V. Pichugina, and Svetlana S. Taldonova

Novosibirsk State University of Economics and Management, Novosibirsk, Russia
s.p.anofrikov@nsuem.ru

Abstract. Today, integration of vocational and higher professional education, a prerequisite for the creation of a continuing education space, is gaining particular relevance. This creates a need for in-depth analysis of how different levels of education are interlinked in the context of the conjugation of universal and professional skills that shape the innovative model of a high school graduate, and a need to highlight the pros and cons of the merger of vocational and higher education. This study uses general research methods: analysis and synthesis, induction and deduction, as well as mathematical and statistical methods. The paper analyzes and evaluates the practices of Novosibirsk State University of Economics and Management that has integrated programs at different levels by restructuring its NSUEM Business College and embedding the vocational education programs in the continuing education model. The study of students' educational paths from vocational education to bachelor and specialist degrees leads to a conclusion that the model has been a success based on statistical data, which indicate that people who graduate from vocational programs are likely to pursue higher education at NSUEM. Program content assessment proves the continuity of the effort to teach future specialists general cultural and professional skills. On the other hand, research also highlights several challenges that jeopardize the strengths of this integration model and call for its further elaboration and modification towards better learning process and greater professional competence of the graduates.

Keywords: Integration processes · Integration in education · Levels of training · Skills · Single educational space

1 Introduction

The quality of domestic education is fundamental to the economic security of the nation. The education system is what drives the development of the country's human resources and directly affects its R&D, social reproduction, and innovative development.

Today's trends in the modernization of Russia's education system concerning the integration of different levels of education bring to the fore the issues of integrating vocational and higher education, as well as challenges of creating and developing skills that form a fairly complex structure that is directly involved in graduate training.

P. V. Trifonov and M. V. Charaeva (Eds.): DITEM 2021, LNNS 380, pp. 698–703, 2022.
https://doi.org/10.1007/978-3-030-94245-8_95

Integration of general secondary education and professional education is well-studied, whereas the design of integrative programs for higher and vocational education calls for more research.

Integrating various stages of education, the duration of studies, and further career are matters that Russian and international researchers have shown interest in. These challenges have been studied in a variety of aspects: adaptation of vulnerable demographics such as people with special needs and immigrants [1, 2]; creating a logically cohesive system for the development of professional skills, methodological coherence of the learning process [3, 4]; the historical aspect [5], etc.; some researchers focus on how the student's race and gender affect their transition between different levels of education [6]. Choice of which path to go after school remains a relevant problem. Russian statistics over past 10 years shows that high school graduates that want to pursue further education have shifted their preferences towards vocational schools [7, 8]; similar issues are being studied by international authors as well [9, 10]. Notably, how far different levels of education are integrated ultimately affects the competitiveness and skills of the graduates [11, 12].

Thus, research towards the creation of an integral system of professional education is of relevance and significance, as it can highlight the existing contradictions that step from the specific development of various levels of education.

2 Materials and Methods

The goal hereof is to analyze the existing model for integration of vocational and higher education at Novosibirsk State University of Economics and Management.

The following objectives stem from such statement of problem:

1. Analyze the flows of students enrolled in vocational education (VE) programs.
2. Review the flow of VE graduates that go on to pursue higher education (HE) at NSUEM.
3. Study the continuity of general cultural and professional skills trained under VE and HE programs.
4. Identify the organizational strengths and weaknesses that relate to the University's restructuring of its VE system.

Methodologically, this study uses general research methods: analysis and synthesis, induction and deduction, as well as mathematical and statistical methods.

3 Results

VE-HE integration at NSUEM began five years ago.

The first step was to disband the independent institution, the NSUEM Business College, and to reorganize it into the Vocational Education Center (VEC), which is a structural unit of the NSUEM Faculty of Basic Training. For the next step, which began in 2019, the VEC was disbanded as a separate institution, and its staff was redistributed among the University's departments.

The VE-HE integration policy arose from a number of factors including the strife for better training of economists on all levels of education as well as the national economy's need for vocationally educated specialists.

Integration of VE programs into NSUEM's continuing education system is expected to improve the quality of training, first and foremost by virtue of continuity between programs, coordination of curricula in VE and HE programs, and unification of learning methodologies.

The authors hereof analyzed the flows of students enrolled in major 38.02.06 Finance as shown in the official reports for 2016–2019, see Table 1.

Table 1. Flows of students in 38.02.06 Finance, persons.

Metric	2016	2017	2018	2019
Total enrolment in the VE program at NSUEM	136	149	235	250
previous education: general basic (9 grades)	58	79	113	150
previous education: general secondary (11 grades)	78	70	122	100
Graduates of the VE program at NSUEM, total	67	117	99	125
previous education: general basic (9 grades)	53	75	56	53
previous education: general secondary (11 grades)	14	42	43	72

Source: [13].

Table 1 shows a year-over-year increase in enrolment.

For the detailed flow of VE graduates majoring in 38.02.06 Finance and pursuing HE programs at NSUEM, see Tables 2 and 3.

Table 2 shows 59.2% of the VE graduates in major 38.02.06 Finance decided to pursue further education at NSUEM in 2019, which is the best result of the three years.

Table 2. Proportion of VE graduates majoring in 38.02.06 Finance and continuing their training at NSUEM.

Metric	2017	2018	2019
VE graduates of NSUEM, persons	117	99	125
VE graduates of NSEUM enrolled in HE programs, persons	63	42	74
Same as a percentage	53.8	42.4	59.2

Source: [13].

Table 3. Distribution of VE graduates majoring in 38.02.06 Finance and pursuing higher education at NSUEM.

Track	Breakdown by year, persons				Breakdown by year, %			
	2017	2018	2019	2020	2017	2018	2019	2020
38.03.01 Economics	30	18	39	36	47.6	42.9	52.7	43.9
38.03.02 Management	5	6	16	10	7.9	4.3	21.6	12.2
38.03.03 HR Management	1	6	3	3	1.6	14.3	4.1	3.7
38.03.04 Public and Municipal Administration	5	3	3	3	7.9	7.1	4.1	3.7
38.05.01 Economic Security	15	5	2	18	23.8	11.9	2.7	22.0
Others	7	4	11	12	11.2	9.5	14.9	14.7
Total VE graduates re-enrolled in HE programs at NSUEM	63	42	74	82	100.0	100.0	100.0	100.0

Source: [13].

As can be seen from Table 3, VE graduates majoring in 38.02.06 Finance find 38.03.01 Economics the most lucrative major. From 2017 through 2020, no less than 42.9% of the 38.02.06 Finance graduates would choose to enroll in it.

However, the authors hereof are most interested in 38.05.01 Economic Security, a specialist degree, since it is administered by the Department of Public Finance. In 2018 and 2019, the numbers of VE graduates in 38.02.06 Finance enrolling Economic Security fell sharply, partly because personalized curriculum is not an option. However, in 2020, enrolment rose back to 2017 levels.

Thus, evidence suggests that VE graduates are loyal to the University; on the other hand, programs within the University compete fiercely for applicants. Departments that administer tracks other than 38.03.01 Economics need to more actively engage students.

Skills that students acquire as part of their vocational training develop further in their HE program, which creates integrity and continuity of the two key stages of education. This is an argument in favor of intra-university integration of different education levels.

Let us trace this continuity on the case of general cultural and professional skills as trained in graduates that proceeded to enroll in 38.05.01 Economic Security (skills as listed in federal state educational standards).

Continuity of general cultural skills and knowledge are of utmost importance. Thus, skill sets OK-1, OK-5, OK-8, OK-9, and OK-10 [14], which are part of the VE training, develop further as skill sets OK-4, OK-5, OK-7, OK-10, OK-9, OK-12, and OK-11 [15] in higher education.

Of special interest is the continuity of professional skills.

FGOS-2018 for 38.02.06 Finance states that a graduate in that major should be able to capably engage in the following core activities consistent with their skillset:

- financial and economic planning, public and municipal administration, organization of budget execution in the Russian Federation's budget system;
- transactions and balance sheet management involving the budgets in the Russian Federation's budget system;
- participation in the organization and implementation of financial audit.

Knowledge and skills acquired further become part of the general professional skillset OPK3.

Skillsets pertaining to Financial Management and Transactions lay the foundation for a whole block of professional skills pertaining to testing and ensuring the economic security of economic actors: PK1, PK2, PK3, PK4.

Skillsets discussed above are nurtured in different courses, some of which are taught by the Department of Public Finance; this opens up opportunities for cooperation.

4 Discussion

The experience of the first years of VE-HE integration at NSUEM shows that the concept of conjugating different levels of education has potential for further improvement.

Evidence suggests VE graduates tend to remain loyal and continue their training at NSUEM under its HE programs; faculty could work on that as well.

This also creates continuity of general cultural and professional skills trained under VE and HE programs. Cooperation between VE and HE teachers at the same department enables them to share experiences and cooperate on the teaching methodology.

5 Conclusion

Despite some positive outcomes, there are still some challenges to focus on. Thus, some VE graduates may find HE programs unappealing if those lack curriculum personalization.

Notably, reorganizing the structure resulted in a weaker focus on adjusting the learning process for VE students of different ages, and abolished the subject-cycle commissions as functional units involved in some professional modules of the institution.

In general, it seems appropriate to continue research into integration of higher and vocational education at NSEUM and to see how it affects the quality of training available to specialists in economics.

References

1. Rowe, D.A., et al.: Career Dev. Transit. Excep. Indiv. **44**(1), 28–46 (2021). https://doi.org/10.1177/2165143420959793
2. Bagga-Gupta, S., Dahlberg, G.M., Vigmo, S.: Learn. Teach. **13**(3), 82–110 (2020). https://doi.org/10.3167/latiss.2020.130306
3. Vaskivska, H.: Sci. Educ. **25**(5), 46–50 (2017). https://doi.org/10.24195/2414-4665-2017-5-9

4. Borlikov, G.M., et al.: Problem of updating universal methodology for integration of teaching units (ITU). In: Social and Cultural Transformations in the Context of Modern Globalism (SCTCGM 2018), Book Series European Proceedings of Social and Behavioural Sciences, vol. 58, pp. 2351–2355 (2019). https://doi.org/10.15405/epsbs.2019.03.02.271
5. Peretyatko, A.Y., Zulfugarzade, T.E.: Euro J. Contemp. Educ. **6**(2), 367–377 (2017). https://doi.org/10.13187/ejced.2017.2.367
6. Robson, K., Anisef, P., Brown, R.S., George, R.: Can. J. High. Educ. **48**(1), 39–59 (2018). https://doi.org/10.47678/cjhe.v48i1.187972
7. Klyachko, T.L.: Educ. Policy **1**(81), 26–42 (2020). https://doi.org/10.22394/2078-838X-2020-1-26-40
8. Vodopianova, E.: Contemp. Europe **98**, 132–141 (2020). https://doi.org/10.15211/soveurope52020132141
9. Maier, R., Robson, K.: Can. J. High. Educ. **50**(1), 82–94 (2020). https://journals.sfu.ca/cjhe/index.php/cjhe/article/view/188609/186389
10. Dorius, S., Tandberg, D., Cram, B.: Educ. Policy Anal. Arch. **25**, 110 (2017). https://doi.org/10.14507/epaa.25.3238
11. Shafranov-Kutsev, G.F., Efimova, G.Z.: Educ. Sci. J. **21**(4), 139–161 (2019). https://doi.org/10.17853/1994-5639-2019-4-139-161
12. Listvin, A.A.: Educ. Sci. J. **1**, 103–119 (2017). https://doi.org/10.17853/1994-5639-2017-1-103-119
13. Federal Statistical Reports HE-1 and VE-1 for 2016, 2017, 2018, and 2019. State University of Economics and Management, Novosibirsk
14. On the Approval of the Federal State Educational Standard for Higher Education in Major 38.02.06 Finance: Order of the Ministry of Education and Science of the Russian Federation No. 65 dd. February 5 (2018). http://www.consultant.ru/document/cons_doc_LAW_291910/
15. On the Approval of the Federal State Educational Standard for Higher Education in Major 38.05.01 Economic Security (Specialist Degree): Order of the Ministry of Education and Science of the Russian Federation No. 20 dd. January 16 (2017). http://www.consultant.ru/document/cons_doc_LAW_212678/

Success Rate of Graduates in the Labour Market: Expectations of Youth from Studying at the University

Inga V. Tsygankova[1,2](✉), Lidiya R. Vakhitova[2], Olga S. Reznikova[3], Dmitry V. Kruglov[4], and Maria P. Soloveykina[2]

[1] North-West Institute of Governance of the Russian Academy of National Economy and Public Administration under the President of the Russian Federation, St. Petersburg, Russia
[2] The Herzen State Pedagogical University of Russia (Herzen University), St. Petersburg, Russia
[3] V. I. Vernadsky Crimean Federal University, Simferopol, Russia
[4] Saint Petersburg State University of Economics, St. Petersburg, Russia

Abstract. The article highlights issues of the quality of higher education in Russia. Both domestic and foreign scientists' approaches to assessing the quality of higher professional education are systematised. It was found that most researchers associate the quality of higher education with the success of a graduate in the labour market although understand the meaning of this term differently. Based on the results of a sociological survey conducted among students at four major universities in St. Petersburg (The Herzen State Pedagogical University of Russia, St. Petersburg State University of Economics, North-West Institute of Management of the Russian Academy of National Economy and Public Administration, National Research University Higher School of Economics), it was concluded that the quality of education in Russian universities is determined by the relevance of theoretical knowledge and practical skills provided by an educational institution; professionalism of teachers, which consists in not only a deep knowledge of their professional area and new development trends but also the ability to transfer knowledge to students; relations between an educational institution and enterprises of the real sector of the economy, since this gives the opportunity for internships in companies thus providing subsequent employment prospects. The authors of the article conclude that the quality of the education obtained is one of the most important criteria for choosing a university by Russian students. Future graduates associate success in the professional field, a decent position in society, and a decent level of income from professional activities with the quality of the education they receive.

Keywords: Higher education · Labour market · Educational service market · Graduate success

1 Introduction

The quality of the professional education obtained largely determines the success of graduates in the labour market. Higher education is no exception. As noted by Dzhurinsky, the coverage of youth in higher education in Russia was 32% in 2014, and domestic

P. V. Trifonov and M. V. Charaeva (Eds.): DITEM 2021, LNNS 380, pp. 704–710, 2022.
https://doi.org/10.1007/978-3-030-94245-8_96

indicators were close to those of the United States (34% in 2014) [1]. Thus, a third of young people in our country strive to get a higher education.

The importance of the higher education system can hardly be overestimated: it educates knowledge workers for various sectors of the economy. University graduates are future professionals in various fields, the quality of their training determines the ability of the Russian society to create innovations, apply new technologies and techniques, implement complex production and management processes, and provide high-tech services.

The labour market is changing, and this transformation results in a change in the university graduate requirements on the part of employers. In the modern world, along with key professional competencies, employers value such personal competencies as the ability to work in a team, initiative and independence, the ability to time management and goal-setting, the ability to predict the development of a situation [2]. Domestic researchers note that such competencies as systems thinking, cross-industry communication, project and process management, work with IT systems, customer focus and communication are essential for Russian employers [3].

The issues of the quality of higher professional education and the success of graduates in the labour market are widely covered in domestic and foreign literature. Dicker, Garcia, Kelly and Mulrooney emphasise that it is difficult to give a clear definition of the term "quality of education"; in general, the quality of education means that a graduate has generated skills and qualities that are valued by employers [4]. Alzafari and Ursin note that at present most of the studies devoted to assessing the quality of the higher education system are focused on national standards, and transnational studies are extremely rare [5]. This is confirmed by the fact that many researchers associate the assessment of the quality of education with completely different indicators. Thus, Mussawy and Rossman say that the quality of higher education in Afghanistan is determined by criteria established at the state level and largely depends on the efforts of the top management of universities [6]. The Chinese scientist Zhang and the German researchers Dorau and Höhns associate the quality of higher education with the success of a graduate in the labour market, but at the same time they understand the term "graduate success" in a different way. Zhang defines the success of a university graduate as the internal satisfaction of a young worker with his/her career choice, the correspondence of achievements in the professional sphere to a certain level according to his/her own scale of values [7]. Dorau and Höhns believe that the success of a graduate is a young worker's social status determined by stable employment over a long period of time and the ability to implement their career plans [8].

When speaking about the quality of higher education, Zborovsky and Ambarova use the term "educational failure" in publications [9–11] and associate it with the low quality of higher education in Russia due to a large number of non-state universities in the country, low motivation of students to study and insufficiently high level of school education.

Demidenko draws the conclusion that the success of a university graduate is not always determined by the quality of the education received: informal employment schemes and "preferential" conditions for obtaining a job contribute to successful employment with a sufficiently low level of knowledge and skills [12]. In some publications, Telyatnikova argues in favour of using the dual system in the field of higher professional education in the process of combining educational and work activities and proves that the quality of education is determined by the existence of practical skills among graduates [13].

The coronavirus pandemic has significantly affected the higher education system including that in Russia. In this regard, domestic and foreign authors' publications on issues of the distance learning in higher education, its advantages and disadvantages were issued [14–17]. They note that the remote learning format significantly changes the activities of a higher school teacher, which will undoubtedly affect the quality of professional training in the following years.

The diversity of opinions, we believe that the quality of education in Russia is determined by the success of graduates in the labour market, which presupposes successful employment in the profession they have acquired; career development opportunities; decent income level.

High competition in the labour market encourages young people to choose universities that provide high quality education. The quality of educational services shapes the attractiveness of the university and determines the confidence of young people who strive to get higher education.

In this regard, it is very important to answer the question: what determines the quality of higher education in Russia in the opinion of university applicants? Why do applicants choose this or that educational institution? What do students value the educational institution for? In order to get answers to these questions, we conducted a survey among students of renowned Russian universities.

2 Materials and Methods

Research methods include analysis and synthesis, comparative analysis, classifications and groupings, sociological methods.

The purpose of the research is to identify factors that influence the choice of a higher educational institution by young people and get an idea of the expectations of applicants from studying at a higher educational institution.

Research hypothesis: for applicants, an important role in choosing a university is determined by the quality and not the availability of the education they receive.

The sample of respondents was 320 people. The survey was carried out among students of The Herzen State Pedagogical University of Russia (244 people), St. Petersburg State University of Economics (50 people), North-West Institute of Management of the Russian Academy of National Economy and Public Administration under the President of the Russian Federation (18 people) and National Research University Higher School of Economics (8 people).

3 Results and Discussion

The majority of the respondents noted that when they enter universities, they expect to gain practical skills in their professional field (230 people, 72.6% of the respondents); actual theoretical knowledge (212 people, 66.9% of respondents); the possibility of building new social ties (182 people, 57.4% of the respondents) (see Fig. 1);

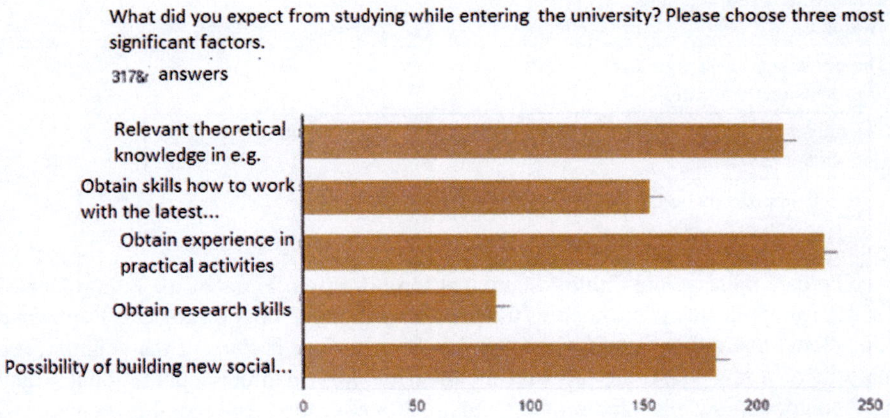

Fig. 1. Applicant expectations from studying at a university upon admission. *Source:* Compiled by authors

Most of the applicants, when choosing a profession, first determine a field of professional activities, and then they chose a suitable university (see Fig. 2).

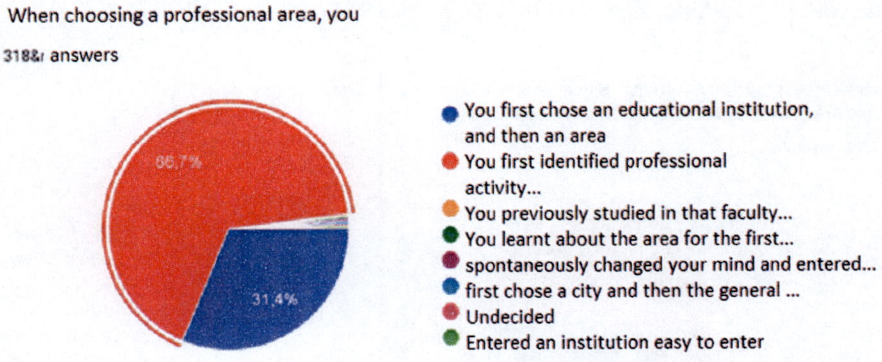

Fig. 2. Priorities in choosing a professional field. *Source:* Compiled by authors

Table 1. Why did you choose the university where you study? Choose the three most important factors.

Possible answer	Number of respondents, persons	Proportion of respondents (of the total number of respondents), %
The key role was determined by the university brand and its awareness	187	60.3
The university has a high level of professional education	134	43.2
The university had state-financed openings	125	40.3

Source: Compiled by authors

The data presented in Table 1 show that top priority in terms of the importance of factors for applicants is attached to the university brand and its awareness. Renowned educational institutions attract young people from various regions of the country. The majority of young people are sure that in universities known throughout the country they provide good knowledge and provide high quality education. In second place in terms of importance is the belief that it is the educational institution, that they have chosen, provides a high level of professional training. The third priority in terms of importance for Russian youth is the availability of state-funded places, which allows them to obtain higher education for free. Other factors in choosing a university were less significant: the traditions of the university and its organisational culture (109 people; 35.2% of the respondents); ample employment opportunities (64 people, 20.6%). Such factors as low tuition fees, geographic location, low passing scores, and the ability to combine work and education virtually play no role in choosing an educational institution.

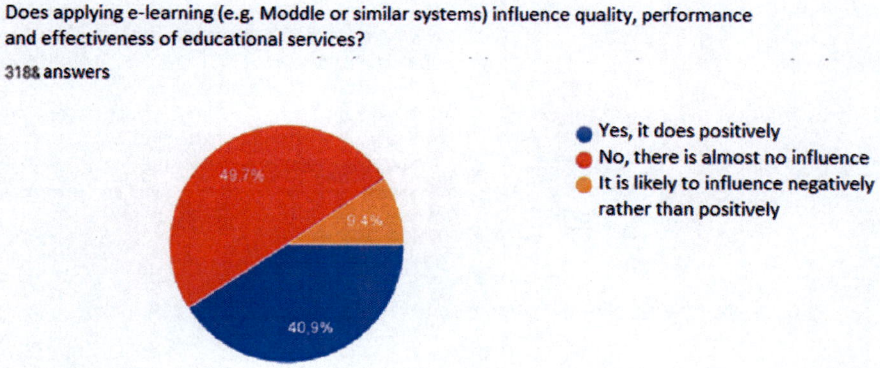

Fig. 3. The impact of e-learning on the quality of educational services. *Source:* Compiled by authors

It was unexpected that the majority of respondents noted that the use of an electronic system in teaching (for example, Moodle) has almost no effect on the quality, results and effectiveness of educational services (see Fig. 3). Only 40.9% of the respondents believe that e-learning systems contribute to enhanced quality of education and acquisition of new knowledge.

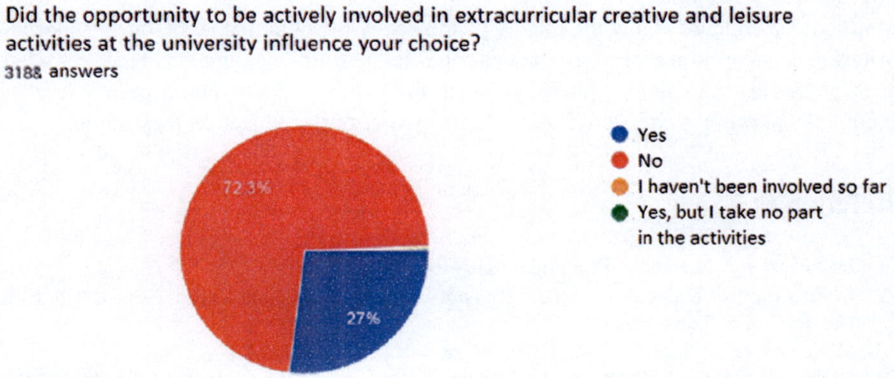

Fig. 4. Influence of active involvement in any extracurricular creative and leisure activities on the choice of an educational institution. *Source:* Compiled by authors

The survey results showed that the choice of an educational institution has little effect on the possibilities of active involvement in extracurricular creative and leisure activities: the majority of respondents enter universities solely for professional knowledge (see Fig. 4).

According to students, the key competencies of a teacher are the ability to transfer their knowledge to students, knowledge of modern trends and problems in the professional field and a deep knowledge of the subject. General erudition, the ability to conduct scientific developments in their professional field, the availability of practical experience and mastery of modern computer technologies are of minor importance.

In answering the question "Why do you get higher education?", the majority of students (57%) answered that the quality of education is the key to successful employment in the profession and specialty acquired and determines a high status in society. At the same time, the majority of respondents noted that the professional competencies generated due to higher education make it possible to ensure a high level of income and a decent quality of life, even if they have to work not in their specialty. Almost 12% of the respondents believe that the level of education changes the interests and lifestyle of a person: the higher the level of education, the more interesting both personal and professional life.

4 Conclusion

According to students, the quality of education in Russian universities is largely determined by the relevance of theoretical knowledge and practical skills that an educational institution provides; professionalism of teachers, which consists in not only a deep knowledge of their professional sphere and new development trends but also the ability to transfer knowledge to students; the links, that exist between the educational institution and enterprises of the real sector of the economy, since this provides an opportunity for internships in companies, which provides for subsequent employment prospects. Students believe that if the chosen institution meets their requirements, their investment in education will pay off.

It can be concluded that the quality of the education obtained is the most important criterion for the choice of higher educational institutions by students. They associate success in the professional sphere, a decent position in society, and a decent level of income from professional activities with the quality of the education they obtain.

References

1. Dzhurinsky, A.N.: Actual Prob. Educ. **1**, 9–19 (2016)
2. Collins English Dictionary – Complete and Unabridged. William Collins Sons & Co. Ltd., Harper Collins Publishers (2012)
3. Mileshina, O.Y.: Econ Soc. Mod. Dev. Models **9**(2), 277–287 (2019)
4. Dicker, R., Garcia, M., Kelly, A., Mulrooney, H.: Stud. High. Educ. **44**(8), 1425–1441 (2018). https://doi.org/10.1080/03075079.2018.1445987
5. Alzafari, K., Ursin, J.: Qual. High. Educ. **25**(1), 58–75 (2019). https://doi.org/10.1080/13538322.2019.1578069
6. Mussawy, S., Rossman, G.: Qual. High. Educ. **27**(1), 99–122 (2021). https://doi.org/10.1080/13538322.2020.1833419
7. Zhang, L.: Educ. Chin. After Sch. **6**, 12–24 (2009)
8. Dorau, R., Höhns, G.: Erfolg der Absolventen auf dem Arbeitsmarkt als Maβstab f¨ur die Qualität der Ausbildung. In: Qualität in der beruflicher Bildung. https://www.bibb.de/veroeffentlichungen/de/publication/download/1178
9. Zborovskiy, G.E., Ambarova, P.A.: Soc. Res. **3**, 17–27 (2021). https://doi.org/10.31857/S013216250012904-7
10. Zborovskiy, G.E., Ambarova, P.A.: High. Educ. Rus. **11**, 34–45 (2019). https://doi.org/10.31992/0869-3617-2019-28-11-34-46
11. Zborovsky, G.E.: Newsl. Surgut. State Pedag. Univ. **3**, 9–20 (2020). https://doi.org/10.26105/SSPU.2020.65.2.008
12. Demidenko, S.: Soc. Res. **46**, 160–162 (2020). https://doi.org/10.31857/S013216250008523-8
13. Telyatnikova, T.V.: Lab. Econ. **4**(1), 51–58 (2017)
14. Vertakova, Y., Plotnikov, V.A.: Manage. Consul. **3**, 54–60 (2020). https://doi.org/10.22394/1726-1139-2020-3-54-60
15. Tsygalov, Y.: Manage. Consul. **10**, 61–73 (2020). https://doi.org/10.22394/1726-1139-2020-10-61-73
16. Masalova, Y.: Lab. Econ. **8**(3), 261–274 (2021). https://doi.org/10.18334/et.8.3.111817
17. Armstrong-Mensah, E., Ramsey-White, K., Yankey, B., Self-Brown, S.: COVID-19 and distance learning: effects on Georgia State University school of public health students (2020). https://doi.org/10.3389/fpubh.2020.576227

Innovative Approaches to the Educational Processes in Higher School in the Context of Digital Transformation

Tatiana P. Maslevich[1](✉), Platon A. Polomoshnov[2], Irina V. Lipchanskaya[2], and Denis A. Zhukovsky[3]

[1] Russian Presidential Academy of National Economy and Public Administration (RANEPA), Moscow, Russia
maslevich@ranepa.ru
[2] Rostov State University of Economics, Rostov-on-Don, Russia
[3] Don State Technical University, Rostov-on-Don, Russia

Abstract. The COVID-19 pandemic has accelerated a transition of enterprises to a digital business environment, restructuring business processes using innovative opportunities. Higher educational institutions in a short time transferred educational processes to an online format using distance learning technologies (DLT) in their work. Today it has become obvious that such a format presupposes not so much a change in the way the participants interact, but the development of new methods for implementing processes using modern technologies of interaction, control and communication of students. The paper provides an overview of the results of a survey of students in connection with their experience in gaining knowledge using digital educational platforms. The survey involved 57 second-year students studying in the field of management. The data obtained allow asserting that a consequence of the change in the format of the educational process was the deterioration of the education quality. Among the main reasons are outdated teaching methods that do not correspond to innovative learning processes, lack of training for participants (both students and teachers), insufficient infrastructure of universities and financial difficulties, personal characteristics of the teachers, peculiarities of intragroup interaction, lack of understanding of the status of a specialist trained in online format and a number of other factors. The results of the survey are recommended to be taken into account by universities when developing innovative educational methods for the implementation of processes using DLT.

Keywords: Innovative educational methodology · Digital educational platform · Digital educational process · Distance learning technologies

1 Introduction

A digital transformation requires the restructuring of the activities of higher education and the creation of an educational process infrastructure capable of solving the problem of integrating resources and services for all participants, on the one hand, and, on the

P. V. Trifonov and M. V. Charaeva (Eds.): DITEM 2021, LNNS 380, pp. 711–716, 2022.
https://doi.org/10.1007/978-3-030-94245-8_97

other hand, not only preserving the level of education, but also raising it to a new modern level. This process is ongoing, however, the forced and rapid transition to a distance format during the COVID-19 pandemic raises a number of issues for universities on the organization of the digital educational process in the near future. Already, 80% of Russian universities are fully or partially working with DLT. Systematic conclusions on the possibility of their full or partial adaptation have yet to be drawn, however, some results are already visible. Results of some conducted research have been published on the preference of students to use various learning formats, the ability of participants in the process to effectively implement them and master disciplines, the success of certification, the level of accessibility of online education, etc. It has been shown that students prefer a mixed approach to learning, and this model does not have a significant impact on overall students' performance [1]. Already today, all participants in the process are fully or partially ready for the use DLT in teaching, and the traditional format of education must be complemented by them without fail [2]. At the same time, university professors question the reliability of the grades received by students during online exams, despite the existing process control system [3]. It is noted that not all educational institutions are completely ready to switch to the distance format, and in the near future the demand for specialists trained online may decrease [4]. The complexity of the implementation of the digital educational process lies in the significant limitation or complete absence of direct communication channels between participants, self-organization skills, as well as the occurrence of cases of imitation by students of educational activities in the absence or decrease of the teachers' control function [5]. A significant problem in the introduction of innovative educational methods is the lagging of the methods of work of universities from the available technological capabilities, when the old methods of education are trying to combine with new technologies. As a result, formalism and bureaucracy grow, hindering the development of innovative processes [6]. It is noted that learning with the use of DLT affects the equality of access to education for students living in different regions of the country or belonging to different social strata due to differences in financial opportunities and differentiation in the provision of technical means [7]. Difficulties in accessing online learning are experienced by students with disabilities, for whom it is necessary to create an accessible environment in the online space [8]. A comparison of the knowledge gained by students as a result of online and offline learning requires special attention. Research shows that students perform better with the traditional approach to education and examinations [9]. At the same time, even before the changes caused by the COVID-19 pandemic, it was noted that an effective transition to a new training format requires preliminary preparation of all participants, including in the development of the teaching methodology, formats for classes, their content and communication methods for participants [10].

Thus, at present, most of the participants in the educational process have formed their attitudes towards distance learning technologies, and they are ambiguous.

2 Materials and Methods

To study the opinion of students about digital educational processes with DLT application, a study was carried out among second-year undergraduate students studying in the field of management. The participants were asked to answer the question of how, in their opinion, the quality of education changed during a transition to the distance format, to indicate the main reasons that influenced this dynamics (the first level reasons), as well as the factors that form them (the second level reasons). The "quality of education" was understood as the possibility of mastering a discipline in full in comparison with the traditional teaching methods. It was proposed to identify up to five first level reasons, the number of second level reasons was not limited. The survey was conducted using the fishbone method, the questions were open-ended, and the opportunity to work both individually and in groups was provided. 57 people took part in the survey.

3 Results

The results of the survey showed that 100% of the respondents noted a decline in the quality of education. The main reason was the discrepancy between the educational methodology and the method of implementing the process (100% of the respondents). The following were named as disadvantages of the method: a large amount of theoretical information, difficult to assimilate as a stream of spoken speech for more than an hour (52%); lack of opportunities for full-fledged teamwork (50%); passive role of listeners, inability to take part in a collective discussion (42%); irrelevant data in the context of rapid changes and lack of connection with practice (38%), lack or absence of electronic educational materials (25%). A number of respondents (23%) drew attention to the fact that theoretical information received online is forgotten faster, since it is perceived only by ear (less often – visually on a computer or phone screen), and is not associated with the teacher's personal characteristics, his/her facial expressions and gestures, the ability to defuse the situation when explaining, use the additional opportunities of the communicative process in the format of classroom work.

About 80% of respondents believe that the deterioration in the quality of education is associated with the personal and professional characteristics of teachers, namely, the inadequacy of qualifications to modern market demands (52%), the lack of motivation to work and the desire for constant self-development (43%), the unwillingness to master new educational technologies and change educational content and teaching style (24%). At the same time, the age of the teachers did not have a significant impact, some of the respondents (10%) noted that too young, as well as older age, is associated with the fact of the lack of modern knowledge in the discipline. A number of respondents (15%) noted that a low level of motivation of teachers is associated, among other things, with the lack of response to the proposed materials from the audience and the inability to assess the degree of mastering the materials in real time. Some factors were named that are not directly related to the qualifications of teachers, but have a significant impact on the content, in particular, an accumulated fatigue of teachers and listeners, which, on the one hand, leads to the reluctance of the former to include in the lectures the issues difficult to explain that require additional physical and emotional efforts for clarification, and on the other hand, the reluctance of students to make additional attempts to master complex materials (10%).

About 65% of the respondents believe that the deterioration in the quality of education is associated with the peculiarities of formation and the composition of student groups and interpersonal relationships in groups. A large number of group members (48%) were named as a factor of influence, as a result of which, in the conditions of limited time and specific communication, an online seminar turns into either individual training of several students, or there is no individual approach at all and training becomes "impersonal", losing personal contact between teacher and student. The process is adversely affected by the stratification of students in groups according to the level of training and desire to acquire knowledge, to work independently, to take part in online discussions (38%), "bullying" of the most successful students (20%). About 20% of the respondents believe that the distance learning format leads to the emergence of personal problems, the solution of which requires, among other things, a consultation of a psychologist.

The next most important factor was the decline in listeners' interest in the online learning process due to the lack of skills to use DLT in both students and teachers, technical imperfections of educational platforms, and lack of communication means (44% of respondents).

Among the significant factors influencing the decline in the quality of education, there were: the lack of a clear understanding of the prospects for being in demand in the profession and for subsequent career growth for those educated online (18%), the absence or high cost of textbooks and teaching aids (15%), and the lack of proper parental control (10%), etc.

4 Discussion

A transition of the educational process to the online format was carried out in such a short time that the technical unpreparedness of universities became obvious already in the first weeks. The Learning Management System (LMS), the platform on which most Russian organizations manage the educational process, only partially fulfills its functions as a tool for creating and providing content, however, it does not have the technical capability to implement the functions of communication and interaction between its participants in full. To fill in the missing options, most teachers use additional means of video communication and information storage – MS Teams and Zoom software, and interact with students using the social media. To date, the question of the possibility and scope of using DLT in the educational process remains controversial.

5 Conclusion

Upon an analysis of the students' opinions, based on the experience of a year of study at a university using DLT, a few conclusions can be drawn that should be taken into account when organizing the educational process using distance technologies.

1. When changing the format of the process, the method of transferring information using DLT was changed, while the design of the discipline, namely, the educational content (the lectures and seminars, methods of intermediate and final certification, colloquium procedures and independent work), developed and tested in class, remained the same. In the future, an increase in operational efficiency and quality

of the process will be directly related to the development of innovative educational methods, improvement of the communication methods between participants, use of new control and measurement tools for checking knowledge.

2. For the full and effective implementation of the digital educational process, a unified digital platform should be developed and implemented, which would allow the communication needs to be fully realized. The variety of digital platforms (in particular, LMS, MS Teams, Zoom) does not allow the formation of a single information space that enables implementing a set of business processes of an educational institution.
3. The problem of financial support of both the institutions and the students with modern technical facilities and communication methods, ensuring the implementation of educational programs, remains important.
4. Already in the near future, the issues of legal support of distance learning for universities and other participants in the process should be resolved.

The experience gained in the implementation of the educational process in the mode of remote communications of participants during the COVID-19 pandemic in the near future should ensure the full integration of DLT into the educational process of universities, replacing the traditional technologies either in part or in full. At the same time, this action should be accompanied by the solution of all problematic issues and the elimination of risks for all its participants.

References

1. Etom, R., et al.: The use of e-learning tools in blended learning approach on students' engagement and performance. J. Phys. Conf. Ser. **1835**(1), 012075 (2021)
2. Shushara, T.V., Delvig, N.A., Abraamyan, S.M., Shushara, S.M.: Digitalization of education: distant learning prospects and realities. CEUR Workshop Proc. **2834**, 526–532 (2021)
3. Al-Karaki, J.N., Ababneh, N., Hamid, Y., Gawanmeh, A.: Evaluating the effectiveness of distance learning in higher education during COVID-19 global crisis: UAE educators' perspectives. Contemp. Educ. Technol. **13**(3), ep311 (2021). https://doi.org/10.30935/cedtech/10945
4. Banerjee, D., Das, D., Pal, S., Paul, S.R., Debnath, A., Reza, M.: Effect of COVID-19 on digital transformations in teaching learning methodology and its consequences in society: a review. J. Phys. Conf. Ser. **1797**(1), 012066 (2021)
5. Frolova, E.V., Rogach, O.V., Tyurikov, A.G., Razov, P.V.: Online student education in a pandemic: new challenges and risks. Eur. J. Contemp. Educ. **10**(1), 43–52 (2021)
6. Shabanov, G.A., Rastyagaev, D.V.: University digitalization: reality and expectations. High. Educ. Today **1**, 2–7 (2020)
7. Agranovich, M.L., Ermachkova, J.V., Livenets, M.A.: Online learning during the COVID-19 pandemic and inequality of access to education. Federalism **3**, 188–206 (2020). https://doi.org/10.21686/2073-1051-2020-3-188-206
8. Russ, S., Hamidi, F.: Online learning accessibility during the COVID-19 pandemic. In: W4A 2021, Proceedings of the 18th International Web for All Conference, Article 8, pp. 1–7 (2021). https://doi.org/10.1145/3430263.3452445

9. Foo, C., Cheung, B., Cho, K.-M.: A comparative study regarding distance learning and the conventional faceoace approach conducted problemased learning tutorial during the COVID-19 pandemic. BMC Med. Educ. **21**, 141 (2021)
10. Jocius, R., et al.: The virtual pivot: transitioning computational thinking PD for middle and high school content area teachers. In: SIGCSE 2021 – Proceedings of the 52nd ACM Technical Symposium on Computer Science Education, pp. 1198–1204 (2021)

Current Trends in Economic Research: Methodology and Practice

Olga B. Ivanova[1(✉)], Tatiana V. Plotnikova[1], Anastasia Ye. Malkhasyan[2], and Vladimir V. Klimuk[3]

[1] Rostov State University of Economics, Rostov-on-Don, Russia
obivanova@bk.ru
[2] Don State Technical University, Rostov-on-Don, Russia
[3] Baranovichi State University, Baranovichi, Belarus

Abstract. Digitalization of the economy and the rapid development of automation systems of various business processes at the macro and micro levels actualize the issues of applying the most modern methods of research during the analysis of economic processes and assessment of individual economic entities, as well as industries, regional or national economies. Despite the possibility of using a wide range of solutions for this purpose, any programs will be based on traditional research methods adjusted to the current economic realities. The article provides an overview of the main methods of economic research. There are examples of symbiosis of several types of methodologies from different fields of knowledge, as well as practical examples of the use of several methods in order to achieve specific economic objectives.

Keywords: Methodological analysis · Expert assessments · Economic research algorithm · Economic observation

1 Introduction

Any economic research is based not on the use of isolated methods, but on the use of a system, the feasibility of which is confirmed not only by theoretical conclusions, but also by practical results. It is often the case that effective research methods come precisely from practical developments in the field of theoretical study. The specific object and subject of observation are determined depending on the goals and objectives of the research. That is why the diversity of economic objects (regions of the country, individual enterprises, industries, population, etc.) determines a large number of existing research methods.

2 Methods

Each of those allows you to achieve certain goals set at the beginning of the study. Methodological analysis of the scientific process allows us to distinguish three types of research methods, the classification of which is shown in Fig. 1:

P. V. Trifonov and M. V. Charaeva (Eds.): DITEM 2021, LNNS 380, pp. 717–725, 2022.
https://doi.org/10.1007/978-3-030-94245-8_98

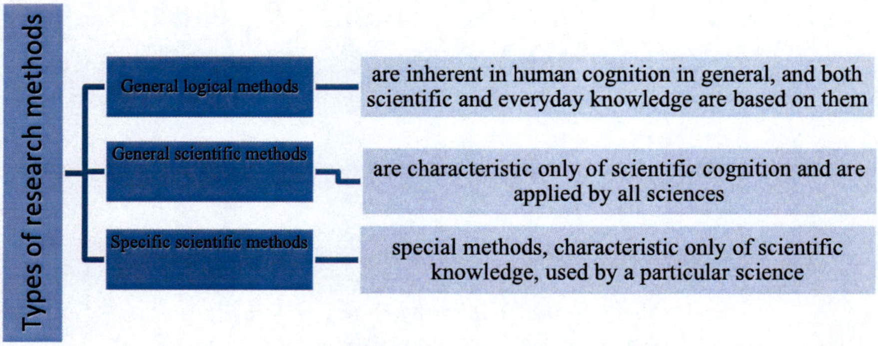

Fig. 1. Types of economic research methods.

General logical methods of knowledge include methods such as analysis, synthesis, abstraction, generalization, induction, deduction, analogy, modeling [1]. If we talk about the peculiarities of the methodology of economic research, their most popular tool is analysis.

3 Results and Discussion

The analysis is a mental or real dissection of a whole subject or object into its component parts. The procedure of economic analysis is widespread in economics. This procedure assesses the financial and economic condition of the enterprise using, for example, methods of horizontal and vertical analysis of the balance sheet, a system of financial ratios, calculating productivity, the state of fixed assets, etc. [2]. Here is an example of such an analysis using the horizontal analysis of Svyaznoy Chain LLC financial results report for 2019–2020 (Table 1).

Table 1. Horizontal analysis of the statement of financial results of Svyaznoy Chain LLC.

Indices	2020	2019	Changes	
	Thousand rubles	Thousand rubles	Thousand rubles	Growth rate, %
1	2	3	4	5
Revenue	95 394 059	48 322 328	47 071 731	197.4
Cost of sales	70 310 145	32 771 371	37 538 774	2.1 times growth
Gross profit (loss)	25 083 914	15 550 957	9 532 957	161.3
Administrative expenses	26 779 754	16 723 759	10 055 995	160.1

(*continued*)

Table 1. (*continued*)

Indices	2020	2019	Changes	
	Thousand rubles	Thousand rubles	Thousand rubles	Growth rate, %
Profit (loss) from sales	−1 695 840	−1 172 802	523 038	The loss increased by 44.6%
Interest receivable	858 676	497 449	361 227	172.6
Interest payable	1 647 933	954 447	693 486	172.7
Other income	3 772 428	3 121 513	650 915	120.9
Other expenses	7 079 435	4 620 585	2 458 850	153.2
Profit before taxation	−5 792 104	−3 128 872	2 663 232	The loss increased by 85.1%
Current income tax	–	–		–
Change in deferred tax liabilities	15 920	28 102	−12 182	56.7
Change in deferred tax assets	961 841	337 339	624 502	2.9 times growth
Other	84 494	88 369	−172 863	95,6
Net profit	−4 898 837	−2 675 062	−2 223 775	The loss increased by 83.1%

The changes reflected in the statement of financial results of Svyaznoy Chain LLC are shown in Fig. 2. From this, we can conclude that the company incurred even greater losses in 2020 than in 2019. The final financial result is a loss of 4.8 billion rubles, which is 83% higher than the loss in 2019. It should be noted that a similar situation has been observed in the company since 2015, when the company's profit amounted to 1.5 billion rubles. Since then, the company has never achieved a positive financial result. Also since 2015, the market value of the company has been falling. If in 2015 it was 11 billion rubles, in 2020 it will be only 3.6 billion rubles.

This example shows us the relevance of economic research by applying economic analysis methods:

- even an outside expert can assess the situation, since the information needed for the study is publicly available;
- in a short period of time you can conduct a full-fledged research to identify strengths and weaknesses in the economic activity of the enterprise;

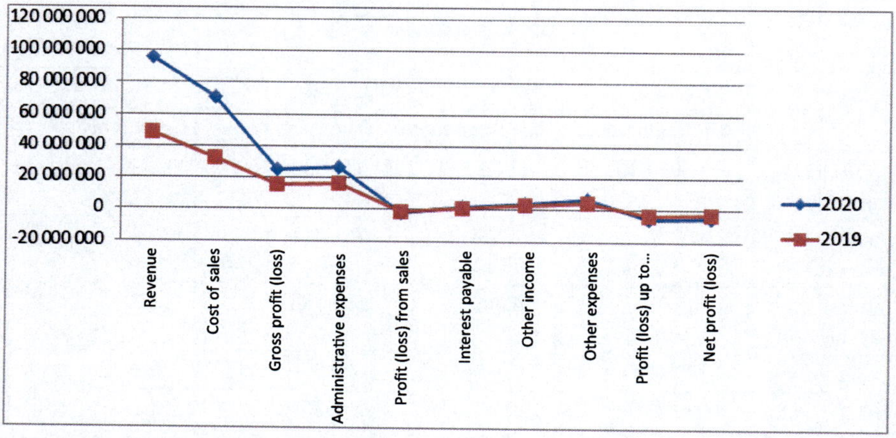

Fig. 2. Changes in financial indicators of Svyaznoy Chain LLC in 2019–2020.

- economic analysis forms the basis for economic monitoring, which allows non-stop tracking of the economic situation in the enterprise and making timely management decisions.

Synthesis is an inverse procedure to analysis, involving the mental or real connection of previously isolated parts of an object or object into a single whole. In economic research, it is most often an interdisciplinary synthesis. Economic theory is expanding into psychology, sociology, law, or vice versa, when economic science borrows methods from other sciences to solve its own problems. Abstraction is another method of economic research. It involves the distraction from a number of properties and relations of the object under study with the allocation of the properties and relations of interest to the researcher. A classic example of economic abstraction is the concepts of perfect competition or pure monopoly, which are often used in economic theory.

Generalization is a method of thinking, as a result of which common properties and features of objects are established. New research methods appear very often in economics, which combine several traditional techniques. A striking example is the method of expert evaluations. This is one of the most popular methods of economic research in recent decades.

Methods of expert evaluations are procedures aimed at the analysis of the problem and the selection of its solution by specialists. Such a system is used if we are talking about a very complex problem, which cannot be solved independently due to various factors. So, for example, it could be a lack of information, low skills, etc. In this case, experts come to the rescue. The solutions proposed by them on the basis of the arguments presented are called methods of expert evaluation. This analysis allows you to identify possible options for business development under different scenarios. Using this methodology, the company can get optimal results as quickly as possible and without unnecessary costs.

Expert evaluation as a method of research can be divided into two types which are the method of collective work of experts, as well as the method aimed at obtaining an individual approach from each of them. Here is a practical example of the method of expert evaluations. Suppose an airline needs to compare the parameters of different types of aircraft. To do this, let us make a table listing these types and common parameters and their weight (Table 2).

Table 2. Example of expert evaluation [3].

N	Parameters	Weight	A	B	C	D	E	F
1	A1	X	10	9	2	5	3	3
2	B1	Y1	9	8	3	5	6	1
3	C1	Z1	8	7	1	7	8	3
Amount								

Next, you will need to write in each of the aircraft type categories (A, B, C, D, E, F) weights (X, Y, Z) according to the parameters (A1, B1, C1). All objects can be rated on a ten-point scale. To calculate the total, multiply the weight by the points. As a result of the calculations, we get the highest sum, which means the highest efficiency of any option.

Next, the evaluation will be carried out by experts, each of whom will analyze a particular object according to all the criteria presented. The data obtained will be recorded in a summary table in order to then come to a consensus.

Induction is the process of deducing a general statement from a series of particular statements, from single facts. Deduction is the process of reasoning, the opposite of induction, going from the general to the particular or less general. The methods of induction and deduction are widely used in economic science. Analogy is a method of cognition in which, based on the similarity of objects in some attributes, a conclusion is made about their similarity in other attributes. Economic analogy is an assumption about the presence of a feature in the studied economic object on the basis of similarity, existing in other features, with an ideal or material analogue of this object. A particular case of analogy is extrapolation. Extrapolation refers to the extension of provisions or conclusions made with respect to any part of objects or phenomena to the totality of these objects or phenomena. Economic extrapolation is most often used in statistical calculations and appears as a spread of established past trends to a future period. Modeling is the study of an object (the original) by creating and studying its copy (model), which replaces the original from certain aspects of interest to the researcher. In the context of economic research, economic and mathematical modeling is of particular interest. This is a type of symbolic modeling, the purpose of which is a mathematically formalized description of economic objects, processes and phenomena [4].

Let us consider an example of the use of a set of general logical methods in the enterprise. For example, the task is to carry out a study of the competitive market of industry. To do this, it is necessary to follow a certain algorithm of research, which includes the stages and methods described below in Table 3.

General scientific research methods can be divided into two main groups: empirical research methods and theoretical research methods. The latter include idealization,

Table 3. Algorithm of economic research with an indication of the methods used (by the example of the study of the competitive market). *Source:* Compiled by the authors on the basis of [5].

Name of the stage	Description	Methods used
Definition of objectives	Gathering information is a time-consuming activity. It will require the mobilization of financial and human resources. Therefore, it is necessary to determine which indicators will be taken into account when preparing a new sales strategy	Description, generalization, synthesis
Indication of the boundaries of the research	It can take a long time to review a niche, but you can save time if you specify a sample of companies to monitor. The study of competitors allows you to know their profits, production volume, characteristics of work, the customer base	Modeling
Setting the time interval	Evaluation is ongoing to ensure that the data is up-to-date. The manager's task is to specify the term that will be optimal for the task and at the same time provide the necessary resources of the company. For example, a comparative study of a competitor's products can be carried out briefly to meet targets and in detail, even with sales forecasts for several years	Modeling, comparison
Preparation of positioning	The experience of colleagues becomes an indispensable tool when a firm brings a product to market or changes its own approach to promotion. For example, if it is known that in the footwear trade in the region most buyers prefer leather goods, then it is better to go to them with such an offer	Analogy

(*continued*)

Table 3. (*continued*)

Name of the stage	Description	Methods used
Planning	The study of competitors will allow the commercial department to forecast volumes for the new period, taking into account the potential for sales	Economic extrapolation
Pricing	The cost of the product to the end consumer determines the demand for it and the amount of net profit. To make it balanced, it is worth taking into account the situation in the industry and assess similar proposals of other organizations	Induction
Development of the product	The goal of any startup is to create a product that is superior to other options and in demand by the majority of the population. A comprehensive review of the existing range, looking for strengths and weaknesses, will help to create it	SWOT analysis

formalization, axiomatic method, hypothetico-deductive method, method of ascending from the abstract to the concrete, historical and logical research methods [6]. Empirical methods relate to those techniques that are the content of practice or its direct result: observation, description, measurement, experiment.

Economic surveillance is a specific type of social observation, which systematically and purposefully collects primary information about various economic events. One form of economic surveillance is economic monitoring. A continuation of economic observation is description. This is the method by which the results obtained during the analysis are processed and reduced to a perceptible form. For example, drawing up an annual report on the work of the company with an explanatory note describing the changes that have occurred in the financial statements of the company are the most common examples of the use of descriptive method in economic research.

Another method named economic measurement is used most often in statistical studies, which can be carried out both at the macro- and micro-level. Those natural, labor and value units are measured. Here is an example of the joint application of general logical and empirical methods. For example, Table 4 shows the factors that, according to the management of trade organizations, significantly limit the activities of their companies. In carrying out such a study, three types of methods were used simultaneously such as expert evaluations (allow to identify factors limiting the activities of the enterprise), economic measurements (allow to make these evaluations statistically reliable and quantifiable), description (provides information in tabular form, convenient for further analysis) and comparison (the information presented over the years allows to analyze the significance of each factor and to identify how the situation has changed over the past five years in the context of each factor) [7].

The application of the above set of methods allows us to conclude that their leaders in the 4th quarter of 2020 indicated insufficient solvent demand of the population as the main factor limiting the activity of their trade organizations (56% of the respondents). Also, if we analyze the dynamics of restrictive factors for the development of trade since 2006. This factor is the only one of them that every year more and more entrepreneurs point to.

Table 4. Factors limiting the activities of retailers (% of the number of respondents). *Source:* Compiled by the author according to [8].

Year	Effective demand failure	Lack of funding	High rent	High transportation costs	High competition	High level of taxes
2016	57	32	27	24	71	49
2017	54	28	28	21	67	46
2018	50	26	27	21	65	44
2019	56	28	30	22	71	48
2020	56	28	29	21	68	45

Another method is the economic experiment. The economic experiment is interpreted as a scientific test of the effectiveness of certain theoretical provisions or economic activities in individual enterprises, economic activities or the entire economy. The purpose of an economic experiment is often to confirm the correctness of a certain hypothesis, put forward on the basis of the study of empirical facts or theoretical provisions.

4 Conclusion

Thus, a comprehensive study of the techniques of economic research shows that empirical methods provide an insight into an economic object or process through a scientific fact, while theoretical methods of research to a greater extent study the world of ideal objects. The problem of correlation of empirical and theoretical research methods is solved in the process of their practical application.

References

1. Barlybaev, A.A., Sitnova, I.A., Yunusova, G.M.: Fin. Credit **32**, 48–59 (2015)

2. Zotova, A., Kandrashina, E., Ivliev, A., Charaeva, M.: Int. J. Environ. Sci. Educ. **11**(14), 6715–6725 (2016)
3. The Method of Expert Evaluations: Types, Criteria and Examples. https://www.kom-dir.ru/article/3450-metod-ekspertnyh-otsenok
4. Charaeva, M.V.: Formation of structural and logical model of company's financial standing analysis. In: 2nd International Conference on Social, Economic, and Academic Leadership (ICSEAL 2018), Prague, Czech Republic, pp. 129, 16th–18th June (2018)
5. Study of Competitors in Trade – Analysis, Methods. https://www.kom-dir.ru/article/3450-metod-ekspertnyh-otsenok
6. Edronova, V.N., Ovcharov, A.O.: Econ. Anal. Theory Pract. **10**, 37 (2013)
7. Barmuta, K., et al.: Entrep. Sustain. Issues **7**(1), 484–497 (2019). https://doi.org/10.9770/jesi.7.1(34)
8. Leading Indicators by Economic Activity of the Federal State Statistics Service. https://rosstat.gov.ru/leading_indicators

Digital Educational Environment for Business: Risks and Trends of Sustainable Development

Elena V. Lyapuntsova[1](✉), Oksana V. Lazareva[2], Ilona I. Drozdova[3], Yulia M. Belozerova[4], and Anna V. Danilina[5]

[1] Bauman Moscow State Technical University, Moscow, Russia
[2] Russian State University of Justice (RPA of the Ministry of Justice of Russia), Moscow, Russia
[3] Russian University of Transport (MIIT), Moscow, Russia
[4] GITR Film and Television School, Moscow, Russia
[5] Institute of Linguistics of the Russian Academy of Sciences, Moscow, Russia

Abstract. The article is devoted to the study and consideration of the features, the main directions of the formation of a digital educational environment for business in a period when the risks and unfavorable conditions are so great that they must be leveled by the presence of effective methods, techniques and technologies. On the example of the new mechanics of using the "case-tech" method proposed by the authors, the urgency of creating educational trends, an effective environment for interaction for subjects of the learning process, taking into account the goals of sustainable development, is substantiated. But, the basis of the method is universal. The main advantage is novelty and formalization. However, a person as an expert and mentor is the main role model in the process both at the start of creating a business and during its development. The important takeaway from this article is that technology does not have to be for technology's sake. But, at the same time, it is necessary to weigh the risks, take into account the goals of sustainable development in the formation of the digital educational environment as an ecosystem.

Keywords: Transformation process · Information educational environment · Non-formal education · Business education · Virtual and augmented reality · Case tech method · Small and medium business · Transformation · Individual trajectory

1 Introduction

Modern educational and social environment dictates the rules for the formation of intellectual activity of various professional groups. Business education is no exception. This is a form of intellectual activity associated with the development of rules, norms, methods, and methodologies of the learning process. The main specific feature of education becomes the thesis of "creative learning" in the formation of entrepreneurial thinking.

The purpose of the research is to create an opportunity to systematize educational elements in the design of digital learning environments (DLE) and scaling.

The prerequisite for research and study of existing methods was the understanding and awareness by the authors of the need to form a new approach, taking into account

P. V. Trifonov and M. V. Charaeva (Eds.): DITEM 2021, LNNS 380, pp. 726–730, 2022.
https://doi.org/10.1007/978-3-030-94245-8_99

the trends in the introduction of numbers in our lives. According to the authors, such a methodology, is based on project activities by means of a case-technology strategy and the use of digital platforms, large databases, datasets, tools, services, services, and many other aspects. The strategy and methodology of case-technology proposed by the authors is based on the study of various existing methods. The authors have deeply studied the case-study method, which emerged in the 1960s at Harvard. In all aspects and at all stages of the formation of an end-to-end digital environment as an ecosystem, the person as an individual is included in the process. The transformational process of personal development is a unity of four elements: education, lifelong learning, upbringing and adaptation of the individual to life in society. Subjects include a speaker, an educator, a mentor, an expert, a mentor, using modern technologies that ensure the process of adaptation of the subjects taking influence to the processes that characterize society [1, 2].

2 Materials and Methods

There are several aspects of doing business: functional, structural, strategic, psychological. It is these factors that form the basis of the "case-technology" or "case-tech" method proposed by the authors [3, 4]. The result of the study is a developed methodology: "case-technology" and includes the development, description of the case, which correlates with the personality of the author or determined by the characteristics of implementation.

It is a role model or example for learning and a technology for implementing the process in a digital environment. The developed methodology includes new approaches: the process of in-depth learning, case study, taking into account the description of the personality itself, the factors in their dynamics and systems, the development of digital aspects of business education, the peculiarities of entrepreneurial education. The method contains not only soft skills, but also elements of hard skills. It is an integrative process of learning as a technology: from the ability to negotiate with partners and investors to a clear business plan, the creation of a specific business, product, facilities or services. Practically from idea to implementation, which is laid out in the methodology as 5 steps to achieve the result.

Trends in digital technology are transforming the way the world's educational system works. It is not necessary to apply the new for the sake of innovative trends. It is reasonable to include technologies such as artificial intelligence, neurotechnology. It is necessary to focus on the individual, who plays the most important role, as noted in the Sustainable Development Goals (SDGs). Two basic approaches to the goals of education as an institution are taking shape: on the one hand, education is a tool for personal development and meeting the intellectual demands of society. On the other, it is a technology for achieving a business result.

3 Results and Discussion

According to experts, automation will eliminate 75 million work places in the world by 2025, but will create even more: 133 work places. They will be taken by specialists with new qualifications and digital-skills with trends in the development of digital-technology. In Russia, according to estimates, in 10 years there could be 20.1 million robotized work places. This is almost half of the places of officially employed citizens. This will form a specific view of the nature of adaptation and retraining programs, if retraining can be talked about at all.

The private sector will need a qualitatively different training of its personnel, both executive and top management level. Then case-tech training becomes one of the main factor aspects and an indicator of the effectiveness of the training of entrepreneurs, managers of all levels and not only. Forbes estimates the global education market at $6.5–7.0 trillion. At the same time, the share of online education is 3% ($165 billion) and analysts predict that it will reach $240 billion by 2023, with an average annual growth rate of more than 5%. This steady growth in demand for educational programs in the online format is due not only to the growth of graduate salaries and the eventual return on their training, but also to transparent career development, engagement in the process. This steady growth is also the result the effectiveness of the professional community and business contacts. The prerequisites for networking and professional communities are created. EdTech is a special global trend in the structure of business education. EdTech is primarily a set of specific techniques and approaches to the construction of the educational process, then in the future it will build into a LMS ("learning management system"). It has a clear investment cycle and life cycle assessment of any educational product depending on the speed of transformation of the business environment, with modern approaches to producing both experts and mentors.

From a business perspective, EdTech's goal is to grow faster than the market. In terms of products, the main purpose of education is the result, not the process. So the factor of reducing entrepreneurial risks is to gain knowledge with effective methods, techniques and technologies [3, 5]. According to RBC rating, the volume of Russian EdTech market reached 30 billion rubles per year. A third of this money (9.625 billion rubles) is in the 35 largest companies. And a third of all case-products presented are sold exclusively online. Domestic development trends for Skolkovo Foundation residents are about 20 EdTech projects. And each year they increase revenue by an average of 2–3 times, and the target audience by 50%.

According to RBC calculations, the solutions of Skolkovo participants could close 80% of the online education market in Russia by the end of 2021. The share of private companies implementing their case-tools will grow. Major players such as Sberbank, Yandex, and Mail.ru are entering the arena and are steadily capturing a significant share of the market. But, the market is currently developing rapidly and rapidly small forms of educational services and services, the so-called information business. Its own ecosystem is being formed and the market and business itself is being structured.

The volume of the global EdTech market is currently $152 billion representing 3% of the total market for educational services. By 2025, it will only grow to 4.4%. Investor interest in the EdTech market is stable. Over the past 5 years, from 2017 to 2021, the volume of investment has increased almost 7-fold. Although the volume of investment in EdTech is significantly lower than in other market niches [6, 7].

4 Conclusion

Risks and challenges dictate new opportunities for the development of business education, a mixed type with the use of virtual and augmented reality technologies, with the application of the innovative educational concept proposed by the authors and the developed method of "case tech". That can be used further in various forms of training business processes, small and medium-sized businesses, applying the rules formed by the information business. It is important to integrate new methods and approaches into the educational process in educational institutions in order to involve the younger generation in the formation of business thinking [8, 9]. The issues of transformation of thinking and the task of psychological adaptation, stability in the formation of an individual trajectory in the education are significant and need to be discussed by professional communities [10]. The implementation of the SDGs on a global scale will require the introduction of new patterns of production and consumption, will lead to changes in the structure of supply and demand [11–13]. The 2030 Agenda for Business with the SDGs in mind means new requirements for environmental and social responsibility of companies and new opportunities for expanding entrepreneurial thinking in view of trends [14]. New niches and competition. However, we must not forget the humanistic and humanitarian aspects for a stable and sustainable development with values and meanings [15].

References

1. Lazareva, O.V., Shuba, V.I.: Modern technologies of organizing work on the formation of a transformational process of personal development of an entrepreneur in the community in times of uncertainty (for example, during the spread of a new coronavirus infection COVID-19). Education **1**(128), 20–23 (2021)
2. A Selection of Articles on the Transformational Process of Personal Development. https://pavelbilskiy.com/lichnostnaya-transformaciya-chto-kak-i-kogda-podborka-statej/
3. Lyapuntsova, E., Belozerova, Y., Drozdova, I., Afanasyev, G., Okunkova, E.: Entrepreneurial risks in the realities of the digital economy. J. MATEC Web Conf. **251**, 06032 (2018). https://doi.org/10.1051/matecconf/201825106032
4. Case Study is Active Learning Method. https://vc.ru/education/39384-case-study-metod-aktivnogo-obrazovaniya
5. Seven Basic Aspects of Business for the Novice Entrepreneur. https://pro-biznes.com/organizaciya-i-upravlenie-biznesom/sem-osnovnyx-aspektov-biznesa-dlya-nachinayushhego-predprinimatelya.html
6. EdTech: Markets, Trends, Investments. https://rb.ru/analytics/pro-edtech/
7. EdTech Company – From Idea to Implementation. https://vc.ru/hr/81641-edtech-kompaniya-ot-idei-do-voploshcheniya
8. Lyapuntsova, E.V., Drozdova, I.I.: Resolution of the Round Table "Student and teacher as key elements of higher education system: directions of interaction and development". Higher Education Didactics from the Positions of Student and Teacher. In: Collection of Articles of the Participants of the Interuniversity Practical and Methodological Seminar (17.03.2020–11.05.2020) MOO "League of Higher Education Teachers". In: Presidential Grants Foundation. RUSAINS, Moscow (2020)
9. Lyapuntsova, E., Vdovichenko, M., Belozerova, Y., Gorbatov, A.: Application of modern modeling methods: virtual technologies in the era of digitalization and their role in modern companies. J. Phys. Conf. Ser. **1425**(1), 012165 (2019). https://doi.org/10.1088/1742-6596/1425/1/012165
10. Lazareva, O.V.: Small Group as a Unit of Organizational Behavior. SKF MTUSI, Rostov-on-Don (2012)

11. The Sustainable Development Goals. https://www.un.org/sustainabledevelopment/ru/sustainable-development-goals/
12. Global Sustainability Trend: Business Opportunities: An Overview of New Niches Emerging from the UN Sustainable Development Goals. https://www.economy.gov.ru/material/file/45e459dca8acad4ecdd396aef4448e10/38526748.pdf
13. Ostrovsky, A., Kudina, M.: New paradigm of education in the era of digital transformation of the state. Public administration. Electronic Bulletin 78 (2020). https://cyberleninka.ru/article/n/novaya-paradigma-obrazovaniya-v-epohu-tsifrovoy-transformatsii-gosudarstva
14. Six Ways to Develop Entrepreneurial Thinking. https://rb.ru/opinion/kak-razvit-myshlenie/
15. Sustainable Development: What It Is and What It Means. https://www.forbes.ru/obshchestvo/425081-ustoychivoe-razvitie-chto-eto-takoe-i-v-chem-ego-znachimost

Education as a Driving Force of the Economy of the Future

Igor N. Molchanov[1,2], Natalia P. Molchanova[1], and Marina V. Charaeva[3](✉)

[1] Lomonosov Moscow State University, Moscow, Russia
[2] Financial University under the Government of the Russian Federation, Moscow, Russia
[3] Southern Federal University, Rostov-on-Don, Russia

Abstract. Education plays a key role in the composition of factors affecting the pace of economic growth. Despite the extensive norm creating and conducted research on the problems of training and accumulation of human potential in the functioning of the industry, there are still issues in the activities of educational organizations that require discussion and decision-making. The purpose of the work is to study the mechanism of managing the development of education in the context of global transformational changes affecting the development of social and economic relations. Using the methods of systemic, monographic and historical approaches, economic analysis and expert assessments, the indicators characterizing the activities of institutions of general and vocational education are considered, the key vectors of training students in undergraduate, graduate and postgraduate studies are identified. The key position in the training of personnel of "basic" universities, as implementing educational programs, taking into account the promising economic specializations of Russian regions and the peculiarities of the spatial development of the country, has been determined. The necessity of priority attention to the training for working in STEM-areas, the use of progressive forms of education in conjunction with the practical activities and research work of leading federal and research universities has been substantiated. Based on the generalization of Russian and foreign experience, recommendations are formulated to improve the training by using the "6i" model to manage the functioning of universities during the transition to the sixth column.

Keywords: Digital transformation · Education system · STEM segment · Competencies · Creativity

1 Introduction

The Russian education system has to solve a number of complex tasks. Along with improving the existing regulatory legal framework, it is necessary to systematize and update the set of documents on strategic management and planning for the development of the industry, which are designed to reflect the targets for the development of general and vocational education in the long term (approximately 20–25 years). This formulation of the question is due to the presence of some economic problems in the field of education and the scale of the tasks solved by the industry, aimed at achieving the most important

P. V. Trifonov and M. V. Charaeva (Eds.): DITEM 2021, LNNS 380, pp. 731–737, 2022.
https://doi.org/10.1007/978-3-030-94245-8_100

national goals outlined by the President of the Russian Federation. It is necessary to create a competitive educational environment and intensify the training of highly qualified experts who are in demand for work not only in the Russian market but in the international markets of the EAEU member states and other integration associations of Eurasia as well. These problems are of high urgency, especially in the field of IT-technologies, and their solution will help to maintain a leading position in high-tech sectors of the national economy, to support the required rates of economic growth. In the works of researchers, the social orientation of education, its key role in creating and disseminating knowledge, developing cognitive and communication skills, motivating students to state their opinions, promoting the ability to think logically, promoting values that are adequate to the needs of society [1].

The peculiarities of the geopolitical and spatial position of Russia put forward the issues of the formation of human potential and the sustainable functioning of the education system for priority positions. In this regard, the development of measures to provide full support to science and education requires priority attention. This is why these spheres of activity include "leading links", based on which it becomes possible to build a socially oriented and innovatively developed national economy. In order to provide the regions with experts of necessary qualification, equal conditions for families to live in urbanized and rural areas will have to be created and to maintain the high quality of the natural environment.

2 Methods

Russian and foreign scientists comprehensively consider the goals of higher education [2–4]. The research carried out on the basis of a systematic approach made it possible to group the entire set of goals and learning outcomes in five areas (groups), each of which has its own characteristic features: creating conditions for productive work (1), providing opportunities for creativity and personal development (2), formation a favorable environment for further training and retraining (3), expanding sources of income (4), consolidating universal values (5) [5].

3 Results

The Russian education system has gone through a difficult period of reform and transition to global educational standards during the 1990s–2010s. The current stage of development is characterized by trends in the transformation of social relations associated with the process of globalization. Features of different countries of the world are pronounced in the indicators of socio-economic development and a variety of approaches to the organization of educational activities. For developed countries, a distinctive feature of investment is investment in human capital, which has a positive effect on the dynamics of government spending on the education and upbringing of the younger generation [6]. However, according to Russian and foreign researchers, the measures applied in practice are not enough. Only a common interest and joint participation in the work to improve the organization of educational activities can bring socially significant results [7, 8].

As you know, in Russia, the education sector is financed mainly with the use of budget funds. In total income of educational institutions, the proportion of extra-budgetary funds fluctuates depending on the organizational and legal form and level of education. Based on official sources of information, it can be concluded that the volume of state financial support is not enough to meet the constantly increasing needs of various groups of the population in obtaining new knowledge, retraining and advanced training. The role of attracting additional non-state, mainly commercial financial resources is vital. Along with the implementation of the main functions, educational organizations can carry out income-generating activities and provide educational services. If the share of extra-budgetary funds in universities is a significant part (up to 50%), then it is relatively small in general education and preschool institutions, which confirms the relevance of finding additional sources of funds and strengthening the resource base as a whole. Leading universities purposefully expand the range of application of the knowledge of teaching staff and practice various forms of paid services. However, the work of regional universities is lagging behind progressive trends: "financial and economic indicators follow the dynamics of development drivers and not *vice versa*" [9].

The variety of models of financial support for education can be illustrated by looking at foreign countries. In the field of higher education in the UK, universities and charitable foundations mainly represent non-profit organizations. The latter assists educational organizations and supports them. Universities receive financial resources from various sources (grant support from the state, other non-profit organizations, and student tuition fees). Such a system allows universities to remain independent from the state and make a significant contribution to public life by providing information about the benefits of education and the educational achievements of students and graduates.

A number of publications emphasize the role of human capital as a factor of economic growth and explain the reasons for regional differences. For example, a study conducted in the UK has shown an uneven geographical distribution of the skills of the population across the country. A characteristic was revealed: for persons with a higher level of qualification, interregional differences in employment levels are less pronounced than for persons with a lower level of qualification [10]. The authors of scientific papers provide evidence that the process of digitalization can initiate an aggravation of the unevenness of regional development and spatial inequality [11].

The digital transformation causes a change in the structure of graduation of students by Russian universities by qualification levels. The partial closure of the specialist's program led to a decrease in the proportion of university students by the level of qualification "specialist" in the entire number from 86.3% (2013) to 11% (2018). This had a negative impact on the number of total output: it decreased by 27.7% during the period under review. At the same time, the training of bachelors (5.5 times) and masters (3 times) has significantly increased, which is associated with the transition to global educational standards of higher education at the bachelor's and master's degrees [12].

According to official data, the shortage of personnel in Russia only in IT specialties currently amounts to more than 700 thousand people, and according to some estimates – up to 1 million people [13]. The number of highly qualified specialists in STEM fields (Science, Technologies, Engineering and Mathematics) in 2018 amounted to 24% of the total output and significantly lagged behind the needs of the economy. According to various levels of qualification, the ratios between the contingent of students in the STEM segment and the total number of students studying at universities are as follows: bachelor's degree – 21%, specialty – 24%, master's degree – 33% [12]. For this reason, the decisions taken by the Government of the country on changing the standards of education concerning the formation of new educational modules in those areas of training that are necessary to ensure the creation of breakthrough digital technologies are of high importance, and should be made before the start of the next training cycle.

According to the research results, "emerging markets of high-tech products and services operate in an environment where digital technologies, smart computers and robots primarily replace workers of medium and medium-high qualifications engaged in routine cognitive work" [14]. In this regard, we can conclude that there is a high demand for highly qualified specialists in the STEM segment to work in the constituent entities of the Russian Federation; accordingly, we can foresee a significant increase in spending on higher education and R & D, which will require an increase in the scale of funding from budgetary and private sources.

The World Bank experts draw attention to the limited financial support for a number of federal programs (for example, "Regional reference universities" and "Universities as centers of innovation creation space"), based on which the development of regional universities is promoted. It is noted that the results of educational and scientific work of universities are not linked to their financial security, but depend on the improvement of such indicators as scientific productivity, cooperation with employers; innovations in teaching in order to cultivate "soft" skills; the availability of innovative educational programs [15].

4 Discussion

Summarizing the opinions of various researchers allows us to formulate several recommendations, answering to which can help to significantly improve some aspects of the work of higher education institutions. First, one of the reserves is to increase research productivity of the faculty by reducing workload and, consequently, increasing the time that is given to scientific work (in an average Russian university, the academic (working) load of an ordinary professor is about 1000 h; in the UK, the maximum annual load is 550 h [15]); the following provision is to strengthen management cooperation between universities and employers; another provision could be the introduction of innovative teaching methods to develop "soft" skills of students. For most universities, the development of project-based learning seems promising. Moreover, improving the quality of teaching and developing the creativity of students is becoming highly relevant.

In the work on personnel training, it is necessary to take into account, on the one hand, the features of the modern spatial development of Russia, which is characterized by objective global trends in the influence of agglomeration construction on accelerating the process of spreading innovations, while increasing differences in income and living standards of the population of nearby territories. On the other hand, attention should be focused on the need to take into account the long-term prospects for the development and functioning of rural areas, which a priori play one of the key roles in the growing geopolitical influence of Russia in the world community. Rural territories as a "special object of state and municipal management" [16] require constant attention due to the need to find solutions to urgent problems at the intraregional level. Despite this, in the current economic climate, due to limited financial resources, priority attention is paid to finding sources of financial support for higher education.

Regional higher education systems should plan their activities taking into account the competitive advantages of the territory. This applies, first, to the "basic" universities, which will have to build training programs for students in bachelor's and master's degrees for working in STEM fields. It is also necessary to strengthen the economic potential of the leading universities in the constituent entities of the Russian Federation – research universities, federal universities, the sustainable functioning of which contributes to the creation of an innovation-oriented academic environment, the involvement of students in scientific research. The national project called "Education" plays a key role in the implementation of the tasks set.

There is another aspect that increased attention to which can increase the competitiveness of the Russian higher education system in the international arena – this is the expansion of the admission of students from abroad. In 2017, foreign students accounted for 4% of the total number of students studying at Russian universities (OECD data). This problem is being solved within the framework of the federal project called "Export of Education", one of the goals of which is to increase the income from the education of foreign students. In 2018, the direct economic effect of this activity in Russia amounted to 30.6 billion rubles (0.5 billion US dollars) [15]. In this regard, it seems a timely measure to create a state strategy for the internationalization of education, in the work on which it is advisable for universities themselves to participate along with the Government. We can recommend the organization of joint activities of all participants in the educational process. We should make informed decisions on the choice of areas that deserve special attention based on interdepartmental approvals for implementation appropriate measures to increase the number of foreign students.

Under the influence of new trends in educational activity and the uneven economic situation of neighboring states in the Eurasian space, competition between national higher education systems is increasing. In this regard, when justifying new vectors of development of Russian education in the future it is necessary to take into account the influence of a number of basic conditions, for example: 1) the availability of a modern educational infrastructure of vocational education, which allows providing training in specialties that are in demand in the branches of the regional economy; 2) promising economic specializations of the regions formed according to the "effective branches" of the subjects of the Russian Federation (their list is developed by the Ministry of Economic Development and is based on maximum consideration of specific spatial features of the distribution of productive forces and other factors of economic development).

For most regional universities, one of the ways to increase the scientific and educational potential can be a system for evaluating the results of work and achievements, acceptable from the point of view of content and gaining recognition in society. Currently, the focus is on the development of research activities of universities. The University of Deusto, Spain, has developed the "6i" model of Caro-Gonzalez et al. [17]. Its authors recommend using this model to carry out "research management for the purpose of developing and implementing integrative institutional strategies by combining six dimensions": "internationalization" – "interdisciplinarity" – "intersectorality" – "impact" – "innovation" – "inclusion" [17]. The developers believe that this model is a comprehensive strategic design "with which universities can steer their priorities, activities and role within local, regional and global ecosystems" [17]. Presumably, the need to apply this model in practice will increase due to the need to expand various aspects of activity: both strategic management of their own (internal) research, and external (international, interdisciplinary and intersectoral) cooperation between universities, as well as value justifications for the effectiveness of their functioning through impact, innovation and inclusiveness.

5 Conclusion

It would be advisable to concentrate further research on studying the possibilities of productive use of existing new sources of funding and attracting new ones to ensure the sustainable position of universities in an unstable macroeconomic environment. When considering issues of improving management and development for the future, priority attention should be paid to the quality of long-term-oriented forecast developments and planning documents of a strategic and tactical nature, containing the justification of the volume indicators and the structure of the required specialists in the context of training forms and areas of training for working in the digital economy.

References

1. Chan, R.Y., Brown, G.T.L., Lidlow, L.H.: What is the purpose of higher education? A comparison of institutional and student perspectives on the goals and purposes of completing a bachelor's degree in the 21st century. https://docuri.com/download/researching-the-aims-goals-and-purposes-for-the-bachelors-degree-_59a7cbcff58171db1d429dc6_pdf
2. Klyachko, T.L.: Univ. Manage. Pract. Anal. **105**(5), 28–35 (2016). https://doi.org/10.15826/umj.2016.105.041
3. Ovchinnikova, O.P., Ovchinnikova, E.V.: Finance **11**, 58–63 (2019)
4. Goksu, A., Goksu, G.G.: Procedia Econ. Fin. **26**, 1152–1158 (2015). https://doi.org/10.1016/S2212-5671(15)00945-4
5. Molchanov, I.N.: Public Admin. **80**, 239–255 (2020)
6. WorldBank. https://data.worldbank.org/indicator/SE.XPD.TOTL.GD.ZS?view=chart
7. Kuzminov, Y.I., Frumin, I.D., Sorokin, P.S. (eds.): How to Make Education a Driving Force of Socio-Economic Development? Publishing House of the Higher School of Economics, Moscow (2019)
8. Andrews, M.: The Limits of Institutional Reform in Development: Changing Rules for Realistic Solutions. Cambridge University Press, Cambridge (2013)
9. Sandler, D.G., Sudakova, A.E., Tarasyeva, T.V.: Econ. Reg. **16**(4), 1087–1103 (2020). https://doi.org/10.17059/ekon.reg.2020-4-6
10. Green, A.E.: Spatial inequality and skills in a changing economy. In: UKCES Briefing Paper Series. UK Commission for Employment and Skills, London (2011)
11. Neugebauer, R. (ed.): Digital Transformation. Springer, Heidelberg (2019). https://doi.org/10.1007/978-3-662-58134-6
12. Bondarenko, N.V., et al.: Indicators of Education: 2020: Statistical Collection. National Research University "Higher School of Economics", Moscow (2020)

13. Lvov, P.: Russia Will Conduct a Comprehensive Assessment of University Rectors. https://ria.ru/20210306/rektory-1600226368.html
14. Akaev, A.A., Desyatko, D.N., Petryakov, A.A., Sarygulov, A.I.: Econ. Reg. **16**(4), 1031–1045 (2020). https://doi.org/10.17059/ekon.reg.2020-4-2
15. Russia: Recession and Growth During the Pandemic. Special Section: Education. World Bank. Report on the Russian Economy 43, July (2020)
16. Lazhentsev, V.N., Ivanov, V.A.: Econ. Reg. **16**(3), 696–711 (2020). https://doi.org/10.17059/ekon.reg.2020-3-2
17. Caro-Gonzalez, A., Ferreira-Lopes, L.: Changes: facts, trends, forecast. Econ. Soc. **13**(1), 217–230 (2020). https://doi.org/10.15838/esc.2020.1.67.13

Reproduction of Human Resources of Academic Organizations as an Imperative of the Future Economy

Aleksander A. Litvinyuk(✉), Pavel V. Zhuravlev, Ekaterina V. Novikova, Ekaterina V. Kuzub, and Felix K. Akhmedov

Plekhanov Russian University of Economics, Moscow, Russia
litvinyuk.aa@rea.ru

Abstract. The purpose of the article is to identify the main factors hindering the rejuvenation of the personnel potential of Russian academic organizations. The scientific novelty of the work lies in the systematic study of all motivational aspects that promote or hinder the attraction of promising young professionals to organizations operating in the academic labor market. The studies were carried out using a modified version of the author's diagnostic technique "Motivation". The respondents were young scientists, postgraduate students and university students. The results obtained showed the low efficiency of administrative measures aimed at rejuvenating the personnel potential of academic organizations in the Russian Federation, since most people entering the postgraduate studies only imitate their interest in the field of activity under consideration. Based on a comparison of the motivational profiles of employment behaviour, the basic scales of the MMPI test of successful young scientists and teachers of higher education and students who wish or do not wish to continue their studies in postgraduate school, the proposals were made for the representatives of talented and creative youth to increase the level of motivation to plan their professional development in science and higher education. They are also aimed at solving the problems of preventive professional selection of talented and creative young specialists in higher educational institutions of the Russian Federation and increasing employment guarantees in the academic sphere of labor.

Keywords: Scientific personnel · Motivation · Talented youth · Academic organization

1 Introduction

The authors believe, that in the reproduction of the personnel potential of domestic academic organizations, a problem has emerged that has not yet been resolved at the proper theoretical and practical level, which actualizes a number of negative consequences for the development of the Russian economy. This is an effective motivation for talented young professionals to plan their careers in academic organizations.

This situation is aggravated by the fact that in Russian science lacks complex and interdisciplinary research aimed at finding a solution to this problem. Basically, the subject area of specific sciences is studied: economics [1], law [2], sociology [3], psychology

P. V. Trifonov and M. V. Charaeva (Eds.): DITEM 2021, LNNS 380, pp. 738–744, 2022.
https://doi.org/10.1007/978-3-030-94245-8_101

[4], etc. This does not make it possible to formulate effective solutions to the problem identified by the authors.

The novelty of the study carried out by the authors is due to consideration of both the material aspects of the motivation of young specialists in the academic labor market and the structure of their motivational complex as a whole. At the same time, a hypothesis was put forward that, relying only on material incentives, the filling of all available vacancies in academic organizations with young specialists cannot be achieved, however, this will not increase the level of their human resources due to the low quality of the labor resources involved. The purpose of the study in this context is to compare the motivational characteristics of university graduates with the reference motivational profiles of successful scientists and teachers. The achievement of this goal presupposes the solution of a multicomponent problem: why do young specialists go to postgraduate school; what master's graduates become postgraduate students and why they do not seek to develop themselves in the academic labor market.

2 Methods

Many options for the classification of labor motives and algorithms for managing motivational processes in an organization built on this basis have been proposed within the framework of management as a science. The most popular scientific literature approach to solving the problem of analyzing labor behavior is the well-known theory of motivation by Vroom (Expectancy theory) [5]. However, it is rather difficult to apply it in practice. It is required to rely on reliable values of Valence (V) and Expectancy (E) indicators to make effective decisions in the field of labor motivation. The meaning of these indicators outwardly seems simple and clear to understand, but the objective quantitative characteristics of these parameters are very difficult to justify.

Proceeding from this premise, the author's tools developed by Litvinyuk [6, 7] and a digital product created on its basis were used in this research.

3 Results

To identify the reasons for admission to postgraduate school, the authors conducted a survey of more than 120 students from a number of Moscow universities, who have identified the continuation of their studies in postgraduate school as a priority task for themselves. The main motivators for such a decision were as follows.

- It is rather difficult to quickly find a promising job in a specialty with a satisfactory starting salary after completion of master's studies. Postgraduate studies make it possible to increase significantly the time range of its search (41%).
- The mode of classes in modern Russian postgraduate studies makes it possible to work part-time. In turn, having an advanced degree often accelerates career growth (28%).
- I enjoy doing research and teaching (15%).
- No personal reasons (9%).
- Postgraduate studies make it possible to avoid conscription for military service (7%).

Consequently, many of the university graduates consider their postgraduate studies as a kind of "alternate airfield" where they can comfortably wait out their temporary difficulties in the labor market.

As a result, all the current attempts of the state to rejuvenate the staff of academic organizations by means of administrative pressure on top management or artificial formation of non-competitive advantages for young teachers and researchers, in our opinion, will not lead to a real result [8]. They only contribute to the temporary improvement of a number of statistical indicators that can "impress" people who are not "immersed" in these problems.

At the same time, the personnel potential is not improving qualitatively, and so far, most of the effective scientific research in Russia is carried out mainly by rather aged specialists.

Based on the foregoing, the authors in 2020 attempted to calculate the characteristic configurations of the motivational profile of successful young scientific and pedagogical workers who linked their professional activities with the academic labor market and achieved, in the opinion of their line managers, noticeable professional success. The authors compared the resulting configuration of the motivational profile with the average motivational profile of those graduate students who deliberately decided to continue their studies in postgraduate school, and those graduate students who definitely do not want to go to postgraduate school and choose a non-academic option for them.

317 scientific and pedagogical employees under the age of 36, full-time postgraduates and undergraduates of the Plekhanov Russian University of Economics, including the branches of the above university stationed in Volgograd, Smolensk and Kemerovo agreed to be voluntary respondents. Moreover, 2-year undergraduates and post-graduate students from a number of other universities in Moscow, as well as in Volgograd, Kemerovo and Smolensk regions, took part in the study. The sample totaled more than 500 people.

For the scientific interpretation of the results obtained, the main provisions of Litvinyuk's motivational theory and diagnostic techniques created on its basis, using modern digital technologies, were applied [6].

The standard motivational profile of successful young scientists and teachers has the following average configuration (Table 1).

Table 1. Average motivational profile of successful young scientists and university teachers.

Basic groups of labor motives and their standard indexing	The strength of the impact of the group of motives on the labor behavior of the respondents (the standard strength of the action is 0)
Safety (MS)	+4
Subordination (MD)	0
Acquisition (MA)	+3
Satisfaction (ME)	+6
Energy saving (MP)	−2

The following conclusions can be made based on the results given in Table 1.

First, the labor behavior of young teachers and researchers is very flexible in relation to the strength of the motives of satisfaction (6 points). In other words, the effectiveness of their professional activity depends a lot on the level of emotional involvement in the labor process.

Second, these respondents are inclined to comply with the regulations for performing labor operations established in the educational organization (4 points). This can be interpreted as unwillingness to receive sanctions from the administration for "improper behavior in the workplace", which indicates their high motivation to keep their jobs.

Third, the nature of their labor behavior is quite dependent on the variation in the size of their material rewards based on the results of their labor activity (3 points).

Fourth, young teachers and researchers are indifferent to the formation of cohesive informal groups (0 points).

Fifth, this group of respondents is focused on the creative and creative nature of work (−2 points). According to the main provisions of Litvinyuk's motivational theory, the negative value of the force of the action of energy conservation motives (MP) on an individual indicates his positive orientation towards the innovative and creative nature of labor behavior and personal development [6].

To assess the real motivational attitudes of Russian students towards professional development within the framework of the academic labor market, it makes sense to compare the average motivational profiles of those students who, in the course of the above survey, expressed a conscious desire to continue their studies in postgraduate school (group A); do not see themselves in the future in science and higher education (group B) with an average profile of successful young teachers and researchers (group C), considered by the authors in Table 1. The comparison of indicators is given below (Table 2).

Table 2. Average motivational profile of work activity of A, B and C groups of respondents.

Standard index of the group of motives for work	Groups of respondents and the strength of their influence on labor behavior of groups of motives		
	A	B	C
MA	+5	+3	+3
MD	+1	+1	0
ME	+3	+5	+6
MP	−1	−5	−2
MS	+3	+4	+4

The calculation of the standard statistical Spearman coefficient (ρ) showed that the degree of similarity of the motivational profiles of group A and group C was (0.675), while the same between the motivational profiles of groups B and C was (1,000). The conclusions drawn from the above data are quite paradoxical. Students who plan to complete postgraduate studies after graduation and continue their professional careers in higher education are much more different in their motivational profile from the profile of successful young teachers and researchers than respondents who consciously do not want to go to graduate school.

In other words, those young people who are less able to engage in scientific and teaching activities, according to the target criteria identified by the authors thereof, turned out to be motivated to develop their professional careers within the framework of the academic labor market. It is they who will "develop" Russian science and higher education in the foreseeable future in conditions when universities, under the influence of administrative pressure, are forced to reduce the average age of teaching staff and researchers.

Similar conclusions were made by the authors earlier [9] based on a comparison of the values of the main classical clinical scales of the MMPI test, which they conducted online with the same respondents.

The studies conducted by the authors in 2019–2020 under the RFBR project 19-010-00168 clearly showed that the first three criterion priorities when choosing a starting and promising profession for further career development for graduates of Russian universities turned out to be: a high level of material rewards (44.2%); the possibility of a fast vertical career (21.3%); good social package (20.5%) [9].

In other words, the planning of their professional development within the framework of the academic labor market is not perceived by this group as promising for the following reasons: low, according to the opinion established in modern Russia, the level of starting wages; a long-term lag in professional development, especially in terms of the implementation of a vertical career.

4 Discussion

The low efficiency of the modern Russian system of training highly qualified scientific and pedagogical personnel is an obvious fact, evident to everyone who works in academic organizations. This problem is not purely Russian. Whatever they say about the effectiveness of the reproduction of scientific and scientific-pedagogical personnel in countries where the development of this sphere of activity is given much more attention than in modern Russia, but they also face very similar problems as our country [10–13]. Most foreign researchers believe that the decrease in the motivation of talented youth for professional development within the framework of the academic labor market is associated, first of all, with insufficient motivation for such a variant of personal development.

A roughly similar situation is observed in modern Russia. This is indicated by the results of studies conducted by Reznik [14–16] and other specialists [17–19]. Problems with the attraction of talented young specialists, similar in content, have arisen in a number of other countries that emerged within the territory of the former USSR [20–22].

The general result of all the above-mentioned studies is the conclusion that the problems of retaining talented young specialists are mainly of a material nature. However, it is possible to attract young people to the academic sphere with the help of financial incentives, but it is impossible to say unequivocally whether those young people will be talented and motivated for development in science and higher education.

5 Conclusion

In our opinion, actual rejuvenation of personnel in academic organizations and the reproduction of human resources require practical solving of the following urgent tasks that naturally follow from the results of our research.

First, a program should be developed at the state level to increase the social prestige of scientific activity. A similar campaign in the media in relation to increasing the prestige of military service has already been successfully implemented. A similar impact on the mass consciousness of youth should be implemented urgently in relation to the field of professional activity under study.

Second, it is necessary to transform radically the system of financial incentives for the work of young specialists in academic organizations by introducing the institution of postdoctoral studies, which is traditional for many countries. This option of additional financial incentives presupposes long-term, usually up to six years, work of a young specialist under the guidance of an experienced scientist. This can be both limited-time scientific research and teaching [23]. This practice allows ensuring the continuity of knowledge and skills in science and higher education and allows a young specialist to quickly complete the process of his professional adaptation.

Third, the practice of short-term competitive selection for filling vacant positions and a project-based approach to the formation of temporary labor collectives should be eliminated in the field of higher education. Young teachers and scientists see this as the main reason for the instability of their position in this segment of the labor market. All the more so as the recently increasing staff reduction in universities is often carried out taking into account such a formal criterion as the age of a teacher or research worker. With this approach, young professionals are given to understand their prospective after reaching adulthood, even if they are fully capable workers.

Finally, when recruiting for graduate school, not only the fulfillment of the plan for budgetary and non-budgetary recruitment, but proactive assessment of the ability of candidates for creative and creative activities, the configuration of their motivational complex are required. The methods of such selection, including the author's one, are available in modern HR-management [6]. They just need to be adapted to the specifics of the scientific specialty in which the university conducts postgraduate studies.

References

1. Ledeneva, S.A., Ivanova-Shvets, L.N.: Bull. Plekhanov Rus. Econ. Univ. **3**(99), 114–122 (2018)
2. Lazarenko, V.A., Tkachenko, P.V., Lipatov, V.A., Severinov, D.A.: Alma Mater. (Bull. High. Sch.) **1**, 31–38 (2020). https://doi.org/10.20339/AM.01-20.031
3. Morozov, A.V.: Qual. Educ. **2**, 23–25 (2019)
4. Terentev, E.A., Rybakov, N.V., Bedny, B.I.: Educ. Stud. **1**, 40–69 (2020). https://doi.org/10.17323/1814-9545-2020-1-40-69
5. Vroom, V.: Work and Motivation. Wiley, NYC (1964)
6. Litvinyuk, A.A., Ledneva, S.A., Kuzub, E.V., Kleschevsky, Y.: Motivation of Youth Employment in the Field of Science, High Technology and Higher Education. Kreativnaya ekonomika, Moscow (2019). https://doi.org/10.18334/9785912922916

7. Certificate of Registration of the Computer Program No. 2020667283 Dated 22.12.2020. https://www.elibrary.ru/item.asp?id=44760498
8. "Science" National Project. https://futurerussia.gov.ru/nauka
9. Litvinyuk, A.A.: Lead. Manage. **7**(3), 489–502 (2020). https://doi.org/10.18334/lim.7.3.110725
10. Shin, J.C., Kim, S.J., Kim, E., Lim, H.: Doctoral students' satisfaction in a research-focused Korean university: socio-environmental and motivational factors. Asia Pac. Educ. Rev. **19**(2), 159–168 (2018). https://doi.org/10.1007/s12564-018-9528-7
11. Litalien, D., Guay, F., Morin, A.J.S.: Learn. Ind. Differ. **41**, 1–13 (2015). https://doi.org/10.1016/j.lindif.2015.05.006
12. London, J.S., Cox, M.F., Ahn, B., Branch, S.: Int. J. Doct. Stud. **9**, 205–227 (2014). https://doi.org/10.28945/2050
13. Cally, G., Asangi, J., Damith, R.: High. Educ. Res. Dev. **34**(1), 89–104 (2015). https://doi.org/10.1080/07294360.2014.934663
14. Reznik, S.D.: Alma Mater. (Bull. High. Sch.) **1**, 37–44 (2016). https://doi.org/10.20339/AM.01-16.037
15. Reznik, S.D.: Univ. Manage. Pract. Anal. **4**, 106–116 (2015)
16. Reznik, S.D., Chemezov, I.S.: Bul. Tomsk State Univ. **430**, 159–168 (2018). https://doi.org/10.17223/15617793/430/22
17. Bednyy, B.I.: High. Educ. Rus. **4**(211), 5–16 (2017)
18. Bednyy, B.I., Sapunov, M.B., et al.: High. Educ. Rus. **28**(1), 130–146 (2019). https://doi.org/10.31992/0869-3617-2019-28-1-130-146
19. Vinokurova, N.A.: Population **4**, 113–121 (2016)
20. Dzianisava, N.F.: Bul. Hrodzien Univ. Janki Kupaly **8**(3), 104–112 (2018)
21. Tkalenko, N.V., Kholiavko, N.I., Hnedina, K.V.: Polissya Sci. Bull. **4–1**(12), 44–49 (2017)
22. Tarvid, A.: Proc. Econ. Fin. **14**, 585–594 (2014). https://doi.org/10.1016/S2212-5671(14)00747-3
23. Miller, J.M., Feldman, M.P.: J. High. Educ. **86**(5), 697–724 (2015)

Author Index

P. V. Trifonov and M. V. Charaeva (Eds.): DITEM 2021, LNNS 380, pp. 745–749, 2022.
https://doi.org/10.1007/978-3-030-94245-8